TROPICAL AMERICAN PLANT DISEASE

(Neotropical Phytopathology Problems)

by

Frederick L. Wellman

The Scarecrow Press, Inc.
Metuchen, N. J. 1972

ISBN 0-8108-0459-X

Library of Congress Catalog Card Number 71-175527

This work was supported in part by the
United States Department of State
Agency for International Development
Grant No. AID/csd - 1149

Dedicated to

JOHN A. STEVENSON

This scholarly gentleman, in addition to his feats as a mycologist, with professional zeal collects and preserves in an organized fashion, plant disease publications from the American tropics. This he has been doing for much more than half a century. His "John A. Stevenson Library" given to the public, is housed at the United States Plant Industry Station in Beltsville, Maryland.

He has responded understandingly, soundly, and unstintingly by every means at his command, whenever asked to make plant disease determinations, to help in preservation of herbarium specimens, to give consultations on control measures, and to send out, as I know, often personally bought pieces of literature to harried field men as well as academicians in every nation in the Neotropics. Dr. Stevenson has collaborated in many, and been joint author in some, of the most important plant disease lists from numerous countries; and wherever I have heard his name mentioned in the tropics, it is with appreciation, respect, and affection.

To quote a Latin American colleague of mine: "I have visited Dr. Stevenson four times, and before and since I have always been helped by him. I write him many times, he answers promptly and often in Spanish. As head of the

United States Mycological Herbarium he has been specially gracious to us. He cooperates with everyone. Compared to the late Professor Spegazzini I consider Dr. Stevenson easily as great. He has gotten together more specimens of plant disease fungi than we have in any collection in this part of the world. I call him a magnificent person." And I have heard this kind of an estimate from many sources. In Brazil a pathologist told me Stevenson has saved tropical American agriculture from so many plant disease losses that he looked upon this one mycologist-pathologist as perhaps the most valuable man in the U. S. government in its relations to its southern neighbors.

[See the following: 1. Fidalgo, O., and M. E. Kaufman-Fidalgo. 1960. "John A. Stevenson (de biologistas do Instituto Botanica)." Secretaria da Agricultura do Estado de São Paulo. Jan. 1960. (Deposited in the John A. Stevenson Library, U. S. National Fungus Collections, Beltsville, Maryland, U. S. A.) 2. Gamundi, I. 1960. "John A. Stevenson". Bol. Soc. Argentina Bot. 8:222. 3. Biographical sketch. 1961. "Stevenson, J. A." American Men of Science. ed. 10, p. 3922.]

ACKNOWLEDGMENTS

I cannot express with best clarity the deep debt I owe to my wife and son. It has been necessary for them to be alone long weeks, and sometimes months, while I have been away and been writing. I recall most appreciatively with what patience my wife has heard my enthusiasms about new observations and my first ideas about principles and characteristics unique to tropical American plant disease.

Literally hundreds of tropical agriculturists have helped me. Some have been non-reading peasants; some, indigenes with languages I knew nothing of; among others were presidents of countries, and many university-trained plantation owners. There were also numerous scientists of 16 countries, working in private and government agricultural research stations, who were solving problems for planters of such crops as cinchona, rubber, coffee, henequen, cacao, banana, sugarcane, tobacco, citrus, forests, coconut, and vegetable gardens.

The time is here to make mention of some special individuals who have given me of their unselfish aid. Not all can be listed, for it would be too long a list. But these have done more for me than being just friendly. I speak first of the continued interest of John A. Stevenson.

Former administrators of mine that were of supreme help have been: Claud L. Horn and Ralph H. Allee, both retired, and the late Ross E. Moore. Of special value in comparing of our observations: Carlos L. Bianchini, Benjamin H. Waite, Marco A. Flores, E. Schieber, Raul Munoz, H. S. Stover, Francisco de Sola, and (the late editor in Costa Rica) Mariano Montealegre. There are some with whom I have long maintained contact: John S. Niederhauser, José A. B. Nolla, David S. Thurston, Eugenio Schieber, Albert S. Muller, Pierre G. Sylvain, and Eddie Echandi. With special gratitude I acknowledge the invaluable challenge and aid from students; the longest times as laboratory assistants of mine were spent by the loyal and dependable Takhar Singh Claire in Honduras and Guillermo Salazar in Costa Rica.

Photographs have come from numerous sources, many in color (It is unfortunate that color film has pre-empted most black and white work which is so much more acceptable in publication). I have given credit where it was due in all captions under pictures, but I wish specially to mention: Francine Fischer, Benjamin H. Waite, Claud L. Horn, Eugenio Schieber, Eddie Echandi, Enrique Ampuero, L. H. Purdy, Marvin Williams, and the late James Mitchell. Outline drawings and maps were made by me, and they may serve the purpose.

Former students of mine whom I talked with about it, were unanimously in favor of such a text as I have prepared; they specified it must not (as some books do) by inference settle questions that are still in abeyance. I discussed the contents and make-up of this book with the late Frank A. Haasis and he helped in its organization and reduction of the number of chapters from 30 to 22. Parts of this book while in manuscript have been read by W. C. Snyder, G. B. Lucas, Don E. Ellis, T. T. Hebert, Ralph R. Shaw, and Eugenio Schieber; and Eddie Echandi has read it all. Suggestions have been helpful (Dr. Shaw's admonition many years ago was to make it the definitive work and I am afraid it misses that) but the last decision has been mine, and anything doubtful is the writer's full responsibility.

In closing, this book would never have come to completion except for Don E. Ellis, head of the Department of Plant Pathology, North Carolina State University; and then the efforts of other University administrators such as J. L. Apple and J. A. Rigney who secured subsidy under which it was written. I want as well to thank all members of my Plant Pathology Department for bearing with my life as a writing recluse, and to Mrs. Joyce Johnson for her pleasant, and untiring helpfulness.

Frederick L. Wellman
Professor Emeritus
Department of Plant Pathology
North Carolina State University
Raleigh, North Carolina

January 1971

CONTENTS

LIST OF TABLES

LIST OF FIGURES

FREDERICK L. WELLMAN

photo by Tom Byrd

FOREWORD: THE AUTHOR

Dr. Frederick L. Wellman was born in Kamundongo, a village in Angola, in 1897. After graduation from Fairmount College in Wichita, Kansas, he went to the University of Wisconsin for further studies. In his professional career he has served well the U. S. Department of Agriculture as a plant pathologist eventually becoming senior pathologist and regional consultant (Office of Foreign Agricultural Relations). The U. S. government chose him for special plant exploration work in Turkey in 1936 and he headed the World Coffee Mission (Europe, Africa, the Orient) in 1952.

He was destined to play an important role in tropical agriculture. Being a pathologist with the United Fruit Company in Central America in 1928-29, having a directorship of the El Salvador Cooperative Experiment Station (U. S. Government sponsorship), teaching at the Inter-American Institute of Agricultural Sciences at Turrialba, Costa Rica, and, finally, becoming the head of the Department of Plant Pathology, University of Puerto Rico Agricultural Experiment Station at Río Piedras, were some ingredients which he molded into a unique experience to serve tropical agriculture. The enormous wealth of information which he possesses on what he terms the Neotropics places him among the foremost living tropical plant pathologists.

He has been a visiting professor at North Carolina State University at Raleigh for several years, and is now professor emeritus there. He has given series of lectures at the University of California (on three campuses); Cornell; the Universities of Florida, Georgia, and North Carolina, and at Purdue.

In the last two years he has travelled again to Brazil and to Central America as a member of a committee surveying the spread of that menace to coffee--the rust disease in Brazil, and the immediate measures which might be taken to stop its northward advance. The disease, hitherto unknown in the Americas, is now a grave menace to the economy of all the coffee-growing countries in Central and South America. Aware of the serious consequences ahead, the U. S. coffee manufacturers are using Dr. Wellman's talents in planning the strategy to follow.

He has contributed numerous research papers on diverse phytopathological subjects many of which have enriched the literature of plant pathology in the tropics. His textbook Coffee Botany, Cultivation, and Utilization, of 488 pages (52 on diseases) and a smaller 130-page one, Plant Diseases; An Introduction for the Layman, provide useful information to a wide range of readers.

A member of many scientific societies, academies, and fraternities, his was the driving force in attaining recognition of the Caribbean Division in the American Phytopathological Society. Fred Wellman became the Division's first president, and later, councilor. In 1967 he received one of two special awards voted by the Division for meritorious service to agriculture in the American tropics.

Students from every country of the Western Hemisphere, and from Asia, Africa and Europe, attest to the kindness, comprehension and selflessness which have marked this respected scientist, who has always found time to spare for those in quest of knowledge. He has never let them down! To them he has been at once teacher, scientist, philosopher and philanthropist. Calm but insistent in truth, cautious when in doubt, never over-excited in an argument, he has been always ready to admit an error and to grant to others the credit owed them.

The present book culminates a lifetime devoted to the science of plant pathology in its ever widening horizons. To him it is a dream come true!

Dr. José Antonio Bernabé Nolla,
University of Puerto Rico at Mayaguez
Councilor, American Phytopathological Society (Caribbean)

SOME HIGH POINTS IN THIS BOOK

This text is neither a manual nor a compendium, but a discussion of problems and progress in the science of plant pathology in the American tropics. That science has a long and colorful history. To the tropical pathologist in the Occident, geography is important. Neotropics include the Americas from south parts of California, Texas, Louisiana and Florida to as far south as just above Tierra del Fuego in Chile, and the Argentine Pampas; the warmest band is the long-known "heat equator" starting well up in northwest Mexico following a curving sloped-line down through Panama and lowered to bisect the mouth of the Amazon eastward to the bulge point of Brazil. The geographic equator is straight across from the Galapagos Archipelago to the Amazon delta, where it crosses the heat equator; ocean currents, prevailing winds, and shielding mountains profoundly affect climates and diseases; elevations as well as exposures make great differences in crops and diseases; in addition to other difficulties of underdeveloped regions, long distances are involved in communications among workers. All usual plant pathology principles are met within the American tropics, along with new ones not encountered in much more studied temperate zone environments.

Parasitic organisms have probably been in the tropics longer than anywhere else on earth. Viruses there are very numerous; many are indigenous, and a number are severe on crops. But there are certain ecologic zones where vectors are inhibited and virus diseases do not exist. A teeming part of tropical life are the saprophytic and other kinds of bacteria, and primitive types of fungi (some rare), among which are plant parasites. There are two distinct kinds of anthracnoses that along with leaf and fruit scabs, molds, and other such blemishes, cause continuous losses from disease. Their seriousness depends on ecologic factors often present in warm-moist tropics. Most white mildews are in cooler-dryer areas; black mildews are in wetter-warmer ones; in the latter are transition types from usual obligate parasites to saprophytes. (Where a Latin binomial of an organism is used without an authority, this latter may be found on referring to the binomial where it occurs in the index.)

Chronic diebacks, woodstains, and like diseases are from invasions by fungi, mostly Ascomycetes, of weakened and susceptible trees; such infections are debilitating, cause losses, and can lead to slow death, but are not spectacular. The more fear-inspiring fusarial wilts cause some of the most notable and commercially destructive tropical plant troubles. Otherwise, fusarial attacks cause twig death, increase in leafspot severity, seedling damping-off, preemergence death of seedlings, stem canker, stunting, sterile buds, and explosive overgrowths from affected flower initials.

No wild or cultivated crop is free from leaf spots (some very serious); symptoms range from speck and weakspot and intermediate kinds to gross leaf-consuming rot. Very many fungus species are involved and not all are known or described. The development of a sclerotium or such fungus structure is typical of organisms in wet-warm climates; it is probably the holdover condition or stage that survives excess rainfall accompanied by bacterial hyperparasitism (in the temperate zone it has been readily adapted to over-wintering).

Rusts or smuts are serious disease organisms on grasses, cereals, and many major tropical crops. Indigenous tropical rusts, such as those on inga and other trees, maize, potato, tomato, tobacco, and many others, will spread to and threaten temperate zone agriculture. There is a smut of beans, and there are smuts also on tropical pasture plants and trees. Certain of the most important crop disease organisms that have come from the Orient have been such as a few smuts of sugarcane and the destructive yellow leafrust of coffee. As is well known, rusts may have complex life histories; the most typical example in the Neotropics is a maize rust, with an alternate host oxalis, which has five spore forms.

There are numerous Basidiomycete parasites with recumbent habit that cause diebacks, foliage rots, leaf spots, and other blights. Most common of all these is rhizoctonial damping-off of seedlings and this occurs in untouched jungle. Some of the recumbent fungi have thread or cord-like travel stages; some release spores (mostly at night) from a thin layer of superficially occurring hymenium; some release spores from gills in caps of typical agarics or "toadstools." One agaricaceous fungus causes cacao witches' broom; has a very narrow host range; infects the youngest cacao tissues with basidiospores. As it is systemic, it causes growth monstrosities and dieback. Another agaricaceous fungus, causing coffee American leafspot, has a phenomenally wide host range of over 500 tree and plant species. Infection is by gemmae, producing leafspots that are luminescent; the spots cause severe defoliation, which results in debility and dieback.

The commonest plant-affecting alga is tropical. It is a typical parasite and the study of it was of great significance in developing the early thinking and philosophy about plant disease. It attacks many hosts and can be a lichen on proper symbiotic union with a fungus, after which it may then differ markedly as to its parasitism. There are about a score of algae that are troublesome to plants in the American tropics.

Exhuberant growths of mosses, orchids, ferns, strangling figs, climbing grasses, bamboos, and many other non-parasitizing plants, through weight and smothering, tear apart and damage trees. However, trees are much parasitized by true flower-bearing bushes and vines, and even by other trees. In all nature there is no series of hyper- and super-hyperparasites that is so extensive as may be found among the complex of diseased and parasitizing bushes and vines on jungle and cultivated trees--this does not include the infections of insects and worms. Among the latter are menatodes, microscopic worms that are absolutely ubiquitous in the tropics; they cause important crop losses and when co-parasitic with fungi may break down disease resistances.

Tropical soils are fragile, easily injured, and need protection. Also in warm and moist areas where crops are grown, the problem of minor element deficiencies is common. Periodic volcanic eruptions spread deposits some of which are good, some bad. In certain diseases there are intimate relations to soils which result for good reasons in some fields being designated as disease-susceptible and some, disease-resistant. There are instances in which there seems to be a "self-healing effect" in soils by which they seem, after a while, to free themselves of a disease. Disease amelioration and control is often accomplished by cultural methods: long distance change in locations of fields; close-by shifting culture (milpa system); stirring of soils; flooding and temporarily holding water over a field to alter conditions; and use of informal exclusion practices or quarantines.

Chemical control methods are widely used in tropical plant pathology. The perpetually warm-moist conditions cause special difficulties (e.g., fast deterioration of certain temperate zone-accepted fungicides). Many spray problems are being given very modern and highly specialized attention; there are eight characteristics of special value that make a spray compound good for our tropics. Sprays almost invariably gain in effectiveness with the use of adjuvants and by alternate applications. Sticking qualities are essential in wet climates, and sprays are not only put on from the ground but also with air-borne equipment.

Plant disease resistance (although it will probably never completely obviate chemical disease control), when incorporated with proper husbandry methods, is the most important method of keeping crops healthy in the American tropics. All our crops have in them natural and inherent resistance to most of the microorganisms that surround them. If resistance is desired to some special organism, the pathologist wisely secures seed, cuttings, or plants of wild types of some plant, generations of which have been growing for mellenia among disease organisms. There, it is probable, that disease resistance has evolved. From such sources, genes may be used by pathologists and transferred to good commercial varieties for exacting, modern field culture. Types of resistances are of much concern against different types of disease. With more slowly reproductive parasites like bacteria, and perhaps Fusarium, adequate resistance may be the rather readily uncovered simple type, also called perpendicular and specific. With rusts and perhaps Cercospora and other parasites which are mutating vigorously and fast increasing, pathologists temporarily use specific resistance, but (though with difficulty) must eventually secure good permanent field resistance, called tolerance and horizontal resistance.

Chapter I

INTRODUCTION. THE TROPICS

Summary -- Introduction -- Scope of the book -- Character of the tropics -- The special Neotropical region -- Important life zones related to elevation -- Crops grown in our tropics -- Certain historical dates of background interest -- The phrase "tropical plant pathology" -- Table of comparisons of tropical American countries -- Literature cited

Summary

Tropical America, also known as the Neotropics, is a large region extending on a biogeographical basis from the lower fringes of continental United States of America down to the colder ends of Chile-Argentina. This tropical land, with its long and at times unsettled history, is physically extremely variable; it has some large cities and numerous remarkable natural features. Through it there slants the thermal equator, line of most heat, starting well up in Mexico and going diagonally southeast, down to the easternmost bulge of Brazil; it crosses the geographic equator at the mouth of the Amazon. Basically the region is agricultural, with various climates resulting from effects of land masses, ocean currents, equatorial winds, and temperature zones of hot-warm, medium, temperate, and cold at different elevations above sea level. The many differences make it possible to grow a wide range of money-making and subsistence crops, all of which have diseases.

Introduction

The American tropics is a large region (see accompanying Map A) and along with its highly variable and extensive territories the people who live in it are different in different parts. The diversity that thus occurs is paralleled by unique types of agriculture and kinds of crops

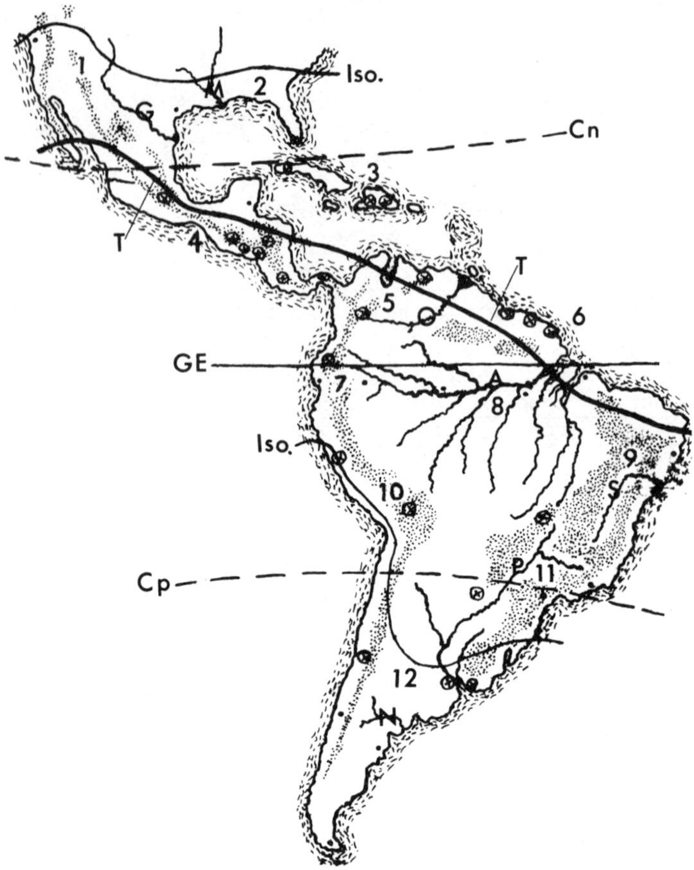

Map A. Sketch of main features of tropical America (Neotropics). Legend: A--Amazon river; Cn--Tropic of Cancer; Cp--Tropic of Capricorn; G--Rio Grande; GE--Geographical Equator; Iso.--Isothermal lines at about 20° C.; M--Mississippi River; N--Rio Negro; O--Orinoco River; P--Paraná River; S--São Francisco River; T--thermal or "heat" equator. Numbered areas to locate a few cities, capitals are encircled: 1. Los Angeles, Chihuahua. 2. Houston, New Orleans, Miami. 3. Havana, Port au Prince, Santo Domingo. 4. Mexico, Belize, Guatemala, San Salvador, Tegucigalpa, Managua, San José. 5. Panama, Bogotá, Caracas. 6. Georgetown, Paramaribo, Cayenne. 7. Quito, Guayaquil, Iquitos. 8. Manaos, Santarem, Belem. 9. Recife, São Salvador. 10. Lima, La Paz. 11. Brasília, São Paulo, Rio de Janeiro, Asunción. 12. Santiago, Buenos Aires, Montevideo, Valdivia, Trelew, Punta Arenas. (Stippling indicates some of highest lands.)

grown. More than this, anyone, be he the indigenous pathologist or the visiting specialist, meets there on every hand new parasites, and unprecedented problems. Here is great and stimulating challenge.

In these tropics there occur, in relatively small proportion, parts where because of elevation, proximity to edges of temperate zones, blasts of occasional frost-carrying winds, and effects of skirting cold ocean currents, temperatures that may go fairly low. However, it is in the vast major portions where it is warm, where there is no winter as understood in the classic sense, that most people live, produce their crops, and are perpetually encountering their plant disease questions.

It is well known that agriculture in the Western Hemisphere tropics, the often-called Neotropics, is difficult. Here the planting, disease control, special art of husbandry, and harvesting processes are adapted to conditions and persons. Much is still pioneer country, hard living, and domesticity of the land is toilsome. While for many the results of farming may be only enough for family subsistence, there are others who find it good monetarily, at times a veritable bonanza; but always, when crop growing settles into permanency, parasites come in, and the single most troublesome limitation to it is plant disease.

From the standpoint of world plant pathology, the work in the American tropics holds a unique place. Over the years, in plant disease study this region emerges as a realm of special concern and interest. This aids in the region, and although taking advantage of standard background and texts of great note in the temperate zones of the Old World and North America, tropical pathology is going beyond that. Researchers in our tropics are now opening new fields of attention. Scientists are realizing that what is now being determined is giving new concepts of global importance in the ever-continuing need for producing crops and controlling disease.

Scope of the Book

This book (first outlined in 1929 and re-worked four times in intervening years) deals especially with plant disease problems of the Western Hemisphere tropics. No attempt has been made to include what is known in the

African, Oriental and South European parts of the world, though in a few cases there has been references, included for their direct value to the Americas.

Much of this volume results from my reviews and studies in connection with my own research, and also from consultation relations, from teaching, from lecturing, from carrying on help in building and promulgating research institutions, and from developing in-service training programs. This book does not cover every tropical disease problem. That would be impossible in the presently fast growing and expanding tropical plant disease field. It is hoped that there may be found in this volume some interesting things, a good deal of them introductory in nature, but parts perhaps, of value now and usable in later developments.

The reader realizes that American tropical plant pathology (Neotropical phytopathology) has been in progress for a long time, but is still in a formative stage. Unfortunately, outside of our region results from the tropics are so little known and read, that only relatively scattering references are seen and very few can be found in such excellent texts as those by Heald (31) Stakman and Harrar (60), Horsfall and Dimond (34), Butler and Jones (10), Leach (44), the United States Department of Agriculture's yearbook of 1953 (61), Large (42), and Walker (67). Not many of the great names of proud and productive workers in the American tropics are commonly known in temperate zone annals, and this is inevitably going to change.

There is also growing up, most encouragingly, a series of books and manuals of and for workers in the Neotropics. Such contributions are of preeminent value, and of special importance in the tropical countries for which they were prepared. These deserve wider acknowledgment, and the care in preparation by the authors bespeaks the pride they have in their accomplishments and realization of the importance to agricultural science of their works. A number of these authors, (not including temperate zone writers translated into Spanish and Portuguese) but not all, are as follows: García Rada (25), Marchionatto (48), Fernandez Valiela (19), de Segura (57), Nowell (49), Galli et al. (22), Chaves (13), Honey (33), Echandi (17), Garcés O. (24), Alvarez García and Chupp (1), and Silveira (59); see also such special bibliographies as those of Otero and Cook (50) and Barnes and Allen (5).

One of the objects of the present text is pointing out problems that are uniquely tropical. Plant pathology is a dynamic study and there are open places, vacuums, left in the whole of this science and these are being filled by efforts in the tropics. As examples, tropical studies will lead to understanding of: algal parasitisms; more intimate relationships between parasitic phanerogams and their hosts; effects of hyperparasitism and super-hyperparasitism; the bearing on sporulation generations that are unlimited by annual winters; decrease or increase in virulence of parasites; basic characters of gemmae (not chlamydospores) in fungi and their relation to the evolution of spores; inter-relations of parasitic and saprophytic organisms under environmental conditions approximately untouched by man; the association of volcanic eruptions and hurricanes and such phenomena with the spread and development of plant disease; the physiology of luminescence in a plant parasite; problems of dust and sprays with additives applied to perpetually warm and moist plant surfaces; and, the place of the tropics as a region where plant variation is rampant as an eventual source of the most satisfactory resistance-tolerance in crop-plants.

In the whole field of phytopathology the tropics is the region which has the most to offer. In the temperate zone great fundamentals of parasite taxonomy as related to pathology have been established (some may be awaiting change). It is believed this book has in it evidence that discoveries and research in the tropics are going to remodel old concepts. In the tropics in some areas still untouched by scientists, parasites-saprophytes have been operating over eons of time; far longer than in the temperate zone. In many cases, as is seen in this volume, what is pointed out is only the beginning.

What is left to us is for others to work at and prepare; some of such a book as this will soon be out of date. This is expected, and immediately acknowledged herewith. Nonetheless, such discussion as appears below may be a possible catalyst in relation to studies in the future. Who knows!

The literature citations are of the special material which the writer has used in preparation of the chapter involved. Many other works of equal value occur, and it is not meant by leaving them out that they are of less account. However, there are space limitations, and over three times as many papers as have been noted could be listed. If all were included these might have been considered an excessive part of the volume.

I believe I discuss in this book some of the most outstanding problems in tropical plant disease study. As the problems are all related to causal organisms, I feel the chapters are best arranged in logical sequence beginning with those that are simplest and primitive and progressing through increasing complexity to those of highest specialization. It is realized that the most important thing to assure any disease control is to have a full knowledge of the infirmity and thus know when to apply remedies. A suggestive portion of knowledge available is given. However, in chapters where it appears feasible, reference is made to controls and to fundamentals upon which we based means of disease control; in the closing chapters special attention is paid to problems dealing with disease control and amelioration. At the end of the volume is a set of appendices of materials that helped me in my work in the tropics.

The first two chapters are fairly long and of background information. Included are tabular histories, briefly itemized, of some things regarding the American tropics-and-man and development of plant pathology in that region. For a pathologist's broader understanding it seems valuable that there should be included brief discussions of geography, climate, and agriculture in these tropics.

Character of the Tropics

The word "tropical" generally connotes a warm winter-free region, conveniently defined as between the imaginary map-drawn lines of the Tropic of Cancer on the north, the Tropic of Capricorn on the south, with the equator running exactly through the middle. It has been described and defined (4, 23, 28, 45, 62 and others) and is given important geobotanical consideration (15, 16, 27, 29, 35, 39, 55).

The geographical equator, considered latitude 0 (zero), is a theoretical straight line around the middle of the earth, half way between the north and the south poles. This line, as may be seen on Map A, in the Western Hemisphere starts in the east across the Gallapagos Islands to the South American continent through the northern part of Ecuador at a point near Quito. It goes on almost to that point where the countries of Ecuador, Peru and Colombia meet, and through relatively uninhabited swamps and flat lands of the southern section (Amazonas) of Colombia. It then cuts through some of the northern Brazil jungle and on eastward across river after river, all flowing south and all subsidiaries of the great Amazon. It meets the Atlantic on the east at Macapa, the city situated on the north bank of the Canal do Norte which is one of the three largest outlets of the Amazon where it reaches the sea. The two great "Tropic" lines are as follows: at 23 degrees and 27 minutes north called the Tropic of Cancer, and the other is at 23 degrees 27 minutes south and is called the Tropic of Capricorn.

It is the pure geographer who considers the broad band between these Tropics as the only true tropical region. These straight lines are accepted by physicists and geographers. However, biologists who are thinking of climate and its effects on life, realize for reasons of ecology that "the tropics" is more difficult to define and is irregular in its borders. It is marked at its center not by the geographic equator but by the thermal- or heat-equator, and on its north and south edges by latitudes where there is sufficient frost to inhibit growth of such a major biological warm country feature as the palm tree.

There has long been argument by botanists, zoologists, and demographers about the true limits of the weather-affected part of the earth that is "tropical." After long research encompassing world-wide experiences of many, Köppen (40, 41), looking at the problem from the standpoint of living organisms, decided the tropics were within those borders north and south where temperatures exceeded 68° F for all 12 months of the year. This is known, now and for considerable time, to be too restrictive. From such works as those of Sapper (54), Cook (15), Grober (28), Garbell (23), von Humboldt (36), Roger (53), Schimper (55), Walter (68), and Supan (62) it appears it best to extend this some. Cook (15) determined: "equatorial" to be 0 to 15 degrees either side of the geographical equator, "tropical" to include 15 to 23.5, "subtropical" 23.5 to 34, and the "warm temperate zone" to start at 34 and extend to 45 degrees. Very roughly speaking I would consider the American tropics as the broad belt of land having as its outer limits the approximate annual isotherms of about 22° C; with a line where it is warmest obliquely running through it, the thermal or heat equator.

This heat equator does not follow the geographic equator. In the Neotropics, it is found far northward up in Mexico and Central America and on the other hand swings over 15 degrees south down far into South America. It is not a stationary or straight line even over the surfaces of weather fronts. These weather fronts waver somewhat and are characterized by low pressures near the ground or water level, with high pressures far aloft in the sky, and are radically tempered by features of geography.

From the standpoint of the plant pathologist, from what he has learned of the characteristics of crop diseases of the region, it is convenient to use a practical biological character limiting the warm or tropical-subtropical part of the earth. This has been suggested (71), but it is here made more specific. This "tropical Line" is irregular, but encompasses the following:

The north border of the American tropics is delimited for this book as including a narrow fringe of the southernmost states in the United States. Starting at the west edge on the Pacific coast, where if there were sufficient moisture coconut palms would grow in the southern part of the state of California (this, for our purposes, is somewhat north of the city of San Diego: latitude close to 35°N) and going eastward, dipping down to the Mexican border, thence crossing Texas from El Paso to a little above Houston. The southern part of Louisiana, mostly the sugarcane growing part is tropical in nature, and the line extends across the Gulf coasts of Alabama and Georgia. This hypothetical line includes most of Florida, and separating it from the continent a narrow band along the Atlantic coast; surprisingly tempered by the Gulf Stream there is a delimitation of a strip of land recognizable as touched by the tropics and found as far north as up into a few miles of North Carolina.

The southern extremity of the American tropics is many degrees further away from the geographical equator than the northern extremity. A glance at a map of South America (Map A) shows that the great land mass tapers down on both sides, abruptly below where the Tropic of Capricorn goes across. This allows the warming effects from the Brazil Current to be felt on the Atlantic or eastern side below Rio de la Plata. On the Pacific coast the climate is more-or-less free from cold, for a distance about half way down the Chilean coastal flood plain. From the standpoint of stricter climatology the southern edge of the Neotropics might be given as starting from the west at about Valparaíso, Chile and across to Montevideo, Uruguay. However, for our purposes plant disease studies made south of this are of great value in Neotropical phytopathology; for this book the southernmost edge is not drawn to exclude any South American plant disease studies.

Some of this region is agriculturally difficult land, but some of it is probably the potentially most productive in the world. During the 1960-1970 decade the population numbered nearly 270 million (mid-America, including some of the south fringe of the United States, being some 67 million and South America, 187 million) with annual increase estimated at from 2.6 to 3.0 per cent. By the year 2000 population is expected to more than double. Increase in food production has been slow, averaging only about 1 per cent, and special efforts will have to be carried on to enlarge crop growing commensurate with the population (51, 60, 63). The difference between population growth and food production is a very serious and dangerous outlook in the Neotropics.

The Special Neotropical Region

The designation Neotropical is not infrequently used in Europe and Latin America. The word is clearly defined in standard dictionaries (e.g., 58 and 69): Neotropical includes that region belonging to, or characteristic of, the subtropical and tropical biogeographical region of the western hemisphere. The understanding is not confined to zoological matters, and the term is used in mycology and pathology; for example see the title of the mycological-phytopathological paper by Bitancourt and Jenkins (8). It is of further interest that a general biological journal appears in South America in which part of the title is "Neotropical."

There are numbers of features (these can be studied in more detail from such maps, geobotanical works, and encyclopedias as are cited in: 2, 4, 6, 11, 15, 16, 18, 20, 26, 28-30, 35, 37, 39-41, 43, 53-56, 62, 65, and 68), many outside the province of plant sciences, that add to background understanding of the American tropics. Below Panama alone, the South American continent itself is a land mass of approximately 6. 8 million square miles (see Appendix 3 at the end of this book for conversion information). To the north of this are about 1. 8 million sq. mi. in Central-Middle America and the subtropical southern edge of the United States and the Caribbean-Bahama Islands. It makes a total of probably 15 or more per cent of all the world's land surface; although some 175, 700 sq. mi. of the whole consist of mountain peaks, cold tips of south Argentina and Chile, and deserts that are agriculturally unusable.

The great region of the American tropics is, as has been suggested, not homogenous as to elevation, neither is it as to having a warm-moist climate, storms or any main ecologic feature. The region, despite impressions often gained from fiction, is neither entirely craggy mountains nor limitless swamps and lowlands. A good proportion of it (63 and 65) is undulating, gentle to steeply sloping hill land. And it has, running through its far western part, in places bare of vegetation, the Mexican-Central American-Andillean Cordillera. It is spectacular, and studded with volcanoes and snow-capped peaks. Descriptions and data about the region and the countries of it are found in encyclopedias and studies as have been cited in the above paragraph.

Geologically, the South American Andes extend up into Mexico and Central America, but in this northern part of the mountain chain there are no peaks of such height as are below the Isthmus of Panama. These, of which only a few are named, are next to the highest in the world, topped only by the Himalayas. In Argentina is the peak of Aconcagua (22, 834 ft.), fourth highest in the world, and also in that country are such other peaks as Pissis (22, 241 ft.) and Bonete (22, 546 ft.). Another great Andean peak is Chimborazo in Ecuador with a height of 20, 561 ft., and at the Argentine-Chile border are peaks of Ojos del Salado, 22, 539 ft. and Tupungato, 22, 316 ft. Many far-up slopes and high flat parts are carpeted with so-called reindeer moss, short grasses, and unusual xerophytes. Somewhat below, there are mountain shoulders that are drizzly-cold where trees still cannot exist, but further down are chilly regions where there are temperate zone-like forests and elfin woodland types of growth.

A few other features of these tropics, indicate a

little of its diversity and interest: In the world, the southernmost town is Puerto Williams, Chile, and the one at the highest altitude is Aconguilcha, 17, 500 ft., in the same country. The largest cities of South America are Río de Janeiro, Brazil, 3.2 million inhabitants; São Paulo, Brazil, 3.15 million; Buenos Aires, Argentina, 3 million; Santiago, Chile, 1.9 million; Lima, Peru, 1.3 million; and Bogotá, Colombia, 1.7 million. Havana, Cuba has 1.2 million. Of all seas the Caribbean is the largest with a surface of 1,063,340 sq. mi. and a depth of 8,172 ft. Near its edge is the abysmal Puerto Rico Trench which is 24,498 ft. deep. Of all rivers the Amazon which runs 3,900 mi. is second longest in the world, the Paraná measures 2,500 mi., the Río Madeira is 2,100 mi., and the Orinoco is 1,800 mi. In Venezuela occurs Angel Falls, a 3,212 ft. drop which is the longest in the world, and the falls of Cuquenán, 2,000 ft. is the fourth. (For comparison, Niagara drops 157 ft.)

Important Life Zones Related to Elevation

In the tropics, specialists know of different climatic effects experienced at sea level and on up to snow covered peaks. Some early information on this in the American tropics was given by von Humboldt (36) and by Bonpland and Humboldt (9). It will soon be 200 years since they made their observations and drew their conclusions. It was not long after that other explorers added to and corroborated the findings and made more detailed studies along these lines (citations to some of which are 7, 11, 16, 32, 40, 65, 66, and 68). What has been learned, and is being further investigated has important bearing on tropical climatology and ecology (e.g., 3, 37, 38, 41, 65). It has become almost a science in itself, and it is admitted as being of fundamental concern to the agricultural scientist.

A portion of quickly felt tropical climatology, is the different "life zones," and is here only briefly discussed. It is of major significance in forestry and botany, and in much understanding of occurrence, spread and severity of various plant diseases; for examples of greater interest in these phenomena see citations 46, 47, 57, 49, 22 and 19. Zones will be referred to numerous times in the following chapters. Figures 1 and 2 give approximations about the "zones." The life zone at sea level and on flood plain is very different from that on foothills, on tablelands, on

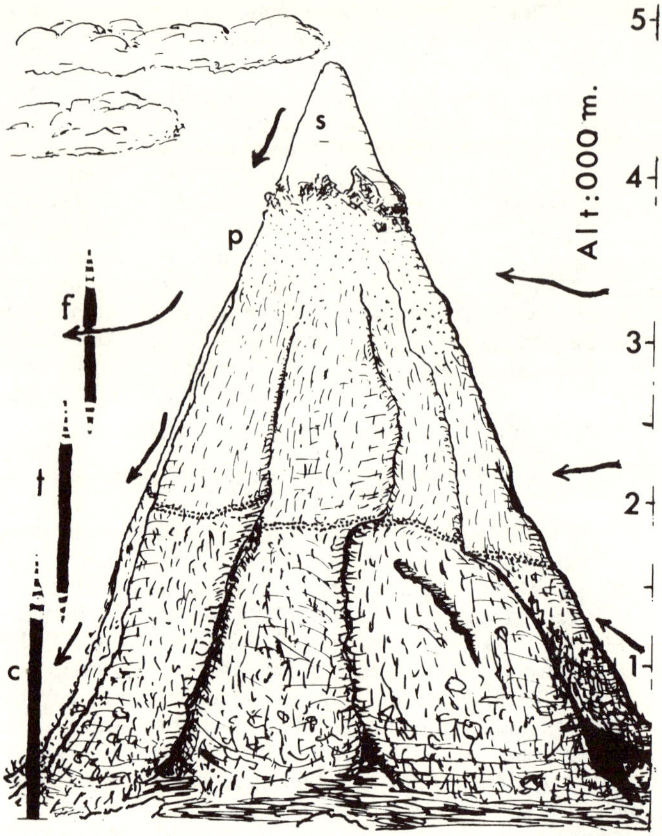

Fig. 1-1. A theoretical mountain, altitude in thousands of meters. Location in tropics at heat equator. Climates encountered*: warm, or caliente, slopes--"c"; intermediate or templada slopes--"t"; cold or fria slopes--"f"; irregular lowest location of occasional frosts indicated by double line of dots; nearly uninhabited paramo--"p"; ice and snow--"s". Directions of prevailing winds suggested by arrows. [*Limits are not exact; depend at fringes between climates on exposures, extent of rainy and wet seasons, amount of winds, and personal reactions of human observers.]

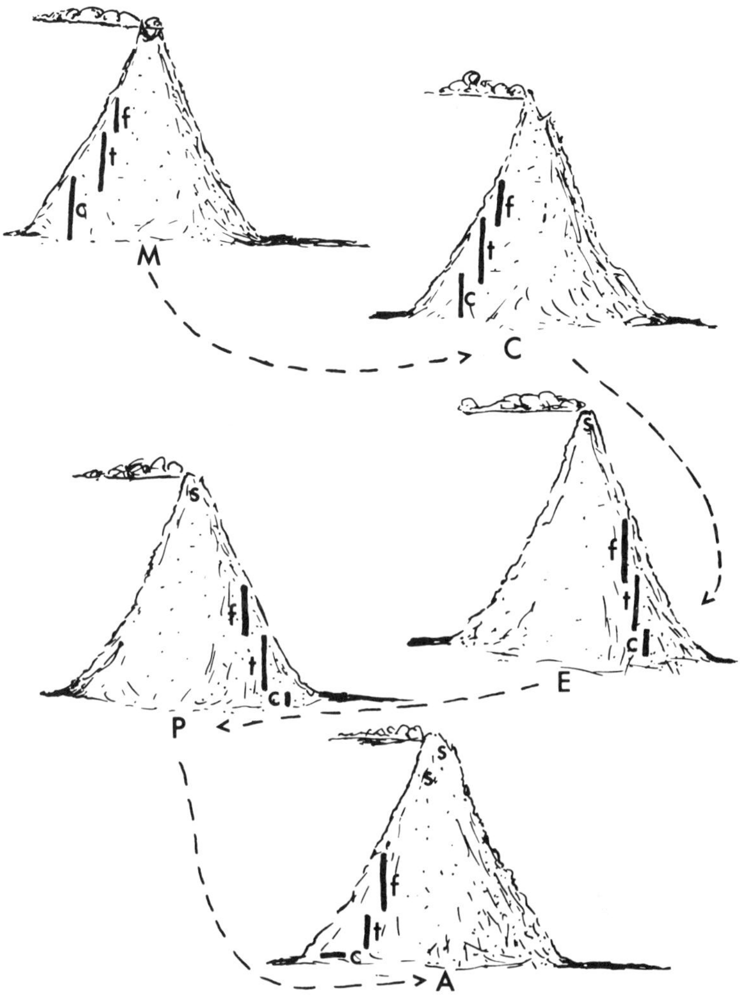

Fig. 1-2. Diagrammatic sketches of relative widths of tropical climates ("c"--caliente; "t"--templada; "f"--fria) on slopes of hypothetical mountains in: M, a south part of Mexico directly on the heat equator; C, in east Colombia some distance from the heat equator; E, in western Ecuador over twice as far from the heat equator; P, in southern Peru; and A, in Argentina at a long distance from the heat equator. The letter "s" indicates snow fields and cloud banners are blown leeward from summits.

mountain shoulders, and on high peaks (see citations especially 35, 66, 37, 36, and 65); in general writers use the Spanish words to describe the presence of the "tierra (land) caliente" which is warm or hot land, the "tierra intermedia" or medium warm land, the "tierra templada" or temperate and cool land, "tierra fria" (cold), and "tierra helada" (freezing).

Authors give various data, but general guides can be seen in them. These, given here are examples only. For average annual temperatures in three zones, von Humboldt (36) published the following: tierra caliente 30°-23°C, tierra templada 22°-17°, and tierra fria 12° to 0.8°. It was shown in a scholarly work of Walter (68) that on an average in the tropics, starting at sea level, the life zones can be looked at in four categories: tierra caliente at elevations of 0-1000 meters; tierra templada, 800-2500 m.; tierra fria, 2300-4200 m.; tierra helada, 3900-6000 m. or more; the upper line of tierra caliente is: 800 meters in Mexico, and 1000 m. in Peru and Bolivia and Colombia; the timber line is: 4600 m. in Mexico, 4200-4500 m. in Peru, and 4300 m. in Colombia; and the approximate snow line is: 4900 meters in Mexico, 4800 m. in Colombia, and 4800-5300 m. in Peru and Bolivia. Also, the effect of altitude on vegetation in equatorial regions is reported by Cook (15) as follows: 0-600 meters, palms and bananas; 600-1250 m., tree ferns and figs; 1250-1900 m., myrtles and laurels; 1900-2600 m., evergreen trees; 2600-3200 m., deciduous trees; 3200-3800 m., coniferous trees; 3800-4450 m., alpine shrubs; 4450-5050 m., alpine herbs; and at 5050 meters: permanent ice and snow.

There have been a few ecologists who have made the laudable attempt to define distinct life zones on bases of simple temperature-moisture-elevation measurements. However, in my own experience, the numerous natural and local physical variations all have to be considered, especially relative to plant disease. Not only must the research man take into account distances from the heat equator which has a marked effect in basic sea level or other temperatures, but also distance from the geographical equator (where for example, all days are about 12 hours long, but at 30°N. which is the northernmost fringe of the American tropics the months of April-August inclusive have days longer than 12 hours and 45 minutes, and at 30°S. which is near the south edge of the Neotropics in the months of the last half of October to the first days of March days are over 12 hrs.

45 mins.; also wind exposures and shieldings, usual relative moisture contents of those winds, habitual cloud cover or lack of it, so-called rain shadows, effects of deep mountain cuts, the orientation and configuration of valleys, and such, all are important. Micro-climates are markedly changed in small areas which are in low spots, close by outcrops of rocks, or are adjacent to jungle, sparse bush, savannah or sands.

In tropical highlands, plantations and farms may perch precariously along sides of deep mountain gulches and on steep slopes that temperate zone agriculturists would view as impossible. There are extensive parts where terrace fields are made by indigenes who seem to prefer them to any they might have on broader, more level places far below; in other places both Indians and peasants farm high quality soils in lands spread on broad, elevated table formations, where the climate is frost-free. In lowlands, bordered with swamps are great areas covered with, depending upon the point of view, the famous (52) or infamous, jungle-and-rainforest (which is no peaceful paradise).

In relation to the future, mention should be made here of the Pampa, or pampas, of Argentina. This is more than a minor, circumscribed, inconsiderable bit of geography. It is about 240 thousand square miles stretched across the central part of that country. The land is productive, under-used, mostly a level plain sparsely grazed, and much of it destined to be as successful agriculturally as places like the fabulous Central Plains of the United States of America, and better than the famed Nile Valley, or the great wheat lands of Canada. The wild pampa is changing, being tested, and becoming increasingly settled for agriculture. Also there should not be forgotten the fertility potentials in little discussed rather broad expanses of tropical deserts that await irrigation, in Mexico, and in Central and South America; not all, but some are even so located that they have seasonal cycles that are not excessively hot. It is of interest that in areas, for example the Atacama of Chile, there has never been even a hint of rain going far back into folk memory, and the weather station records of Calama there are utterly free of any sign of precipitation. It and a few other such parts of South America are more desiccated than the torrid sands of Egypt.

Crops Grown in Our Tropics

There are a large number of economically important crops produced in the Neotropics. The variations in growing conditions met in American tropics are so great that except for some plants that require especially long day lengths, or definite periods of dormancy it appears almost any crop can or could be grown in this vast region. There continues to be the greatest need for plantation crops, since economically these tropics are first and foremost agricultural; the most important, widespread resource. It is by plantation cultures of products specially adapted to the tropics that all nations of the Caribbean archipelagos, of Central America, and of South America have become stabilized and have kept going.

In some situations, what may have been done in the name of developing profitable plantations now seems harsh with regards to either nature or people. Advances of military men, occurring long before agricultural determination resulted in the exploitation of yellow gold, but this proved limited. The army men moved on. Agricultural possibilities were believed promising but indigenous people under foreigners were not a tractable labor source. Nevertheless the first European settlers with almost superhuman strength nestled into groups of farms. Often they brought with them or imported their own labor, cleared land, burned it over, and planted tropical crops. Those who succeeded were able to remain and adapt to, and adopt, the land which by hard work they won from the wild.

A successful tropical plantation crop--an example such as coffee (70)--is of somewhat special nature. It must flourish under the conditions of the region, and produce well with the labor available. Often produce is of a kind that can be shipped long distances. If the harvest is more perishable, such as fruit, it must be handled under a sophisticated system of rapid gathering and shipping; if a dried bark, spice, grain, nut, or other stable material it can be taken from the field, processed and shipped in a slower, more simple fashion.

There are a number of permanent to semi-permanent plantation crops that come to the mind of workers in the Neotropics: coffees (*Coffea arabica*, *C. canephora*, *C. excelsa*); sugarcane (*Saccharum officianarum*); banana (*Musa* spp.); plantain (*M.* spp.); manila hemp or abacá (*M. textilis*); pineapple (*Ananas comosus*); the cacao or chocolate tree (Theobroma cacao); coconut (Cocos nucifera); orange (Citrus sinensis); African oilpalm (Elaeis guineensis); Pará rubber (Hevea brasiliensis); sisal (Agave sisalana); henequen (Agave fourcroydes); quinine (Cinchona ledgeriana); tung (Aleurites fordii); vanilla (Vanilla planifolia); black pepper (Piper nigrum); various spice trees; eucalyptus (Eucalyptus spp.); Pyrethrum (Chrysanthemum cinerariaefolium); maté (Ilex paraguayensis), and of course there are others.

There is an article by Fosberg (21) listing many crop plants of the American tropics and all are given concise treatment: making specific mention of 31 native food plants, 19 introduced food plants, 20 spices and flavors and perfumes, 20 that are beverage, narcotic, stimulant and pharmaceutical plants, 31 sources of insecticides, oils and waxes, 41 fiber, cork and pulp plants, 22 timber and fuel sources protected and collected for special purposes, 11 dyestuffs and tan materials, and 17 that yield latex, gum and resins. A total of 216 species are in his condensed discussion and there are over a hundred more used and valuable species. There are many others that are important as uncultivated plants, employed by indigenous people and peasants.

All these crops and useful plants have their diseases. No species is without, and as they gain in importance troubles increase for good reasons: The numbers of a valued plant multiply through cultivation; as the population increases there is more of a stable source of the host upon which parasites may live and reproduce; additional amounts of infective spores develop and are released to infect the host. Disease plants are moved by man, and with them he carries crop sicknesses to new regions. Any new region has its own, different plant diseases, which, if of the right character, become added enemies of the introduced crop.

It is to be noted that there are exotic crops grown successfully in the tropics that were at one time brought out of, and adapted from the temperate zone. Tropical resident agriculturists, who on moving to the tropics, took favorite plants to a warmer locale, and by selecting and adapting methods, there grew them. The seeds and transplants of these also had in them temperate zone diseases. More parasitic organisms were introduced into the already rich parasite flora of the tropics. At the same time, the temperate zone parasites on their old temperate zone hosts quickly escaped into the tropics onto new, tropical, hosts.

On these appeared, thus, new diseases and in a relatively short time after arrival they had to be accepted as tropical disease problems.

Certain Historical Dates of Background Interest

There are main points of general interest in the annals of the American tropics, and are found in historical and encyclopedic reports (2, 12, 14, 18, 30, 43, 50, 56, 65). A selection of materials from them aid in understanding developments in the region of the Neotropics. For matters of more specific concern in plant pathology the reader will refer to the next chapter with the list there.

Dates presented directly below suggest the flow of events. It is now evident that indigenes and people of the republics knew agriculture was the most important activity in our tropics. They fully realize the increasing population will require, in the near future, more and more modernization and specialization in crop-plant production.

Table 1. SELECTED DATES OF GENERAL HISTORICAL INTEREST in the American tropics, obtained from various sources.

Date		Historical occurrence
2500 B. C. (approx.)		Early Inca remains in the Andes
1000	"	Inca weaving, pottery, culture of maize and manioc
300 A. D.	"	Start of Inca "classic period"
300-900	"	Flowering of the Mayan culture, Mexican-Guatemalan
1000	"	Close of Inca "classic Period"
1200	"	Rule of Manco Cápac, "the Great Inca"
1325		Founding of Mexico City by Aztecs
1440-69		Rule of Huehue Moctezuma (Montezuma I) first Aztec king

1492-98	Columbus from Spain discovers West Indies and South America
1494	Pope Alexander VI in Treaty of Tordesillas divides New World between Spain and Portugal
1495-7	Banana and sugarcane introduced to Neotropics
1499-1502	Amerigo Vespucci from Spain and Portugal explored and defined a "New World"
1500	Cabral from Portugal lands on and claims Brazil for Portugal
1500-01	De Bastidas explores Colombian coastline
1502-20	Rule of Moctezuma (Montezuma II) last king of Aztecs
1504	Cortés settles in Hispañola
1507	New World named America after Vespucci, by Waldseemüller of Lorraine
1508-09	Ponce de León conquers San Juan Bautista island (Puerto Rico) for Spain
1513	Ponce de León discovers Florida; Balboa in Panama sights Pacific Ocean
1517-21	Spanish expeditions from Cuba explore Mexico and lower Mississippi Valley
1518	Diaz de Solis travels region of Argentina
1520	Planting of sugarcane in Brazil and Argentina

1521	Cortés conquers Mexico City
1524	Alvarado conquers Guatemalan region
1526	Death of Huayana Cápac "the Last Great Inca"
1544	Orellana crosses Andes, descends Amazon to Ocean
1547	Death of Cortés
1551	University of San Marcos founded in Lima, Peru
1573	University of Bogotá founded in Bogotá, Colombia
1584	First western hemisphere printing press in Lima, Peru
1585	First population count in Brazil
1613	University of Córdoba founded in Córdoba, Argentina
1621	Dutch West Indies Company formed
1629	Spain's Mexico Fleet destroyed
1630	Silver mining by Spaniards started in Peru
1664	French pirates establish Port-du-Paix in Hispañola
1676	University of San Carlos de Borromeo founded in Antigua, Guatemala
1693	Gold discovered in Brazil, followed by gold rush
1715	Coffee trees brought by French to American tropics

1728	Diamonds discovered in Brazil
1746	Earthquake destroys Lima
1763	England exchanges city of Havana with Spain for Florida
1765	England's New England colonists manufacture chocolate
1783	Birth of Simón Bolívar
1787	University Centrál founded in Quito, Ecuador
1791	Slave insurrection in Hispañola
1810-28	Era of independence: Argentina 1816, Bolivia 1825, Brazil 1822 (-29), Chile 1818, Colombia 1821, Ecuador 1822, Guatemala 1821 (subsequently El Salvador, Nicaragua, Honduras, Costa Rica separated from it), Mexico 1810, Paraguay 1814, Peru 1821, Uruguay 1828, Venezuela 1813(-28).
1812	Earthquake destroys Caracas, Venezuela
1817-23	O'Higgins rule in Chile
1821	Founding University of Buenos Aires in Buenoe Aires
1823	Monroe Doctrine stated
1826	Bolívar's Panama conference of American nations
1829	Bolívar's proposal of League of Hispanic-American States
1830	Death of Simón Bolívar
1840-89	Emperor Dom Pedro II of Brazil

1850	Death of San Martín
1861	Juárez enters Mexico City
1867	Río Amazon opened to world trade
1867	Maximilian of Mexico executed
1869	Founding of newspaper La Prensa in Buenos Aires
1872	Death of Juárez
1880	Oil production started in Mexico
1889-90	First International Conference of American States
1889	Brazil becomes a republic
1910	International navigation of Río de la Plata
1914	Panama canal opened
1914	Discovery of oil in Maracaibo, Venezuela
1915	Famous model labor code of Mexico enacted
1916	Adoption of civil law code in Brazil
1922	Founding of National Historical Museum in Brazil
1928	Meeting of Pan American Conference
1936-38-40	Conference of American States on collective defense
1945-47	Chapultapec and Río de Janeiro treaties of Inter-American Reciprocal Assistance in case of outside aggression

1948	Organization of American States treaty signed in Bogotá, Colombia
1949	Iron ore mining started in Venezuela
1960	Model land reform bill passed in Venezuela
1961	Alliance for Progress, signed at Punta del Este, Uruguay

The Phrase "Tropical Plant Pathology"

The words "tropical plant pathology" appear often in this book, and it seems there is a good reason for discussing this usage briefly. When "Neotropical phytopathology" appears it is recognized as referring to the western hemisphere part of tropical plant pathology. These words encompass a part of the whole field of plant pathology (phytopathology). It is of course that special sector of applied botany which has been most fully developed in the temperate zone.

There is much precedent and well recognized usefulness in setting apart tropically centered scientific developments. For example there is tropical herpetology, tropical entomology, and tropical botany. In the world-wide study of soils there is the special study of them in the warm, circum-equator regions and these are called tropical soils. In weather and climate study there is tropical meterology, just as in human disease problems there is what is known as tropical medicine. Thus there is, in the over-all study and control of plant diseases, a specialization that we speak of as tropical plant pathology.

All plant pathology is of international concern, and it has no political boundaries. As it was developed it was affected by using great biological concepts from the germinal thinking of numerous great pioneer workers in science and practically all were of different nationalities.

The realization that plant pathology is much needed and has to be studied as a profession separate from botany or human medicine or horticulture, came first in the temperate zone. It developed because of catastrophic crop failures and famines, and the belief that these had to be

Table 2. BACKGROUND FACTS[a], in brief about tropical American countries.

Country	Area in Sq. km. (in thousands)[b]	Certain important crops	Remarks on climate
Argentina	2, 794	Cereals, citrus, other fruits, sugarcane, maize, potatoes, alfalfa, grazing grasses	In Tucumán; Chaco subtropical; sunny, mild in Pampa; Patagonia cold and windy
Bolivia	1, 099	Potatoes, grains, alfalfa, cotton, sugarcane, cacao	Andean altiplano dry and cold, eastern lowlands warm, in mountain valleys ("yungas") cool but not freezing
Brazil	8, 514	Coffee (largest producer in world), forest products, rubber, nuts, cacao, cotton, sugarcane, tobacco, jute, manioc, rice, bananas, oranges, beans	Amazon basin warm and wet, being hot in the north, highlands delightfully subtropical
Chile	742	Grazing and soiling crops, potatoes, grains, sunflowers, fruits, winegrapes	Deserts in north; central more moist and dry lowlands irrigated, south chill forests and grazing

Colombia	1, 138	Coffee, sugarcane, bananas, cacao, maize, grains, subsistence crops	Lowlands warm, steamy; Andes ridges cold, dry; tablelands, valleys cool with 2 rainy, 2 dry seasons annually
Costa Rica	51	Coffee, rice, vegetables, cacao, bananas, rubber	Meseta Central subtropical, Caribbean lowlands warm-wet, Pacific slopes dry-warm
Cuba	115	Sugarcane, maize, beans, rice, fruits	Warm, humid; hurricane effects frequent
Ecuador	271[c]	Bananas, balsa, cacao, coffee, rice, potatoes, forest products	Has one of world's wettest areas, Humboldt current tempers Pacific coast, Andes highlands cool, drier
El Salvador	21	Coffee, rice, beans, maize, grazing and hard fiber crops	Warm; Pacific slopes dry; mountain sides, high valleys cooler
Guadeloupe	1. 5	Bananas, subsistence crops, sugarcane	Insular climate, warm tempered by trade winds

Guatemala	109	Coffee, fruits, rubber, bananas, cinchona, maize, beans, potatoes	Pacific side of volcanoes dry, Caribbean exposure wet, mountains cool
Guiana, British Guiana, Dutch[d] Guiana, French	215 143 91	Sugarcane, forest products, rice, vegetables, citrus, hardwood logs, grazing crops	Low lying countries with remarkably uniform high temperatures, 2 long, rainy seasons per year broken with short dry seasons
Honduras	112	Bananas, tobacco, sugarcane, maize, beans, coffee, oilpalm, grazing grasses	Warm, moderately dry interior lowlands and in cooler highlands
Honduras, British	21	Forest products, beans, maize, cabinet woods, subsistence crops	Warm to hot in lowlands, moist with long rainy season
Jamaica	12	Coffee, bananas, spices, coconuts, hard fibers	Insular climate that is warm, but cool in mountains
Martinique	1	Coconuts, bananas, vegetables	Warm climate, tempered by trade winds
Mexico	1,969	Cereals, maize, beans, fruits, hard fibers, cotton, tobacco,	Continental effects felt from north, part of it

		vanilla, vegetables, grazing	dry, Yucatan desert-like, cool central table lands without excess rain, warm and some parts moist at sea level
Nicaragua	187	Tobacco, coffee, sugarcane, cotton, maize, timber	Low parts warm cooler in mountains, 6 months wet--6 months dry
Panama	74	Coffee, tobacco, rice, oil-palm, bananas, coconuts	Hot and warm lowlands, mountain slopes cooler and more moist
Paraguay	407	Cotton, tobacco, grains, oranges, yerba maté, quebracho (tan bark), oil of petigrain	Dry season June to September, hot and humid December through March
Peru	1, 285	Sugarcane, tobacco, tea, potatoes, cotton, flax, rubber, coffee, peanuts, forest products, unique native Indian crops	Andean part sunny days with chilly nights, in Selva (tropical jungle) warm and wet all year Pacific coast dry with some desert cooled by periods of hazy sky
Puerto Rico	9	Coffee, sugarcane, vanilla,	Warm to hot, tempered

		vegetables, citrus, ñame(yam), pineapples	by trades winds, occasional hurricane effects
Republica Dominicana[e]	48	Oranges, warm-country fruit crops, grazing crops, subsistence crops, hard fibers	Warm to hot, tempered by trades winds, occasional hurricane effects
Trinidad and Tobago	9	Sugarcane, cacao, henequen, rice, jute, bananas	Hot and wet, remarkably even high temperature
Uruguay	187	Grazing crops, cereals, flax, rice, maize, fruits	Subtropical, frosts occasional at high elevations, rains moderate all year
Venezuela	912	Sugarcane, maize, cotton, beans, cacao, coffee, peanuts, sesame	North coast at sea level hot and humid, dryer and chilly in Andes, mountains valleys and table lands subtropical

(a) Authorities are from specialists and encyclopedias (2, 18, 20, 43, 56, 63-65) and my own observations.
(b) Areas in square kilometers are approximate. For comparison's sake: area of Continental United States of America ia less than Brazil.
(c) Area includes Gallapagos Islands.
(d) Also known as Surinam.
(e) Often given as Dominican Republic; area and information includes Haiti.

reduced or averted by the use of scientific methods.

In tropical plant pathology in the Western Hemisphere, where the population of scientists is thinner than in Europe or North America, the work has been difficult. It has to cover many more kinds of ecology, of disease organisms and susceptible crop-plants than ever before. Far more than are commonly encountered in the temperate zone.

In addition, tropical plant pathologists, often with only moderate to minimum means of support, have training that fits them to aid crop-disease-harrassed growers with the greatest of ingenuity. These specialists (at this writing, dangerously few in the American tropics) carry on scientific work that is arduous, sometimes dirty but always interesting, with a notable dedication and confidence based on years of detailed training in a vigorous and complex profession. Their advisements on plant disease controls to eliminate famine and crop failures have to be economical, and within what is available to the grower in the particular area where he is located; and they must be easily carried out. It is not always the case, but the plant pathologist in the tropics deserves from the public at least as much appreciation as if he were a dentist, a lawyer, or a human doctor.

Tropical plant pathology is a specialization in the larger sense of general phytopathology. It has its limitations, for it deals mostly in warm-country plant disease problems. It is granted that there is no sharp dividing line, that can be drawn upon a biological basis, between plant pathology of the temperate zone and the tropics. There is, however, a considerable difference in all that needs to be known to be a competent pathologist in either of the main portions of the two great regions. It is likewise a great convenience to be able to indicate semantically a recognizable, worldwide, professional difference in the two categories.

Literature Cited

1. Alvarez García, L. A. and C. Chupp. 1957. Enfermedades de las hortalizas en Puerto Rico y medios de combatirlas. Puerto Rico Univ. Agr. Exp. Sta., Dep. Plant Pathology, Typescript 164 p.

2. Americana Corporation. 1969. The Encyclopedia Americana, International Edition. New York. 30 vols.

3. Arias, A. C. 1942. Estudios climatológicos: áreas geográficas de dispersión Parthenium argentatum, Hevea brasiliensis, Castilla elastica. Secc. Agr. Fom. (Mexico). 112 p.

4. Aubert de la Riie, E., F. Bourlière, & J.-P. Harroy. 1957. The tropics. Knopf, New York. 208p.

5. Barnes, H. V., and J. M. Allen. 1951. A bibliography of plant pathology in the tropics and in Latin America. U. S. Dep. Agr., Bibl. Bulletin 14. 78 p.

6. Bartholomew, J. 1950. The advanced atlas of modern geography. McGraw-Hill, New York. 108 p.

7. Beard, J. S. 1949. The natural vegetation of the Windward & Leeward Islands. Clarendon Press, Oxford. 192 p.

8. Bitancourt, A. A. and A. E. Jenkins. 1950. Estudos sôbre as Miriangiales-II. Vinte novas espécies de Elimoaceae neotropicais. Arq. Inst. Biologico, 20:1-28.

9. Bonpland, A. and A. Humboldt. 1825. Nova genera et species plantarum quas in peregrinaciones ad plagam aequinoctialem orbis novi collegerunt, descripserunt, partim adumbraverunt, Amat. Bonpland et Alex. Humboldt excludis autographis Amati Bonpland in ordinem digessit Carol Sigismund Kunth accedunt tabulae aeri incisae, et Alexandri de Humboldt notationes ad geographiam plantarum spectantes. Lutecia Parisorum 7. (Rare book; excerpts consulted, not seen as a whole. F. L. W.)

10. Butler, E. J., and S. J. Jones. 1949. Plant pathology. Macmillan, London. 979 p.

11. Cárdenas, M. 1945. Aspecto general de la vegetación de Bolivia. 312-313, <u>in</u> F. Verdoorn (ed.), Plants and plant science in Latin America, Waltham, Massachusetts. (<u>Cf.</u> citation 65.)

12. Chardón, C. E. 1949. Los Naturalistas en la America Latina. Torno 1. Los Siglos XVI, XVII y XVIII, Alejandro Humboldt, Carlos Darwin, La Española, Cuba y Puerto Rico. Ciudad Trujillo, Rep. Dominicana, 386 p., 27 pl.

13. Chaves, G. M. 1966. Fitopatologia Geral. Univ. Rural, Est. Minas Gerais, Brasil, (mimeographed). 148 p.

14. Colon, E. D. 1930. Datos sobre la agricultura de Puerto Rico antes de 1898. Cantero, Fernandez & Co., San Juan, Puerto Rico. 302 p.

15. Cook, R. 1947. The geography of flowering plants. Longmans, Green, London. 403 p.

16. Dugand G., A. 1934. The transition forests of Atlántico, Colombia. Tropical Woods 40:1-14.

17. Echandi, E. 1967. Manual de Laboratorio para Fitopatología General. Herrero Hermanos, Sucs., Mexico, 59 p.

18. Enciclopedia Universal. Eds. and Contribs. 1930. Enciclopedia Universal Ilustrada, Europeo-Americana. Espasa-Calpe, Madrid. 70 vols.

19. Fernández Valiela, M. V. 1952. Introducción a la Fitopatología. Gadola, Buenos Aires. 872 p.

20. Foreign Agricultural Service Committee of Associate Authors (W. Anderson, G. Burmeister, N. P. Guidry, R. G. Hainsworth, A. G. Kevorkian, C. M. Purves, R. P. Schottroff, and K. H. Wylie). 1958. Agricultural Geography of Latin America. For. Agr. Serv., U. S. Dep. Agr. Miscl. Publ. 743. 96 p. and 85 maps.

21. Fosberg, F. R. 1945. Principal economic plants of tropical America. Pp. 18-35, *in* F. Verdoorn (ed.), Plants and Plant Science in Latin America. Chronica Botanica Co., Waltham, Massachusetts. (*Cf.* Citation 65.)

22. Galli, F., H. Tokeshi, P. de C. Torres de Carvalho, E. Balmer, H. Kimati, C. O. Nogueira Cardoso, and C. L. Salgado. 1968. Manual de fitopatologia.

Doenças das Plantas e seu Controle. Ceres, São Paulo. 640 p.

23. Garbell, M. A. 1947. Tropical and equatorial meterology. Pitman Publ., New York. 237 p.

24. Garcés O., C. 1954. Control de las Enfermedades de las Plantas. Editorial Bedout, Medellin. 381 p.

25. García Rada, G. 1947. Fitopatología agrícola del Peru. Est. Exp. Agr. La Molina, Peru. 423 p.

26. Good, R. A. 1947. The geography of the flowering plants. London. 403 p.

27. Greenman, J. M. 1925. The age-and-area hypothesis with special reference to the flora of tropical America. Amer. J. Bot. 12:189-193.

28. Grober, J. 1954. Triennium bioclimaticum Tropicale. Forsch. u. Fortsch. 28(6):176-178.

29. Guppy, H. B. 1917. Plants, seeds, and currents in the West Indies and Azores. Williams & Norgate, London. 531 p.

30. Hanson, E. P. (ed.). 1950. The New World guides to the Latin American republics, 3d ed. Duell, Sloan and Pearce, New York. 3 vols.

31. Heald, F. D. 1926. Manual of plant diseases. McGraw-Hill, New York. 891 p.

32. Holdridge, L. R. 1941. Determination of world plant formations from simple climatic data. Science. 105:367-368.

33. Honey, E. E. 1930. Phytopatologia Geral. Esc. Superior Agr. Luis Queiroz. (Mimeographed) 650 p.

34. Horsfall, J. G. and A. E. Dimond. 1959. Plant pathology; An advanced treatise. Academic Press, New York. 3 vols.

35. Hopkins, A. D. 1938. Bioclimatics: a science of life and climate relations. U. S. Dept. Agr. Miscl. Publ. 280. 188 p.

36. Von Humboldt, A. 1817. De distributione geographica plantarum secundum coeli temperiem et altitudinem montium, prolegomena. Lutetiae Parisorum. 249 p.

37. (James, P. E.) 1963. Physical geography, geology and structure of Ecuador. Encyc. Brittanica. 7: 947-950.

38. Johnson, J. 1947. Water-congestion and fungus parasitism. Phytopathology. 37:403-417.

39. Klages, K. H. W. 1942. Ecological crop geography. Macmillan, New York. 615 p.

40. Köppen, W. 1900. Versuch einer Klassifikationder Klimate vorzugsweite nach ihren Beziehungen zur Pflanzenwelt. Geogr. Zeitschr. 6:593-611; 657-679.

41. _______. 1918. Klassifikation der Klimate nach Temperatur, Niederschlag und jahreslauf. Petermans Mitt. 64. 193 p.

42. Large, E. C. 1940. The advance of the fungi. Henry Holt, New York. 488 p.

43. Larousse . (Eds. and Contribs.) 1960. Grand Larousse Encyclopédique. Libraire Paris, 10 vols.

44. Leach, J. G. 1940. Insect transmission of plant diseases. McGraw-Hill, New York. 615 p.

45. León, J. 1950. (Personal communications.)

46. Marchionatto, J. B. 1932. Les enfermedades parasitarias de las plantas y sus relaciones con el medio ambients. Agronomía. 25:463-486.

47. _______. 1934. Influencia del medio sobre la predisposición de las plantas a las enfermedades. Rev. Argentina Agr. 1:111-114.

48. _______. 1948. Tratado de fitopatología. Editorial Sudamericana, Buenos Aires. 537 p.

49. Nowell, W. 1923. Diseases of crop-plants in the Lesser Antilles. West India Committee, London. 383 p.

50. Otero, J. I. and M. T. Cool. 1937. A bibliography of mycology and phytopathology of Central and South America, Mexico and the West Indies. Puerto Rico Univ. J. Agr. 21:249-486.

51. Pendleton, R. L. 1955. El Papel de los suelos tropicales en la alimentación del mundo. Turrialba. 5:6-16.

52. Richards, P. W. 1964. The tropical rain forest; An ecological study. Cambridge Univ. Press, Cambridge, England. 450 p.

53. Roger, L. 1951. Phytopathologie des pays chauds. Encyc. Mycologique. P. Lechevalier, Paris. 1126 p.

54. Sapper, K. 1902. Mittleamerikanische Reisen und Studien aus den Jahren 1888 bis 1900. Braunschweig, Germany. 419 p.

55. Schimper, A. F. W. (trans. W. R. Fisher). 1943. Plant geography upon a physiological basis. Oxford Univ. Press, Oxford, England. 839 p.

56. Schweizer Lexikon. 1945-1948. Encyclios-Verlag, Zürich. 7 vols.

57. de Segura, C. B. 1965. Enfermedades de cultivos tropicales y subtropicales. Editorial Jurídica, Lima. 439 p.

58. Shorter Oxford English Dictionary, The. 1955. Clarendon Press, Oxford. 2515 p.

59. Silveira, V. D. 1949. Elementos de fitopatologia. Agronomía. 8:1-42, 87-130, 189-247, 281-328 (pub. in fascicles, chap. 1 to 9 seen).

60. Stakman, E. C. and J. G. Harrar. 1957. Principles of plant pathology. Ronald Press, New York. 581 p.

61. Stefferud, A. (ed.) 1953. Plant diseases; The yearbook of agriculture 1953. U. S. Dep. Agr., Washington, D. C. 940 p.

62. Supan, A. 1903. Grundzuge der physischen Erdkunde. Von Veit, Leipzig. 852 p.

63. Turk, K. L. and L. V. Crowther (eds.). 1967. Rural development in tropical Latin America. Cornell Univ. Press, New York. 446 p.

64. United States Dep. Agr. Economic Research Service. 1967. Indices of agricultural production for 20 Latin American countries (plus Guyana, Jamaica, and Trinidad and Tobago). West. Hem. Branch For. Reg. Anal. Div., publ. 44, 49 p.

65. Verdoorn, F. (ed. 85 cooperators). 1945. Plants and plant science in Latin America. Chronica Botanica Co., Waltham, Mass. 381 p.

66. Wadsworth, F. H. 1960. Records of forest plantation growth in Mexico, the West Indies, and Central and South America. Caribbean Forester 21 (suppl.)

67. Walker, J. C. 1969. 3d ed. Plant pathology. McGraw-Hill, New York. 819 p.

68. Walter, H. 1964. Die Vegetation der Erde in Okophysiologischer Betrachtung. Bd. I: Die tropischen und subtropischen Zonen. Fischer, Jena, East Germany. 592 p.

69. Webster's Third New International Dictionary of the English Language, Unabridged. 1961. Merriam, Springfield, Mass. 2662 p.

70. Wellman, F. L. 1961. Coffee botany, cultivation, and utilization. Leonard Hill, London; and Interscience Publishers, New York. 488 p.

71. ________. 1969. More diseases in the tropics than in the temperate zone. Ceiba. 14:1-12.

Chapter II

THE TASK OF TROPICAL PATHOLOGY

Summary

About half or more of harvests from agriculture in our tropics is lost due to disease in field or afterwards; food has to be imported; there is local starvation and death because of reduced crop yields from disease. Fungi are known from early geologic ages to have arisen under warm tropical-like conditions. In parts of the earth all organisms were destroyed by deep freeze--excepting in the great circum-equatorial region; tropics is now where most fungi and remaining representatives of early organisms live. Aside from most temperate zone kinds of typical diseases, the tropics is specially notable for abundance of plant diseases. There is "nudging about" and over-running of some organisms by others. Altitudes are important in disease occurrence. There are about five levels from lowest and warmest through medium situations up to highest and coldest. Pathologists become aware there are climax-conditions of disease and typical disease climates. History of tropical pathology starts from time of the American indigenes. Determination and research were begun by great nomenclatural scientists, but as early as 1727 the first catastrophic plant disease was reported. In the next two centuries, at first developing slowly, then with more assurance, tropical plant pathology became what it is now, a modern, greatly rewarding profession.

Introduction

Through the use of the phrase "tropical plant pathology" the audience is directed to a special aspect of plant disease study. In some ways the information about it is slightly diffuse, but nevertheless it is developing around an increasing body of published work in several languages. In the American tropics it is mostly, but not entirely, in Spanish, Portuguese, and English. It should be repeated that there is no strict demarcation line where the temperate zone meets the tropical region. An approximation has been made of how much the American tropics (Neotropics) comprise (see Map A in Chapter I).

In this region, plant diseases include a wide range of types of parasites and physiological disturbances. Although fungus distribution is worldwide, such as Bisby (18), Ainsworth and Bisby (1) and many others have amply pointed out, there are a large number of parasitic fungi that are limited to tropical conditions. In addition many organisms found in both the tropics as well as under temperate conditions, are by far the most active and parasitically important in the tropics.

As pathology progresses it is found that there are special husbandry relationships that affect disease occurrence as well as disease control. In the tropics may be found areas in which conditions are quite primitive, in some instances agriculture is carried on in about the same manner as was done by stone age man. Popenoe (84) has listed seven types of the "pre-industrial" means of growing crops in the tropics. These are all the way from small, haphazard home gardens to extensive shifting plots in jungle and irrigated permanent terraces.

Impact of Tropical Crop Disease

Crop growing is not easy in the tropics. Often farm labor saving devices are unavailable or are far from perfected for the conditions. Agriculture requires skill, knowledge and energy and when tropical harvests that are gathered are very little or are failures because of disease, the result is more distressing than equal reduction in the temperate zone. It means much to lose 10 or 20 per cent of the bean crop where the work involved has been done using oxen to plow, sowing by hand, weeding with machete, and harvesting depending on help from the grower's wife and small children. In comparison where mechanization does practically everything, and financing can be secured from state or federal assistance, a 10 to 20 per cent loss may be felt, but it has much less emotional and exhausting impact.

This has been a natural concern of mine (116), and I have obtained through personal contacts some over-all, very conservative percentages of crop loss through disease, figures reasonably accurate under the conditions. These are expressed as follows, in percentages: sugarcane, 20-30; bananas, 20-29; wheat, 10-12; oranges, 28-34; maize, 20-23; dry beans, 20-35; seed sorghum, 25-35; rice, 35-50; tobacco, 17-25; white potatoes, 20-30; tomatoes, 35-50; carnations, 15; pasture lands, 17-30; coffee, 20-27; cacao, 27-35; roses, 60-80; and green cut hay, 15-20. These figures are not exact as there is no crop in which disease losses are reported from every country in which it is grown. The fact is, crop disease losses are nothing to be proud of and are not given much publicity, especially in the tropics.

However, reviewing loss estimates published in the last 25 years by 72 pathologists on 18 crops has led me to conclude that more is untaken than my estimates indicate. This became more apparent on discussing it with other pathologists and counting not only field losses but also those of aboriginal and peasant farmers living along rivers who wait for the long, low water seasons and grow their crops in river banks and sandbars, and, last, in some parts the losses of growers who plant rice in wetlands. In tropical America the indigenes and illiterate peasants still make up a large proportion of the agricultural population. However, others, such as the planters who grow the money-making crops in their countries, gain the most attention. There are also the smaller growers, apart from the primitives but not in the industrialized planter class who produce specialized and lesser crops sometimes under intensified types of agriculture.

I am aware than under many tropical conditions, in large part little or no disease controls are attempted on locally used food and fiber crops. There is a good deal done on plantation cash crops, and in more recent years considerable attention is starting to be given to studies leading to disease control of maize, beans, rice, potatoes, and cucurbits. Up to the present, where spraying and soil treatments are applied the costs of these controls are some 4 to 6 times more than in the temperate zone. The total disease loss, per annum, especially in the Latin American tropics, must cost that vast and developing, but not rich, region at least over $4 billion (in U. S. currency). This is more a guess than a close estimate, such as was made (37) for the United States, but I consider it probably fair.

Billions of units of any currency of Latin America seems astronomical in numbers. However, to make it more understandable it is well within the truth to say that if even half of the disease losses could be saved in the American tropics it would be enough to carry on all costs of the Alliance For Progress and implementation of all of the official dealings of the Organization of American States.

But giving percentages and money losses is not enough for presenting the true impact of plant disease in tropical America. So long as populations of rural dwellers in the moist, warm tropics remain as widely spread jungle inhabitants, their plant diseases are not publicized and they subsist by agricultural activity that is both hard and dangerous, in unsettled regions. Crop diseases appear soon in their clearings, inevitably, and more trees must be cut and burned. This shifting culture is exhausting work, hard on life. With skill and luck a clearing is good for about three years--but it is often good for only one or two. Under some good conditions, where diseases are less severe, agriculture is more settled.

If this greater stability comes about, the flimsy huts that used to be common are no longer seen. Widely separated shelters become clusters of better built dwellings. As this occurs, there is more economical use of the land; crops are grown in weeded and mulched beds or rows; what is cleared is used longer than before (in fact, may never be returned to the wild); and wilts, rusts, rots and spots of crops are thought about and control measures are attempted.

As we bring to bear more scientific study on agricultural problems in the tropics, it becomes clear that after disease difficulties are solved, from the standpoint of plant production the tropics is the most favored part of the earth. It has the greatest potential of any place; because of soils, rainfall, sunlight, and the lack of killing cold. However, in the warm, moist tropics there is a special impact from parasitic and other plant diseases.

> [Some of the paragraphs below are based in large part on an article (116) published in connection with a symposium on the problems and future of the hot and humid tropics. See also such works as those by Ordish (77) and Padwick (79) and articles from bibliographies (9, 78).]

Commercial garden locations for producing cut flowers are usually at some distance from markets, and growers and sellers have learned many effective tricks of the trade. These are devised by small holders whose flowers are their only means of income. When diseases appear on flowering plants the owners may be more than usually affected because of the more primitive conditions.

Small fields of roses were studied on the slopes of a volcano, where they were profitable sources of income for small farmers for decades. The old style so-called "creollo" or "Spanish" roses produce in a relatively short season a great profusion of salable, moderate- to small-sized flowers. One year an innovative grower obtained rooted cuttings of longer blooming, large flowered, tea roses. By use of these plants the black spot, *Diplocarpon rosae*, disease was introduced where it had never been before. In three years all his roses and those of neighboring growers, were infected with the disease. He and his family were ruined, and when seen again he had become a poorly paid garbage collector in the city.

In one old market, a flower stall was suddenly closed that always was filled with healthy zinnias, grown from local seed. When an outside supply of seed was brought in, *Cercospora zinneae* became an exceedingly serious leaf and petal spot and the flowers produced were completely unacceptable. These are only two examples, out of hundreds. Plant diseases, for small florists of the tropics, are painful. Such a man and his family often live on so restricted an amount of money that a single failure is a major catastrophe.

In Latin America there are over 50 minor subsistence crops grown in edges of clearings and in waste corners of plantation fields. For protection against disease, peasants rogue out wilted plants carefully and pick off infected leaves.

Unless plots are moved around promptly, season after season usually to new soil, the much-valued food plants will not continue to thrive. They are nearly always being troubled with root infections from tropical *Rosellinia*, *Rhizoctonia*, *Phytophthora*, nematodes, and occasionally bacteria. On the leafy vegetables, in addition to the viruses, there are attacks by hundreds of leaf infecting fungi.

The greatest impact from these diseases comes from the fact that the native grower is deprived of, or restricted in, his favored foods. When sources of these are destroyed by disease then wheat, maize, rice or something foreign has to be brought in. It may keep him from starving but it reduces in the peasant his sense of pride and independence.

Every year in many parts of the Neotropics there are imports of large amounts of rice, wheat flour, increasing quantities of canned and other processed food, maize, beans, other kinds of eatables, and even tobacco, spices, tea, and medicinals. The fearful consequences of failures from diseases in the great food crops are in some cases extreme.

The common white potato came originally from tropical America and its most serious disease is, and always has been, Phytophthora blight. This has, from long ago, killed crops especially in cultivated conditions. It is believed by some to be the basic cause of ancient food shortage among mountain living people in our tropics, and of resulting weakness in highland populations.

There is no doubt (see, e.g., Fernandez Valiela (42), Galli et al. (44), de Segura (99), Nowell (76), and Spegazzini (101)) of the impressive impact of this blight in the tropics. Cox and Large (27) give in classic manner some of its inter-hemispheric importance. In Central America I have seen areas of a hundred or more hectares at a time, planted solidly to potatoes, completely destroyed by the "gotera" blight. Farming stopped and populations moved.

Over three decades ago I once rode by what was left of a maize-growing community in a remote part of Central America. The Indians in it were a tribe religiously antithetical to most white men. When the trail reached what was once a flourishing village, only a few inhabitants remained. Half of those living were sick, in filthy rags and hardly able to move. Scattered about were many shallow graves, some recently opened by wild animals, and remains from cadavers could be seen fed upon by buzzards. From inquiries it was learned this was the closing scene of famine that occurred because of maize disease, probably a combination of *Puccinia* rust, *Helminthosporium* leaf blast, and root rots.

Beans are a basic source of important food in Central America. In Guatemala and El Salvador it was found that in some of the best regions fields had Pacific slope exposure where there is a comparatively short wet season during which the crop grows to near maturity. Harvest was timed to take place in the ensuing dry season. About every third year there might be a slightly lengthened period of rains when the dried and temporarily dormant blight and anthracnose diseases would have a chance to develop. Fields that ordinarily give a 40-bag harvest did well to produce 10 or 15. Small farmers can never do more than work harder at the kind of labor that shortens life.

When crop diseases become serious on plantations, losses are felt in national economies. The waste and reduction strike at the monetary and even social foundations of tropical America. In the beginning cacao was grown and used by pre-Columbian Americans. Commerce developed in it when it was found that it could be profitably produced in plantations in the Guianas, Ecuador, Colombia, Trinidad and other like countries. Port cities grew up, settlers were attracted, and a country had something to finance civilization. However, as has been shown by Stahel (102), in 1895 the witches' broom disease, *Marasmius perniciosus*, was found in Dutch Guiana cacao plantations. In a few years it was in other countries, and in Ecuador it quickly reduced production as much as 60 and 80 per cent, so harvest was stopped. Chocolate companies in Europe and North America had to go elsewhere for the product. European ships had their courses changed, for there was no longer business in certain former cacao ports.

About this time growers in West Africa had obtained the chocolate tree and found that it developed well there, free from American disease. A large amount of African cacao now takes the place of what was an exclusive New World product. However, science was brought in by Latin American plantation owners to help them solve their problems and put cacao growing on a modern basis; not only in horticulture and processing but most of all in disease control.

Another example of the impact of plant disease on a plantation crop is found in sugarcane. There is no more important world food staple than crystalline sugar. Commerce in it is not only of large international scope, but it also reaches intercontinental levels and this product knows no true ideological limitations. Sugarcane has, (see Edgerton (35)), over 440 diseases. This is partly because it has been taken to grow in so many parts of the tropics, and has then been sent as living cuttings or setts to other parts. It is cultivated and becomes infected under very many different conditions, next to so many different kinds of diseased crops, and next to so many different wild areas with their diseases. With some variations, this kind of disease history is true of all great world crops.

It is well known that in the American tropics there is no tropical crop with such financial and international repercussions as coffee (116). For years coffee expanded rapidly in Brazil, Colombia, Central America, Cuba, Puerto Rico and other parts of Latin America. However, a disturbing leaf spot was discovered in a special region in Colombia. It was not a rust, which was and is so much feared by coffee men, but it was severe and it was killing coffee plantations. In 1876 it was described by Saenz (96). Soon after that it was found in many other countries. It is caused by an organism given one name after another, but now known as *Mycena citricolor*. It is especially severe in its defoliating effects. Research introduced spraying and other control methods, and where coffee had once been abandoned, producing it is once more back on a stable basis. It is disease effects, such as have been mentioned above only as examples, that have made a great difference in the history of plant pathology in the Americas. For further discussion and illustrations, see Chapter XV.

Tropical Origin of Parasitic and Other Fungi

There is a great deal published, mostly from the temperate zone, about plant diseases and fungi that cause them. An impression gained is that these came from the region. However, numerous mycologists know, as Stevens (103) points out, how impressive is the fungus flora in the ever-warm parts of the world. These organisms seem everywhere in rainforest areas.

Among pathologists there is an understandable preoccupation with disease-causing fungi and minor concern with saprophytes. This preoccupation has hindered almost to the point of preclusion giving attention to the epiphytic, macroscopically invisible, non-parasitizing forms. Hand lense and microscope studies show that they are to be found on almost any of the older foliage on non-herbaceous plants in moist-warm conditions. Very young stems and leaves are free from these epiphytes. Hyphae seen by me often are septate, variable in dimensions and conformation, and when isolated on potato dextrose agar grow into both white and tawny to brown mycelial mats. None was black. In one culture microsclerotia developed but it was not related to *Pellicularia.*

In intriguing reports of Ruinen (94), on what she called the "phyllosphere," it is shown that leaves in the tropics, both in the Orient and Occident, have over and around them as a natural occurrence, non-parasitic bacteria and fungi. Her studies open up interesting aspects that should be given further planned investigation. I made observations in 1949 and 1960 on leaf-surface-fungi of 3-year-old coffee leaves and on 6-year old orchid leaves in moist forest conditions. It is possible to find on such leaves not only non-parasitic fungi but also superficially growing algae, lichens, and leafy liverworts. But the most frequently met organisms were fungi. In Costa Rica a few instances were seen in which gemmae of *Mycena citricolor* were apparently kept by a physical barrier of epiphytic organism growth in the "phyllosphere" from reaching, and thus causing attack on, coffee leaf tissue. In some cases epiphytic fungi, although invisible to the naked eye, were found to form a web of sparse filaments on leaf petioles and slender twigs. Studies and observations such as the above emphasize the common occurrence and good habitat in which fungi subsist in the tropics--it is believed they have been there a very long time.

It seems to me that parasitic microorganisms evolved under conditions of what we call tropic warmth. It is probable that at one time there was plant life on all continents of the earth; even at the poles it was warm. The ice ages came later, and it is only after comparatively recent geologic times that the so-called temperate zones came into existence. The north and south ice caps killed parts that were once always warm (though the ice masses retreated toward those poles, uncovering and letting thaw some of these parts); annually, periods of snow and ice then became interspersed by warm seasons. Plants and their diseases became adapted to the winter-summer rhythm.

The advent of temperate zone summers was when erstwhile tropical organisms once again throve. In the tropic zone is where they are most unhindered in life. Thus we can assume all temperate zone pathogens come from tropical-conditioned ancestry. In fact, it has been proved, Tyler and Barghoorn (113) and Barghoorn and Tyler (8), that filamentous organisms that could be either fungi or algae were actually present in the Precambrian era. This was a long time ago when it was moist and warm, going back to about half the commonly accepted present age of the earth.

Archeologists have found that during the Eocene epoch, about 55 million years ago, plant leaves were infected with a highly specialized tropical fungus of the genus *Asterina* (see Fig. 2-1 for example). Fossilized specimens, when examined microscopically, appear as well organized morphologically as are Asterinaceae of the present day. These highly developed, singular tropical black mildew parasites, after extended geologic time of evolving into their parasitic character, were then recovered from sediments deposited under tropical conditions 50-60 million years ago. This was much before the Pleistocene epoch and the Ice Ages which ended in making the temperate zones. It is only in the circum-Equatorial region around the globe, where anything like the primordial moist-warm conditions still remain (8, 15, 37, 70, 95, 98, 113).

Change and variation helped in adaptation for survival in the temperate zone. In it, however, where there are so few generations annually as compared to the tropics, there cannot help but be, where winter intervenes, less chance for mutations and gene admixtures. There is reason to expect, and here is where we find, more types and races of fungi in the tropical world. Here, it is believed, is their more natural habitat.

Some Typical Tropical Diseases

It is often held by botanists and pathologists, that

Figure 2-1. Symptoms on Agave sp. leaf found in Mexico by Kern and Trelease, of specialized fungus Asterina agaves Ell. & Ev. (Photo from material in U.S. National Fungus Collections.)

plant diseases under tropical conditions are mostly a regular continuance of problems in the temperate zone. There is even a comforting thought by some of those working in the tropics that with the strenuous efforts going on in the temperate zone, the tropical problems will eventually be solved by work in that presently wealthier region, the temperate zone where there is a larger population of plant pathologists. For some things, this may be true, and some of it is coming about.

However, there are parasites and diseases unique to the tropics. Sometimes ecology is such that certain of the parasitic organisms are so well adapted to a warm environment that they can only survive in it. Among such are those that are sensitive to cold; or are dependent on warm and heavy rains. Some are so specialized they are only able to attack hosts that are confined to the tropics. Others survive only because of the primitive means of agriculture (72) which allow their perpetuation.

Certain parasites, like *Puccinia*, *Alternaria*, *Oidium*, and *Cercospora*, may not flourish in certain tropical regions where rains are frequent and prolonged. Spores of some kinds are actually washed away. On the other hand some disease organisms are splashed about and depend very especially for their perpetuation on drops of rain. This is the case for example with members of the genus *Phyllosticta* (conidia), *Fusarium* (chlamydospore stage), *Rhizoctonia* (sclerotia), and *Mycena* (gemmifers).

A short list of some well known typical tropical American plant diseases is as follows: banana fusarium wilt or Panama disease (*Fusarium oxysporum cubense*), cacao witches' broom (*Marasmius perniciosus*), coffee leafspot (*Mycena citricolor*), rubber leafspot (*Dothidella ulei*), algal leafspot (*Cephaleuros virescens*), maize rust (*Puccinia polysora*), milky or lichen spot (*Strigula complanata*), black mildews (*Asterina* spp. and *Meliola* spp.), citrus tristexa disease (virus), grass downy mildews (*Sclerospora* spp.), black and white horse-hair or wire thread blights (*Marasmius* spp.), web blights (*Pellicularia* and *Corticium* species), black rootrots (*Rosellinia* spp.), red rootrots (*Ganoderma pseudoferreum*), and leafgall (*Exobasidium* spp.), an example of which is shows in Fig. 2-2. There are, of course, thousands more.

In the tropics, certain diseases of major importance

Figure 2-2. Example of a leafgall, in this case Exobasidium lauri, on sweetbay tree, Magnolia virginiana. Picture from photograph in files of Plant Pathology Department, University of Florida, Gainesville, Florida.

are also common in the temperate zone. Examples: wheat black stem rust (Puccinia graminis triciti), classic potato late blight (Phytophthora infestans), common main causes of seedling damping-off (Rhizoctonia solani and Pythium debaryanum), cucumber and tobacco mosaics (common viruses), maize leaf-blight (Helminthosporium turcicum), bean anthracnose (Colletotrichum lindemuthianum), tomato fusarium wilt (Fusarium oxysporum lycopersici), corn brown spot (Physoderma zea-maydis), beet leafspot (Cercospora betae), tomato leafmold (Cladosporium fulvum), bean angular leafspot (Isariopsis griseola), and soil rot (Sclerotium rolfsii).

Abundance of Tropical Diseases

[This section is based in part on my paper (116) recently published in the Journal Ceiba, from the Escuela Panamericana Agricultura, in Honduras. See also such other examples in works in citations 2, 6, 7, 9, 16, 35, 40-42, 45, 68, 76-79, 85, 89, 95, 99, 101, 103, 104, 116.]

From studies it is evident that diseases are more numerous in the tropics than in the temperate zone. See Table 3. It is perhaps conservative to estimate that for one important disease in the temperate zone that 10 (116), or even 20 may be found in the tropics. For every 60 or 65 diseases being given exhaustive attention in the temperate zone, there may be only 4 or 5 being given equally intensive study in the tropics; annual losses may be only 4 to 6 per cent in the temperate zone but 30 to over 50 per cent in the tropics.

While it is true that plant disease research is increasing in the tropics, and control measures are continually being improved, it is also true that more research discloses many new problems. More than ever there is an increasing need for trained workers as crops become modernized and more and more important. As was pointed out by Yarwood (117), in the temperate zone those crops that are most valuable, consequently most intensively and longest cultivated, are those most heavily diseased. Lists show that already in the tropics, on many most-studied crops about 300 diseases are reported. New parasites and variations of them are constantly being discovered.

In spite of the large numbers of diseases reported, it is fortunate for growers that only about 5 per cent of them are of prime importance. Even this relatively small per cent means that some 15 severe troubles out of a total of around 300 are found. For purposes of simplification, in addition to the most severe and important there are some 50 per cent which are well scattered infections, and while fairly common may not cause much actual losses. There is a third group of about 40 per cent in it in some cases, as is seen in Table 4, the parasites are sometimes "nudged out" by others. They are often somewhat scattered in occurrence and a few may be even rare. The fourth group is small, about 5 per cent, and are the hyperparasites attacking main disease organisms.

In reviewing the progressing history of tropical plant diseases it is apparently difficult to predict what may happen to certain of them: The sugarcane rind disease, was at one time a definitely serious limiting factor but it is now almost unheard of. The pink disease of such trees as rubber, citrus, and coffee under certain conditions becomes epidemic rapidly. In some old locations where it was once prevalent it can hardly be collected. On the other hand the chlorotic streak of sugarcane, the hoja blanca of rice, and the cacao cushion gall complex all seemed of minor importance, at one time almost like curiosities, but they are all increasing in severity and deserve every attention now being given them.

Unfortunately, I was never able to carry out well planned, paired experiments on abundance of fungus spores in the tropics compared with the temperate zone. However, a few simple assays were made that show tendencies, but not enough to give final numerical data.

In the first place spore catches made during dry seasons in the tropics (El Salvador) on glass slides treated with alboatrum, showed there were not likely to be large numbers of spores from a single species, but many different species of spores in a slide. In the temperate zone (Washington D. C.) during the summers it was not at all uncommon to find hundreds of individual spores, but all of a spore catch, to be of only one or two species of fungus.

When hard agar (PDA and WA) plates were exposed, outside city limits in El Salvador, in the dry season in the tropics, from 20 to 38 colonies were secured on plates

Table 3. NUMBERS OF DISEASES ON CERTAIN IMPORTANT CROPS compared, tropics to temperate zone.

Crop	In the tropics	In the temperate zone*
Brassica oleracea capitata, cabbage	43	36
Citrus spp., orange and relatives	250	110
Cocos nucifera, coconut	35	1
Coffea spp., coffees	385-400	10
Cucurbita pepo, pumpkin-squash	110	19
Ipomoea batatas, sweetpotato	187	17
Lycopersicon spp., tomatoes	278	32
Musa acuminata, bananas	200-220	8
Nicotiana tabacum, tobacco	221	30
Oryza sativa, rice	550-600	54
Phaseolus vulgaris, bean	253-280	95
Saccharum spp., sugarcane	450-460	56
Solanum andigenum-tuberosum, potato	175	91
Theobroma cacao, cacao	52	1
Zea mays, maize	130	85

*Includes observations on greenhouse-grown naturally infected exotic specimens, and on growths under subtropical conditions bordering on the temperate zone.

Table 4. PRIMARY PARASITES BEING "NUDGED OUT"[a] by secondary or parallel attacks of other(s) not considered hyperparasitic.

Primary parasite	Being "nudged out" by:	Disease involved
Cephaleuros minimis	Diplodia theobroma	Cacao, twig damage
Cephaleuros virescens	Pyrenomycete, undetermined	Lagerstroemia, leaf attack
Cercospora coffeicola	Valsa-like organism	Coffee, twig dieback
Cercospora hibisci	Diplodia sp. & bacteria	Jamaica sorrel, leafspot
Colletotrichum sp.[a]	Colletotrichum sp.[b]	Hevea, leafspot
Colletotrichum sp.	Tryblidiella rufula	Bougainvillea, stem disease
Dothidella parryi	Gloeosporium agaves	Agave, leafspot
Erysiphe polygoni	Diplodia sp. & Pestalotia sp.	Bean, mildew
Fusarium oxysporum cubense	Soil bacteria	Banana, wilt (root attack)
Fusarium solani	Diplodia sp.	Sweetpotato, rootrot
Helminthosporium sp.	Xylaria (apiculata?)	Guineagrass, culm attack
Hemileia sp.	Penicillium sp.	Orchid, petal attack
Marasmius perniciosus	Schizophyllum commune	Cacao, witches' broom in aging condition

Melanconium sacchari	Nectria sp.	Sugarcane, rind disease
Mycena citricolor	Pestalotia funerea	Cinchona, leafspot
Nigrospora oryzae	Cladosporium zeae	Corn, dry earrot
Pellicularia koleroga	Fusarium (roseum?)	Bungalan, thread blight
Sclerotium rolfsii	Bacteria and fusaria	Squash, soil rot
Septobasidium sp.	Capnodium sp. & Pestalotia sp.	Inga, felt disease (aging)

a. Slow growth type.
b. Fast growth type.

opened for 20 seconds. These colonies were all, apparently, different species of fungi. Where the experiment was repeated in the temperate zone, in the country outside Washington, D. C., in a dry summer season, several exposure times were tried. No spores were caught in 20 seconds, and only two or three spores in 45 seconds. The best exposure time was found to be about 3 minutes. After such a period about 21 to 30 colonies grew from spores on the agar and these were from four species of fungi.

In the wet rainforest, there are periods at least, when air-borne spores do not float at all. In Costa Rica, agar plates were exposed many times under moist jungle growth in connection with an attempt to recover spores of disease organisms. There were no air-borne contaminants. To study this further, plates were carefully opened and exposed, especially in order to collect spores, under rainforest jungle canopy. These were left open 20 and 30 minutes, of course during periods when there was no rain. Out of 20 exposed plates, one rainless morning, and 25 exposed plates on another such morning, four contaminating spores were thought to be caught. These were, all four of them, spores of a *Curvularia* and probably came from the laboratory where plates were taken to be incubated.

Altitudes and Disease

It is often misleading to make generalizations in agricultural science, and this is in particular true in tropical plant pathology. However, researchers in this specialty know that disease occurrences are affected in profound manner by elevation. Indeed this is of great importance, as on mountainsides; with the same crops growing at different elevations, certain diseases may only be present at certain levels. An example of this is well illustrated by clear findings on beans in Central America. On that crop not only are the occurrences of some diseases pretty well confined within certain limitations as to elevation, but also the severity of some diseases is definitely connected with the altitude at which the beans are grown. Wellman (116) has given further indications of effects from elevation on numerous other crop diseases.

In the tropics, average temperatures (given in this book mostly as "C" or centigrades of Celsius (106)) are reduced about 0. 3 C. to 0. 9 for every 100 m. elevation.

(For discussion and citations about this effect see Chapter I.) In both Central America and in South America, in work on diseases affected by elevation, I have concluded that in relatively moist conditions temperature reduction is a little more than half a degree C. for every 100 m. in elevation above sea level. It is also a remarkable feature that diurnal temperatures in the tropics in general vary only a few degrees from the mean; in the so-called temperate zone such variations are 10 to 50 times of what we experience in the tropics.

Where disease conditions are studied more in detail, and judging from known disease temperature relationships (6, 11, 20, 27, 30, 42, 44, 57, 68, 76, 99, 111, and 116), the cooling-from-elevation effect is often more complex and less clear-cut on irregularly lying tablelands, along rounded ridges, and on tops of hills with ragged outlines. However, it is more distinct in effect when diseases are followed upwards on more even and broader mountainsides which progress higher and higher in wide sweeps up long distances. It is also evident that wherever fields are on a mountainside that faces tradewinds, temperatures are perceptably lower than in fields at the same zone level but on the opposite exposure, the other and leeward side, where there is no tradewind.

It is always hot and steamy at the heat (or thermal) equator at sea level, and to the worker in it the heat is only relieved by ascending a hillside; it bears repeating that the farther one goes north or south of the heat equator the less elevation is required to reach cooler areas. Wherever you are in the tropics, crops and diseases change with elevation. The different effects of elevation on plant diseases are more striking in open fields than on crops under planted shade or under part forest cover.

In my experience there appear to be about five levels respecting disease, as follows:

1. A first level, using the heat equator as the base, is from at or near sea level to approximately 300 meters up. It (in tradewind region, the windward side) is hot, humid, and it has a mean temperature of about 28° C (82.5 F). There is commonly about six months of rains and about six months of dry. With mountains intervening, at the same level on the leeward side the climate is dryer and in some areas almost desert-like. Where there is a fair

amount of rainfall there are still long dry periods, and this is where there is an abundance of virus diseases. These have a chance to be insect-spread, for the extended dry periods provide good weather for sucking insect vectors.

There are also severe attacks on leaves of crops from such parasites as *Cercospora* and *Helminthosporium*. Wilts of tomato and banana from *Fusarium* are more likely to abound here, the root and stem attacks on tomato from *Pythium* are seen, the *Phytophthora* gummosis and foot rot of citrus is severe at this level, and it is here where the wilt-producing *Pseudomonas* bacteria on many hosts are the most devastating in wet periods. It is in low areas on the wet side where relative air humidity is high, especially in places protected by shade, that one finds the wiry true thread- or horsehair-blights from *Marasmius* spp. It is in this low country that the non-thread-forming *Marasmius sacchari* is most frequently collected in old, poorly kept sugarcane fields and the witches' broom-forming *Marasmius perniciosus* is most often found causing its swellings on cacao. Here *Tryblidiella* spp. are often seen associated with dieback of woody plants.

During dry seasons the sooty molds build up following abundant sucking insect infestations, and the *Septobasidium* felt-blights nurtured by, and symbiotic with, certain branch and twig sucking insects are also here where they serve to cause disturbance to tree health. Species of *Colletotrichum* may invade after excessive sooty mold and felt-blight occurrences. The anthracnoses of anona, cacao, mango, coffee and other tree crops due to *Colletotrichum* spp. are common at the low level. In fact actual killing out of mango plantations has been seen at this level, due apparently primarily to *Colletotrichum gloeosporioides* attacks that weakened trees which become seriously invaded later with numerous secondary parasites.

Nematode redring disease in coconut and African oil palm appears to require the conditions of heat and moisture that occur at the low levels where there are good rains but interspersed with dry seasons. *Phytophthora parasitica* on peppers is more severe in fields at low levels than where they are grown at higher much cooler locations; the same is true of *Pythium aphanidermatum* on beans. Dodders such as *Cassytha filiformis* and *Cuscuta americana* are more frequent at the low level.

2. The second level going up is at medium-low elevations. This is about 500 meters above sea level (considered as measured at the heat equator). Here it is still warm with an annual mean of near 26° C, which is not quite so hot and steamy as at the lowest locations. Here it is often, especially where tradewinds occur, more moist than below. Dews are long lasting and occur during the short dry season periods, and it often rains 200 or more days in the year. In some locations where cultivation is carried on there are more or less three to four months of a dry season that is occasionally broken with scattered showers.

It is under such conditions that many of the true fan-type or pellicle-forming web-blights from *Pellicularia* (= *Thanatephorus*) spp. develop. Here web-blights are vigorous, and occasionally cause great damage to beans and other crops. Where there is well distributed rain fall the koleroga blight, *Pellicularia koleroga*, may occur profusely on coffee and other hosts. At this level *Fusarium* wilts are seen, as are other wilts and blights from bacteria. At this level occur many leaf and stem injuries attributable to species of usually weak parasites like *Pestalotia* and *Xylaria*. A miscellaneous lot of serious grass leafspots are more common here than at lower or higher elevations.

Diebacks of trees related to *Colletotrichum*, *Diplodia*, *Schizophyllum*, *Hypoxylon*, and *Lenzites* occur at this second level (although these parasites are found as well below and above) where moisture is plentiful. It is in the second level that legume, maize and grass tarspots from *Phyllachora* are often found, and it is this zone in which many of the Dothideaceous fungi are richly developed. This also seems a specially important level for dispersion and occurrence of black mildews; *Meliola* spp., *Asterina* spp., *Amazonia* spp. and relatives.

From my experience it appears that if agarics can be searched for with promptness after rainy periods that are then immediately followed by dry, the collector finds at this level more sporocarps of wood-rotting fungus genera such as *Polyporus*, *Fomes*, *Poria*, *Schizophyllum*, and *Hydnum*. Here also are found certain but ephemeral, fleshy species of Agaricaceae. The leafblast of maize, *Helminthosporium turcicum*, is most severe at this medium low level. This is also true of the maize rust *Puccinia sorghi*. Rice blast from combined *Piricularia* and *Helminthosporium* attacks vigorously at this medium low level, but here, of the two, Piricularia is the more vigorous in its pathogenicity.

In tree crops, wood blue-stain and branch-dieback from combined borer and Ceratocystis attack seem readily encountered at the medium lowlevel. Some very serious coffee dieback is frequent at this level, at first considered specially due to Colletotrichum. However, on closer study with nematologists it becomes clear this dieback is secondary upon root injury by both sting and rootknot nematodes.

3. What is termed medium elevation is about 500 to 1000 meters. At its coldest edge it has an average temperature of about 24° C (75° F), and where there is much cloud cover the temperatures remain on the cool side. On the windward side rains are likely to be frequent, and the wetness does not dry off rapidly. Dry seasons usually last only about three months, but during this period the relative humidity of the air averages about 75 per cent. There may be winds, but they are not drying winds, and when clouds are absent the sun is brighter than at lower elevations.

It is at this zone level in fields played upon by ocean winds, that there are fewer viruses; in fact in some mountain valleys it may be that there are no virus troubles as recurring periods of heavy rainfall make it impossible for insect vectors to survive in numbers.

The Cercospora coffeicola leafspot (see Fig. 2-3) of coffee that is so destructive at low elevations is more severe at the medium high level if the crop is grown without shade. However, observations indicate that when put under shade this coffee leafspot is more effectively reduced in plantings at the higher elevations than at the lower. In banana, the same is true with the Cercospora musae leafspot (Sigotoka disease). At medium high elevations it can be severe but here shade has a good ameliorating effect on Sigotoka. Certain other leaf-inhabiting fungi like Isariopsis spp. and Cladosporium herbarum grow vigorously at this level. The Cladosporium fulvum leafmold of tomato makes that crop almost prohibitive to produce in the medium high zone unless it is sprayed. This is also true of certain leaf-spotting bacteria. Where it is not too wet there are still a few of the insect-associated Basidiomycetous felt blights, Septobasidium spp., and at this level outstanding effects are observable from pink disease attack, Corticium salmonicolor on sun-exposed coffee, Hevea rubber, orange, and various other

Figure 2-3. Symptoms on coffee, *Coffea arabica*, of leafspot, *Cercospora coffeicola* Berk. & Curt., from studies in El Salvador. *Leaf on left* from tree *under medium shade*, on right in light shade.

plantation tree crops. Tomatoes are severely attacked at the medium high level by races of Phytophthora infestans. The stem swelling and splitting and leafspot diseases from the algal organisms, Cephaleuros spp., on several hosts may be severe at this level.

The rice leafblast, Piricularia and Helminthosporium complex, often are found on upland plantings of so-called "dry" rice grown at this level. However, up here the Helminthosporium part of the complex is the most profuse in its development. Up further at an even cooler level grasses are more affected with Balansia spp. and Claviceps spp.

Mycena citricolor = Omphalia flavida which is the American leafspot of coffee (it attacks some 500 hosts in the Neotropics), can be specially prevalent and severe in wet parts of the medium high level zone. There is occasional devastation from cool-temperature species of Pythium and from certain tropical rusts on trees and miscellaneous weed hosts. Here occur also severe potato tuber storage rots, including specially those from Pythium and bacteria. The branch-wood-stain fungi, like species of Ceratocystis also occur frequently, and it is here that may be found fairly commonly the slow-working tree collar-infecting fungi such as Nectria spp. and as well Rosellinia spp. It is in the higher parts of this level, but where it is not too wet, that there is a considerable profusion of many genera and species of the tree-top infecting Loranthaceae.

4. The fourth level is termed high or templada and is approximately 1000 to 2000 meters above sea level. At this level on mountain sides facing prevailing winds it is generally never dry, and in some locations, relative air humidity may be almost constantly close to saturation; rains and fogs are common. The cold downpours may be light, but they are slow in finishing and there is a lengthy period when the drops are sparse and cold. At this high level are grown certain special mountain crops, and in this and the level below one often sees nostalgic attempts by settlers at planting temperate zone flowers, vegetables, and fruits, far from their normal environments. At this level, plant virus troubles are rare to absent. There are pockets of moist land where there are seen such troubles as: those caused by Phytophthora spp. which causes cankers; certain species of Pythium, especially on the sugarcane which struggles to grow in these colder regions and what appear to be specialized races of Colletotrichum (Glomerella cingulata).

This is ideal country for fruit and vegetable rots (*Sclerotinia* spp.), and for powdery mildews from *Oidium* spp., *Erysiphe* spp., and special kinds of cool-country herbaceous weed and tree rusts. There are also numerous examples of leafspot from *Septoria*, *Phoma*, *Chaetoseptoria*, *Albugo*, and *Ascochyta*. The *Mycena* coffee leafspot, that is also found on many other hosts, while dangerous at slightly lower levels is much more severe in the higher locations. In addition, it produces the agaric stage more freely higher up, whereas the miniature yellow toadstools are less common below. The algal leafspot, *Cephaleuros virescens*, is found many places, but is most severe on citrus and guava in higher locations, and it has been seen as severe on trees in high-grown coffee, although only mild in effect at levels lower down the mountainside. Here the indigenous maize rust, *Physopella zeae*, is more common, is severe, and somewhat confined to the cool heights.

5. There is also the fifth and highest plant-disease level as one goes up in the mountains. This is neither the altiplano nor the snow fields, but is a little below them and is the level known as fria in Latin America. Here temperature may approximate "temperate zone conditions." However in the tropics, with the uniformly short days and much reduced diurnal fluctuations in temperatures, the "conditions" are actually far from those of the temperate zone. The elevation at this fifth level is usually around 2000 meters to something over 2200 meters elevation which is at, and just below, the line where occasional frosts may occur.

Unpleasant as it sounds, level five is undoubtedly an important part of the Neotropics. Here in dryer parts is where some of the temperate zone cereals have been introduced, along with their rusts, and where the white potato (*Solanum andigenum-tuberosum*) is grown as a staple and, after fermenting and drying, stored in that condition. At this high level the winds blow, it is fairly misty at times, with clouds, and yet the sky may be quite clear in certain periods with occasional frosty nights. At this level there are no plant virus diseases, and bacterial plant pathogens are practically non-existent. In sheltered locations, especially on the lee of mountains, several food crops grow well and in these locations are seen diseases that develop from such fungi as species of *Chaetoseptoria*, *Sclerospora*, *Phyllosticta*, special *Phytophthora*, special *Sclerotium*, and *Phoma*.

Powdery mildews are not common, but the downy mildews become dangerous. Some of the mycologically speaking more complex Ascomycetes and Basidiomycetes that are never seen in most of the tropics, are found in these high cold regions. Among the parasitizing phanerogams, in addition to the large population and numerous species of the Loranthaceae that can be found, there are more of the Balanophoraceae found at the colder levels, and where there is shade and good tree cover many more Orobranchaceae.

Before leaving this subject, it is well to note that some tropical plant disease organisms show a much more extensive range of occurrence than others. For some examples, there are species of Botrytis, Torula, Penicillium, Alternaria, Colletotrichum, and Cephaleuros. These commonly attack the ever renewed supply of senescent leaves that are found perpetually at any level where plants grow without much seasonal rhythm in the tropics.

While the Helminthosporium turcicum is apparently most severe at the medium low level, it occurs all up and down the levels from the lowest to almost the highest at which maize will grow. The cucurbit mildew, Pseudoperonospora occurs at all elevations where rains are not excessive and where the cucurbit crops fruit well. Tobacco mosaic is also found at all levels where the crop grows except in some wetter areas. Algal leafspots, Cephaleuros virescens, are found from low to high. Cucurbit viruses attack plants like the cucurbits, bananas and relatives, grasses and Commelinas only at lower elevations under natural conditions. Valsa spp. (see Fig. 2-4) and valsa-like fungi are common at low elevations but are found in medium cool locations as well.

"Disease Climates"

There are situations in the American tropics in which certain fungi reach a practically stable or climax condition as to prevalence and relative severity. For such climax situations there are definite requirements, which are about as follows: obviously, both the fungus and the host must be present for the disease to occur. However, it appears ecologic conditions must be such that the fungus multiplies and spreads easily and it is more likely to be at climax stage if its propagules are mostly vegetative in

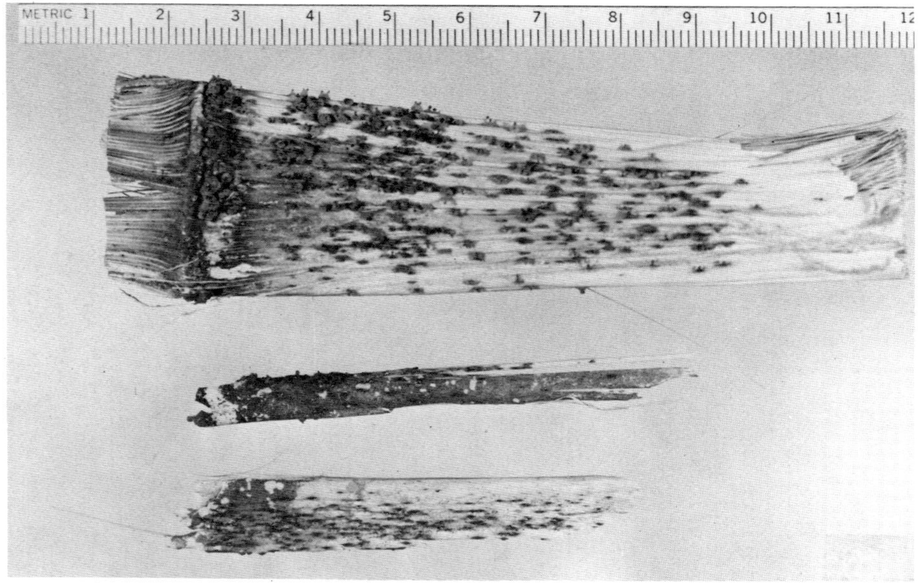

Figure 2-4. Valsa sacchari Stevenson on old bases of sugarcane stalks, collected by Bourne in Florida; material in herbarium of U.S. National Fungus Collections. Valsa spp. and valsa-like fungi infect numerous hosts and disease old and dying tissues, and occur under a wide range of climatic conditions.

character such as the soil-inhabiting actively growing hyphae of a Pythium, the fragile aerial Oidium spores, the water-dispersed soft gemmae of Mycena, and the like. If there are occasional unusual weather fluctuations the holdover phase of the fungus must be such that it will preserve the parasite. There must be consistent preponderance, but not necessarily replacement, of lesser fungus pathogens by the dominant disease producer.

Examples of fungi observed as established in a climax condition are as follows: Balansia on grass observed in Costa Rica; Cercospora on coffee in Costa Rica; Cladosporium on tomato in Nacaragua; Colletotrichum on lime in Paname; Diplodia on henequen in El Salvador; Dothidella on rubber in Costa Rica; Fusarium on banana in Honduras; Marasmius on cacao in Ecuador; Meliola on sweetpotato in the Virgin Islands; Mycena on coffee in Costa Rica; Oidium on orange in Panama; Phytophthora on potato in Guatemala; Puccinia on wheat in Mexico; Pythium on sugarcane in Colombia, Rhizoctonia on various seedlings in El Salvador; and Rosellinia on potato in Colombia.

It is possible for experienced tropical pathologists to recognize, almost intuitively, that there are distinct environments that are what might be called disease climates. A pathologist may speak with considerable certainty of areas as "black mildew country," "good for Fusarium wilt," "where you can expect Rosellinia root rots," "Rhizoctonia damping-off soils," "downy mildew type of exposure," "where white powdery mildews develop," "specially favorable for aerial Phytophthora," "a Mycena leafspot zone," "the woodstain canker part of a slope," "Cercospora leafspot land," and similar phrases.

I have often observed that where the climate is dry and irrigation has to be introduced to produce a crop, there is a somewhat different disease-fungus flora from that in a wet region. In the dry regions the smuts and rusts may be common on crops. There are many troubles from viruses, as the insect vectors flourish more readily under these dryer conditions, and from insect presence the sooty molds, caused by species of Capnodium, are seen. There are leaf injuries from sun scorch and there may be the killing of plants in a round area in a field which can be traced to centers where lightning has struck. In some regions where there are short rainy seasons, observations show that it is at the time when rains die down and the drought intensifies that the powdery mildews develop from species of *Oidium*, *Erysiphe*, and *Sphaerotheca*. As the rain-free days lengthen into months, powdery mildews increase on irrigated fields.

In a different climate, where rains are common and frequent, where air humidity is high with its accompanying swamp and jungle growth, there develop certain kinds of leafspots. These are those that depend on spread by the dashing of water drops; they develop in their most flourishing manners if leaf and stem surfaces are never completely dry and much of the time there is a continuous coating of liquid water. Specialized tropical types of wet-country organisms belong to species in a very long list of genera of such as *Mycena*, *Glomerella*, *Cephaleuros*, certain nematodes, *Marasmius*, *Corticium*, *Meliola*, *Phyllosticta*, and *Phoma*.

There are some organisms, for example like those belonging to species of leaf-infecting *Helminthosporium*, root-infecting *Fusarium*, certain large-spored species of *Cercospora* and *Cladosporium*, and wood-stain fungi such as *Ceratocystis* and *Diplodia*, that develop better in an intermediate type of environment. These do well surrounded by an ecology of a median kind that is typical of a more nearly true forest region. Either excessive rains or extremes of dryness reduce diseases that come from infections by those organisms that are found most frequently in less drastic environment.

During one season in El Salvador a hillside planted to beans, facing towards the northeast so the moist tradewinds came directly to it, apparently was especially suitable for rust disease, *Uromyces appendiculatus typica*. On the same mountain, at the same elevation, but around on the other side, where the air was dryer and tradewinds were not felt, beans grew comparatively free from the rust. Another typical instance was that of a series of hillsides, that were planted to banana. These fields were in Costa Rica and were grown at about 1500 feet elevation and bore profitable crops although the leafspot, *Cercospora musae*, was present but not excessive. There was a small, narrowly cleft valley along with the hillside plantings, and it lay at such an angle that, although it was at the same elevation, the banana plants growing in it were in shade for about half of each day. In the plantings in this narrow and partly shaded valley the bananas bore well, and had only a few scattered leaf spots. "Heads" or corms from this area were used as "seed," and planted in a warmer new location of low flat land near the Caribbean coast, and were not sprayed. In two years the lowland planting was abandoned because of leafspot.

A large experimental plantation of cinchona was grown in a highland area of Guatemala. Its most persistent problems were stripe cankers, *Phytophthora* spp., that destroyed planting after planting. One slope faced the northeast where the cinchona trees grew well and remained vigorous and free from canker. This was not because of resistance, as the genetics of the trees on these slopes was identical. In the healthy field the soil was comparatively solid, had good percolation, was well drained and fairly neutral. The soil of the nearby west slope where cinchona repeatedly succumbed to *Phytophthora* infections, was poorly drained, slightly acid, and spongy wet and had mosses on it before it was cleared. On both of these slopes workers were frequently criss-crossing from one to the other field so there must have been equally good chance of contamination with the disease fungi.

The sole effect of shade, no other factors involved, is at time striking in relation to plant disease. In work in Costa Rica where there was an ample natural supply of inoculum, leaf infection occurrence was studied on coffee *Cercospora coffeicola*. Unsprayed seedlings were grown in nursery beds under slat houses with four different amounts of shade, described as: full sun, light shade, medium shade, and heavy shade. Actual counts of leafspots from *Cercospora* revealed that by far the most severe attack occurred in full sun-exposed leaves. The least spotting was in leaves under heavy shade, under the medium shading there was a medium number of leafspots, while next to the most spots were on leaves that had only light shade.

In studies on the same disease made ten years later in Puerto Rico, conditions were more controlled. Trees were grown in an inert medium, regularly watered and flooded with nutrient solution every two weeks. These trees, never sprayed, were about three feet tall and grew in a series of shade houses in which the amount of shade was regulated with plastic screening. There were many old spot-diseased coffee trees around the shade houses. The shade was so arranged that there was ample chance for fungus spores to be carried inside and reach all coffee leaves. In three months time all the experimental trees had some leafspotting. The trees in full sunlight averaged about 800 spots per tree, while trees under 90 per cent shade averaged about nine spots each. At 50 per cent shade the average was 312 spots per tree, at 27 per cent shade there were 460 spots per tree, and at 12 per cent shade there were about 660 spots per tree.

The bean angular leafspot, Isariopsis griseola, is much more serious in the tropics than it is in the temperate zone; in the tropics presence or absence of shade appears to have a distinct affect on it. Beans planted between rows of dead but standing corn stalks, with old dry leaves, are often shaded to almost a 40 per cent condition. Under these conditions there is about half the number of angular leafspots as those in full sunlight. Where edges of bean fields are planted so they come under the influence of shade from trees or buildings, it has been repeatedly observed that angular leafspot is much reduced on the bean leaves.

In connection with work on diseases of citrus nurseries, one of the problems in El Salvador, Florida and Puerto Rico has been the common withertip, Gloeosporium limetticola, of lime trees. Conditions that are best for fruit production are also good for the disease production. In both El Salvador and Puerto Rico sun scorch and withertip occurred on lime seedlings at the beginning of the wet season. Comparative freedom from withertip, however, could be maintained beneath a canopy of trees. Growth was tested in lath houses: in both El Salvador and Puerto Rico, seedlings under lath shade grew almost completely free from withertip infection, while seedlings grown in full sunlight showed over half of their twigs and leaves badly diseased.

In the relatively wealthy temperate zone regions, where the sprays and other means of disease control can be afforded and readily applied, disease reduction under shade is not so important. In the tropics, where agriculture is a less technological art than in the temperate zone, growers rely as much as possible on any "natural means" of disease control that they can discover. The use of shade, and its manipulation is important in this regard and in some coffee regions it has become an art. Observations indicate that both cacao and coffee shading, as will be discussed elsewhere, is not so much a matter of horticultural practice affecting tree growth as a horticultural practice that has its most important effect on amelioration of some of the serious diseases.

Winds and Disease

In some of the American tropics we are blessed with trade-winds. Generally, north of the geographical equator the trades come from northeast (averaging about 18 km. per hour). Those living in the region recognize the winds for their "steady on course" character, so valued by sailors. These winds (see 10, 37) travel long distances and undoubtedly at times bring with them fungus spores. When the trades reach land their moistened air strikes foliage cooled by self-shade and respiration. This moist air condenses and hygroscopic leaf and stem surfaces absorb the moisture, and chance parasite spores imbibe water for germination and host infection.

In the Caribbean the strong east-northeast "sea breeze" starts over the land at, say, 10 A. M. This moving air is 3° to 5° C cooler than still air, and by about 4:30 P. M. the breeze dies away. As night falls the air out at sea, but close to land, has cooled a few degrees and the gentle, night off-shore breeze begins. Thus there is diurnal interchange and this furnishes water clouds for the rains during the rainy season, and the humidity that affects and moistens plant surfaces in the dry season.

A very brief account of mountainous land regions played upon by the northeast trade-winds is about as follows. Because these trade-winds are dryer, cultivation of crops is often more advanced on the leeward (western) slopes than on those at the wetter of windward (eastern) side. Clouds coming in off the sea discharge rains on flood plains, and even more heavily a little higher up, where a distinct coolness is encountered from elevation. By the time clouds reach mountain summits they are fairly spent and there is only moderate rain fall. As the air mass is forced over and down on the other side, by compression the air is warmed but not before some rain falls. The clouds more readily disperse and the result is a comparatively dry climate. From what has been outlined it is clear why low temperatures, at an elevation up-hill on the windward (east) exposure, are warmer at the same elevation on the leeward (west) side. Diseases are very different on windward and leeward slopes on the same mountain.

The play of winds must have a profound effect, sometimes indirect, on crop disease occurrence and severity. Observations show distinct differences: The sea level lands to windward, sultry and treated to tepid sea breeze, are favored parts for such diseases as *Fusarium* or bacterial wilts, *Cercospora* leafspot, *Helminthosporium* blasts, and *Sclerotium* soilrots. Further up where winds have been slightly cooled and rains are more frequent, the diseases are more likely to be black mildews, *Albugo* infections, *Leptothyrium* flyspeck of fruits, and *Triblidiella* dieback. Where winds have shoved the clouds up to even colder slopes, where more rain falls and drizzly fog is common, the diseases are often such as *Mycena* leafspot, foliage attacking *Phytophthora*, and certain black *Phoma* spots. After the breezes are over the top, still cold but dryer, and have dropped down on the other side there may be more cereal and other grass rusts, smuts, ergots, and white powdery *Sphaerotheca* or *Oidium* mildews. On sunny leeward slopes still farther and dryer, *Phyllachora* tarspots, *Diplodia* charcoal fruit rot, and anthracnoses are more common.

The tracks of western hemisphere hurricanes (the most common "seeding grounds" are off the coast of Africa) are no doubt much affected by crop disease. Winds of hurricane force are about 150 to 400 km. per hour, accompanied by heavy, driven rain. Once started by a complex of weather factors, whirlpools of air form, spinning ever more rapidly. These make up, with their surrounding clouds, the body of the hurricane and this body moves slowly at first while the cyclonic winds within are extremely fast. The body of a hurricane picks up speed as it rushes, quite often in a northeasterly direction, towards the Bahamas-Antilles archipelagos. Sometimes a hurricane goes more nearly due west towards Central America. Hurricanes are almost certain to strew disease along the broad paths they follow. I think, although absolute proof is lacking and would be difficult to secure, storms probably brought west from the Orient such tropical fungi as *Pellicularia koleroga* and *Ganoderma pseudoferreum*. Hurricanes have within them the possibilities of very fearsome plant disease distribution. To name just one such possibility: either or both of the rusts, *Hemileia vastatrix* and *H. coffeicola*, might have been picked up on the African coast and brought over to coffee in Central America or the West Indies. From there it would be a short move to South America. (See Appendix I for discovery of Hemileia in Brazil.)

History of Tropical American Phytopathology

[I wish to express here my great appreciation to John A. Stevenson for aiding me in gathering together historical facts (12). Any features that are incorrect in this historical summary or the table at the end of this chapter are entirely my responsibility.]

In the period after the discovery of the New World the great navigators and explorers were exceedingly busy in travel thrusts into what they called terra incognita. This preceded the settling of the American tropics, and some of it is indicated in Chapter I. Expeditions brought to Europe fabulous reports that stirred scientific interests, as they had never been moved before. Europeans had just finished laying some of the foundations of botany and mycology. The amount more to learn seemed now to be almost inexhaustable. It was a symptom of the stimulation going on, that when Spanish or Portuguese ships took expeditions to the Americas they always returned with all kinds of specimens, and not the least of such materials were preserved plants and fungi for study in the established centers of learning in the Old World.

It can be stated with absolute confidence that primitive man in the Neotropics was mostly agricultural. He produced his maize in plots known as milpas, in which were diseases. These small and shifting fields were (some still are) carved out of jungles by hand, using archaic tools and fire to make the clearings. This was going on a thousand years and more before the time of Columbus. When cultivated plants in exposed and eroded soils showed weakness and disease, a new plot was readied and planted, and the abandonment was reabsorbed by the surrounding jungle. Such rotation, although unplanned, halted intensification of crop plant diseases.

Indigenous peoples long ago and even now carry on magic and religious practices to avert famines and crop failures in the West Indies, the Central American countries, and in South America.

From the very beginning of tropical American modern years, the Columbus expeditions (1492-96) took from the Old World to the New many European plant parasites. For centuries explorers and colonists were earnestly commissioned by their governments to bring back, or send back, plant materials from the New World for testing in the Old.

In the meanwhile plant diseases were being studied in Europe. It is significant that they were also realized as occurring in the Americas. From old records it is known that in 1571 a blight of potatoes (Phytophthora?) was causing serious rot and short supply in Peru. Numbers of dried disease plant specimens were being sent to Europe by adventuring botanists in the West Indies.

Some of the first plant disease specimens ever examined by microscope were seen in England in 1660 by Robert Hooke using recently invented lenses from Holland. Historical accounts show that as early as 1691 wheat rust was being found by growers in Mexico. In 1727 a destructive disease of cacao (probably pod rot) was observed in Trinidad. The first collection of fungi from Argentina was secured in 1767 by the naturalist Padre Filiberto Commerson. By 1772 wheat rust, some 80 years previously just beginning to be seen, had become fully established in Central Mexico and was spreading into Baja California and northward; by 1790, it was in California Norte (the present State).

There came to the hands of Linnaeus parts of dried more-or-less saprophytic fungi from the Americas, but there are three points of special interest with respect to parasites: he named and described two of the most common and destructive dodders of the western hemisphere tropics, i.e., Cassytha filiformis in 1735 and Cuscuta americana in 1753, while in 1740 he published about the American tropical mistletoes in the genus Loranthus.

While the Linnean system of Latin binomials used in classification was being developed it was adapted to mycology. The mycologists of that distant period devised their own system of organization and used among their examples numbers of tropical fungi. In 1788 fungi were described by O. P. Swartz (1760-1818) from Cuba, Jamaica and Haiti. Many of these were from collections by naturalists and botanists such as Spruce, traveling in the Americas. Many shipments of fungi were given to men we now recognize as premier mycologists of the time. Four of the names that for us have great prestige, are Persoon (1761-1838), Fries (1798-1878), Montagne (1784-1866), and Léveillé (1796-1870). The meticulous labors of such men and their colleagues are still referred to after two centuries. These are some of the bases upon which depended the beginnings of disease control measures, for tropical planters and growers.

Valliant described a South American wood-destroying Clavaria in 1727 and in 1736 Linneaus named a Hydnum reported to be killing trees in Cuba and Brazil. Persoon discovered from Brazilian material the wood-decaying Odontis in 1796 and a Stereum in 1797. There was after that an increasing interest in Neotropical fungi and plant diseases.

When the main organization of the fungus world had been decided upon by the premier mycologists, later work developed. Carlo G. Bertero (1789-1831) collected in 1818 in Colombia, Chile, Puerto Rico, and Hispaniola, and the material was determined mostly by the famed A. DeCondolle (1778-1841). Among several others an outstanding mycological project was on Cuban fungi from several sources which Montagne published about in 1845-1850. There were in such years newer concepts evolving among mycologists, even such matters as variations in pathogenicity were used in differentiations, and such applications made mycology more dynamic and for pathologists more satisfactory. Indeed many mycologists welcomed the chance to work with pathologists. This brought to them specimens of multitudes of unknown organisms from struggling plant researchers in the Neotropics. Some of the later mycologists, to mention only a few who worked with tropical materials, were such well known names as M.A. Curtis (1808-1872), C.E. Broome (1812-1886), M.C. Cooke (1825-1914), M.J. Berkeley (1836-1884), N.T. Patouillard (1847-1928), G. Bresadola (1847-1929), G.E. Massee (1850-1917), and J.C. Arthur (1850-1942). Men like these in many countries have willingly allowed themselves to be imposed upon by pathologists from our tropics. They felt the responsibility of being of service to anyone requesting it.

There was a stimulated interest at about this time in pathology in the Neotropics, because of several circumstances. A serious, so-called "curl disease" of potatoes was found in 1847 in Bermuda. In 1869 F.M. Draenert published, from Brazil, studies on a sugarcane disease bacterium, the first report in the world of a bacterial disease of plants. There was a great stir in all the Latin American coffee countries when in 1868 coffee rust was found in Ceylon. The potential of losses that might come from it soon publicized and it made the tropical American people anxious as never before. Further, in 1870 it was seen how the hollyhock rust, long confined to Chile, suddenly spread to Europe and after that to the United States. It was a shock when Saenz in Colombia described the devastating coffee leafspot in 1876, and by 1878 Ernst had described other coffee diseases in Venezuela. In Brazil the physician J. I. Pugiarri collected fungi and plant disease specimens from 1877 to 1892. In that same country, A. Goeldi worked on coffee diseases during 1890 to 1896, and during this time described for the first time Meloidogyne, the root knot nematode. Berkeley and others identified large numbers of fungi from Brazil from about 1877 on for some years. One name of fame from those days is that of Padre Rick.

During the years about 1876 to 1880, the lively wine industry of Chile discovered in its vineyards a serious disease. At this time LeFeuvre proved there was a grape mildew in Chile and worked out control measures for it. The first official phytopathologist for Brazil was F. A. Dafert, who determined the dread Phytophthora blight of potato there in 1896. The pathologist-mycologist F. Noack worked in Brazil on diseases of coffee in 1896 to 1898. Training of Brazilians was also in progress and courses in plant pathology had been started in 1893 by Prof. García Redondo at the Polytechnical School in São Paulo. He taught the subject until 1899. A year before that, A. Puttemans (1892-1937) had started his vigorous life of work in Brazil, and taught plant pathology from 1903 to 1910. In that latter year he became the first Chief of the Phytopathological Laboratory started in Río de Janeiro.

In 1879, Carlos Spegazzini (1858-1926), the star student of the great mycological compiler Saccardo, had arrived in Argentina. Young, well-trained, energetic, with an insatiable mycological interest, he enjoyed helping solve tropical plant disease problems. He became an outstanding Argentine citizen and served the country, soon earning the confidence of specialists not only of nearby Paraguay, Chile, and Brazil, but of the U. S. A. and Europe too. During the years 1880 to 1923 he described and made world recognized lists of thousands of tropical American fungi, a large proportion of which caused diseases.

He started teaching in the 1880's and soon workers were giving him queries and specimens from subtropical parts of the United States and from the West Indies Islands, from Central America, and from northern countries of South America. Meanwhile Argentine growers sent him problems, and questions came in as well from Peru, Paraguay, Chile, and Brazil. During his life he was a dominant influence in mycology and pathology in much of the American tropics. His name still remains one of the brightest for us in our tropics, and in Argentina a research institute bears his illustrious name.

Trained workers came to or were trained in Argentina. Names of a few of those notables, besides Spegazzini, are Hauman, Sanzin, Huergo, Renaco, Zeman, Baez, Fernandez-Valiela, Singer, and Marchionatto. These and their colleagues make up a select army of diagnosticians and specialists in pathology. Illustrations of a number of interesting early examples of Argentine disease work are: Huergo, during 1897 to 1902, made many general surveys of plant diseases; in 1897-1903, he did special work on cereal diseases; in 1899, on a severe grape anthracnose; in 1902, on orange diseases; and on the peach, in 1908. In addition to many other duties, from 1880 to 1887, Spegazzini made detailed studies on diseases of tobacco, tomato, sugarcane, and coffee. In 1897 a plant disease laboratory was established for him in the Oficina Químico of Argentina.

There are outstanding early examples of plant pathology in other parts of the American tropics of which only three will be mentioned: A devastating colocasia tuber rot was causing wide spread hunger in Jamaica when a commission studied it and suggested control measures in 1887. Severe maize rust was investigated in Venezuela by Ernst and his results were published in 1891. There was an increasing interest in the large number of tropical American plant rusts along with other leaf-spotting organisms, and two mycologists, Paul Sydow (1851-1925) and his son Hans Sydow (1879-1946) worked on collections and described such organisms during the years 1884 to 1941.

The Oriental coffee rust was accidentally sent to Puerto Rico in 1903, was eradicated, and a year later scientists certified its absence from the Western hemisphere. From about 1895 to 1920 many official pathologists were hired for long and short terms by governments and planters' organizations throughout the Neotropics. In 1904 the U. S. government formed the agricultural experiment station in Mayaguez, Puerto Rico; around that time the Imperial Department of Agriculture for the West Indies was instituted, and one began also for Guiana. Pathologists were in all these organizations. Long before 1905 the fatal redring disease of coconut was seen in the West Indies, but its nematode causation was not determined for many years.

From about 1890 to 1915 occurred a flourishing exchange of preserved plant disease specimens among American workers in the tropics. As well, much was sent to Europe and the U. S. A. In connection with tropical rusts, Arthur (1850-1942) was specially helpful. From 1913 to 1928 he clarified questions from many rust workers in the Americas; from his own collections and research he determined that certain rusts were specifically adapted to tropical conditions.

Tropical plant pathology, with its past and present debt to tropical mycology, has continued to be a much needed and valuable profession. A listing of brief items about it indicates early and modern developments, for which see the special chronological review in table form that appears at the end of this chapter.

Not all of the Neotropical countries had plant pathologists at work for them all of the time. There had developed from 1930 to the present, laboratories and departments of pathology in many countries hitherto lacking in plant disease specialists. Training and assistance in starting these centers of research and service were secured through the help of numerous neighboring tropical countries; some help came from the United States, through land grant colleges of that country, and some from many workers in Europe.

Table of History of Phytopathology in The American Tropics

In the American tropics--Neotropics--plant pathology for a long time has required attention and dedication from specialists. As in any great science that is so critically important to food production as is tropical plant pathology, progress is marked through the years by the events and works of man. There are many things that have not been included below that should be, but what is given indicates most of the general flow of circumstances.

For a time the most impressive plant disease items

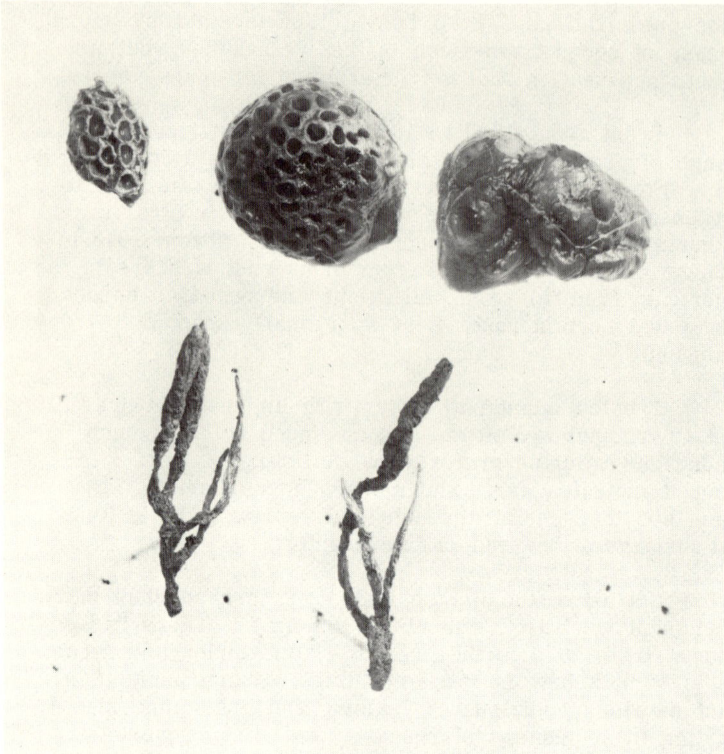

Figure 2-5. Fruiting structures from South America, of Cyttaria spp. Material preserved in the Lloyd Collection deposited in the herbarium of the U. S. National Fungus Collection. (Photo courtesy of John A. Stevenson.)

in the Americas has been those of taxonomic botanists and mycologists. However, this changed. Early scattering disease experiences became important and mycologists and pathologists from both the past and the present combined forces. As plant disease has increased, research attained a more and more sophisticated technological stage and is presently moving forward in a modern and intensive fashion.

There has been a rapid improvement in communications between pathologists, there are active professional societies now operating, and the researchers and practitioners are appreciated and valued as never before by a public perpetually up against additional crop disease problems. The training of men in the profession is progressing.

In the table of items below citations have been compressed, so that each author has one citation number. Successive titles within the citation number can be skimmed for the source relevant to the subject carrying the citation in the table.

Table 5. CHRONOLOGY OF TROPICAL AMERICAN PHYTOPATHOLOGY.

Date	Item	Literature citation numbers
--	Fossil organisms, filamentous, bacteria-like	8, 113
--	Fossil organisms, black mold infecting leaves, fungus saprophytes, etc.	14, 37, 70, 98, 113
400-800 A.D.	Maize decline in Maya country (diseases contributing?)	84
Pre-Inca	Ceramic representation, diseased potato tuber	25
1660	South American dried specimens of fungi were first of any plant disease materials examined microscopically by R. Hooke	42
1727	deVerteil report of cacao blast in Trinidad	110
1730-53	Linnaeus, working with specimens sent to him from the American tropics named parasites such as _Loranthus_ spp. in 1730, _Cassytha filiformis_ in 1735, the genus _Hydnum_ from Cuba and Brazil on dead material in 1736, and _Cuscuta americana_ in 1753	66

Date	Item	Literature citation numbers
1767	Collection in Tierra del Fuego on dwarf beech, later found again by Darwin, of parasite _Cyttaria_ (see Fig. 2-5)	42, 13
1794-1801	Fungi from tropical America, e.g., _Odontis_ and _Stereum_ in Brazil	81
1799-1804	Humboldt and Bonpland explorations and collections	20, 37, 50, 54, 57, 63
1806	Tropical fungi studied, Persoon	81
1830-51	Plant disease notes and collections, St. Hilaire	105
1831-36	Voyage of Beagle collections and notes, Darwin	42, 68, 114
1835-92	Fungi studied of Guyana, Chile, West Indies by Montagne	74
1840-60	Sugarcane degeneration in Brazil	44
1841-57	Descriptions of fungi from Chile, Guyana, Venezuela, etc., Berkeley	13
1850 approx.	_Phytophthora infestans_ on potato in Brazil	85
1853	Puerto Rico fungus work, Klotzsch	62, 104
1860-80	Sugarcane leafscald ruinous in Brazil	2, 44
1865	Wheat rust common in Chile	46
1868	French Antilles fungus studies, Roussel	93
1869	First bacterial disease in world described in plants by Draenert on sugarcane in Brazil	32, 85

Date	Item	Literature citation numbers
1875-1888	Tropical American plant disease and wood rot fungi, Cooke	26
1875	Orange and olive diseases in California, Farlow	39
1876	Coffee leafspot described in Colombia, Saenz	96
1876	*Phytophthora* footrot on orange in Florida	41
1876-97	Rusts on barberry and wheat in South America, Magnus	67
1877	Brazilian fungi described, Berkeley and Cooke	13
1877-78	Grapevine *Oidium* and anthracnose studied in Chile, LeFeuvre	64
1878	First book on plant disease in Peru, García Merino	99
1878	Coffee nematodes in Brazil	60
1878	Coffee leafspot from *Cercospora* in Brazil	48
1879	Spegazzini arrival in Argentina	42, 68
1880-91	Coffee diseases and maize rust studies, Ernst	38
1880-1923	Tropical fungus studies, Spegazzini	95, 101
1881	Coffee *Cercospora* in Jamaica, Ernst	38
1882-1925	Appearance of Sylloge Fungorum, many tropical American plant diseases, Saccardo *et al.*	95

Date	Item	Literature citation numbers
1882	Cape Horn fungi studied, Heriot	52, 68
1883	Florida orange blight	41
1883	Florida fungus diseases, Ellis and Martin	36
1884-1930	Rusts and other fungi from Neotropics, the Sydows	42, 109
1887	South American sugarcane and coffee diseases, Spegazzini	101
1887	Colocasia disease in Jamaica	69
1887	Florida fungus studies, Calkins	22
1887-1903	Neotropical (Venezuelan Orinoco, Antilles and other areas) disease fungi described, Patouillard *et al*.	80
1887	First description of *Meloidogyne* nematode cause of coffee root knot in Brazil, Goeldi	48
1889	Tropical American fungi studied, Heriot	52
1889-1905	Rehm's consultations and mycological descriptions	87
1890	Fusarium banana wilt in American tropics	107
1890-91	Maize rust studies in Venezuela, Ernst	38
1890	Bourbon sugarcane stemrot in West Indies	110
1893-97	Disease fungi described from California and other parts of American tropics, Ellis and associates	36

Date	Item	Literature citation numbers
1893	Fungi described from Chile, Philippi	82
1893-1905	Hennings extensive publishing on tropical fungi	51
1893-99	First teaching of phytopathology in Brazil by García Redondo	85
1893-95	Rusts of South America studied, Möller	73
1894-96	Dafert's plant disease work in Brazil	29
1894-97	Citrus diseases studied, Webber	115
1895	Cacao witches' broom in Surinam	102
1895-1920	Increased hiring of pathologists in tropical countries and planters associations	
1895	Dafert in Brazil first official South American country phyto-pathologist	85, cf 29
1895	Sugarcane red rot in West Indies	2
1896	Incidental fungus collecting in Yucatan, Millspaugh	71
1896-99	Rusts from Chile	29, 31
1896	Special fungus collection studies from Brazil	21, 68, 95
1896	Florida citrus disease contributions	41, 108
1896	Costa Rica fungus studies published	19, 95, 101, 112

Date	Item	Literature citation numbers
1897	Coffee disease studies in Paraguay, Bertoni	17
1897	Oficina Químico-Agricola in Buenos Aires with work on diseases of tobacco, alfalfa, and other crops	42
1897-1904	Rust materials studied from Mexico	53
1897-1907	Huergo official Argentine nation phytopathologist	42, cf. 56, 68
1898	Coffee rootrot in Brazil	75
1898	Tobacco bluemold found and described in Argentina	42, 68, 101
c. 1899	Sugarcane mosaic in West Indies and Tucumán	
1899-1908	Early researches in Chile on diseases of cereals, potato, bean, watermelon, grape, pear, apple	56, 65
1899	Oficina de Botanica y Fitopatología, Argentina	68
1899-1901	Earle in disease fungus studies--South America, Florida, Puerto Rico	34
1900	Coffee nematode disease studies in Brazil, D'Utra	33
1901	Bean _Colletotrichum_ study in Brazil	85
1902	Costa Rican fungus studies, Pittier	83
1903	Coffee rust accidentally shipped	116

Date	Item	Literature citation numbers
	to Puerto Rico but completely eradicated	
c. 1903	*Sclerotium rolfsii* in Neotropics	5
1904	Special formal plant disease publication in Argentina	47
1904-05	Brazilian fungus studies, Rick	88
1904-1906	Establishment of experiment stations in Puerto Rico, British West Indies, and Guiana	
1908	Coconut budrot in Cuba	55
1910	West Indies plant disease handbook, Bancroft	7
1910-15	Researches on tropical rusts, Arthur	4
1910	Cacao *Phytophthora* diseases in Trinidad, Rorer	91
1911	Bacterial disease of banana and plantain, Rorer	91
1911	Citrus canker from Japan to Texas	41
1912-13	Banana *Fusarium* wilt in Cuba and Neotropics	100, 107
1913	Cacao witches' broom in Trinidad	91
1914	Diseases of: potato in Brazil (97), coffee in Colombia (23), cacao in Mexico (24), citrus in Cuba (41), cotton in West Indies (76), various crops in Brazil (49), various crops in Puerto Rico (104), probably sugarcane mosaic in Tucumán,	

Date	Item	Literature citation numbers
	Argentina (92), algal cacao disease in Trinidad (43), citrus canker in Cuba (61), banana wilt in Jamaica (58), and cacao spraying in Trinidad (91).	
1915	Diseases of: coconut in Jamaica (4), coffee in Puerto Rico (40), limes in Trinidad and Tobago (91), papaya in Barbados (76), pigeon pea in Brazil (86), cassava in Trinidad and Tobago (91), cacao in Jamaica (4), coconut budrot in Cuba (61), potato blight in Cuba (59), and studies on citrus canker control in Florida (90), spraying to control diseases on cacao, coconut, and cassava in Trinidad and Tobago (91).	
1915	Publishing on rusts in Puerto Rico, Arthur	3
1915	Cacao witches' broom fungus described in Surinam, Stahel	102
1916	Amazon Valley fungi described	109
1917-19	Mottling disease (mosaic) in Puerto Rico, Stevenson	104
1918	Brazil established Laboratorio Patalogía Vegetal	
1923	Text on diseases of the Lesser Antilles, Nowell	76
1926	Introduced mosaic resistant POJ sugarcane into Neotropics	
1926	Pittier's work on 298 spp. of Venezuelan fungi	83
1927	Descriptions of Costan Rican fungi	109

Date	Item	Literature citation numbers
1927	Citrus canker eradicated from Florida	41
1930-1950	Acknowledgment of lessons from crop failures and need in our tropics for better manning of the profession of phytopathology; training of official heads and leaders for Departments or Sections in country Ministries; teaching of the science in selected institutions	
1936	Meeting of Phytopatologistas do Brasil	
1937	Sugarcane smut in Argentina	28
1937	Publication of mycology and pathology bibliography for the American tropics, Otero and Cook	78
1940-50	Renewed emphasis on tropical plant disease research	
1949	Meeting in Guatemala of Central American pathologists	
1950	International meeting of Latin American pathologists in Mexico	
1951	Bibliography of tropical plant pathology, Barnes and Allen	9
1951-54	Publication of 3 volumes on pan-tropic plant diseases, Roger	89
1952	Cacao witches' broom in Peru (illustrates disease spread)	99
1954	University of Recife establishes Instituto de Micología	

Date	Item	Literature citation numbers
1960-61	Caribbean Division of American Phytopathological Society	
1962	Symposium on tropical plant diseases	116
1962	World adoption of Celsius (C = centigrade temperature scale)	106
1964	In Argentina founding of Instuto Spegazzini	
1964-66	Asociación Latinoamericana de Fitopatología	

Literature Cited

1. Ainsworth, G. C. 1961. Ainsworth and Bisby's dictionary of the fungi. 5th ed. Commonwealth Myc. Inst., Kew (Surrey) England. 547 p.

2. Arruda, S. C. 1941. A historia das grandes epifitias da cana de açúcar. Biologico. 7:313-318. --. 1944. A escaldadura das folhas, doença da cana de açúcar, nova no Brasil. Arq. Inst. Biol. (São Paulo) 15:142-196. --. with J. Franco do Amaral. 1945. Leafscald of sugar cane in Brazil. Phytopathology 35:135-137.

3. Arthur, J. C. 1915. Uredinales of Porto Rico based on collections by F. L. Stevens. Mycologia 7:186-196. (Numerous other publications.)

4. Ashby, S. F. 1915. Cacao diseases. J. Jamaica Agr. Soc. 19(1):8-11.

5. Aycock, R. 1966. Stem rot and other diseases caused by Sclerotium rolfsii; or, the status of Rolf's fungus after 70 years. North Carolina Agr. Exp. Sta. Tech. Bulletin 174. 202 p.

6. Baker, R. E. D. 1940. Distribution of fungus diseases of crop plants in the Caribbean region. Trop. Agr. (Trinidad) 17:90-94.

7. Bancroft, C.K. 1910. A handbook of the fungus diseases of West Indian plants. Putnam, New York, 40 p.

8. Barghoorn, E. S. and S. A. Tyler. 1965. Microorganisms from the gunflint chert. Science 147:563-577. --. with J. W. Schopf. 1965. Microorganisms from the late Precambrian of Central Australia. Science 150:337-339. Also see Schopf, J. W., E. S. Barghoorn, M. D. Maser, and R. O. Gordon. 1965. Electron microscopy of fossil bacteria two billion years old. Science 149:1365-1367. (Many citations in connection with these articles.)

9. Barnes, H. V. and J. M. Allen. 1951. A bibliography of plant pathology in the tropics and in Latin America. United States Dep. Agr., Bibl. Bulletin 14. 78 p.

10. Bartholomew, J. 1950. The advanced atlas of modern geography. Meikeljohn, London; and McGraw-Hill, New York.

11. Bateman, D. F. and A. W. Dimock. 1959. The influence of temperature on root rots of poinsettia caused by <u>Thielaviopsis basicola</u>, <u>Rhizoctonia solani</u>, and <u>Pythium ultimum</u>. Phytopathology 49:641-647.

12. Benjamin, C. R. 1963. The national fungus collections. Plant Sci. Bull. 9:1-6.

13. Berkeley, M. J. 1842. An edible fungus from Tierra del Fuego and an allied Chilean species. Trans. Linnean Soc. (London) 19:37-43. --. 1852. Enumeration of some fungi from Santo Domingo. Ann. Mag. Nat. Hist. II, 9:192-203. --. 1860-1869. On a collection of fungi from Cuba. Pt. II. J. Linnean Soc. (London) 10:341-392. --. 1874 and 1876. Enumeration of the fungi collected during the expedition of H. M. S. Challenger. J. Linnean Soc. (London) 14:350-354; 15:48-53. --. with M. C. Cooke. 1877. The fungi of Brazil including those collected by J. W. H. Trail, Esq., M. A., in 1874. J. Linnean Soc. (London) 15:363-398. --.

with M. A. Curtis. 1869. Fungi Cubensis (Hymenomycetes). J. Linnean Soc. (London) 10:280-341. (Numerous other reports on fungi from such other countries as Surinam, Mexico, and additional from Santo Domingo, Cuba, Brazil, and South America.)

14. Berry, E. W. 1916. Remarkable fossil fungi. Mycologia 8:73-79.

15. Berry, J. A. and C. A. Magoon. 1934. Growth of microorganisms at and below 0° C. Phytopathology 24:780-796.

16. Bertero, C. 1828-1829. Lista de las plantas de Chile. El Mercurio 4:194-195; 12:558; 13:596-605; 14:613; 16:742.

17. Bertoni, M. S. 1897-99. Una nueva enfermedad de cafe. Rev. Agr. (Paraguay) 9. 7(R32):211-223.

18. Bisby, G. R. 1943. Geographical distribution of fungi. Bot. Rev. 9:466-482.

19. Bommer, J. E. and M. Rousseau. 1896. Fungi in Primitiae; Florae Costaricensis. Bull. Soc. Bot. Belgium 35:151-166. --. 1905. Champignons, resultats du voyage la S. M. Bélgica en 1897-1899. 15 p.

20. Bonpland, A. and A. Humboldt. 1825. Nova genera et species plantorum quas in peregrinaciones ad plagam aequinoctialem orbis novi collegerunt, descripserunt, partim adumbraverunt, Amat. Bonpland et Alex. Humboldt exchedis autographis Amati Bonpland in ordem digessit Carol Sigismund Kunth accedunt tabulae aeri incisae, et Alexandri de Humboldt notationes ad geographiam plantarum spectantes. Lutecia, Parisorum 7. (Not seen in original.)

21. Bresadola, J. 1896. Fungi brasiliensis lecti a Dr. Alfredo Möller. Hedwigia 35:276-302.

22. Calkins, W. 1887. Notes on Florida fungi. J. Mycol. 3:33-34.

23. Carradine, E. G. 1914. Enfermedades del cafeto.

Colombia Rev. Nac. Agr. X, 9(123):93-97.

24. Chevalier, A. 1914. Algunos enemigos y enfermedades del cacaotero. Bol. Soc. Agr. Mexicana 34:662-663.

25. Christiansen G., J. 1967. El cultivo de la papa en el Peru. Lima, Peru. 351 p.

26. Cooke, M. C. 1875. Mycographia seu Icones Fungorum. I. Discomycetes. London. 120 p. --. 1880. Exotic fungi. Grevillea 9:10-11. --. 1881. Some exotic fungi Grevillea 9:97-101. --. 1882. Exotic fungi. Grevillea 10:123. --. 1884. Notes on Hypocreaceae. Grevillea 12:77-83. --. 1886. Praecursores ad monographiam Polypororum. Grevillea 15:19-27. --. 1888. Some exotic fungi. Grevillea 16:121.

27. Cox, A. E. and E. C. Large. 1960. Potato blight epidemics throughout the world. U. S. Dept. of Agr., Agr. Handbook 174. 230 p.

28. Cross, W. E. 1946. La actuación de la estación experimental frente a la crisis producida por el carbone de la cana de azucar. Estac. Exp. Agr. Tucumán, Circular 136. 7 p.

29. Dafert, F. W. 1896. Molestias de planta observados no estado em 1894 e 1895. Relatorio An. Inst. Agron. (est. São Paulo 1894-1895) 7 & 8:317-326.

30. Diehl, W. W. 1937. A basis for mycogeography. J. Washington Acad. Sci. 27:244-254.

31. Dietal, P. and F. W. Neger. 1896, 1897, 1899. Uredinaceae chilensis. Engler Bot. Jahrb. I, 22: 348-358; II, 24:153-162; III, 27:1-16.

32. Draenert, F. M. 1869. Bericht über die Krankheit des Zuckerrohres. Zeitschr. Parasitenk. 1:13-17; 212.

33. D'Utra, G. 1900. Molestias vermiculares do cafeeiro. Bol. Agr., (São Paulo) L:1-16.

34. Earle, F. S. 1899. Some fungi from South America. Bull. Torrey Bot. Club 26:632-634. --. 1900. Some Florida fungi. Bull. Torrey Bot. Club 27:120-123. --. 1901. Some fungi from Puerto Rico. Muhelenbergia 1:10-17.

35. Edgerton, C. W. 1915. Effect of temperature on Glomerella. Phytopathology 5:246-259. --. 1958. Sugarcane and its diseases. Louisiana State Univ. Press, Baton Rouge. 297 p.

36. Ellis, J. B. and B. M. Everhart. 1893. New California fungi. Erythea 1:145-147. --. 1894. Description of a new fungus. J. Trinidad Field Nat. Club 1:1. --. 1896. New species of tropical fungi. State Univ. Iowa, Lab. Nat. Hist. Bull. 4:62-72. --. with F. D. Kelsey. 1897. New West Indian fungi. Bull. Torrey Bot. Club 24:207-209. --. with T. H. MacBride. 1896. Nicaraguan Hymenomycetes. State Univ. Iowa, Lab. Nat. Hist. Bull. 3:190-194. (And others.)

37. Encyclopaedia Brittanica. Eds. & Contribs. 1963. Chicago. 23 vols.

38. Ernst, A. 1880. Coffee diseases in New Granada. Nature 23:292. --1891. El tizón de maíz. Boletin Min. Obras Publ. Caracas (Dec. 26, 1891) 4 p.

39. Farlow, W. S. 1876. On a disease of olive and orange trees occurring in California in the spring and summer of 1875. Bussey Inst. Bull. 1:404-414.

40. Fawcett, G. L. 1915. Fungus diseases of coffee in Porto Rico. Porto Rico Agr. Exp. Sta. Bulletin 17. 28 p.

41. Fawcett, H. S. 1936. Citrus diseases and their control. McGraw-Hill, New York. 656 p.

42. Fernández Valiela, M. 1955. Introducción a la fitopatología. Talleres Gráficas Gadola. Buenos Aires. 872 p.

43. Freeman, W. G. 1914. An algal disease of cacao, Cephaleuros virescens. Bull. Dep. Agr. Trinidad and Tobago 13:263-264.

44. Galli, F. et al (vide supra, Chap. I). 1968. Manual de fitopatologia, doenças das plantas e seu controle. Editoria Agronomica Ceres, São Paulo. 640 p.

45. García Rada, G. 1947. Fitopatología agricola del Perú. Estac. Exp. Agr., La Molina, Peru 423 p.

46. Gay, C. 1865. Historia física y politica de Chile; Agricultura [vol. 2]. E. Thunot Paris. 450 p.

47. Girola, C.D. 1904. Investigación agrícola en la República Argentina. An. Min. Agr. Argentina 1: 217-238.

48. Goeldi, E.A. 1892. Relatorio sobre a molestia do cafeeiro na provincia do Río de Janeiro 1887. Arch. Mus. Nac. Río de Janeiro 8:8-123.

49. Guimaraes, R. F. 1914. Molestias e parasitas das plantas e seu tratamento. Bol. Agr. (São Paulo), 15:445-460; 773-776.

50. H B K. (Humboldt, Bonpland, Kunth. See citation also of Hooker.)

51. Hennings, P. 1893. Fungi brasilienses. Engler Bot. Jahrb. 17:523-526. --. 1895. Fungi americano-borealis, jamaicenses, centro-africani, turkestanici. Hedwigia 37:267-295. --. 1897. Beiträge sur Pilzflora Südamerikas II. Hedwigia 36:190-246. --. 1899. Fungi chilenses a. cl. F. Neger collecti. Hedwigia 38:71-73. --. 1899. Uredineae aliquot brasilianae novae a. cl. E. Ule lectae. Hedwigia 38:129-130. --. 1900. Fungi paráenses I. Hedwigia 39:78-80. --. 1902. Fungi paráenses II. Hedwigia 41:15-18. --. 1902. Fungi costaricenses I. Hedwigia 41:101-105. --. 1902. Fungi S. Paulensis I and II. Hedwigia 41:104-118; 295-311. --. 1904. Fungi Amazonica I, II and III. Hedwigia 43:154-186; 242-273; 340-351. --. 1904. Fungi S. Paulensis Hedwigia 43:197-209. --. 1905. Fungi Amazonica IV. Hedwigia 44:57-71. --. with P. Magnus and J. Bresadola. 1893. Die von Herrn P. Sintensis auf der Inseln Portorico 1884-1887 gesammelten Pilze. Engler Bot. Jahrb. 17:489-501.

52. Heriot, P. 1889. Champignons en mission scientifique du Cap Horn, 1882-1883. Botanique 5:173-206.

53. Holway, E. W. D. 1897. Mexican fungi. Bot. Gaz. 27:23-38. --1901. Mexican fungi III. Bot. Gaz. 31:326-338. --1904. Mexican Uredineae. Ann. Myc. 2:391-394.

54. Hooker, W. J. in Kunth, C. S. (q. v.) 1822. Synopsis plantarum quas in plagam acquinoctialem orbis novi collegerunt Al. de Humboldt et Am. Bonpland, Fungi. pp. 7-15.

55. Horne, W. T. 1908. The bud rot and some other coconut troubles in Cuba. Estac. Centr. Agron. Cuba, Santiago de las Vegas, Boletín 15. 43 p.

56. Huerto, J. M. 1897. Enfermedades criptogámicas en la zona norte y oeste de la provincia de Buenos Aires y el broussin. Ofic. Nac. Agr. Argentina 21:575. --. 1899. Antracnosis de la vid. Min. Agr. Buenos Aires, Special publication. 22 p. --. 1903. El lino en 1902 en Santa Fe y Cordoba. Las prejuicios, las enfermedades y las causas. Bol. Agr. Ganad. Argentina 3:854-861. --. 1904. Enfermedades de los tomates. Bol. Min. Agr. Argentina 2:4-5. --. 1908. El enrulamiento de las hojas del duraznero. Min. Agr. Argentina, Circula 3. (cited by Marchionatto.)

57. Humboldt, A., A. Bonpland, & C. Kunth. 1815-1825. Nova genera et species plantarum quas in peregrenaciones ad plagam aequinoctiolem Orbis Nova collegerunt, descripserunt, partim adumbraverunt, Amat. Bonpland et Alex. Humboldt exchedis autographis Amate Bonpland in ordineim digessit C. S. Kunth accedunt tabulae cricisae, et A. de Humboldt notaciones at geographiam plantarum spectantes. Lutecia Parisorum, 7 folios, 2872 p.

58. Jamaica Committee. 1914. Report of special committee on banana disease made to Board of Management. J. Jamaica Agr. Soc. 18:450-455.

59. Jehle, R. A. 1915. A common but very serious potato disease in Cuba. Modern Cuba 3(4):46-48.

60. Jobert, C. 1878. Sur une maladie du caféier observée au Brésil. Paris Acad. Sci. Compt. Rend. 87: 941-943.

61. Johnston, J. R. 1914. The citrus canker. Modern Cuba 2(11):63. --. 1915. A remedy for the coconut bud-rot. Modern Cuba 3(3):76-80.

62. Klotzsch, J. 1853. Fungi Portoricensis quas Schwanecke hort. collegit. Beitrag zur Flora der Insel Portorico. Wernigerode, Germany. (Excerpts seen.)

63. Kunth, C. S. 1822. Synopsis plantarum quas in itinere at plagam aequinoctialem orbis novi collegerunt Al. de Humboldt et Am. Bonpland. Paris. (Pp. 7-15) vol. 1: 491 p.

64. Le Feuvre, R. 1877. El Oidium tuckeri, enfermedad de las viñas y medio de curarlas. Imprenta Nacional, Santiago. Cortilla 22 p. --. 1878. La antracnosis de la vid. Chile Bol. Soc. Nac. Agr. 9:490-495.

65. Lavergne, G. 1899. Una enfermedad criptogámicas de la papa. Chile Bol. Soc. Nac. Agr. 30:667-674. --. 1900. Una enfermedad de los papales y sandiales en las provincias de Santiago y Coquimbo. Chile Bol. Soc. Nac. Agr. 31:6. --. 1901. Observaciones sobre algunas enfermedades de la vid. Rev. Chile Hist. Nat. 5:114. --. 1901. La antracnosa de la vid (Gloeosporium ampelophagum). Chile Estac. Pat. Veg. Cartilla 8. 8 p. --. 1901. A propósito de una enfermedad de los papas y los sandías. Actes Soc. Sci. Chile 11:43-50. --. 1902. Enfermedades de los perales y manzanos. Rev. Chile Hist. Nat. 6:195.

66. (Linnaeus, C.) von Linné, C. 1735. Systema naturae. Holmiae, L. Salvatii --. 1753. Species plantarum Holmiae, L. Salvatii. --. 1754. Genera plantarum. Holmiae, L. Salvatii. --. 1803. Système sexuel des végétaux. Bertrand, Paris.

67. Magnus, P. 1876. Uber Aecidium magellanicum Berk. Hedwigia 15:2-4. --. 1892. Uber einige in Südamerica auf Berberis-arten wachsenden Uredineen.

Hedwigia 10:319-326. --. 1897. Ein auf Berberis auftretendes Aecidium von der Magellanstrasse. Deut. Bot. Gesell. (Berlin) 15:270-276. --. 1897. Uber das Mycelium der Aecidium magellanicum Berk. Deut. Bot. Gesell. (Berlin) 15:148-152. --. 1897. Uredo Goebeliana n. sp. auf Parietaria sp. bei Macuto in Venezuela. Flora (Marburg) 84: 176-177.

68. Marchionatto, J. B. 1932. Las enfermedades parásitarias de las plantas y sus relaciones con el medio ambiente. Agronomía 25:463-46. --. 1934. Influencia del medio sobre la predisposición de las plantas a las enfermedades. Rev. Argentine Agr. 1:111-114. --. 1937. Las especies de Cyttaria y Cyttariella en la Argentina. Darwiniana 4:9-32. --. 1944. Manual de las enfermedades de las plantas. Editorial Sudamericana, Buenos Aires. 368 p. --. 1948. Tratado de fitopatología. Editorial Sudamericana, Buenos Aires. 537 p. --. 1951. Los parásitos de las plantas. Acme Agency. Buenos Aires. 118 p.

69. Massee, G. 1887. Remarks on the disease of Colocasia in Jamaica. London. (Quoted by West Indies workers). --. 1891. Mycological notes II. J. Myc. (London) 6:178.

70. Meschinelli, A. 1898. Fungorum fossilium omnium hucusque cognitorum iconographia. Vicetiae [Vicenza, Italy], Aloysii Fabris 144 p.

71. Millspaugh, C. F. 1896. Contribution II to the coastal plain flora of Yucatan. Field Columbian Mus. Publ. 15 Bot. Ser. 1:281-339. (Fungi p. 285-286.)

72. Mohr, E. C. J. (Trans. and ed. R. L. Pendleton). 1944. The soils of equatorial regions with special reference to the Netherlands East Indies. Edwards, Ann Arbor, Mich. 766 p.

73. Möller, A. 1893. Die Pilzgärten einiger südamerikanischer Ameisen. Bot. Mitt. Tropen. 6:1-127. --. 1895. Brasilische Pilzblumen. Bot. Mitt. Tropen 7:1-152.

74. Montagne, C. 1835. Prodromus florae fernandezianae,

pars prima, sistema enumerationem plantarum cellularium quas in Insula Juan Fernandez a Cl. Bertero collecto describi, edique curavit. Ann. Sci. Nat. Bot. (Ser. 2) 3:347-366; 4:86-99. --. 1837. Centurie de plantes cellulaires exotiques nouvelles. Ann. Sci. Bot. (Ser. 2) 8:345-370. --. 1839. Sertum potagonicum. Voyage dans l'Amérique meridionale par A. D'Orbigni. Paris. --. 1842. Fungi cubensis. Ann. Sci. Nat. Bot. (Ser. 2) 17:119-128. (and other works on material from Chile, Guayana, West Indies.)

75. Noack, F. 1899. A podridão da raiz mestra do cafeeiro. Bol. Inst. Agr. Campinas (São Paulo) 10:224-229.

76. Nowell, W. 1914. Two physiological affectations of Sea Island cotton in the West Indies. West Indian Bull. 14:304-317. --. 1914. Diseases of the cotton plant in the West Indies. Imperial Dep. Agr. West Indies Pamphlet 74:94-105. --. 1915. The internal disease of cotton balls. Barbados Agr. News 14:222; 238-239. --. 1923. Diseases of Crop-Plants in the Lesser Antilles. West India Committee. London. 383 p.

77. Ordish, G. 1952. Untaken harvest, man's loss of crops from pest, weed and disease. Constable, London. 171 p.

78. Otero, J. I. and M. T. Cook. 1937. A bibliography of mycology and phytopathology of Central and South America, Mexico and the West Indies. J. Agr. Univ. Puerto Rico 21:249-486.

79. Padwick, G. W. 1956. Losses caused by plant diseases in the colonies. Commonwealth Myc. Inst., Phytopath. Papers 1, 60 p.

80. Patouillard, N. 1887. Contribution a l'étude des champignons extra-européens. Bull. Soc. Myc. France 3:119-131. --. with A. Gaillard. 1887. Champignons du Venezuela et principalement de la région du Haut-Orínoque, rícoltés en 1887 par M. A. Gaillard. Bull. Soc. Myc. France 4:7-46; 92-127. --. 1891. Quelques espéces nouvelles des champignons extra-européens. Rev. Mycol. 13:135-138.

--. with G. Lagerheim. 1895. Champignons de l'Equateur, Pugillus IV. Bull. Herb. Boiss. 3: 53-74. --. 1902. Champignons de la Guadeloupe, recueillis par le R. P. Duss. Soc. Myc. France 18:1-16. --. 1903. Enumération méthodique des champignons recueillis à la Guadeloupe et a Martinique. Lons-Le-Saunier, France. 94 p.

81. Persoon, C. H. 1801. Synopsis methodica fungorum sistens enumerationem omnium hucusque detectarum specierum, cum brevibus descriptionibus nec non synonymis et observationibus selectis. Göttingen. 2 vols.

82. Philippi, F. 1893. Die Pilze Chiles, sowert dieselben als Nahrumsgsmittel gebraucht werden. Hedwigia 11:115-118.

83. Pittier, H. 1902. Fungi costaricenses. Hedwigia 41:101. --. 1926. Manual de Plantas Usuales de Venezuela. Caracas. 458 p. (Given therein are 298 species of fungi.)

84. Popenoe, H. 1963. The pre-industrial cultivator in the tropics. Proceedings Internat. Union Cons. Nat. Res., Nairobi, Kenya, Proc. Sept. 17-20, 1963; 66-73.

85. Puttemans, A. 1906. Sobre uma molestia dos feijoeros (Isariopsis griseola e sus synonymos). Brazil Rev. Agr. 130. 7 p. --. 1936. Alguns dados para servir a historia da phytopathologia no Brasil e as primeiras notificações de doenças de vegetaes neste paiz. (Primera Reunião de Phytopathologistas do Brasil, Rio de Janeiro) Rodreguésia 2:17-36. (Trans. A. E. Jenkins and A. D'A. Marchant. 1940. Univ. Puerto Rico J. Agr. 24:77-107.). --. 1936. Reivindicação visando a denomição scientifica da doença da batatiera (Phytophthora infestans (Mont.) deBy. Rodriguésia 2:341-350.

86. Rangel, E. 1915. Fungos parasitas do guando (Cajanus indicus Spreng.) Bol. Agr. (São Paulo) 16 (2):145-156.

87. Rehm, H. 1889. Exotische Ascomyceten. Hedwigia 28:292-303. (Authored several other publications

on South American fungi 1897 to about 1905).

88. Rick, J. 1904. Fungi austro-americani exsiccati. Fasc. I. Annal. Myc. 2: (quoted by Puttemans). --. 1904. Fungos do Rio Grande do Sul (Brasil). Broteria 3:276-295. --. 1905. Fungi austro-americani. Fasc. II. Annal. Myc. 3:15-18.

89. Roger, L. 1951-54. Phytopathologie des pays chauds. Lechevalier, Paris. 3 vols.

90. Rolfs, P. H. 1915. Seeking adequate means of control of citrus canker. Florida Grower 11(16):20-21.

91. Rorer, J. B. 1910. Pod-rot, canker, and chupon-wilt of cacao caused by Phytophthora sp. Bul. Dep. Agr. Trinidad and Tobago 9:79-103. --. 1911. A bacterial disease of bananas and plantains. Phytopathology 1:45-49. --. 1913. The Surinam witch-broom disease of cacao. Board Agr. Trinidad and Tobago Circ. 10. 13 p. --. 1915. Fungous diseases of cassava. Bul. Dep. Agr. Trinidad and Tobago 14(2):36-38. --. 1915. Spraying of cacao, coconuts and cassava. Bul. Dep. Agr. Trinidad and Tobago 14(2):62. --. 1915. Fungous diseases of limes. Proceedings, Agr. Soc. Trinidad and Tobago. 15:14-15.

92. Rosenfeld, A. H. 1914. Las manchas blancas en las hojas de caña. Rev. Indus. Agr. Tucumán 5(4): 158-159.

93. Roussel, M. 1870. Enumération des champignons recoltés par M. Husnot aux Antilles Françaises en 1868. Paris.

94. Ruinen, J. 1956. Occurrence of Beijerinckia in the phyllosphere. Nature 177:220. --. 1961. The phyllosphere; I: An ecologically neglected mileu. Plant Soil 15:81-109. --. 1963. The phyllosphere; II: Yeasts from the phyllosphere of tropical foliage. Antonie van Leeuwenhoek J. Microb. Serol. 29:425-438.

95. Saccardo, P. A. 1882-1928. Sylloge fungorum omnium hucusque cognitorum. 25 vols.

96. Saenz, N. 1895. Memoria sobre el cultivo del cafeto. 3d. ed. Casa Editorial Perez, Bogota. 185 p.

97. São Paulo Secretaría. 1914. Cancro das batatas inglezas. São Paulo Bol. Agr. 15:390-391; 701-704; 214-221.

98. Scott, R. A. 1956. Cryptocolax--a new genus of fungi (Aspergillaceae) from the Eocene of Oregon. American J. Bot. 43:589-593.

99. de Segura, C. B. 1952. Escoba de brujas del cacao en Bagua. Cent. Nac. Invest. Expt. Agr. La Molina Boletin 44. 9 p. --. 1965. Enfermedades de cultivos tropicales y subtropicales. Editorial Jurídica, Lima. 439 p.

100. Smith, E. F. 1905. Bud-rot of the coco-nut plam in the West Indies. Science 22:500. --. 1910. A Cuban banana disease. Science 31:745.

101. Spegazzini, C. 1880-1929. (This great mycologist-pathologist contributed numerous mycological papers in many countries; some, especially pathological, are listed here.) --. 1896. Hongos de la caña de azúcar. Rev. Fac. Agr. Vet. La Plata 1896: 227-258. --. 1896. Las enfermedades del cafeto en Costa Rica. Rev. Fac. Agr. Vet. La Plata 1896:339-346. --. 1898. Sobre una nueva enfermedad del tabaco. Ofic. Quím. Agr. Buenos Aires Boletin 4. 14 p. --. 1908. Una nueva amenaza para los durazneros. Argentina Chron. Agr. II (2):38. --. 1909. El polvillo de la caña de azúcar en Tucumán. Div. Agr. Buenos Aires (quoted by Marchionatto). --. 1925. La piptostelechia del alamo blanco. Physis 8:1-11. --. 1925. Uredineas Argentinas nuevas o criticas. Rev. Argentine Bot. 1:93-174.

102. Stahel, G. 1915. Marasmius perniciosus nov. sp. de verovzaker der Krullotenziekte van de cacao en Surinam. Dep. Landb. Surinam. Bulletin 33. 27 p.

103. Stevens, F. L. 1932. Tropical plant pathology and mycology. Bull. Torrey Bot. Club 59:1-6.

104. Stevenson, J. A. 1914. (This illustrious mycology-pathology worker, whose great talent and knowledge has enriched his friends in the American tropics, has published much in mycology by himself and with others. A very few things of historic interest given.) --. with R. C. Rose. 1917. Vegetable diseases. Porto Rico Insular Sta., Ann. Rept. 1916-1917, p. 83-88. --. 1918. A check list of Porto Rican fungi and a host index. Porto Rico J. Dep. Agr. 2(3):125-264. --. 1919. The mottling or yellow-stripe disease of sugarcane. Porto Rico Dep. Agr. Labor J. 4:3-76. --. 1951. A resume of the activities of the Mycological Collections of the United States Department of Agriculture, with a phytopathological slant, 1885-1950. Plant Dis. Reptr. Suppl. 200:21-29.

105. de St. Hilaire, A. F. C. P. 1830-1851. Voyages dans l'interieur du Brésil. 8 vols.

106. Stimson, H. F. 1962. Celsius versus centigrade: the nomenclature of the temperature scale. Science 136:254-255.

107. Stover, R. H. 1962. Fusarial wilt (Panama disease) of bananas and other Musa species. Commonw. Myc. Inst. Phytopath., Paper 4. 117 p.

108. Swingle, W. T. and H. J. Webber. 1896. The principal diseases of citrous fruits in Florida. U. S. Dep. Agr. Div. Veg. Phys. Path. Bulletin 8. 42 p.

109. Sydow, P. and H. Sydow. 1900. Fungi aliquot novi a F. Stuckert in Argentina lecti. Mém. Herb. Boissint. 2 p. --. and --. 1901. Fungi novi brasiliensis a cl. Ule lecti. Bull. Herb. Boiss. 2 ser 1:77-85. --. and --. 1916. Fungi amazonica a cl. I. Ule collecti. Ann. Myc. 14:64-97. Sydow, H. [only]. 1927. Fungi in itinere costaricensi collecti. Ann. Myc. 25:1-160. --. 1930. Fungi Venezuelani. Ann. Myc. 28:29-224.

110. Tempany, H. and D. H. Grist. 1958. An introduction to tropical agriculture. Longmans Green, London. 347 p.

111. Togashi, K. 1949. Biological characters of plant pathogens temperature relations. Meibundo, Tokyo. 478 p.

112. Tonduz, A. 1897. La fumagina del cafeto. Ann. Inst. Fisico-Geogr. Iac. San Jose, Costa Rica. Publication 7. 39 p.

113. Tyler, S. A., and E. S. Barghoorn. 1954. Occurrence of structurally preserved plants in pre-Cambrian rods of the Canadian Shield. Science 119:606-608.

114. Von Hagen, V. W. 1945. South America called them; explorations of the great naturalists. Knopf, New York. 311 p.

115. Webber, H. J. 1897. Sooty mold of the orange and its treatment. U. S. Dep. Agr. Div. Veg. Phys. Path. Bulletin 13. 34 p. --. 1897. Notes on orange diseases. Florida State Hort. Soc. Proc. 10:69-73.

116. Wellman, F. L., J. C. Walker, A. A. Cook, and M. Gallegly, Jr. 1951. Effects of temperature on vegetative growth of five coffee disease fungi. Mexico Sec. Agr. Gan., Ofic. Estud. Esp. (Prim. Asambl. Latinoamer. Fitop.), Miscl. Fol. 4:251-257. --. 1961. Coffee botany, cultivation, and utilization. Leonard Hill, London; Interscience Publ., New York. 488 p. --. with T. J. Grant, L. Sequeira, L. Calpouzos, J. A. B. Nolla, and H. A. Rodenhiser. 1962. Symposium on tropical plant pathology. Phytopathology 52:928-953. --. 1967. Impact of plant disease on crop production. p. 296-302; also remarks, p. 329-330; in Turk, K. L. and L. V. Crowder., eds., 1967. Rural development in tropical Latin America. Cornell Univ. Press, Ithaca, New York. 446 p. --. 1969. More diseases on crops in the tropics than in the temperate zone. Ceiba 14:1-12.

117. Yarwood, C. E. 1962. Crop value and disease. Plant Dis. Reptr. 46:755-757.

Chapter III

NEOTROPICAL PLANT VIRUS PROBLEMS

Summary

Plant viruses are second in importance as the cause of losses to crops in the tropics. Tropical viruses are those that cause disease on tropical plants; a good many such viruses are pantropical, some have first arisen in the Americas and some are still confined to this hemisphere. The places where a specific tropical virus disease may be found are often remarkably far apart. Because of conditions injurious to insects that carry viruses, there are places in wet jungle areas where no viruses are found. When studied by electron micrographs, there are numerous shapes and sizes of viruses encountered in the American tropics, varying from very small spheroids, through rods, needle-like, and finally filamentous particles.

Introduction

The most serious loss from tropical crop diseases is due to fungus attacks, but second in importance is that of viruses. As is well known viruses are ultramicroscopic infective particles; many have been passed through filters that hold back the smallest bacteria. The particles can be seen only by use of electron microscopy, and pictures can be taken of them. They have characteristic sizes, shapes, structures, modes of occurrence in tissues, and their occurrences result in varying disease symptoms.

In the American tropics there are areas where virus diseases are plentiful (65, 75, 82-85, 94, 95, 102, 121, 140, 141, 149), some with comparatively restricted, and others with wide, host ranges. The mounting presence of crop virus troubles give evidence of many kinds of insect vectors. In spite of severity in some crop-growing areas, virus diseases are of little concern under certain tropical conditions, either because the viruses have not been introduced or because there is no dissemination of them. There are also areas where, although viruses are present, small fields are managed under a system of rapid change to new locations (such as the "milpa") and it keeps virus infections on subsistence crops at a low degree.

There is a large body of well illustrated publications (nearly a thousand in the Neotropics) on plant disease viruses. The grand total is about two thousand including findings in the American tropics, and related studies in Africa, the Oriental tropics, and fundamental discussions and descriptions from temperate zone workers. In this chapter there is no attempt to give a full coverage. This is neither a manual about viruses nor a compendium of them; only main points regarding tropical virus problems are suggested, and the relatively small number of articles cited have in them other references valuable in pursuing more details.

(Until an effective work is accomplished, I have had to use several references in attempting to give names to viruses (23, 25, 26, 35, 84, 97, 101, 141). Many cooks have compounded no broth, the aroma of which satisfies everyone. There is almost a tradition of wrangling unsettlement about virus names; certainly a flavor of ineluctability and jealousy. I have used a combination of names, some from Holmes (84), some from those employed in the Review of Applied Mycology (35), and some from Thornberry (141). A promised "New System" (101) has not appeared and it may be necessary to wait for it for years. Meanwhile, it is believed no one can mistake what I have used for names of the viruses I mention.)

The Designation of "Tropical Virus"

There are virus affections that are ordinarily spoken

of as "tropical virus diseases." However, the basis for this category is not on some clear cut distinction that characterizes the infective agents themselves. The thing that makes them fit into this is that they are found on crops in the tropical part of the world.

While it is evident that in the decade of the 1960's there are many more virus studies being carried on in the temperate zone than in the tropics, this preponderance must eventually swing in the other direction. Indeed, work on plant viruses in the tropics has lagged, and plant men in this region have relied in great measure on temperate zone findings for their explanations and results. It has been valuable and right that the tropical workers have adapted to their use temperate zone results, but this will probably not last. For, from what has been seen, it is doubtless that there are more different viruses and more different virus problems in the tropics than in the temperate zone.

There are types of diseases that are discussed by many, and unquestionably are from tropical viruses. This group, so called, is encountered on tropical crops, and some are specially good examples.

Possibly the most well known of these is the common sugarcane mosaic, Marmor sacchari Holmes. Its origin is the Orient (24, 53, 84, 85) and was brought from the Far East to such places as Argentina and Puerto Rico in seed pieces. From these areas, it has spread in other directions as indicated in Fig. 3-1. Another typical tropical virus causes the ratoon stunt of sugarcane (57, 60, 64, 141, 145, 149). It has been seen by the writer in isolated places in Central America, obviously along field edges occurring in such a pattern as to suggest mechanical transmission from naturally infected perennial week grasses, to be discussed below.

Pathologists know that plant viruses are not necessarily affected in adverse manner by being subjected to the warm (not killing) temperatures that occur so commonly in the tropics. In fact Yarwood and Holm (150) and Joshi and Holmes (88) have demonstrated that viruses may be adapted to heated conditions.

It seems well recognized that common tobacco mosaic, Marmor tabaci Holmes, is perhaps the classic kind of plant virus disease (3, 65, 82, 83, 85, 142). It is of special interest and importance that this virus in all probability first

Figure 3-1. Sketch map to indicate (dashed lines) the coming of sugarcane mosaic to the Americas. It was brought in dormant and healthy-appearing seed pieces, or setts, to plant for testing as a new crop in the Western tropics. These setts were from far away sources like the old Dutch East Indies; came by ship around the Cape of Good Hope; and were planted in such locations as the Argentine and the West Indies. Tests proved good and setts from New World plantings moved (arrows show probable directions) to other parts of the American tropics. From the beginning, there were undoubtedly some mosaic infected plants throughout, taken along in these many moved.

occurred in the middle eastern part of South America; Holmes (83 and 85) considers the center of origin to be in the Bolivian-Peru area. In any case it has spread from this tropical region so that it is now world wide in distribution (see Fig. 3-2).

The common sugarcane mosaic (24, 42, 53, 65, 84, 117, 136, 141) is a grass-infecting virus. Symptoms of it are leaf

Figure 3-2. Sketch map indicating probable center of origin (large star) of tobacco mosaic virus in South American tropics and some of the directions (arrows) in which it has been spread. It is now worldwide in distribution.

streaks and plant degeneration, and it is so usual in the Neotropics that it is looked upon as a ubiquitous sugarcane disease in the Americas. The maize dwarf virus (see such works as Cervantes et al. (32), Forbes and King (57), Galvez and Thurston (60), Thornberry et al. (142), Holdeman (82), and Wellman (149)) appears to have been first observed from the American tropics. During the period of many years observation (149) it was found by me first in 1944 in regions of Central America. Since that time I have followed literature reports of its rather rapid spread northward (12, 44, 45, 57, 82, 141, 142) into the large maize-growing areas of continental United States of America where it is destructive, and has been given various names, and is being much studied.

In cacao, which is a typically tropical crop in the Americas, are viruses in at least both Colombia and Trinidad (11, 34, 46, 89, 137). Their presence has been a fearsome threat as all planters know of the millions of trees lost in Africa from the swollen shoot virus. Fortunately this does not occur in the western hemisphere. Quarantine regulations adopted by the Organización Inter-Americana Regional Sanidad Agropecuario (OIRSA), scrupulously backed and carried out by cacao growers themselves, have kept cacao viruses localized and up to the present there is no movement of viruses on this crop from one continent to another.

On citrus crops there are virus diseases that appear to be truly tropical. The most serious of them is tristeza of sweet orange on certain root stocks (14, 16, 20, 34, 37, 40, 55, 65, 67-71, 106-108, 111, 133). While it is found at the cool edges of the Neotropics it is common in more central areas and, although not in Central America, it is especially a problem of citrus growers. The citrus psorosis, *Rimocortius* (*Citrivir*) *psorosis* (Fawcett) Holmes, which causes severe bark loosening and mild leaf symptoms, has been under study many years. Apparently closely related is concave-gum psorosis, *R. psorosis* var. *concavum* Fawcett and Bitancourt. These and other such diseases (27, 58, 62, 66, 84, 100, 102, 141) are of true tropical nature. One of these causes the so-called leprose spots on sweet orange fruit, illustrated in Fig. 3-3.

The coconut and the oil palm are killed by another tropical virus known as the lethal yellowing disease (34, 96, 124). Its violent action and final killing is striking, reminding the observer of effects from a virus that is a comparative newcomer to a crop, in which evolution has not run long enough for the host(s) or the virus to have become mutually accommodated so that the host reaction is lessened to the point that both host and virus exist together.

There are two coffee viruses of different types in Costa Rica (149): one is blister spot, found in three limited areas; found on Arabica coffee; aphid and graft transmissible; not seed-borne; and probably came from some wild tree but this is not proved. Another, causes leaf rugosity; found on Canephora (Robusta) coffee; graft transmissible; seed-borne; and seems originally introduced through imported seed.
There is no more typical tropical fruit than papaya and on the tree the virus mottling mosaic, probably *Marmor* sp., is known in the Caribbean part of the Neotropics. It is

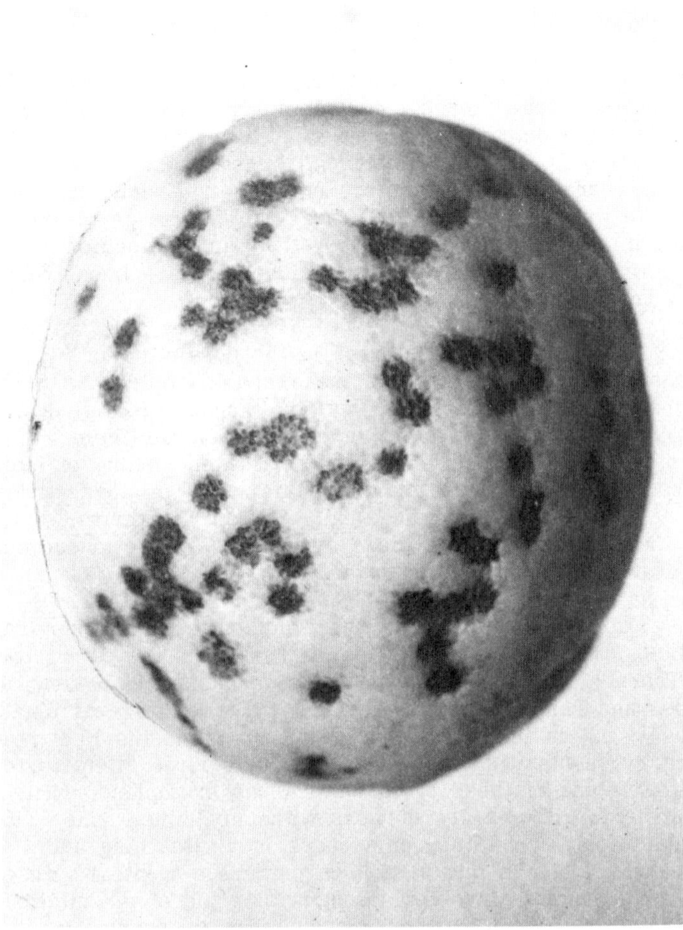

Figure 3-3. Symptoms of slight malformation and fruit spots, called leprose, of an orange; caused by systemic infection by a virus. Photo taken in Brazil, courtesy of A. S. Muller.

aphid carried and causes general stunting and mosaic patterns on leaves. A virus called bunchy top is also on papaya and disseminated by leaf-hoppers; the symptoms are shortened internodes along with leaf chlorosis, and according to the Holmes scheme, it is probably _Chlorogenus_ sp. (2, 18, 84, 99, 122).

In 1945, I observed in fields of _Cajanus cajan_ Millsp., jungle pea (gandul or pigeon pea), in _El Salvador_, what appeared to be a virus-caused witches' broom. Use of microscope, and laboratory isolation and inoculation studies proved the insects, fungi, and bacteria found on these brooms did not cause the disease. Leaves from a broom were small and misshapen, had faint mottling, were abnormally numerous and were borne on clusters of stiff and shortened stems. Absolute proof of its probable viral nature was never obtained since the field was destroyed by an uninformed administrator. However, this is believed to have been from a truly tropical virus.

There is special interest in the possibility of virus diseases from wild and shade trees in the tropics, as they may spread their troubles to coffee, cacao, spice trees, ginger, vegetables and other crops grown under them. I have seen not only leaf mottling but also shortening of internodes on new growth forced after trimming, on shade trees among species of _Gliricidia_, _Inga_, _Croton_, _Erythrina_, and _Spondias_. Suggestive as these symptoms are they were not duplicated on crops grown under them.

Lists such as that by the Commonwealth Mycological Institute (35), the work by Holmes (84), and the compilation by Thornberry (141) are of utmost importance in reviewing names and research results on the numerous tropical plant viruses. It is impossible for a pathologist in the Neotropics to carry on his responsibilities without serious attention to at least some of these diseases. As an example of what occurs: in a few years of work with only minor time available for expenditure on viruses in my teaching and research, I had to take action on numerous tropical viruses. The crops mainly involved, along with some of the citations for reference which I used, are as follows: banana and relatives (4, 19, 29, 34, 105, 129, 144, 149); celery (149); cereals (2, 84, 141); citrus tristeza (14, 16, 20, 37, 40, 55, 65, 67-71, 106-108, 111, 123, 133); citrus other than tristeza (16, 27, 54, 58, 66); maize (28, 59, 135, 149); malvaceous plants including cotton (13, 37, 38, 43, 56, 131, 141); orchids (110, 119, 141);

papaya (2, 18, 99, 122); pepper (116); potato (1, 10, 37, 39, 41, 112, 113, 116); rice (9, 51, 61, 72, 94, 98); strawberry (30, 84); sugarbeet (22, 53, 84, 141, 142); sugarcane (53, 57, 60, 64, 84, 136, 141); sunflower (143); tobacco (48, 53); tomato (7, 53).

The typically tropical banana and its relatives, plantain and abaca, are occasionally rather severely damaged by infection with the Cucurbit virus *Marmor cucumeris* var. *commelinae* Holmes. These plantation hosts, and the usual reservoir weed hosts *Commelina* spp. are especially warm country plants and when infected are illustrative of effects of a virus disease under tropical conditions. For fruit symptoms on banana see Fig. 3-4.

Indigenous Plant Viruses in the American Tropics

There are probably more plant viruses in the American tropics than in all of the temperate zone. In the tropics there are many more species of plants. Most plant viruses are intimately associated with insect vectors, and insects are much more numerous in the tropics than in the temperate zone. This insures the greater expectation of virus occurrence, spread, and evolvement under tropical conditions. It is safe to prophesy that once a much greater amount of virus study becomes established in a number of places in the western hemisphere tropics, fundamental information not now known or possible to secure, will come from such research.

There is no comprehensive coverage of all tropical crops affected by viruses in the Americas. It is hoped this lack will be somewhat helped in a volume I have in preparation listing at least a large number of the tropical American plant diseases. There is clear suggestion in reports on virus studies (e.g., host ranges: 21, 39, 40, 65, 72, 83, 84, 94, 131, 140, 141, 149) that probably all crop-plant species grown in our tropics are susceptible to virus injuries, and that viruses are serious and indeed widespread.

The viruses that are in the solanaceous crops grown for commerce in the tropics are very numerous. There are probably more tobacco viruses in the tropics than in the temperate zone. Many are still to be carefully studied, but it is likely that all those that are known in the temperate zone are present in the tropics and with them probably a number never seen before. A situation that disturbs growers

Figure 3-4. Banana fruit with symptoms of necrosis (blackening of fruit skin and dark islands in meat of fruit) from natural infection of probably the commelina strain of the cucumber mosaic virus complex. Photo taken in Ecuador, courtesy of B. H. Waite.

of tobacco in Central and South America is that certain of the viruses cause more severe necrosis than is found in the temperate zone. Many of the tomato viruses are also serious in the tropics and it is not only the common tobacco mosaic virus that attacks them but there are many others. In South America there occurs the dwarf-yellow disease which is vicious in its effect. A common complaint was that it was found more in small fields, rather than in large plantings; sometimes over 80 per cent of plants were diseased, and fruits from these are unusable even for cooking. The virus comes from nearby escoba, *Sida* sp., plants brightly marked with chlorotic, yellow-splashed, leaves. Moreover, the disease is more common in dry seasons (56, 131, 141).

Work in Brazil and Puerto Rico has shown that the virus is disseminated by whitefly, *Bemisia* spp. This fragile and small insect cannot flourish during wet seasons or in permanently moist localities. However, where it is adapted to the conditions it will take the virus to tobacco, tomato, possibly cotton, and many other plants. The occurrence of chlorotic *Sida* plants extends from the Gulf parts of the United States, southward through the West Indies, Central America and South America. In fields of tobacco in El Salvador, on its dry side where diseased escoba plants occurred abundantly, the writer was struck by irregularity of growth of the tobacco crop, the curled leaves, and the stunting and splitting of stems.

There is evidence from its exceedingly widespread occurrence, that the common tobacco mosaic, *Marmor tabaci* Holmes (also popularly called TMV, for tobacco mosaic virus), although known everywhere that tobacco is grown, is probably originally from the American tropics. It was shown by Holmes (83, 85) that disease resistance to this virus came from plant progenies out of the Neotropics. Certain wild peppers the writer has studied appeared only mildly mottled when infected with TMV. Other peppers have the habit of forming necrotic protective areas around leaf infections and of prompt dropping of infected leaves. Wild, small fruited tomatoes seldom show more than mild effects when inoculated. The famous Ambalema tobacco will become infected systemically in the leaf inoculated, but the virus does not move very far into the rest of the plant. It is evident that resistance-tolerance occurs where crop varieties have evolved along with the occurrence of *Marmor tabaci* in the wilds of the American tropics.

A virus, that enhances the horticultural attractiveness of plants in the genus Abutilon, is definitely known to have come from the American tropics. It is a striking example of such a disease, and Fig. 3-6 illustrates leaf symptoms of it. Plant explorers continually search for and obtain valuable food and ornamental plants for use in the metropolitan temperate zone. In one shipment of ornamental Abutilon spp. secured in 1858 from the West Indies, a single seedling was discovered that was exotically marked with "variegated" foliage and, when produced by cuttings from it, other plants retained this characteristic. It was at first considered at least a new variety and by some even given a special specific ranking as a plant. However, in 1906 there was published in Germany the work of Bauer (13) who proved by grafting procedures, done under controlled conditions, that the "variegation" was transmissible; since it was not of bacterial or fungus nature it was recognized as a filterable virus. It, Marmor abutilon Holmes, is known to have many malvaceous hosts, and to produce striking symptoms; it occurs not only in the American tropics but now in cultivated plants in Europe and in North America as well.

Some of the very least observed symptoms of viruses are those of wild grasses, mild streaks and stunts invisible to the general observer looking at pastures or along roadsides. Once in a long while (72, 149) workers may connect trouble on introduced crop plants with virus infection from native sources. Usually common grasses are completely disregarded as virus sources. There must be many such sources still awaiting detection. For example, when years after introduction to our tropics the much valued pangola grass, Echinochloa colonum, had won its much deserved popularity as pasture, it began to exhibit virus symptoms (49, 86). This has resulted in considerable consternation in the expanding animal husbandry of the West Indies.

Early in the 1950's rice plants were seen with leaves that exhibited chlorotic streaking that bleached and was apparently from a virus; it was called "hoja blanca" or "white leaf." It is severe in Cuba and in the northern part of South America and has been, and is being, studied in the American tropics (9, 52, 61, 72, 94, 98). This occurrence is of supreme importance as rice has become an increasingly needed food staple of our tropics. History is known of a rice famine in the Orient some 200 years ago caused by a contagious stunt, probably a virus disease.

Plants that have the Neotropical hoja blanca disease exhibit leaf bleaching that accompanies a little less striking symptom of streaks and broken lines of yellow. In seedlings that are young when infected, plants are small with complete abortion of ovules, and in some cases the fruiting panicle is arrested before it emerges from the spiga or boot. If infected after grain has set, kernels shrivel and are ruined. Hoja blanca plants often are seen first at and near field edges, with infected individuals soon appearing farther in; symptoms then spread from these in directions all around.

Research proves (94) that hoja blanca is carried by leaf hoppers, *Sagota* spp., feeding first on virus-diseased wild grasses such as *Echinocloa* sp., and then moving to rice. Seldom do the wild grasses look severely affected. The chlorotic markings are more clearly seen in young plants of wild grasses; as these approach maturity there is a tendency for the mild yellowing to disappear, and grass seed stalks develop without much hindrance. Hoja blanca appears to be a tropical American wild grass virus of systemic habit and of the group to which McKinney (84) gives the generic name of *Fractilinea*. (Until distinguished panels of virologists finally work out an acceptable (to them) means of classifying and naming viruses, I will continue with satisfaction to use as I have for years the descriptive binomial *Fractilinea oryzae* for the virus of this highly destructive rice disease.)

Wide Gaps in Geographical Occurrences

From studies at the northern border of the American tropics (Florida in the U.S.A.), a celery virus was worked with and named. It not only injures celery there but numerous other crops and has one of the widest plant virus host ranges (149). It is a specially virulent race of the common cucumber mosaic virus, *Marmor cucumeris* Holmes var. *commelinae* Holmes; it resides in weed hosts, very often in the monocotyledonous *Commelina*. In this low growing plant it has been seen in states of subtropical North America (Florida, Louisiana), in Mexico, in the West Indies, in Central America, and in South America. It is of interest to note both the almost continuous occurrence of this virus. However, it disappears in regions deeply covered with jungle (4, 19, 34, 84, 115, 144, 149).

A remarkable spread between occurrences is that of

two beet curlytop viruses. One of these is centered in the provinces of Tucumán, Río Negro, and Chubut of Argentina and is known as "encrespamiento de la hoja" or crisping of leaf. It is specially notable but not extremely severe on sugarbeet, beans, and some other wild cultivated plants (it does not infect potato or tomato). Dissemination is by Agalla ensigera, a well distributed leaf-hopper in dry areas.

At a distance of about halfway across a continent north, with the Amazon basin, much of the Andes, the northern Pacific littoral, and Central America intervening, is found the North American curly top virus, Ruga verrucosans Carsner & Bennett. It occurs in Mexico and in the hot-dry irrigated parts of California, Texas and New Mexico. It is different from the encrespamiento virus: it is disseminated by Eutettix tenellis, has a very wide host range, is specially severe on sugarbeets and causes disease in tomatoes and potatoes. Every practical means is being used to keep the South American virus where it is, and although the curly top of North America was introduced into Brazil some time ago, it is being energetically studied and quarantines are set up hopefully to keep it from other parts of the American tropics (22, 53, 55, 84, 141).

There are interesting instances of wide spread gaps among occurrences in certain still not fully studied viruses that may originally be of tropical origin but that have serious effect in the temperate zone. The radish mosaic virus, Marmor raphani Holmes (RMV), first described from material in California, U. S. A. (84), causes a rather mild chlorotic mottling, but does not induce severe plant malformations. The exact symptoms of this disease were seen by the writer on radish plants in Panama and in highland Indian gardens in Guatemala.

A typical seed-carried virus is Marmor lactucae Holmes (84) of lettuce, that causes stunting, leaf mottling and malformation, necrotic speckling and scorched leaves, and a bitter flavor. It has been known a long time in Europe and the United States. For many years I watched for it in Central America without success. However it was eventually found, in the West Indies (Puerto Rico and the Virgin Islands).

A disease, at present being questionably included among the viruses, is aster yellows, Chlorogenus callistephi Holmes (50, 80, 84). It is well known in the North American temperate zone, from the eastern seaboard to the west coast. It is likewise known in Mexico and is reported to be in Bermuda. The writer found typical symptoms of it in lettuce in Guatemala, and in New Zealand spinach on the west coast of El Salvador.

Many years apart, urgently requested consultation trips were made to visit two different growers with remarkably similar problems. One was in a Central American country and the other in an island in the West Indies. Both finca owners had made clandestine shipments of strawberry plants from California and Oregon and were growing them in cool hill country. The beds of plants developed individuals, scattered here and there, showing symptoms typical of *Marmor fragariae* Holmes (30, 84). Since a nematode causes a somewhat similar effect of crumpling and stunt, in one case (the island material) microscopic examination was used to be certain it was virus, or at least not a nematode disease. In both, on advice, the diseased beds were pulled and all plants immediately burned, closing the incidents permanently. A virus closely related to strawberry crinkle is *M. rosae* Holmes that causes the common rose mosaic. It is most commonly accepted as of temperate zone origin, but it has been long known in Brazil, and I have observed roses of the criollo type, progeny of very old "native" stock, with typical rose mosaic virus symptoms in Costa Rica.

An instance of a subtle, tissue-inhabiting virus is that of the orange tristeza (sadness) disease. Under certain conditions, the virus, in its citrus hosts remains as a latent entity and may cause little or no effects, especially on sweet orange seedlings. It probably occurred invisibly but commonly in nature first in either South Africa or southwestern Asia. There was no development of it as a serious problem until after horticulturists were selecting superior varieties of citrus that had to be multiplied by grafting or budding methods. When this was done, the virus which was ordinarily benign, would cause in certain graft combinations (see Bitancourt (20)) cellular disturbances at the union between stock and scion. This causes disease and rot of rootlets, yellowing of foliage and death of trees.

Since about 1930 the tristeza virus has been a major citrus disease problem in such parts of the Neotropics as Louisiana, Florida, California and Arizona in the U. S. A., and in South America at least in Uruguay, Argentina, Venezuela, and Brazil. In the latter country, in the state of São Paulo alone, during the decade 1936-1946 citrus growers lost over 7 million trees from this virus trouble (16, 34, 37, 55, 65-71, 84, 107-108, 148).

An example of occurrence at far distant places and of an apparent introduction and disappearance of a virus disease was experienced in Florida in 1931 to 1939. Literature reviews had suggested a potential threat to Florida growers from a virulent maize stripe virus, Fractilinea zeae Holmes, known in the Pacific Islands tropics, in Africa, and reported in Trinidad and in Cuba in the Neotropics (28, 82, 84, 135, 149). On a trip to compare possible virus symptoms on maize in Florida, maize fields were seen in two Cuban provinces where the maize stripe virus was causing farmers up to 40 per cent losses. Various aspects of the disease were noted, herbarium specimens were made, and the leaf-hopper vector, Peregrinis maidis was caught, killed, and identified (135). When I returned to Florida, the comparisons of symptoms on maize both from stripe and from Commelina virus, again added to the information and different character of symptoms from only the latter virus on maize. Search was instituted and no stripe, as seen in Cuba, could be detected in Florida. However, subsequently, in the years 1933 and 1934 I found true typical maize stripe symptoms on a few plants in maize fields in Florida near Lake Okeechobee the first year, and the next year, although causing little damage, it was again positively seen near Homestead, Chapman Field, and west of the Everglades. The vector was collected and identified from comparison with dead Cuban specimens. However, and surprisingly, in 1935 no maize stripe was encountered in any of the previous Florida locations where I saw it before, and subsequent visits a few years later have never disclosed its presence again.

Wet Jungle and Plant Virus

An impressive feature noted in wet, rainforest-type jungle is the very large number of different genera and species of trees that are free from virus symptoms. These trees are well mixed in the jungles and there is no solid stand of any species. Numbers are great, as is indicated by the true and several times repeated saying that "There are more different kinds of trees and woody plants on a 400-foot hill in the tropical rainforest than in a whole continent in the temperate zone."

In making virus observations the only thing that can be done is to take advantage of clearings and edges of tree growths on river banks and lakes. To report in the most concise manner: from examinations of tree foliage in hundreds of rainforest locations, covering many thousand specimens in the countries of Costa Rica, Ecuador, and Guatemala, at no time was it possible for the writer to see any symptom of virus on such jungle trees.

It is believed that this probably holds good in all regions that are wet, relatively wild, and seldom touched by man. There may be exceptions, but where there is enough wetness that the ecologist calls the region a moist-to-wet rainforest, certain of the trees have become adapted to conditions in which some plants would not survive. Insects of the kinds that act as vectors of viruses seemingly cannot multiply and develop normally under those conditions.

Another feature of the undisturbed rainforest is that both bushes and lesser herbaceous plants may live but do not thrive in those difficult conditions. If such lesser plant life is involved in carrying and spreading of viruses, these, along with insect vectors, can be eliminated as a factor in the jungle tree virus disease complex.

In certain areas, crop plant varieties well known to be virus-susceptible, if kept free from bacterial and fungus troubles, can also be virus free. Standard, old kinds of beans, tomatoes, sugarcane, rice, and pepper were seen in jungle clearing where, from a superficial observation, viruses should attack. However, mosaics, streaks, stunting chloroses, and similar diseases were absent. One of the contributing factors to this freedom from contracting expected virus is that the insect vectors, many of which are rather delicate, cannot withstand the attrition from constant moisture and heavy rainfalls. Deluges (I have personally worked in 275 mm. [11 in.] of rain in one ten-hour period, and 140 mm. of rain in one four-hour period) simply chill and wash off all kinds of sucking insects. Rains have been known to occur daily in a farming district in Costa Rica for over 40 days. In the northern part of Ecuador there is no real dry season, and it is one of the heaviest rainfall areas on earth. These are examples. Rains of consistency and great magnitude, wound already moist insects, tearing at wings and soft bodies, beating under scale edges, bruising and watersoaking the hapless virus vectors.

Another reason certain sucking insects do not flourish under conditions in parts of the tropics is from the larger numbers of parasitic fungi, bacteria, and viruses that may attack them. This phase of plant pathology, the control of insect vectors by diseases, deserves much more attention than it has had to the present. (For examples of some findings in the past see citations: 15, 54, 95, 125, 126, 134, 138.) Insect diseases are much more common under tropical conditions, and here is where best studies may be made on them.

All tropical areas probably have microorganisms that attack insects, but under rainforest conditions it seems they are more prevalent. This has an important bearing on plant virus diseases and bacterial troubles as well; abundance and mobility of vectors govern not only local occurrence, but breadth of dissemination and pattern as well as rate of disease spread in the field. Where certain types of insect populations are held to a small number from infections of them, the plant diseases they spread are also reduced.

A very brief list (see also Leach (96), Fawcett (54) and Sweetman (138)) of examples of organisms attacking tropical insects that are likely to spread disease on plants is as follows (those I have personally identified are marked with an asterisk *): apparent virus disease* of army worms on vegetables and undetermined bacterial diseases* of aphids on cucumber and on cotton; *Aegerita webberi* Fawcett* of whitefly on orange; *Aschersonia aleyroides* Webber*, the red fungus of whitefly on citrus, mango, and other tree crops; *A. goldiana* Sacc.* the yellow fungus of scale on citrus; *Aspergillus* spp.* of aphids on various herbaceous crops; *Cephalosporium* sp.* of rust mite on coffee; *Calonectria* sp.* of scale on citrus; *Cordyceps* sp.* of grubworms and grasshoppers on various hosts; *Pellicularia* sp.* of tree cricket on coffee; *Empusa* spp.* of flies and of small moths; *Fusarium* spp.* of scale insects on rose, and overgrowth-smothering (?) of aphids on cucurbits; *Colletotrichum* spp.* smothering of aphids on maize; *Isaria*-like fungi of various sucking insects on citrus; greenish mass of *Muscadine* fungus* of mealybug on coffee root that had been abandoned by caretaker ants; and a white *Verticillium*-like fungus* hyperparasitizing *Aschersonia goldiana*.

Of course, in some cases entomologists are called in answering anguished requests from growers before the natural fungal, or bacterial insect enemies have had a chance to become established. Unless the insect parasites have been given protection and insecticides are withheld, and fungicides have not been applied, there may be little possibility of good natural biological control of virus-carrying insects.

Some Cucurbit Virus Diseases

Symptoms of complex virus attacks on squashes and close relatives are often seen in Central America and the West Indies tropics. Studies (84, 115) have shown that squash mosaic, melon mosaic, cucumber mosaic, and others are distinctly different and can be distinguished seriologically and by other means. I have seen three different viruses in one field of cucumbers and in another of squash. As cucurbit crop varieties are being purified genetically and forced to ever higher production horticulturally, viruses are becoming more and more important pathologically.

Huisquíl, or chayote, is an important subsistence and fresh market crop in all Central America and the West Indies. I have observed in El Salvador and Honduras that the crop is widely affected by two viruses: in a few instances the common cucumber mosaic virus, *Marmor cucumeria* Holmes, which is the most destructive and severe, in some cases even killing plants; and a much milder virus disease, common everywhere but never satisfactorily identified, that in some fields will be evident in every plant but result in practically no injury.

M. cucumeris becomes very severe on cucumber in certain special farming areas. Observing the crop in Cuba, when it was being grown for U. S. trade, fields planted to a susceptible variety produced only part of one picking. In Florida in "winter-grown" cucumbers, squashes, and melons a similar thing occurs. Numerous races or varieties of *M. cucumeris* are in Florida and Cuba, and even more in other parts of the American tropics. Holmes (84) presents some suggestion of this in his Section I under the cucumber-mosaic virus group, which comprises those viruses infecting both dicotyledon and monocotyledon hosts. The latter are often the perpetuating reservoirs of these viruses. I have seen mosaic-infected *Commelina* in such widely separate regions as Florida in the U. S. A., in the islands of St.

Figure 3-5. Sketch map indicating (within dotted outline) where wild and crop plants with symptoms of infection by the cucurbit virus complex are very common, and possible area of origin of the virus; arrows show some of the directions of spread.

Thomas, St. Croix, Puerto Rico, Trinidad and Cuba in the West Indies, and in Colombia, Brazil, Mexico, Guatemala, El Salvador, Honduras, Costa Rica, and Panama. This widespread area (Fig. 3-5) may be where the virus originated; at least indications are that it probably spread in all directions from there. Not only do representatives of the cucumber viruses infect such a perennial as Commelina, but they also cause disease in grasses, maize, and, interestingly, in species of Musa.

In my first years of work, banana was artifically infected (149) with Marmor cucumeris var. commelinae

Holmes. This was thought to be a possible but more or less artificial circumstance, since it was under greenhouse conditions, and possibly of no practical moment. However, later interest was shown by banana companies (e.g., 144) and it became evident that the virus diseases of this nature had serious potential. Studies (4, 19, 29, 34, 35, 115, 129, 144, 149) in various countries now show that bananas, plantains, and abaca are all susceptible and disease symptoms vary a great deal from acute effects giving shock symptoms such as heart rot, aborted fruiting stalk, severe necrosis accompanying spectacular leaf mottling and streaks, and stunted bunchy-top-like growth (140), to lesser symptoms of mild mottling, apparent almost-complete recovery, and the phenomenon of shoots from some plants remaining virus-symptom-free. On giving the problem of virus source in nature special attention, it was learned that in a field of plantains, _Musa_ sp., there were large numbers of individuals with virus-shock symptoms. Between the rows of these comparatively giant plantain plants were cucumber vines growing close to the ground and all were severely mosaicked. It is to be recalled that the virus, Holmes's _Marmor cucumeris_ var. _commelinae_, which I found (149) to have a phenomenally wide host range, is common in many plantation inhabiting tropical weeds, such as species of _Bidens_, _Commelina_, _Physalis_, _Convolvulus_, _Ipomoea_, _Datura_, _Cucurbita_, and _Emilia_. When diseased weeds are cut in plantations the aphids leave the wilting source-hosts, crawl onto _Musa_ plants and by even short feeding immediately infect them (18).

Growers who have virus disease in plantain or banana plantations avoid, as much as possible, using virus-affected plants in their fields as sources of "seed heads" (corms) for multiplication. In this way clone transmission is kept to a minimum but I have seen it in two instances when "seed" was taken from mats of plants with such mild disease symptoms that it escaped observation by trusted laborers.

As shown by Bird and Wellman (19) the banana aphid, _Peregrinis nigronervosa_, because of its abundance, its adaptability to living on _Musa_ spp., and its comparative toughness, was used in transmission studies. From one to five viruliferous aphids are enough to transfer the virus, from one plant to another, and from plants of one _Musa_ species to another. Both this aphid and _A. coreopsidis_ are capable of transmitting the _Musa_ virus between plantain and banana. However, neither are successful in infecting a cucumber plant if they are first fed on virus-diseased banana or plantain. Colonies of insects the pathologists originally identified as *A. gossypii*, were found to be good in transmitting the virus from weeds (*Commelina*, *Bidens*, and *Physalis*) to crop-plants of *Musa* spp., and from diseased *Musas* to *Angura*, *Luffa*, and *Vinca*, and from *Musa* (banana) to *Musa* (plantain) and vice versa.

An interesting series of circumstances given careful attention are as follows. In fields, aphids multiply rapidly on such plants as angura, squash, melon, cucumber, and similar cucurbits. After some time the aphids become crowded, the colonies become restive; winged forms develop and both winged and wingless members leave their old hosts. Plants of *Musa* are not favorite food for aphids, but feeding of only one or two viruliferous individuals is sufficient to induce virus infection. In Puerto Rican studies it was found through work with *Luffa* and *Vigna* that there are at least four races of the virus *Marmor cucumeris* on *Musa* hosts.

Observations show that although cucurbits and certain other dicotyledonous hosts often exhibit the most spectacular of symptoms, this CMV, the so-called cucumber virus complex, is no doubt best thought of in warm countries as most adapted to and coming from monocotyledonous hosts.

Sugarcane Mosaic and Ratoon Stunt

Every great tropical crop is susceptible to virus attack, and to the threat of severe losses from mosaic or similar disease. Perhaps the classic example for the Neotropics is the common sugarcane mosaic.

In its early decades, after introduction to the American tropics, sugarcane became highly profitable as a plantation crop. Possibly nowhere in the world is it so well adapted to the conditions it found as it was to the islands of the Caribbean. In fact, it soon brought so much wealth to the developing Antilles that what came from it far outdistanced, in money returns, the wildest dreams and speculations of gold-seeking adventurers of the old, fabled Spanish Main.

Small islands, once nothing but ports of call for potable water and sources of the scurvy-fighting lime fruit, became settled. Their red latertic soils were cleared of jungle and planted to bananas, coffee and sugarcane. Sugar was the best product. These islands became worth fighting for by maritime powers, and were fortified. They shipped to mother countries in Europe increasing cargoes of sugar, first in individually wrapped loaves of crystals, and later in bags, and more recently in even greater bulk.

As is well known, sugarcane is a giant tropical grass; the most efficient sugar producer of any plant. At least most of the first sugarcanes brought to the American tropics were healthy. The plantings grew well, produced exceptional crops, and the future looked bright indeed. Sometime in the decade following 1890 the planters of the West Indies and of Brazil began to have trouble with a degeneration disease. This degeneration of "yellow stripe" along with the rind disease combined to make losses so severe that the whole sugarcane production faced immediate ruin (for a few citations see: 24, 53, 55, 84, 102, and Stevenson 136).

The rind disease was studied and soon was found to be caused by the fungus, _Melanconium sacchari_ Mass., which could be held in check by various cultural practices. But the other disease was a more difficult problem. It spread continuously, and with movement of the crop from area to area, "yellow stripe" was eventually everywhere its host was in the western hemisphere; from the north border of the tropics, in Louisiana and in Florida in the U. S. A., south to and including Argentina.

By 1915 it was believed the "stripe" was mosaic, later named _Marmor sacchari_ Holmes (84), and undoubtedly the same as that known in the Dutch East Indies. It menaced all the sugarcane in the American tropics and subtropics, caused 60 to 80 per cent losses, and it was determined (24, 33) that it was transferred by aphids (_Aphis maidis_) from cane plant to cane plant in the field. It was then proved that it was a mosaic and that the virus was carried from country to country in living sugarcane stalks or setts used for "seed." Where mosaic was severe, it caused so much damage it seemed inevitable that it would drive out the much needed sugar business.

The highest yielding sugarcanes are rank-growing plants, giant grasses, with thick juice-filled stalks whose internodes sugarcane men call "barrels," and are classed as "noble" canes; and were in the beginning from the species

Saccharum officianarum. When mosaic struck, specialists in the Neotropics secured varieties of sugarcane from Java, known as the P. O. J. 's (initials denote sugarcane experiment station work in Java) that had great tolerance to mosaic. They are hard and waxy canes, and more difficult to run through the old-fashioned mill than the softer varieties like "crystallina," "Louisiana purple," "estrella," and such. However, the wonderful P. O. J. canes although tolerant were not mosaic-immune, but in spite of disadvantages they saved the tropical American sugarcane industry from complete failure. Mills were modernized to accept harder canes. It became necessary to take the step of looking for even more resistant canes. Such work is dependent on controlled means of inoculation (17, 25, 53, 55) allowing more exact tests. Wild canes, S. barberi and S. spontaneum, secured in the Orient were brought to the American tropics. Crossing these with S. officinarum, with the P. O. J. 's, and then back crossing and studying disease reactions, has resulted in superior resistant types that have revolutionized and made permanent sugar production (8, 55, 84, 121).

The second most important tropical American virus disease of sugarcane is ratoon stunt. Known first in Queensland and other parts of the Orient, it was recognized in the western hemisphere, in the U. S. A., in 1952. Symptoms of this disease are exactly what the name implies: severe stunting. There is no leaf mottling or plant malformation. Cane sections are small, and distance between nodes is shorter than normal. When dissected, vascular elements are seen to be darkened and in some cases reddish brown, especially at the nodes.

On request, a considerable amount of my time was spent surveying for this disease as there was the possibility that it might have been introduced to Latin America by importations of improved sugarcanes from the United States. After two years of attention in consultation visits in El Salvador, Panama, Costa Rica, Nicaragua, and Guatemala, it seemed evident that the virus had been in Central America long before the arrival of experimental canes from U. S. government sources. Many private and local government, and veteran cane specialists, were interviewed and their fields were studied.

In both Nicaragua and El Salvador there were specialists who believed the stunt was most severe at edges of plantings, and "traveled along rows." In El Salvador it was seen in a two manzana test field, in an area where no cane had ever been planted before. After two years the first 8 to 15 mats of plants at row ends were stunted. In another two years it had spread irregularly in the other parts of the rows. Further observations on distribution showed that ratoon stunt occurs in areas where no new planting materials have been brought in for over a century. In one mountain-side plantation it was seen where it was believed no planting had ever been made of anything but crystallina, originally brought in over 60 years before. A field of "P. O. J.," believed to be "number 31" and from one of the oldest introductions, was seen in Panama and it had much stunt in it.

As a matter of curiosity, attention was given in Costa Rica to the possibility of the virus symptoms being present in gramineae other than sugarcane. Both from the standpoint of general stunting, and from occurrence of typical vascular discoloration at nodes, symptoms were found in some collections of the perennial grass species of *Panicum*, *Cynodon*, *Coix*, *Paspalum*, and *Echinocloa*. Inoculation tests to prove the virus nature of the symptoms were unfortunately not completed. It is significant, to me, however, that apparent stunt effects were never seen on any annual grasses.

In reviewing the work of so many, and examining my own experiences with tropical plant viruses, the conclusion I reach is that the only sure and practical method of control is by tolerance or resistance. It seems certain that working long enough, and with proper backing, this can be accomplished. Some examples of what has been done along these lines are evident on reading numerous reports; examples of some are found in citations 5, 8, 9, 20, 32, 51, 67, 68, 71, 108 and 121. Probably there is no agricultural region in the world where more is to be gained through plant improvement and breeding for tolerance and resistance to virus diseases of crops, than in the Neotropics.

Shapes and Sizes of Tropical Virus Particles

[This discussion comes, in part, from an unpublished paper by F. L. Wellman and T. T. Hebert. It was prepared in answer to a request from pathologists of the West Indies and Central America. I presented it in December of 1967 in Honduras at the annual meeting of the Caribbean division of the American Phytopathological Society.]

To carry on adequately verifiable researches of plant viruses, including its isolation, requires absolute surety of the identity of the virus in question. One of the major problems is that of transmission of a virus under controlled conditions. Much has been done on this in the Neotropics (see: 17, 24, 33, 45, 46, 53, 56, 89, 95, 99, 103, 106, 111, 147, 149) and it is part of the background upon which details of present and future findings about particle characters depend.

Since about 1940 the study of especially well-known and isolated plant viruses has been much improved through electron microscopy. Photographs, so-called micrographs, are made of virus particles using electronic equipment. The result is that virus particles that are invisible with the finest of conventional microscopes, now can be seen, measured, and their relation to parasitized tissues determined.

Pathologists of the American tropics have been well aware for a long time of the fact that viruses differ detectably as to susceptible hosts, distribution, transmission, serology-cross protection, thermal inactivation, and induced disease. It has not been until after the use of micrographs that much has been known about structure. Numerous plant viruses that from early electron micrographs were described as round bodies, have been shown by Caspar (31) to be polyhedra; sphere-like particles bounded by many small and flat surfaces.

In the center of such a polyhedral particle is a core of ribonucleic acid, RNA, surrounded by an envelope of protein molecules or sub-units. Some virus morphologists call a whole virus particle, whatever its shape, a virion. The protein envelope is known as a capsid and the molecular sub-units of which it is composed are called capsomeres. A rod-like virus particle is composed of capsomeres that are in a helical arrangement about a hollow tube-like center. It seems a single strand of RNA is also wound round such tube-like center, and embedded in the protein sub-units. Among the viruses are phages that attack bacteria (139) and the particles have spheroidal, or rounded, "heads" to which are attached rod-like "tails" with a fringe of appendages at the distal end. It is with these latter that the phage particle attacks a host bacterium.

It is not the purpose at this point to review the many highly specialized methods and refined techniques required in some of the ultimate findings in plant virology. However, the reader may refer to citations 6, 17, 36, 47, 64, 74-76, 81, 91, 92, 104, 109, 117, 120, 128, 130, 132, 133 for some of the methods used to obtain some of the more exciting, modern findings about plant viruses.

The mechanics of parasitism by a virus particle follows a much more intimate association between host and parasite than one finds in parasitism by fungi, bacteria, dodders, and algae. Electron microscope examinations of ultra-thin sections of tropical-virus, diseased cells, reveal (74, 77, 78, 85, 90, 123) that virus particles, in parasitic position, show no film or other fine structure between them and the host protoplasm. It is now generally agreed that after a particle enters the host cell it somehow sheds its protein coat and the freed nucleic acid then acts as genetic material; it changes the character of the cell protoplasm and it "directs" production of more virus particles; this is a complex act as RNA and protein are synthesized separately and then assembled into the _virions_.

In our tropics advantage is being taken of a few techniques making simplest and most direct preparations. A virus particle may easily contaminate the sap stream and be exuded in leaf guttation drops, and these can be used for mounts on grids. Sap smears from cut disease tissue are also a source of mounts, and, in some cases, plant juices forced from cut vasculars by water pressure can be a source. There is also the "leaf dip" method, and as well, standard precipitation. Samples, preserved as mounts, are then used for micrograph purposes (Fig. 3-6). When dried they can be taken or sent long distances. It is through such methods that virus discoveries are progressing in the American tropics, see such examples as Perez _et al._ (117, 119) in Puerto Rico; Grogan and Kimble (73) in Mexico; Herold _et al._ (77, 78, 79) in Venezuela; Kitajima _et al._ (90-93) in Brazil; Herbas (76) in Costa Rica; Forbes and Ling (57) in the American tropics; and others.

It has been shown (50, 80) that there are virus particles, mycoplasma-like, of a wide range of sizes and pleomorphic in character. One of these is from aster yellows, _Chlorogenus callistephi_ Holmes. It is known in Bermuda, Mexico, the southern part of California, and other places in our tropics.

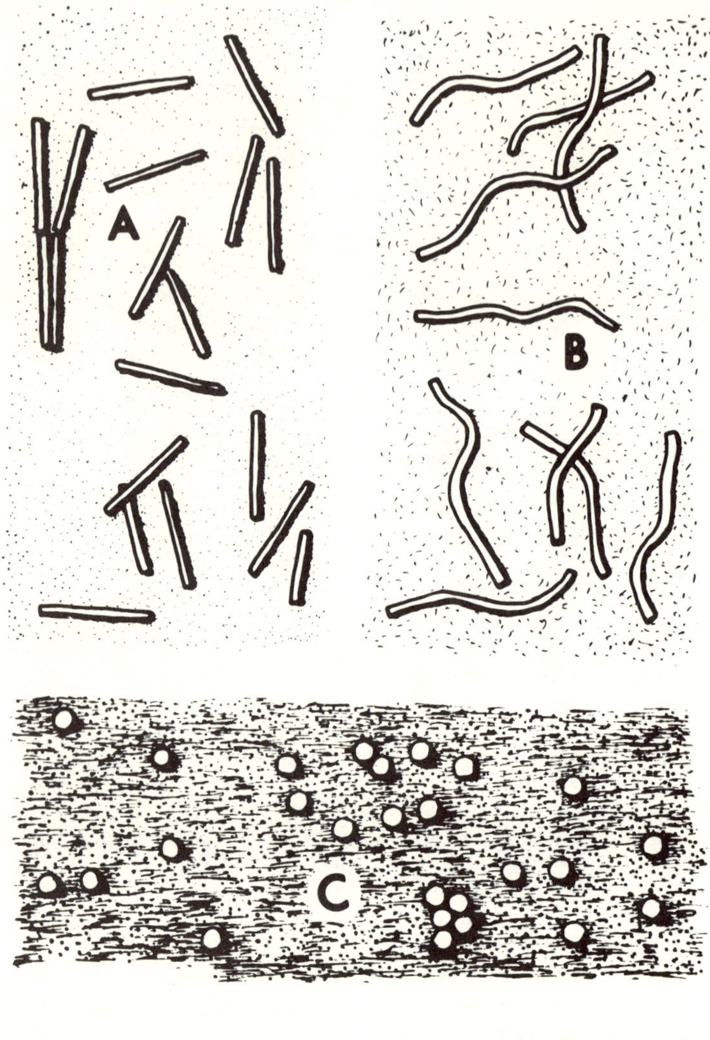

Figure 3-6. Sketches of virus particles: A. rod shaped particles; B. sinuous particles; and C. rounded or spheroidal particles. From films of electron microscope photographs made by T. T. Hebert, using the leaf dip method to secure the particles for his studies.

Particles that are small polyhedra, 25 to 35 millimicrons in diameter, are known all over the Neotropics. The Marmor cucumeris Holmes group, of this small polyhedra type, is serious at times on many subsistence crops and is a cause of disease in musaceous (e. g., banana) plantations. The sugarcane ratoon stunt (57, 64) is also from small spheroidal viruses, and another such common tropical virus is a bean mosaic (73). Relatively large polyhedral particles, about twice as large as those just mentioned, cause such diseases as the Brazilian vira cabeca (90). Another particle shape is the short rod, sometimes called "bacilliform." This is found as the cause of alfalfa mosaic in South America (63). Longer rod-like virus particles are found in two orchid virus diseases known as cymbidium mosaic (36) and odontoglossom mosaic (29, 87, 110). The virus of common tobacco mosaic, Marmor tabaci Holmes, is a needle-like particle and its long list of hosts and stability as a contaminant in dead tissues assures its widespread occurrence in the American tropics (6, 146). In this connection, it was found in diseased orchids in Puerto Rico (118) and the occurrence gave rise to considerable surprise, and almost unbelief.

Perhaps the most unusual shaped particle is tadpole-like, characteristic of at least certain species of the genus Holmes calls Phagus. Viruses of this form (139) are still to be given electron microscope attention in the American tropics. It would seem to be a particle combining characters of both the spheroid (so-called "head") and the rod (so-called "tail").

Thread-like and very long particles are another kind. Examples are the orange tristeza (93, 123) most studied in Argentina, Brazil, and in the U. S. (California and Florida), the amaryllis mosaic virus (76), and the papaya ring spot (79). Apparently the filamentous condition is found also in the tobacco mosaic virus sometimes in its attack of orchids (118, 119).

Micrographs show that rod-like particles, both short and long, vary in respects other than measurement. Some types have characteristically rounded ends, in some the ends seem "broken off" and not rounded, in some rods there is bending, in others, the rod is always stiff and straight, and in long filamentous types in some cases they are straight and in others flexuous.

Table 6. DIFFERENCES of certain Neotropical plant viruses, from micrographs

Shape (approximate)	Size (millimicron)	Induced Disease (and Citation Number)
Spheroid	26 dia.	Cowpea mosaic (73)
(small)		severe bean mosaic (73)
	32 dia.	sugarcane ratoon stunt (57, 64)
Spheroid	80-120 dia.	vira cabeça (90)
(large)		
Head-with-tail	----	bacteriophages* (139)
Rod	46 X 17	Lucern mosaic (63)
"	50 X 25	orchid virus (118)
"	45-55 long	tobacco rattle (127)
Slender	175-190 long	tobacco rattle (127)
"	90-126 X 21-24	" " (152)
"	224 X 50	maize mosaic (78)
"	242 X 48	maize raya blanca (77)
"	280 X 24	odontoglossum mosaic (25, 87, 110)
---	250 X 90	gomphrena virus (91)
"	480 X 18	cymbidium mosaic (36)
Rod-like	553 long	papaya mild mosaic (21)
"	750 X 13	maize dwarf (12, 142)
"	750 X 36	maize mosaic (128)

Filament	780 long	papaya distortion mosaic (21)
"	795 long	sugarcane mosaic (25)
"	690-770 long	" " (120)
Needle-like	300 X 15	tobacco mosaic (6, 146)
"	360 X 25	amaryllis virus (76)
Filamentous	715 X 25	amaryllis mosaic (76)
"	800 X 12	papaya ringspot (79)
"	2000 X 12	tristeza (93, 122)
"	1080 long	tobacco mosaic in orchid (118, 119)
Pleomorphic	----	aster yellows* (50, 80)
mycoplasma-like		

*Up to the year 1970, not studied under the electron microscope in material from the American tropics.

Above is presented a brief review in tabular form of the appearance of some of the virus particles of importance that infect crops in the Neotropics. As has been noted above, but deserves repeating here, electron microscopy studies show that the basic virus particle structure is of protein sub-units combined with ribonucleic acid. A particle, when it infects, acts practically like a gene in its extremely intimate parasitism of the host cell protoplasm it attacks.

Newly simplified techniques and increasing availability of electron microscopes with trained technicians, is making it easier for pathologists to secure good micrographs of hitherto unknown viruses. The viruses listed in the table vary much in shape and size, and represent a series that is going to be much added to in the coming years.

Literature Cited

It is repeated here that this chapter is not a manual-like coverage and the literature list is by no means an exhaustive presentation of all plant virus publications that have appeared from the American tropics. Various special occurrences, several host ranges, many works indicating insect vectors, and other matters are not included. Below, I give citations to works I have referred to specifically in preparing this chapter on the problems I encountered. There are over a thousand publications that might be listed at the close of this chapter.

1. Accatino L., P. 1966. Virus desconocido determinado en dos variedades de papa (*Solanum tuberosum* L.) autóctonas de Chile. Agr. Tec., Chile 26:85-86.

2. Acuña, J. and F. DeZayas. 1946. El mosaico y otras plagas de la fruta bomba (*Carica papaya* L.). Estac. Agr. Cuba Circular 85, 32 p.

3. Allard, H. A. 1914. The mosaic disease of tobacco. U. S. Dep. Agr. Bulletin 40, 33 p.

4. Allen, R. M. 1957. A virus-type disease of gros michel bananas in Costa Rica. Turrialba 7:72-83.

5. Alvarez, A., M. 1965. Determinación de la resistencia por hipersensibilidad al mosaico común del frijol Phaseolus virus 1. Agr. Tec., Chile 25:114-119.

6. ________. and P. Accatino L. 1965. Preparación del virus del mosaico del tabaco para observaciónes al microscopio electrónico. Agr. Tec., Chile 25: 28-35.

7. Alvarez G., L. A. and J. Adsuar. 1943. Studies on tomato mosaic in Puerto Rico--a new mosaic disease of tomato. Univ. Puerto Rico J. Agr. 27:141-148.

8. Arruda, S. C. and H. C. Arruda. 1949. Comportamento e producão da sã da variedad Co. 290, no Estado de S. Paulo. Arq. Inst. Biol. (São Paulo) 19:175-201.

9. Atkins, J. G. and C. R. Adair. 1958. More about hoja blanca. Authorities report on discovery of hoja blanca in Florida; also resistance tests in Cuba, Venezuela. Rice J. 61:10; 13-14.

10. Bagnall, R. H. 1963. A virus from _Solanum phureja_ Juz. et Buk. Phytopathology 53:546-547.

11. Baker, R. E. D. and W. T. Dale. 1947. Notes on a virus disease of cacao. Ann. Appl. Biol. 34:60-65.

12. Bancroft, J. B., A. J. Ullstrup, M. Mersieha, C. E. Bracker, and T. E. Snazelle. 1966. Some biological and physical properties of a meristem isolate of maize dwarf mosaic virus. Phytopathology 56:474-478.

13. Bauer, E. 1906. Weiter Mitteilunger über die infectiose chlorose der Malvaceen und über einige analoge Erscheinungen bei _Ligustrum_ und _Laburnum_. Bericht. Deutsche Bot. Ges. 24:416-428.

14. Bennett, C. W. and A. S. Costa. 1949. Tristeza disease of citrus. J. Agr. Res. 78:207-237.

15. Berger, E. W. 1908. White fly control--introducing the friendly fungi. Florida Agr. Exp. Sta. Press Bulletin 82, 2 p. --. 1921. Natural enemies of scale insects and whitefly in Florida. Florida State Plant Bd. Quarterly Bul. 5:151-154. --. 1932. The latest concerning the natural enemies of citrus insects. Florida State Hort. Soc. Proc. 45:131-136.

16. Bertelli, J. C. 1947. Primer agregado al estudio de la etiologia de la podredumbre de la raicilla o tristeza de los citrus. Uruguay Min. Ganad. Agr., Dir. Agron. Publicación 91, 16 p.

17. Bird, J. 1961. Inoculation of sugarcane plants with the mosaic virus using the airbrush. Univ. Puerto Rico J. Agr. 45:1-7. --. 1961. Further studies on transmission of the causal agent of chlorotic-streak disease of sugarcane. Univ. Puerto Rico, J. Agr. 45:8-18. --. 1962. A white fly-transmitted mosaic of Rhynchosia minima and its relation to tobacco leaf curl in Puerto Rico. Phytopathology 52:286.

18. _______. and J. Adsuar. 1952. On the viral nature of papaya bunchy top. Phytopathology 42:3. --. and --. 1952. Viral nature of papaya bunchy top. Univ. Puerto Rico, J. Agr. 36:5-11.

19. _______. and F. L. Wellman. 1962. A mosaic disease of musaceous crops in Puerto Rico. Phytopathology 52:286.

20. Bitancourt, A. A. 1940-41. A podridão das radicelas dos citrus na provincia de Corrientes, Argentina. Biologico 6:285-288; 356-364; 7:62-69. --. 1943. (Portuguese translation, of work by H. J. Webber, 1947.) Biologico 9:345-355. --. 1944. Um teste para a identificação precoce da tristeza dos citrus. Biologico 10:169-175. --. 1951. Estudos sôbre a tristeza dos citrus II. Susceptibilidad das diversas combinacoes de enxertia de laranjevia doce e langeira azêda. Arq. Inst. Biol. (São Paulo) 20: 39-52.

21. de Bokx, J. A. 1965. Hosts and electron microscopy of two papaya viruses. Plant Dis. Reptr. 49:742-746.

22. Boncquet, O. A. 1923. Discovery of curly leaf of sugarbeets in the Argentine Republic. Phytopathology 13:458-460.

23. Bos, L. 1963. Linneaus en de plantevirologie. Vakl. Biol. 43:116-128.

24. Brandes, E. W. 1920. Artificial and insect transmission of sugarcane mosaic. J. Agr. Res. 19:131-138.

25. Brandes, J. 1964. Identifizierung von gestreckten pflanzenpathogen Viren auf morphologischer Grundlage. Mitteil. Biol. Bundesanst. Land-u. Forstw., Berlin-Dahlem 110:1-130.

26. ________. and C. Wetter. 1959. Classification of elongated plant viruses on the basis of particle morphology. Virology 8:99-115.

27. Brieger, F. G., and S. Moreira. 1945. Experiências de cavalos para citrus II. Bragantia 5:597-658.

28. Briton-Jones, H. R. 1933. Stripe diseases of corn (Zea mays L.) in Trinidad. Trop. Agr. Trinidad 10:119-122.

29. Cardenosa-Barriga, R. 1962. La rayadilla del platano en Colombia. Turrialba 12:118-127.

30. Carvalho, A. M. B., and A. S. Costa. 1961. Ocorrencia do virus do mosqueado do morangueiro no estado de São Paulo. Bragantia 20:563-578.

31. Caspar, D. L. D. 1964. Structure and function of regular virus particles. (pp. 267-291). In Plant Virology. (M. K. Corbett and H. D. Sisler Eds.) Univ. Florida Press, Gainesville, Florida, 527 p.

32. Cervantes, R., J., A. Rodriguez V., and J. S. Niederhauser. 1958. Resistencia al virus causante del achaparramiento del maiz. Mexico Ofic. Estud. Esp., Tec. Folleto 29, 18 p.

33. Chardon, C. E., and R. A. Veve. 1923. The transmission of sugarcane mosaic by Aphis maidis under field conditions in Porto Rico. Phytopathology 13: 24-29.

34. Ciferri, R. 1948. Una virosis del cacao en Colombia y en la República Dominicana. Rev. Fac. Nac. Agron. Medellin 8:79-84. --. 1949. Infectious chlorosis of banana in Colombia. Nature 153:175. --. 1950. A decline of citrus plants in Venezuela. Nature 165:32. --. 1951. Bronze leaf of coconut palm in the Dominican Republic and Venezuela. Phytopathology 41:756-766.

35. Commonwealth Mycological Institute. 1957. Common names of plant virus diseases used in the Review of Applied Mycology. Rev. Appl. Myc. 35. Supplement, Aug. 1957, 78 p.

36. Corbett, M. K. 1960. Purification by density-gradient centrifugation, electron microscopy, and properties of cymbidium mosaic virus. Phytopathology 50:346-351.

37. Costa, A. S. 1950. Mancha anular de amendoim causada pelo virus de vira-cabeca. Bragantia 10: 67-68. --. 1951. Observaciones sobre la tristeza de los naranjeros Argentina Dir. Gen. Agr. Bol. 3:27-28. --. 1954. Identity between ordinary cotton mosaic and abutilon mosaic. Bragantia 13, nota 9. --. 1956. Anthocyanosis, a virus disease of cotton in Brazil. Phytopath. Zeitschr. 28:167-186.

38. ________. and A. M. B. Carvalho. 1960. Comparative studies between abutilon and euphorbia mosaic viruses. Phytopath. Zeitschr. 38:129-152. --. 1962. Molestias de virus do algodoeiro. Bragantia 21:45-62.

39. ________. and R. Forster. 1942. Lista de hospideras do virus de vira-cabeca. Bragantia 2:83-91.

40. ________., T. J. Grant, and S. Moreira. 1949. Investigações sôbre a tristeza dos citrus. II. Conceitos e dados sôbre a reacao das plantas citricas a tristeza. Bragantia 9:59-80. --. 1954. Reação da laranjeira azêda à tristeza. Bragantia 13:199-216.

41. ________. and J. Kiehl. 1938. Una molestia batatinha --necrose do topo--causada pelo de vira-cabeça. J. Agr. Brazil 1:193-200.

42. ________. and M. P. Penteado. 1951. Corn seedlings as test plants for sugar-cane mosaic virus. Phytopathology 41:758-763.

43. ________. and H. F. G. Sauer. 1954. Cotton vermelho or anthocyanosis. Bragantia 13:237-246.

44. Dale, J. L. 1965. Additional data on corn virus in Arkansas. Plant Dis. Reptr. 49:202-203.

45. ________. and L. Anzalone, Jr. 1965. Infection of sugarcane with mechanically transmissible corn virus. Plant Dis. Reptr. 49:757-760.

46. Dale, W. T. 1953. Further notes on the spread of virus in a field of clonal cacao in Trinidad. Imp. Coll. Trop. Agr.; Rept. Cacao Res. 1945-51:130-131.

47. Delgado-Sanchez, S. and R. G. Grogan. 1966. Purification and properties of potato virus Y. Phytopathology 56:1397-1403.

48. Delle Coste, A. C., and S. Zabala. 1946. La peste negra del tomate o corcova del tabaco en la Republica Argentina. Argentina Min. Agr. Inst. San. Veg., Publ. Ser. A., vol. 2:1-23.

49. Dirven, J. C. P. and H. A. van Hoof. 1960. A destructive virus disease of pangola grass. T. Pl. Ziekten 66:344-349.

50. Doi, Y., M. Teranaka, K. Tori, and H. Asuyama. 1967. Mycoplasma- or PLT group-like microorganisms found in the phloem elements of plants infected with mulberry dwarf, potato witches' broom, aster yellows, or paulowinia witches' broom. Ann. Phytopath. Soc. Japan 33:259-266.

51. Drummond, O. A. 1949. Notas sôbre o compartamento do limão rosa em face da tristeza, verrugose e gomose em Vincosa. Lilloa 21:1-5.

52. Elias R., G. and O. A. Gonzalo. 1962. El estado actual de la hoja blanca en Mexico. Agr. Tec. Mexico 2:2-11.

53. Fawcett, G. L. 1923. La transmisión del mosaico de la caña. Rev. Ind. Agr. Tucumán 13:129-131. --. 1925. El mosaico o enfermedades de las rayas amarillas de la caña. Rev. Ind. Agr. Tucumán 15:103-111. --. 1925. Notas preliminares sobre una enfermedad del tabaco. Rev. Ind. Agr. Tucumán 15:5-17. --. 1926. Encrespamiento de

las hojas de la remolacha azucarera. Rev. Ind. Agr. Tucumán 16:39-46. --. 1927. El enrulamiento de las hojas de la remolacha y el insecto transmisor. Rev. Ind. Agr. Tucumán 18:61-68. --. 1929. El enrulamiento de las hojas de la tomatera. Rev. Ind. Agr. 20:49-54. --. 1940. La peste negra de los tomates y lo corcova del tabaco. Rev. Ind. Agr. Tucuman 30:221-226.

54. Fawcett, H. S. 1908. Fungi parasitic on Aleyrodes citri. Univ. Florida, Special studies 1, 41 p. --. 1910. An important entomogenous fungus. Mycologia 2:164-168. --. 1910. Webber's brown fungus of the citrus whitefly (Aegerita webberi n. sp.). Science 31:912-913. --. 1911. Whitefly fungus from India. California Cult. 36:165-166. --. 1936. Citrus diseases and their control. McGraw-Hill, New York. 656 pp.

55. Fernandez Valiela, M. V. 1948. Informe preliminar acerca de la etiología de la podredumbre de las raicillas del naranjo agrio injertado. Rev. Invest. Agr., Argentina 2:139-146. --. 1951. Tristeza o podredumbre de las raicillas de los citrus en la Republica Argentina. Argentina Cent. Nac. Invest. Agr. Pampeana, Publicación Tecnico 1, 63 p. --. 1952. Introducción a la Fitopatología. Segunda Edicion. Gadola, Buenos Aires. 872 p. --. 1958. Tristeza de los citrus en Tucumán. IDIA 124:34-37.

56. Flores, E. and K. Silberschmidt. 1963. Ability of single whiteflies to transmit concomitantly a strain of infectious chlorosis of Malvaceae and of Leonurus mosaic virus. Phytopathology 53:238.

57. Forbes, I. L. and K. C. Ling. 1960. Virus-like particles associated with the ratoon stunting disease of sugarcane. Phytopathology 50:635.

58. Frezzi, M. J. 1940. La lepra explosiva del naranjo. Bol. Frut. Hort., Min. Agr. Nac. Argentina 5:3-15.

59. Frieda H. 1963. Estudios sobre dos enfermedades virales del maiz en Venezuela. Acta Cient. Venezolana 14:221-227.

60. Galvez, G. E. and H. D. Thurston. 1961. Ratoon stunting disease of sugarcane in Colombia. Plant Dis. Reptr. 45:954-956.

61. Garces-Orjuela, C., P. R. Jennings and R. L. Skiles. 1958. Hoja blanca of rice and the history of the disease in Colombia. Plant Dis. Reptr. 42:750-751.

62. García Rada, G. 1946. La psorosis del naranjo. Estac. Exp. Agr. La Molina, Peru. Informe 62, 19 p.

63. Gibbs, A. J., H. L. Nixon and R. D. Woods. 1963. Properties of purified preparations of lucerne mosaic virus. Virology 19:441-449.

64. Gillespie Jr., A. G., J. E. Irvine and R. L. Steere. 1966. Ratoon stunting disease virus. Assay technique and partial purification. Phytopathology 56: 1426-1427.

65. Grant, T. J. 1934. The host range and behavior of the ordinary tobacco-mosaic virus. Phytopathology 24:311-336. --. 1952. Evidence of tristeza, or quick decline, virus in Florida. Florida State Hort. Soc., Proc. 65:28-31. --. 1959. Tristeza virus strains in relation to different citrus species used as test plants. Phytopathology 49:823-827.

66. ________. and M. K. Corbett. 1960. Mechanical transmission of the infectious variegation virus of citrus. Nature 188:519-520.

67. ________. and A. S. Costa. 1949. A progress report on studies of tristeza disease of citrus in Brazil. I. Behavior of a number of citrus varieties as stocks for sweet orange and grapefruit, and as scions over sour orange rootstock when inoculated with tristeza virus. Florida State Hort. Soc., Proc. 61:20-23.

68. ________., A. S. Costa and S. Moreivia. 1950. Studies of trasteza disease of citrus in Brazil. III. Further results on the behavior of citrus varieties as rootstocks, scions and seedlings when inoculated with the tristeza virus. Florida State Hort. Soc. Proc. 62:72-79. --. 1951. Studies of tristeza disease of citrus in Brazil. V. Further information on the

reactions of grapefruits, limes, lemons, and trifoliate hybrids to tristeza. California Citrograph 36:310-311; 324-329.

69. ________. and R. P. Higgins. 1957. Occurrence of mixtures of tristeza virus strains in citrus. Phytopathology 47:272-276.

70. ________., L. J. Klotz and J. M. Wallace. 1953. The tristeza disease of citrus (pp. 730-734). _In_ Plant Diseases; The yearbook of agriculture 1953. U. S. Dep. Agr., 940 p.

71. ________., S. Moreira and A. A. Salibe. 1961. Report on general aspects of tristeza and stem pitting in citrus varieties in São Paulo, Brazil. Florida State Hort. Soc. Proc. 73:12-16.

72. Greene, V. E. and J. R. Orsenigo. 1958. Wild grasses as possible alternate hosts of hoja blanca (white leaf) disease of rice. Plant Dis. Reptr. 42:342-346.

73. Grogan, R. G. and K. A. Kimble. 1964. The relationship of severe bean mosaic virus from Mexico to southern bean mosaic virus and its related strain in cowpea. Phytopathology 54:75-78.

74. Hall, C. E. 1964. Electron microscopy: principles and application to virus research (pp. 253-266). _In_ Plant Virology (M. K. Corbett and D. H. Sisler eds.) Univ. Florida Press, Gainesville, Florida, 527 pp.

75. Hebert, T. T. 1963. Precipitation of plant viruses by polyethelene glycol. Phytopathology 53:362. --. 1967. (Unpublished Ms, observations from consultation trip on plant virus problems in South America.)

76. Herbas A., R. 1964. El mosaico de los amaryllis y su agente causál. Turrialba 14:140-146.

77. Herold, F., G. H. Bergold and J. Weibel. 1960. Isolation and electron microscopic demonstration of a virus infecting corn (_Zea mays_ L.). Virology 12:335-347.

78. ________. and K. Munz. 1965. Electron microscopic

demonstration of virus-like particles in Peregrinis maidis following acquisition of maize mosaic virus. Virology 25:412-417.

79. ________. and J. Weibel. 1962. Electron microscopic demonstration of papaya ringspot virus. Virology 18:302-311.

80. Hirumi, H. and K. Maramarosch. 1969. Mycoplasm-like agents of aster yellows in the salivary glands of leafhopper vectors. Phytopathology 59:399-400.

81. Hitchborn, J. H. and G. J. Hills. 1965. The use of negative staining in the electron microscopic examination of plant viruses in crude extracts. Virology 27:528-540.

82. Holdeman, Q. L. 1965. Virus diseases of corn--tentative key for preliminary field and laboratory decisions. Plant Dis. Reptr. 49:654-655.

83. Holmes, F. O. 1938. Taxonomic relationships of plants susceptible to infection by tobacco-mosaic virus. Phytopathology 28:58-66.

84. ________. 1948. The filterable viruses. pp. 1127-1281. In Bergey's Manual of determinative bacteriology. Williams & Wilkins, Baltimore.

85. ________. 1951. Indications of a New World origin of tobacco-mosaic virus. Phytopathology 41:341-349. --. 1968. Trends in the development of plant virology. Ann. Rev. Phytopath. 6:41-62.

86. van Hoof, H. A. 1959. The delphacid Sagota cubana vector of a virus of Echinochloa colonum. Tidschr. Plantez, Suriname 65:188-189. --. 1962. Een Virusziekten van Bacove en Banaan. Surinaam Landb. 10:47-52.

87. Jensen, D. D. and A. H. Gold. 1951. A virus ringspot of Odontoglossum orchid: symptoms, transmission, and electron microscopy. Phytopathology 41:648-653.

88. Joshi, L. and F. O. Holmes. 1968. Induced tolerance of tobacco mosaic virus to heat. Phytopathology 58:60-61.

89. Kirkpatrick, T. W. 1950. Insect transmission of cacao virus disease in Trinidad. Bull. Ent. Res. (England) 41:99-117.

90. Kitajima, E. W. 1965. Electron microscopy of vira-cabeca virus (Brazilian tomato spotted wilt virus) within host cells. Virology 26:89-99.

91. _______. and A. S. Costa. 1966. Morphology and development of stages of Gomphrena virus. Virology 29:523-539.

92. _______., A. S. Costa and A. M. B. Carvalho. 1963. Detecção de partículas de vira-cabeça ao microscópio electrônico em preparaçoes feitas pelo método do dipping. Bragantia 22:35-38.

93. _______., D. M. Silva, A. R. Oliveira, G. W. Müller and A. S. Costa. 1965. Electron microscopical investigation on tristeza. Internat. Org. Citrus Virologists, Conf. 3:1-10.

94. Lamey, H. A., W. W. McMillan and R. D. Hendrick. 1964. Host ranges of the hoja blanca virus and its insect vector. Phytopathology 54:536-541.

95. Leach, J. G. 1940. Insect Transmission of Plant Disease. McGraw-Hill Book Co., Inc., New York and London, 615 p.

96. Leach, R. 1946. The unknown disease of the coconut palm in Jamaica. Trop. Agr., Trinidad 23:50-56.

97. Lwoff, A. and P. Tournier. 1966. The classification of viruses. Ann. Rev. Microbiol. 20:45-74.

98. Malaguti, G. 1957. La hoja blanca extraña enfermedad del arroz en Venezuela. Agron. Trop. (Venezuela) 6:141-145.

99. _______., H. Jiménez and S. Horovitz. 1957. Pruebas de transmissión del mosaico de la lechoza a otra especies de *Carica*. Agron. Trop. (Venezuela) 7:23-31.

100. _______., and W. N. Stoner. 1954. Psorosis de las citricas en Venezuela. Agron. Trop (Venezuela) 4: 127-149.

101. (Maramorosch, K.) 1965. A new system of classifying viruses. Phytopathology 55:1158 (news item 13).

102. Marchionatto, J. B. 1935. Nueva estudios sobre la lepra explosiva del naranjo. Mon. Internat. Def. Plant. 9:137. --. 1948. Tratado de Fitopatologia. Editorial Sudamericana, Buenos Aires, 537 p.

103. Martorell, L. F. and J. Adsuar. 1952. Insects associated with papaya virus disease in the Antilles and Florida. Univ. Puerto Rico J. Agr. 36:319-329.

104. Matthews, R. E. F. 1957. Plant virus serology. Cambridge Univ. Press, 128 p.

105. Medeiros, A. G. 1963. Nota sôbre a ocorrencia de mosâico em banameira, no Estado de Pernambuco. Inst. Pesq. Agron. Pernambuco, Boletin Tecnico 4, 13 p.

106. Meneghini, M. 1946. Sôbre a natureza e transmissibilidade da doença tristeza dos citrus. Biologico 12:285-287. --. 1947. Reacão de amido nas enxertias de seedlings de laranjeira utilizadas em experiencias de transmissão de tristeza. Biologio 13:91-92.

107. Moreira, S. 1942. Observações sôbre a tristeza dos citrus, ou podridao das radicelas. Biologico 8: 269-272.

108. ________., A. S. Costa and T. J. Grant. 1954. Metodos para identificação e controle da tristeza dos citrus. Bragantia 13:223-236.

109. Murakishi, H. H. 1958. Serological and morphological relationships among orchid viruses. Phytopathology 48:137-140.

110. Nobrega, N. R. 1947. Uma doença de virus em orquidea. Biologico 13:62.

111. Norman, P. A. and T. J. Grant. 1956. Transmission of tristeza virus by aphids in Florida. Florida State Hort. Soc., Proc. 69:38-42.

112. Obregon, G. 1954. Enrollamiento de las hojas en el papa. Grave enfermedad virosa para los cultivos en Colombia. Agr. Trop., Colombia 10: 29-31.

113. Offermann, A. M. and E. Vitoria. 1941. Estudio sobre un virus productor de marchitamiento apical de la papa. Rev. Argentina Agr. 8:105-113.

114. Paris N., R. 1958. El enanismo del trigo y la cebada en Narino (Colombia). Reunion Interam. Fitogen. Fitopat. Ent. Edaf. 3:206-297.

115. Pérez, J. E. 1963. A strain of squash-mosaic virus and other cucurbit viruses found in Puerto Rico during 1958-1962. Univ. Puerto Rico J. Agr. 47: 141-153.

116. ________. and J. Adsuar. 1955. Antigenic relationship between Puerto Rican pepper mosaic virus and a strain of potato virus Y. Univ. Puerto Rico J. Agr. 39:165-167.

117. ________., J. Adsuar and E. E. Padilla. 1957. A survey for common tobacco- and pepper-mosaic viruses in the filler-tobacco area of Puerto Rico. Univ. Puerto Rico J. Agr. 41:20-21.

118. ________., J. Adsuar and O. Sala. 1956. Tobacco mosaic virus in orchids in Puerto Rico. Phytopathology 46:650-654.

119. ________., and A. Cortes-Monllor. 1960. Survey for the presence of tobacco mosaic, cymbidium mosaic, and other viruses in orchids. Univ. Puerto Rico J. Agr. 44:138-149.

120. Pirone, T. P. and L. Anzalone Jr. 1966. Purification and electron microscopy of sugarcane mosaic virus. Phytopathology 56:371-372.

121. Pompeu de Amaral, A. 1926. O mosaico. Historico do seu apparecimento e extensão-caracteres da doença-meios practicos de combatel. Bol. Agr., Sao Paulo 27:147-156.

122. Pontis Videla, R. E. 1953. Las virosis de la lechosa

(Carica papaya L.) en Venezuela. I. Transmision del mosaico. Agr. Trop., Venezuela 2:241-251.

123. Price, W. C. 1966. Flexuous rods in phloem cells of lime plants infected with citrus tristeza virus. Virology 29:285-294.

124. ________., A. P. Martinez and D. A. Roberts. 1967. Mechanical transmission of lethal yellowing pathogen to young coconut palms. Phytopathology 57: 826.

125. Quayle, H. J. and A. R. Taylor. 1915. The use of the fungus Isaria for the control of the black scale. California Comm. Hort. Monthly Bul. 4:333-338.

126. Rolfs, P. H. and H. S. Fawcett. 1908. Fungus diseases of scale insects and whitefly. Florida Agr. Exp. Sta. Bulletin 94, 17 p. --. and --. 1913. Fungus diseases of scale insects and whitefly. Florida Agr. Exp. Sta. Bul. 119:69-82.

127. Semancik, J. S. 1966. Purification and properties of two isolates of tobacco rattle from pepper in California. Phytopathology 56:1190-1193.

128. Shepherd, R. J. 1965. Properties of a mosaic virus of corn and Johnson grass and its relation to the sugarcane mosaic virus. Phytopathology 55:1250-1256.

129. Silberschmidt, K. and N. R. Nobrega. 1941. Sôbre una doença de virus da bananeira. Biologico 7: 216-219.

130. Silberschmidt, K. and R. H. Rimpan. 1963. Cross protection tests with related and non-related viruses in Nicandra physaloides (L.) Gaertn. An. Acad. Brasil. Cienc. 34:391-403.

131. Silberschmidt, K. and L. R. Tommasi. 1955. Observações e estudos sôbre espécies de plantas suscetíveis à chlorose infecciosa das malváceas. An. Acad. Brasil. Cienc. 27:195-214. --. 1956. A solanaceous host of the virus of infectious chlorosis of Malvaceae. An. Appl. Biol. 44:161-165.

132. Silva, D. M , M. Meneghini and H. L. de Souza Santos. 1958. Purificação e propriedades do virus do mosaico do quenopodio. Bragantia 17:167-177.

133. Silva, D. M., A. R. Oliveira and E. W. Kitajima. 1965. Partial purification of tristeza virus. Conf. Internat. Organ. Citrus Virologists, Proc. 3:10-13.

134. South, F. W. 1910. The control of scale insects in the British West Indies by means of fungoid parasites. West Indian Bul. 11:1-30.

135. Stahl, C. F. 1927. Corn stripe disease in Cuba not identical with sugar cane mosaic. Trop. Pl. Res. Found. Bul. 7:3-11.

136. Stevenson, J. A. 1917. Enfermedad de la caña de azúcar en Puerto Rico. Mundo Azuc. 5:19-24.
--. 1919. The mottling or yellow-stripe disease of sugar cane. J. Porto Rico Dep. Agr. 4:76.

137. Swarbrick, J. T. 1961. Cacao virus in Trinidad. Trop. Agr., Trinidad 38:245-248.

138. Sweetman, H. L. 1936. The Biological Control of Insects. Comstock Publ. Co., Ithaca, New York, (cited by J. G. Leach which see).

139. Takeya, K. and K. Amako. 1964. The structure of mycobacteriophages. Virology 24:461-466.

140. Tinsley, T. W. and A. L. Wharton. 1958. Studies on the host ranges of viruses from *Theobroma cacao* L. Ann. Appl. Biol. 46:1-6.

141. Thornberry, H. H. 1966. Index of Plant Virus Diseases. Agr. Res. Serv. U. S. Dep. Agr., Agr. Hdbk. 307, 446 p.

142. ________., A. G. Otterbacher and M. R. Thompson. 1966. Gama grass, *Tripsacum dactyloides*: a new perennial host of maize dwarf virus. Plant Dis. Rptr. 50:65-68.

143. Traversi, B. A. 1949. First study about a disease of *Helianthus annuus* in Argentina. Lilloa 21:271-278.

144. United Fruit Company Research. 1958. Problems and progress in banana disease research. United Fruit Co., Boston, Massachusetts (review bulletin), 36 p.

145. Viega, F. M. 1956. Ratoon stunting disease in Brazil. Cane Grower's Quart. Bul. 20:3-6.

146. Warmke, H. E. and J. R. Edwardson. 1966. Electron microscopy of crystalline inclusions of tobacco mosaic virus in leaf tissue. Virology 30:45-57.

147. Weathers, L. G. 1965. Transmission of exocortis virus of citrus by _Cuscuta subinclusa_. Plant Dis. Reptr. 49:189-190.

148. Webber, H. J. 1943. The tristeza disease of sour-orange rootstock. Amer. Soc. Hort. Sci., Proc. 43:160-168.

149. Wellman, F. L. 1932. Celery mosaic control in Florida by eradication of the wild host _Commelina nudiflora_. Science 76:390-391. --. 1934. Identification of celery virus 1, the cause of southern celery mosaic. Phytopathology 24:695-725. --. 1934. A disease of bananas, remarkably similar to bunchy top, produced by celery virus 1 in U. S. A. Phytopathology 24:1032-1034. --. 1936. The host range of southern celery-mosaic virus. Phytopathology 25:377-404. --. 1936. Dissemination of southern celery-mosaic virus on vegetable crops in Florida. Phytopathology 26:289-308. --. 1937. Control of southern celery mosaic in Florida by removing weeds that serve as sources of mosaic infection. U. S. Dept. Agr., Technical Bulletin 548, 18 p. --. 1949. A list of maize diseases from a limited area in Costa Rica. Plt. Dis. Reptr. 33:81-85. --. 1956. (Notes from 1944 to 1956 on sugarcane and maize dwarf virus diseases in widely separated areas in El Salvador, Guatemala, Nicaragua and Costa Rica.) Unpublished, typescript, 14 p. --. 1957. Blister spot of arabica coffee from virus in Costa Rica. Turrialba 7:13-15. --. 1961. Coffee botany, cultivation, and utilization. Leonard Hill, London; and Interscience Publ., New York, 488 p. --. 1961. Enfermedades de algunas plantas ornamentales

de Puerto Rico, con recomendaciones para contrarrestarles. Rev. Agr. Puerto Rico 48:179-213.

150. Yarwood, C. E. and E. W. Holmes. 1962. Heat adaptation in a rust and a virus. Phytopathology 52:709-712.

151. Zoeten, G. A. 1966. California tobacco rattle virus, its intracellular appearance, and the cytology of the infected cell. Phytopathology 56:744-754.

Chapter IV

BACTERIAL AND PRIMITIVE-FUNGUS DISEASES

Summary -- Introduction -- Tropical origin of Pseudomonas solanacearum -- Some other bacterial diseases -- Certain primitive fungus diseases -- Diseases from Pythiaceae and close relatives -- Table of differences in primitive-type fungus diseases -- Literature cited

Summary

Saprophytic bactera in soils, debris, and on living plant surfaces are common and aid in the decay and therefore biologic destruction, of plant disease fungi. Bacterial slime-rot, *Pseudomonas solanacearum*, is highly pathogenic on many crop plants; probably it is indigenous to the tropics. It is the focus of much research, perhaps more than any other plant disease bacterium. Numerous bacteria of various genera and species attack many plantation and subsistence crops. Some have been and are being investigated. Various parasitic primitive-type fungi, low in the phylogenetic scheme, are encountered on numerous hosts. Among many of these parasites (e.g., *Physoderma* of maize and *Spongospora* on potato) are those that have originated in the Neotropics. Some of the most unusual parasites in plant pathology are the protozoans that cause disease in coffee and sugarcane. *Pythium* disease affects more plants than a disease from any other fungus genus. Diseases from it are especially common in the tropics, and there are no control measures; amelioration is by expensive replanting practices.

Introduction

No one needs any statistical proofs to appreciate that in the warm, moist tropics there is an ubiquitous occurrence of abundant, naturally occurring saprophytic bacteria of all sorts on the foliage of plants and in the soils. They are found also in drainage waters, on the outside of all kinds of living and dead vegetable matter, and the rot complexes in jungle floors are full of them. The occurrence of so much bacterial life is taken for granted in the tropics but has never been critically studied; and no generalizations, except for the presence of large numbers, can be made about the populations of saprophytic soil bacteria in the tropics.

Crude observations give evidence that soil bacteria in the moist, warm tropics are very much more numerous than in the cool, dry parts of the temperate zone. A forester or naturalist quickly learns that in the warm, moist tropics of the lowlands there never forms the deep, felty blankets of so-called duff known on forest floors of the temperate zone. Something similar to these duffs can be found in small amounts in specially favored forested areas on cold mountain sides in the tropics. But the deposits are comparatively rare. Where great quantities of organic matter are produced per year in the tropics, in amounts far above the growth in the temperate zone, the microorganisms consume it at a high rate.

Another evidence of great bacterial effects on tropical soils is in the occurrence of the widely encountered laterite formations. These soils are red and not easily cultivated with profit, because they are old soils containing large amounts of aluminum and iron. These are what is left from effects of intensive bacterial action along with rains and warmth that consume and leach away the soil components most needed agriculturally. Laterites are not farmable in the same manner as a temperate zone loam. They have to be protected by cover crops and mulches to cut down bacterial action, and must be disturbed as little as possible in the cultural processes. Tropical soils (see Mohr and Pendleton (41)) are lacking in organic content. Wellman (75) and others have repeatedly pointed out that these are relatively fragile and easily injured soils; as their high bacterial content keeps the organic matter at such a low ebb.

The bacterial content of tropical soils can be of use in plant pathology. There is speculation that bacterial populations have in them types that must depend on fungus hyphae in soil for their existence. At least soil inhabiting parasitic fungi being studied in unsterilized soil have been found (Wellman (75), Marshall and Alexander (38)) several times obviously being attacked by soil bacteria as hyphae weaken. There are certain types of fungus propagules, such as oospores, chlamydospores and sclerotia, which are protected with sufficiently resistant coverings to escape a certain amount of bacterial action in soil. However, it seems probable that such husbandry practices as mulching, growing cover crops to be turned under, crop rotations and similar methods have been notable in taking advantage of bacterial action to reduce the amount of parasitism from soil fungi.

Not only are bacteria numerous in the tropics. There are many fungi, the Archimycetes and relatives, of primitive nature in the warm, moist regions. Here such fungi have been in existence continuously without being halted by the advent of ice ages and subsequent annual winters or cold periods. This means that not only are the more original types of primitive organisms found in the tropics, but also the variants from them are also to be discovered there. While at present they are somewhat neglected because so many other serious disease problems require attention, the occurrence and parasitism of the more primitive type parasitic fungi are beginning to be studied more. It is in the tropics where they are more common. There is every reason to expect discoveries in the tropics that will clarify knowledge about theories respecting the species concept and evolving of the parasitic habit. Among microorganisms, compared with a higher plant, species are less rigidly defined. The parasites, because of the economic reasons for their study, may one day give much fuller information about what a species really is in the plant world.

Tropical Origin of *Pseudomonas Solanacearum*

Fresh, untreated soils on being examined microscopically are found to contain large quantities of fungus threads, algal growths, and bacteria. These organisms appear to act as adhesive agents in holding soil particles together. Such soils have two invariable characters: they will be found to have an extremely steep angle of repose, and cast off vegetation that comes into contact with such soils is rapidly decomposed.

Soils in the tropics are well adapted to bacterial habitation as is evidenced by the fact that leguminous trees and lesser plants, which are so common in the tropics, never seem to lack for root nodule (nitrogen fixing) bacteria. This is also the case with bacteria that cause disease. Three examples can be given here. Once *Pseudo-*

monas solanacearum becomes established in a region it becomes a permanent part of the soil. This I have found is also true of so-called bacterial blight (Xanthomonas phaseoli) on bean and the sorghum bacterial stripe (Pseudomonas andropogoni). There are in the Neotropics about 20 species of bacteria in 6 genera that are fairly common causes of plant diseases.

A common finding is that plant diseases arise in the region of origin of the most important host plant. In certain cases there seem to be exceptions, as for example (24, 25, 27) the more exclusive oleander tubercle, Pseudomonas tonelliana (sometimes considered a strain of the olive bacterium P. savastanoi). It has been found only on oleander, and the bacterium occurs in Italy, Europe, the U. S. A. and in South America. However, it does not seem to have been reported from Turkey which is the country of origin of its host.

Where there is such a wide host range (see Kelman (30) and Smith (61)) as in the bacterial slime-rot, Pseudomonas solanacearum, the origin may not be easily traced. While the organism was so named to indicate its favoring Solanaceous hosts, there are many others involved. Organisms like this are spread with seeds and other planting material, and this occurs with P. solanacearum. This bacterium is transmitted with seeds such as soybean and peanut, and the numerous planting stocks such as tomato and tobacco seedlings and the corms of banana that through the usual horticultural procedures take the bacterium from one place to another. It affects potato seed tubers, and invades manioc and ñame (yam) by way of their tubers too. It is occasionally severe in such other crops as sunflower, castor bean, vanilla, abaca (hemp), lima bean, soybean, chayote, watermelon, sweetpotato, Heliconia, Java jute, field bean, sugarcane, canna, lleren, and a number of others. It has had a chance to be disseminated very widely.

It can be seen from literature that it is found in both the Occident and the Orient. Pathologists in the Neotropics many times have found the disease developing in crops in newly planted clearings of virgin jungle and forest land. Early in this century it was in Florida (56), and it has been observed by me to develop in new banana plantings in Honduras, using corms known to be absolutely free from the organism, and in tomato and eggplant fields in Central America in lands never cultivated before and seeded directly to those crops. Researchers have found (9) that P. solanacearum is insect carried, and with banana and plantain crops this adds to the serious problem of control.

This disease is in the parts of the United States bordering on and influenced by the adjacent Neotropical region, but also occurs far north (13, 23, 30, 72). Between the years 1881 to 1898 there were wilt troubles on potato, tomato and tobacco in the southern part of the USA that we know now were all due to P. solanacearum.

It appears this bacterial disease was recorded first from the Eastern tropics, and the very earliest published notice of its occurrence was from the warm parts of Japan in 1683. It has long been common and severe also in Australia, Indonesia (where it had early serious attention), in Sumatra, Africa, India, Ceylon and such countries (24, 30, 56, 57). Significant studies have been accomplished to make it possible to identify and work with the organism in pure cultures (9, 28, 30, 31, 45, 46, 68) for most detailed research.

With respect to early reports of P. solanacearum in the Neotropics a few notices are as follows: in Florida and Mississippi, in the U. S. A., it was reported in 1898 and 1892 respectively. The disease was known in British Guiana about 1840 to 1844 in bananas. Rorer (57) showed it was probably present in Trinidad as early as at least 1890. It is now known as very severe in certain weed hosts (8) as well as such crops as bananas, tobacco, tomatoes, eggplant, plantain, and potato at least in the following countries: not only in the U. S. A., but also in Mexico, Guatemala, Costa Rica, Honduras, Puerto Rico, Trinidad, Guadeloupe, Ecuador, Panama, Colombia, Peru, Uruguay, and Brazil (1, 4, 10, 13, 16, 18, 20, 21, 23, 30, 31, 32, 43, 46, 50, 55, 56, 57, 61, 63, 68, 69, 73).

The optimum temperature for the bacterium, in culture or as a pathogen, is at a warm condition, that of about 27° C. In the tropics it remains in the soil from one growing season to the next. Once established in a field the problem of avoiding further damage from this bacterium is difficult. Research (68) has shown it can be controlled in bananas by careful farming and technological means used in plantations. While the organism is not ubiquitous, it is well spread and Pseudomonas solanacearum probably originated in the tropics. It is certain that tropical pathologists do well in their work to assume that, and to plan their studies and control measures with this as a basic consideration.

Some Other Bacterial Diseases

In the genus *Pseudomonas* there are a number of parasitic species of special note in the Neotropics. One of these is *P. garcae*, the coffee aureolada or halo leafspot in Brazil (4, 5, 15, 25, 55). It caused consternation and disturbance when it was first encountered in the Brazil coffee industry. However, as studies on it have progressed pathologists have learned ameliorative and control measures so that its once great potential for drastically limiting the crop is no longer to be feared. It has a host range that includes at least potato, tomato, bean, and citrus (5, 15).

The tobacco leafspot bacterium, *P. angulata*, is a common organism and it causes many angular-shaped spots on leaves and fruits, malformations, defoliations and discolorations. It is carried in seed and can become quite widespread. Bean, halo blight, *P. phaseolicola*, a most serious disease (Fig. 4-1), is probably in all bean-growing areas of the Neotropics, except, perhaps where it is very dry (7, 17, 24, 25, 45, 59). The olive knot organism, *P. savastanoi* (27), comes from the hot, dry eastern tropics and is a bacterium that causes the production of swollen protective layers of host tissue about invaded parts, and thus it is able to withstand long periods of the very dry and hot, near-desert, conditions of the tropics. It is well established in the so-called tropical parts of California, and in quasi-desert areas in Chile and Mexico.

The disease from the bacterium causing tobacco wildfire, *Pseudomonas tabaci*, is known in several Neotropical countries. It has been seen as serious on tobacco in El Salvador in dry areas, occurs in the West Indies islands, but although present it was of less severity in some of the more moist parts of Costa Rica. Whether such differences could be observed in other years is still to be learned. In Costa Rica, El Salvador, and Puerto Rico I have found occasional fruits of tomato showing speck, *P. tomato.* This bacterium has never been considered as a serious rotting organism in the tropics. Its effects as a blemish have not been much of a deterrent in salability of tomato fruits, and small fruited tomatoes have seemed to be free from the

Figure 4-1. Leaf symptoms on bean of halo blight, Pseudomonas phaseolicola. A. young lesions, B. late stages. (From Department of Plant Pathology, University of Florida, Gainesville.)

speck disease. P. syringae attacks black pepper in Brazil (55) and is transmissible to lemon and beans.

A remark should be interjected here about the woefully unknown bacterial leaf infections on grasses in the Neotropics. Few people worry about them, because you walk on grasses, animals graze on them, and they are neither selected for growth characteristics nor are many cultivated. I did not have sufficient time to investigate the evident bacterial leafspots I saw in Central America. However, I saw in many places leaf symptoms typical of Pseudomonas andropogonis (broad, dark-colored stripes) and of Xanthomonas holcicola (spotting and streaking). I believe that in El Salvador, Puerto Rico, and Guatemala I found undoubted occurrence, based on symptomatology, of Pseudomonas coronofaciens var. purpureum on a number of wild and pasture grasses. In several areas it was causing severe early dying of leaves (24, 25, 43, 55, 64).

Bacteria of the genus Xanthomonas occasionally cause serious crop losses in the tropics. The cabbage blackrot organism, X. campestris, may not often be serious, but if it becomes established in a growing region it is devastating in effect unless carefully controlled. The citrus canker, X. citri (for symptoms on leaf see Fig. 4-3) occurs sporadically but compared with P. solanacearum, is of little importance in the Neotropics. In some lowland areas the cotton crop is attacked by bacteria causing both boll rot and angular leafspot from X. malvacearum (47, 24, 25). Under conditions I observed, there has occurred severe defoliation, and of course considerable reduction in harvest in normally planted fields. However, where field planting is mixed in with other crops, as in Indian culture, the leafspot is usually mild in effect.

On the other hand, the bean fields of many countries, become seriously attacked with common bacterial blight, X. phaseoli (7, 17, 45, 59). This disease often becomes intensified in severity in a few years. In El Salvador, starting with a few plants in a field affected, in three years time the crop was practically destroyed by this blight. Attempts to change the cultural practices of the Indian farmers were unsuccessful. The Indians insisted on keeping the kind of beans they had eaten for generations and any sort of crop disease control was resisted. The village subsequently had to move to another side of the volcanic slope. The sugarcane redstripe, X. rubrilineans, sometimes occurs so com-

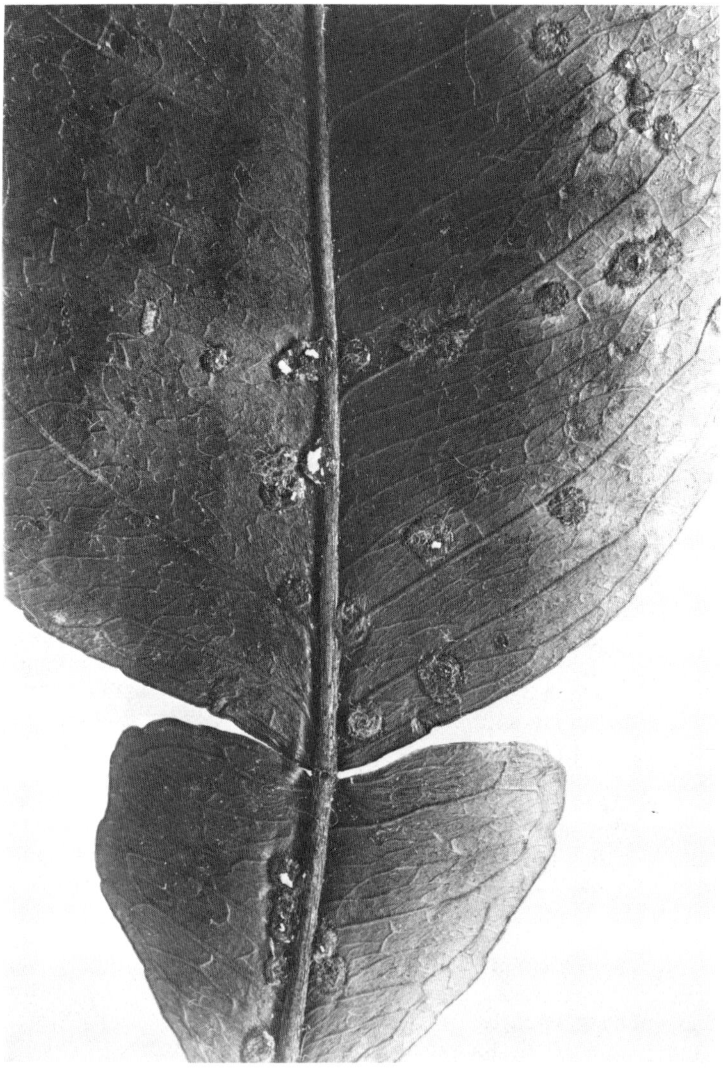

Figure 4-2. Symptoms on leaf of citrus canker, Xanthomonas citri, showing slightly raised lesions; on twigs infected places are scabby and raised. Photograph courtesy of Department of Plant Pathology, University of Florida, Gainesville.

monly in fields it becomes accepted like a botanical leaf character. On cleaning up such a field, by cultural and exclusion methods, cane men have claimed over 40 per cent increases in production (6, 54, 55).

In an experimental planting of sweet corn in Costa Rica, undoubted symptoms were seen of bacterial wilt, X. stewartii. The few plants diseased were immediately rogued out and the experimental field changed to a new location. Ten years of subsequent search did not disclose more of this disease in nearby fields in the locality where it was first observed. Certain cereals such as wheat, barley, and rye are grown in the Neotropical highlands, and have been reported to show symptoms of bacterial stripe, X. transluscens, which is known to be severe but numerical data on losses are seldom expressed about such diseases.

However, another graminicolous bacterium, the one causing sugarcane gumming, X. vasculorum, has been at times very severe in Brazilian, U. S., West Indian, Central American, and Colombian cane crops. As a matter of fact, and of further interest, the disease was first described by Draenert (19, 49) in 1869, the result of 6 years of study in Brazil. This, although seldom acknowledged and never given wide acceptance, is actually the first proof of a bacterium causing a plant disease. This has been discussed above in Chapter II. It had appeared as an announcement in 1869, which is eight years before Burrill's (11) statement of the bacterial nature of pear blight in 1877. Fields of cane growing in Cuba, before newer methods of control were developed, when diseased by the gumming organism, were said to have less than half the usual yield (personal discussions with Cuban authorities in 1933 and 1945). Such losses were also felt in early plantations in El Salvador, and the same kind of reduction was also true in many other sugarcane countries.

Tomato fruits are occasionally diseased with the common bacterial spot, X. vesicatoria. It is a severe problem and spreads rapidly causing major losses in a short time. This tomato fruit spot has been of some concern in the tropics because not only is the good quality tomato hard to produce, but its presence for use in the human diet is of special importance in the warm countries where foods rich in vitamins are so greatly needed.

The most easily found of all plant bacterial diseases

is the common softrot, Erwinia carotovora. Its presence seems everywhere, and in warm countries where refrigeration is more of a problem than in the temperate zone, this trouble is extremely important as a market and storage problem. All markets, under normal tropical conditions, discard large amounts of rotted fresh vegetables (18, 20, 24, 25, 31, 33, 47, 55, 58, 63, 72). Percentages have not been secured during enough times and in sufficient places to obtain an estimate of the total economic loss from this. But it is large. I was convinced that at a minimum, and regularly, 10 to 30 per cent losses were incurred week in and week out, from one season to the next.

There are a handful of diseases from bacteria of lesser concern that should be mentioned. Crowngall, Agrobacterium tumefaciens, has been reported (e. g., 24, 25) at times from Latin America. This disease may be of historic interest but mostly of little economic importance. I have never seen it in the tropics, but there is no doubt of its somewhat rare occurrence. The Corynebacterium insidiosum causing alfalfa wilt, and the C. michiganense that results in tomato canker are found in the tropics and have been seen when on trips to cool, mountainous areas. Tobacco hollowstalk, Erwinia aroideae is of occasional occurrence, and has added to confusion in diagnosing the Phytophthora black shank symptom in tobacco.

Tropical pathologists are often confronted with bacterial diseases, on ornamentals. Two examples are a leaf and stem rot of certain aroids (37, 44) and bacterial leaf spotting of many ornamentals, one of which is begonia (36).

Certain Primitive Fungus Diseases

In the western tropics, where there are so many ancient-type vascular plants such as cycads, clubmosses, tree ferns, hosts of aerial algae, and relatives of these--direct descendents from forms that grew in ancient geologic ages--it is not surprising to find remaining representatives of primitive fungi. By far the largest number are saprophytic in habit but others are parasite. Without making time-consuming identifications, thallophytes have been seen invading: weak root hairs, algae, the ooze of myxomycetes, water-soaked pollen grains, and the hyphae and spores of higher fungi. Often slime molds are found (Fig. 4-3), and while they may be superficial in the temperate

Figure 4-3. Stalks of celery with masses of the slime mold Physopella sp. on surfaces; reticulations from climbing slimy growth up from the soil extending to profusion of sporulating structures above. Photograph, taken in Florida, courtesy of Department of Plant Pathology, University of Florida, Gainesville, Florida.

Figure 4-4. Leaf infection symptoms on maize, from attack by Physoderma zeae-maydis; occurs throughout the Neotropics. Photograph courtesy Plant Pathology Department, University of Florida, Gainesville.

zone, growth is sometimes so profuse in the tropics as to cause serious blemishes and even smothering of plants.

There are a great many organisms other than bacteria, phylogenetically speaking of a lowly character to be found causing disease in the tropics. Doubtless there are many more of them in this part of the world than have been discovered in the temperate zone. For example there is the organism that is very well known in the southern part of the United States and practically all the American tropics (22, 26, 35, 65, 66) that causes brown spot of maize. This organism is adapted to warm, moist (tropical) conditions (29) and it is called Physoderma zeae-maydis, and symptoms of its attack are illustrated in Fig. 4-4. It invades many of the different kinds of maize crops and other relatives of maize. The relatives of this organism also attack quack grass (65) and other grasses such as Bermuda grass and there are species that infect trees such as citrus (14). These organisms need much study. While at present citrus growers may consider the infections as having only benign effect, in years to come with further selections of varieties more severe disease may be expected from these primitive fungi.

The individuals of Physodermas are remarkable parasites. The resting spores remain in the crop tissues until the cells decay when spore masses are released into soil where the spores germinate and from these develop the zoospores. These zoospores are furnished with a single flagellum; with it they move around rather rapidly in water, and by it they are splashed from place to place and from plant to plant.

There occur quite commonly in the tropics and subtropics conditions specially favorable for other primitive forms of fungi. Yeasts are frequent, as anyone knows who has dealt with problems of keeping fresh fruits and vegetables for tropical marketing. It is among the so-called yeasts that organisms of such undoubted primitive types as Spermophthora, Nematospora, and Eremothecium are sometimes included among the fungi.

These are considered by Ainsworth and Ashby (2) as the most primitive of ascomycetes, and Gäumann (26) places them here likewise. They, along with the yeasts like Endomyces are without doubt closely related to the Ascomycetes, but do not produce truly perfect asci. Some of these unusual fungi are found in such out-of-the-way places as stagnant, moist areas protected by leaf scales of bromeliads and in "leaking" tree crotches of such crops as cacao, sweet orange, and avocado (unpublished observations by me). Some also are parasitic on agricaceae (see Gäumann 26), and produce spores between the gills of their host fungi.

All through the Neotropics, except where rains are excessive and the vectors cannot live, Nematospora coryli Pegl. is found causing seed damage in such crops as grapefruit, occasionally in coffee, cotton, pecan, and pomegranate. The closely related N. phaseoli Wingard, attacks lima bean, and I have seen pods of the inga tree with seeds affected in the same way. For typical internal symptoms of Nematospora attack of grapefruit, see Fig. 4-5.

There are those who believe, and with reason, that perhaps the most unusual plant parasite organisms that occur may be those in the tropics. It is clear from what is known in the temperate zone that studies in the tropics are far from complete and have not uncovered, long and varied as the list is, all the types of tropical diseases that attack plants. The observer is then not greatly shocked when he discovers, as Matz (39) and other have, that a vascular infection of sugarcane producing what is known as dry top rot, is caused by the protozoan organism, Amoebosporous (Plasmodiophora) vasculorum (Matz) W. R. I. Cook.

Dry top rot is now rare but may be encountered in old plantings of small farmers, where there has been little attention given to weeding, and where replanting is neglected. The symptoms of attack on sugarcane by a protozoan is, as I saw in the West Indies, first even before wilting--a sudden darkening of the base of the tip roll of leaves. The darkening of the tissue increases in the tender parts, and soon dries out, at which time the above top leaves wilt and die. On dissecting the diseased part, there is seen in the stem, darkened vasculars; and the apical bud shows discoloration and distinct constriction prior to its death.

In plant pathological literature Plasmodiophora brassicae Wor., is considered a type of the so-called primitive fungi that causes disease in plants. It is most common in cool lands of Europe and Russia, and in the northern part of North America. Thus it has the reputation of being inhibited by warm conditions. However, it is known to cause disease at the wide range of from 9° to 30° C, and is found,

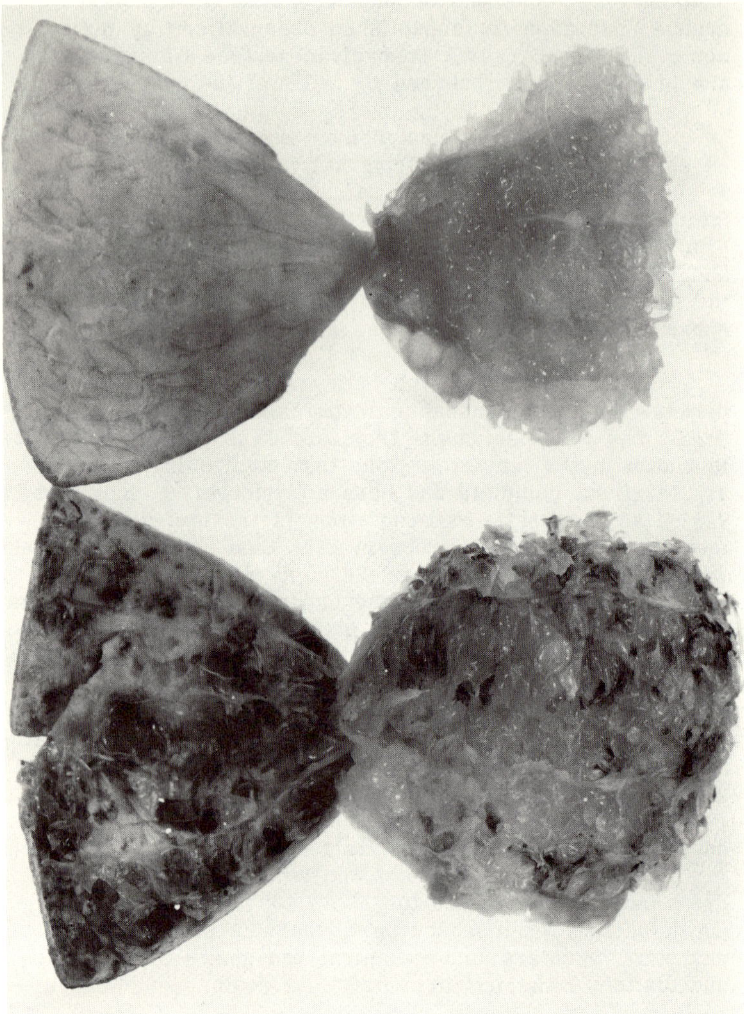

Figure 4-5. Infections by Nematospora: grapefruits, healthy above, diseased below; skins peeled off from pulp which is at right. The disease symptoms consist of blackening which is more notable in skin tissue than in pulp--seeds are also affected. Photograph courtesy of Department of Plant Pathology, University of Florida, Gainesville.

although rarely, in Brazil, Argentina, Mexico, and Peru. It seems probable to have been taken to the Neotropics as contaminations from subterranean parts of such planting materials such as: growing ornamental plants and cuttings, on seed tubers, roots, and bulbs (24, 25, 26, 72, 75).

There is likewise the organism, Spongospora subterranea Lagerh., that came originally from the Peruvian-Bolivian and Ecuadorian Andes region. In 1966 (12) the fungus has been found to transmit a potato virus disease. Years ago, in the last century, it was a severe disease in coolest potato growing regions of North America and in Europe. Indeed, at one time it was thought to be a native of the temperate zone. On giving the disease closer attention in the American tropics, it was found on wild potatoes, Solanum andigenum, and as well on Nicotiana tabacum, Physalis nicandra, and other Solanaceae, and on other types of crops like Ullucus tuberosus Caldas of the Andes and of the family Basellaceae, and on the mountain vegetable Tropaeolum tuberosum Ruiz & Pavón of the Nasturtium family. There is no doubt of its Neotropical homeland (1, 10, 24, 26, 42), and of its primitive nature. A relative, Synchtryium endobioticum Percy, is also found in South America, and is another spectacular potato disease causing Sarna or Verrugosa (24).

A remarkable disease in this category is that which develops systemically in coffee, most seriously in Coffea liberica and C. canephora (the so-called liberica and robusta types respectively). In liberica coffee it was studied and described by Stahel (62) as causing severe losses in Surinam. It is called a phloem necrosis, and was reported in Central America (74) on Arabica coffee. The causal organism is the flagellate Phytomonas (Leptomonas) leptovasorum Stahel. For some time the presence of such an unusual kind of disease-causing plant parasite was questioned by temperate zone pathologists. However, repeated work on it by such a careful scientist as Stahel (13, 62), its description in detail, and (75) its being diagnosed and found in Colombia and Brazil attacking Coffea arabica (the coffee in which it causes the so-called red disease) leaves no doubt but that the Phytomonas-Leptomonas invades coffees. It is distributed by coffee root mealybugs of at least three genera: Rhizoecus, Pseudorhizoecus, and Linus.

Symptoms of the disease first seen are sudden wilting and tree collapse, accompanied by leaf yellowing and darkening and hypertrophy of vascular elements under the bark.

Leaves die so suddenly that they are not cast off the trees, and when the soil is removed at the base, diseased trees show remains of root-inhabiting mealybugs and the rootlets are killed and decayed. The disease does not cause above-ground symptoms in young seedlings, and the killing is most common after trees are over two years of age. Once it appears in a plantation the disease spreads readily from tree to tree as the ants which attend the mealybugs dig galleries and carry the root-inhabiting insects from tree to tree and from one plantation to another.

Diseases from Pythiaceae and Close Relatives

There is some argument among plant pathologists about which fungus genus causes most different kinds of diseases on plants. There is however no question that damping-off diseases and related seedling rots attack by far the largest numbers of crop plants in the tropics, and if careful calculations could be secured, it is believed the diseases from *Pythium* spp. probably would be found to be the most frequent and the most destructive type of injury of any in all of plant pathology. *Pythium* spp. mostly attack subterraneously in the tropics. There is probably no crop plant immune from injury by the genus, and it is one of the most potent causes of the necessity for rotations and other field changes. It may be a great, if not the principal, reason that the shifting agriculture of the milpa system had to be evolved in the Neotropics (3, 6, 17, 24, 25, 26, 40, 43, 47, 48, 51, 52, 53, 60, 63, 70, 71, 72).

If searched for there are many of the Pythiacea commonly encountered in the tropics. Because of the nature of things, however, most of the time these organisms are given little attention by tropical pathologists. The reason for this is that largely there is little special premium given in pioneer-type cultivations, for good stands of seedlings. When crops are sown directly into the field, or in seed beds, planters and peasants expect and plan for either much replanting or expect to use whatever comes up from the seeding and from what survives the damping-off and rot disease. Replanting to replace "normal" seedling deaths, is almost a foregone part of the system of field culture carried on by a large proportion of tropical farmers. As our tropical agriculture progresses further, in its rapidly developing, modern, and more intensive fashion, it will soon be necessary for the blights, wilts, and damping-off diseases from *Pythium* to be given most careful attention.

As this is being written no one has a satisfactory idea of the number of species of the genus *Pythium* there are in the Neotropics. In this region, most pathologists commonly do not attempt specific determinations but lump together what they find of these organisms as simply *Pythium*. There are, indeed, times when culture material in the tropics is practically impossible to distinguish as *Pythium* instead of *Phytophthora*. Until simpler means of species differentiations are devised, even when there is easy access to a rare mycologist interested in the genus, the best method seems to be to work with these parasites as *Pythium* sp. Up to the present the complexities wrought by the specialists dividing *Phytophthora* from *Pythium*, make the species concept in *Pythium* difficult for the pathologists who are the ones most urgently in need of rapid means of determination. The most modern information about the still unknown kinds of *Pythium* in the American tropics is still to appear. In connection with this there will have to be presented some method of easier, and ready, assaying of species differences. This, probably, will be through disease reactions on standard hosts under standard conditions.

Symptoms of *Pythium* attack on warm country-grown tomatoes are of varied characters under different conditions. Whether this is due to some differences in ecology or in the species of the organism is not known. An example of what may be found is as follows. From my determinations the causal organism was called *P. ultimum*, in all cases, but this was recorded without absolute certainty. Tomatoes grown at about sea level in Costa Rica were affected in the field by a rapid collapse and a complete involvement of the root and collar region of the plants. Roots were not much discolored. The fungus isolated was consistently a *Pythium* from lesions in these plants. At near 400 meters above sea level there was a seriously cankered condition on tomatoes. This took place at the soil line and the cankers were darkened and slightly sunken, which was accompanied by hardened growth of the plants. Canker tissue again yielded *Pythium* isolations, but some appeared close to *Phytophthora*. At about 600 meters elevation a gardener had old vines of tomatoes with almost blackened side roots. Isolations from these seemed altogether of *Pythium*.

Pythium diseases on grasses show a wide range of symptoms. In an undetermined, bamboo-like perennial grass in a cool mountain region of Guatemala, there was a distinct decay of haulms in spots between the nodes. On microscopic examination of cultures it was a Pythium and with field materials typical oogonia of Pythium were seen in fresh tissue. In a coarse-growing grass, Paspalum sp., in a pasture complex at about 400 m. elevation in El Salvador, there occurred a severe killing effect near a spring. Whole plants had darkened bases, and the disease was localized in the wet place close to the spring. On microscopic examination the causal fungus was seen to be a Pythium. In a lawn grass garden in Puerto Rico Stenotaphrum (St. Augustine grass) was found by microscopic examination of disease material to be rotting from Pythium sp. This rotting could be readily seen on spreading apart the grass leaves making up the turf. Although weak, when transplanted, the sod in starting to grow often recovered from the alarmingly rotted appearance.

A severe sugarcane rootrot has been known in certain parts of the Oriental tropica for years. Several organisms have been considered at one time or another as the main causes--for example, species of Marasmius and Rhizoctonia. It remained for Rands and Dopp (51) to show that in such places as the island of Barbados and in southern Louisiana the undoubted cause is Pythium arrhenomanes Drechs., "a highly variable species" as they describe it and emend it in their publication. It appears that in the Neotropics this rootrot is extremely severe only in sugarcane cultivated at the coolest edges of its best climatic zone. For example in Louisiana, U. S. A., the disease causes special damage following long cool winter seasons, and at the highest and therefore coolest levels of cane growing in West Indies islands, and in such countries as El Salvador and Costa Rica where plantings are in the mountains, this rootrot is severe in old varieties such as rayadilla and crystallina. It is of interest that Rands (51) reported he had isolated and had determinations made of 13 different species of the fungus Pythium from sugarcane. The genus attacks the roots of the crop in all countries where the crop is grown in the western hemisphere. At the very least, it causes dark lesions and strangulations in sugarcane roots. It has not been reported from every sugarcane country in the world at this writing, but it is probable that it occurs in all of them and is being overlooked because so many of the modern cane varieties are now more highly tolerant to the fungus.

I have found that, over a number of years, with Phytophthora and Pythium specialists who are interested in the relationships between the genera, and with the taxonomy of the two genera, and Pythium especially, in its present state, the best I can do is be unspecific relying only on the use of the genus name, Pythium. Species of it are at home in the tropics growing saprophytically in soils, and it is possible they are not pathogenic until they inadvertently encounter higher plant tissues. If a crop plant is past the weak seedling stage and is growing vigorously it will often be able to escape Pythium attack.

The soil conditions most favorable to Pythium are water soaking, a clayey consistency, with often a covering of a great deal of rotting vegetative matter. Should conditions be adverse to the best root growth of a plant and there is Pythium present, the fungus soon penetrates hairs of weakened roots and thus enters the main part of the root. As Pythium parasitism advances there is further breakdown of underground stem, and also root tissues. The hyphae from the fungus grow out and conidia are produced on mycelial tips. As is well known from literature, from these organs there germinate free-swimming zoospores. In addition to conidia I have observed apparently resistant propagules that are densely filled with protoplasm, are about the same size as conidia, but do not produce zoospores. These are killed on drying, but would seem to be something akin to gemmae and germinate in moist conditions to penetrate directly into crop plant root tissue. If kept moist these gemma-like organs do not require being near host tissues to germinate and produce a mycelium that continues its growth among the soil particles.

The gemma-like propagules of Pythium have been seen under the microscope when rootrot specimens of kenaf, tomato, papaya, pasture grass, maize, and watermelon are being examined. On the other hand, I have not observed at any time in the tropics what I was certain was the formation of zoospores. It is possible these develop more commonly in cool country, and in temperate zone regions. Oogonia, the round bodies seen imbedded in tissues decaying from Pythium rootrot, are readily found. The bodies are protected with a thickened wall and are apparently impervious to bacterial dissolution. This is probably another type of propagule that withstands drying and most other adverse conditions for Pythium.

Cultivation of maize by mountain Indians of Guatemala and El Salvador is not only for food but also for construction, fuel and religious purposes. At high, misty, and cold elevations the large seeded maize requires as much as 11 to 14 months from seed planting to maturity. When the crop is planted in the folk system, the soil selected for it is deep volcanic sand preferably on a mountain shelf. Seed hills are staked off at widely spaced intervals; over 2 meters apart in both directions. At the edge of an Indian holding, a sparse hedge of bushes of *Sambucus* sp. is maintained to furnish "fertilizer" and "disease control" sticks.

Holes are dug at hill sites, and several maize seeds are planted over 20 centimeters (8 inches) deep in the sand. They are first loosely covered. When seedlings emerge they are thinned to three or four and stocks about a span long are cut from stems of *Sambucus*, put in a circle around the new maize plants, and covered with sand. As the "mats" of plants grow, more fertilizer sticks are added at a little distance and the sand is again drawn farther up around the stalks, and this is repeated. The prop roots of maize are kept covered, and in a few months the growth is about knee high. It develops even more slowly from then on.

This kind of cultivation is thought of by some to be cumbersome and wasteful of energy. However, when well-meaning agronomists have planted maize in a so-called "normal" fashion, seed depth about 5 centimeters, in rows and with side dressings of commercial fertilizer, the field gives a poor stand, in many cases because of apparently *Pythium* rootrot and basal decay. I believe the folk system is something that has evolved by trial and error over countless generations in amelioration mainly of the *Pythium* disease of maize.

There are other organisms closely related to *Pythium* considered to be some of the most typical causal agents of tropical disease. These are fungi of the genus *Sclerospora*, and they attack plant tops. They often do not cause actual soft or blackened rot and as a consequence have not the serious impace on some observers that other diseases may have where discolored rots and wilts occur. The plants attacked by the *Sclerospora* downy mildews belong to the Gramineae, such as sugarcane, maize, sorghum, teosinté, and many pasture grasses (24, 25, 53, 54). Losses from crop infections are at times up to 20 to 40 per cent in maize and sorghum and 30 per cent in pasture grasses, and these diseases are found in numerous parts of the American tropics.

Table 7. Differences in primitive type fungus diseases

Fungus	Host	Disease	Environment
Physoderma zeae-maydis	maize	Brown spot, stalks break at nodes	Warm and moist
Nematospora coryli	Coffee, cotton, pecan, grapefruit, pomegranate, inga	Punctured fruits, injured seeds	Not too wet (must not injure bugs that are vectors)
Nematospora phaseoli	Lima beans	Injured seeds	As above
Plasmodiophora brassicae	Cruciferous crops	Malformed roots that decay easily	Moist and slightly acid soil
Spongospora subterranea	Potatoes and certain other solanaceous plants, ullucus tuber, and *Tropaeolum* tuber	Spongy scab readily infected with other rots	Cool and moist
Phytomonas leptovasorum	Liberica coffee Canephora (robusta coffee Arabica coffee	Phloem necrosis (called "red disease" on Arabica)	Moist and warm (not wet enough to injure root mealybug vectors)
Amoebosporous vascularum	Sugarcane	Dry top rot	Relatively dry and neglected fields

The disease caused in its most spectacular form is a stunting of stalks and multiple, witches' broom growths of the leafy parts. Closer examination reveals chlorotic and transluscent-appearing leaf streaks on the upper surfaces. On turning these over, or in picking apart the witches' broom-like "heads," there is found a downy growth, from which the disease takes its common name. The white and downy growth on the affected leaves is from the numerous conidiophores. These are the fungus organs that grow into the air from the sickened leaf, and on which are borne the multitudes of spores.

Sclerospora graminicola is pan-tropical in its occurrence on pasture grasses and it has had special attention in the Philippines and other parts of the Oriental tropics. While it was described in detail from the Orient, it is seen as well in the Neotropics where it is known in Mexico and in northern areas of Central America and in Florida (26). It appears that it has spread into these northerly parts from the warm tropics and is even reported in places as far away as in Iowa, where it probably lives over the winters in the oögonial condition.

Literature Cited

1. Abbott, E. V. 1931. Further notes on plant diseases in Peru. Phytopathology 21:1061-1071.

2. Ainsworth, G. C. 1961. Ainsworth and Bisby's Dictionary of the Fungi. (5th ed.). Commonwealth Myc. Inst., Kew (Surrey) England. 547 p.

3. Alvarez G., L. A. 1951. A new disease of orchids in Puerto Rico: *Pythium* black rot. Univ. Puerto Rico J. Agr. 35:126-135.

4. do Amaral, J. F. 1945. Doenças vasculares das plantas causadas por bactérias. Biológio 11:250-253.

5. ________., C. G. Teixeira and E. D. Pinheiro. 1956. O bactério causador da mancha aureolada do cafèeiro. Arq. Inst. Biol. (São Paulo) 23:151-155.

6. Anon. 1957. Diseases. Sugar Manuf. Assoc. Res. Dept., An. Rept. 1956:27-28.

7. Bonilla, H. A. 1958. Algunas enfermedades del frijol (Phaseolus vulgaris L.) en el Valle del Cauca. Acta. Agron., Palmira 8:1-75.

8. Buddenhagen, I. W. 1960. Strains of Pseudomonas solanacearum in indigenous hosts in banana plantations of Costa Rica and their relationship to bacterial wilt of banana. Phytopathology 50:660-664.

9. ________., and T. A. Elsasser. 1962. An insect-spread bacterial wilt epiphytotic of bluggoe banana. Nature 194:164-165.

10. Burbano, J. and C. Ruiz. 1943. Monografía agricola de la Provincia de Imbabura. Dep. Agr. Ecuador, Boletin 26, 119 p.

11. Burrill, T. J. 1877. Pear blight. Trans. Illinois State Hort. Soc. 1877:114-116.

12. Calvert, E. L. and B. D. Harrison. 1966. Potato mop-top, a soil-borne virus. Plant Path. 15:134-139.

13. Carter, W. 1962. Insects in Relation to Plant Disease. Interscience Publ., New York and London, 705 p.

14. Childs, J. F. L., L. E. Kopp and R. E. Johnson. 1965. A species of Physoderma present in citrus and related species. Phytopathology 55:681-687.

15. Costa, A. S. and D. M. Da Silva. 1960. (Studies on halo blight of coffee.) Bragantia 19:LXIII-LXVIII. (Rev. Appl. Mycol. 42:684-685.)

16. da Costa Neto, J. P. 1941. A murcha bacteriana da batata. Sec. Estado Negoc. Agr. Indust. Comer., Rio Grande do Sul, Circulár 47, 4 p.

17. Crispín M., A., A. Ortega C. and C. C. Gallegos B. 1964. Enfermedades y plagas del frijol en México. Ser. Agr. Ganad. México, Divulg. Folleto 33, 40 p.

18. Deslandes, J. A. 1938. Doenças da bananeira. Min. Agr. Dep. Nac. Prod. Veg., San. Veg. Publicacao 10, 93 p. --. 1944. Doenças da batata. Min.

Agr. Dep. Nac. Prod. Veg., San. Veg. Publicacao (3d. ed.) 4, 37 p.

19. Draenert, F. M. 1869. Bericht über die Krankheit des Zuckerrohes. Zeitschr, Parasit. 1:13-17; 212.

20. Drummond-Goncalves, R. 1953. A bacteriose e a mandioca guaxupe. Biológico 19:114-117.

21. ________. and J. F. do Amaral. 1945. Murcha bacteriana (Phytomonas solanacearum) da batatinha e de outras Solanáceas. Biológico 11:139-140.

22. Eddins, A. H. 1933. Infection of corn plants by Physoderma zeae-maydis Shaw. J. Agr. Res. 46:241-253. --. 1936. Brown rot of Irish potatoes and its control. Florida Agr. Exp. Sta. Bulletin 299, 44 p.

23. ________., G. D. Ruehle and G. R. Townsend. 1946. Potato diseases in Florida. Florida Agr. Exp. Sta. Bulletin 427, 96 p.

24. Fernandez Valiela, M. V. 1952. Introducción a la Fitopatología. Gadola, Buenos Aires. 872 p.

25. Galli, F., H. Tokeshi, P. de C. Torres de Carvalho, E. Balmer, H. Kimati, C. O. Nogueira Cordoso and C. Lima Salgado. 1968. Manual de Fitopatologia; Doenças das Plantas e seu Controle. Biblioteca Agronomica Ceres, Sao Paulo 640 p.

26. Gäumann, E. A. (Trans. C. W. Dodge). 1928. Comparative morphology of fungi. McGraw-Hill, New York, 701 p.

27. Halperín, L. 1942. La tuberculosis del olivo. Almanaque Min. Agr., Argentine 16:128-129. --. 1942. La quemazón bacteriana del tabaco. Agronomia (Buenos Aires) 31:33-38.

28. Hayward, A. C. 1963. A method for characterizing Pseudomonas solanacearum. Nature 186:405-406.

29. Hebert, T. T. and A. Kelman. 1951. The germination of sporangia of Physoderma zea-maydis as influenced by environmental factors. J. Elisha Mitchell Sci. Soc. 67:178-179.

30. Kelman, A. 1953. The bacterial wilt caused by *Pseudomonas solanacearum*, a literature review and bibliography. North Carolina Agr. Exp. Sta. Tech. Bulletin 99, 194. --. 1933. The relationship of pathogenicity of *Pseudomonas solanacearum* to colony appearance in tetrazolium medium. Phytopathology 44:693-695.

31. Kramer, M. and J. F. do Amaral. 1944. A identificacão da murcha bacteriana presente em culturas de batatinha do Estado de São Paulo. Biológico 10:199-207.

32. Leach, J. G. 1940. Insect transmission of plant diseases. McGraw-Hill, New York, 615 p.

33. Loos, C. A. 1962. Factors affecting incidence of rhizome-rot caused by *Erwinia carotovora* of Gros Michel and Lacatan bananas. Phytopathology 52: 110-114.

34. ________. and S. B. Loos. 1960. The black-head disease of bananas (*Musa acuminata*). Proc. Helm. Soc. Washington 27:189-193.

35. Lyman, G. R. 1918. Plant disease survey work on the *Physoderma* disease of maize. J. Washington Acad. Sci. 8:43-44.

36. McCulloch, L. 1937. Bacterial leaf spot of begonia. J. Agr. Res. 54:583-590.

37. McFadden, L. A. 1961. Bacterial stem and leaf rot of *Dieffenbachia* in Florida. Phytopathology 51: 603-668.

38. Marshall, K. C. and M. Alexander. 1960. Competition between soil bacteria and *Fusarium*. Plant & Soil 12:143-153.

39. Matz, J. 1920. J. Dep. Agr. Porto Rico 4:45 (and personal communications 1946-48).

40. Middleton, J. T. 1943. The taxonomy, host range and geographic distribution of the genus *Pythium*. Torrey Bot. Club, Memoir 20, 171 p.

41. Mohr, E. C. J. (Trans. and ed., R. L. Pendleton). 1944. The soils of equatorial regions with special reference to The Netherlands East Indies. Edwards, Ann Arbor, Mich. 766 p.

42. Mujica R., F. 1942. Susceptibilidad de variedades de papas a la sarna polvorienta causada por la _Spongospora subterranea_ (Wallr.) John. Dep. San. Veg. Bol., Chile 2:17-19. --. 1953. Outbreaks and new records: Chile. FAO Plant Prot. Bull. 1:106-107. --. 1954. Determinaciones micologicas. VIII. Agr. Tec., Chile 14:110-118.

43. Muller, A. S. 1941. El reconocimiento de las enfermedades de las plantas cultivadas en Venezuela 1937-1941. Bol. Soc. Venezuela Cienc. Nat. 7: 99-113.

44. Munnecke, D. E. 1960. Bacterial stem rot of _Dieffenbachia_. Phytopathology 50:696-700.

45. Muntañola, M. 1955. Parásitos criptogamicos de arvejas (_Pisum_ spp.) y poroto (_Phaseolus_ spp.) en el provincia de Tucumán. Rev. Agron. Noroeste Argentina 1:283-299.

46. Nolla, J. A. B. 1931. Studies on the bacterial wilt of the Solanaceae in Puerto Rico. J. Dep. Agr. Puerto Rico 15:287-308.

47. Nowell, W. 1915. The internal disease of cotton bolls. Agr. News, Barbados 14:222; 238-239. --. 1923. Diseases of Crop-Plants in the Lesser Antilles. West India Committee, London, 382 p.

48. Pontis Videla, R. E. 1951. Una podredumbre del tallo del maíz (_Zea mays_ L.) en Venezuela, causada por _Pythium aphanidermatum_. Agron. Trop., Venezuela 1:13-28. --. 1952. Una podredumbre de la caraota (_Phaseolus vulgaris_ L.) en Venezuela, causada por _Pythium aphanidermatum_. Agron. Trop., Venezuela 2:207-209.

49. Puttemans, A. (trans. A. E. Jenkins and A. D'A. Marchant.) 1940. Alguns dados para servir a historia da phytopathologia no Brasil e sa primeiras notifiçacões de doenças de vegetaes neste paiz.

Primera Reunião de Phytopathologistas do Brasil, Río de Janeiro, Rodreguésia 2:17-18-36. In Univ. Puerto Rico J. Agr. 24:77-107.

50. Ramirez, R. 1921. Plagas de la agricultura en el Distrito Federál. Rev. Agr., México 5:662-663.

51. Rands, R.D. 1926. Root disease of sugar cane in Louisiana. U.S. Dep. Agr. Circular 366, 20 p.

52. ________. and E. Dopp. 1938. Pythium root rot of sugarcane. U.S. Dep. Agr. Tech. Bulletin 666, 96 p.

53. Rangaswami, G. 1962. Pythiaceous Fungi. Indian Council Agr. Res., New Delhi, 276 p.

54. Revilla, V.A. 1952. Algunos resultados de las investigaciones sobre la nueve enfermedad de las caña de azúcar en el país. Peru Dir. Gen. Agr. Bol. 3-4:11-18.

55. Robbs, C.F. 1946. Contribuição ao estudo das bacterias que atacan plantas no Brasil. Bol. Soc. Agron. Brasil 9:195-221. --. 1959. Uma doença bacteriana da piminteira do reino (P. nigrum), nova o Brasil. Rev. Agr., Piracicaba 34:263-266. --. 1960. Bacterias fitopatogênicas no Brasil. Inst. Econ. Rural Rio de Janeiro, Divulg. Publicacõ 2, 63 p.

56. Rolfs, P.H. 1898. Diseases of the tomato. Florida Agr. Exp. Sta. Bull. 47:128-136.

57. Rorer, J.B. 1911. A bacterial disease of bananas and plantains. Phytopathology 1:45-49.

58. Ruehle, G.D. 1940. Bacterial soft rot of potatoes in southern Florida. Florida Agr. Exp. Sta. Bulletin, 348. 36 p.

59. Shands, H., V. Clibas and W.J. Zaumeyer. 1964. Observations on dry bean diseases in Brazil. Plant Dis. Reptr. 48:784-787.

60. Sideris, C.P. 1931-32. Taxonomic studies in the family Pythiaceae. I. Nematosporangium. II.

Pythium. Mycologia 23:252:295; 24:14-61.

61. Smith, T. E. 1939. Host range studies with Bacterium solanacearum. J. Agr. Res. 59:429-440.

62. Stahel, G. 1917. De zeefvatenziekte (phloem necrose) van der Liberica koffie in Suriname. Dep. Landb. Suriname, Bulletin 12, 2 p. --. 1932. Zur kenntnis der Siebröhrenkronkheit (Phloem-nekrose) des Kaffeebaumes in Suriname. II. Phytopath. Zeits. 4:539-544.

63. Stevenson, J. A. 1917. Diseases of vegetable and garden crops. J. Porto Rican Dep. Agr. 1:93-117. --. 1918. A check list of Porto Rican fungi and a host index. J. Porto Rico Dep. Agr. 2:125-264.

64. Tessi, J. L. 1949. Presencia de Psuedomonas coronafaciens en la Argentina y reacción de algunas variedades de avena frente a este parasito. Rev. Invest. Agr., Argentina 3:319-334.

65. Thirumalachar, M. J. and J. G. Dickson. 1947. A Physoderma disease of quack grass. Phytopathology 37:885-888.

66. Tisdale, W. H. 1919. Physoderma disease of corn. J. Agr. Res. 16:137-154.

67. Tocchetto, A. 1943. Peronospora da parreira. Rev. Agron., Brasil 7:253.

68. United Fruit Company, Research Department. 1958. Problems and Progress in Banana Disease Research. United Fruit Co., Centr. Res. Lab. Booklet 10, 36 p.

69. Velasco Llano, V. 1944. La enfermedad (Bacterium solanacearum) del platano es muy grave. Estac. Agr. Exp. Palmira, Colombia Boletin 5, 2p.

70. Verona, O. and T. Benedek. 1959. Iconographia Mycologica I. Mycopathologia 11: (Supplement) 105 p.

71. Viegas, A. P. and A. R. Teixeira. 1943. Alguns

fungos do Brasil (Phycomycetos). Bragantia 3: 223-245.

72. Walker, J. C. 1969. Plant pathology (3d ed.). McGraw-Hill, New York. 819 p.

73. Wardlaw, C. W. Diseases of the Banana. MacMillan, London, 615 p.

74. Wellborn, V. 1940. La necrosis de los cafetos. Café de El Salvador 10:484-500.

75. Wellman, F. L. Clubroot of crucifers. U. S. Dep. Agr. Tech. Bulletin 181, 31 p. --. 1961. Coffee Botany, Production, and Utilization. Leonard Hill, London; and Interscience Publ., New York, 488 p. (--. 1929 and 1950. Unpub. notes on _Fusarium_ relations to soil components.)

Chapter V

TROPICAL PHYTOPHTHORA AND LEAF-CURL DISEASES

Summary

Phytophthora fungus diseases strike fear into a grower from the long and famine-noted history of the classic potato blight. Its causal organism was indigenous to the Neotropics, and it is now spread world wide. Impacts from tropical *Phytophthora* diseases are great and have had effects on settlements, great cities, and civilization. Observations on the tropical members of the fungus (there are said to be about 70 species) indicate many problems of occurrence, dissemination, optimum conditions for disease, and pathogenicity. There are many workers investigating along these lines in the tropics, and it may be from them that world knowledge about these diseases will have its next best impetus. The long decades of unfruitful work on oospores in Europe and northern U. S. A. were suddenly released to a new life when studies were centered in the tropics. There seems to be no symptom type to characterize attacks from species of *Phytophthora*. There is recent work that indicates crossing between species; significance with respect to pathology is still to be determined. White rusts (*Albugo* spp.) known well in cold countries are present but not serious in the Neotropics. There may be possible exceptions to this in the future. There are several tropical leaf-curls (*Taphrina* spp.). Some were imported from the temperate zone but others have originated in the tropics.

Introduction

The plant diseases caused by fungi belonging with Phytophthora and Taphrina are of great concern in the Neotropics. One of the reasons is that so much of the tropics has climatic conditions specially adapted to the increase and development of the organisms. Another reason is that there are so many more species of host plants, both uncultivated and as crops, that are susceptible to infection by these primordial types of organisms. A third reason is that in spite of the known severity of diseases they cause, the most common treatment is ignoring the occurrence of these troubles and replanting, or changing to another area, or just living on less and doing the best that can be done with poor productivity.

There is another reason these plant diseases are not given the control attention that should be expended on them. Plant pathologists and mycologists are in some places only recently working with more modern equipment and ideas in the Neotropics. It is with sufficient permanency, and long time effort, that some of the extremely difficult problems of pathology, physiology, genetics, and species relations can be solved. Even though studies on parasitic phytophthoras have been in progress for over a century most of it has centered on problems in the temperate zone. Good and unquestionably valuable as all of it is, it will not be until more of the problems are attacked under tropical conditions down in the real tropics, that the tropics, where it all began, will answer the most subtle questions, and it is here that the most important discoveries regarding control measures will be devised.

When newly found, aside from rusts, there is no disease of crops that merits so much fear as that said to be from a fungus species of the genus Phytophthora. This is understandable, for it is continuously remembered that a famine, killing millions of people, occurred in Ireland in 1845-1846 from a potato blight caused by a Phytophthora. Farming has never been the same since. From what it has been possible to deduce, indications are that in about 1833 diseased tubers from the tropics were carried in a merchant ship's galley. The tubers seem to have been picked up by a sailing craft which on its weary way to Ireland had stopped for some fresh food supplies at Chiloe, a cool island at the edge of the Neotropics in the southern part of Chile. (See also discussions in such citations as 6, 9, 20, 27, 35, 38, 40, 50, 54, 55, 57-59.)

Differing Impacts from Phytophthora Diseases

It is not recorded in any history that a city was settled at some specific location chosen for the single fact that the area around was free from one dangerous and important crop-plant disease. However, the reverse has been the case many times; tropical cities and, as well, certain loosely gathered population concentrations within circumscribed rural areas, have disappeared because a crop became diseased. At numerous points in the great region of the Neotropics, principal populations have grown up and settled around crops. On the other hand, some places after attaining fame and wealth dropped by the way and were deserted when a disease entered and destroyed a once dependable agricultural product. A few different examples of such disruptions and ruinations are given here, especially in the case of diseases from parasitic *Phytophthoras.*

The largest city of Ecuador is on its west coast, and is built around the port that lies upstream a short distance from the bay of Guayaquil. Up until the time that cacao was found to be of such importance in world commerce, the village that later became the city of Guayaquil was as quiet as a port city could be, but it became wealthy and notorious when its fame spread abroad from handling the chocolate bean. Great colonial structures were built, a fine cathedral came into being, and in typical equatorial style stucco houses and the cheap bamboo and log shelters that went for residences, spread around quickly. The latter of flimsy make were used as if they were permanent houses and the population moved into all it could get to take advantage of the chocolate boom.

Guayaquil became the home base of wealthy people, as its worldwide contacts were established for chocolate business; connections for them were in about all the cosmopolitan areas of the temperate zone. For some decades in the last century the future for Guayaquil seemed great and secure. It was at about the end of that century that pod rot, caused by *Phytophthora palmivora*, became a problem (1, 48). By the 1920's, this disease, combined with two others, witches' broom and monilia pod rot, brought production down to such a low ebb that businesses closed in Guayaquil, banks no longer operated, ships did not visit the port, the place was ruined, and a great migration took place away from the city. The main trouble was the *Phytophthora* disease. It was not until pathological studies were instituted that sprays and other means of amelioration brought back money-making productivity to the great cacao region of Ecuador.

The nearest to a tropical city on the mainland of the U. S. A. is Miami in Florida. It is a flourishing winter resort, world renowned, and, as almost a trademark, pictures show friendly coconut palms growing in parks and along avenues. The tree is naturally an exotic to most of temperate zone United States, but it is found in south Florida and its scenic and attraction value to Florida can hardly be overestimated. During the early part of this century budrot, again caused by *Phytophthora palmivora*, of coconut was seen (Fig. 5-1) and studied in the West Indies (1, 14, 15, 55, 56, 59). It was also found in alarming amount in Florida, and the budrot disease was proved to be from *Phytophthora palmivora*. A rigorous program of eradication was instituted by the state of Florida, and after several years, by 1928, through hard work and at great expense, people of the area were repaid by the complete eradication of the trouble.

As a postscript: when I was in Florida, especially during the years 1931 and 1932, but also in 1934 and 1935, I made a search every time I saw coconut plantings for budrot, and no examples were ever seen. I had become familiar with the symptoms from seeing them in Honduras, and later on, a good deal of the disease was seen in coconut groves in El Salvador and a few in Costa Rica. In 1963 on being called for consultations about palm diseases in the American Virgin Islands I found budrot from *Phytophthora* on palms growing around the picturesque winter resort Island of St. Croix; from which it was subsequently eliminated by eradication procedures. Species of this fungus also attack roots of trees.

For a long time oranges have been a dependable, money-making, agricultural product in Argentina, and an area of big production is in the Province of Cordoba. The fruit is shipped long distances, it has entered into international trade and is an important part of the province's income. Many growers have a substantial yearly outlet in Europe. Large holdings of orange groves had developed and a large population of growers, laborers, and transporters lived on the proceeds. In about 1945 (see Fawcett (11)) it became apparent that over half of the sweet orange trees of that Province had a disease of the base called

Figure 5-1. Coconut palm budrot, Phytophthora palmivora, causing death of three palms in foreground, those in back show no symptoms. Picture taken 1924 near Miami. Photograph courtesy of Department of Plant Pathology, University of Florida, Gainesville.

foot-rot, caused by Phytophthora citrophthora Groves (and other species as well), were sold or allowed to deteriorate and return to bush. The situation increased in severity and the orange men appeared headed for bankruptcy. It was then that researchers discovered (11) resistant root stocks and these were used sometimes to preserve still standing growing trees by inarching techniques. Amelioration methods were instituted, and farmers partially exposed roots (Fig. 5-2) to the air and improved drainage. The one most important thing accomplished has been the studied replacement of old languishing trees with superior types of sweet orange tops grafted onto Phytophthora-resistant sour orange root stocks.

It is well known to students of World War II, that a serious problem had arisen in the American tropics because of increasing malarial infections in indigenous populations. Quinine had been relied upon during more than a century for keeping populations free of chills and fever with their extremely weakening effects. Because of military reasons, normal sources of quinine brought from the Orient were cut off from the Western Hemisphere. A so-called "crash program" of planting the cinchona tree was instituted in South and Central America, and the trees were cultivated where no crop-plant had ever grown under the hand of man before. Progress was good and results were such that the success of its production in the Western tropics looked assured. The army planned confidently on obtaining its needed malaria control medicine; even the plastic industry, which had learned of the value of quinine in making certain of its compounds, looked expectantly for the cinchona bark supply required.

All went well and millions of dollars were spent on the program. Without its being expected, there were sudden occurrences of leaf-, stem-, and root-infecting organisms. The most devastating were found to be species of Phytophthora. These caused havoc in seed beds, nurseries and newly planted fields. They were organisms that were natural inhabitants of primeval jungle. Those who worked on them (e.g., 7) realized that unless controls could be devised a promising and essential new industry in the American tropics was doomed; and that, of more importance the health of the populations in the tropics, who could be given treatment and safe guard by medication with quinine, was endangered. Control methods were earnestly sought, and varietal tolerance of tree varieties to Phytophthoras, and ecological studies related to moving plantations out of wet

Figure 5-2. Base of old susceptible orange tree attacked by gommosis or footrot, Phytophthora spp.; soil excavated from around roots to expose them for examination and treatment. Photograph courtesy of Department of Plant Pathology, University of Florida, Gainesville.

rainforest areas were being started. This was never completed. There is no doubt that the result after some years would have been successful. (Fortunately, just when things seemed most serious, chemical researchers put onto the market improved insecticides for control of the malaria mosquito, and synthetic anti-malarials to take the place of quinine. If these had not come when they did, Phytophthora infections on cinchona would have made the world war more prolonged, more deadly, and more difficult than it was.)

Observations of Tropical Phytophthora Diseases

There are many more Phytophthora diseases of plants in the warm, but not hot, tropics than in the temperate zone parts of the Americas. For an idea of the numbers, hosts and distribution, the reader should see the volume List of Tropical American Plant Diseases. There are reports (15, 31, 35, 55, 59) of at least 70 species of this fungus parasitizing plants. When more is known about the variability of those parasites and the susceptibilities of the hosts, some species may lapse into synonymy (3, 10, 14, 15, 20, 23-25, 31, 35, 39, 50, 54, 55, 59). On the other hand, it is possible, and even probable, that in years to come even more disease-producing species may be discovered by tropical workers.

From the standpoint of diseases caused, tropical Phytophthoras are quite often limiting factors in crop production until control measures are devised. If searched for, the parasites usually are found to be present in moist areas where farming has become established and carried on for some years. Sometimes where jungle has been recently destroyed and plantations have been newly started with susceptible crops such as cacao, coconut, and oranges, the phytophthora diseases seem to be absent. Almost invariably, with intensification of agriculture over many years, and the coming of observant growers, and with them pathologists, sooner or later agriculturists are confronted by phytophthora problems. These the pathologists are challenged with on all sides.

Conditions recognized as good for such diseases are long rainy seasons, coupled with abundant and persistent dews in intervening dry seasons. Temperature is also important. From a review of literature on the effect of temperatures on phytophthora diseases it is evident that warmth, from about 30° to 35° C, is rather inhibitive to the causal organisms. Such as these highest temperatures exist in special, rather restricted parts of the tropics where it is relatively dry, practically at sea level, and near the heat equator (see Map A in Chapter I). By far the majority of the crop-producing areas have, either or both, soil and air temperatures that are within the range of 9° to 28° C at which these diseases flourish (4, 6, 11, 12, 20, 31, 38, 44, 55, 59). As is discussed in Chapter I, in the tropics cooling is quickly felt from elevations on hillsides, exposures to prevailing winds, shading effects, and at distances away from the heat equator.

There are some growing methods observed in the tropics that I know are favorable to phytophthora diseases. Special shading of cacao is commonly employed in Central America and podrot, cherelle wilt, and canker from P. palmivora are more prevalent in shade than in sun. Where heavy mulching is used with potatoes in the West Indies islands, P. infestans has been found as more severe than where clean culture is carried on. Where avocado trees have been unexpectedly unfruitful in Guatemala, unlettered Indians have been seen to expose the roots around the tree crown and beat the main roots with a heavy cudgel. Of seven such treated trees, in two years six succumbed to typical carbonaceous-type feeder-root disease from P. cinnamomi. In a fruit tree huerta (mixed planting) of orange, guava, kakee, and ladyfinger banana in Honduras, the peasant owner had scraped up dirt from between the trees and heaped it around the base of all his trees, three years before my visit. Of 26 orange trees 20 were severely attacked by gomosis from P. citrophthora. Lesions of P. parasitica have been seen as occasional in Puerto Rico on ornamentals such as vinca and cordyline. Where grown without much cultivation, except weeding, of these plants about 8 per cent were damaged. In plots with overhead sprinkle irrigation, both of these ornamentals showed over 40 per cent damaged by the disease. The five species of Phytophthoras named here are, from all that can be learned, the most widely disseminated in the American tropics.

Spread of a foliage or fruit disease, whatever its symptoms and whatever the species involved, if caused by Phytophthora, is likely to progress most rapidly downwind. This is specially true in such as P. palmivora on cacao or coconut and P. infestans on potato or tomato. In the case of P. parasitica as a cinchona rootrot, P. citrophthora as sweet orange gomosis or footrot, or P. cinnamomi as avocado rootrot, spread is faster downhill if the grade is steep enough that water will flow with reasonable ease in a definite direction. The footrot diseases are more severe in trees grown on poorly drained and heavy soils, as is certainly to be expected.

In the case of P. capsici, notable on a wide range of hosts and under many different ecological environments in the Neotropics (22, 31, 32, 35, 50, 59), in many cases injury symptoms are in the nature of girdling and stem lesions. Some of these if on sufficiently lush growth encircle stems causing the part above this infection to die. When this occurs, recovery of the main causal Phytophthora is difficult, and sometimes impossible in old material. The best lesions to use for isolations of this fungus in the tropics are from edges of comparatively new infections, when the centers are beginning to turn dark in color.

Samples of known P. palmivora used for inoculum, have been observed making saprophytic growth in cacao leaf debris on paper toweling in a moist chamber; when a diseased sector of a cacao pod was wrapped in moist filter paper, hyphae grew into the paper and lived where sufficient juice from the pod, and moisture, leached from the pod. Isolations from these kinds of fungus growths on paper were cultured and all were shown to cause lesions on young cacao leaves in moist chambers.

Seedling damping-off has been observed personally, and positively (by microscope) identified as caused by Phytophthora sp. in seed beds of kenaf, rubber (Hevea), pre-emerged seedlings of arabica coffee, cucumber, tobacco, an undetermined kind of grass, tomato, and sesamum. In both kenaf and sesamum the first attack was by Pythium sp. followed and overrun by Phytophthora sp. A moist type of soupy appearing (sancocho) type of damping-off occurred on seedlings of lettuce, cucumber, tobacco, grass, and tomato grown in moist chambers. In rubber and coffee seedlings the hypocotyls were darkened, softened, and then collapsed.

Rots that are more or less solid occur from Phytophthora attacking such things as orange fruit pedicels, sugarcane setts, cacao pods, and buds of coconut, areca and other palms. These rots are rather firm, and not unpleasant to smell. In the coconut bud disease especially, when bacteria are involved as well as the fungus, the rot tissues may soften and develop a discoloration and bad odor. In Honduras, budrot of coconut was given some attention in 1929. In it was found both phytophthora and bacteria. The latter had apparently been carried to the tops of the palms by scavenger buzzards that, after after feeding on decayed materials and village sewerage, sunned themselves on coconut tree fronds. Diseased coconut buds from such trees were most disagreeable to smell and no phytophthora could be isolated from dissected material.

In the temperate zone, the potato disease (from P. infestans*) is called late blight as the blight does not appear on young foliage when it first sprouts up in newly planted rows. Indeed, there is no observable trouble from it until spring has gone and summer is well along. In the tropics where there is no true winter, the name "late" in relation to the all-over growing season, has very little meaning. In many parts of the Neotropics the potato blight is more reasonably called such names as mildew, mildíu, míldio, heilo, gotera, trizón, rancha, mancha, Phytophthora blight, the potato blight, and enfermedad de papa. Some of the life history and symptoms of this disease are indicated in Fig. 5-3.

In general the most critical condition for a severe potato blight epiphytotic in the tropics is for the crop to be growing well, abundant inoculum to be present, and cool weather come, accompanied by about 15 days of successive rains. This means there will be heavy dews at night. Temperatures should be, on the whole from 22° to 25° C. In Peru, research has found the special conditions that enhance epidemics of rancha (4, 50) and the same problem is under past and future study in Chile (38). Of course, the so-called "blight weather" complex so much watched for and feared in temperate zones (6) is well known, and it occurs every year in some parts of the tropics.

One of the less publicized symptoms in the Neotropics is the tuber infection. (Tubers stored some months for food, when affected with Phytophthora infestans, show increasing amounts of brown to reddish-brown and firm decay under

*It is of interest to note here that Puttemans (46) held till his last days that another binomial, antedating that of de Bary and worked out in the American tropics, should have been used in our literature on the disease. According to accepted rules of nomenclature he was correct.

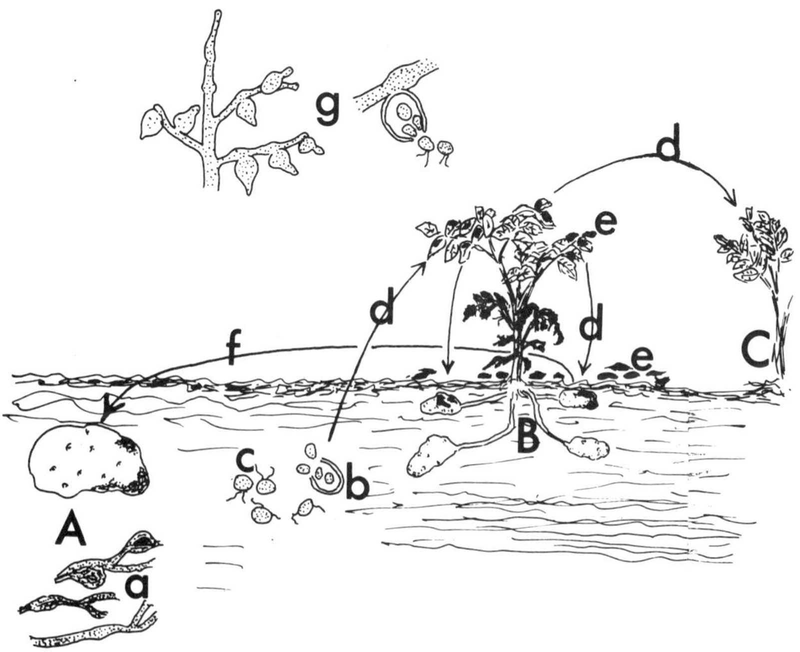

Figure 5-3. Diagram of stages in attack of potato by Phytophthora infestans: A. diseased seed tuber, B. growing infected plant, C. neighboring plant. Quiescent mycelium (a) buried in tuber, sporangium (b) develops from resting mycelium and produces swarm spores (c) splashing of rain (d) distributes spores from ground to foliage, and from foliage to foliage, and in the end from diseased leaves (e) to maturing and matured underground tubers (f). On leaves aerial sporangia (g) are developed in profusion under best conditions which are several successive days of rain, heavy night dews, and cool temperatures at about 22° to 25° C; sporangia may germinate and infect directly or develop swarm spores that infect. In the tropics holdover stage of the organism is encountered in A (a), and this readily adapts itself to over-wintering possibilities in the temperate zone.

unbroken skin). This is known as chocolate (pron. cho-co-lah-tey), but in early stages it is practically invisible in its effect on the surface of the tuber (4, 6, 12, 35). During tuber examinations for seed sources the early stages are easily unrecognized. Diseased tubers have been shipped long distances.

Where the potato blight has occurred so severely in many tropical countries the course of events is practically always as follows. Seed stock is brought in with invisible internal infections. On planting, tubers germinate and shoots appear above ground in a normal manner. Growth is good until usually or at least quite often after blossoming of the vines, when the first blight spots are found on the lower leaves. The infection may remain at a minimum unless a long stretch of cool and moist weather is encountered. When this happens the blight quickly spreads from a few leaf spots at ground level to the whole plant, and then rapidly from plant to plant.

I have seen many times tropical hillside fields of both potato and tomato crops completely and severely covered with this blight. Sometimes the disease was most spectacular during periods when no rain or very slight amounts had fallen, but when studied it was learned there were heavy dews consistently present at nights. It has also been noted that in fields nearer sea level there is at times an anamoly found with respect to effect of temperatures. Given almost continuous 100 per cent air humidity and fine drizzle of rain, severe epidemics will occur during periods in fields where air temperatures seldom touched as low as 25° C and go to a high of about 28.5°; temperatures a few degrees warmer than temperate zone workers think of as critical for epidemics. With showers every day or every other day, *Phytophthora* spreads freely and leaf kill is severe. Field effects from the *Phytophthora* causing stripe canker of cinchona trees and blight of beans were the same as to the conditions just mentioned above. In the West Indies the footrot of mature sweet orange trees, and effects of *Phytophthora* causing leaf and stem cankers of seedlings occurred with almost constant temperatures at 28° C and above.

In their careful studies on Phytophthora, Misses Waterhouse and Blackwell (59) showed that of 29 species the minimum temperatures for fungus growth was 2° to 12° C, optimums were about 25° to 28° and maximums from 30° to

36° C. There is a small group of five *Phytophthora* species that appear adapted to slightly cooler conditions. Minimum growth temperatures of this group is about the same as the rest but the optimum appears to be nearer to 20° C, with the vegetative condition of the fungus killed at about 25°. It is to be repeated, however, that these are temperatures well within the regular situation in tropical mountains and are of common occurrence in even low lying lands that border on the north and south edges of the Neotropical region.

There is an often overlooked feature about tropical soil-inhabiting and leaf-killing fungi that is a great hazard to their existence. This is the abundance of bacteria that dissolve decaying plant tissues and weak fungal hyphae. This can be observed to occur in petri dish cultures, and I have seen it in both cacao podrot and cinchona root-attacking *Phytophthoras*, when soil and debris from a disease-free field were sprinkled over colonies of the fungus growing on cornmeal agar. In both the treated cultures and in the selected bits of field debris, it was found that the only thing that remained of the *Phytophthora* fungus, after bacterial action had progressed, was round oogonia that had formed thickened protective walls. In some cases in both *P. cinnamomi* and *P. palmivora*, especially dense chlamydospores were found undissolved in month-old bacteria-infested fungus cultures.

Various Symptoms from Phytophthora Attacks

There are no ways to demonstrate, exactly, whether members of the genus *Phytophthora* might have been and then left, a primordial condition of independence as a saprophyte to become specialized as a parasite. Some workers believe there is no saprophytism in the genus, but non-pathogenic cultures have been found in tropical material by me. In one study, instituted in Guatemala, a *Phytophthora* was recovered on water agar from soil of a mountain jungle where no symptoms of this kind of fungus attack was seen on any plant in repeated visits. The isolate looked exactly like *P. palmivora* when cultured in the laboratory and studied under the microscope. However, when tested on leaves of cacao, tomato, bean, cinchona, and kenaf it caused no disease.

In Costa Rica, soil and debris specimens were secured and observations made looking for such fungi as *Fu-*

sarium, Rhizoctonia, and Pestalotia which were on volunteer seedlings of cacao. Soil and debris smears were made on slides and stained with eosin and in some cases with cotton blue. Under the microscope there appeared to be many hyphae with chlamydospores and oogonium. Over 600 bits from these soil samples were placed on water agar, and 19 showed phytophthora-like growth. In pure culture, on maize meal agar, these isolates were even more like the organism in question. No isolate caused disease on young cacao leaf in moist chamber. In such cases as noted in the above two paragraphs, Phytophthora spp. appear to be capable of good growth in a saprophytic condition.

Among notable symptoms are fruit rots that, although primarily are caused by phytophthoras, are also those complicated with secondary invaders. It is remarkable to the observer how well P. infestans-infected tomato fruits stay mostly firm yet characteristically, and spectacularly, ruined as edibles. The fruit softening if it occurs with P. palmivora on cacao or on hevea rubber is softened because of bacteria, yeasts and weakly parasitic fungi. Phytophthora spp. attack fruits of lemon and orange, and may enter through the pedicels. The pedicels may not be softened, but in many instances the secondary organisms, when they act in a mixture with Phytophthora, cause softened rot. In most cases phytophthora fruit rots are dark-colored, and, if uninvaded by secondaries, are more-or-less firm. (See 11, 15, 16, 20, 45.)

Wilt has been seen induced in both sugarcane and maize by a combined(?) Phytophthora-Pythium (sometimes it was not possible to be certain which) infection of roots. It was seen in poorly drained soil, and roots were the only parts invaded. In El Salvador while studying wilted plants of tomato known to be of a Fusarium-resistant variety it was found that the unexpected wilt was not from fusarium but was the result of strangling, firm lesions at the collar. When isolations were made from the lesions Colletotrichum, Rhizoctonia, and Phytophthora were all recovered.

Rounded to elliptical stem cankers, caused by Phytophthora, are fairly common in the tropics in such crops as cacao, citrus, and papaya. In both citrus and cacao the cankers dry up if the place of infection is fairly matured and the conditions are right. In some cases the cankers are rather shallow and are surrounded with a roughening of cortex. In papaya occurs a firm basal rot, but if infection is up on the trunk, the stem canker spreads downward mostly. With age the decay penetrates into the central lumen of the trunk and, with secondary infections, finally results in tree death. The same kind of symptoms are seen in the herbaceous tobacco (13, 20). A stripe canker going from roots to the first branches at first, and finally farther up, is caused by Phytophthora cinnamomi on trees of cinchona, avocado and of macadamia nut. (See also 7, 8, 11, 14, 15, 20, 34, 35, 42, 43, 47-50.)

There is a tree footrot, a basal infection called such names as gomosis (gummosis), and brown rot that is caused by P. citrophthora. It is probably one of the most well-known, serious pathological problems of the Neotropics. Symptoms of such a footrot are shown in Fig. 5-2. In all places where the susceptible grapefruit, lemon, tangerine, and sweet orange are grown on their own root stocks the trees are in imminent danger of footrot. It is known that sour orange is resistant and the other types can be grafted onto resistant roots. This control is successful unless the tristeza virus has to be reckoned with (see the chapter on tropical virus above), for the tops of such grafted trees succumb to root weakening that is caused by the virus. As another example of footrot, it should be noted that a basal rot from Phytophthora has been found attacking papaya. In this there is no gumming reaction, and the disease is a firm, almost dry rot. Numerous other phytophthora footrots are found on herbaceous annuals, such as on various solanaceae (14, 22, 50), and even on spice or black pepper (26, 30). A different type of firm-rot occurs if P. infestans invades potato tubers. When this occurs the water-gorged tissues lose crispness, darken and become leathery (4, 12).

From the standpoint of recognizable appearances of different types of diseases and of diseases on different hosts, one phytophthora species may be involved with 20 or more maladies. An example of this is Phytophthora palmivora. On cacao alone it causes: 1. leaf blight; 2. twig dieback; 3. branch canker; 4. seedling damping-off; 5. pre-emergence decay of seed in seed bed; cherelle wilt; and podrot. On coconut: 8. bitten leaf; 9. young frond disease or wilt; and 10. budrot (see Fig. 5-1). On rubber it causes: 11. leaf spots; 12. early defoliation before normal season drop; 13. twig disease; 14. dieback; 15. stem canker; 16. tapping panel disease; 17. fruit rot. On castor bean it causes: 18. seedling damping-off; 19. seed pre-emergence rot in planted rows; 20. leaf blight; 21. capsule rot; 22. wilt of bearing plant. When this Phytophthora infects pineapple it causes: 23. fruit rot. On okra it causes: 24. seedling damping-off; 25. leaf disease; 26. a stem rot and collapse; 27. pod rot. In various ways this same fungus parasite attacks other crops such as black pepper (Piper), red pepper (Capsicum) and eggplant. It is also at times a problem causing diseases on numerous ornamentals as well as spice crops.

The Life History and Species Problem in Phytophthora

The organism, P. palmivora, exemplifies certain usual developmental stages of phytophthoras in the field. The usual experience in the tropics with P. palmivora (9, 16, 20, 31, 35, 43, 48, 49, 50) is that in spite of abundant bacteria on all sides it readily persists in nature from rainy-season to rainy-season. In cacao it remains as semi-dormant, heavy-walled and nutrient-filled hyphae in old disease lesions. In some cases dormant oospores are preserved buried in diseased tissues. On the advent of rains the semi-dormant hyphae revive and send into the air sporangiophores. The oospores, as well, on being released by decay of surrounding diseased host cells, imbibe water, germinate, and develop sporangiophores. Whichever happens, sporangia are carried away by a breeze or a raindrop which serves to deposit them in a susceptible host. It seems all phytophthoras act in the same fashion.

On examination of any phytophthora material under the microscope, the ovoid, aerially borne sporangium appears as a fragile organ. It is easily dislodged from the sporangiophore, is thin walled, absorbs water quickly, and germinates from its upper (distal) end. If there is not direct growth and penetration there emerge from a sporangium several two-tailed rapidly swimming swarm spores; they settle quickly in dew or rain drops and are ready to penetrate. On piercing the host epidermis, the slender germination thread, whether all from a sporangium itself or from a biflagillate swarm spore, appears to ramify directly both between cells and through them. After development and maturation of the disease condition, the time arrives for reproductive spore production.

Judging from special studies on life history problems (2, 14, 16, 17-21, 23-25, 28, 29, 31, 38, 40-43, 51, 52, 60), if there has been no sexual fusion between compatible Phytophthora strains, no germinable oospores are produced. This has been the case in *P. infestans* as seen in the temperate zone (5, 6, 9). On the *other hand,* once research on *P. infestans* was taken into the tropics, workers found there *occurred* in nature hitherto unknown strains of that species that were compatible, proceeded to fuse sexually, and developed fertile oospores that germinated (see especially 17, 19, 36, 40, 51, 52). Of further interest and still to be more completely studied, researches of Haasis and Nelson (23-25, 35) show very significantly that there is capacity for ready interspecific crosses in the phytophthoras. The question inevitably rises, what is species in this fungus genus? In addition, what can this permissiveness among the phytophthoras portend in the future as respects changes in virulence, variation in host attack, and similar problems? Also, what should a pathologist use for a species name in *Phytophthora*? Where does all this lead?

A scientific name used by the struggling tropical pathologist is the handle with which he has to refer to a special organism that is his adversary in crop disease. There is no question that the academic requirements of the needed mycology must be sufficiently satisfied that a new name or disposition will have world-wide acceptance, both in cloistered places of literature reference and herbarium collection, and in research laboratory and field plots for control study. It is of interest that the genus *Phytophthora* in the tropics is apparently in an unfinished *state with respect* to its mycology. Concepts respecting it are well advanced--the ground work is done; what is required are more years of dedicated effort. Preferably this should be carried out in the Neotropics.

As years have progressed, to this genus are attached many named species, most of which were species based on several features such as: host range, growth in culture, effects of media, temperature relations, and morphology. A large number of what we know of as species in *Phytophthora* came from isolations secured in the tropics. Tucker (55) as a result of his and others' experiences in the American tropics, paid especial attention to facts of occurrence and in his taxonomy of the genus. This was given further attention by several, among whom are Leonian (31) and Frezzi (14, 15).

The delimitation of species in *Phytophthora* as understood at present depends partly on morphology, partly on symptoms of the disease caused by the isolate, partly on host range, and partly on culture appearance and compatibility reactions. It is apparent that, as results now stand, judging from host reactions, there are wide differences in pathogenicity within the effects of different isolates from what are considered the same species. In some cases, as in P. infestans, the species has been demonstrated to vary its vigor of pathogenicity in connection with ability to overcome resistance characters in potatoes. It was shown by Mills (36) that passing of races of this potato fungus to tomato started with practically no infection in tomato. However, after it was once established on tomato the race increased its adaptability to its new host and eventually became a severe pathogen of the new host.

A point of significance to all of the phytophthora disease researchers is that P. infestans, the classical strain that is of worldwide temperate zone occurrence but has reproduced all these years asexually, has remarkable pathogenic malleability. Complexities of findings have marked Phytophthora as one of the most difficult of the genera respecting the definition of species. Variability is great in the fungi, and may be specially so in dealing with the usual collections of isolations of Phytophthora.

The Disturbing So-Called White Rusts

There are many plant parasites that cannot subsist on decayed vegetable matter and these survive only as obligate parasites by their attack on living hosts. Some of the best known of these are true rusts, species of such genera as Puccinia, Hemileia, and Physopella. Far removed in relationship from the true rusts are certain parasitic species nearer to the Pythiaceae, of the genus Albugo, which likewise are obligate in habit and to further confuse matters are called white rusts in the temperate zone. In translation this name is used in the tropics, too, where these diseases are also known as white blister, whitening, white disease, white cloud, blister, whiteness, and even fungus leaf miner. While Albugo is a genus that is a member of the same large family to which Phytophthora and Pythium belong, it is much more restricted in its existence. In species of Albugo there are probably numerous well-defined physiologic races.

Since most of the critical studies on Albugo spp.

have dealt with the diseases from it in such places as Canada, northern Europe, the U. S. A., and Japan, the idea is prevalent that the parasite is a native inhabitant of the cool to cold temperate zone. However, it is notable that in the United States infections by *A. ipomoeae-panduranae* (Schw.) Swingle on sweetpotato are most abundant in the southeastern, warmer and moister, parts of the United States; and in Argentina, from such a report as that by Spegazzini (53) and Fernandez Valiela (14), there is little doubt that this *Albugo* disease is more common in the northern, warmer parts of that country. It is common throughout the West Indies, Brazil, parts of Mexico, and Central America. Invariably, where I have seen it, it has been in warm, moist climate surroundings. In Central America it was found that there was no such chilling needed for spore germination as many text books state is necessary. Gardens were observed where temperatures averaged 25° C with only 5 or 6 degrees variation from that at the most, over periods of more than a year. At no time could the temperature possibly have fallen to 20°, 15° or less. Yet this occurred on *A. portulacae* (DC.) Kuntze on cultivated purslane. *A. bliti* (Biv.) Kuntze was identified on beet and an amaranth, alternanthera and on salsify. The white spotting on sweet potato foliage appeared and spread without the remotest benefit of chilling.

The conidium of any *Albugo* is a sporangium. A conidium starts as an end of a fertile hypha that is pinched off into this rounded organ. It soon is pushed on and another conidium forms but is retained by connection with the older organ. This repeats itself until rather long chains are formed. When dislodged and caught in water the conidia germinate, releasing biflagellate, rapidly swimming zoospores. If any of these reach stomata the zoospores infect and thus enter the leaf tissues. The moisture for this is common in occurrence. Where observed, diseases from *Albugo* are not usually encountered in tropical places where irrigation is required to grow crops. However, it is of interest that with *Albugo* if there is an extreme amount of rainfall throughout the year, such as in wet, rainforest areas, the white spot-blister diseases disappear. If there is a situation in which part of the year is wet season and part dry, the disease easily persists over the drought period. In the tropics, where, it is to be remembered, temperatures are free from great fluctuation, spread and severe development of the whitening diseases from species of *Albugo* are most effective during such seasons as when there occur light rainfalls and nightly but not heavy dews.

Some crops known to me as specially affected by white rusts are sunflower, salsify, radish, cabbage and relatives, all attacked by A. candida (Pers.) Kuntze; the beets and the Andean or Peruvian grain Amaranth by A. bliti (Biv.-Bern.) Kuntze; New Zealand spinach by A. portulaceae(?); the sweetpotato by A. ipomoeae-panduranae (Schw.) Swingle; and cultivated purslane by A. portulacae (DC.) Kuntze. Many weeds and ornamentals are also susceptible to species of Albugo. Morning glory vines are often spotted by not only A. ipomoeae-panduranae but also by A. convolvulacearum Speg. There is a possibility that these two species may be synonymous, although I consider them probably different: convolvulacearum in my work is characterized as causing smaller but thicker spots that soon break through the epidermis of morning glory leaves, while it seems the ipomoeae-panduranae species is of a type that makes lighter, less severe-looking spots, tends to be confluent, and takes much longer to break through leaf epidermises.

The red bedding plant, alternanthera, is much blemished by A. bliti and under some conditions is not used in grounds beautification because not only are the diseased leaves unpleasant in appearance, but also it is more difficult to grow plants from infected cuttings than from those where white rust does not occur. In some areas the common pigweed, the closely related spiny amaranth, and the weed lambsquarter are all commonly diseased with an Albugo. It is often believed that species of the genus have quite narrow host ranges. However, these studies have not been given much attention in the tropics. It may very probably be found that host ranges under those conditions are fairly wide. I have found in several cases where no crop of any sort had ever been grown before, white spotting in crops of sweet potato and domesticated purslane planted with healthy starts.

In the case of both sweet potatoes and purslane, which are commonly replanted vegetatively, the use of vine cuttings that may not show lesions even though from diseased mother plants will carry infections from field to field. Some indigenous housewives, it has been fascinating to learn, and cooks, pick out white spotted purslane as being the better and more flavorful type. There may be some variation in resistance to Albugo in sweet potatoes. Certain of the stringy Indian-grown sweet potatoes have been seen that appeared highly tolerant; although their leaves were well spotted with the disease they produced good crops of roots. Next to them were grown introduced varieties of sweet potatoes with plump roots full of sugar, and the foliage they bore was much reduced in its growth by severe spotting, and so were the roots.

A Few Taphrina Leaf-Curl Diseases

Several well-recognized introductions of diseases into the American tropics have occurred. One of these is the leaf-curl, *Taphrina* (Exoascus), of peaches and relatives. The causal fungus may be carried long distances as dormant conidia adhering to the bark of seedling trees or budwood. This is probably the manner in which *Taphrina deformans* was introduced on peaches from Europe to Argentina, Chile and Brazil. It may occur in Central America and in the West Indies but I have not collected it nor has it been brought to my attention.

There is an interesting historical feature about *Taphrina* identifications in the Neotropics. The peach leaf-curl fungus had been well studied in Europe before the end of the last century. In 1901 the witches' broom of Surinam cacao with symptoms similar to peach leaf-curl was described by Europeans as caused by *Taphrina theobromae*. Some years later another disease was described as caused by a fungus of the same genus, *Taphrina bussei*. True witches' broom was, of course, shown later to be caused by *Marasmius perniciosus*.

While remarkable malformations and bright discolorations are caused by infections of *Taphrina* on various species of fruit trees, mostly *Prunus* spp., it has never become in the Neotropics the very severe disease that was so long known in Europe and North America. One reason, perhaps, is that the fungus *T. deformans* demands for best conditions long periods of crop growing at temperatures below 20° C. In dry areas where the air is low in moisture and where fruit trees can only be grown under irrigation, the fungus does not survive. It does, nevertheless, become of importance in some areas in our tropics (as can be seen in such references as 14, 20, 33, 35, 37, 50 and 53).

Knowing its temperature requirements would give pathologists some assurance that *Taphrina* diseases would never cause much difficulty in the Neotropics. It is certainly true that *Taphrina* is not reported from the warmer parts, and is not found even in such places as South Florida, the West Indies, and most if not all of Central America, Panama, and the north coast of South America. However, in a country like Peru, those peaches grown in its cooler parts are susceptible to infection with *T. deformans.* This kind of cool country distribution is likewise the case in such countries as Brazil, Chile, Colombia, Argentina, and Uruguay.

There is much still to be done with other species than *T. deformans,* especially in the Neotropics. With the need for ever greater amounts of fruits of various kinds as dietary supplements in a wildly expanding population, more plums, peaches, cherries and relatives are daily being grown and multiplied in the Neotropics. Not only will the commonly known fruits of the *Prunus* group, but there are in our tropics native, wild *Prunus* species that will be affected by diseases such as those caused by *Taphrina.*

There has been a considerable drop in interest in studies on *Taphrina* in the temperate zone, where economics are such that spraying is so easily used to control tree diseases. In the tropics such means are not so easy and are at times almost prohibitively expensive--they would cost more than the comparatively modest money from harvests will often allow. Such means for control such as resistance and systems of planting, pruning, locations of orchards, and the like will have to be used.

To those not reading much in tropical pathology, it is sometimes surprising to learn that there are several indigenous species of *Taphrina* in the American tropics: for example *T. andina* on the capulin cherry in Colombia; *T. mexicana* on the wild plum *P. microphylla* in Mexico; and *T. cerasi* on sour cherry in Chile. Undoubtedly there are numerous *Taphrina* species in the tropics that have not thus far been found by mycologists or pathologists. No one, anywhere, has a definitive notion about the variabilities in species of the genus *Taphrina* nor knows much about its races or strains.

A few species of trees other than those of Prunus are also attacked by leaf-curl fungi in the Neotropics. It is known that in Chile the highly valued black poplar is severely attacked by *T. aurea,* the pear is affected by *T. bullata,* and an undetermined species of *Taphrina* diseases azaelia. At the very tip of the South American continent the dwarf beech, *Nothofagus,* which was so impressive to Charles Darwin in his studies during the famous voyage of the Beagle, has not only a leaf-curl but also a witches' broom caused by *Taphrina entomospora.* On that handsome shade tree, so much known in the farthest-south edges of the continental United States of America, the live oak, there occurs the leaf blister disease caused by *Taphrina caerulescens.* In many parts of South America there is found a true witches' broom of birch caused by *T. betulina.* These, and many other kinds of *Taphrina* that are probably still to be discovered in the Neotropics, must form the basis for very rich researches by tree specialists.

Table of Some Important Phytophthora Diseases

Many of the diseases caused by species of the fungus genus *Phytophthora,* are known in parts of both the Neotropics and the eastern tropics (or Paratropics). The diseases included below are of special importance and best known in the Americas since they are the *Phytophthora* infections causing largest economic losses. In a few cases--marked with an asterisk in the table below--special emphasis has made them studied, as the threat from their occurrence was of paramount effect on human beings. The troubles were so difficult to work with, and the damage caused was so serious and of such long duration, that past growers sometimes lost hopes for the cultivated crop that was attacked. However, trained plant pathologists were brought in to meet the challenge of the diseases, and satisfying control measures were devised.

Controls for *Phytophthora*-caused diseases are of different types. Some are, at best, temporary ameliorations. However, in some instances the problem is being taken care of very successfully. Where possible, high tolerance, resistance, or even immunity is being sought. Results are progressively better as the decades of plant breeding continue under the pathologists. These things, and the significance of tolerance and immunity, are discussed more fully in the last two chapters of this book. Tolerance or resistance may never completely replace a certain amount of control by chemical applications.

Table 8. WELL KNOWN TROPICAL PHYTOPHTHORA DISEASES in a large part of the American tropics.

Common name of host (and part involved)	Phytophthora species	Disease
Coconut (bud)	*P. palmivora*	Budrot*
Tomato (fruit)	*P. infestans*	Firm textured decay
Hevea rubber (fruit)	*P. palmivora*	Dark-colored decay
Orange (fruit)	*P. citrophthora*	Brown rot
Cacao (fruit)	*P. palmivora*	Pod rot*
Potato (leaf)	*P. infestans*	Blight*
Tomato (leaf)	*P. infestans*	Blight*
Citrus species (leaf)	*P. citrophthora*	Blight
	P. palmivora	Blight
	P. parasitica	Blight
Hevea rubber (leaf)	*P. palmivora*	Leafspot*
Cinchona (seedling tops)	*P. parasitica*	Killing
Hevea rubber (seedling tops)	*P. palmivora*	Killing
Potato (seedling tops)	*P. infestans*	Blasting*
Tobacco (stem)	*P. parasitica*	Black shank*
Papaya (stem)	*Phytophthora sp.*	Canker
Orange (stem)	*P. citrophthora*	Canker
	P. palmivora	Canker

Cinchona (stem)	P. cinnamomi	Stripe canker
Macadamia nut (stem)	P. cinnamomi	Stripe canker
Cinchona (stem)	P. parasitica	Girdle canker
Orange (stem)	P. citrophthora	Girdle canker
Cacao (stem)	P. palmivora	Girdle canker
Orange (ground line)	Phytophthora spp.	Gomosis or footrot*
Cinchona (ground line)	P. cinnamomi	Footrot (feeder roots black)
Potato (storage root)	P. infestans	Chocolate rot
Mandioc (storage root)	Phytophthora sp.	Dark colored rot
Hevea rubber (tapping panel)	P. palmivora	Blackening*

Literature Cited

1. Ashby, S. F. 1929. Strains and taxonomy of Phytophthora palmivora Butler (P. faberi Maubl.). British Myc. Soc., Trans. 14:18-38.

2. Bartnicki-Garcia, S. 1966. Chemistry of hyphal walls of Phytophthora. J. Gen. Microbiol. 42:57-69.

3. Buddenhagen, I. W. 1958. Induced mutations and variability in Phytophthora cactorum. Amer. J. Bot. 45:335-365.

4. Christiansen G., J. 1967. El Cultivo de la Papa en el Perú. Editorial Jurídica, Lima, Peru. 351 p.

5. Clinton, G. P. 1910. Oospores of potato blight. Connecticut Agr. Exp. Sta. Ann. Rept. 1910:753-774.

6. Cox, A. E. and E. C. Large. 1960. Potato blight epidemics throughout the world. U. S. Dep. Agr., Agr. Handbook 174, 230 p.

7. Crandall, B. S. 1947. A new Phytophthora causing root and collar rot of Cinchona. Mycologia 39: 218-223. --. 1954. The diseases of the cinchona tree. F. A. O. Plant Prot. Bull. 3:33-37.

8. ________. and J. C. Dieguez. 1948. Phytophthora stem canker of sesame in Peru. Phytopathology 38:753-755.

9. Duggar, B. M. 1909. Fungous diseases of plants, with chapters on physiology, culture methods and technique. Ginn & Co., Boston, 508 p.

10. Erwin, D. C., G. A. Zentmyer, J. Galindo, and J. S. Niederhauser. 1963. Variation in the genus Phytophthora. Ann. Rev. Phytopath. 1:375-396.

11. Fawcett, H. S. 1936. Citrus diseases and their control. McGraw-Hill, New York, 656 p.

12. Fawcett, G. L. 1922. La gomosis de los naranjos. Rev. Indust. Agr. Tucumán. 12:149-151. --. 1938. (See in citation 35, footnote page 225, work on potato tuber disease.)

13. Fernandez R., M. 1957. La enfermedad prieta del tabaco en Cuba. Agrotecnia 12:32-36.

14. Fernandez Valiela, M. V. 1952. Introducción a la Fitopatología (2d. ed.). Gadola, Buenos Aires. 872 p.

15. Frezzi, M. J. 1937. Podredumbre del pie de los citrus. Mem. Labor. Fitop. Bella Vista, Argentina 1937:15-22. --. 1940. La Phytrophthora citrophthora, causante de la podredumbre del pie del naranjo y la gomosis del tronco del limanero, en Corrientes. Rev. Argentina Agr. 7:165-170. --. 1942. Podredumbre morena de las frutas

cítricas y parásitas que producen en Corrientes, Argentina. Rev. Argentina Agr. 9:217-220. --. 1950. Las especies de Phytophthora en el Argentina. Rev. Invest. Agr., Min. Agr. Ganad., Buenos Aires. 4:47-133.

16. ________. and T. Mácola. 1943. Phytophthora palmivora, causante de la prodredumbre morena de las frutas cítricas en Córdoba. Rev. Argentina Agr. 10:227-233.

17. Galindo, J. and M.A. Gallegly. 1960. The nature of sexuality in Phytophthora infestans. Phytopathology 50:123-128.

18. ________., and G.A. Zentmyer. 1964. Mating types in Phytophthora cinnamomi. Phytopathology 54:238-239.

19. Gallegly, M.E. and J. Galindo. 1958. Mating types and oospores of Phytophthora infestans in nature in Mexico. Phytopathology 48:274-279.

20. Galli, F. et al. (see Lit. Cited, Chapter II). 1968. Manual de Fitopatologia Doenças das Plantas e seu Contrôle. Ceres, São Paulo 640 p.

21. Gómez-Miranda, B. and J.A. Leal. 1965. Influencia del esterol, luz y fuente de nitrógeno en la reproducción de Phytophthora y Pythium. Microbiología 18:235-245.

22. Godoy, E.F. 1940. El mildew o tizón del pimiento producido por la Phytophthora capsici en la Republica Argentina. Rev. Fac. Agr. La Plata 24:235-280. --. 1943. Epifitología del tizón de la papa. Rev. Fac. Agr. La Plata 25:97.

23. Hassis, F.A. 1962. Qualitative and quantitative genetic variation among isolates of Phytophthora parasitica. Phytopathology 52:12.

24. ________. and R.R. Nelson. 1963. Studies on the biological relationship of species of Phytophthora as measured by oospore formation in intra- and interspecific crosses. Plant Dis. Reptr. 47:705-709.

25. ________., ________., and D. H. Marx. 1964. Morphological and physiological characteristics of mating types of Phytophthora cinnamomi. Phytopathology 54:1146-1151.

26. Holliday, P. and W. P. Mowat. 1963. Foot rot of Piper nigrum L. (Phytophthora palmivora). Commonw. Myc. Inst. Phytopath. Paper 5, 62 p.

27. Jehle, R. A. 1915. A common but very serious potato disease in Cuba. Modern Cuba 3:46-48.

28. Klemmer, H. H. and J. F. Lenney. 1965. Lipids stimulating sexual reproduction and growth in pythiaceous fungi. Phytopathology 55:320-323.

29. Leal, J. A. and B. Gómez-Miranda. 1967. Effect of amino acids on the sexual reproduction of species of Phytophthora and Pythium. Trans. British. Mycol. Soc. 50:77-84.

30. Leather, R. I. 1967. The occurrence of a Phytophthora root and leaf disease of black pepper in Jamaica. F. A. O. Plt. Prot. Bull. 15:15-16.

31. Leonian, L. H. 1925. Physiological studies on the genus Phytophthora. Amer. J. Bot. 12:444-498. --. 1935. Identification of Phytophthora species. West Virginia Univ. Agr. Exp. Sta. Bulletin 262, 35 p.

32. Lindquist, J. C. 1932-1935. Sobre la presencia de la Phytophthora capsici en la Republica Argentina. Physis 11:170-174.

33. Macola, T. 1929. Sobre la presencia del torque o enrulamiento de los durazneros en la provincia de Mendoza. Rev. Memo. BAP., Argentina. 1929: 42-46.

34. Malaguti, G. 1951. Epifitias de tizón en plantas cítricas causadas por Phytophthora parasitica. Agron. Trop., Venezuela 1:213-226.

35. Marchionatto, J. B. 1928. La gomosis o pie podrido del naranjo. Physis 9:94-98. --. 1937. El torque o enrulamiento de las hojas del durazuero.

La Chacra (Argentina) 84:32-34. --. 1948. Tratado de fitopatología. Editorial Sudamericana, Buenos Aires, 537 p.

36. Mills, W. R. and J. S. Niederhauser. 1953. Observations of races of Phytophthora infestans in Mexico. Phytopathology 43:454-455.

37. Mix, A. J. 1936. The genus Taphrina. II. A list of valid species. Bull. Univ. Kansas, U. S. A. 24: 151-176.

38. Müller, K. O. 1952. Potato blight in Chile. FAO Plt. Prot. Bull. 1:7-8.

39. Nelson, R. R. and F. A. Haasis. 1964. Interspecific compatibility patterns among 6 species of Phytophthora. Phytopathology 54:902.

40. Niederhauser, J. S., J. Cervantes, and L. Servin. 1954. Late blight in Mexico and its implications. Phytopathology 44:406-408.

41. ________., and W. R. Mills. 1953. Resistance of Solanum species to Phytophthora infestans in Mexico. Phytopathology 43:456-457.

42. Nolla, J. A. B. 1928. The black-shank of tobacco in Puerto Rico. J. Dept. Agr. Puerto Rico 12:185-215.

43. Orellana, R. G. 1953. Infection and tissue changes of Theobroma cacao L. by Phytophthora palmivora Butl. Turrialba 3:167-172. --. 1954. Contribución al estudio de la sobrevivencia, diseminación y combate de la Phytophthora del cacao. Cacao, Costa Rica 3:10-11. --. Estado de las investigaciones sobre la enfermedad del cacao por Phytophthora. Cacao, Costa Rica 3:10.

44. Perlasca, G. and G. Malaguti. 1954. Ensayos de cultivo de tomate en la epoca de lluvias. Agron. Trop., Venezuela 4:95-97.

45. Pontis Videla, R. E. and A. R. Landaeta. 1953. Una podredumbre de los frutos de la berenjena (Solanum melongena L.) en Venezuela causada por

Phytophthora capsici Leonian. Agr. Trop., Venezuela 3:117-119.

46. Puttemans, A. 1936. Reivindicação visando a denominação scientifica da doença da batatiera (Phytophthora infestans (Mont.) deBy.) Rodreguésia 2: 341-350.

47. Rands, R. D. 1922. Streepkanker van Kaneel, veroorzaakt door Phytophthora cinnamomi n. sp. Meded. Inst. Plantenziekten, Netherlands Indies 54:5-53.

48. Rorer, J. B. 1910. Pod-rot, canker, and chupon-wilt of cacao caused by Phytophthora sp. Trinidad Dep. Agr. Bull. 9:79-103.

49. Sanchez Potes, A. 1966. Enfermedades algodonero, del cocotero y de la palma africana en Colombia. Acta Agron., Palmira 16:1-13.

50. de Segura, C. B. 1965. Enfermedades de Cultivos Tropicales y Subtropicales. Editorial Jurídica, Lima, Peru. 439 p.

51. Smoot, J. J., F. J. Gough, H. A. Lamey, J. J. Eichenmuller and M. E. Gallegly. 1958. Production and germination of oospores of Phytophthora infestans. Phytopathology 48:165-171.

52. ________., J. L. Jinks and M. Grindle. 1957. Oospore formation in Phytophthora infestans. Phytopathology 47:33.

53. Spegazzini, C. 1908. Una nueva amenaza para las durazneros. Crónica Agr., Argentina 2:37.

54. Thomas, K. M., T. S. Ramakrishnan, C. K. Soumini and M. S. Balakrishnan. 1947. Studies in the genus Phytophthora I. Oospore formation and taxonomy of Phytophthora palmivora Butler. Indian Acad. Sci., Sec. B., Proc. 26:147-163.

55. Tucker, C. M. 1931. Taxonomy of the genus Phytophthora de Bary. Univ. Missouri Agr. Exp. Sta. Bulletin 153, 208 p. --. 1933. The distribution of the genus Phytophthora. Univ. Missouri Agr. Exp. Sta. Bulletin 184, 80 p. (--. 1926. Phyto-

phthora bud rot of coconut palms in Puerto Rico. J. Agr. Res. 32:471:498.)

56. Turner, P.D. 1961. Strains of Phytophthora palmivora (Butl.) Butl. from Theobroma cacao L. II. Isolates from non-African countries. British Myc. Soc. Trans. 44:409-416.

57. Vargas C., C. 1950. Algunos resultados de investigaciones con papas nativas Peruanas. Mexico Ofic. Estud. Esp.; Fol. Misc. 3:187-200.

58. Viirosoo, E.V. 1954. Observaciones en papas Andinas. Rev. Agron. Noroeste Argentina 1:87-98.

59. Waterhouse, G.M. and E.M. Blackwell. 1954. Key to the species of Phytophthora recorded in the British Isles. Commonwealth Myc. Inst., Myc. Paper 57, 9 p.

60. Zentmyer, G.A. 1957. The search for resistant rootstocks in Latin America. California Avocado Soc. Yrbk. 41:101-106. --. 1959. Spore stages of the avocado root rot fungus, Phytophthora cinnamomi. California Avocado Soc. Yrbk. 43:110-112. --. 1960. Phytophthora canker of macadamia trees in California. Plt. Dis. Rptr. 44:819. --. 1965. Bacterial stimulation of sporangium production in Phytophthora cinnamomi. Science 150:1178-1179.

61. ________., A.O. Paulus, and R.M. Burns. 1962. Avocado root rot. Univ. California, Berkeley, Ext. Serv. Circular 511, 18 p.

(N.B.: The above 61 numbers represent a fraction of the studies carried on with these problems. I have referred to materials I have been specially interested in, for reasons of consultation experiences. The workers in the southern part of South America are poorly represented in the above; I have followed a skewed responsibility towards Central America and the West Indies. The serious student on the subject of this chapter needs about 300 citations, and special attention to laboratory researches in Europe and the United States of America--F.L.W.)

Chapter VI

ANTHRACNOSES AND SIMILAR BLEMISHES

Summary

Leaf spots that are raised to slightly-raised, some with severe effects, are plentiful in the tropics. A group there, common anthracnoses (*Colletotrichum* = *Glomerella*) is nearly ubiquitous. These cause not only leaf spots, but also have a lengthy incipient stage in fruit attack, result in branch or twig dieback, some of the organisms acting as hyperparasites, and in certain instances there are systemic infections. The latter may be benign to eventually severe in shock effect. Quite different are the spot anthracnoses (*Sphaceloma* = *Elsinoë*), which are serious on some fruits and other crops. The spots from them are often small, raised, with corky surfaces. Tropical leaf crusts are found widely; perhaps the most studied is the South American leaf disease (*Dothidella ulei*) on rubber. It has a complex life history, as do other such organisms. There are some 600 described species of tarspot organisms (*Phyllachora* spp.) that infect grains, grasses, and dicotyledonous hosts. The symptoms are shiny black lesions. Leafmold spots, largely *Cladosporium* spp., comprise about 260 named causal fungus species; *C. fulvum* of tomato is an example, and, as with others, the seriousness of attack is dependent on ecological factors often present in warm-moist tropics.

Introduction

During years of diagnosing plant diseases in the Neotropics, a worker finds he often discusses his conclusions not only with growers who have the infected specimens but also with students and curious or friendly onlookers. It is evidently a common reaction for a great many people to include together as a natural group all of the very many raised leaf lesions. In a distinctly separate group are the rotted tissues, smudgy carbon black marks, specks, rusts, death where fungus threads like spider webs touched, and such. They definitely are not included in the raised-spot category. The raised spots make an excrescence on the leaf surface and have various textures: rough, wrinkled, smooth, pubescent, and in some cases the appearance of scabs.

Materials examined of such nature fall into assemblages based on symptoms. There is a certain logic in such association and the present chapter discusses a number of such leaf-excrescence diseases. They are related as to effect, but sometimes only remotely as to mycological connections; a practical recognition as to appearances is followed rather than an association on the basis of classification of causal parasites. In the Neotropics there are numerous scabby leaf spots from different anthracnoses, crusts, tarspots and molds. In this discussion only a few are to be mentioned, but those given attention are of particular interest and have special problems.

Common Anthracnose

Study of common anthracnoses leads into various problems and phases in pathology of these troubles. As is seen below at the close of the section "Colletotrichum, Gloeosporium, or Glomerella?", although much of anthracnose is thought of as a spot, there is a great deal more to it than that. In average phytopathological usage, anthracnose is characterized mostly as a lesion with delimited edges, dark-colored with a slightly raised rim but slightly sunken towards the center, within it a concentric arrangement of a peppering of small black acervuli (plate-like fungal structures bearing spores in a creamy to reddish mass). Such stippled lesions on foliage and fruits in tropical crops are caused by species of Colletotrichum = Glomerella; they are almost everywhere and in some instances are actually hyperparasites (57).

While some authors, even in modern reports, include both of the fungus genera of Colletotrichum and of Gloeo-

sporium as causes of common anthracnose, it can be seen from the definitive work by Von Arx (54) how arbitrary and how artificial is the division between these two.

Blemishes from common anthracnose are readily recognized when they occur on surfaces of foliage and fruit. These spots have characteristics that once seen are remembered. Infections also develop into both dry and at times soft rots. Under certain conditions the rots may proceed to a point at which they become dark mummifications. In some instances primary infections become courts of entry for secondary, otherwise weak, parasites that ordinarily could not invade hosts. There are times when the anthracnose fungus is systemic and results in no observable surface change in health of the infected plant. However, at other times the systemic condition becomes acute and causes shock effects.

Typical anthracnose lesions most commonly are furnished with a sticky mass of spores found in the shallow, saucer-shaped acervulus. Spores in this matrix are single-celled, transparent, more or less reniform to cylindrical, and at the edges of the acervulus there are often but not invariably spike-like, short, fungal hairs. On the advent of water, whether from rain, dew, or hydathode leaf-exudations, the glue-like matrix is softened, releasing the spores. Without losing their sticky quality, the spores are splashed or otherwise spread to hitherto healthy plant surfaces.

The researcher carrying on general plant disease work, or the consultant who is sought out to diagnose plant infections, becomes impressed with the large numbers of times he finds common anthracnoses. These troubles, most correctly reported as due to fungi of the genus _Colletotrichum_, are remarkably widespread in the Neotropics. Widespread, from the standpoint of geographical areas, from the standpoint of the large numbers of hosts, and from the standpoint of the breadth of environmental conditions in which these anthracnoses occur. (For some more inclusive and general sources of information on common anthracnose occurrences and its developments reference can be made to citations 1, 3, 7, 19, 21-23, 25, 34, 44, 45, 48, and 54.)

Infection and Incipiency

One of the characteristics of the species of _Colletotri-_

chum is that when spores germinate, the germ tubes generally have appressoria: on the end of the germ tube there develops a swelling of special nature--an organ for attachment--to assure early infection. The course of infections by Colletotrichum spp. have been studied in some detail (6, 21-23, 25, 34, 44, 45) and in general follow a sequence. So long as the spores remain buried in the glue-like matrix on the acervulus they are quiescent and do not germinate. When they are released from the mass by water and are thus carried off and stuck to a healthy host surface, the spores imbibe moisture. These spores have a hygroscopic character absorbing moisture from the air, swelling but not germinating until there is liquid water. This is important to the tropical pathologist. In most of the agriculturally used parts of the Neotropics there is enough relative humidity in the air that Colletotrichum spores, which are rather resistant anyway, do not become quickly desiccated and die. Of course in deserts with a consistently low moisture content of air, the anthracnose organisms neither thrive nor long exist.

Once out of the acervulus, one of these hyaline conidiospores sticks temporarily to its host, absorbs moisture and in a film of it pushes out a germ tube. On its contact with a congenial surface there grows the appressorium (mentioned previously), a special organ. It is fixed by its growth so strongly that usual, tropical heavy rains will not wash the infective organism away. Appresorium walls are heavy and fortunately resist bacterial action (which is common under tropical conditions). Proof of this has been observed in moist chamber studies. Coffee leaves inoculated with spores of Colletotrichum (coffeanum) gloeosporioides and bean leaves inoculated with C. lindemuthianum were incubated at 24° C for eight days and appressoria were found to be developed on both coffee and bean leaves. During this period no excess water was on leaf surfaces. The moist chambers were then moved to 29° and water was increased keeping leaf surfaces undisturbed and wet, but at no time was there enough to wash the leaves. In 3 weeks bean leaves were browned and coffee leaves were starting to discolor, and after another week bits of epidermia were examined microscopically. Bacteria were profuse. Where mycelium of either Colletotrichum had extended over a leaf surface hyphae apparently had been consumed by bacteria, but the "shells" of appressoria were still intact. No attempts were made to prove if they contained living protoplasm.

Soon after the appressorium is attached to host epidermial cells, it pushes down a sharp peg that pierces plant tissues. This is direct penetration into cells: invading hyphae are much larger in diameter than the sharp peg from the appressorium, and the fungal threads grow inter- as well as intracellularly in the host.

In the beginning the hyphae may grow rapidly. However, many laboratory observations were carried on, on free-hand sections from spore-inoculated outer tissues of fruits of banana, mango, plantain, and coffee. At first no discoloration or other symptom of disturbance is macroscopically seen. In green banana fruits on the plant, disease infections may be invisible for weeks. It is only when banana fruits start to ripen that symptoms appear. This is also true in fruit of mango, although sometimes coffee cherries will show anthracnose blemishes before the fruits become red.

The fungus *C. gloeosporioides* is abundant in many tropical areas. An example is in two experimental shaded coffee plots in Turrialba Valley in Costa Rica. Trees in the plots were fertilized, pruned, weeded, and in one lightly shaded. Here, when examined, the trees showed no anthracnose. In the contiguous plot without shade, otherwise handled the same, all trees had anthracnose spots on a few leaves, and fruit bearing was followed by dieback in about 1 per cent of the branches. Isolations were made from trees in these two plots with the following qualitative results. In 100 isolations from coffee dieback branches in the unshaded plot, 62 yielded *Colletotrichum*, along with other fungi that were miscellaneous and no attempts were made to identify them. In 100 isolations from partly denuded, but not killed, branches in the shaded plot, 18 yielded *Colletotrichum*, and 9 were found to have a *Fusarium* growing with the *Colletotrichum* and in 4 isolations of the *Colletotrichum* there grew out with it a *Cytospora*. The three different fungi were considered as probably parasitic; they were cultured and an attempt was made to prove their pathogenicities. Results of inoculations (see Table 9) are that the *Colletotrichum* caused disease, whereas the *Cytospora* and *Fusarium* gave uncertain reactions.

In the softer lamina of young but fully expanded leaves of coffee, anthracnose infection follows the usual course: spore germination, appressorium attachment to the epidermal cells, piercing of the outer cuticle by its infection peg and thus direct entrance to laminal tissues.

Table 9. INFECTION STUDIES OF FUNGUS CULTURES from diseased coffee branch isolates

Inoculum used	Inoculated[a]	Infected
Colletotrichum sp.	42	22
Cytospora sp.	52	1[b]
Fusarium sp.	50	2[b]

a. Inoculation: Coffee tree in field with "spent" apparently disease-free fruit branches was used at end of harvest. Branch tips were clipped, split, inoculum from agar culture inserted, covered with filter paper and bound with adhesive tape. Abundant moisture supplied by rains.

b. Considered questionably positive as wood and cortex showed discoloration. Microscopic examination gave uncertain results, but small bits of fungus mycelium were seen. Isolations from the wood on water agar gave negative results.

The infection may require as much as two weeks before it becomes macroscopically visible. In leaves of such trees as coffee, inga, magnolia (as in illustration, see Fig. 6-1), rubber, mango, and lime, the *Colletotrichum* fungus species infection often spreads in all directions. If leaves of coffee or rubber have reached a point of maturity in which there is a hardening of cells, the anthracnose infection may develop in such a way that main veins stop its spread and irregular to "piece-of-pie-shaped" lesions form. In lime leaves the anthracnose often spreads in an ever-increasing round spot, including main veins as easily as the rest of the leaf (Fig. 6-2). In some leaves, as in beans and in trees like *Bauhinia,* anthracnose lesions are of two types. There are typically rounded and irregular dark decayed leaf spots, and there are also elongated lesions on the underside of the larger veins. In the bean the causal organism is *C. lindemuthianum.* Whether the same species also attacked certain leguminous tree leaves was not proved. However, in leaves of these hosts as well as others there may be a long incipient periods between the infection and the moment a spot

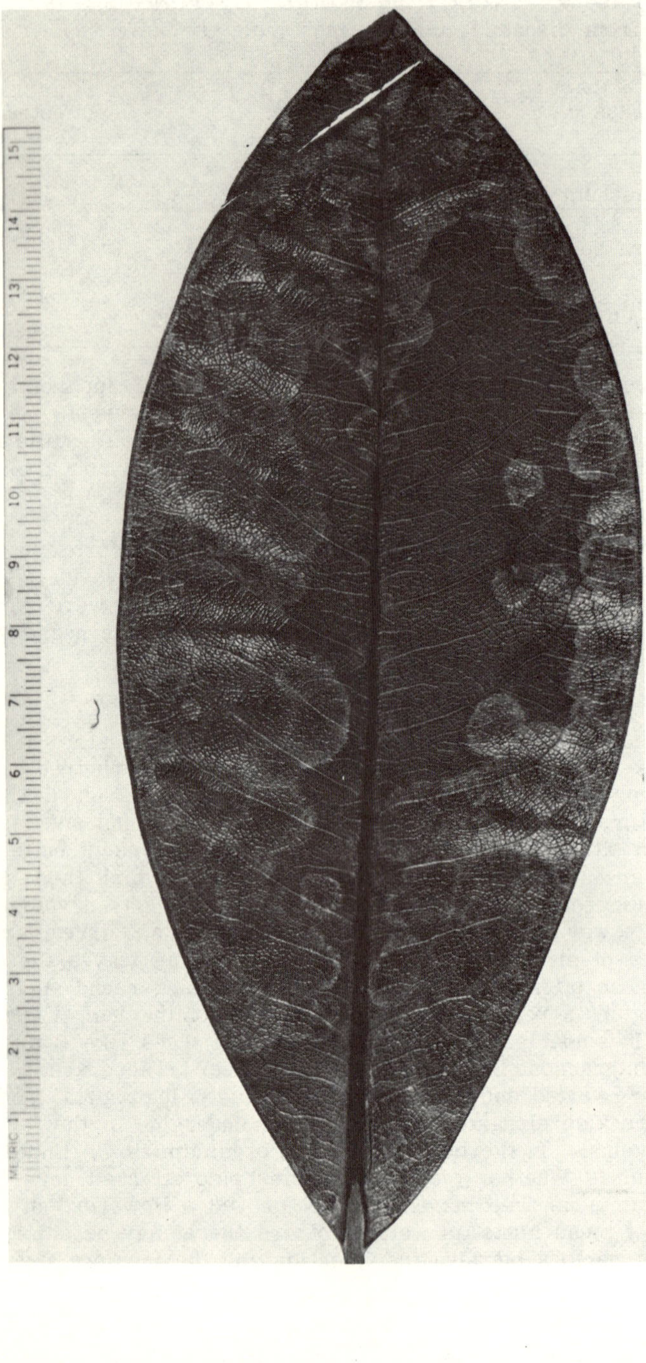

Fig. 6-1. Systemic invasion, following natural infection, of *Colletotrichum gloeosporioides* in leaf of magnolia.

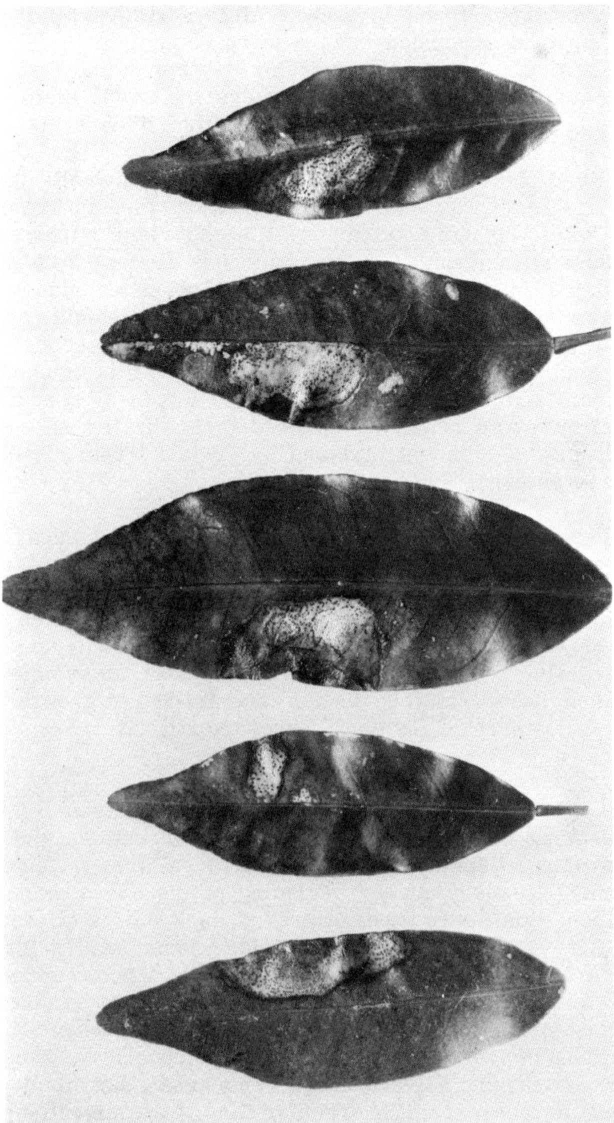

Fig. 6-2. Leaves of the lime tree exhibiting discrete, old lesions of anthracnose, *Colletotrichum limetticola*. Photograph courtesy of Plant Pathology Department, University of Florida, Gainesville.

from it becomes visible (6, 21, 40, 41, 45, 46, 54, 56).

Perhaps the ultimate in Colletotrichum invisible incipient infection is found in seed. After certain fruit infections such as in garden bean, coffee, and lime the fungus gains entrance to the vascular system of the fruit, and in a small number of cases invades the seeds themselves. This occurs in such a manner that it does not kill the embryo, and the seed is able to germinate. The fungus revives when the beans germinate, causing no notable disease. In some instances, as has been seen and studied in arabica coffee, Colletotrichum (coffeanum) gloeosporioides is completely internal and under some conditions seems to cause no injury. It has been recovered by isolation techniques from stelar tissues of apparently normal experimental seedlings that were one and one and a half years old; seeds were planted originally after smearing with Colletotrichum spores from agar cultures. In some 20-month-old seedlings Colletotrichum was readily recovered from the fibrous central cylinders of both epicotyl and hypocotyl. There had never been external symptoms.

In somewhat the same way, C. falcatum of sugarcane, under certain poor conditions of cane growing, will progress long distances or even remain alive but without much growth, almost invisible entirely within sugarcane stalks. When the cane is harvested, if the disease is serious it is found. Sometimes slowly growing canes with no external symptoms will suddenly show stunting disease and irregular growth if the field is subject to flooding, excess fertilization or other such accident.

C. gloeosporioides isolations from the externally healthy-looking coffee seedlings grew well in culture and produced disease spots with spores from cultures on coffee cotyledonary leaves. At the beginning of the dry season, spores from this source were sprayed upon old, apparently disease-free coffee foliage in the field. Four months later, typical "weak spots" were found on many of the old sprayed leaves. Many of these leaves with the weak spot symptoms were taken to the laboratory for studies. Microscopic examinations of free-hand sections gave inconclusive results. However, on making isolations we recovered: ordinarily very weak or non-parasitic organisms such as miscellaneous sarpophytic bacteria, Diplodia sp., two Valsa-like organisms, a white hyphal basidomycete, Pestalotia sp., Fusarium spp., Cercospora sp., Helminthosporium sp., Nigrospora sp., the alga Cephaleuros virescens, and an unidentified organism that had brown, yellow, and gray mycelial mats on PDA but never produced spores. It appeared these all entered weakened leaves that had typical weak spot lesions generally started by Colletotrichum, except in a few cases where Cercospora sp. infected first.

Fruit Drop and Fruit Rot

A problem well recognized by tropical fruit growers (exception is banana producers who do not have this trouble) is the matter of unseasonal fruit drop. This occurs in practically every dicotyledonous tree crop, and it has been observed especially on the following: lime, orange, coffee, Hevea rubber, cacao, mango, peach, zapote, jocote, anona grande, níspero (sapodilla), avocado, marañón (cashew), and wild fig. In many hundreds of dropped fruits taken to the laboratory and isolations made from peduncles, the organism recovered is Colletotrichum.

In citrus, often in the lime, in the West Indies and Central America this unseasonal early fruit drop can be quite serious (21). Growers of only a few isolated trees have discussed this with tropical experiment station specialists and it was noted that the maiden (first) crop was usually heavy. However, often the next year the unseasonable drop of immature fruits is serious. It has been found that on West Indian Islands and keys, and in South America there are places where, when lime was at first introduced, it had no particular enemies and the crops continued to be good year upon year. However, many times this kind of history was not the experience, and after one heavy maiden crop and perhaps a following medium good crop, drop occurred and fruit production then decreased. These smaller crops were blamed upon what was commonly called normal drop.

This drop came when fruits were still very small, and it was something believed by some temperate zone-trained horticulturists to be analogous to the so-called "June drop" of apples in the temperate zone. It seemed the phrase "normal drop" was a misnomer, since relative amounts fluctuated and often trees diagnosed as in a "hard growth condition" such as should have presaged normal drop did not always drop small fruits. Practically the identical history of "normal drop" also called "purga," occurs with coffee trees.

As has already been mentioned, on making simple isolations using acidified potato dextrose agar, the organism almost invariably recovered is Colletotrichum gloeosporioides. In a special test of organisms that might be found occurring naturally in peduncles of "normal drop" fruits, on healthy leaf, stem, and branch surfaces, using a coffee tree, a cacao tree, a citrus tree and two unidentified jungle trees, most of the isolates were Colletotrichum. While this was the only isolation study run in this manner, it convincingly suggests how common is the anthracnose organism in fruit peduncles and on healthy surfaces of the trees tested for presence; and furthermore indicates something of its ubiquity in this situation and how its spores are naturally available, and ready to attack where weakened tissues are present (see accompanying Table 10.)

The mango is another fruit tree sometimes severely afflicted by the so-called normal drop. In this crop, it seems fruit drop has an effect on tree size; after a number of years of repeated light crops from disease causes, the result is stronger tree growth. The mango bears large numbers of flowers in panicles at tips of branches. With a good year from two to five fruits may set per panicle. When this occurs season after season the trees may grow well but not excessively. In some areas almost all fruits contract early anthracnose of the peduncle and consequently drop early. There are certain moist locations in which mango trees produce excellent flowering but Colletotrichum gloeosporioides is so abundant that flowers are destroyed so quickly that no fruits ever set. In either situation where anthracnose is so severe and fruit production is practically non-existent, the vegetative growth of trees is much more than where heavy fruit crops are the general rule.

In the case of cacao there is much evidence, see Chapter V, that the once mysterious, so-called cherelle wilt of fruits, may start with Phytophthora, but at least in some cases it is exaggerated in its effect by early fruit infection with C. gloeosporioides. In some ways the cacao tree appears fairly tolerant of anthracnose. The parasite has been isolated from darkened internal vasculars of young pods that showed no outward sign of disease, except for dwarfing and hardened growth. However at other times it causes a severe rotten spot.

There is a condition during enlargement of some fruits when Colletotrichum will infect, cause a scab-like

Table 10. RESULTS FROM ONE SERIES OF ISOLATIONS[a] to assay for fungus propagules (spores, chlamydospores, hyphal fragments) in dropped fruits and on healthy tree surfaces growing in moist tropics[b]; scrapings[c] were used for isolations.

Kind of tree	Recovery of *Colletotrichum* sp. from:						
	Early drop fruits	Old[d] leaves	Recovered	Old[e] stems	Recovered	Old[f] bark	Recovered
Coffee	30	9	7	9	2	9	9
Cacao	10	9	6	9	5	9	7
Orange	20	9	4	9	9	9	6
Tree No. 1[g]	--	9	9	9	3	9	9
Tree No. 2[g]	--	9	8	9	6	9	9
Lime	18	--	--	--	--	--	--

a) Acidified potato dextrose agar in large test tubes. b) Turrialba, Costa Rica at close of rainy season 1960. c) Scrapings made with flamed scalpel. d) More than a year old, many with miscellaneous liverworts and lichens but no disease of leaf. e) Leaf-bearing stems past juvenility with about half of leaves dropped. f) Bark bits taken including cortex with cork, within one meter of the roots. g) Unidentified jungle tree.

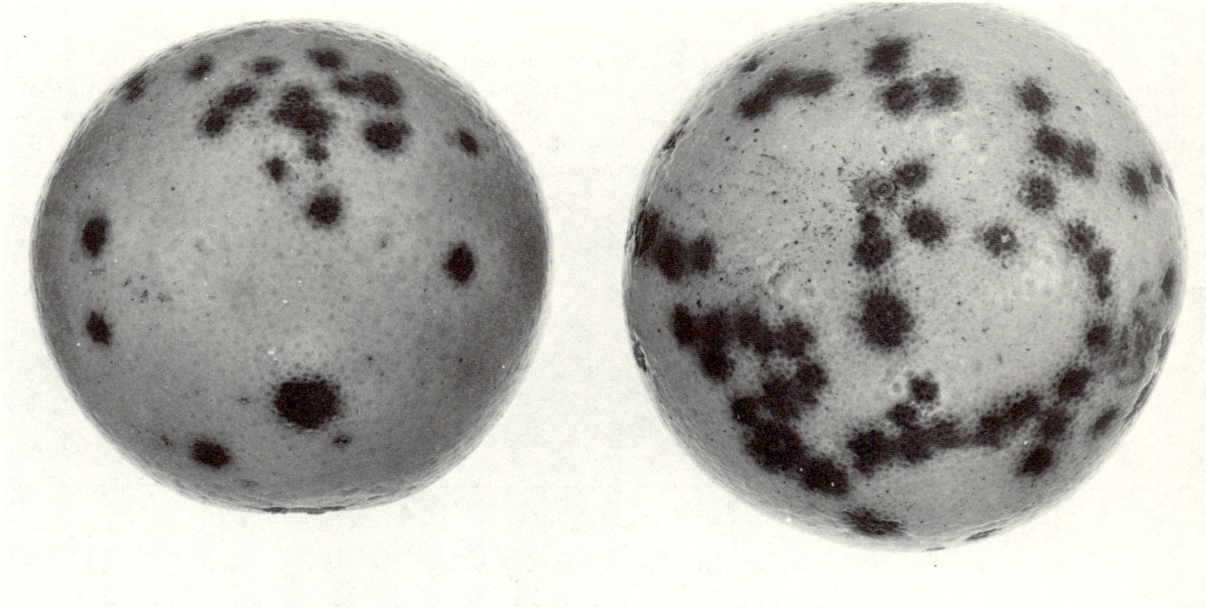

Fig. 6-3. Fruit spotting symptom on orange fruit, from slow spreading, shallow infections by *Colletotrichum*. Photograph courtesy of Plant Pathology Department of University of Florida, Gainesville.

growth but no soft rot. An example of this is seen in the medium small but dark-colored lesions illustrated in Fig. 6-3 on orange fruits. In lime the fruit may become irregular in shape but the organism never penetrates below the skin and the injured part heals. In some citrus fruits the fungus is superficial in its effect and makes only a discoloration of the skin (the so-called tear stains). In grapefruit the organism at times attacks the outer peel and shallow scab-like lesions develop but do not soften or penetrate to the pulp. With mango, if the organism infects the fruit skin, just before time for picking, it may not do much more than produce small, dark surface spots that are surrounded with a dark-green halo. These spots do not soften and the fungus does not penetrate. In banana and in plantain once the fruit skin is infected it finally darkens but unless kept for an abnormally extended period the anthracnose does not penetrate to the pulp.

Colletotrichum, Gloeosporium, or Glomerella?

Fungi have been found associated with common anthracnoses of plants for a long time. During about a century, over a thousand different Latin binomials were given to organisms studied in connection with these diseases (54): many times the basis was a host plant, sometimes the basis was the commemoration of colleagues, often some morphological quirk of the fungus as seen under the microscope guided the naming, and in a few cases names indicated cultural appearances. The result from all these namings was the increasing difficulty for a plant pathologist to be sure just which disease-causing organism he had. Generic names for the organisms were varied, and there grew up several schools of thought about where these *fungi imperfecti* were to be placed.

In the meanwhile there were more and more collections and reports about the fungi with different names that caused anthracnose troubles. The anthracnose problem was recognized as of major importance and studies absorbed the lives of numerous competent specialists in Europe and the United States of America. Probably the most complete and widespread confusion came about when the anthracnose troubles began to gain prominence in the tropics and in the subtropics (Neotropics), see such citations as: 1, 5, 7, 10-14, 16, 17, 21, 23, 25, 40, 41, 43-45, 48, and 54. It became a difficult thing for pathologists who were led to believe in a complex situation in which many different species of organisms were involved.

From the standpoint of strict mycology, for a long time two generic names were used (7, 25, and others): Colletotrichum was described first in 1837 by A. C. J. Corda. His organism he considered one of the fungi imperfecti, with, on the whole, unicellular spores, but all hyaline, kidney-shaped, and borne on a more or less flattened fungal plate, with stiff setae and spine-like fungus spikes at the edges. The erection of this genus helped a great many and soon a hundred or more species were described, many of them given specificity solely because of the host. During this period a group of these organisms were seen which on being studied disclosed no presence of the setae so prominently a part of the description of Colletotrichum. This added confusion to the problem for pathologists, and the two mycologists, Desmazieres and Montagne, described in 1849 another genus, Gloeosporium, to include these anthracnose organisms without the spike-like growths. Quickly thereafter a number of organisms then became included under Gloeosporium. Along with these two, several other generic names came into use for anthracnose fungi (see review, 54).

It is pertinent to show some of the confusions that came about in the tropics in the choosing of generic names for anthracnose fungi: coffee anthracnose, given the name Colletotrichum coffeanum by Noack, was named Gloeosporium coffeanum by Delacroix but Spegazzini named it G. coffeicola and Averna-Sacca called it G. coffeigenum. Anthracnose of a tropical vine was called C. philodendri by Hennings and the name was changed to G. philodendri by Spegazzini. An orchid had anthracnose from, according to Allescher, G. maxillariae; Spegazzini named the genus Hainesia. One anthracnose was named in succession C. piri Noack, Colletotrichopsis piri (Noack) Bubak, and Hainesia piricola Oudem.

Working with the anthracnose disease of vanilla the fungus names, all for the same organism, have been Trullula vanillae P. Henn., G. vanillae Cke., and C. vanillae Verpl. and Cloess. Anthracnose of castor oil plants have been designated as G. ricini Maubl. and C. ricini Bub. and Frog. On rose Bresadola called the anthracnose Glomerella and Hennings called it Colletotrichum. There is G. similacis Elisei and C. similacicola Speg.; G. eucalypti McAlp. or C. eucalypti Bitancourt; and G. bignoniacearum Speg. or C. bignoniae-ingeae Rangel. Delacroix named on breadfruit both

G. artocarpi and C. artocarpi. Such are a few of the fungus names in the tropics that lead us to uncertainties.

Complications and comparisons, once understood, have been illuminating in the works on names (see the two paragraphs directly above). This has been aided by extensive studies on host ranges and genetics, and detailed cultural work. Use has been made of isolates from so-called Gloeosporium spp. compared with Colletotrichum spp. Only a moderate amount of the literature regarding these researches is indicated at the close of this chapter but see especially: Andes and Keitt (4), Ashby (5), Burges (11), the team of Edgerton, Lucas, and Chilton (15, 18, 19, 33), Shear and Wood (45), and Von Arx (54). Results, after many years of work, have clarified many problems--not all by any manner of speaking--and have solved points that were once mysteries.

It is now known that the matter of presence or absence of setae to "split" between Colletotrichum and Gloeosporium is artificial as either may occur; sometimes it is apprently wholly dependent on host, culture medium or other environment. While the latter name has been given, on good evidence, to many isolates and field collections, when cultured and worked upon further the production of setae occurs, and Von Arx (54) presents this quite effectively. He lists 735 Gloeosporium species, of which he includes 288 under the imperfect species Colletotrichum gloeosporioides Penz., whose ascosporic or perfect form is Glomerella cingulata (Stonem.) Spauld. and Schrenk. Among the Colletotrichum species Von Arx placed about 600 of them in synonymy with C. gloeosporioides. Most of the time, then, these anthracnose organisms are Colletotrichum in the imperfect stage, and the majority with which pathologists deal are not many different species but only C. gloeosporioides.

However, it is clear from the analyses by Von Arx (54) that there are a number of valid species other than C. gloeosporioides: on beans and certain relatives there is C. lindemuthianum (Sacc. & Magn.) Bri. & Cav. If anthracnose is on cucurbits the fungus is usually C. orbiculare (Berk. & Mont.) Von Arx. As seen on banana, although it also is capable of infecting apples, oranges, and tomatoes, there is C. musae (Berk. & Curt.) Von Arx. When various members of the malvaceae are diseased with common anthracnose the fungus is C. malvarum (A. Br. & Casp.) Southw. On certain of the pasture and cover crop legumes such as Medicago, Melilotus, Vicia, and Trifolium the anthracnose is most often caused by C. trifolii Bain & Essary although C. destructiva O'Gara also is sometimes present. Common anthracnose on passion fruit, sansevieria, agave, avocado, certain orchids, and tomato fruit, is usually C. crassipes (Speg.) Von Arx, an especially interesting organism, well set apart from the usual type of Colletotrichum since its appressoria are very irregular to dendritic in form. Anthracnose affects dahlia, certain cucurbits, cabbage and relatives, clematis, and potato and on study the organism is C. atramentarium (Berk. & Br.) Taubenh. The well known redrot of sugarcane is caused by C. graminicola (Ces.) Wilson (also called Gloeosporium tucamensis (Speg.) Von Arx & E. Müll.) its perfect stage being Physalospora tucamensis Speg., and it is the causal agent on several other crops such as sorghum, maize, and pasture grasses like Andropogon and Lolium. Work on anthracnoses of species of Lobelia, Pithecolombium, Ranunculus, Iris, Lilium, Helianthus, Aralia, Dianthus, Capsicum, Zingiber, and Smilax resulted in designations and descriptions of 86 different organisms all of which are now under the name of the perfect stage Sphaeria dematinna Pers.

Part of the great contribution of Von Arx in clarifying some of the mycological problems respecting anthracnose organisms, is that he has given attention to many other generic determinations and has placed genera in synonymy. For species in the following genera: Gloeosoprium, Macrophoma, Tubercularia, Phyllosticta, Vermicularia, Septogloeum, Volutella, Hainesia, Phoma, Collestroma, Fusoma, Marsonia, Myxosporum, Chaetostroma, Discella, Discula, Fusisporum, Piggotia, Fellmenia, Sphaeropsis, Hypodermium, Colletotrichopsis, Cryptosporium, Gloeosporidium, Septoria, Ascochyta, Monostichella, certain "species" of Fusarium, and Trullula, all are now under Glomerella cingulata (= Colletotrichum gloeosporioides).

As seen in field and moist chamber studies especially, during eight years in Costa Rica, a brief summary of growth and morphology of the asexual stage, Colletotrichum gloeosporioides, and its sexual stage, Glomerella cingulata, is as follows. In all cases Colletotrichum appeared first with the perfect stage Glomerella much later. This was always the situation on live material in moist chambers with anthracnose diseases of old coffee leaves, cacao pod, and mango flower. It was confirmed during wet seasons with the same diseased hosts in debris-protected field material put on the ground under trees.

Conidia of C. gloeosporioides are kidney shaped, hyaline, and in the material described measured 6-10 X 3-3.5 microns (μ). Setae are black and pointed, and those seen in these studies measured 75-100 X 2-6μ. In plant parts dead from anthracnose, if kept moist, often the G. cingulata stage could be found in a week or 10 days. Bodies characteristic of this stage were the dark-colored and flask-shaped perithecia that measured 85-300μ in diameter. A mature perithecium has a pore (ostiole) at its tip, and on crushing the body, hyaline paraphyses and asci are seen within. The asci in the Costa Rican material measured 35-80 X 8-14μ, each ascus had the expected eight spores and they measured 9-30 X 3-8. All dimensions were within variations that might be anticipated for the organism (4, 11, 45, 54).

Spot Anthracnoses

The name anthracnose for a plant disease was once loosely applied. It covered almost any acute malady. However, with greater understanding of the part special fungi played in plant infections, such designations as rusts and mildews and anthracnoses were given to denote different distinct kinds of diseases. The general term anthracnose, as given in the Fungus Dictionary (1), refers to "a plant disease having characteristic limited lesions, necrosis, and hypoplasia, generally caused by one of the Melanconiales." However, spot anthracnose is more specific in that the causal fungi are of the asexual genus Sphaceloma (in perfect stage, the ascomycete Elsinoë). It is well known (1, 9, 26, also general texts such as 21, 22, 23, 25, 34, 48) that spot anthracnose fungus species, of which there are at least 40 of Elsinoë and over 50 of Sphaceloma, are much the most common in the tropics. Indeed, many are confined in occurrence to warm regions of the earth. In spot anthracnoses the lesion is often well described as a scab, or at least as a scab spot.

Lesions in the tropics from infection by the fungus Elsinoë ampelina (de Bary) Shear that are dark and discrete on such a fruit as the grape, are not always scabs and are not startlingly different in their superficial aspects from fruit spots on mango caused by Glomerella cingulata. However, further study discloses that in most of the lesions caused by Elsinoë spp., they develop, at least in part, as what may be described as typically raised scabs. Scabs grow out

Table of anthracnoses and certain other leaf spots

Table 11. Simple comparisons between examples of two anthracnoses and three other somewhat similar leaf blemishes, all causing severe defoliation and other leaf injury.

Common names of Diseases	Causal Fungus	Host Attacked	Leaf lesions usually seen in field
Common anthracnose	*Colletotrichum gloeosporioides*	Sour orange	Not numerous on a leaf, spread rapidly in lamina, a rounded spot peppered with small acervuli
Spot anthracnose	*Elsinoë fawcettii*	Sweet orange	Often numerous, spread slowly, irregular raised spots with corky surfaces
Crusty leaf spot	*Dothidella ulei*	Rubber (hevea)	Numerous crusty spots, often in groups, some become confluent
Tarspot	*Phyllachora maydis*	Maize	Slightly elongated, shiny black spots
Leafmold	*Cladosporium fulvum*	Tomato	Medium sized, somewhat diffuse with dark powdery surface

especially on the large veins on the undersides of leaves and on fruits as well as green stems. The raised scabs, or "sarnas" of the Latin American tropics, many times do not cause frightening losses. However, the insidious manner in which these infections cause early leaf cast, rots, disfigurements and effects from absorption of toxins, and regularly reduce productions, often means all the farmer can do is learn to work that much harder and to live with the disease rather than try to control it (8, 9, 20, 21, 24, 26-28, 30, 34, 37-40, 42, 49, 52, 53.)

Scabs may seem on first encounter to be superficial, as if they might be picked off. The surface of the spot anthracnose in many instances becomes eruptive, broken and dry, almost bark-like in appearance. (Symptoms on an orange fruit of spot anthracnose are shown in Fig. 6-4). From watching it in nature on several hosts in Central America, there seems to be no great amount of progression of infection backwards into the vascular tissues. Spot or scab anthracnoses appear to remain fairly well contained within somewhat circumscribed areas of the affected part on the host.

Climatic conditions for development of spot anthracnose disease are limited, but well within the scope of effects that are recognized as tropical in nature. Temperatures especially good for many of these anthracnose organisms range between about 15° or 25° C or a little higher; often as much as 28° although this may be excessive if maintained a long time. Temperatures of 28° to 30°, if constant, are too warm but this is also a condition not common in most of the cultivated lands of the Neotropics. Long exposure to 30° or 32° C may actually kill _Elsinoë_ spp. (See 21-23.)

Shade is of considerable importance in field husbandry in the American tropics, and Briton-Jones (10) working in Trinidad found that well-shaded sour orange seedlings showed 17 per cent of disease whereas they were 73 per cent diseased in full sunlight. I, making similar observations on sour orange seedlings in El Salvador, found in one year 22 per cent disease under lath house shade and nearby 82 percent disease in seedlings that had no shade protection. The next year, sun-grown seedlings were all discarded as unfit for graft use, while all the shaded seedlings had been budded and grafted and were growing well. In Costa Rica where in a few places _Elsinoë_ attacks coffee causing roña, or scab,

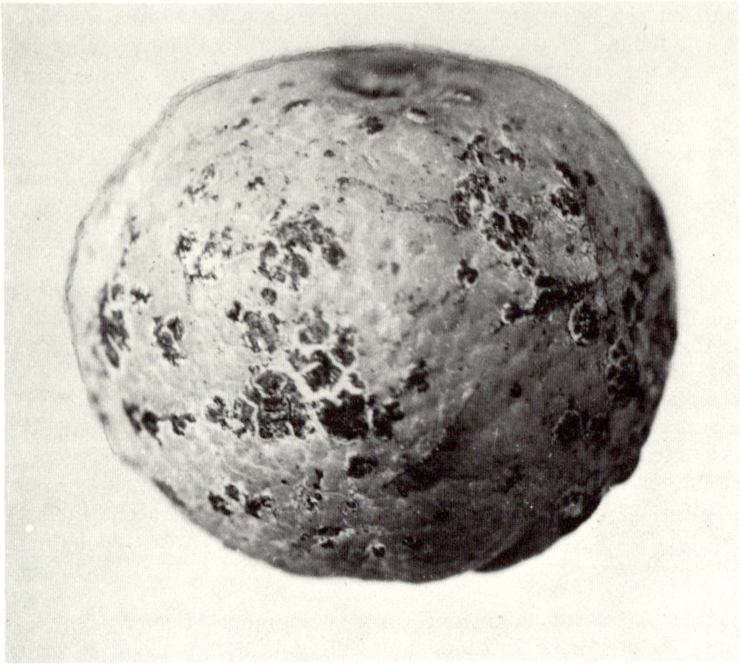

Fig. 6-4. Spot anthracnose, Elsinoë sp. (probably australis), or scab on fruit of sweet orange in South America. Photograph taken in state of Minas Gerais, Brazil. Courtesy of A. S. Muller.

it was invariably found on sun-grown trees. In Puerto Rico it is also clear that this anthracnose, from Elsinoë, on both avocado and on mango was most common and most severe on the more sun-exposed sides of the trees; very much less in well shaded parts. No counts were ever made, but differences in infections from scab were so distinct in mango that the peasants claim the sun causes the scab, and on some trees outside fruits on high branches become a mass of scab and are left unpicked. Lower fruits in shade may have only a few scattered lesions in occasional cases.

The fundamental reasons for this difference in disease in shade and in sun has not been determined, but this type of tolerance or resistance is unquestionably a shade effect. The basic reason awaits, as it must, dedicated research in

Fig. 6-5. Numerous black lesions and malformation of leaf from mango tree, caused by spot anthracnose Sphaceloma perseae. Photograph taken in Guatemala, courtesy of A. S. Muller.

the Neotropics that may add another good chapter to the information so much needed on the physiological nature of resistance and tolerance to diseases in plants.

With respect to the so-called spot, or scab, anthracnose on citrus, it is of practical interest that there is considerable difference in the natural tolerances of resistance (and susceptibilities) of certain species and varieties. Types that are most attacked are the common lemon, the tangelo, and the sour orange, and the most severely injured is the rough lemon. In some regions the parasite Elsinoë fawcettii Bitanc. & Jenk, has been so severe that rough lemons are no longer grown. On the other hand, in some of the citrus families, there is high resistance.

I have seen in both Central America and the West Indies many mixed plantings in which certain citrus types will remain absolutely free from Elsinoë infections. They may even be growing underneath or closely adjacent to, touching, and even grafted onto, severely diseased rough lemon trees. The key lime and Tahiti lime are apparently both immune from scab anthracnose, so is the cumquat and some, but not all, of the sweet oranges. There are other kinds of citrus intermediately classifiable as to susceptibility such as the tangerine, sweet lemon, and the Rangpur lime. Spot scab anthracnoses disease a large number of the major agricultural crops in the world. Some of these that are attacked in the Neotropics are coffee, guava, lemon, erythrina, manioc, soybean, orange, sweetpotato, peanut, grape, mango, avocado, forage legumes, grapefruit, and pecan. There are, as well, a great number of ornamentals and weeds that are affected. The latter need especial attention since these organisms multiply rapidly on wild hosts and are apparently in a continual state of mutation and other changes. (For symptoms on mango see Fig. 6-5.)

There are probably at least 100 of these spot or scab anthracnoses that have been described on as many different crops. It is significant that, so far as the tropical world is concerned, the largest percentage of these parasitic species come from the Neotropics. While some are described from the Philippines, Guam and such other places as the Dutch East Indies, the larger proportion of those described up to date have been found in the tropics of the American continents (8-10, 13, 14, 20, 21, 24, 26-28, 34, 37-40, 42, 51-53).

Cladosporium, Sphaceloma, or Elsinoë?

In the above section the spot and scab anthracnoses have been indicated as caused by Elsinoë or Sphaceloma. In literature, there is still some confusion in this, and it is interesting to have for the record the fact that at one time authors described the causal fungi under the name of the imperfect fungus Cladosporium, and even Sporotrichum (10). This is easily understood as the spore characters of the conidial stage are irregular in shape, often two-celled and are not necessarily hyaline but dark in color (21, 25, 48).

A fungus was described in 1886 by Lamson-Scribner, who called it Cladosporium sp., as the cause of these scab anthracnoses on citrus, and in 1896 it was considered to be

Cladosporium citri Penz. It had resulted in such widespread damage in those days that it was thought of as one of the most important citrus diseases. It was then given a great deal of study as to its epidemiology; resulting in changes in field culture methods that greatly reduced its severity in citrus groves (21).

It was a very interesting organism to pathologists, for in some characters it appeared different from usual Cladosporium species. The occasional increased severity of attack observed warranted closer research. It became evident through studies by Jenkins to be a species of Sphaceloma. Fawcett (21) gives a concise review of the history of naming the organism and shows the genus of the fungus was called, in turn, Gloeosporium, Fusarium, Cladosporium, and Sphaceloma so long as only its imperfect stage was known. During special studies in Brazil, the ascogenous stage was discovered and in 1936 (9) the citrus organism was renamed Elsinoë fawcettii Bitanc. & Jenk.

At the present time it is acceptable to use the generic name Sphaceloma for the conidial forms of these spot or scab anthracnoses. In fact, for some this is a common practice especially when occurrence of the perfect stage has not been proved. However, the microscopist who makes special study of the spot (or scab) anthracnose organisms has clear evidence from what he sees that the spores of Sphaceloma, although they are more or less irregular in shape and often dusky or slightly colored, are not true Cladosporium and also are much different from the more regular and reinform Colletotrichum spores that are also hyaline in character.

Among species of Elsinoë (Sphaceloma) there are various dimensions for the irregular conidia (Sphaceloma stage): they are 1 to 3 celled and may be small as in E. ampelina which are about 5-6 X 2-3 microns in size, to the slightly larger E. fawcettii spores that are about 5-10 X 2-5μ. E. piri Wor.) Jenkins has conidia that are irregular but mainly ellipsoidal and measure about 4.5-6 X 3.5-4μ.

Crusts on Leaves

There are certain leaf spots that appear neither like woolly growths nor like scabs, and are not flat-shiny black spots. These are lesions that push out and break the epidermises of leaves. They have a rough and brownish color, and look as if they can be scraped off with a dull knife. In conformation they are circular to slightly irregular, warty, and are called crust, or crusty, spots. Although not well studied from the standpoint of pathology and crop loss, these may deserve much more attention in the future than they have been given in the past or up to the present. In the tropics, where most broad-leaved trees are evergreen, some may hold onto their leaves three and four years and some leaves live longer and even have perennial increments in growth of petioles (e.g., on the Castilloa elastica); a type of slowly developing disease organism has a much greater chance to prosper in the tropics on long-held leaves than it does in the temperate zone where broad-leaved trees defoliate naturally each year. In reality, only during a few months of the year are their leaves present in a condition to become infected, in the temperate zone.

Miscellaneous observations have been made on development of leaf crusts on certain hosts, from material collected in El Salvador. This was done in 1945 and diseased leaves were dried between newspapers. Hand-cut cross sections were made of crusts to observe internal structures. On the host Inga laurina (Swartz) Willd., which casts its leaves readily, a crust fungus determined as a species of Trabutia was found attacking leaves both young and old. However, asci were observed in the ascomas formed only on well-matured leaves. Very young leaves produced crusty spots, but no sporuliferous growth. On another host, Inga spuria Humb. & Bonpl., whose leaves may hang on two years and are not easily cast, two kinds of crust spots were found, determined as Catacauma sp. and Trabutia sp. What seemed of special significance in these was that with both fungi the larger crusts were at places where there was a confluence of spots, where these confluences were located on main leaf veins or towards leaf centers. Crust spots did not show signs of producing good asci at locations in thin lamina, or along the sides of leaves where there were no main veins.

It appears from literature (e.g., 7, 13, 14, 43) that some of these crust fungi perhaps are more interestingly mycologically than pathologically. They are found frequently and develop vigorously in the tropics, and there is concern over the possibility that some crops that are intensively produced may show at times more and more losses from these leaf infections. Leaf drop is induced by crust fungi, and sometimes there is likewise attack on petioles and stems. In any case branch tips are weakened. With unusual leaf fall from infection, fresh leaf scars with unmatured vasculars are presented for secondary invasion by other parasites, such as Colletotrichum sp., and consequent dieback often develops. With severe dieback there is always serious root shock, often leaving roots susceptible to infections by parasites that cause root and trunk-base diseases.

In the Neotropics the growing of crops in certain conditions under shade trees is an important agricultural technique. The farmer using it, finds that not only must there be protection of his shorter statured plantation tree crops, or other crops, under shade, but that he must also protect the leaves and foliage that gives him the shade that makes it possible, easier and less expensive, to grow his cash and food crops. Important agricultural shade trees, a few genera of which are Eugenia, Inga, Gliricidia, Erythrina, Bauhinia, Leucena, Albizzia and Grevillea all have crust-like leaf diseases. This is also true of many fruit and plantation trees such as mango, Hevea rubber, bread fruit, members of the genus Prunus, cacao, and numerous others that may be affected by fungus-caused leaf crust diseases. The generic names of many of the leaf-crust spot organisms are: Dothidia, Catacauma, Dothidella, Trabutia, Robelia, Ophiodothella, Dothidina, and Dothiorella. Mostly these cause blackish, irregularly raised, spectacular crusts on leaves. The leaf epidermis, above the internally diseased mesophyll, loosens and sloughs away, exposing the raw crust.

Probably the best studied of all these, the South American leaf disease of rubber, is the one that attacks the latex-producing tree (Hevea), seringueira or hule. It is well known as the limiting factor for the securing of natural latex rubber in the western hemisphere; it is the main cause of rubber plantation failures in South America. Long ago this resulted in the loss of the so-called "India Rubber" trade, to the Malayan and African growers who introduced the tree from Brazil and perfected its technology. (The name "India Rubber" is old in European commerce and refers to the fact that it had been brought a long time ago from West India voyages returning to the Old World.) The fungus is known (e.g., 35) to have various races, and has been studied by a researcher of Malaya (50). However, by luck the leaf disease has not been transferred across the seas from the Americas to the Orient. This, nevertheless, remains an ever present threat in that region.

The scientific name of the causal organism of this leaf crust disease, is most commonly given as *Dothidella ulei* P. Henn. Mycologically the generic name of *Dothidella* was erected by Spegazzini and is classed by some mycologists as a "nomen ambiguum" and is thus questionable. In some writings it is given perhaps more correctly as *Melanopsammopsis* (*Dothidella*) *ulei* (P. Henn.) Stahel (47). However, in the Neotropics it is still mostly called *D. ulei* by pathologists there.

The disease was reported in 1901 by the German botanist-mycologist Ernst Heinrich Ule. He collected it in South America on more than one species of wild rubber trees, among them being not only *Hevea brasiliensis* that was tapped in the jungle by the serengueros who then took the smoked latex to market, but he also found the disease on related unused rubber tree species such as *H. guayensis* and *H. confusa*. When the mycologist P. Hennings examined the fungus, he based the name he gave on leaf spots he obtained in 1904 that had pycnidia on them. This he called *Aposphaeria ulei*, and this pycnidial stage is still the most common stage. When the conidial stage was discovered in 1911 it was given the name of *Fusicladium macrosporum*. It required several years before the complexities of these findings were sorted out and it was realized that there are three stages in development; the two imperfect forms namely the conidial and the pycnidial, and the perfect ascomycete form that is regularly given as *Dothidella* = *Melanopsammopsis*.

Studies (36) have shown that with *D. ulei*, the conidia (from the *Fusicladium* stage) penetrate only very young tissue. In fact susceptible leaves are those less than seven days old. Conidiospores germinate quickly, in 10 to 12 hours, in dew. It furthermore requires only five to seven days before an infection which at first is absolutely invisible to the naked eye, produces a spot on which many new spores are found on the surface of the lesion.

It is very interesting that the pycnidial stage develops on the nearly full grown, but not old, leaves. The pycnidia have spores in them, but curiously enough, these pycniospores do not seem to have the capacity to infect Hevea leaves. It is the fully-grown leaves upon which the perithecial stage or the *Dothidella* that we now speak of, is developed. The ascospores from Dothidella are formed only on the leaves of Hevea that are not too heavily infected. This is a situation of considerable interest. When a leaf is too thoroughly invaded with Dothidella, spots interfere with each other and the organism does not seem to have enough available nutrients to proceed to form its perithecial, or ascogenous, stage.

Ascospores are rather delicate and do not withstand more than four to six hours of exposure to drying conditions. This is important from the standpoint of the spread of the organism, as for this reason viable spores cannot be blown on dry winds for very great distances. Ascospores if they are fresh, will germinate in about two hours. However, the mere germination is not all that as important. It requires an extra 14 to 16 or more hours before an appressorium is formed at the end of the germ tube. After the appressorium develops it then is able with this organ, to penetrate into the tissues of the leaf.

Tarspots on Grain and Pasture Crops

Closely related to the crust-like organisms are the so-called tarspots. The tarspots are an important group and they produce a symptom of which the name is very descriptive as it looks like a spot of shiny black tar that has splashed onto a leaf surface. It is interesting that, as has been said before, in the case of the crust forms of spots, they seem to be of a kind of blemish that might be picked off with the thumb and finger. The shiny tar-black surface is due to the fact that the black disease area underneath retains over it the host epidermis.

Tarspots (for example on maize is found Phyllachora maydis) commonly affect pasture grasses, grain crops and pasture legumes, although there are also numerous kinds on many trees. Most tarspots encountered by pathologist are species of Phyllachora. Over 600 have been named, although some day some may be recognized as synonyms. Probably at least 200 species of Phyllachora are recognizable as notable pathogens. Phyllachora diseases result in serious defoliation problems and are more or less of cosmopolitan occurrence; there are many more in the tropics than in the temperate zone and they are much more severe in their pathological effects in the warmer climates (7, 13, 14, 31, 43, 55).

The tarspot fungus appears to be confined to two stages of spore forms. In the imperfect or conidial stage there are several types, the Cylindrosporium, the Polythrincium, and the Septoria spore types all readily classified under the fungi imperfecti. In all cases conidia are many-celled, produced in abundance, readily released from spot surface, and then carried by various means to other host-plant surfaces.

The perfect, or ascogenous, stage is Phyllachora, which is characterized by the shiny black lesion appearance on the host leaf surface. These tar spots are made up of stromatic masses--the perfect stage, and contain inside them many asci. These asci are buried in the black stromae, which although rather thin have in them the locules in which the numerous minute sacs are buried that contain the perfect ascospores. These are shot out during moist periods and germinate after a few hours of continuous moisture. A spore forms a germ tube, which grows out on to the leaf surface where it develops an appressorium which in turn becomes very firmly appressed to the host surface. From the heavy walled appressorium a strong, sharp peg emerges next to the host epidmis which it penetrates.

Leafmolds

The leafmolds are caused largely by what we now know as Cladosporium species; there are at least 260 species of this genus affecting an exceedingly wide range of tropical hosts from delicate and lowly herbaceous crops to some of the longest lived, most vigorous, and tough appearing of trees. One of the best known of these leafmold fungi is Cladosporium herbarum. This and all other species cause brownish mold spots and are imperfect or conidial stages of ascomycetous organisms. There are many ascomycetes, for example, Mycosphaerella and Elsinoë, that produce as their imperfect stages what is or has been interpreted as a Cladosporium.

In the Neotropics leafmold diseases are characterized by the following: on the surface growth they are woolly organisms, olivaceous in color, and when there are enough on a leaf in a spot, the dark color is predominant and readily recognized. Another characteristic is symptoms of toxin production for after these moldy spots develop there also often occurs rapid yellowing of the leaf and soon premature dropping. There is another point of interest and that is that leafmolds have odor which becomes quite distinctive. Some of these molds also attack old and aging leaves in association with "weak spot." These molds also change the taste of stored grain products on which they are growing. They have also been implicated in their effects on coffee flavor from occurrence in the harvested product (29).

Something should be said about the ecology under which I have observed that Cladosporium molds flourish. They are apparently most vigorous where it is not extremely hot, but where there are heavy night dews that occur over a long period of weeks. The dry seasons may last for only a few weeks in those areas with warm temperatures especially at nights. The conditions of moisture and temperatures at night that are specially good for leafmold, are recognizable for a normal person as when he feels close to, but not quite at, the point of sweating.

In parts of the tropics seasonal rhythms often include periods when there are unusually heavy rains for a lengthy time. When these rains occur, if they are strong enough, spores of all kinds are washed off and the crops may grow away from attack by Cladosporium. These molds, however, will finally recur when the rains lessen or cease followed with little wind but night dews. Fields of crops grown in the sun are likely to be more infected with mold than those in shade. It is also true that small cleared fields surrounded by thick growths of jungle are more truly susceptible to leafmold than where there is a better drainage of air over the area.

One of the most interesting and pressing problems about leafmold has been the question of how the leaf parasites survive the extremes of wet seasons and exist from one crop season to the next. From field observations it appears that there are probably two possibilities; neither has been proved in detail but are working hypotheses. In thicker leaves on trees there are stromatic masses that form under one to three years leaf spots that have developed through Cladosporium growth. These stromatic masses are probably the portions of leaf tissues in which the organism must reside from one season to the next. In trees and plants with comparatively lush vegetative leaves there is another problem, for these leaves decay rapidly. For example in tomatoes, the Cladosporium attacks the leaves, causes them to die, drop, and soon the whole plant collapses to rot on the wet ground. Bacteria consume the fungi as well as the dead host tissues. It seems probable that there is a saprophytic stage of this fungus, or that parts of old, dead, diseased leaves escape some of the vicious bacterial decay so common to all things in the tropics.

In preliminary studies, never completed, I have found spores of *Cladosporium herbarum* germinating and infecting aged tree leaves of species of *Eugenia*, *Coffea*, and *Castilloa*, which are dicotyledonous. However, in a species of *Canna* and one of *Cordyline* both being monocotyledonous herbaceous plants, old leaves were also thus affected. This growth on old leaves is superficial, mostly at edges of the lamina where it is collapsed. In old leaf tissues of *Coffea liberica* when sprayed with a fungus suspension, the spores on germinating thrust out germ tubes that entered old opened stomata. The fungus grew a short distance in the substomatal cavity but the invasion did not go farther. So long as the leaves were green there was little killing of tissues, although at times in the center of leaf lamina, small, chlorotic flecks could be seen from *Cladosporium* infection.

Such fleck spots do not develop a progressive rot, and there is very much reduced fungus growth, with no spores from them. It requires great care to recover *Cladosporium* from such spots. Taken to the herbarium these lesions are invariably classified as "sterile spots." On orange, attack by a *Cladosporium* may be common under unusually dry season conditions following many months of rain.

In citrus (sweet orange, lemon, lime, grapefruit) a species of *Cladosporium* sp. was found in Costa Rica that caused greasy appearing spots on old leaves. Leaves with such spots put into moist chambers, produced upon incubation *Cladosporium* spores and these spores were used to test their infectivity on tomato, beans, and maize. In leaves of all these herbaceous hosts, spores germinated and entered stomata, but there was no infection of cells.

In the open field in the temperate zone, disease from *Cladosporium fulvum* on tomato is found occasionally but it is not considered any problem (see Heald (25), Walker (58) and experience of others quoted by them). However, in the glass house environment, with its high humidity, slow moving air, and temperatures specially good for the organism, tomato leafmold can be extremely severe, and unless controlled causing complete crop failures (58). The environmental complex so ideal in the temperate zone tomato greenhouse, or glass house, is almost exactly what is duplicated in tropical fields of tomatoes. Cladosporium fulvum is a very severe and limiting factor in tomato production in many parts of the Neotropics (e.g., 23).

Tomato leafmold occurs very widely up and down the slopes of tropical mountains where tomatoes are grown. The Cladosporium fulvum investigations that have been started in the tropics have not gone very far and there has not been enough time for study. However, it is known that there are some very good resistant types to be found in both introduced and native materials, and it is these that are to be, and are being, used for developing tolerant and resistant tomatoes especially adapted to tropical conditions.

There are other leafmolds in the tropics of very serious consequence. These occur on such crops as agave, onion, cucurbits, many grasses, orchids, species of tropical plums, tropical walnut, sweet potato, maize, tomato, rice, and grape. The main injury from Cladosporium infections, of course, is defoliation. When severe, this defoliation progresses into a dieback and after the dieback starts secondary organisms enter to increase its severity.

There are minor fruit rots, also caused by species of Cladosporium in the tropics; on the whole they have not been much investigated and are generally of only superficial note. They are a danger that should not be overlooked, however. Races of the organisms might turn up having special virulence on fruits. Cladosporium spores have been recovered from diseased mango, genipe, marañón, jocote, and anona. Isolations from especially severe rot of mamey, grown in the higher plateau regions of Central America, yielded Cladosporium sp. on acidified agars. Spores of Cladosporium sp. have been recovered three different times from dry rot spots on dropped apples grown in high locations in El Salvador, Guatemala, and Colombia.

Literature Cited

Herewith are presented some of the titles of publications I have used in my consultative work on the subject of this chapter. Only a few of the many mycological works are indicated. For example reports from Spegazzini, Saccardo, many lists of occurrences, and the like are not included. The reader will find in them suggestions about and likewise drawings of, pathological features in these diseases. There have been numerous contributions from Peru (44), many from Brazil (23), and from any tropical country in South America perhaps the most from Argentina (22).

The subject of this chapter is a large one, is sufficient for a book in itself, and the wide ranging substance of titles given indicates some idea of the wealth of informa- that has not been any more than touched upon in the above discussion of the problems. Numerous works listed below have many citations which deserve examination. The reader is well aware that existing abstract journals are obtainable from which can be secured coverage of the subjects in this present chapter.

1. Ainsworth, G. C. 1961. Ainsworth and Bisby's dictionary of the fungi. Commonwealth Myc. Inst., Kew (Surrey) England, 547 p.

2. Alvarez G., L. A. 1949. Anthracnose of the annonaceae in Puerto Rico. Univ. Puerto Rico J. Agr. 33:27-43.

3. Abrahão, J. 1964. Mangueira com antracnose murcha. Biologico 30:120.

4. Andes, J. O. and G. W. Keitt. 1950. Variability of Glomerella cingulata (Stonem.) S. and v S. from apples. Phytopathology 40:915-925.

5. Ashby, S. F. 1937. Gloeosporium strains. Trop. Agr., Trinidad 8:322-325.

6. Baker, R. E. D. 1938. Studies in the pathogenicity of tropical fungi. II. The occurrence of latent infections in developing fruits. Ann. Bot. 2:919-931.

7. Bessey, E. A. 1935. A Text-Book of Mycology. P. Blakiston's Son & Co., Ltd., Philadelphia. 495 p.

8. Bitancourt, A. A. 1936. A verrugose do abacateiro. Biológico 2:157-160. --. 1936. Doenças do abacateiro. Biológico 2:385-389.

9. ________. and A. E. Jenkins. 1936. Elsinoë fawcetti; the perfect stage of the citrus scab fungus. Phy-

topathology 26:393-396. --. and --. 1939. Ciclo evolutivo de Elsinoë australis Bitancourt e Jenkins, agente da verrugose da laranja doce. Arq. Inst. Biol. (Brazil) 10:129-146. --. and --. 1942. New discoveries of myringiales in the Americas. Amer. Sci. Congr. 8th. Washington, U. S. A., 1940. Proc. 3:149-172. --. and --. 1946. A verrugose de mangueira. Arq. Inst. Biol. (Brazil) 17:205-228. --. and --. 1956. Estudo sôbre as Mirangiales VIII. A anthracnose maculada da seringueira causada por Elsinoë. Arq. Inst. Biol., Brazil 23:41-66.

10. Briton-Jones, H. R. 1931. Trinidad plant diseases. Notes on some diseases of main crops in Trinidad. Trop. Agr., Trinidad 8:300-302. --. 1933. The control of scab disease (Sporotrichum citri Butler) in the British West Indies. Trop. Agr., Trinidad 2:40-42.

11. Burger, O. F. 1921. Variations in Colletotrichum gloeosporioides. J. Agr. Res. 20:723-736.

12. Cardeñosa-B., R. 1963. La antracnosis del platano cachaco en el Tolina. Turrialba 13:88-95.

13. Chardon, C. E. 1921. A contribution to our knowledge of the Pyrenomycetes of Porto Rico. Mycologia 13:1-15.

14. ________., and R. A. Toro. 1930. Mycological explorations of Colombia. J. Dep. Agr. Porto Rico 14: 195-369. (See especially Dothidias with relatives and Phyllachoras and relatives pp. 243-266 and illustrative plates.)

15. Chilton, S. J. P., G. B. Lucas and C. W. Edgerton. 1949. Genetics of Glomerella. III. Crosses with a conidial strain. Amer. J. Bot. 32:549-554.

16. Condado, C. 1938. La antracnosis de Melilotus albus en la Republica Argentina. Rev. Argentina Agr. 5:240-248.

17. Cruz, B. P. B. 1958. Podridão peduncular e antracnose do mamoeiro. Biológico 24:113-114.

18. Edgerton, C. W. 1914. Plus and minus strains in the genus Glomerella. Amer. J. Bot. 1:244-254. --. 1915. Effect of temperature on Glomerella. Phytopathology 5:246-259.

19. ________., S. J. P. Chilton and G. B. Lucas. 1945. Genetics of Glomerella. II. Fertilization between strains. Amer. J. Bot. 32:115-118.

20. Fawcett, G. L. 1920. La verruga o scab de los citrus. Rev. Ind. Agr. Tucumán (Argentina) 21:124-128.

21. Fawcett, H. S. 1936. Citrus Diseases and Their Control. McGraw-Hill, New York, 656 p.

22. Fernandez Valiela, M. V. 1952. (Ed. 2.) Introducción a la Fitopatología. Gadola, Buenos Aires 872 p.

23. Galli, F. (et al., see "Lit. Cit." in Chapter II.) 1968. Manual de fitopatológia, doenças das plantas e seu controle. Ceres, São Paulo. 640 p.

24. Conçalves, R. D. 1935. Verrugose e mancha das folhas do abacateiro. Biológico 1:443. --. 1937. Antracnose do abacateiro. Biológico 3:158-159.

25. Heald, F. D. 1926. Manual of plant diseases. McGraw-Hill, New York, 891 p.

26. Jenkins, A. E. and A. A. Bitancourt. 1939. Illustrações das doenças causadas por Elsinoë (sphaceloma) conhecidas na América do sul até Janeiro de 1936. Arq. Inst. Biol. São Paulo 10:31-80. --. and --. 1941. Antracnosis de la vid en Chile. Dept. San Veg., Chile Bol. 1:19-53. --. and --. 1941. Revised descriptions of the genera Elsinoë and Sphaceloma. Mycologia 33: 338-340. --. and --. 1946. Elsinoë on Randia. Mycologia 38:65-68.

27. ________., ________., and V. Rosetti. 1954. Uma anthracnose da seringueira. Biológico 20:107.

28. Jenkins, A. E. and G. D. Ruehle. 1942. A new species of Sphaceloma on poinsettia. Proc. Biol. Soc. Washington, U. S. A. 55:83-84.

29. Krug, H. P. 1941. Cafés duros. IV. Relacão entre zonas, qualidade de café e porcentagem de microorganismos. Rev. Inst. Café, Sao Paulo 16:288-295.

30. Leach, C. M. 1958. A disease of dodder caused by Colletotrichum destructivum. Plant Dis. Reptr. 42:827-829.

31. Lopez-Rosa, J. H. and R. T. Sherwood. 1966. Symptoms and host-parasite relations in the tar spot disease of lespedeza caused by Phyllachora lespedezae. Phytopathology 56:1136-1142.

32. Lucas, G. B. 1946. Genetics of Glomerella. IV. Nuclear phenomenon in the ascus. Amer. J. Bot. 33:802-806.

33. ________., S. J. P. Chilton, and C. W. Edgerton 1944. Genetics of Glomerella. I. Studies on the behavior of certain strains. Amer. J. Bot. 31: 233-239.

34. Marchionatto, J. B. 1934. Las formas de conservación invernal de las sarnas del peral y del manzano. Rev. Argentina Agr. 1:245-247. --. 1948. Tratado de fitopatología. Editorial Sudamericana, Buenos Aires. 537 p.

35. Miller, J. W. 1966. Differential clones of hevea for identifying races of Dothidella ulei. Plant Dis. Reptr. 50:187-190.

36. Rands, R. D. 1924. South American leaf disease of para rubber. U. S. Dep. Agr., Bulletin 1286, 19 p.

37. Rubin, H. L. 1961. Poinsettia scab--a new report for Puerto Rico. Plant Dis. Rptr. 45:375.

38. Ruehle, G. D. 1932. Life history and control studies for melanose and scab of citrus. Citrus Industry 13:8-34.; --. 1941. Poinsettia scab caused by Sphaceloma. Phytopathology 31:936-939.

39. ________. and R. B. Ledin. 1955. Mango growing in Florida. Florida Agr., Exp. Sta. Bul. 574, 90 p.

40. ________. and D. A. Wolfenbarger. 1949. Diseases and pests of the mango in Florida. Florida Subtrop. Exp. Sta. Rptr. 13. 6 p.

41. Sanchez P., A. 1953. La antracnosis foliar del cacao. Acta Agron., Colombia 3:41-64.

42. Schiel, E. 1942. La lucha contra Sphaceloma australis, parásito del mandarino en la provincia de Santa Fe. Rev. Argentina Agr. 9:19-27.

43. Seaver, F. J. and C. E. Chardon. 1926. Mycology. Botany of Porto Rico and the Virgin Islands. Scientific survey of Porto Rico and the Virgin Islands. Vol. 8, Part 1. New York Academy of Sciences, 208 p. (Dothideaceae and Phyllachoraceae pp. 48-56; Colletotrichum and Gloeosporium pp. 85-87.)

44. de Segura, C. B. 1965. Enfermedades de Cultivos Tropicales y Subtropicales. Editorial Jurídica, Lima, Peru, 439 p.

45. Shear, C. L. and A. K. Wood. 1913. Studies of fungous parasites belonging to the genus Glomerella. U. S. Dep. Agr., Bur. Pl. Indus., Bulletin 252, 110 p.

46. Simmonds, J. H. 1963. Studies in the latent phase of Colletotrichum species causing ripe rots of tropical fruits. Queensland J. Agr. Sci. 20:373-424. --. 1966. A study of the species of Colletotrichum causing ripe fruit rots in Queensland. Queensland J. Agr. Anim. Sci. 22:437-459.

47. Stahel, G. 1917. De zuid-amerikaansche hevea-bladziekte veroorzaakt door Melanopsammopsis ulei, nov. gen. (= Dothidella ulei P. Hennings). Dept. Landb. Surinam, Bulletin 34, 111 p.

48. Stevenson, J. A. 1926. Foreign plant diseases. A manual of economic plant diseases which are new to or not widely distributed in the United States. U. S. Dep. Agr., Government Printing Office, Washington, D. C. 198 p.

49. Tartakowsky H., S. J. and S. T. Arentsen S. 1941.

La antracnosis de la vid. Bol. San Veg., Chile, 1:6-18.

50. Taysum, D. H. 1966. Pathological division. Report Rubber Res. Inst. Malaya 1965:55-65. (R. A. M. 46:202.)

51. Viegas, A. P. 1955. Manchas de folha da *Hevea brasiliensis.* Bragantia 14:63-69.

52. ________. and A. E. Jenkins. 1943. Stem and foliage scab of sweet potato. J. Washington (D. C.). Acad. Sci. 33:244-249.

53. ________. and H. P. Krug. 1934. Desenvolvimento de uma espécie de *Elsinoë.* J. Agr. Piracicaba 2: 277-284.

54. Von Arx, J. A. 1957. Die arten der gattung *Colletotrichum* Cda. Phytolog. Zeitsch. 29:413-468. --. 1957. Kultur- und infektions - versuche mit einigen *Colletotrichum*-arten. Tidschr. Plantenz. 63:171-190. --. 1957. Revision der zer *Gloeosporium* gestellten Pilze. Noord-Hollandsche Uitgevers Mattschappij. Amsterdam, (Special publ.) 51, 153 p.

55. Wallin, J. R. 1946. Prevalence of *Helminthosporium turcicum*, *Angiospora zeae*, and *Phyllachora graminis* on corn in Guatemala. Phytopathology 36: 412-413.

56. Weir, J. R. 1926. A pathological survey of the para rubber tree (*Hevea brasiliensis*) in the Amazon Valley. U. S. Dep. Agr. Bulletin 1380, 130 p.

57. Wellman, F. L. 1964. Parasitism among Neotropical phanerogams. Ann. Rev. Phytopath. 2:43-56.

58. Walker, J. C. 1969. Plant pathology (2d. ed.). McGraw-Hill, New York, 819 p.

Chapter VII

MILDEWS AND CARBON-SOOT DISEASE PROBLEMS

Summary

Both the black and the white mildews that are obligate parasites are found in the tropics. Probably the most common and numerous of these diseases are the black mildews. Although notable, and disturbing, the black mildews on the whole are not severe types of disease on tropical crops. These causal organisms seem to flourish in some parts of the wet tropics. Black mildews are in some cases hyperparasitic on a number of plant disease fungi. Transition types between parasitism and epiphytism occur in black mildews; epiphyte-saprophyte, carbon-soot fungi seem closely related to black mildews. The white powdery mildews, quite different from the black mildews, are widely distributed in the American tropics. Some are serious diseases. The white mildews are not well adapted for the most flourishing development, under the rigors of climates marked by heavy and long continued rainfall in the wet-warm tropics.

Introduction

Both the black and the white mildews occur with such prevalence in the tropics, that, among agriculturists, there is a tendency to accept these diseases as if they were botanically a healthy part of the foliage of plant or tree on which they prey. Mildews are not spot diseases. Mildews, be they black or white, are from disease fungi that are seen as more or less thin and diffuse surface coatings. Many healthy leaves are pubescent, and to farmers where mildews abound, the "hairyness" from fungus infection may be readily considered by them as a normal condition.

In the temperate zone the name "mildew" was used first as referring to a powdery white parasitic growth on crop plants. In long-ago peasant Europe, as grains were ground to flour in stone grist-mills, there escaped white dusts that the wind deposited on plant leaves. A disease seen that resembled this mill dustiness became "mill-dew," simplified to "mildew." This has been transliterated in Spanish to "mildieu," and "mildú." Other names Latin-Americans use are such as "anublar," "polvillo" and occasionally "tizón."

Any mildew disease, and the organism that causes it, of whatever color it may be, grows on the surface with various kinds of hyphal projections that sink into the parasitized leaf, and organs known as haustoria develop which absorb nutrients from the host. As all pathologists know, a distinct character of the mildews, whether they are the white and powdery kinds or the black sorts, is that these fungi require living tissues on which to grow.

Difference Between Mildew and Mold

It is well recognized by pathologists that there is a tendency for some growers in the tropics to lump together molds and mildews. There has to be a clear distinction between them, for their means of control are very different. Molds have a very large part of their growth inside leaf, stem and fruit tissues, with a relatively small part of its woolly structure outside the host. Mold organisms that cause acute disease not only attack and grow on and in fresh living host tissues, they are also able to grow vigorously in aged, moribund, and often in decaying tissue. Mildews only live through attacking fresh living tissues. Molds are known on roots of plants as well as other parts. Mildews, the fungus structures of which grow in large part on top of foliage surfaces, are aerial only and for that reason are on foliage.

The word "mold," also spelled "mould," was taken by pathologists as referring to the dark-colored organisms on leaf and other plant surfaces. Many times in popular writing any fungus is called a mold, but in the more technical articles the word has to be restricted.

The black mildews are one of the most remarkable groups of parasitic fungi in the tropics. They are rather common, are truly black fungi, and are specialized in their modes of parasitic habits in the tropics or temperate zone. It was pointed out by Toro (51) that John A. Stevenson gave the name black mildews to these important tropical fungi. Some of the genus names among the black mildews are very well known in literature from the tropics, and with increased study there they will be found to be as important, as the white mildews are considered in the temperate zone. However, though the present day black mildews have been studied more by mycologists than by pathologists. In the case of *Meliola*, is a genus that has names of well•over 1000 species under it. In spite of all that is known about the differences between species, very little has been done to determine for the tropics the amount of losses sustained by crops infected by them. Nowhere are these black mildews being made a subject of major pathological study, although agriculturists who observe their crops well, know that at times these fungi are very damaging in their effects. There is another black mildew genus, *Asterina*, which has at least 250 species under it, of which not one species is a subject of pathological investigation. In the case of *Amazonia* there are also a large number of species, described by mycologists. Some of the other genera are *Irenopsis*, *Parodiopsis*, *Lembosia*, *Ireina*, *Parodiella*, and *Scolecopeltis*. Some of the greatest students and collectors of these fungi have been F. L. Stevens, A. C. Batista, A. Gaillard, R. Ciferri, H. Sydow, I. J. Puiggari, C. Spegazzini, J. A. Stevenson, R. A. Toro, and C. G. Hansford. (1, 6-9, 14, 20-22, 24, 25, 27, 37, 43-51.) Symptoms of a black mildew disease are represented in accompanying Fig. 7-1.

As early as 1828 Elias Fries stated that fungi of the genus *Meliola* and its close relatives, were typically tropical in distribution (24, 51). Since that time it has been amply confirmed. There are, of course, a few rare exceptions. According to details discussed by Roger (37), Toro (51), Hansford (24), Stevenson (47, 48), and others, there is no question that the ideal climate for the true black mildews is moist tropical. These obligate parasites require high humidity around foliage surfaces for best development. In my experience in climates where there is heavy rainfall that will not allow white mildews to become established at all, these conditions do not prevent the development of black mildews. Not only are they able to infect hosts in such places, the diseases also develop well.

Black mildews, although of largely tropical distribu-

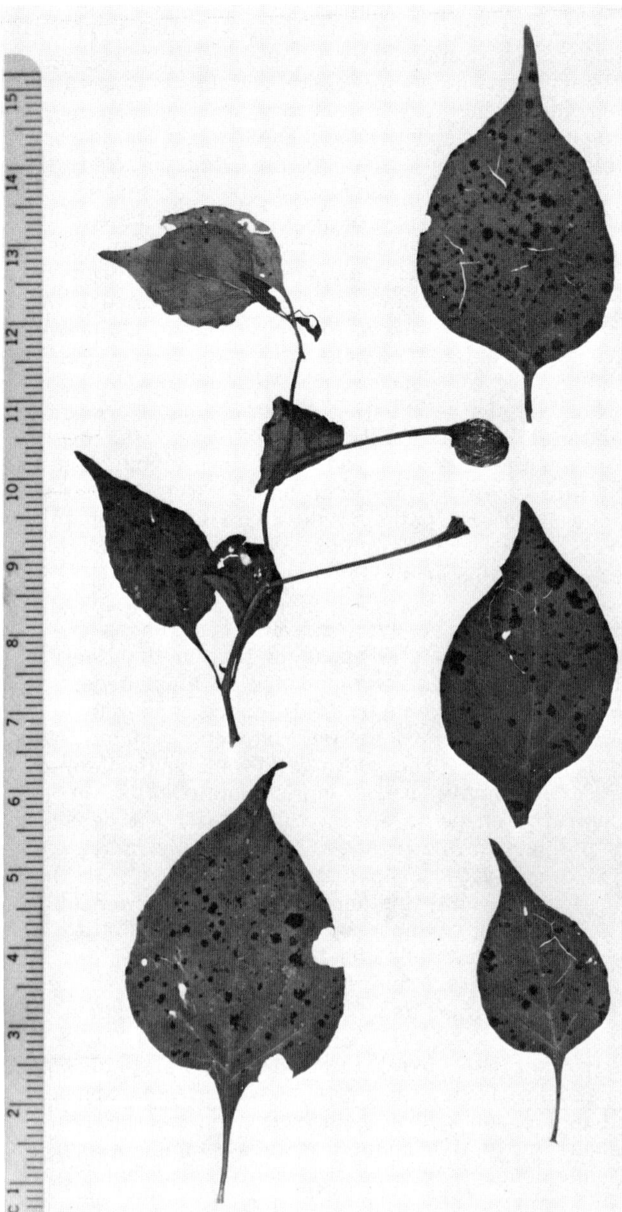

Fig. 7-1. Symptoms of black mildew, *Meliola capsicolum* F. L. Stevens, developing as characteristically rounded, black patches with irregular edges; disease on a wild pepper, *Capsicum* sp., in Costa Rica.

tion, while they require good warmth such as 24° to 28° C mean temperatures (common in our tropics), are stopped in their growth by mean temperatures much higher than that. From my observations it seems clear that these tropical fungi are dependent upon a year-round growing season; they require a constantly renewing supply of host leaves in the vicinity of old infected leaves. This situation, of course, occurs only in the tropics. There are ecologic zones specially suitable (51, 56). Shade also seems to be a condition that favors attacks by black mildew. Most species of it are readily killed by frost; they are severely injured when dried, so that both drought and excessive sun exposure inhibit their development. This is evident even in the distribution of the more resistant species.

There are some interesting comparisons between occurrences and field conditions that seem to favor the black mildews more on the one hand and those that favor most the powdery or white types on the other hand. It is believed that with close attention, pathologists could soon learn to prophesy what may be "black mildew country" or "where white mildews are going to be the most serious."

In my experience comparing field occurrences of black and white mildews, it is apparent that in the Neotropics the white mildews, such as *Erysiphe* spp., flourish best where the wet season is comparatively short, and where during the dry season conditions are characterized by periods of light nightly dewfall. The white mildews, in the tropics, are found most commonly in cool midmountain belts, and in locations where the shade is absent or at least not dense. These white powdery mildews seem definitely the most devastating in a cultivated crop that is not shaded. In the Caribbean region, in many instances white powdery mildews are worse if the fields of the crop are on gently sloping plateau land or on mountain sides that face the west and south. The black mildews may be more severe in sheltered locations or on mountain sides facing the east and north.

On the other hand black mildews such as species of *Meliola*, *Amazonia*, *Irene*, *Irenopsis*, and *Asterina* appear well adapted to regions with a relatively long wet season. These black parasites also appear at an advantage where, if wet seasons are long, there are two wet seasons annually, and where in intervening dry months there is heavy dew at night. The black parasites develop most commonly in warmer, low mountain belts, and flourish as noted above, in protected areas as under shade.

To repeat in another way: heavy rains, by attrition, wash away white mildews, such as Erysiphe, Oidium, and Sphacelotheca. White mildew spores do not germinate in drops of water. But the black mildews seem different. Species of Meliola and Asterina are found in rainy regions where Erysiphe species probably never can occur.

Parasitic Habits Among Black Mildews

The most attentions given to the black mildews have been those studies by mycologists and as Hansford (24) points out these organisms, especially in the genus Meliola, have species that seem markedly restricted to a host-of-choice by the fungus species. This, Stevenson (personal communication) has agreed with. There are not many species of these dark-colored mildews that have several host suscepts. Indeed, mycologists and pathologists realize that without knowing the species of the host from which the black mildew is collected, it is almost impossible to make an accurate specific identification of the fungus in question.

It is of interest to discuss some of the parasitic habits of true black mildews. It has been shown by Hansford (24), Gaillard (19) and others that when the spores of some kind of Meliola germinate on a leaf surface, the rapid growing, far-reaching germ tubes develop special organs, or hyphopodia, that attach themselves closely to the leaf surface. These so-called feet are hyphal outgrowths from which slender infection threads pierce the cuticle of the host and send very small, rounded or irregular haustoria into the epidermal cells. The morphology of parasitism is thus known, but not a great deal has been done otherwise, especially as to physiology. It is probably not the same kind of thing as occurs with certain of the powdery mildews. On the whole what little has been demonstrated thus far is not as clear as it should be, and not fully satisfactory for pathologists. A beginning on it was made many years ago in France (19, 20, 37) and in Argentina (6-9, 17), but our techniques and methods are so much more improved now that there is no doubt that there will be discoveries of hitherto unexpected and new features. This work when done, is going someday to be of great benefit in science and for the advanced recognition of the specialist who accomplishes it.

Although I have made only superficial observations,

they may indicate some of the possibilities in store for real investigation. Certain species of Meliola do not attack the younger leaves of a preferred host. Why this is, is not clear. Very young tissues are resistant to infection by some of the black mildews. This was observed especially in the case of sweetpotato, coffee, and avocado that were infected respectively with Meliola malacotricha Speg., Meliola sp., and Irene perseae (Stevens) Toro. There is sufficient difference in leaf type in these three hosts that it seems likely this habit of infecting only the fully-matured epidermises may be true in other cases; possibly, all cases.

To the majority of coffee growers, black mildew is of relatively minor concern. It is not a frequent affection, and when it does cause disease it is on old, so-called "spent" leaves, and in comparatively restricted, moist regions. However, with M. malacotricha of sweet potato on old vines, there is a much wider ecological occurrence than occurs with black mildew of coffee. The same appears to be the case with the black mildew, Irene perseae, on avocado. The latter is at about all the elevations where the host is grown, in Puerto Rico and in El Salvador. Black mildew on monocotyledonous hosts such as cooking banana and sugarcane are most vigorous on their oldest leaves, and occur most commonly in moist regions.

Observers of black mildews do not seem to be fearful about economic losses from these diseases. Apparently the reason is that the kind of diseases these organisms cause result mostly in a slow and chronic loss of phytosynthetic surfaces rather than any occurrence of a crisis like quick-rot, wilt, cancerous swelling, or exudation of life-giving sap. This has seemed specially so on wild trees in moist, but not wet, jungle regions. There are certain black mildews, however, that at times grow so vigorously on leaves of cultivated crops such as mandioca and sweetpotato that there is definitely severe leaf destruction. The leaf drop that comes from severe infection can, in mandioca at least, keep green leaves reduced to only the youngest leaves towards ends of branches and the result is tuberous roots that are small and of poor quality. In certain cases of Meliola sp. on inga trees used as coffee shade, it was found that a definite chlorosis developed although infected leaves were not dropped. Such host reaction indicates a toxic effect, since uninfected leaves remained green. On the other hand, if black mildew infections are on leaf petioles or covering bases of leaves, they often cause unseasonal drop.

Hyperparasitism by Black Mildews

One of the things that interests pathologists in the tropics is the large number of hyperparasitic organisms that are encountered. They are found in many fungus groups and this is true among the black mildews. There are certain tropical fungi that are especially known as hyperparasites, and generally include such genera as Calonectria, Trichoderma, Verticillium, and Darluca. Generally these are about all that are encountered, although in the genus Meliola there are species which are also vigorous hyperparasites (e.g., 24, 25, 43, 50). A plant clinician sees many of them, some of which probably have never been described. It is believed that some of these may be of considerable economic importance in reducing sporulation of primary parasites. It is likewise interesting that there are some fungi that parasitize the black mildews themselves. In some cases this latter, in my categorization, becomes classed as superhyperparasitism and deserves more attention in the future. It is in such study that some of the most interesting subtleties are involved in infection habits and toxic effects of fungi on fungi.

It is known from mycological work (for example Hansford (24), Stevens (43), Stevens and co-workers (44, 45) and Toro (50)) that in the tropics certain black mildews, mostly species of Meliola, are hyperparasites on several parasitic fungus genera such as: Helminthosporium, Dimerium, Pellicularia, Calonectria, Fusarium, Colletotrichum, Trichothyrium, Glomerella, Arethrobotryum, Erysiphe and relatives of it, Nectria, Gibberella, Nigorospora, Pestalotia, and others. It is worth repeating that even Meliola itself and its close relatives are at times hyperparasitized by species of Meliola.

I have seen, contrariwise, hyperparasitism of Meliola by a Rhizoctonia. It is accomplished by the hyperparasitic organism wrapping itself around, and absorbing nutrients from, the vegetative hyphae of the black mildew fungus. This phenomenon was observed in my laboratory under tropical conditions in El Salvador. In 1946, isolates from stem canker-diseased seedlings of tomato and a weed (Purslane) yielded a mixed culture of a pythiaceous fungus, probably both Phytophthora sp. and Pythium sp., and a light-colored (in culture) isolate of Rhizoctonia (solani?). When held for three weeks growing on PDA petri plate cultures, interactions were seen. The more slender hyphae of Rhizoctonia proceeded to wrap themselves around the coarse hyphae of the pythiaceous fungi. In moist chamber cultures this same Rhizoctonia was used as inoculum on black-mildew diseased leaves of sweet potato. In these chambers, the black mildew fungus was also hyperparasitized by the basidiomycete. (This is super-hyper-parasitism.)

Five years later, microscopic examinations in Costa Rica of anthracnose-infected material of morning-glory that also had black mildew, Meliola malacotricha Speg., hyperparasitism was observed. On vines near ground level, a Rhizoctonia was found attacking the Meliola. In addition acervuli of Colletotrichum gloeosporioides were being engulfed by hyphae of M. malacotricha, and those spore-bearing anthracnose spots were covered with a profusion of black mildew growth; meanwhile, and this was all on the same vines in the field, the black mildew was causing new black mildew growth on the morning-glory vines. In the case of the infected Colletotrichum hyphae, closer high-power microscope examinations showed the diseased fungus threads appeared disorganized internally and not so dense in content as the hyphae of unattacked healthy anthracnose fungus.

In Puerto Rico in 1964 while examining diseased mango leaves another effect was seen of hyperparasitism by a black mildew. Here, the black mildew of mango, Meliola mangiferae, developed as mats and showed thin, gray areas on leaves. On careful microscopic examination it was found that M. mangiferae was being in turn parasitized by another Meliola(?) of smaller dimensions. The black leaf parasitizing mycelial strands of M. mangiferae, which were ordinarily tightly held onto the leaf surface, became loosened when they were hyperparasitized. It was also seen that where the hyperparasite infected a foreward edge of M. mangiferae the parasitic black mildew fungus growth was effectively stopped in its march into new uninfected leaf territory.

In connection with certain special studies on bean diseases a hyperparasitizing Meliola sp. was found attacking two bean leaf spots. In the case of spots from Chaetoseptoria wellmanii, it was found that where the hyperparasite attacked the fungus that was the primary cause of the round spots on leaves, the setae in the Chaetoseptoria lesions were completely inhibited from forming. This also occurred on the bean anthracnose spot from the fungus, Colletotrichum lindemuthianum. The host fungus in both cases was markedly weakened, very rapidly after the hyperparasitizing Meliola started its infection. This was amply indicated by the growth of the Colletotrichum hyphae. In them, if the Meliola attacked when the growth was young, and the anthracnose organism had just formed its spore-bearing surfaces, the acervuli, no spores were produced on the structures.

One added point should be made about Meliola hyperparasitism. From observations it appears that if one of these hyperparasites gains access to a well-developed fungus lesion, such as that of a Colletotrichum anthracnose spot with many spore-bearing acervuli, the Meliola, on parasitizing the anthracnose acervuli, quickly grows over these fungus structures. It no doubt draws nutrients from the sporulating fungal organs. Where seen on anthracnose it was over the lesions that Meliola perithecia were formed. Where seen on scattering vegetative threads of another black mildew, the hyperparasitic Meliola did not develop perithecia. Studies of black mildews have remarkable phases still wanting serious and adequate attention. These are organisms with a special challenge because of their subtle host and hyperparasitic relations.

Superficiality of Carbon-Soot Diseases

The somewhat more delicate and fine-grained appearance of certain black mildews, such as that of the genus Asterina, is overshadowed by the coarser and denser black-appearing mildewed blotches from infection by many species of Meliola. Both of these so-called black mildews and their relatives are mostly obligate parasites closely associated with preferred hosts; a few attack several crops. There is also not one of these obligately parasitic black mildews that is characterized by the notably heavy pellicle of carbon-black-like skin of fungus growth that is produced by the carbon-soot, fumago, or fumagina fungi, such as belong in the epiphytic and saprophytic black-fungus genus Capnodium.

However, there appears to be a bridging step among these black fungi, from especial parasitism to true epiphytism. Probably all species of Meliola are parasitic and possibly no types saprophytic, but in the genus Asterina we find (as Hansford (24, 25) has shown) that although many are typically black parasites, some are truly epiphytic in habit. This is likewise true of species in the black mildew genus Atichia. The parasitic types of this genus all are strictly epiphytic in habit, while all the true epiphytic species depend entirely for growth on insect secretions. Thus there seem to be examples of a transition between the more gross and probably more primitive saprophytic habit, to the more specialized parasitic existence. In any case along with true black mildews, the carbon-black fumagos or sooty disease organisms are often thought of, incorrectly, by tropical crop growers as seriously parasitic organisms.

The importance of fumago or fumagina disease, frequently brought to crop disease clinics for determination, is not so much in the parasitism as it is in the discoloration of the foliage and fruit surfaces on which it occurs. One of the most common of the fumago genera is *Capnodium,* and there are hundreds of species of it. Many of them in the Neotropics. A few of these have been grown in pure culture on agar. In nature they always develop on insect honeydew. In my laboratory, in Puerto Rico, in far from extensive trials which were prematurely stopped on account of a change in location, difficulty was encountered in culturing *Capnodium coffeicola* on agar; doubtless because little is understood about the amounts and kinds of sugars and vitamins the fungus prefers. In studies with *C. coffeicola* numerous unusual contaminating organisms were found that grew much faster than *C. coffeicola.* An illustration of the spectacular symptoms of coal-black growth of this organism is shown in accompanying Fig. 7-2.

Capnodium spp. are most common in those parts of the Neotropics where sucking insects flourish. Some of the insects with this habit that likewise secrete honeydews in sufficient amount to feed the epiphytic fumagina organisms are: mealybugs such as *Dactylodius,* any or all the plant lice--many of which are species of *Aphis,* white fly such as *Aleyrodes,* many kinds of plant scales, and numerous Fulgorid leafhoppers. Most of the honeydew-secreting insects feed on the undersides of the leaves but the excretions fall in small sticky, sweet drops deposited by gravity below the locations of the insects on leaf and fruit surfaces. It is in the honeydew that spores of the superficial fungi germinate and grow; there is actually no infection of host tissues in these black pellicles that flourish so well on honeydew.

These black epiphytic fungi grow so rapidly that they quickly disfigure leaves and fruits, and, in some bush and tree crops such as tea, coffee, cinnamon, nutmeg, citrus, bay rum tree, tamarind and rain tree, if the tree's new

Fig. 7-2. Black mat of the fungus Capnodium sp. causing coffee fumagina; the black organism is an epiphyte completely blanketing upper surfaces of these leaves. Collected in Colombia on Arabica coffee growing under shade trees harboring multitudes of such sucking insects as aphids, fulgorids, and coccids during the dry season.

growth is most involved, the black-velvety to harsh-black fumago membranes and pellicle-like sheets may smother growing tips. There is no question that the black growth interferes with photosynthesis and where there is a great deal of the fungus on a tree or bush it is no doubt injurious. Observant planters realize that both lower yields and poor quality of the fruit or other product is a result of abundant presence of carbon black epiphytic growth. Spores from these growths, whether of the asexual stage or those from asci, and as well bits of black mycelium itself, are widely spread by breezes, insects, and the meteoric water that is rain or dew. (2, 3, 5-7, 12, 16, 17, 21, 22, 29, 31, 37, 42, 44, 48-50, 55.)

Where rainy seasons are long and precipitation is heavy, sucking insects do not thrive in the tropics. In some regions there is a timing of rains so that, even if rains are not heavy in total amount, they may come down at such intervals and at such time that the slight washing force suffices to keep sucking insects few, and the fumago fungi do not become a menace. From the above it is clear that it is almost a natural law that can be stated as a principle--where sooty molds abound there are likewise many sucking insects. Where tree leaves are subject for long periods to insect honeydew deposits, they often show varying amounts of fumagina or sooty mold. In the accompanying Fig. 7-3, fumagina is illustrated as a relatively thin coating after about six months with the leaf green and turgid; deposit black but leaf still living although weak, after one year; and after a year and a half, deposit very heavy and leaf dead.

The highly ornamental Hibiscus rosa-sinensis, common in so many tropical countries, is grown throughout the West Indies. I found a remarkable occurrence in Puerto Rico when I noted tufts of a sooty mold, Capnodium hibisci, developing at the base of the underside of mature leaves, on each side of the main vein. The perfect stage of this fungus is of the genus Microxyphiella. A good deal of attention was given for several months to its occurrence, searching for the presence of sucking insects any place on the sooty mold-diseases bushes. None was found.

The confirmed observations made years before in Costa Rica where, in a rainforest region, under conditions where no Capnodium coffeicola could develop on coffee because of the heavy rains, I found tufts of a Capnodium

Fig. 7-3. Increasing severity of "fumagina disease" on leaves from a wild citrus tree studied in Puerto Rico. At left, fumagina deposit light and thin after six months of development; center, more severe covering of fumagina after 12 months of exposure (patch of loose black fungus growth removed to show the turgid and living but yellowed leaf underneath); right, a heavy carbon black cover of fumagina fungus on a leaf exposed more than 18 months and now dead. Healthy leaves were found on the tree that were up to at least four years of age.

(microxyphiella?) growing out of leaf invaginations, or domatia of a large leaved species of Inga. There were no sucking insects ever found near the infected leaves.

White Powdery Mildews

The white powdery mildews are caused by species of such fungus genera as Erysiphe, Sphaerotheca, Oidium, Uncinula, Podosphaeria, Microsphaera, and Phyllactinia. They belong under a single fungus family, the Erysiphaceae, and one of their distinguishing characteristics, as with the black mildews, is that they are obligate parasites. The white parasites are spread widely in the temperate zone, causing powdery mildew attacks in locations far north in that region, where summers are short, relatively dry, and cool. They also occur south, into the subtropics where the climate is somewhat dry. These white mildews may also occur in some places in equatorial regions. While they are not thought of as major causes of crop disease losses in the Neotropics they become of considerable concern at cool elevations in dry seasons.

The white powdery mildews seem to be much better known in phytopathological literature than black mildews. This is because the white fungi have been noted and studied for a long time in the temperate zone which has been the source of much literature, and thus publicity, about these diseases. It is in temperate zone publication sources that the most information has appeared in the early years of plant pathology. Tropical sources are recently important in this regard.

The powdery mildews are said to be practically cosmopolitan in occurrence and there are hundreds of species. It seems possible that there are many kinds now given specific ratings that when studied under a variety of tropical conditions will prove to be races or strains.

Symptoms of the white powdery mildews are very characteristic. They are diffuse lesions of superficial white growths, and these white powdery coverings on leaves are sometimes accompanied by malformations, aborting of flowers and fruit, and growth reduction of the plant that is attacked. These symptoms are well known from such descriptions as those seen in the temperate zone (15, 28, 44, 54), and are identical in the tropics (for examples see 17, 21, 31,

37, 42). Symptoms in the tropics are especially severe in crops that are grown on hillsides in dryer cool regions of equatorial America. On the whole it appears that both in the wettest and in the hottest parts of the tropics there is the least damage from the powdery mildews. However, powdery mildews of several species cause much concern to tropical pathologists as is seen in such citations as 3, 10, 11, 13, 18, 23, 32, 34-36, 38, 39, 41, 48, 52, and 53. Symptoms of effects from attack by a powdery mildew, in this case *Erysiphe polygoni* DC., on garden bean are shown in Fig. 7-4 A and B.

A common powdery mildew, brought to me from cooler elevations of Central America and Mexico, is one that occurs on several plant species, *Erysiphe polygoni* DC. It is found on various introduced and native tropical legumes, of which there are numerous species in the fields, and where found it appears towards the middle to the end of the dry seasons. As suggested, it is most frequent in occurrence on hosts growing at fairly high, cool elevations, such as mountain table lands and shallow valleys on dry mountain sides. The disease, furthermore, is apparently most frequent in the Trade Winds areas on west slopes; probably because this location is less subject to the heavy, white mildew-damaging rains. In the dry season the rains are relatively light, and in wet seasons severe washing rain storms seldom last for more than a day.

It has been shown (e.g., reports 4, 18, 22, 26, 33, 52) that white mildews attack grasses and small cereals such as wheat, oats, and others. These are found where the fields are in high mountain locations as in the Andes, Guatemala, Mexico, cool parts of southern Chile, and where there is a prolonged dry season. Under such ecology may be, for example, severe trouble from *Erysiphe graminis* DC. before the rains commence. The mildew, where severe, causes leaf tips and edges on plants to wither, discolor, and die which is of great concern both in fields of small grains and in grass pastures. The most severe effect is often increased in prolonged dry conditions, when the rainy season is slow to start. After rains commence it is only a short while, often within a few days, before the mildew is practically gone.

An early observation on diseases on roses that are grown exposed to full rainfall, out in the garden, under rainforest conditions in Costa Rica, is that there was no white powdery mildew. (A black mildew, *Meliola* sp., was

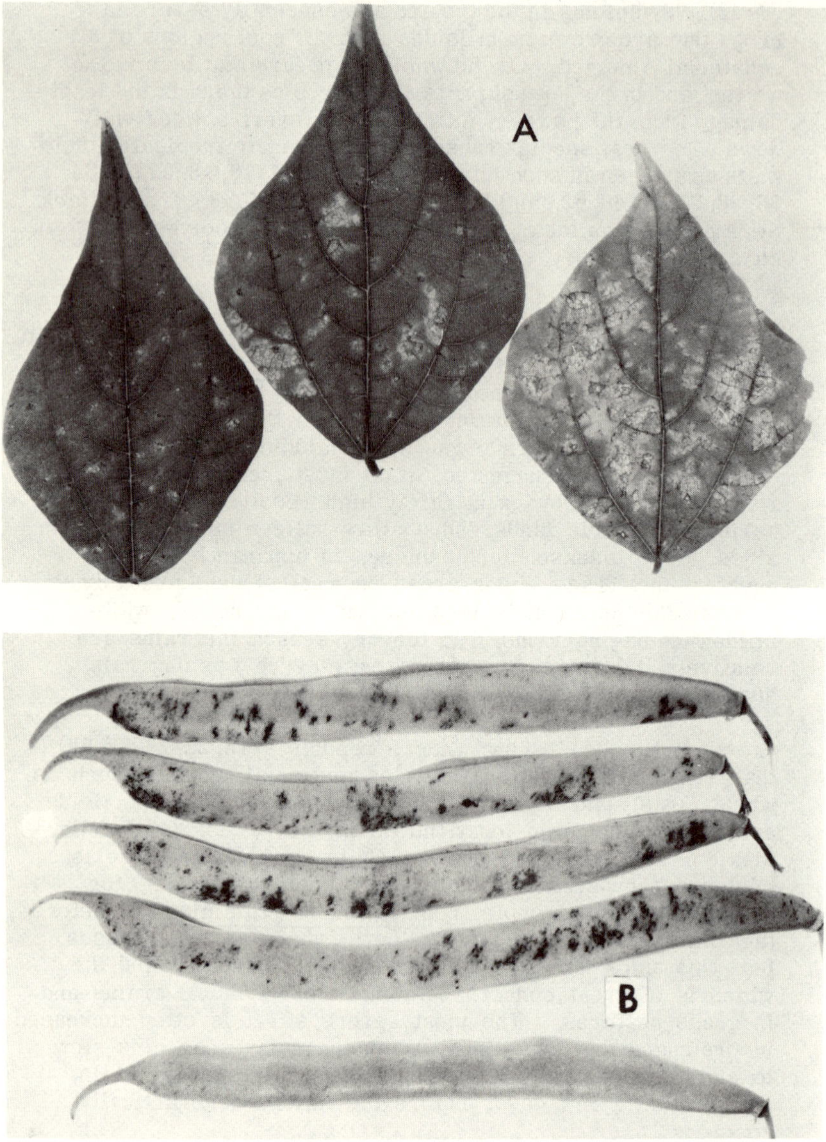

Fig. 7-4. Powdery mildew, *Erysiphe polygoni* DC., on garden bean. Advanced symptoms on leaves A and on pods B. Photographs courtesy of Plant Pathology Department, University of Florida, Gainesville.

present.) Other rose bushes were planted, according to a gardener, "to keep them healthy" and well protected from rains. This was done with narrow thatch roofs held high by bamboo poles, and arranged so as to cut off as little light as possible, but at the same time protect flowers on plants from the long periods of strong rains. In some other cases the rose bushes were grown next to house foundations where rainfall did not reach, under the very wide eaves of single story, bungalow-type, residences. Where roses were under either of these two kinds of rain protection the white mildew, in this case identified as *Sphaerotheca pannosa* (Wallr.) Lev., was severe. Bushes with much mildew on them were transplanted to the open field, and when they produced new foliage in the rains it was found to be free from white mildew.

Not every white powdery mildew is as important in the Neotropics as it is in the temperate zone. All of the genera occur in the tropics but some, for example *Podosphaeria* while it can be found in our tropics, is generally not important there and occurs only under specialized conditions. On the other hand the genus *Erysiphe* is widespread in parts of the Neotropics and at times may be a serious disease-causing fungus genus on several important crops.

Primordial Oidium Mildews

The plant disease diagnostician in the Neotropics soon learns to recognize symptoms of attack by white powdery mildews. Often, knowing the host that is infected and knowing from literature what mildew attacks it gives the specialist confidence so he can usually make a close guess as to genus and species of the parasite. Most of the time, however, spore scrapings yield no perithecia (ascocarps), and the mycelial characters are not strikingly different among many species. It is possible for tropical pathologists to work with a microscope in some regions a full lifetime and see only the many elliptical, egg-shaped, hyaline and one-celled conidia. These spores, characteristic of the genus *Oidium*, occur in such large numbers that, seen on top of the cobweb-like mycelium which bears them, they are the fungal stage that gives the white powdery appearance. For an example of the appearance of *Oidium* sp. disease symptoms, see Fig. 7-5. In most cases and from many hosts, the spores have remarkable similarity, although in some microscope mounts where conidiophores have been included, and slight differences in these observed, there may be con-

Fig. 7-5. Example of disease effect from one of the primordial types of Oidium in El Salvador. Typical leaf malformation; the parasite, producing extreme numbers of spores, develops the white-appearing growth on the leaf surface. (Oidium bixae Viegas on Bixa orellana L.)

firmation as to suspected causal organism specificity.

However, when spore traps yield the typical conidia of a white powdery mildew the only thing a cautious tropical pathologist or mycologist usually concludes is that it belongs to the genus _Oidium_. When, as happens so many times in the tropics, pathologists encounter a severe white powdery mildew on a host but without perfect stages, the best and only unquestionable disgnosis is that the parasite belongs to the genus _Oidium_. According to the Dictionary of the Fungi (1) there are 25 species of the genus in the tropics on various important hosts, but in reality there must be many more. Species designations with _Oidium_ are employed freely for the whole gamut of white powdery mildews on papaya, hevea rubber, avocado, citrus, mango, tropical peaches, tropical plums, tomato, and other such hosts. These _Oidium_ spp. will undoubtedly be given more detailed study in the future under wide ranges of ecology. It is believed there are numerous white powdery mildews in our tropics never before described, and that do not occur in the temperate zone.

The spore type characteristic of _Oidium_, is one-celled, transparent, thin-walled, almost weightless, very numerous, and short lived. It is possible for them to be considered a primordial kind especially adapted to rapid and wide local spread of the disease they cause. Their habits fit in well with the ecological conditions of the dryer tropics, and the seasonal development of their hosts in such regions. The small size of oidium spores is a great advantage in carrying on the species, and the weeks and months of time over which the conidia are produced, indicates adaptation to survival that surmounts very long periods of geologic time. If these fungi, when they became parasites, were never under difficult temperate zone or other conditions, they never needed to develop the perfect stage or the various handsome-appearing perithecial organs on which mycology depends as its last resort for genus and species names.

Occasionally a single species of a white powdery mildew attacks many crops. As work is still in progress on it in the tropics, a completed host list of weeds, crop plants, and ornamentals that are known as suscepts of _Erysiphe polygoni_ has not been made. This organism in the past has been designated in its imperfect stage as _Oidium balsalmi_. It is very destructive on over 300 species of hosts in the tropics and attacks at least half that many in the temperate zone. Another of these mildews with a long host list is

Erysiphe cichoracearum, an organism that was known at first by many in the past as Oidium ambrosiae, causes a great deal of trouble as a white powdery mildew. This organism may not have so large a number of host species in the tropics as O. balsami. The attack by O. ambrosiae covers perhaps a wider range of genera than the other. For example, this Oidium is found on tobacco as well as on certain of the weeds that are in the aster family, species of verbena, many cucurbits, and other types of plants.

There are a large number of genera and species of the important cucurbitaceae, that occur both in the wild and as cultivated crops in the Neotropics, and many of them have been taken to the Old World either as curiosities or for their agricultural worth. In tropical America a species, or perhaps more than one species, of Oidium is found on the plants in this family of squashes, melons, pumpkins and the like. Powdery mildews seem especially adapted to the relatively dry climates where some of the cucurbits abound naturally, and is also found in some areas where many are grown for food. When these mildews are brought to pathologists and the organism determined, frequently it is designated as Oidium communis and also as Erysiphe cichoracearum DC.

Unfortunately E. cichoracearum has not been given much attention as to details of many of the hosts parasitized in the American tropics. It can be confidently expected that when more is known about it, more is going to be learned about races of the species, how these mildews are held over from season to season, where best to search for tolerance, and what weeds should be eliminated as perennial sources of the parasites. There are certain other powdery mildews that are known to have special races in the tropics, one example is Erysiphe polygoni var. calendulae which is apparently confined to the ornamental flowering calendula in South America. Incidentally the determination of this organism in such countries as Chile and Argentina is sometimes given as E. communis var. calendulae but more frequently as the imperfect form Oidium erysiphoides var. calendulae.

One of the widely encountered diseases from the tropical Oidium monilioides (Nees) Link (in its imperfect, primordial stage) is mildew found generally on wild, and also cultivated, gramineae. This fungus is mentioned above under the binomial Erysiphe graminis. At about the beginning of the present century it had caused some problems, resulting in European and North American researches which demonstrated that in the temperate zone there were distinct physiological races of the parasite. Since then these have been given designations according to preferred hosts: *E. graminis tritici* on wheat, *E. graminis hordei* on barley, *E. graminis poae* on bluegrass, *E. graminis avenae* on oats and others. As is shown by such workers as Fresa (18), Sarasola (41), Vallega (52), and others, these are also found in South America. It is of additional interest that a mildew specifically on mountain-planted sugarcane in the Neotropics, is given the trinomial *E. graminis sacchari.*

It seems significant that the occurrence of the mildew, *E. graminis*, has been on the rise. In 1909 there appears in a discussion by Duggar (15): "*Erysiphe graminis* ... is seldom a fungus of any consequence in most sections of the United States." Some 50 years later Walker (54) gives a different view: "Powdery mildew of cereals is important ... [and] is most destructive on barley and wheat but occurs on other cereals and grasses." Apparently there has been an obvious increase in the U. S. A. in the importance of powdery mildew on cereals. It seems logical to expect that the causal organism at first was probably a relatively mild parasite in the tropics; it probably had a generalized spread. As small cereals were grown in greater concentrations in the temperate zone, spores of *Oidium* drifted up from the tropics and infected plants. The most virulent variants multiplied, and some were races more adapted to growth on vast fields of special hosts. This kind of disease increase will undoubtedly be evident with other tropical white mildews as their host crops become intensively grown.

Some tropical mildews have not, through the present day, displayed presence of ascocarps, and therefore still come under the imperfect form genus *Oidium*. A few of these in the Neotropics are as follows:

Oidium hevea has been known (37) for a long time as a limiting factor in rubber latex production from trees of *Hevea brasiliensis* in such places as Ceylon, Malaya and other regions of the Orient. In 1958 the mildew was reported from Brazil (38), where vigorous action eliminated its presence. No perfect stage of the fungus has been discovered.

On citrus trees there occurs a mild powdery mildew, *Oidium tingitaninum* Carter (16). It was apparently first seen in the East Indies by Petch (32) and later in California.

Although the distances between the places of collection are very far, Fawcett (16) concluded the fungi were of the same species. Citrus was found with a powdery mildew in Brazil that seemed definitely to be a different species, which was given the name O. citri Bondar (21).

There is a powdery mildew on avocado, that Ruehle (39) has reported from Florida as occasional, and since no perfect stage has been seen, is listed commonly as Oidium sp. and also as O. persicae.

Papaya is occasionally severely attacked by the mildew known as Oidium caricae Noack, which was described a long time ago in Brazil, is found in much of South America, and of more recent years (35) has become a serious problem in Venezuela.

Table Comparing Mildews

In connection with observations on certain tropical plant mildews (obligate type fungus causal organisms), it becomes evident that there are some distinct differences. As the causal organisms have not been studied in the Neotropics under controlled conditions, and attempts at growing them on artificially inoculated hosts are studies I have not carried out on which to base my findings, results, such as given below are subjective and in some ways superficial. A longer lifetime of detailed study would give data and objective conclusions.

However what is given in Table 12 is suggestive and is presented after some years of seeing both the black and white mildews. The conclusions, as tabulated below come from numerous visits to widely separated agricultural areas: 1. In the West Indies islands and Nicaragua when observations were made on sweet potato. 2. In El Salvador, Puerto Rico, and Guatemala in coffee areas. 3. In Puerto Rico and the west coast of Central America in consulting work on mango. 4. In Puerto Rico and El Salvador to study diseases on mountain grown sugarcane. 5. In highland areas of El Salvador, Puerto Rico and Guatemala visiting pastures with special Desmodium diseases. 6. In four places in mountainous parts of Guatemala where farmers attempt to grow apples for a profit.

Table 12. COMPARISONS OF DISEASE REACTIONS from certain tropical mildews caused by fungi with obligate-parasite habits.

Fungus	Mildew type	Host	Remarks on localities
Meliola malacotricha	Black	Sweetpotato	At medium to low elevations on moist exposures, 6 months wet season and 6 months dry
Meliola sp.	Black	Coffee	Low mountain valleys in shaded fincas, moist exposures, long rainy seasons
Meliola mangiferae	Black	Mango	Low foothills, moist exposures, long wet seasons
Erysiphe graminis sacchari	White	Sugarcane	Medium high mountainsides, relatively dry but ample fogs in long dry season
Erysiphe polygoni	White	Desmodium sp.	High mountain pastures, on dry west slopes, short rainy season
Sphaerotheca pannosa	White	Apple	Plantings on cold mountain shoulders with Pacific side exposure, relatively short wet season, dry period long with much fog and clouds

Literature Cited

1. Ainsworth, G. C. 1961. Ainsworth & Bisby's dictionary of the fungi. Commonwealth Mycol. Inst., Kew (Surrey) England, 547 p.

2. Alvarado, J. A. 1935-36. Tratado de caficultura práctica. Tipografía Nacional, Guatemala. 2 vols., 1227 p.

3. Alvarez-G., L. A. and C. Chupp. 1954. Enfermedades de las hortalizas en Puerto Rico. Univ. Puerto Rico, Agr. Exp. Sta. (Unpub. MS dated March 28, 1954; 164 type. p.)

4. Arentsen S., S. 1954. Novedades micologicas. I. Agr. Tec., Chile 14:70.

5. Barr, M. E. 1955. Species of sooty molds from western North America. Canadian J. Bot. 33: 497-514.

6. Batista, A. C. 1951. Alguns fungos de fumagina de Pernambuco. Mycopath. Mycol. Appl; Italy 5: 147-172.

7. ________. 1959. New considerations about the genus Caolthyriolum Speg. Soc. Biol. Pernambuco, Brazil, An. 16:19-22.

8. ________., A. F. Vital, and J. N. da Silva. 1953. Novos ou interessantes fungos da familia Meliolaceae. Soc. Bot. Brasil, Engr. Nac. 4:88-114.

9. ________. and T. J. Gayão. 1953. Alguns fungos da familia Hemispheriaceae. Soc. Bot. Brasil, Congr. Nac. 4:114-134.

10. Botero, R. O. 1936. El mildew de los rosales. Bol. Agr., Colombia 210:492-493.

11. Campacci, C. A. 1951. O míldio da cebola. Biológico 17:191-196.

12. Carter, W. 1962. Insects in relation to plant disease. Interscience Publ., New York, 705 pp.

13. Christiansen G., J. 1967. El cultivo de la papa en el Perú. Editorial Jurídica, Lima, Peru 351 p.

14. Ciferri, R. 1954. Meliolae of Santo Domingo (W. I.). Mycopath. Mycol. Appl., Italy (special publ.) 131 p.

15. Duggar, B. M. 1909. Fungous diseases of plants, with chapters on physiology, culture methods and technique. Gin & Co., Boston, 508 p.

16. Fawcett, H. S. 1936. Citrus diseases and their control. McGraw-Hill, New York, 656 p.

17. Fernandez Valiela, M. V. 1952. Introducción a la fitopatología (2d. ed.). Gadola, Buenos Aires, 872 p.

18. Fresa, R. 1943. Oidio o blanco de las plantas cultivadas. Argentine Min. Agr. Almanaque 18:103-106.

19. Gaillard, A. 1891. Les hyphopodies myceliennes des Meliolas. Bol. Soc. Myc. France 7:99.

20. ________. 1892. Le genre Meliola. (Paris, France) 164 p.

21. Galli, F. et al. (see "Lit. Cit.," Chapter II). 1968. Manual de fitopatologia. Doenças das plantas e seu contrôle. Ceres, São Paulo, 640 p.

22. Garcia Rada, G. and J. A. Stevenson. 1942. La flora fungosa Peruana. Lista preliminar de hongos que atacan a las plantas en el Perú. Est. Exp. Agr. La Molina, Lima, Peru (special publ.) 112 p.

23. Godoy, E. F. 1939. Oídio del tomate. Bol. Frut. Hort., Argentina, Vol. 4, and Rev. Argentine Agron 6:49. (quoted by Fernandez Valiela.)

24. Hansford, C. G. 1961. The Meliolinae. A Monograph. Ann. Mycol., England, Ser. 2, (Special publ.) 806 p.

25. ________. 1963. Iconographia Meliolinearum. Ann. Mycol., England. volume, no pagination of 284 plates.

26. Heald, F. D. 1926. Manual of plant diseases. McGraw-Hill, New York, 891 p.

27. Marchionatto, J. B. 1943. La contribución de Carlos Spegazzini a la fitopatologica Argentina. Rev. Fac. Agron., Argentina 25:11-20.

28. Miller, P. R. 1953. Plant disease situation in the United States. FAO Pl. Prot. Bul. 1:104-105.

29. Montagne, C. 1849. De Capnodio, nov. gen. Ann. Sci. Natur., France 11:233-234.

30. Mujica R., F. 1957. Determinaciones micologicas. IX. Agr. Tec., Chile 17:106-107.

31. Nowell, W. 1923. Diseases of crop-plants in the Lesser Antilles. West India Committee, London, 383 p.

32. Petch, T. 1915. Citrus mildew. Phytopathology 5: 350-352.

33. Pittier, H. 1926. Manual de plantas usuales de venezuela. Caracas, Venezuela, 458 p.

34. Pontis-Videla, R. E. 1953. Outbreaks and new records. Venezuela. FAO Plant Prot. Bul. 1:91-92.

35. ________. and A. R. Landaeta. 1952. El oídio o blanco de la lechoza (Carica papaya L.) en Venezuela, causado por Oidium caricae Noak. Agron. Trop., Venezuela 2:55-57.

36. Richardson Jr., R. W. and O. Brauer H. 1954. Verduras en el huerta familiar. Mexico Ofic. Estud. Espec. Folleto 16, 79 p.

37. Roger, L. 1953. Phytopathologie des pays chauds. II. Etude descriptive des maladies parasitaires des plantes des pays chauds. Ency. Mycol, France XVIII:1127-2256.

38. Rossetti, V. 1958. O oídio da seringueira. Biologico 25:266-272.

39. Ruehle, G. D. 1955. A note on powdery mildew of mango. Florida State Hort. Sci., Proc. 68:277-278.

40. ______. and D. A. Wolfenbarger. 1949. Diseases and pests of the mango in Florida. Florida Subtrop. Exp. Sta. Rpt. 13, 6 p.

41. Sarasola, A. A. 1943. El oídio o mal blanco que ataca a las zapallos y melones. An. Rur. Dir. Agr. Ganad. Ind., Prov. Buenos Aires, Argentina 1:105-107.

42. de Segura, C. B. 1965. Enfermedades de cultivos tropicales y subtropicales. Editorial Jurídica, Lima, Peru, 439 p.

43. Stevens, F. L. 1916. The genus _Meliola_ in Porto Rico. Illinois Biol. Monograph 2, 87 p.

44. ______. 1925. Plant disease fungi. Macmillan, New York, 469 p.

45. ______. and M. H. Ragan. 1939. The Microthyraceae. Illinois Biol. Monograph 17, 138 p.

46. ______. and L. R. Tehon. 1926. Species of _Meliola_ and _Irene_ from British Guiana and Trinidad. Mycologia 18:1-22.

47. Stevenson, J. A. 1918. Citrus diseases of Porto Rico. J. Dep. Agr. Porto Rico 2:43-123.

48. ______. 1918. A check list of Porto Rican fungi and a host list. J. Dep. Agr. Porto Rico 2:125-264.

49. ______. and F. L. Wellman. 1944. A preliminary account of the plant diseases of El Salvador. J. Washington Acad. Sci. 34:259-268.

50. Toro, R. A. 1930. Hemisphaeriales and Perisporiales (pp. 227-240). _In_ Mycological explorations of Colombia. J. Dep. Agr. Porto Rico 14 (no. 4).

51. ______. 1952. A study of the tropical American black-mildews. Univ. Puerto Rico J. Agr. 36:24-87.

Chapter VIII

CHRONIC TREE CROP DISEASES

Summary -- Introduction -- Prelude to Dieback -- Isolations from Dieback -- Inoculations Using Cultures of Dieback Parasites -- Woodstain Diseases -- Limitations as to Field Location and Woodstain -- Ceratocystis Organisms -- Dissemination in Woodstain and Like Diseases -- Chronic Woodstain Effect Enhanced by Dieback -- Table of Chronic Diseases -- Literature Cited

Summary

Slow-acting, chronic diseases of trees and other plants are common, and result in such effects as dieback, basal rot and woodstain. Poor growing conditions are often the prelude to diebacks; various organisms are involved: *Colletotrichum* (*Glomerella*), *Diplodia* (*Physalospora*), *Pestalotia*, *Cytospora*, *Fusarium*, and others. Woodstain infections are mostly from the many species and races of *Ceratocystis* also known as *Ceratostomella*. In some cases these have characteristic ecological limitations. Chronic woodstains may be severely intensified when diebacks also develop.

Introduction

Trees in the great region of the American tropics on the whole are the most important and significant form of plant life. This everyone knows. Individual trees do not become as large or as old as trees do in the temperate zone. However, they are of prime importance in agriculture and in the tropics they grow more rapidly and there are many more species than in the colder lands. In numerous parts tree agriculture has been forced on man (because annual crops are so laborious to produce) and trees furnish him everything: human food, browse for animals, spices, medicines, pleasure-giving narcotics as well as beverages, clothing, shelter; trees are worshipped by jungle-inhabiting South American and Antillean indigenes, and to this day we find such things as the veneration and respect paid to the shade under wide spreading branches of the ceiba (*Ceiba pentandra*) in the West Indies and Central America.

Farming is necessary but difficult in the hot humid tropics, and it bears repeating that in a great deal of it agriculture has become, more than anything else, a tree-oriented development. In both wet and dry areas the most important plantation crops are tree crops; it is likewise appreciated that the very jungle that lies between plantations and that covers all of some landscapes, is thick, heavy, long-existent tree growth that should never be lightly destroyed. Some information on this, from several sources, has been reviewed by Wellman (96).

It is indisputable: that there is comparatively little desert in the Neotropics where trees cannot grow; that jungle regions abound and selective jungle clearing is absolutely essential if cultivated crops are to be grown; that some jungles have to be kept for land cover to preserve sources of irrigating and drinking waters; that some tropical forests and jungles are the only repositories of natural renewable resources such as specially valuable woods, medicinals, nuts, wild game, water sheds, places for recreation, and are worth more for such things than if the lands are cleared and, under difficulties, indifferent crops grown on the soils. Some of the arborescent covers have flourished vigorously through the millenia by trees growing mutually protected in very shallow top soils (laterites, sands, clays, and volcanic tuffs), that are many times poor and fragile, and from which when tree cover is removed the lands degrade and deteriorate so rapidly that they are useless agriculturally. In its natural state the protected jungle maintains a normal relative humidity in the atmosphere that helps in cultivated fields nearby and even at some distance. Factory gasses are not in the jungle which gives off, to circulate long distances away from it, a rich share of unpolluted life-giving oxygen from its photosynthesizing green tree and plant population. However, plantation trees as well as wild and jungle trees have a full complement of diseases. What is being learned about cultivated tree crops is, with slight modifications as applicable to understanding the problems of tree health of the jungle as it is to the cultivated trees that make up our plantations.

Among numerous tropical agronomists, horticulturists,

and pathologists, conditions of tree dieback, branch rots, sapwood stains, and the concomitant gradual root failures often are not given any of the intensive research they deserve. It appears that the chronic, slow-acting, dieback diseases have been little studied. Far back in the minds of many growers and some scientists there is the feeling that the chronic tree diseases are not to be feared. Trees are thought of as tough and woody and therefore are not killed quickly. The deliberate course of events that occurs goes along with debilitation, and some consider the presence of chronic troubles inevitable as the effects are not marked by excessively spectacular symptoms.

These are different from the more publicized acute types of diseases such as leafspots, foliage rots, blights and rusts which come quickly to a maximum, devastating effect. In acute diseases the plant tissues die rapidly and the plantation owner applies ameliorative or control measures and he can guide his operations accordingly. With the slower acting, or chronic, troubles, the expert plantation man has a much more serious and insidious problem to circumvent. He has to train himself as a scientifically minded person and a knowing, successful technician, in preventing, ameliorating and controlling the chronic tree diseases.

Prelude to Dieback

A plantation starts as a piece of wild and tree-covered land, often densely grown, extensive and difficult to travel through. Knowledgeable planters understand that the superficially vigorous appearance of the wild bush cover can not be taken as an indication that there is under it either good or bad soil. The primeval tree growth seen is something that has been there, perhaps for millions of years. The soil problem is a highly complex one and it is very important that it be appreciated in the tropics. For soil is the main basis upon which the planter must depend for continued health and productive vigor of his tree crops. A plantation owner's earliest success is in full measure subject to how well he is able to adapt himself and his chosen crop to the soil and the growing conditions of the place he has selected to cultivate (7, 16, 19, 30, 31, 33, 50, 63, 64, 66, 71, 76, 92, 93, 96).

The planting of trees or bushes where they are to grow profitably for years takes much more soil judgment

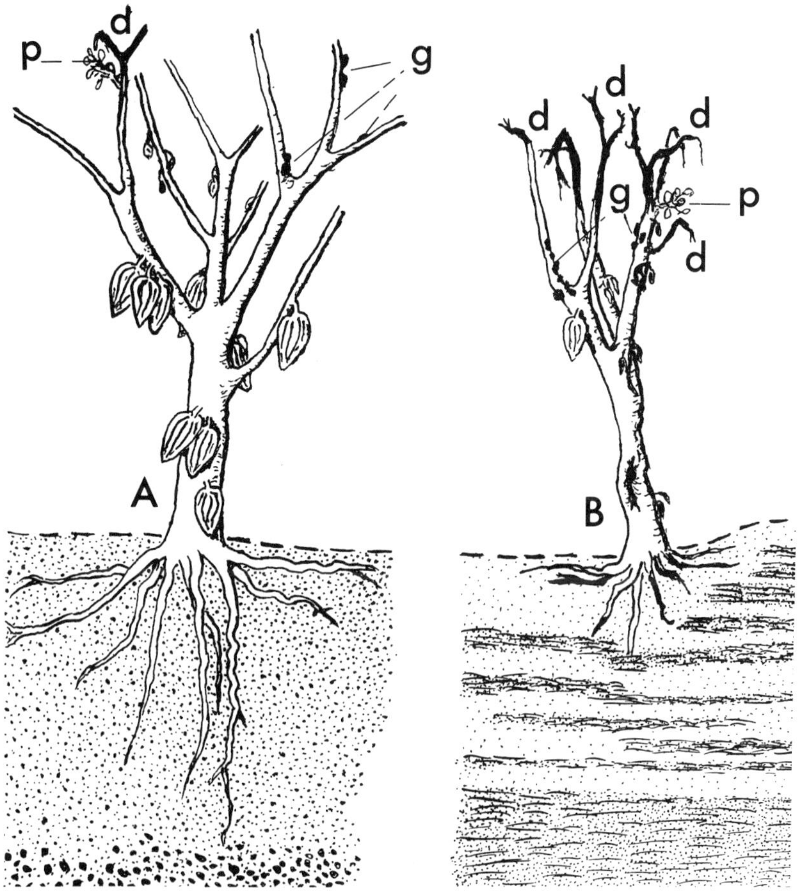

Fig. 8-1. Dieback (d) occurrence on cacao. Field sketches of 12-year-old trees grown 200 m. apart, dug out in road making. Indicated are branch scaffolding, fruits, but no leaves, and soil condition affecting roots: A. soil friable, underlaid with gravel, good root system; dieback (d) on one branch possibly induced by parasite (p); cushion galls (g) present but not killing branches; production is 21 large pods ready for harvest and seven younger and healthy fruits. B. soil thin, underlaid with clay strata, roots shallow growing, weak and some darkened from poor drainage; dieback (d) on every main branch; galls (g) present; also parasite (p); production is one medium-sized pod and nine shrivelled pods that will never grow.

than the planting of maize, beans, rice or other such annual crops. Fields of these can be sown often three or more times a year. The grower can see effects and easily change methods and types of crops to suit what the situation warrants. This is not so simple with the perennial tree crops like coffee, rubber, tea, coconut, rose, or orange; also, there are the more or less permanent herbaceous plantings too, such as sugarcane, banana, plantain, and certain tuberous crops.

The prelude to dieback in crops, is when they draw meager sustenance from fields with short-lived soils (Fig. 8-1). From this knowledge has come a widely recognized principle among planters in the tropics: trees on naturally poor soils are doomed to early chronic infections, unless supplementary fertility is furnished. Soils for the more permanent crops must be good, and enough must be known about managing and protecting them that they are not despoiled in a short period of time. Plantation crops are not replanted at short intervals and therefore have roots that remain for years to occupy and draw nutrients from one location. The nutrients must be in available form at the beginning, but if the plantation is to last, the ground must have in it the potential of bound-nutrients that will be released a little more slowly over long periods. Where this is not the case the tree crops of coffee and cacao, for example, sooner or later show degeneration and become subject to the chronic diseases.

The usual situation is that in which the first diseases are the acute types: rusts, thread blights, rots, leafspots, mildews, wilts and such. These quickly become spectacular. However, on the whole, they may not be nearly so costly in the long run as the chronic troubles that cause slow weakening (6, 15, 31, 33, 38, 39, 57, 66, 88, 95, 96). The acute diseases in many cases are the kinds that yield to chemical control measures, but chronic debility is harder to cure. It is in later years in older plantations that the chronic troubles are more common. Diseases of this kind intensify with passage of seasons and more work is required in connection with curative measures. Methods used for amelioration and for control, become routines established along with husbandry techniques; of course through control applications growing costs become much increased even though diseases are reduced. Along with this the grower develops, invariably, a better understanding and care of a plantation. Those who through the years have watched disease effects know that all plantation crops must be given disease controls if they are to be permanent. Disease amelioration and control is always expensive, but there is a bright side. As he becomes an expert on disease problems the planter inevitably becomes more of an expert on his crop, and he modernizes and changes his horticulture, always for the better. The result is gain, any way it is considered.

Isolations from Dieback

There are numerous kinds of chronic tree diseases, and the most common trouble is the one in which occur dead branches and dead twig ends. Such diseases have not been given as much attention as some others, which is one of the reasons I spent some time on it, so there would be more than guess work to discuss with growers. Diebacks disturb not only the agriculturists who want harvest, but as well the very great body of ornamental growers, professional and amateur, who maintain trees for appearances.

Isolations from dieback (94, 95, 97) were not a main project, but they have been carried on many years. Results have confirmed the belief that many, but not all, residents in dieback twigs are secondary decay organisms. An example of this is the following. In isolations from tissues of coffee dieback twigs, some results of which are indicated in accompanying Fig. 8-2, main fungi recovered have been first the parasitic organisms like *Colletotrichum*, *Diplodia*, and *Fusarium*. These invade the weakened stem tissues, hasten decay, and apparently after that such saprophytes as *Nigrospora*, *Cladosporium*, *Tryblidiella*, and *Aspergillus* gain a foothold.

In 212 isolations from old, "spent" fruiting branches dying from dieback after harvest, all showed presence of fungi. These included species of *Diplodia*, *Pestalotia*, *Botrytis*, *Fusarium* (*roseum*?), a *Phoma*-like isolate, some undeterminable because in the time they were maintained in culture they were non-sporulating fungi, although from half-buried faggots of dead branches kept over a year in moist soil under a greenhouse table, some developed *Xylaria* (*thwaitesii*?) fruiting bodies, which are common in the tropics as is indicated in work by Dennis (18).

Coffee trees occur with very severe decline and dieback. Several such were dug in El Salvador to study the

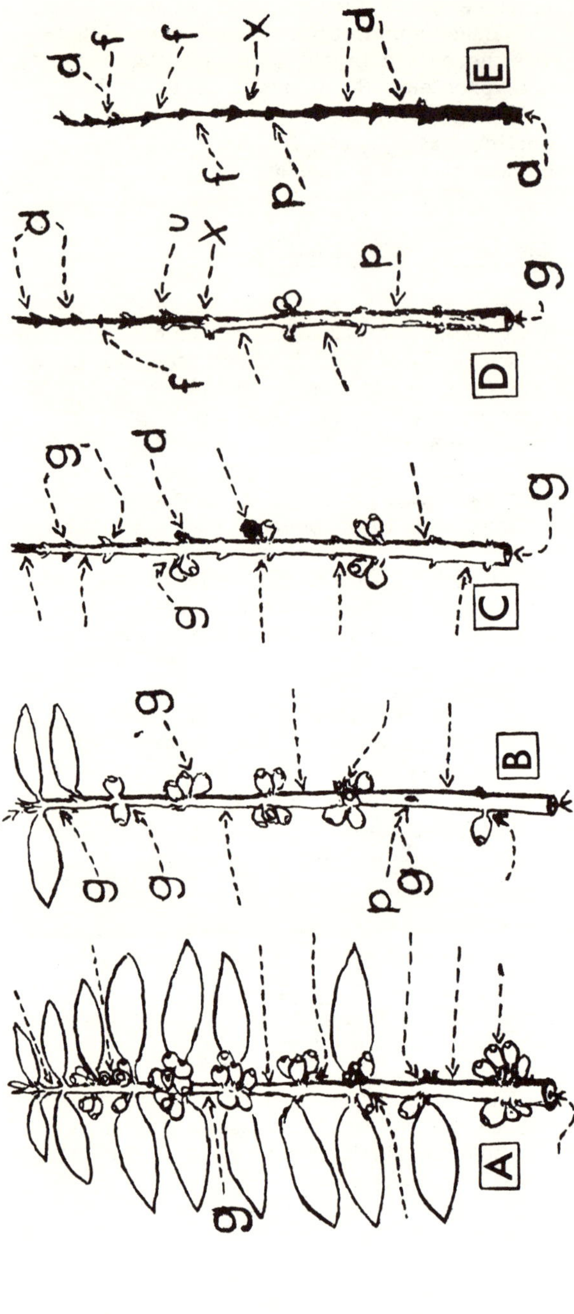

Fig. 8-2. Results secured in Costa Rica by isolations (arrows indicate where internal stem tissues were dissected out and plated on PDA) from fruiting branch dieback of Arabica coffee from an abnormally severely affected field. Increasingly diseased: A. dieback started, B. C. D. intermediate conditions, E. all of branch dead. Where tissue isolations yielded fungus, marked with letters: (d) *Diplodia*, (f) *Fusarium*, (g) *Glomerella* = *Colletotrichum*, (p) *Pestalotia*, (u) undetermined fungus with white mycelium, (x) *Xylaria*-like. Drawings from laboratory-clinic sketches; branches averaged 28 cm. long.

possible effect on roots, and these were proved to be seriously affected by collapse (96, pp. 105-112). Seldom it is possible for pathologists either in the temperate zone or in the tropics to correlate damage to tree tops with corresponding collapse of roots. However, this has been followed in coffee (3, 12, 88, 93, 96). When tree-top damage is severe enough from overbearing and roots are affected because of the lack of nutrients elaborated and received from the tops, the tissues that are subterranean are adversely injured from starvation. Therefore, root deterioration, as indicated by the sketch in Fig. 8-7, is something that in many cases may be wholly dependent upon top degenerescence. I made studies of such root deterioration for a number of years. Much of the samples of collapsed coffee roots, when isolations were secured from stelar tissues, showed no fungi. Bacteria were obtained, but were of saprophytic types. In the year 1945 an old experimental field of coffee was dug up in El Salvador, and samples of diseased roots were secured. From 300 groups of isolations in which there were 10 to 15 isolates in each group, a few fungi were recovered and 108 colonies were kept three months for study. None was a Basidiomycete, and there were no Phycomycetes. Isolates that were finally determined came from shallow roots and were species of *Cladosporium*, *Cylindrocladium*, and *Curvularia*. There were also numerous isolates of a fast growing, fluffy white fungus with single celled, hyaline spores.

A gradual weakening of large specimen trees of *Samanea saman* occurred in the American Virgin Islands and in Puerto Rico. It was found that a *Fusarium* was in their roots and in the trunk sapwood. Trees also showed dieback. Drying branch ends of Saman were collected and isolations made from them. Many saprophytic organisms were recovered, but only a few that might cause disease. Possible parasites were *Physalospora*, *Pestalotia*, and *Glomerella*.

Studies were made on three coffee trees with top weakening, known to be the result of both poor soil conditions and from tissue exhaustion following overbearing. This always ends (96) in severe dieback. Where sufficiently intense, this disease above ground is accompanied by subterranean root collapse. In the particular study here reviewed many cultures were made from the collapsed roots, but no parasitic organisms were recovered. The tops had been sun-exposed and had occasional dark lesions in young bark, typical of the first appearance of dieback. It will be recalled that darkening of diseased tissues is commonly known, see for example work by Idrobo & Naundorf (46) and the review article by Rubin & Artsikhovskaya (73). Tissue discoloration does, at times, extend well beyond where the actual parasite may have reached in a plant. In my study, from 63 newly darkened lesions only three yielded *Colletotrichum*.

Dieback material from *Anona* trees in El Salvador yielded 2 weak parasites: *Colletotrichum* and *Diplodia*, which were then obviously being followed by a saprophyte, *Nigospora*. Collapse of foliage occurred in Puerto Rico at tips of mahogany branches. Nothing that was certainly a parasite could be recovered. However, using moist chamber incubation for two weeks, a small spored Moniliaceous fungus was seen. On growing cultures on PDA the fungus displayed brown mycelium. Isolations from collapsed mahogany twigs yielded two different fungi, neither one producing spores in agar slants, but both having mycelium that displayed clamp connections, evidently Basidiomycetes.

In Costa Rica observations were made in connection with earliest studies of cacao cushion gall (the disease, caused by *Calonectria*, is discussed in Chapter IX) and it appeared that a condition of dieback might be related to it. In Fig. 8-1 is shown a suggestion of some of the occurrences which drew attention to this possibility. For this reason it was necessary to carry on isolation studies of dying branch ends close to and at a distance from galls on long-infected cushion gall trees. Laboratory investigations disclosed nothing consistent. Weakly parasitic fungi recovered were species of *Diplodia*, *Pestalotia*, *Valsa*, *Fusarium*, *Colletotrichum*, and through moist chamber work it was possible to secure a superficially growing fungus (*Acrostalagmus*?), a species of *Aspergillus*, and an organism with no spores but angular rhizomorphs in either liquid or agar cultures. One isolate produced vegetative growth superficially similar to *Armillaria mellea* in culture (69), but which proved nonpathogenic.

In many parts of the tropics a large fig tree is grown for ornament and is known as "laurél." In El Salvador and in Puerto Rico it is realized that it has a useful life of about two decades after which it gradually dies over a period of two or three years. During the terminal aging process, twig dieback from which many weakly parasitic fungi can be

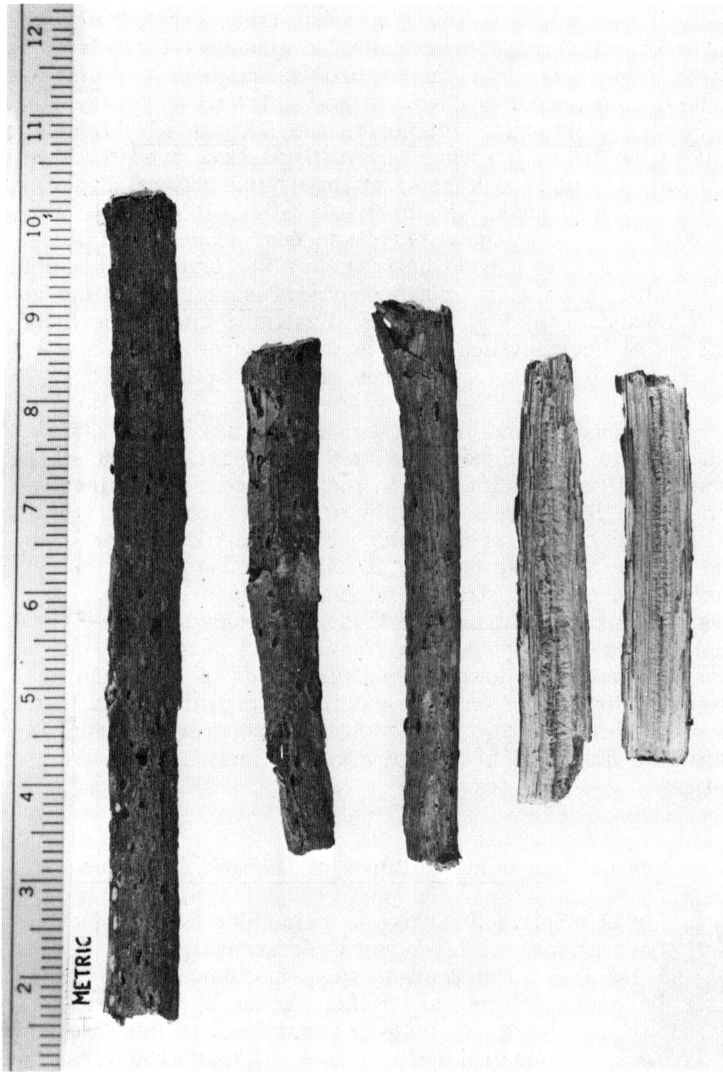

Fig. 8-3. Branch tips of *Bougainvillea* vine affected with dieback; fungus fruiting structures of *Triblidiella* sp. on surfaces of dead stems.

recovered, is the first symptom of the end. The fungus most consistently found on dead branches is usually a species of Nummularia or Hypoxylon sometimes, as Shear mentions (80), difficult to classify. However, there is no invasion by the organism(s) in young vigorous trees growing next to, and even with branches intermingled among, the old branches of old dying trees that display profuse numbers of Nummularia-Hypoxylon fruiting structures. Flowering, woody vines of Congea, Allamanda, Bignonia, and Bougainvillea flourish well in the tropics. When any of these have aged the vine tips begin to show collapse and finally this natural condition results in dead ends. In the West Indies all of these four kinds of vines often have old dead ends invaded by Triblidiella, which has a wide host range (79), and it is possible to isolate Diplodia and a saprophytic Fusarium from dieback in Allamanda. In Fig. 8-3 are illustrated specimens from dieback of bougainvillea, and on the surfaces of the dead branches are fruiting structures of Triblidiella.

When the new dieback twigs from any woody plant are cut into, it will be found that these parts exhibit symptoms of slight discoloration in the sapwood region directly underneath the bark of the twig or the branch itself. If this darkened region is peeled back by scraping the upper bark and cutting away the cortex, it will be found that in many cases dark colored strands extend into wood bundles. A first judgment might be that these are from invasion by Diplodia which is very often common and in addition has black mycelium. However, these strands of stain are not always indicative of any one particular organism. In fact, in many cases the organisms that are recoverable from the darkened vasculars are fungi with relatively colorless mycelium.

Inoculations Using Cultures of Dieback Parasites

In attempting to arrive at something more objective than observations and mycological determinations on miscellaneous isolated dieback organisms, inoculation studies were instituted in greenhouse and field. In these, the results were so irregular as to belie observational isolated evidence of common pathological occurrences. It was finally realized that under the existing conditions, more depended on the vigor of tree growth than on almost any other consideration.

Isolations of Fusarium for example were divided into

two groups: those isolated from the bark and cortex, and those isolated from the dead but once sappy inner wood, of dieback twigs. This group, nearly all of them, are large spored, with fast growing mycelium. The deeper wood invaders mostly are quite different. In them spores are smaller, the colonies do not grow rapidly on acidified nutrient agar, and the odor of the culture is commonly like some kind of an alcoholic compound mixed with a musty smell. Inoculations, using organisms from bark and cortex, were never successful in causing infections in *Inga* and *Erythrina* twigs. Companion twigs were inoculated with *Fusarium* isolates from inner wood, and out of 110 isolations, 27 caused local infection.

Many isolates of typical *Glomerella cingulata* were obtained from trees having dieback. These isolations were, of course, from the *Colletotrichum* stage. When they were used to inoculate vigorously growing *Inga* or *Erythrina* twigs, there was never any infection. However, infection occurred when the twigs were on an old *Inga laurina* tree that had been partly uprooted.

Attempts were made, using miscellaneous isolations, to infect old side-fruiting branches of arabica coffee. The terminal young portions of old branches were cut away, and at the tip of what was left a split was made and a bit of the pure agar culture inserted (and bound) in the split. Out of 52 tests with pure cultures of *Cytospora* sp. ? (a brownish *Valsa*-like growth on PDA), infection was positive in one case. In this case re-isolations from the newly diseased twig were of the *Cytospora*. This indicated it had successfully invaded the old coffee branch. It was common to secure infections in coffee using *Glomerella cingulata* material and the anthracnose organism was easily re-isolated. From 50 cultures of a small spored *Fusarium* from dieback of diseased jungle trees, 3 isolates infected old coffee fruit branches from which cherries had been harvested the year before.

A significant difference between the woodstain disease and the twig dieback trouble, is that the former is so much more invasive of trunks and is a potent factor in the infection proceeding downwards eventually and often well into the roots. This is true in either coffee or cacao but is apparently more rapid in its progress in the latter; in all hosts dieback of twigs and tops progresses the most rapidly outwards towards the tip of branches. There is invasion of wood in a direction back towards the tree trunk, but with dieback fungi this is slow in occurrence. The most rapid of the effects from dieback is deliberate enough that leaves have time to go through the process of dehiscence. Leaf drop and death is brought on by poisons (for example, see such studies as 26, 32, 46, 73) that exert toxic effects from organisms living in the stem tissues, on which the leaves are borne. It should be re-emphasized here, that a common character found in large numbers, in fact in most of the slow dieback and woodstain diseases, is the discoloration of the host's wood elements.

Chemical changes in cellulose and wood components made by fungus infections from Basidiomycetes have been given more study than the chemistry of darkening tissues that result in chronic tree diseases. Little, or nothing, has been done in the Neotropics on tissue discolorations that are often characteristic in dieback. A review of some suggestive findings in the United States, Russia, Japan and other parts of the temperate zone is given by Rubin and Artsikhovskaya (73). It seems fairly certain that in our tropics in the dark decays of dieback there is oxidation along with an accumulation of at least phenolic compounds, tannins, anthocyanins, and concentrations of carotinoids and chlorogenic acids.

Along with the oxidative processes in invaded cells there develop chlorogenic acids. These are substances readily induced in plants and under some conditions such an acid may be a material that by reacting in infected tissue aids in disease resistance. It is one of the protective compounds that is in plants before diseases enter. However, this chlorogenic acid does not exist in very large quantities in a great many cases and so the organism that is parasitizing can enter and start its decomposition work well. After they have been invaded, certain plants manufacture and lay down a great many different compounds and among them chlorogenic acid is perhaps increased. The presence of such an acid, in itself, results in a considerable amount of discoloration of tissues where it is concentrated.

In addition to the tree crops mentioned, there are several other plantation and tropical subsistence crops diseases by fungi of the kind making diseased tissues black and that cause woodstain and such in trees (5, 6, 8, 13, 14, 17, 20, 22, 25, 36, 42, 45, 48, 50, 52, 54, 56, 57, 59, 62, 68). The herbaceous crop plants most commonly recognized as susceptible are: sugarcane, sweetpotato, crotalaria and pineapple. Symptoms of disease very similar to the typical so-called black rot, Ceratocystis fimbriata, in roots and stems of sweetpotato have been seen in stems and tubers of the ñame (Dioscorea), and in roots of yautía (Xanthosoma).

There appear to be a number of undetermined kinds of tropical American timber trees that when made into boards show woodstain effects. The stained wood is strong and the stain is considered a natural characteristic; such wood is given special names and sometimes carefully selected and sold at a high price as decorative material. The stained lumber, with an accenting of already handsome graining, is employed by artists for carving, and in making fancy boxes and fine cabinets.

Indian baskets from South America have been seen, in fact purchased as unique and of special beauty, which were made by weaving palm wood strips having handsomely stained strands running in the material. In the West Indies, peasants and artists weave mats and containers from leaves of sabutan (Pandanus) that in some cases are beautifully marked with stained vascular strands; probably, as we know from Philippine work (72), this marking is from infections by the same fungus, Ceratocystis sp., that causes woodstains. Fancy Arundinaria walking canes are cut, with specially attractive staining patterns, which find their way into the hands of prosperous men in the temperate zone, and the mouth pieces of some wood-wind instruments made from Arundinaria "reed" are said to have special tone quality inherent in the degree of wood tissue staining they have acquired.

Woodstain Diseases

There is a large group of chronic, that is to say slow-acting, diseases, that eventually kill. The effect is more systemic than in dieback, and is seen in main branch and trunk attack. These are the woodstain diseases, already mentioned. They became of serious concern to me in experiment stations where work was being carried on with tree crops. In addition many planters require answers to critical disease problems of woodstain, most generally in plantations of cacao, eucalyptus, hevea rubber, and coffee. As the common name of the disease suggests, the most easily seen symptom, of clearest diagnostic value, is

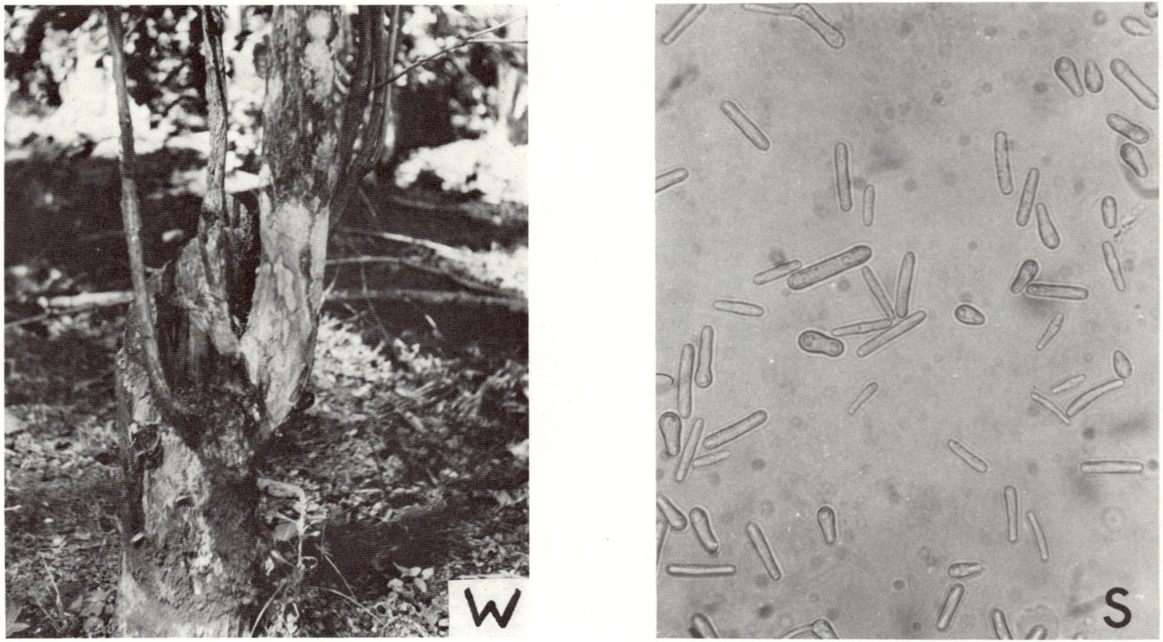

Fig. 8-4. W: Woodstain disease killing once vigorous coffee tree; S: Spores of the causal organism, Ceratocystis fimbriata (coffee race), from photomicrograph. Pictures courtesy of E. Echandi, University of Costa Rica, San Pedro del Monte de Oca, San Jose, Costa Rica.

the presence of stained wood graining in branches and trunks of diseased trees. External symptoms which draw attention to the inner condition are leaf chlorosis and defoliation of tree branches and their wilting and then death, loosening of bark, and finally the succumbing of the whole tree. (This is illustrated in Fig. 8-4).

In citrus (see Fig. 8-5) there occur chronic troubles and sometimes they may be due to such a parasite as <u>Fomes</u> rather than a woodstain fungus. It is probable that nearly all tree crops eventually may be found with this kind of disease. As years of growing continue, chronic conditions develop even where original plantings were known to produce well, with acceptably healthy growth in them. In some instances certain individuals or varieties show resistance (20, 44). This the intelligent grower appreciates and it stimulates his interest in the possibilities of new and improved types of his trees. On learning of different varieties he often brings in examples for his own private and limited tests. If they look good the few trees are increased to a thousand or more. After some growth, it is about this time when the pathologist is asked to come to advise about two or three diseased specimens. These have not actually frightened the grower and his foreman, because tree death is usually slow and chronic disease requires years to become serious. Even if eradication that is advised is carried out, the woodstain disease becomes established and in a few seasons trees with the killing woodstain disease are found scattered in many parts of the holdings; it becomes a permanent part of the plantation.

Limitations as to Field Location and Woodstain

It is significant, and it is moreover a dependable principle in tropical pathology, that woodstain diseases in the field may be limited in extent by ecological factors. The limitations are not on a simple temperature basis, as temperature range is wide at which the causal organisms themselves grow in pure culture. For example the usual strains of <u>Ceratocystis</u> that attack sweetpotato, and cause woodstain disease in trees, in pure cultures grow from about 5° to 34° C. It is well known that these organisms withstand freezing but are killed at 35°, the latter a very high temperature for most fungi to withstand for a long time. The apparent optimum for these woodstaining organisms in pure culture is 25° to 27° C; temperatures quite

Fig. 8-5. Fomes applanatus causes concentric canker, an occasional chronic disease and finally death of aging citrus trees. Photograph courtesy of F. E. Fischer, Department of Plant Pathology, University of Florida, Gainesville.

commonly present in the warmer parts of the tropics and not very often exceeded in tree wood and in the soil.

It is of interest to note from my careful observations and from detailed studies by Echandi and Segall (27), that the woodstain and canker of coffee, caused by Ceratocystis fimbriata coffee race, developed on hillsides between the isotherms of about 18° and 23° C. No stain troubles were found in the colder much higher-grown coffee trees. This was followed through quite carefully and it was also seen that below the 24° C isotherm the disease did not occur. However, the stain disease was killing and causing much destruction in plantation coffee only slightly above, and in therefore cooler area. This is a comparatively narrow band for a disease occurrence, and it was of great interest that the disease lines seem quite definite and the effects sharply limited, from infections by C. fimbriata coffee race. It was found in these first experiments that there appeared to be a good reason to refer to race differences in Ceratocystis species. More careful mycological investigations on this are in order, but for convenience at least race designations seem not out of place.

In cacao the woodstain organism, Ceratocystis fimbriata cacao race, although found in many countries is of special interest where it occurs in the littoral of Ecuador. This region is bathed by the cool Humboldt current and the temperatures are generally evenly maintained at around 22-26° C. Thus it is not a hot climate, although it is athwart the geographical equator, and field men with light tasks often work comfortably wearing a coat or jacket rather than in shirt-sleeves. In this area also it is of interest that there is very little strong sunlight because the clouds form a haze and cut out much of the sun's rays, except for a short while of less than half an hour after sunup and a short period before sundown.

In cacao the Ceratocystis woodstain disease in its chronic form causes tree wilt (47-49) is distributed by the Xyleborus beetle (10, 35, 37, 41, 43, 67, 77, 81) and areas where this insect is able to breed and work are the apparent ecological confines in which the disease occurs. The slow acting woodstain disease in Arabica coffee seems to be carried by insects (27), and several kinds of flies and moths may be involved. In the case of rubber tapping panel disease from Ceratocystis, flies (Diptera) have been observed by myself, feeding on and crawling over, spore exudations and then visiting newly bled rubber tree panels nearby. Moreover, the fungus of the tapping panel disease was isolated in my laboratory from the feet of flies caught on tapping panels.

In Hevea rubber we have an interesting, slightly different situation with respect to occurrence of disease from infections by Ceratocystis. In the rubber tree, it is well known that the product of commerce is from the latex which the tree exudes in its sap. This sap comes from the latex-bearing part of the cortex immediately under the corky bark. The portion of the tree that is cut by special knives and used for securing the latex, is known as the "panel," and the panel has a good many fungus diseases that attack it. One of the important fungi, Ceratocystis fimbriata rubber race, that infects it causes a black injury known as panel disease. At the higher point in altitude of the warmer belt where Hevea trees grow, just above the foothills in the tropics, one may find this rubber panel disease. Here it is warm but not hot, about 24° C. From up here the occurrence of panel disease extends down to the flood plain and sea level, and temperatures down there are from 29 to 30° C. It has not been studied long enough so that we are certain about its absolute limits of occurrence with respect to temperatures, altitude and rainfall in the tropics. However, enough is known to realize that the isotherm of disease occurrence reported for woodstain in coffee caused by C. fimbriata coffee race is distinctly different from the rubber panel disease from C. fimbriata rubber race. It seems certain that ecological differences, even as wide as occur between the coffee and the rubber diseases, will prove to be related to the mode of field distribution.

It has been found (62, 99) that Ceratocystis fimbriata will cause infection of carrot roots, and such roots can be used in testing for its presence in questionable diseased tissues. The C. fimbriata which was isolated from sweet-potato black rot was not able to infect cacao under experimental conditions in Costa Rica. In Puerto Rico and other parts of the West Indies, C. adiposa (Butler) Moreau is a different species that is the causal organism of the so-called pineapple disease or black rot of sugarcane. This disease, in Puerto Rico, is most severe in the cooler months of the year, and is more of a problem in hillside fields where temperatures are always less than in the warm areas near sea level.

Ceratocystis Organisms

The reader has seen above that there are a number of woodstain disease-producing fungi that cause chronic ailments of trees. Some have been discussed briefly but for fuller details it is necessary to go to the works of many authors. In fact there is enough published material on the subject to fill a small book. The woodstain types are mainly from fungi in one genus, although there are numerous other types of organisms that are primary causes of chronic disease.

It seemed convenient to present here a number of the most commonly encountered *Ceratocystis* disease organism types that are encountered in the Neotropics. Information on these is presented here only in outline form, and is based largely on the classic memoir of the late J. Hunt (42). The order of listing the species but not the same numbering, is as arranged by Dr. Hunt.

1. *Ceratocystis adiposa* (Butler) C. Moreau (Syn.: *Sphaeronema adiposum* Butl.; *Ceratostomella adiposa* (Butl.) Sart.; *Ophiostoma adiposum* (Butl.) Nannf.; and *Endoconidiophora adiposa* (Butl.) Davids.). It is the cause of sugarcane black rot in the Caribbean and West Indies. The conidial stage is endoconidial and these conidia are halfmoon or hat-shaped. Ascospores are halfmoon-shaped. Perithecia have long necks, and develop superficially with ornamented black globular base.

2. *Ceratocystis fimbriata* Ell. & Halst. (Syn: *Sphaeronema fimbriatum* (Ell. & Halst. Sacc.; *Ceratostomella fimbriata* (Ell. & Halst.) Elliott; *Ophiostoma fimbriatum* (Ell. & Halst.) Nannf.; and *Endoconidiophora fimbriata* (Ell. & Halst.) Davids.). See also the description by Elliott (29). This disease is widespread, pan-tropical, and I believe certain hosts have races that are associated and probably especially adapted to them, which for convenience may be designated as "races," examples being *Ceratocystis fimbriata* coffee race (coffee race or strain also causes canker and tree collapse), *C. fimbriata* cacao race (besides woodstain it causes tree death), *C. fimbriata* yautia race (black root and stem rot of *Xanthosoma*), *C. fimbriata* hevea race (a severe tapping panel disease characterized by rot and discoloration of tapping panels of *Hevea* rubber tree), and *C. fimbriata* ñame race (black tuber and black stem of *Dioscorea*). These parasites cause some of the worst of the woodstain and black

rot diseases. They occur throughout the Neotropics, but as was pointed out a few paragraphs above, in some places certain of the diseases appear confined to occurrence within rather specific ecological limitations. Diseases from some of the races of this one species are found on susceptible crops in the Oriental tropics and in the temperate zone. The ascospores are of the typical so-called hat shape. Perithecia have long necks, and the brown to black globular bases are found superficial as well as immersed in host tissues.

3. Ceratocystis paradoxa (Dade) C. Moreau (Syn: Ceratostomella paradoxa Dade; Ophiostoma paradoxum (Dade) Nannf.; Endoconidiophora paradoxa (De Seynes) Davids.; and Sporoschisma paradoxum De Seynes. Also this is the perfect stage of Thielaviopsis paradoxa (De Seynes) Von Hoehn). It is pan-tropical on various hosts, including old sugarcane, husks of cacao and coconut, and found in subtropical regions. Ascospores have unequally curved sides. Perithecia are partly to completely immersed and are ornamented and with long necks that are comparatively stout in structure.

4. Ceratocystis radicicola (Bliss) C. Moreau (Syn: Ceratostomella radicicola Bliss and Ophiostoma radicicola (Bliss) von Arx). It causes root disease, and in consequence wilt, of date palm in southern California and northern Mexico. It has elongate-ellipsoidal shaped ascospores, and is also characterized by the long, slender-necked perithecia that have globe-shaped bases partly to completely immersed in host tissue.

5. Ceratocystis obscura (Davids.) Hunt (Syn: Ceratostomella obscura Davids. and Ophiostoma obscura (Davids.) von Arx). This causes stain of pine logs in southern U.S.A. and in Central America. The ascospores are hat-shaped to quadrangular, while the perithecia are short necked.

6. Ceratocystis ips (Rumb.) C. Moreau (Syn: Ceratostomella ips (Rumb.) Nannf. and Grosmannia ips (Rumb.) Goid.). Its ascospores are hat- to quadrangular-shaped, from long-necked perithecia. The organism causes severe pine log stains in southern and subtropical U.S.A. and probably in Mexico and Central America.

7. Ceratocystis pluriannulata (Hedgc.) C. Moreau (Syn: Ceratostomella pluriannulata Hedgc. and Ophiostoma pluriannulata (Hedgc.) H. & P. Sydow). It causes light stain of oak in southern U.S.A. I have seen a tapping panel

of Hevea rubber that was streaked and stained with a Ceratocystis that answers closely to the description of this species. Perithecial necks were very long with several whorls of fimbriate annuli, and there were globular bases, which were superficial, unornamented, and dark-colored. Ascospores that were exuded were kidney-shaped.

The worker, or team of specialists, which studies the tropical species of Ceratocystis is going to have to make his own decisions as to whether all the techniques now in use by research people in the temperate zone will satisfy his requirements. The standard types of agar gave me difficulties with C. fimbriata races until I made the agar more solid than common PDA and in some cases fortified it with maize meal. It was also difficult to maintain and study some Ceratocystis cultures without frequent re-isolation from diseased material. Growing of woodstain organisms on sterilized leaf petioles and branch tissues gave more long-lived cultures than did refined agars. C. fimbriata grew on autoclaved twigs of coffee, mango, cacao, avocado and crotalaria. Preliminary work indicated considerably better growth of the presumed cacao-adapted race on cacao twigs than on coffee twigs, and the rubber isolate grew better on sterilized Hevea tree twigs than on twigs of coffee, cacao, or crotalaria.

Use of some of the roots or tubers from tropical sources will being immediate and interesting results. As is known, Yarwood (99) was the first to use carrot discs for isolations from soil. This has been of great convenience. This technique, with adaptations (Moller and DeVay (62)) is a standard system in connection with diagnoses made in the plant clinic at the North Carolina State University. There are numerous different tropical roots and tubers that when inoculated will certainly give results never believed possible with the present systems in the temperate zone.

Dissemination in Woodstain and Like Disease

Insects and diseases, with reason, are traditionally linked together in the study of Neotropical plantation problems. It becomes significant in the researches of the chronic tree diseases, that planters have by themselves come to the conclusion that cankers, woodstain, and black rot diseases can readily "come out of the soil." Knowing that the genus Ceratocystis has had its major impact as a parasitic genus, it becomes of special interest to learn that there are many non-parasitic forms of it, and that it is a member of a family noted for its adaptability to saprophytism (42, 97). Spores of these Ceratocystis saprophytes appear to remain long viable in the soil, or in dead and decaying plant tissues. There is no reason to doubt that the same is true of species members of Ceratocystis, and observations as well as collections made indicate the probability that the genus is as much saprophytic as parasitic.

It is apparent that conidia of such chronic disease-producing fungi as have been diagnosed as both Thielaviopsis basicola and C. paradoxa (11, 13, 23, 33, 35, 39, 58, 85-87, 90, 99) remain alive in soil and/or decaying plant tissue in the soil, for it has been seen many times that citrus plants, cacao cuttings, sweetpotato and ñame tubers, and sugarcane setts from undoubtedly disease-free plants contract infection from the chronic-type fungi when planted in old fields sometimes fallow for more than a year. In one sugarcane field that the planter insisted was of virgin soil, I found over 2 per cent of plants of first ratoon infected with C. paradoxa. Contaminated cane knives (machetes), are also a potent means of spreading the disease. This is a common experience with C. fimbriata on coffee and on cacao.

It is significant to note that with Ceratocystis paradoxa the "seed pieces" of sugarcane taken at the tips are quickly infected as well as those from the basal parts. One of the features that very much strikes the person who comes down to the tropics from the temperate zone, is that when this disease is watched for year after year, it will be found that it is not so common in the wetter parts of sugarcane regions. Why this is so is something that has not been determined but it may be in connection with insect relationships. In any case if the sugarcane is grown in dryer parts the organism can cause a great deal of loss from infections that begin at the base in plants from setts. There is an increasing need to understand the epidemiology of these chronic disease organisms. In a few cases (27, 28, 37, 49, 67) some background has been established. However, more has to be known of spore dispersal in fields. This is seldom attempted in the tropics with these blackening, chronic, disease organisms; although Meredith (61) worked out methods and made successful studies on air-borne spores of Botryodiplodia and Nigrospora in banana plantations.

In the case of Ceratocystis fimbriata sweetpotato race, it is well known (33, 35) that much of the time the disease resides in the soil. The spores also spread readily on roots that are contaminated externally, handled and kept under storage conditions. In the Neotropics where I made some studies of the sweetpotato black rot, there is no formation of the perfect stage of the parasite. It is therefore safe to judge that in the tropics in spite of extremes of bacterial decay and other natural antibiotic activities of one sort or another, C. fimbriata conidia encounter little difficulty in surviving in sufficient numbers to carry on parasitically.

The phenomena are of a different sort, when it is the ascospore that must be relied upon for distribution and dissemination to keep the species alive. A review of the literature such as that in Hunt (42) Elliott (29) and Galli et al. (35) shows that the ascigerous stage of Ceratocystis consists of ascospores produced in evanescent asci in the globular basal part of a perithecium, with an opening or ostiole, at its apex. Most often there is, in the nature of things, a neck to the globular perithecium. Ascospores are released from the thin-walled asci in the bottom of the perithecial globe, and are surrounded with a semi-liquid matrix. This, with spores in it, develops pressure inside the perithecium and is exuded through the natural opening. There it is seen as a cloudy-colored drop full of ascospores.

A point should be made here about one of the control measures that is most often used in connection with these diseases from the woodstain organisms, and, as well, others of similar chronic action such as Botryosphaeria (34, 35), Diaporthe-Melanose (31, 74, 98), Diplodia (1, 32, 61, 75, 84, 94, 97), Glomerella (2, 24, 31, 66), Physalospora (28, 53, 82, 83, 91), Rosellinia (7, 16, 30, 31, 66, 70, 76, 89), Sphaceloma-Elsinoe (31, 33, 51, 66), Sphaeropsis (31, 35, 40), Thielaviopsis (13, 52, 85, 99), and Xylaria (31, 35, 40). Some general discussions of such troubles (4, 35, 66, 68, 78, 86-88, 90, 92) are continuously appearing and in some cases specific means for control of chronic diseases have been recommended (4, 20, 21, 23, 31, 33, 35, 41, 44, 63, 66, 75, 88, 96). In some of my work I have dealt with growers of many kinds asking searching questions. Planters are often advised to carry out careful pruning of the diseased trees, cutting back as far as the operator possible can so that he is going to take all of the affected wood. What has been cut off is heaped and these prunings are buried (and rapidly decay). Some people have said that the only way to dispose of such infective prunings is to burn these heaps of diseased stems and branches, but in some parts of the tropics burial may be

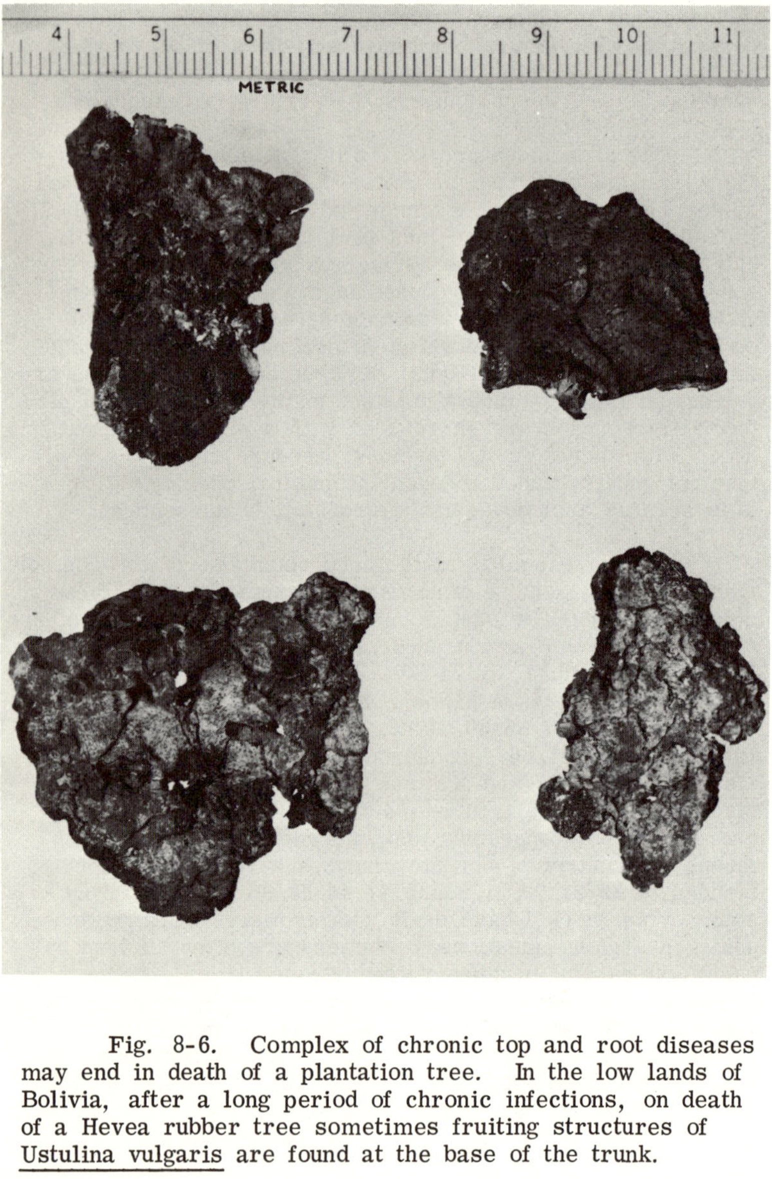

Fig. 8-6. Complex of chronic top and root diseases may end in death of a plantation tree. In the low lands of Bolivia, after a long period of chronic infections, on death of a Hevea rubber tree sometimes fruiting structures of Ustulina vulgaris are found at the base of the trunk.

the only practical way of disposal, moisture sometimes being a serious limitation to burning.

Chronic Woodstain Effect Enhanced by Dieback

There are secondary organisms such as *Colletotrichum* spp. that add to the severe effects following primary invasions from woodstain fungi (*Ceratocystis* spp.). The *Nummularia*-*Hypoxylon* complex apparently does not invade until weeks or months after the woodstain fungi have diseased the tree branches. *Valsa* as an associate organism has been noted in work of Wellman (94, 95, 97) especially in woodstain isolations from ficus, rubber, coffee, tea, cacao, and crotalaria. In this same relation a *Diaporthe*-like organism was recovered as well as *Botryosphaeria* and *Daldinia*. It has also been found that one of the common invaders is a *Chaetomium*-like organism. Various genera and species of Basidiomycetes develop in sapwood, apparently following invasion by bark beetles and with them *Ceratocystis* infection. I observed cases of coffee woodstain in which primary sapwood invasion was followed by development of species of *Ustulina* and *Lenzites*. These came in after there had been killing from *Ceratocystis*. In a few instances *Ustulina* (in Fig. 8-6, fruiting structures are shown of *Ustulina* from severe dieback effects on Hevea rubber trees) and *Lenzites* appeared to follow attack of tree bases by *Polyporus* sp. The *Ustulina* and *Lenzites* secondary decays are dark with distinct irregular bands of black in the rotted wood, and are the results of successive secondary, tertiary and even quaternary seasonal waves of invasions which are common in the tropics.

In numerous instances, the tropical phytopathologist is confronted by a major challenge to identify the main cause of slowly-acting, but eventually sure, chronic failures in diseased plantation trees. The woodstain symptom, if seen, can give an acceptable temporary answer to a planter's query. When clear-cut, the stain shows up well. However, the researcher knows that this is only a partial answer, because both woodstain and many diebacks in the final stages consist of a complex. The pathologist, who is trying to help the troubled planter, must search for a recognizable earlier disease condition in a tree, amenable to amelioration, so that his client may attempt to re-invigorate his trees to "throw off" the otherwise surely fatal chronic diseases.

Fig. 8-7. Symptoms of tree root disease (example Rosellinia bunodes on arabica coffee). Infection may require a few years of increasing penetration before it causes death of a plantation tree, and during the pre-collapse period on such a tree a chronic top disease is more severe than if the roots were not infected.

Table of chronic diseases

Table 13. Comparisons of certain relatively slow-acting chronic diseases affecting crops growing in the tropics of the Western Hemisphere

Host	Organism(s)	Disease	Control	Country[a]
Ananas comosus	*Ceratocystis paradoxa* (pineapple race[b])	Fruit rot, leaf base disease	Field hygiene, rogueing and rotation	Puerto Rico and subtropical U. S. A.
Anona cherimolia	*Diplodia*[c] *Colletotrichum* complex	Twig and small branch dieback, fruit charcoal rot	Clean-up pruning, field hygiene, spraying	Guatemala, El Salvador
Cassia grandis[d]	*Cytospora Diplodia*[c] complex	Dieback at ends of secondary branches	--	Costa Rica
Citrus sinensis	*Sphaeropsis tumefaciens*	Branch swelling and knot, terminal weakening	Pruning, hygiene	Subtropical U. S. A.
Citrus limon	*Botryosphaeria ribis*	Branch canker, gomosis, and fruit stem-end rot	Pruning with improved fertilization	Nicaragua, Puerto Rico, El Salvador

Host	Organism(s)	Disease	Control	Country[a]
Coffea arabica	*Ceratocystis fimbriata* (coffee race[b])	Woodstain in main branches and trunk, slow death of tree where sun grown	Careful pruning, hygiene, added fertilization, shade regulation, move field to new location	Costa Rica
Coffea canephora[e]	*Physalospora rhodina*[c]	Dieback on spent fruiting branches in sungrown trees	Pruning and heavy fertilization with shade regulation	Costa Rica
Ficus carica	*Ceratocystis fimbriata*	Woodstain, wilt of branches, weak tree	Pruning, added fertilization	El Salvador
Ficus nitida[f]	*Ceratocystis fimbriata* followed with *Nummularia-Hypoxylon*	Woodstain, old trees dying, dead sapwood invaded with *Nummularia-Hypoxylon*	Pruning, replacement of old with young trees	Puerto Rico
Hevea brasiliensis	*Ceratocystis fimbriata* (rubber race[b])	Black rot and decay of tapping panel	Field hygiene, possible sprays to repel insects	Costa Rica
Inga vera	*Pestalotia* sp.	Twig dieback	Pruning	El Salvador Puerto Rico

Saccharum officinarum	*Ceratocystis paradoxa* (sugarcane race[b])	Rot of sett (seedpiece)	Using setts from healthy and vigorous "seed" field	Puerto Rico, El Salvador
Theobroma cacao	*Ceratocystis fimbriata* (cacao race[b])	Woodstain, causing weakness, wilt and tree death	Resistant varieties	Ecuador
Theobroma cacao	*Diplodia theobromae*[c]	Severe dieback of twigs and branch ends	Pruning, regulating shade	Ecuador,

---

a) Where I gave the disease some special attention.

b) In my experience there appeared to be special adaptations of *Ceratocystis* species. In determinations and recommendations these were designated in preliminary fashion as "races" depending upon the host involved.

c) In by far the majority of cases isolations of *Diplodia* remained the imperfect stage. In the case of isolations from *Coffea canephora* in 17 cultures out of 42 the perfect stage *Physalospora rhodina* was developed.

d) Jungle tree, near plantation of sungrown coffee with severe dieback.

e) Robusta strain. For botanical classification see Wellman (96).

f) Tree grown for ornament.

It is imperative to make certain if roots are in healthy condition, and not attacked by nematodes, or exhibit symptoms of attack (Fig. 8-7) by such a root disease fungus as Rosellinia, or growing in a poor soil or in an inadequately drained piece of land. It is in such situations that tree growths, even in trees planted to replace the originally chronically-diseased plantation, are weakened and more rapidly than ever they succumb to the chronic woodstain and dieback complex. An inherently weak top growth cannot help but be on a declining root system, and it is often the case that horticultural practices such as soil protection through use of "green manures," mulching, pruning, fertilizing, and proper shading are introduced to make a more vigorous tree crop. These on analysis are as much pathological as horticultural treatments, and their understanding helps in a full realization of the nature of chronic tree declines.

The truth is that infections by the dieback organisms which are classed as producers of primary tree decline or dieback, would not go far without the inevitable accompanying occurrence of secondary complications after primary infections. For orientation purposes I classified as primary those infective agents bringing on in their wake chronic decline. These are species of such fungi as Ceratocystis, Glomerella (Colletotrichum races of high pathological potency), Physalospora (specially pathogenic Diplodia types), foliage infecting Phytophthora, certain rare twig-blight-causing Fusarium spp., two aerial Corticiums, and occasional cord blights from Marasmoid fungi.

Literature Cited

1. Alicbusan, R. V. and L. A. Schafer. 1952-1959. Diplodia rot of mango. Philippine Agr. 42:319-322.

2. Alvarez-G., L. A. 1940. Die-back of cassava caused by Gloeosporium manihotis. Agr. Exp. Sta. Puerto Rico An. Rpt. 1939-40:42-43.

3. ________. 1952. Estudios de enfermedades de la raîz del cafeto en Puerto Rico. Rev. Café (Puerto Rico) 8:5-6.

4. Amos, E. R. and J. A. Barañao. 1946. Resumen de

las observaciones sobre cancrosis, en distintas variedades de álamos. An. Rural Prov. Buenos Aires 14:297-307.

5. Anon. (Trinidad). 1958. Ceratostomella canker causing death of cacao trees. FAO Pl. Prot. Bull. 6:189.

6. Arbelaez G., E. 1957. La llaga macana del tronco del cacao. Acta. Agron. (Colombia) 7:71-103.

7. Averna-Sacca, R. 1926. Algumas molestias cryptogámicas novas do systema radicular do cafeeiro. Sec. Agr. Com. Obras Publico, Brasil Publ. 17, 12 p.

8. Barba D., C. and A. J. Hansen. 1962. Un hongo semejante a Ceratocystis moniliformis en cacao en Costa Rica. Turrialba 12:46-47.

9. Bateman, D. F. and A. W. Dimock. 1959. The influence of temperature on root rots of poinsettia caused by Thielaviopsis basicola, Rhizoctonia solani, and Pythium ultimum. Phytopathology 49:641-647.

10. Benavides G., M. 1955. Insectos pasadores del tronco del cacao. Cacao en Colombia 4:43-45.

11. Berkeley, M. J. and M. A. Curtis. 1869. Fungi cubensis. J. Linn. Soc. 10:280-392.

12. Bianchini, C. L. 1955. Las llagas del café en Costa Rica. Biol. Trop. (Costa Rica) 3:203-235.

13. Casas-C., C. 1951. La mancha negra de la piña originada por Thielaviopsis paradoxa. Mexico Ofic. Est. Esp., Fol. 4:335-342.

14. Chevaugeon, J. 1957. Ceratocystis fimbriata Ellis et Halstead. Rev. Myc. Paris (France) 22, Suppl. colon 2:45-60.

15. Childs, J. F. L. and E. West. 1953. Butt rot of queen palms in Florida associated with Ganoderma sulcatum. Plant Dis. Reptr. 37:632-633.

16. Choussy, F. and J. C. T. Uphof. 1938. Las enfermedades de la raíz en los cafetales. La podredumbre negra y la podredumbre blanca de las raíces. Asoc. Cafetaleros El Salvador, Publ., 32 p.

17. Costa, A. S. and H. P. Krug. 1935. Eine durch Ceratostomella hervorgerufene Welkekrankheit der Crotalaria juncea in Brasilian. Phytopathol. Zeit. 8:507-513.

18. Dennis, R. W. G. 1959. Further notes on tropical American Xylariaceae. Kew Bulletin 1957:297-332.

19. Dennis, R. W. G. and D. A. Reid. 1957. Some Marasmoid fungi allegedly parasitic on leaves and twigs in the tropics. Kew Bulletin 1957:287-392.

20. Desrosiers, R. 1956. Diferenciación entre variedades de cacao con base a su susceptibilidad a la infección con Ceratostomella fimbriata (E. & H.) Elliot, en el Ecuador. Turrialba 6:48-52.

21. ________. 1956. Progresos en el combate de la escoba de bruja, Monilia y Ceratostomella del cacao. Cacao 3:11.

22. ________. 1960. El problema de la Ceratostomella en el Ecuador. Inter-Amer. Cacao Conf. (Colombia) 7:67-70.

23. ________. and A. von Buchwald. 1955. Fungicidal treatments for control of cacao cutting disease in Ecuador. FAO. Pl. Prot. Bull. 4:17-21.

24. Earle, F. S. and J. M. Rogers. 1915. Citrus pests and diseases at San Pedro in 1915. San Pedro, Isle of Pines, Citrus Path. Lab. An. Rpt. 1:5-41.

25. Echandi, E. 1955. Trunk and stem canker of coffee trees. Biol. Trop. (Costa Rica) 3:237-241.

26. Echandi, R. and E. Echandi. 1958. Influencia de vitaminas y aminoácidos en el crecimiento y sporulación de Mycena citricolor, Phoma costarricensis, Cercospora coffeicola, y Ceratostomella fimbriata. Biol. Trop. (Costa Rica) 6:103-111.

27. Echandi, E. and R. H. Segall. 1956. Trunk, branch, and stem canker of coffee trees. Plant Dis. Reptr. 40:916-918.

28. Eddins, A. H. and R. K. Vorhees. 1933. Physalospora zeicola on corn and its taxonomic and host relationships. Phytopathology 28:63-72.

29. Elliott, J. A. 1923. The ascigerous stage of the sweetpotato black-rot fungus. Phytopathology 13: 56 (Abstr.).

30. Fawcett, G. L. 1916. Enfermedades del café causadas por hongos en Porto Rico. Est. Exp. Agr. Porto Rico, Bul. 17, 31 p.

31. Fawcett, H. S. 1936. Citrus diseases and their control. McGraw-Hill, New York, 656 p.

32. ________. and O. F. Burger. 1911. A gum-inducing Diplodia of peach and orange. Mycologia 3:151-153.

33. Fernandez Valiela, M. V. 1952. Introducción a la fitopatología. Gadola, Buenos Aires, Argentina, 872 p.

34. Frezzi, M. J. 1952. Presencia de Botryosphaeria ribis en la Republica Argentina y su importancia economica. Rev. Inst. Agr., Buenos Aires 6: 247-262.

35. Galli, F., H. Tokeshi, P. de C. Torres de Carevalho, E. Balmer, H. Kimati, C. O. Nogueira Cardoso, and C. Lima Salgado. 1968. Manual de fitopatologia, doenças das plantas e seu controle. Ceres, São Paulo, Brasil, 640 p.

36. Gómez L., C. 1962. Infección y síntomas progressivos occasionados por Ceratostomella fimbriata en el cacao. Turrialba 12:216-217.

37. ________. 1962. Relación de los coleopteros Xyleborus sp. con la infección por "Ceratostomella fimbriata." Turrialba 12:218-220.

38. Grimm, G. R. 1957. Factors affecting dieback and

growth response of young transplanted citrus trees. Florida State Hort. Soc. 70:82-85.

39. Grodsinsky, L. 1941. La podredumbre seca de la espiga del maíz. Argentina Almanaque Min. Agr. 16:153-154.

40. Hedges, F. and L. S. Tenny. 1912. A knot of citrus trees caused by Sphaeropsis tumefaciens. U. S. Dep. Agr. Bur. Plant Ind. Bul. 247, 74 p.

41. Henao C., R. 1958. Represión de los insectos taladradores del tronco de cacao (Theobroma cacao L.) Acta Agr. (Colombia) 8:167-180.

42. Hunt, J. 1956. Taxonomy of the genus Ceratocystis. Lloydia (US) 19:1-58.

43. Idrobo M., S. 1960. El complejo Xyleborus-Ceratostomella en Colombia. Inter-Am. Cacao Conf. (Colombia) 7:73-79.

44. ________. 1960. Una prueba posible de resistencia a Ceratostomella fimbriata (E. & H.) Elliot. Inter-Am. Cacao Conf. (Colombia) 7:149-151.

45. ________. and R. Cardenosa B. 1956. Grave epifitotia en Colombia causa el Ceratostomella fimbriata (E. & H.) Elliot en cacao, Theobroma cacao L. Cacao en Colombia 5:25-27.

46. ________. and G. Naundorf. 1956. Producción de toxinas por el Ceratostomella fimbriata en cacao. Cacao en Colombia 5:37-39.

47. Iton, E. F. 1959. Studies on a wilt disease of cacao at River Estate. Trinidad Imper. Col. Trop. Agr., Cacao Res. Rpt. 1957/57:55-64.

48. ________. 1961. Ceratostomella wilt in cacao in Trinidad. Inter-Am. Cacao Conf. (Trinidad & Tobago) 8:201-207.

49. ________. 1961. Studies on a wilt disease of cacao at River Estate. III. Some aspects of wind transmission. Rept. Cacao Res. 1959-1960. pp. 47-58. Univ. Coll. West Indies, Imper. Coll. Trop. Agr., Trinidad.

50. Jauch, C. 1961. La viruela de la púa en los duraznεros y almendros del nordeste bonaerense. Rev. Argentina Agron. 8:206-215.

51. Jenkins, A. E. and A. A. Bitancourt. 1941. Antracnosis de la vid en Chile. Chile Dep. San. Veg. Bul. 1:19-53.

52. Johnston, J. R. and J. A. Stevenson. 1917. The pineapple fungus (Thielaviopsis paradoxa). J. Agr. Univ. Porto Rico 1:208-210.

53. Leach, R. and J. Wright. 1930. Collar and stem canker of pigeon pea (Cajanus indicus), caused by a species of Physalospora. Trinidad Imperial Coll. Agr. Mem. 1, 4 p.

54. Malaguti, G. 1952. Una podredumbre del tallo de Crotalaria juncea causada por Ceratostomella fimbriata. Agr. Trop. (Venezuela) 1:287-292.

55. ________. 1952. Ceratostomella fimbriata en el cacao de Venezuela. Acta Cient. Venezolana 3:94-97.

56. ________. 1956. La necrosis del tronco del cacao en Venezuela. Cacao (Costa Rica) 3:17.

57. ________. 1960. Observaciones sobre la enfermedad necrosis del tronco de cacao por Ceratocystis fimbriata en Venezuela. Inter. Am. Cacao. Conf. (Colombia) 7:80-85.

58. ________. and A. V. Rodriguez. 1954. La pudrición de los esquejas de la caña de azucar y posibilidad de su control con fungicidas. Agr. Trop. (Venezuela) 4:61-67.

59. Marchionatto, J. B. 1938. Argentine Republic: fermented maize ears and control measures. Internat. Bulletin Pl. Prot. 12:25-26.

60. ________. 1949. Hongos exoticos. Lilloa 21:135-153.

61. Meredith, D. S. 1961. Botryodiplodia theobromae Pat. and Nigrospora sp. in the air of a Jamaican banana plantation. Nature 190:555-557.

62. Moller, W. J. and J. E. DeVay. 1968. Carrot as a species-selective isolation medium for *Ceratocystis fimbriata.* Phytopathology 58:123-124.

63. Montealegre, M. R. 1954. Cafetales a pleno sol, versus cafetales a la sombra. Suelo Tico 7: 263-275.

64. Nowell, W. 1917. *Rosellinia* diseases in the Lesser Antilles. West Indian Bulletin 16:31-71.

65. ________. 1919. A root disease of cacao in Trinidad --*Rosellinia pepo.* Trinidad & Tobago Dep. Agr. Bulletin 18:178-199.

66. ________. 1923. Diseases of crop-plants in the Lesser Antilles. West India Committee, London, England, 383 p.

67. Naundorf, G., S. Idrobo M., and M. Sandemente. 1956. Transmisión y diseminación del *Ophiostoma fimbriata* causante de la pudrición azul en el cacao. Cacao en Colombia 5:29-33.

68. Pontis, R. E. 1941. El mal de la tinta del nogal en la República Argentina. Rev. Argentina Agr. 8: 317-325.

69. Raabe, R. D. 1966. Variation of *Armillaria mellea* in culture. Phytopathology 56:1241-1244.

70. Rangel, J. F. 1935. Roselliniose, doença do cafeeiro. Brazil Min. Agr. Serv. Defesa Veg., Publ. 5, 23 p.

71. Ribeiro Filho, J. 1958. Estudo preliminar sôbre a incidência do secamento de ponteiros ou dieback em algumas variedades de cafeeiro (*Coffea arabica* L.) Rev. Ceres 10:413-421.

72. Roldan, E. F. 1962. Species of *Ceratocystis* (*Ceratostomella*) causing stain in rattan. Philippine J. Sci. 91:415-423.

73. Rubin, B. A. and E. V. Artsikhovskaya. 1964. Biochemistry of pathological darkening of plant tissues. Ann. Rev. Phytopath. 2:157-178.

74. Ruehle, G. D. and W. A. Kuntz. 1940. Melanose of citrus and its control. Florida Agr. Exp. Sta. Bul. 349, 54 p.

75. Ruehle, G. D. and S. J. Lynch. 1940. Copper sulfate as a corrective for dieback, a new disease of the avocado. Florida State Hort. Soc. Proc. 1940: 152-154.

76. Saccas, A. M. 1956. Les Rosellinia des caféiers en Oubangui-Chari. Agr. Trop. Nogent 11:551-595, 687-706.

77. Saunders, J. L. 1965. The Xyleborus-Ceratocystis complex of cacao. Cacao (Turrialba) 10:7-13.

78. de Segura, C. B. 1947. Nuevo tipo de gomosis de los naranjos en el Perú. Est. Exp. Agr. La Molina, Lima, Peru, Bol. 32, 7 p.

79. Shear, C. L. 1933. Life histories of Tryblidiella species. Mycologia 25:274-285.

80. ________. 1945. Studies of types and authentic specimens of Hypoxylon-I. Lloydia 8:245-262.

81. Siller, L. R. 1960. La Ceratostomella fimbriata en el cacao en Centroamerica. Inter-Am. Cacao Conf. (Colombia) Proc. 7:95.

82. Spencer, J. L. and W. C. Kennard. 1955. Physalospora perseae on mango. FAP Pl. Prot. Bulletin 4:43.

83. Stevens, N. E. 1926. Two species of Physalospora on citrus and other hosts. Mycologia 18:206-217.

84. ________. 1941. Host relations in species of Diplodia and similar genera. Mycologia 33:69-73.

85. Stover, R. H. 1956. Effect of nutrition on growth and chalamydospore formation in brown and gray cultures of Thielaviopsis basicola. Canadian J. Bot. 34:459-472.

86. Trinidad Department of Agriculture. 1950. Agriculture in the colony. Trinidad Dep. Agr. Rept. 1950:2-10.

87. Trinidad and Tobago Department of Agriculture. 1953. Work of the department of agriculture. Trinidad and Tobago Dep. Agr. Rept. 1951:16-39.

88. Uphof, J. C. T. 1938. Parásitos en las heridas de los cafetos y manera de controlarlos (mal de macana--mal de irigoyen). Asoc. Cafetaleros El Salvador, 16 p.

89. Viegas, A. P. 1939. Ocorrência de Rosellinia bunodes em Ubatuba, Estado de São Paulo. J. Agr. Piracicaba 2:327-328.

90. Vizioli, J. 1923. Some Pyrenomycetes of Bermuda. Mycologia 15:107-119.

91. Voorhees, R. K. 1942. Life history and taxonomy of the fungus Physalospora rhodina. Univ. Florida Agr. Exp. Sta. Bul. 371, 91 p.

92. Weir, J. W. 1926. A pathological survey of the para rubber tree (Hevea brasiliensis) in the Amazon Valley. U. S. Dep. Agr. Bul. 1380, 129 p.

93. Wellman, F. L. 1946. Observations on coffee root rot in El Salvador. Plant Dis. Reptr. 30:247-252.

94. ________. 1948-1953. Unpublished results secured: isolations and inoculation studies with primary and secondary causal organisms of tree-crop dieback disease. Inter-American Institute of Agricultural Sciences, Turrialba, Costa Rica.

95. ________. 1954. Fungi isolated most frequently from dying coffee fruit branches. Phytopathology 44: 509.

96. ________. 1961. Coffee botany, cultivation, and utilization. Leonard Hill, London; and Interscience Publ., New York, 488 p.

97. ________. and G. Salazar. 1956. Unpublished information: isolations from tree branches with typical dieback symptoms from untouched jungle growths in region of Turrialba, Costa Rica.

98. Wolf, F. A. 1926. The perfect stage of the fungus

which causes melanose of citrus. J. Agr. Res. 33:621-625.

99. Yarwood, C. E. 1946. Isolation of *Thielaviopsis basicola* from soil by means of carrot disks. Mycologia 38:346-348.

Chapter IX

FUSARIAL DISEASES OF LOCALIZED AND SYSTEMIC NATURE

Summary

Fusarium diseases, numerous in the American tropics, may be of major importance; symptoms range from twig die-back, damping-off and cankers to systemic wilt and abnormal production of sterile buds. Cacao cushiongall affects flower initials so crop is a failure; such abnormal growths are not confined to cacao trees. Fusarial wilts include, among others, diseases on a few trees like coffee and saman and on sugarcane, maize, tobacco, plantain, banana, and abaca. Last 3 crops are infected by the same parasite; were brought to the Neotropics centuries ago from the India-Malaya region. The susceptible Gros Michel banana is commercially popular and every possible effort has been made to grow it in spite of wilt, inspiring much mycological, physiological, and soil research. Studies have disclosed: soils of short-, medium-, and long-life characterized by chemical and physical differences. Means of combatting the disease have all eventually been failures, except use of resistant bananas. While fusaria are the imperfect stages of disease pathogens, with a few exceptions including all the many subspecies of *Fusarium oxysporum*, the following have ascosporic stages: *Hypomyces*, *Calonectria*, *Neonectria*, *Pleonectria*, *Nectria*, and *Gibberella*.

Introduction

There are rusts, rots, blasts, leafspots, and swellings but one of the most frightening symptoms of a disease in plant or tree is drooping of leaves and final death. There are many diseases such as these latter that occur in the Neotropics. In truth, the wilts are among the most publicized and spectacular of all plant diseases. The average plantation owner, or lesser grower, suddenly confronted with three wilting trees is frightened and much impressed. For 15 years without much comment he may have suffered 20 per cent annual losses from leafspot on his crop. This he has learned to live with and considers it almost a normal condition. But a year in which three trees die in a month's time is likely to cause more concern than any growing problem he has thus far encountered.

After the human race had suffered long ages of recurrent starvation and death from famine on account of crop diseases, professional plant pathologists began to study them with great care. For the classic plant plague of potato late blight, _Phytophthora infestans_, spraying with chemicals was devised to control it; with the equally classic black stem rust of wheat, _Puccinia graminis_, pathologists selected and bred resistant varieties. Even if some progress was apparent in understanding, and there was at least some reducing of disease losses from mildews, blights, blasts, and rusts, there was, with reason, a long standing fear in man of crop disease. And when the systemic wilts caused by species of the fungus _Fusarium_ appeared in valued crops, the pathologists found themselves faced with another major challenge. Intensive work was instituted on these wilts, some of it well before the beginning of the present century. In the first three or four decades of the 1900's there was much fervor about the wilt fusaria and they became one of the major interests of the profession.

It was pointed out in its history (Duggar (28), Wollenweber (102, 103, 104), Synder and Hansen (85), and others) that certain systemic wilts were certainly related to roots and the soil--the trouble began with a few diseased plants at first but soon spread--and that after a number of years what were once flourishing farms became dismal failures.

Perhaps the most consistent phenomenon noted with respect to the systemic fusarial diseases (for examples: Galli _et al._ (32), Beckman _et al._ (10), Wollenweber and Reinking (104), Carrera (17), Gordon (36), Risbeth (72), and Wardlaw (97)) is the fact that wilt occurrence is dependent on warm soil conditions. There has been at times a tendency for workers to regard, as separate from others, so-called Tropical Fusaria (36, 70, 71, 102, 104) which implies that the place of origin is the temperate zone. This has been due to traditional views and conclusions of temperate-zone-based researchers and research. The more work is carried on in the Neotropics, the more it is realized that the conditions of life for these organisms is nowhere better than in our tropics. Repeated finding of so many species and forms of *Fusarium* that have occurred spontaneously under tropical conditions (2, 6, 7, 15, 18, 24, 26, 31, 32, 34, 36, 40, 45, 46, 51, 52, 55, 59, 68-71, 77, 84-86, 102-104) is overwhelming evidence that these fungi are not only at home in the tropics and subtropics but probably originated there. Types of disease from organisms of the genus cause an exceedingly broad gamut of troubles, and a list of some of them is represented in Table 14.

It seems significant to point out here that from the American tropics, fusaria are most commonly reported infecting crops belonging to the Leguminoseae, Gramineae, Rosaceae, Annonaceae, Solanaceae, Euphorbiaceae, Sterculiaceae, Musaceae, Rubiaceae, Moraceae, Orchidaceae, Cucurbitaceae, and Myrtaceae. However, numerous other families of plants have fusarium attacks, and it almost seems that whenever a crop has very special economic value and thus requires greatest intensive plant disease scrutiny that some *Fusarium* sooner or later is found to be important as a limiting infection.

The fusaria are not only the systemic wilt producers but also they cause rots, leaf and fruit drop, bud proliferation; they increase dieback, and both the collapse and stimulation of tissues; and they include an innumerable lot of saprophytes found in great numbers in the tropics. These fungi have been so common in my isolations and culture studies in the Neotropics, that it is impossible for me to make a fair estimate of the number of different sources of *Fusarium* isolates, or of prepared dried specimens contaminated with *Fusarium* that have been discarded in some 35 years. These fungi are many more times as common in the tropics than in the temperate zone. By one way or another they must have been disseminated from tropical haunts to the temperate zone where certain of them with their disease propensities stirred scientific interest. Thus they were discovered towards the end of this last century and interest in them came to fruition; a newly awakened

(Cont. p. 348)

Table 14. A LIST OF TYPES OF DISEASES FROM FUSARIUM encountered[a] in crops in the American tropics.

Type of injury	Causal Fusarium	Host
External presence on seed	F. sp.	Coffee[b]
	F. (solani?)	Garbanzo
	F. sp.	Dodder
Preemergence seed rot	F. moniliforme	Maize
	F. sp.	Sorghum
Seedling damping-off	F. solani	Tobacco
	F. roseum	Petunia
	F. roseum	Tomato
Storage root dry rot	F. oxysporum	Sweetpotato
	F. solani	Sweetpotato
	F. (solani?)	Mandioc
Tuber dry rot	F. solani	Potato
Corm dry rot	F. oxysporum	Gladiolus
	F. roseum	Gladiolus
	F. solani	Gladiolus
	F. oxysporum	Banana
	F. (solani?)	Banana
	F. oxysporum	Plantain
Feeder root tip rot	F. oxysporum	Coffee
	F. sp.	Avocado
Main-root non-systemic disease	F. solani	Bean
	F. oxysporum	Coffee
	F. roseum	Orchid
	F. oxysporum	Ñame
	F. oxysporum	Cabbage
Systemic wilt, annual	F. oxysporum	Gladiolus
	F. oxysporum	Tomato
	F. oxysporum	Cotton
	F. oxysporum	Tobacco
	F. solani	Dry bean[c]
	F. solani	Green bean[c]
	F. oxysporum	Watermelon

Systemic wilt, plantation crop	*F. oxysporum*	Coffee
	F. oxysporum	Banana
	F. oxysporum	Plantain
	F. oxysporum	Abaca
	F. moniliforme	Sugarcane
Systemic wilt, shade tree	*F. oxysporum*	Saman
	F. perniciosum	Albizzia
Branch-stem canker	*F. roseum*	Gandul
	F. lateritium	Crotalaria
	F. roseum	Snapdragon
	F. lateritium	Coffee
	F. lateritium	Clusia
Fruit peduncle failure[d]	*F. roseum*	Breadfruit
	F. sp.	Jakfruit
	F. roseum	Coffee
Stalk and leaf disease	*F. moniliforme*	Maize
	F. roseum	Alpinia
	F. moniliforme	Sugarcane
	F. sp.	Pasture grasses
	F. (*tricinctum*?)	Zoysia lawn
Head-ear blight	*F. moniliforme*	Maize
	F. spp.	Grain sorghum
	F. moniliforme	Paspalum
Twig dieback	*F.* spp.	Cacao
	F. roseum	Coffee
	F. roseum	Grapefruit
	F. (*roseum*?)	Orange
	F. spp.	Jungle trees
Flower infection	*F.* sp.	Orchid
	F. sp.	Mango
	F. roseum	Cacao
Flower cushion proliferation	*F. rigidiuscula*	Cacao
Flower cushion knob	*F. rigidiuscula*	Cacao
Cushions of buds on trunk	*F.* (*rigidiscula*?)	Lagerstroemia tree

Witches' broom complex	*F. roseum*	Oleander
Leaf spot, secondary invasion of	*F. roseum*	On coffee *Mycena*
	F. solani	On coffee *Mycena*
	F. spp.	On rubber *Dothidella*
	F. (roseum?)	On cacao *Phytophthora*
	F. tricintum	On grass leaf decay
	F. roseum	On grass stem rot
	F. spp.	On grass after invasion by web blight

---

a) There are, of course, many other fusarial diseases. These have been those dealt with in connection with research, training, and consultative services; all from the West Indies and Central America. According to botanists and collectors there occur in the American tropics thousands of plant species in about 200 phanerogamic families on which fungi are known, and of these about 60 families are affected by fusarial troubles (*Fusarium* spp. and perfect stages). Of the thousands of species of plants Viegas (94) in 1961 lists 193 phanerogamic families in South America on which fungi have been reported, and of these 57 are attacked by *Fusarium* or an ascosporic type like *Calonectria*, *Nectria*, *Gibberella*, or *Hypomyces*.

b) No actual disease caused, but result of its presence is a definite change in flavor of the coffee bean contained inside the dried parchment shell (the so-called "Santos flavor" of roasted coffee is seemingly associated with superficial *Fusarium* contamination).

c) In these plants the vascular systems not markedly discolored.

d) In some cases fruit drop is considered a "normal" condition early in the season. The disease indicated here is more than the usual fruit cast from an over-loaded but healthy tree system.

---

surge of phytopathological interest in root disease resulted. Studies by researchers in the temperate zone proliferated rapidly. The number of publications from Europe and the United States on Fusarium problems at first soon far out numbered such reports from the tropics. Of later years this is not so true.

As any pathologist is well aware, fungus diseases of plants caused by species of Fusarium are a major problem in continued crop production. There are numerous plant disease specialists who have spent the largest part of their life's work on study of such diseases. The genus has been a matter both of profitable interest and careful study--and at times even extended controversy--among pathologists and mycologists. When it comes to the physiology of parasitism, many of the troubles caused by fusaria have lent themselves to important studies of complex and revealing results. Some of the most successful researches on the fundamentals of host and parasite relations in plants, and findings on the true basis of resistance, have been accomplished by work on fusarium diseases.

It seems probable that there are various phases of fusarium (including ascosporic forms) disease research that may always be under study. One feature, thus far given only the most superficial attention, is that of hyperparasitism and even super-hyperparasitism of these organisms and what they do in reducing disease in the tropics under natural conditions. It is encouraging to learn that in South America (see Viegas' Portuguese work (94) "Index of South American Fungi") there are at least 49 widely different kinds of plant-affecting fungi that are attacked by fusaria or perfect forms related to them. Some of these are of unusual interest; two special examples are Nectria erinacea Starb. that infects Nectria albicans Starb. and Nectria calonectricola P. Henn. that in turn diseases Calonectria sp.

Explosion-of-buds Type of Fusarium Tree Disease

It is stimulating to watch, over an extended period,

while an unusual type of disease becomes increasingly prominent in an expanding, world-important, tropical tree crop--in this case cacao or the chocolate tree. In January of 1949 a group of pathologists saw in a few cacao trees in Guatemala the occurrence of a condition growers called maleness (macho). Although flowering was excessive no pods were set, and in some cases no normal flowers developed. It was a confusing situation, since horticulturists had sure knowledge that occasional trees, otherwise healthy, did not set fruit because of a flower component incompatibility.

From 1950 to 1954 agriculturists sent in frequent reports of so-called "male" cacao trees from South and Central America. In a few more years, in Costa Rica, in addition to known "male" trees, an increasing number began to show flower abnormalities, and the latter was considered a disease called buba or cushiongall. Symptoms of this are developments of pads of tightly-grouped, small, closed buds on otherwise smooth flower cushions. This follows infection of flower initials. Where a normal flower cushion may produce several, spaced, pastel-colored flowers, about two to four of which will set a pod, on a buba-diseased cushion there grow literally hundreds, up to over a thousand, closely crowded little buds and no fruit is produced. See accompanying Fig. 9-1 for drawings of this condition. On old diseased flower cushions, rounded knobs of woody tissue develop under the pads of the multitudes of buds doomed never to grow into fruit (35, 41, 42, 48, 80).

This disease goes from tree to tree, and in its first years of study it was found to spread up and down branches in the direction of vascular connections between flowering cushions. In several ways it resembled what might be expected of a witches' brooming virus. Cacao growers associations, in their concern for improvement of cacao varieties, became disturbed about the possible transfer of the buba disease from country to country in bud wood and other cacao planting material, and pathologists began intensive study of the problem. Early isolations from buba tissues, made as a side line in my coffee disease laboratory, gave no consistent results. A total of 27 different species of bacteria and fungi were isolated, including three kinds of Fusarium, but these latter were found in only a few cases. At about this time some major attention was turned to the buba problem. Surveys were made and literature was searched for information about it. A full time pathologist, L. M. Hutchins, was secured and through cacao interests given help and a laboratory in Costa Rica.

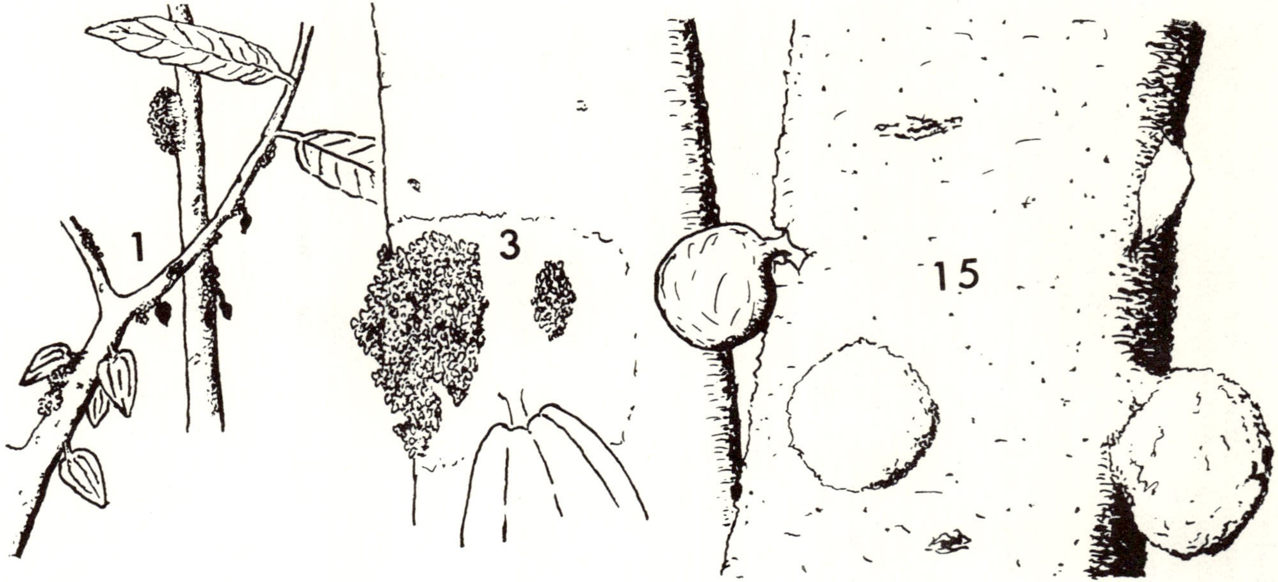

Fig. 9-1. Cacao cushiongall disease, Fusarium rigidiuscula: One-year-old (1) galls, some evidently following up vascular systems of branches, and three killed cacao pods; galls about three years old (3) on tree trunk with fruit cushion maturing a pod; knobs remaining from approximately 15-year-old (15) cushiongall infections, the smooth knob at left in profile had bark-like covering removed. (In Nicaragua, bush children used such knobs for playthings.)

From the possible standpoint of cushiongall being viral in nature, many kinds of vegetative connections were made between diseased and healthy materials. Buba was found to be graft transmissible. Meanwhile hundreds of isolations from diseased tissues resulted in cultures of possible fungi that might cause the disease, and among the cultures was a group of fusaria. The apparently most important of these was identified as *Fusarium decemcellulare*. It was studied in culture; after at first confusing results Hutchins obtained infection when pedicels of young developing pods of cacao were inoculated. Soon thereafter he and others, using special techniques, secured transmission of the disease by inoculating seeds and young seedlings (3, 20, 41).

Meanwhile typical cultures of *F. decemcellulare* were brought to the attention of W. C. Snyder and sent to his fusarium laboratory in California. There the specificity was confirmed, and the perfect stage, the ascosporic *Calonectria rigidiuscula*, was determined. The more acceptable pathologists' usage for this organism appears now to be *Calonectria* (*Fusarium*) *rigidiuscula*. To find that a *Fusarium* causes such a severe and remarkable disease reaction as develops in cushiongall of the cacao tree is sufficiently surprising. To learn, as Hutchins did, that cultures from the *Fusarium*, or imperfect stage, are the ones that produce the explosion-of-buds type of disease, but that single spore cultures from the *Calonectria*, or perfect stage, do not produce buba, is nothing less than an exceptional discovery in plant pathology. In tropical work it is the new and unusual that quite often comes to light (30, 38, 41, 49, 85).

It is gratifying to find that the thorough research on cushiongall of cacao continues. More work is being conducted on various phases of the disease, and doubtlessly much is soon to be published about the host-parasite relationships in this unique tree infection. The physiology and morphology and the fine-structure studies on cushiongall are going to bring great strides in information to the pathologists. It is already known that there are evident sources of resistance-tolerance in cacao to buba (1 and 43).

The studies on cushiongall have revealed that other hosts (such as Malaguti (49) has seen in mango in Venezuela, and I found in Puerto Rico that large trees of astromero (*Lagerstroemia speciosa*) in plantings along streets in Rio Piedras) have the same kind of disease. The researches emphasize the fact that some of the older settled concepts so long taught and known in the temperate zone about how the fusaria act, in general, as disease organisms do not always hold in the tropics. Work in the tropics is serving to round out a more complete idea of the range of diseases that may be due to Fusarium. (44, 50, 65)

Vascular Fusarium Wilts

There is a long list of major crop plants susceptible to attacks by Fusarium spp. in the temperate zone. The list is likely to be even longer in the American tropics (e.g., 17, 32, 36, 55, 59, 77, 94, 102, 104). It behooves the professional worker to be able to make an authoritative statement about the special wilt Fusarium with which he may be forced to deal. This is not always easy with a fungus genus so malleable and unstable in cultural behavior (66, 88, 100) as is the case in Fusarium. However: knowing the disease symptoms and the host, having key literature (see directly above), using simple moist chamber and isolation-culture methods, and having available a microscope for spore studies, can give good answers.

The fusarial wilts of the three closely related crops: banana, plantain, and abaca, all caused by Fusarium oxysporum f. cubense (e. F. Sm.) Snyder & Hansen*, are widespread throughout the Neotropics and also the Oriental tropics (6, 12, 13, 23, 32, 37, 70, 72, 77, 81, 84, 86, 87-91, 94, 96-98, 100, 104).

There is a widespread dangerous rootrot and wilt of common bean from F. solani f. phaseoli (Burk.) Snyder & Hansen (17, 26, 32, 36, 51, 59, 60, 86, 93). It is usually most serious in the low elevation tropical fields of slightly acid, friable soils, especially in regions on the leeward slopes if

---

*Use of some of the trinomials from the works of Snyder and Hansen (85) has not been adopted by everyone. Studies by Gordon (36), Carrera (17), Armstrong (4, 5), and co-workers, indicate that in some cases there are instances when a certain forma specialis may successfully attack several hosts. It is probable that controversy on this may continue for years before all is settled, but for the convenience of a recognizable label, I retain the usages commonly given in the phytopathological literature at the time of this writing.

there is not escessive rainfall. Where cleared fields were far apart, sown only once or twice in Honduras and Guatemala and then in typical shifting culture method allowed to return to second growth jungle and not replanted for about seven years (or more), bean crops could be secured. Where good bean land is so much at a premium as in the more densely populated country of El Salvador, I observed some peasants who took a chance on diseases and repeatedly replanted their fields. About the time the fourth crop was harvested the returns did not pay for the seed used, and the main cause was the *Fusarium* disease. The experience of indigenous growers in wooded tropical country is often about as follows: year 1, jungle trees felled and debris partly burned, the crop seeds planted grow well with no weeding of field; year 2, same crop replanted more closely, a few weeds and some disease but harvest good with a second crop replanted the same year, closely grown with good harvest but more weeds and disease; years 3 and 4, same crop planted two and three times a year, weeding much more troublesome, diseases more severe, lighter to very poor harvests; plans made to move to other nearby selected location.

For some years I had reported to me a wilt of ñames (yams), *Dioscorea* spp., used both in food and as sources for certain medicinal products. Dry autoclaved specimens from Mexico, Colombia and Central American locations were sent to me. All had vascular darkenings, typical of systemic fusarial infections, and the wilt effects described were similar to such diseases in other crops. In 1960 persistent requests from farmers growing *D. alata* L. in Puerto Rico were answered to solve what was called rust and wilt disease. It proved to be *Fusarium* wilt. This was proved when isolations were made and cultures studied in laboratories in Rio Piedras, Puerto Rico and in Berkeley, California. In California, W. C. Snyder concurred in the judgment that the isolates were all *F. oxysporum*. On testing pathogenicity of pure culture material on lengths of the base of surface sterilized ñame vines in test tubes in the laboratory in Puerto Rico, it was found the organism entered the vascular elements, discolored them, and in two weeks' time could be re-isolated 15 cm. away from the original point of inoculation. No cross inoculations have been made on other plants. However, many times in informal reports, using the trinomial system, I have identified the causal organism as *F. oxysporum* f. *dioscorea.*

Diseased tobacco has been seen and examined in several Central American locations. This has been in small groups of older plants after priming or leaf harvest was well advanced. It is at this period in the crop progress that wilt occurs and with it vascular discoloration symptoms. Isolations yielded typical F. oxysporum f. nicotianae (J. Johnson) Snyder & Hansen. In many parts of Central America and the West Indies occur typical wilt-diseased tomatoes. Isolations, in all cases (except from a few plants in Panama areas known to have bacterial wilt) diseased stems yielded F. oxysporum f. lycopersici (Sacc.) Snyder & Hansen. In both tobacco and tomato the fusarial wilts develop in warmer locations but are of less and less concern as newer, more resistant strains of the crops are being introduced.

A very large and beautiful tree, Pithecolombium saman (Jacq.) Benth. (= Samanea saman), is found in many parts of the Neotropics where it is grown for shade and ornament and is given such names as raintree, cenicero, saman, and campano. In Puerto Rico and the Virgin Islands occasional specimens are seen that show what has been called branch decline, ans finally the tree dies. Studies gave no clear answers to the primary pathogens causing top disease, until root crowns were dug out and wood cut open. Here, and below and above in the trunk, were seen discolored vasculars accompanied with a pungent odor not found in healthy wood. Repeated isolations yielded (unpublished) a Fusarium which I considered deserved subspecific ranking as F. oxysporum f. samanea. Cultures of it were studied in California by Snyder who confirmed it as an oxysporum. The disease is found in wilted trees on at least four islands of the Caribbean.

There are other shade and ornamental trees wilted and killed by Fusarium: e.g., wilt of Albizzia (1). A systemic wilt of Crotalaria sp. was brought to the plant pathology laboratory in El Salvador and from it were obtained pure cultures of F. roseum. This is a singularly variable species and it infects many hosts, often is found following or along-with invasion by another parasite, and may some day be recognized as having serious impact in the tropics (in a report of 1934, Wellman (100) described it as F. retusum developing in wilted tomato in the temperate zone). in 1962, in connection with some routine isolations from wilted tomatoes from the south coast of Puerto Rico I secured both F. oxysporum and F. roseum. It is also known from material sent to my laboratory from Paraguay that the latter Fusarium is at least part of the disease complex in wilt, a stunt, and a decay of broad bean.

A fusarial wilt of coffee occurs in the Neotropics (2, 34, 94, 100). This is due to a systemic invasion and a disease quite separate from the Nectria canker (40, 46, 69). The latter, which attacks trunk injuries, often eventually kills badly diseased trees, and may lead to confusion in identifying Fusarium disease on coffee in the field. The coffee fusarial wilt does not sweep through plantations rapidly. Generally this is because coffee is grown at elevations where soil temperatures are below the ideal for Fusarium. However, there are occasional low locations where it has been seen in severe form. It appears from field and greenhouse study that the organism is at first held in check by the host; in Puerto Rico it was found that seedlings often do not succumb for 12 to 14 months after infection of feeder root tips. Large and well established plantation trees show signs of the following course of events: the first infection is in the smaller feeder roots, then the larger roots are infected, and finally the main root and stem axis. Where this tree wilt has been watched most closely, it appears that it is only after the main part of the tree is invaded that sudden collapse occurs. It is estimated in some instances that trees may have been infected for two or more years before final collapse. The coffee tree wilt is not asymetrical in effect, as is seen in many fusarial wilt diseases in herbaceous hosts. In coffee, the whole tree goes down at once. (The cause: F. oxysporum f. coffeae Wellm.)

There are a large number of systemic wilts from different fusaria in the American tropics. For a listing of many of these, in addition to those mentioned in this chapter, the reader may also see the volume entitled "List of Tropical American Plant Diseases" now in preparation. An important help to Fusarium studies in our tropics has been the influence of such men as Wollenweber (102), Carrera (17), Stover (88), Viegas (94), Snyder (85), Gordon (36), Mujica (59), and de Segura (77), with their co-workers.

The Fusarial Wilt of Banana

When a banana plant* has its huge, lush (cont. p. 357)

*The common banana of commerce and subsistence agriculture

Fig. 9-2. Effect of fusarial wilt, Fusarium oxysporum cubense on giant susceptible banana Musa AAA, two years from planting (sketched in Honduras). A. Growing in long-lived, "resistant" soil, mat of six healthy plants and a new sprout or peeper all of which would have fruited; all have good root systems determined by pushing pseudostem and feeling solid anchoring in soil. B. Growing in short-lived, "susceptible" soil, mat of two plants and peeper, all wilted, leaf sheathes splitting, some dropping away from pseudostem; root systems dead and dying and some plants easily pushed over.

is a giant herb, not a tree or a palm, which it resembles. It is a plant in the same sense as either the crocus or the iris is. These much smaller herbs of course are neither trees nor palms, and it is incorrect to call the banana plant either a tree or a palm. For the botany and anatomy of this giant herb see Simmons (81). See also Appendix 2 on Musa at the end of this book. It is best that scientifically our common banana, depending on its genes, be designated as either Musa AAA or as Musa AAB. Abaca or hemp is Musa textilis.

leaves wilted and dropping from its upright stem, it is a sad and a spectacular condition (Fig. 9-2) and it is killed. Sometimes there is a yellowing of the great leaf blades prior to wilt, and always at the base of the pseudostem, that part of the plant that looks like the "trunk of the tree," there is hardening and splitting of the enwrapping leaf sheaths. When slashed open, vascular bundles display reddish or purplish discolorations; the sap has a pungent odor reminiscent of a weak alcohol-formalin mixture. Banana men know these symptoms well, and a pathologist can, with certainty, name the causal organisms as *Fusarium oxysporum* f. *cubense* (E. F. Sm.) Snyder & Hansen; there is for conservative reasons some persistence in the use of the old binomial "F. cubense" as more succinct (12, 84, 84).

In any case, this disease, one of the greatest all-time plagues of agricultural man, has been known a long time. Where growing of susceptible banana varieties is still persisted in, it is an absolute, limiting factor in their productivity. It has cost hundreds of millions of dollars (U. S.) to continue the use of old banana varieties. The hope has been to circumvent the inevitable wilt effects by delaying its spread through soil treatments to kill the fungus, and searching out, and planting in soils that restrain the pathogenicity of *F. oxysporum cubense*. (6, 37, 71-73, 88, 95, 98).

There is reason to believe this especial banana *Fusarium* is not indigenous to the Neotropics. From carefully documented sources, it is concluded by Stover (89) that probably this banana wilt organism originally came from the India-Malaya region. There is always the unlikely possibility that it might either have evolved independently in the American tropics or was brought in by rare chance transport of its spores by storms from the Orient to the Occident. It more probably came over as infestations in or on old importations of plantain or banana corms, which can be shipped great distances by ocean craft in semi-dormant condition for long periods. Many corms were brought to the New World tropics by early slave merchants and peasant peoples. A common and popular, heavy-producing sweet, eating banana of those days is the old wilt-susceptible seda or silk variety, and old kinds of cooking bananas or plantains are also susceptible; many of such century-old types are still found in the mixed gardens, or huertas, around homes in the Neotropics.

However, there is also little doubt, which should be

noted, that there are some wilt fusaria that seem definitely to originate in our western tropics. An example is one with which I have had some responsibility. The coffee fusarial wilt, F. oxysporum coffeae, does not occur in coolest mountain soils but in warm to warmer soils at lower levels. It was found in absolutely virgin land planted with uncontaminated coffee seed: in two fields in Costa Rica; two reportedly virgin fields in Puerto Rico, and in one such field in El Salvador. It gave every evidence of being indigenous in these instances. From these five fields isolations were made from over 80 diseased roots on wilted trees. All yielded typical coffee Fusarium. In studies on F. o. coffeae spore germination phenomena, attempts were made to discover if chalamydospores of this fungus required host presence as is known with other fusaria on suscepts (14, 60, 61, 88, 93), of coffee roots, root tips, peeled seed, or seedling exudates, for germination. Under the conditions in the Puerto Rico laboratory (it may have been too warm) coffee root presence or leachates had no appreciable effect on germination of coffee Fusarium chalamydospores. Macrospores and microspores were readily germinated, while chlamydospores remained dormant.

The very earliest, well-substantiated reports (89) of what must have been banana fusarial wilt was in Australia in 1876, but it is probable the disease was common in many parts of the Oriental-African tropics, long before its report in Australia. Records show that by 1890 it was first found in the Neotropics in a few plants in both Costa Rica and Panama. In 1903 it was reported in Jamaica, in 1906 in Surinam, in 1907 in Trinidad, a year later in Cuba, and in 1910 in Guatemala as well as in Puerto Rico. It has spread steadily since that time.

By the year 1940 it was in the additional Caribbean Islands of Dominica, Hispañola, Grand Cayman, Grenada, Martinique, Guadeloupe, St. Lucia, and St. Vincent. By then it had been reported as well from mainland countries of British Honduras, British Guiana, Colombia (the Pacific side, which is well separated from the east), Venezuela, Mexico, Nicaragua, Brazil, and Ecuador. In more recent years, 1940-1965, it has been found in El Salvador, in Colombia (Atlantic side), and I identified it in both Peru (1950) and in the island of St. John (1963). It is still spreading.

The banana is a preeminent tropical crop and was

introduced (so some believe) to the Neotropics by the intrepid Ponce de Leon in about 1493-1496. But the crop in its new habitat did not succumb for centuries to the now infamous wilt, Fusarium oxysporum f. cubense. The parasitic fungus was described by Smith (84) in diseased banana material from Cuba, and first given major pathological study by Brandes (12). This seems to be a western hemisphere race of the subspecies. It is of great interest to us that a banana wilt from a second race of this Fusarium occurs in the Orient (see Stover (89) and Wardlaw (97)), and is distinctly different in its host specificity from the one we know so well and that is of such great concern in the American tropics.

Habits of the Banana Fusarium

As soon as banana planters began to realize how important wilt disease was in their crops, they watched closely for field reactions. They discovered by observation that in banana wilt, there were some areas, sometimes quite localized, where plantings rapidly succumbed to disease. In some fields the soil was short-lived with respect to banana growing, and it required only a few months for everything to wilt. In some cases--sometimes only a few kilometers away from the bad or susceptible fields--there were locations where bananas could be grown for 30 or more years. These latter became for growers the long-lived, the good or resistant fields. In very admirable researches (37, 54, 57, 76, 89, 95, 98) it was likewise found from reviewing cultivation histories and technical studies that there were also types of fields that were what banana men called intermediate in resistance and lasted perhaps 5 to 10 years before disease destroyed them. Other fields grew bananas 20 or more years (37, 95, 98).

Soils and Banana Wilt

Fusarium oxysporum cubense where found, subsists in largest numbers mostly in the first few inches of field soil. Where recovered at much depth partly decayed, diseased host tissue is present (33, 54, 71, 74, 75, 87, 89, 97). Relying on fusarium plate-count assays (for example, see Risbeth (72) and Stover (89)), it has been found that heavy soils, of high clay content, do not harbor large numbers of the fungus, but lighter textured loams and sandy loams retain them in profuse amounts. It is known furthermore that not only are acid (pH 3. 8 to 5. 0) soils likely to allow more colonization of the Fusarium, but they also are conducive to production of abundant spores under those conditions. However, in most alkaline (ph 7. 2 to 8. 0) soils, less of the Fusarium is found and apparently the organism does not sporulate well under alkaline conditions.

Conidial spores perish, according to Sequeira (as quoted by Stover (89)), on being buried a few weeks in clay, while sandy loam soils with considerable organic content are not so deleterious to the spores. Fusaria do not penetrate much into slightly alkaline clay soils but go considerably more deeply into the acidic and more friable types. All these findings on F. o. cubense are of much value in understanding activities of other soil fusaria.

Amounts and sources of calcium, and supply of soil nutrients had much to do with soil "life expectancy" if planted to wilt-susceptible Gros Michel banana. Volk (95) found that soils are most likely to be "resistant or long-lived" if they have in soil solution about 1000 or more parts per million of total salts, and that the "susceptible or short-lived" soils have less than 400 ppm. Amounts of active calcium are also important and so-called long-lived soils have over 0. 48 per cent, while the short-lived have less than 0. 32 per cent.

To further understand the soil-wilt relation, studies were carried out on additional soil components, such as magnesium, potash, clay, humus, and sand. Some of the conclusions are: that potash is of more importance than was once realized in wilt amelioration; that magnesium serves at times to complicate matters when it in part replaces calcium; that humus and sand make wilt more severe; and that clays are important sources of nutrition and buffering action in good wilt-resisting soils. The types and amounts of clays in the soils are of extreme importance in forecasting how long fields may be expected to continue to produce the Gros Michel banana in wilt country.

It is clear from Stover's review (89) that: in the presence of Fusarium, if grown in high calcium conditions, Gros Michel is most resistant to wilt; high potash reduces wilt in many, but not all, cases; an abundance of nutrients significantly ameliorates all kinds of fusarial wilt; wilt disease is least injurious where there is a balanced nutrient condition optimum for the host; and excess of a salt may be deleterious to wilt resistance.

At one time, banana men operated on the premise, never proved scientifically, that no matter in what class of soil, the wilt *Fusarium* was ubiquitous (and certainly in all tropical soils). However, in some fields it seemed that certain subtle components, akin to vitamins or other such compounds, kept the crop from being diseased, or at least that the nutrients were such that diseased fields lasted longer. This was a most intriguing plant disease problem; if true and "good" soils were plentiful, all growing could be geared to the engineers' special expertise; if not true, the pathologist had to solve it.

During 1929-1930, after learning how to carry on special *Fusarium* isolation methods, I found *F. o. cubense* in severely diseased field soils and debris, and searched for the organism in other places. It was recovered by a few crude tests from laboratory sweepings, and dirt from boots worn after a weekend in a wilt-diseased field. The *Fusarium* was not isolated in jungle less than a kilometer from a wilting field of bananas. Debris and soil from walks of the Lancetilla Botanical Garden (near Tela, Honduras)--frequently visited by workers who had contacts with wilt-diseased bananas--yielded fusaria but not *F. oxysporum.* Search for the organism was also made on an expedition working back and forth in the west coast of Guatemala. A few thousand samples were secured from long stretches of untouched jungle and in these no *F. o. cubense* was demonstrable by isolation methods. However, it was isolated from soil in three out of seven peasant clearings where plants of both ladyfinger banana and a large fruit-type plantain showed disease. Results from these studies seemed controversial at the time and were never published. Since those years others such as Risbeth (72) and Stover (88), using more perfected techniques, published information containing much data that show *F. o. cubense* in the Neotropics is evidently always an introduced organism. The *Fusarium* has been multiplied so much in the years 1920 to 1970, that permanent growing of wilt-susceptible bananas is not possible in the best, highest producing, so-called banana soils; whether in the West Indies, Central America, or South America.

Working in Puerto Rico with parallel cultures of *F. oxysporum coffeae* and *F. oxysporum cubense,* some soil effects were noted. A heavy clay soil and a light sandy loam, from two different parts of a botanical garden (believed to be free of the banana *Fusarium*), were used as test materials. Soils were sieved through common window screening, moistened to about 30 per cent water-holding capacity, and put about 30 mm. deep in 250 ml. flasks. Suspensions of vigorous week-old cultures of the two fusaria were used to inoculate the soils, and incubated at 26° to 29° C for 96 days. Soil smears were stained on glass slides, using the Nash (60, 61) technique for direct observation. With both fusaria, results were closely similar. In the clay only a few chlamydospores were found; in the loam, chlamydospores had developed well, and were especially numerous with the banana fungus. No *Fusarium* chlamydospores were found in non-inoculated flasks of clay or loam soil.

Studies by several researchers (56, 57, 62, 73, 87, 89, 91) have been made on the effects of soil moisture on *F. oxysporum cubense.* The organism in its mycelial and/or conidial condition survives for only short periods in very dry soil. Chlamydospores survive in air-dry soil for 30 or more years. In general, the organism lives in the surface part of field soil submerged in water for at least 2 or 3 months; after long periods the fungus is destroyed at depths much below the surface. In soils of different moisture contents, the quantity of *Fusarium* decreases rapidly with large increases of water in the soil; at times there is good survival in soils with moisture contents of about 30 per cent of water of saturation.

In the case of the banana *Fusarium* it has been found (see Stover (89) for literature) that *F. o. cubense* under proper conditions is a good colonizer of banana roots (11, 72, 75, 90, 97). As is well known from other fusarium studies, host exudates have a stimulatory effect on *Fusarium* chlamydospores; in some soils it has been seen by me that infectivity of susceptible banana will be retained for 20 to 40 years of interim cultivation to such crops as sugarcane, cacao, coffee, and rubber. It appears that some *F. o. cubense* propagules must be assumed to remain dormant during long periods in soil. This indicates a rather complex life history of the banana wilt *Fusarium.* In the absence of host roots, chlamydospores may remain intact, alive but inactive, possibly for half a century. However, this fungus is likewise able to subsist on a few other hosts than banana (89, 96).

At one time it appeared that F. o. cubense was killed in a soil after 18 months of dyking around it and keeping it soaked with water. (Known as flood-fallowing, this is an old system used by Dutch farmers in Surinam and Holland to control certain other plant diseases, and in Florida it has been employed to reduce nematodes in celery seed beds.) Several difficulties were soon evident in Central American and Panamanian banana soils. Medium to heavy clay retained the flood waters best; these are the longer-lived soils and the last to really need control treatments. The more easily tilled, slightly acid, friable, highly productive, and more readily managed banana soils require extreme amounts of water to keep them well flooded. Fields that are not level are impossible to keep inundated at not too great an expense. Many variations were tried in managing this system of soil treatment, and a comprehensive review of this has been made by Stover (89). Perhaps his estimate of the trials is the best conclusion: "In spite of improvements in flood-fallow disease control, the procedure has become uneconomic due to the rising costs and failure to extend the average production life of flood-fallowed areas." [See Fig. 9-3.]

Certain banana planters saw the once promising flood-fallow failing, and viewed with utmost alarm the slow and certain fashion that, in short-lived and even in long-lived soils the wilt progressed soon to make banana production impossible. Pathologists redoubled their work on soil treatments that would still allow susceptible bananas to be grown in diseased land. Very briefly an indication of some of what was done is as follows. For the reason that short-lived soils were acid in reaction (pH of about 5.6 to 4.0), and the long-lived were neutral to slightly alkaline (pH of about 6.6 to 7.2), Scarseth (76) in a bold experiment attempted to change soil reaction. He selected a typical, seriously diseased, acidic, abandoned field of short-lived soil and applied quantities of lime. He grew cover crops on it and plowed them under, put on more lime, and after a time tested the soil. It had lost some acidity, and it was replanted to susceptible bananas. They grew some, but finally wilted. In a small way this has been confirmed by others.

In the banana growing tropics, one of the abundant, naturally available, possible soil amendment materials are trash, leaves, and other organic constituents. Examples of results from incorporations of such in soil of wilt

Fig. 9-3. Results from flood fallow of Fusarium wilted banana plantation. Aerial views showing disease control. Plantation area surveyed and dyked had over 50 per cent dead plants, many more wilting; Top: before flooding with water from nearby river, flooding maintained probably a year and a half; bottom: after flooding, area was drained, cultivated, replanted and produced good crops for a few years but was soon severely diseased. Photographs courtesy of The United Fruit Company, Tela Honduras.

Fusarium-infested fields are in researches reported by *Sequeira* (78). Organic matter, along with bacteria and other soil organisms, does reduce *Fusarium* counts in soil, but the use is not enough to eliminate wilt of bananas. In microbiological research in connection with fusarial growth in soils (33, 53, 54, 72, 74, 78, 87, 88, 91, 100, 101) it is known that bacteria, certain fungi and especially Actinomycetes have a counteracting effect on *F. o. cubense*. Great efforts were made to harness anti-fusarial soil organisms in the control of Gros Michel banana wilt, but after much study (54, 89) it was found that although theoretically feasible, from the standpoint of the banana planter it was practically impossible to eliminate, in sufficient measure, the banana *Fusarium* by antibiotics. To culture and multiply antagonists, such as actinomycetes, in the vast quantities needed for large field applications apparently cannot be achieved on an economic basis.

On the other hand, it should be suggested here that it is reasonable to conjecture that as intensive tropical food production agriculture progresses, due attention will be given to aiding parasite antagonists. In decades and centuries to come inoculations of soil with antagonists may well become common practice in many of the annual grains and other food crops. Biotic means of keeping spore loads of *Fusarium* (or other fungus genus) parasitic species at a moderate level, will allow of using inheritantly disease-tolerant crops. In many cases this is going to become a desirable change, and an advance over the search for high types of resistance or immunity of crop strains that sooner or later always do succumb to a new race of a parasite.

Banana wilt pathologists have, at one time or another, tried various chemical treatments, even including insect extractives (92), to disinfest *Fusarium*-filled soil. For example, chlorine, mercurials, and formaldehyde are all potent cleansing agents against *Fusarium* in the laboratory. When used in the field, the amounts needed to reduce fungus growth are too costly for practical purposes, and even where such a disinfestant has some effect it is for only a short time. Stover (89) has reviewed results on some of the chemical soil disinfestants; whether by themselves or in connection with flood-fallow treatments results with chemicals have been inadequate. It is worth stating that there is no soil-borne disease in the annals of all plant pathology that has had the same close attention by tests in vast field areas to the possibility of practical control through chemical amendments in the field than has banana fusarial wilt. At no time has a chemical given an adequate long-time control that under crop growing conditions has repaid for its use (19, 54, 73, 88, 89, 92).

In any country where fusarial wilt disease is found, wilt-beleaguered banana growers have become well informed about the *Fusarium*, and how it acts in the soil. They have tested many a *Fusarium*-resistant banana which was at one time strange to them and failed in some way in its production, in a shipping requirement, or in a marketing quality. Their engineers had adopted as their own the giant Gros Michel variety, and if it could be retained they were determined to keep it. For generations, it had been studied and tried out, found good in those days, and shipped all over the world. With application of proper tropical engineering management, it was profitable. It should be emphasized that the great plague of wilt drove out of a region companies of planters who had to move to new jungle lands or go bankrupt. However, virgin areas, adaptable for banana culture on a grand scale, after several decades were no longer available. Occasionally companies then returned to old once-used, excellent but long abandoned land that years ago grew fields which had succumbed to wilt. In replanting they used the susceptible and old favorite Gros Michel. However, the long-ago discarded land, soon again succumbed to wilt, often without harvests even starting to repay reclaiming costs. Eventually the planters had to cast about for resistant varieties. These have long been available, they have been much studied and chemical bases of their resistances determined in some cases, and they are being used (8-11, 21, 22, 25, 32, 39, 64, 66, 81, 89, 97, 98).

Host-Parasite Relations in Banana Wilt

When it became accepted that disease resistance--tolerance--was the only dependable chance to make banana production a permanent business, work was carried on towards it; many types and varieties of species of *Musa* were tested, and breeding programs were begun (9, 25, 32, 81, 89). In the interim, the long-life soil phenomenon gave promise of showing a hitherto unknown relationship between host physiology and disease tolerance, and much effort was spent upon the subject. The companies still hoped to continue some growing of the susceptible crop (Gros Michel banana) in the continuing presence of the parasite (*F. oxy-*

sporum cubense). Through many years of hard work at considerable expense and with great interest, a large number of investigators have probably learned more about intricacies of host-parasite relationships in the fusarial wilt of banana than that in any other Neotropical plant disease.

One of the questions not easily answered has been, what is the mode of entry of the Fusarium into the banana plant? Root infections have been well demonstrated in the temperate zone in several fusarial wilts of plants and study in the tropics has proved repeatedly (12, 37, 54, 75, 87-89, 95, 97, 100, 104) that the banana wilt comes from the soil and that artificial soil inoculations give positive results. However, it is of special interest, as several have shown (75, 79, 89 see review 97, 98, 100) that the intact large, vigorous and fleshy main roots*, as Stover (89) in his review states, "cannot be directly infected."

Some kind of injury has to occur to allow infection of a fleshy main root. Stover observed the pathogen growing over tips of healthy main roots without penetration. If the cortical cells of main roots are badly injured or even killed in soil, by drought, by high water content or by other effect they sometimes are infected and the stele reached by the Fusarium. I induced artificial infection of main roots. These were grown into flasks of three-month-old sand-rice-Fusarium cultures. Root injury when the root grew into the flesh appeared to be caused by pressure of the root growing against the glass of the flasks, and possible damage to root ends, by the old autolytic juices from the Fusarium cultures.

In Stover's (89) definitive review of studies it is shown that Fusarium oxysporum cubense does not go directly into cut surfaces of the banana corm, or "head" used as the seed piece. When dug and the main roots trimmed off even with the corm surface, the openings that are left would seem ideal courts of infection. Inoculation and microscope studies reveal that wounds secrete Fusarium-inhibitory substances,

---

* In Gros Michel banana for example, these roots may be almost as large in diameter as a pencil, have a rounded end, and come out of the corm in groups of four. I found that in many cases corm infections (100) occurred at points where the true roots pushed through corm tissues out into the Fusarium-infested soil.

and that host cells are stimulated to grow tyloses within them, to close the open ends of vascular elements. In addition the cut surfaces of rhizomes develop a cambiform barrier and enough suberization that even if it starts in, the fungus cannot progress. Insect borings, it is surprising to note, in rhizomes do not in themselves open the plant to successful infection. There does occur, however, much more effective fusarial invasion of banana plants if both Fusarium oxysporum cubense and various nematodes of both ecto- and endoparasitic habits are present and at work together in the Gros Michel banana rhizosphere, a situation that is found in other wilts as well (examples of literature are 29, 47, 63, 82, 83, 89, 99).

It seems that, from observation of effects in field and laboratory, there is no doubt that Gros Michel banana although susceptible has a small amount at least of natural tolerance to primary F. oxysporum cubense attack. In comparison with certain other plants hosting fusarial vascular diseases, such as the rapidly Fusarium-wilted Bonny Best tomato, the highly susceptible White Burley tobacco, or the easily Fusarium yellows-diseased Danish Ballhead cabbage (even considering the banana plants larger size in comparison with susceptible tomato, tobacco, and cabbage) the Gros Michel banana root system is more slowly attacked by its Fusarium to the extent of causing its plant to wilt and collapse.

By far the most common route of Fusarium entry into the banana plant is by way of the slender and fragile secondary and tertiary rootlets that grow on the main roots. Once a rootlet is infected the parasite grows rapidly in the stelar elements releasing toxins and causing discoloration. This is readily observable by dissection studies, and it is known that progress of the organism in main roots is much more rapid backwards towards the rhizome than forwards. Stover (89) reports a primary infection can travel back as much as two meters inside a main root to enter the rhizome, and only one or a few root infections are enough to result in a large amount of rhizome involvement.

Once within the vascular system of the subterranean rhizome or corm (Fig. 9-4), which is true morphological "stem" and "branch" in banana, the disease progresses readily inside (100). The organism kills side buds on rhizomes, and it moves along the vasculars extending through their connections to heads "branched-off" of "followers-from"

Fig. 9-4. Some internal symptoms from Fusarium in corm of susceptible banana, Musa AAA. Disease darkens tissues (healthy corm is nearly white when cut across). In this picture can be seen: blackening and rot of roots at sides and bottom, dark vascular strands leading up into leaf sheathes of pseudostem, accumulation of disease in base of corm, and killing of side buds on left. Photography courtesy of B. H. Waite.

old originally diseased corms. It is from such a diseased base that the aerial, above-ground part of a standing banana plant is invaded by the wilt organism. The *Fusarium oxysporum cubense* cannot infect cuts on leaves or pseudostems of banana, although interestingly Newcombe (62) grew the organism on leaf tissue that was gas sterilized, and Deese and Stahmann (21) grew it on aseptically removed and kept-living stem material. Invasion of the standing part of the plant is rapid when it once gains entrance, and the fungus enters leaf sheaths while the rhizome is only partly attacked.

The fungus hyphae are found within the xylem of the vascular system, and there they produce large numbers of microconidia which move upward with the sap. A notable feature of most vascular invasion in fusarial wilts is that there is commonly discoloration of the invaded elements. This is true of the banana disease, but in it the darkening of the xylem strands is sometimes discontinuous; a phenomenon of much interest and investigated by Stover (89), Drost (27), and Risbeth (72). Generally, when the main, long sap-carrying cells are invaded, they react to the organism inside. The vascular cells become toughened, discolored fibers. Where the discoloration has "skipped" no *Fusarium* can be found; microconidia have moved through cells without any harm. A little distance upwards, in an again darkened vessel, spores have been stopped, germinated and grown; this induces the dark color. Some vascular strands leading to diseased leaves are not discolored, and to me this suggests an inherent tendency towards at least some kind of partial resistance in Gros Michel.

The discoloration of vascular strands results in the stain of fiber cells. This is spectacular in fusarial wilt of abaca when the stalks of diseased plants are used for so-called manila hemp fiber. Soft tissues are stripped off fibers, and from diseased material the hemp is blackened (see Fig. 9-5) and the tensile strength materially weakened.

Work has been carried out to understand something about banana resistance mechanisms. There are resistance reactions to be found in Gros Michel, if inoculum is not excessive. This disease in Gros Michel operates as much more than a mere wilt. According to studies by such as Deese and Stahmann (21, 22), and Page (66) and others the harmful toxins appear to be a complex of at least fusaric acid and enzymes, among them pectic enzymes. It is coupled with longtime establishment of their poisoning actions

Fig. 9-5. Discoloration of "hemp" fibers obtained from fusarial wilt susceptible abaca. Upper is white and clean and strong from healthy plants; lower is dark, dirty-looking and somewhat weakened from diseased plants. Photograph from B. H. Waite.

that there occurs the effect from stopping sap flow. In diseased plants the sap stream is very severely interfered with, even before *Fusarium* can be isolated from the vasculars, since very soon the rate of transpiration from green leaves on a diseased plant is reduced to half that of similar leaves on healthy plants.

Because of space limitations at times, the banana plant which is large is not the ideal experimental subject for growing in environment chambers. Its gigantism has an advantage, however, in study of individual cells. Experimenters can and do probe the inside of xylem vessels with small glass rods, and the leaf-sheath vasculars can be treated *in situ* as individuals while alive. In plant pathology, study of fusarial wilt disease in this unique herb is resulting in increasing understanding of plants in health and in disease. Some of the greatest intricacies of systemic *Fusarium* parasitism are related to healthy plant life processes in very intimate fashion. In certain ways, some of these things are more difficult to learn about than any other phase of plant pathology, but it is probable that tropical pathologists, in the next decades, will make some of our most telling contributions to information about the inner workings of plant disease from research with bananas.

When effects of invasion of main roots of susceptible and resistant banana plants are studied side by side, microconidia at first are found in the xylem of both. These are the smallest of the minute spores and they move readily in the sap of any xylem vessel, although they are trapped at the sieve ending of the vessel. In both susceptible and resistant roots, microspores germinate where they are trapped in the sieve-end structures, and grow through them, producing more microconidia in the xylem cell that "fits" above (or on its distal end). The great difference is that in resistant xylem the microconidia that are usually set loose to travel further on are held back. The resistant plant secretes gels that surround and then immobilize the microconidia so there is no further spread.

An unusually interesting feature in the fusarial wilt problem relating to host-parasite relationships is that when microspores are injected into susceptible Gros Michel leaf-sheath xylem cells, the spores are recoverable a number of feet above the point of injection; during this time, internal presence of spores causes no systemic disease (Stover (89), Brandes (12), Stover and Waite (90)). Another phenomenon about this (Beckman *et al.* (11)), is that in comparing the rapidity of travel of microconidia a difference is found in the xylem sap of susceptible Gros Michel and resistant Lacatan banana plants. Movement is much slower in the resistant plant, even if either the resistant or the susceptible plant is not diseased. Moreover, this slowing down of spore movement for example in Valery or in Lacatan xylem, appears to be a specific reaction to *Fusarium oxysporum cubense* since small-size spores of other kinds of fungi move in xylem elements without much if any differential rapidity of travel. It appears that in the resistant plant, major occlusion develops rather rapidly in reaction to the superspecific parasite. It seems probable that the much slower passage of the microconidia up the xylem of a resistant variety is a habit that helps to retard the fungus; it gives time for the resistance mechanisms and chemical changes to become operative.

Ascigerous Stage in Study of Fusarium

Intensive work has been done in connection with many fusaria in the parasitic stage. This point in the organism's life habit is mostly involved with the asexual mycelium, micro- or macroconidium, and in some cases in either the intercalary-formed or terminally-borne chlamydospore. These three types of spores, and bits of food-gorged thick-walled, dormant mycelium, make up the populations of oft-mentioned propagules in fusaria. The chlamydospore is the type of fungus condition that is most likely to remain quiescent but viable under the rigorous biotic conditions of tropical soils. Antagonism results from numerous mycophagus nematodes, large amounts of *Fusarium*-inhibiting fungi of several kinds, many Actinomycetes and relatives that thrive on other types of fungi, and abundant bacteria with antibiotic habits in their contacts with fusaria; effects of physical nature such as drought and extended periods of extremely moist conditions in the soil are not likely to destroy chlamydospores.

A question that arises in *Fusarium* studies is presence or absence of the perfect stage. An ascospore would furnish another kind of propagule to the already formidable group of kinds upon which the fusaria depend for persistence in nature. Numerous kinds of *Fusarium* isolations have been secured from diseased plants in Florida, El Salvador, Costa Rica, Guatemala, Honduras, Nicaragua, and Puerto Rico. In these studies (1931 to 1962) greatest attention has been given to wilts, and most of them are from F. oxysporum of various special physiologic forms. Never, in all my experience, have I ever found an ascigerous stage from F. oxysporum. This, it seems, is characteristic of the species (85).

Work with oxysporum cultures used various kinds of mixings of isolates (as is known (66, 87, 88, 97, 100, 102-4), isolates often vary in culture characters), but also those growing on numerous types of agar media, in moist chamber studies, and on many kinds of sand and debris and vegetative tissues. Cultures were given "shock treatments" such as flushing with water of different pH reactions, artificial chilling, subjecting to intense light, alternating between bright light and darkness at short intervals, and attempting to infect living host tissues after freezing and alcohol injuries. No ascigerous stages were ever encountered with any method of crossing oxysporum races.

However, in one non-systemic Fusarium disease of coffee in Central America there has been a different result. In this disease, side branch stubs and upright growths are infected after pruning. The stubs die back several centimeters, and this kills leaves attached to infected tissue. The disease inhibits sprouting of adventitious buds that develop in healthy and normal upright (orthotropic) branches. The organism, as isolated from Costa Rica material, is a typical, light-colored Fusarium growing on solid agar (PDA). Along with the host tissue killing in the field, there are, in moist locations, occasional small numbers of ascocarpic bodies answering the description by Picado (68) of Nectria anisophilum Picado. These perfect, ascigerous stage components seem to be the main way in which the organism lives over from season to season in the plantation. I have seen, in four plantations, complete eradication of this disease through the careful pruning away of all diseased parts that contain the ascocarps and destroying them.

There is a tree canker damaging coffee in Guatemala (Heiser (40), LeBeau et al. (46), Wellman (100)), and which has spread south into El Salvador. In some fincas it is known to infect 95 per cent of the trees; it may not kill them but it causes a weakened condition. It appears to be associated with weeding operations and cuts are made, accidentally, with machetes. Under the canker is dark sap wood, from which it is easy to isolate a Fusarium that proves pathogenic when cultures of it are used to inoculate cuts in bark. Old cankers are characterized by cortex decay and a type of wrinkling of diseased tissue at its juncture with healthy bark. It is in this wrinkled part that the red- to purple-colored kernels develop. When dissected and examined microscopically, they are seen to be the organism *Nectria dodgei* Heiser. In isolation tests for parasite viability in my laboratory, cankers of young to old types were used. The sap wood under cankers from one to five years old readily yielded cultures of *Fusarium*. Heart wood was sterile. Cankers, one estimated to be eight and one ten years old, had no longer viable *Fusarium* in either heart wood or sapwood. However, from each, cortex just under the bark, with the *Nectria* ascocarps still in it, was dissected out and a few asci squeezed from the reddish fungus kernels. From 60 ascospores obtained, five germinated: two from the 8-year-old and three from the 10-year-old cankers.

The phenomenon of the apparent long life in ascogenous growths of fusaria has not been given much study anywhere. It would appear that in the tropics there are compelling reasons for the lifetime to be determined. In the case of maize, in some of the dryest years in Central America there are serious attacks by *Fusarium moniliforme* Sheldon. No adequate figures on losses are available, but native growers have told me that when maize crops follow successively, it is in the second and third dry year that fields are most subject to this disease, and a series of comparatively wet years greatly reduces damage. In observations it has been of interest that the ascogenous stage, *Gibberella fujikuroi* (Sau.) Wr., of this fungus is not common in my experience in wet country or in wet seasons. This may be sheer lack of sufficient observations or other accident, but after several years it remains to me a clear conclusion although based upon nothing more than field findings.

A common phenomenon in the tropics has been the encountering of saprophytic fusaria of different kinds in certain isolation studies. This is specially true when the subjects for isolation are such materials as: twig dieback from coffee, hevea rubber, cacao, lichi, anona, and also mango; root dieback of coffee; and old slow-spreading undetermined leafspots of coconut, breadfruit, citrus, coffee, oil palm, mango, and tree tomato. In a large number of instances, among the fungus organisms most commonly recovered appear to be *Fusarium roseum*. Occasionally short-spored fusaria are encountered, which on culturing give occasional sectors that produce longer spores, but in past experience these fitted well into the *F. solani* species requirements. In long extant, non-spreading leafspots of oil palm and mango grown in dry regions, the isolates obtained had among them unusually slender and long macrospores similar to F. moniliforme. Fusarium isolations from such sources as the above have not been pathogenic when used as inoculum on healthy, growing twigs or on vigorous leaves of host species from which they were originally secured.

At several times, unused remains from tissues collected for isolation studies have been kept for periods of one to several months in large, moist glass chambers in front of a window. In some of the chambers, on dead twigs or on old leaves thus treated, have grown the small, somewhat delicate and variously colored fungus fruiting bodies that appear to be ascogenous stages of fusaria. Dark-colored fungus bodies, determined as Gibberella sp., were found on old coffee twig dieback material. From coffee leaves with non-spreading spots on them that were over two years old, and on coffee dieback twigs from exhausted fruiting branches that had been dead 18 months, creamy to brown colored fungus bodies developed that were determined as possibly Calonectria sp. On dieback twigs from cacao and hevea rubber that had been originally killed by pink disease (Corticium salmonicolor), when kept for more than three months in a moist chamber, there developed red to pinkish- and dark-purple ascocarps of Nectria sp. At another time, in a moist chamber containing jungle debris, yellow and brown ascogenous bodies, Hypomyces sp., grew from a piece of a long dead, wild, and undetermined palm frond. In several instances numbers of undetermined ascus-bearing perithecial growths were seen being destroyed by mites and nematodes feeding on evident Fusarium hyphae growing on dead tissue surfaces incubated in the moist chambers.

Nematodes with Fusarial Wilts

A recognized technique among nematologists in certain types of study is feeding of their eelworms on pure cultures of various kinds of fungi. In the moist tropics, both fungi and nematodes may be found in more abundance than in the temperate zone. The worms are very common in soils that are not subject to long droughts; the plant parasitic nematodes of crops may multiply with extreme rapidity under moist tropical conditions. Nematodes can cause a plant to wilt through their own feeding actions on the host root system, but sometimes when parasitic fusaria are present along with root-attacking nematodes, the wilt disease is much more severe.

With the root knot nematode *Meloidogyne*, it is often found that normally expected *Fusarium*-resistance reactions of cotton and other crops may be much reduced and greatly complicated by the derangement introduced by nematode infections.

When there is presence of many root knot nematodes, tobacco is most likely to succumb to fusarial wilt from *F. o. nicotianae*. Resistant varieties developed in nematode-free soil can be induced to wilt from the fungus if planted in a location where the parasite occurs and at the same time the roots are heavily attacked by *Meloidogyne* sp. It can be readily believed that in many parts of the tropics, fusarial wilts are now, or have been in the past, made much more severe from the effects of their combination with nematodes.

In confirmation of long-time observations, Montealegre (58) pleaded, and presented cogent reasoning, for most of the coffee growers to preserve the natural shade of their plantations, as it decreased disease. It has been observed by me that coffee fusarial wilt, *F. oxysporum coffeae*, and coffee root knot, *Meloidogyne* sp., increase when shade is removed. In attempting to find a reason for this, it was finally concluded that shade trees not alone shade but also furnish extra leafy soil mulch that has its advantages under certain conditions. This increased ground cover appears to retard certain diseases. When a coffee planting, with its shade taken away and showing declination from *Fusarium* and *Meloidogyne* diseases, is replanted to shade, the diseases are reduced. This is experienced not only among coffee growers. Heavy loads of weed and grass mulch materials are cut and carried in to their fields by peasant and Indian growers of Central America. The mulches are arranged around hills of beans, maize, cucurbits and tomatoes, and such treatment may or may not reduce fusaria, but in any case it improves production. Field visits and collecting trips have shown by examination that root troubles, especially from fusarium, are more prevalent in unmulched than in mulched plants of beans and tomatoes.

An Addendum

In spite of all the excellent agricultural engineering

and plant disease research expended upon it, *Fusarium* destruction continues of the desirable but wilt-susceptible Gros Michel banana. The final and only permanent solution remains in securing banana varieties with high tolerance-resistance to *Fusarium oxysporum cubense*. There are modern moves in that direction and some dozen varieties are successful and finding their way into the market. At the present time of writing the so-called Valery variety, a selection out of Cavendish, is growing wilt-free, producing double the usual crops in old wilt disease fields in Honduras.

Banana breeding is slow and expensive but research in Jamaica (see reviews in Wardlaw (97), Dodds (25), Stover (89), and Osborne (64)) has given results usable in both eastern and western tropics. For the future, resistant bananas tailored to exacting requirements, and changing with each new variation in disease, will be grown in the very susceptible but excellent (now completely abandoned) fields of soils of banana-producing countries. For all of the Neotropical population, resistant bananas as a source of food for local populations in the near future may become of even greater importance than ever before. It should be being worked upon perpetually by a team of international experts, under such auspices as the Food and Agriculture Organization of the United Nations with a guarantee of being carried on for centuries.

During some years in the temperate zone, working with wilt of tomato, *Fusarium oxysporum lycopersici* (see (100), three publ. in 1942 and 1943), studies were pursued concerning growth changes by mutations in the fungus in pure culture--both test tube slants and petri plates of PDA. Variation in strains was marked and demonstrable in culture appearance, physiology, and pathogenicity. An important phase, started but incomplete because of assignment change, was effects from sectoring of growing fusarium mats. In one series of 1000 mat sectors, three were proved as somewhat more virulent than were the cultures of origin; in a second series of 1000 mutant sectors, two were more virulent. Meanwhile (100) under controlled conditions, virulent and mild isolates mixed in equal proportions were used as inoculum of susceptible tomato seedlings in greenhouse bench and pot soils. The avirulent and virulent races both invaded hosts, but the virulent types outgrew and outmultiplied the mild races.

Certain related *Fusarium oxysporum* studies were

Table 15. A FEW UNUSUAL TROPICAL *FUSARIUM* DISEASES brought or sent to the writer in Puerto Rico during one month of 1960.

Disease	Organism	Control Measure
Vanilla plant failure	*Fusarium oxysporum vanillae* (in part)	Sanitation, resistance
Orchid root disease	*Fusarium roseum*	Sanitation, sprays
Banana fruit stalk disease	*Fusarium solani* (in part)	Fruit washes in chemical solution
Gandul canker on immature branches	*Fusarium roseum*	Sanitation, crop rotation
Snapdragon stem disease	*Fusarium roseum*	Sanitation, rotation
Clusia stem disease	*Fusarium lateritium*	--
Grain sorghum head blight	*Fusarium* spp.	Use of varieties producing open heads
Oleander witches' broom	*Fusarium roseum* (in part)	Pruning, sanitation
Mycena leafspot on coffee, enlarged	*Fusarium roseum* (secondary)	--
Lagerstroemia tree trunk gall	*Fusarium* (*rigidiuscula*)	Cutting away, burning

carried on at intervals in the tropics on banana wilt (for partial results, see (100), three publ. during 1930, 1931). With the help of assistants, widely varient isolations were secured of F. o. cubense from both diseased hosts and soil: some of the parasites were highly virulent while others were F. oxysporum but could not infect host root hairs. At another period (see (100) during 1949-1951) numerous isolations from soil and banana hosts affected with F. o. cubense yielded many mutants and both pathogenic and non-pathogenic races of the parasite; in connection with making a diagnosis of tomato wilt in plants and soils from the south coast of Puerto Rico in 1961, using special methods (67) in Fusarium recovery from soil, isolates yielded virulent cultures of F. o. lycopersici and as well as avirulent Fusarium oxysporum.

From experience, as given in the above two paragraphs, there is a suggestion of how increasingly virulent races of a Fusarium can arise through rare mutations, and be multiplied in susceptible hosts. Propagule populations of the more virulent are probably left in overwhelming numbers in soils, and it is probable this multiplication may be more in the warm tropics than in cooler countries.

Table of Unusual Tropical Fusarium Diseases

From the standpoint of a better and fuller concept of all that the fusarium troubles can be, and the problems that will be more serious in the future, Table 15 on page 379 gives a suggestion of some of the disease types not commonly treated in literature

Literature Cited

It is solely the literature, and work consulted and discussed in preparing this chapter, that is included here. At one time a total of 882 citations had been gathered for reference to fusarial problems; over two thirds of the items dealt with western hemisphere tropics.

The appended list, therefore, is far from exhaustive. This is illustrated in two examples: in covering literature on the fusarial wilt of tobacco, Fusarium oxysporum nicotianae, I at one time collected over 200 citations and in the present chapter only 11 are used; in connection with the classic monographic treatment of banana fusarial wilt, Fusarium oxysporum cubense, Stover (89) in his citations, including sub-reports, has a total of 358 items. Even with the aid of Stover's list, in this chapter I have given only 70 citations hereunder.

1. Alvarado, R., E. Ampuero, and K.D. Doak. 1961. Reacción de differentes clones y progenies hibridas a la buba o cushion gall del cacao. Inter-Amer. Cacao Conf., Trinidad & Tobago Proc. 8: 290-301.

2. Alvarez-G., L.A. 1945. Studies on coffee root disease in Puerto Rico. I. A coffee Fusarium wilt. Univ. Puerto Rico, J. Agr. 29:1-29. --. 1952. Estudios de enfermedades de la raíz del cafeto en Puerto Rico. Rev. Cafe, Puerto Rico 8:5-6.

3. Archibald, J.F. 1961. Transmission of gall-diseases of cacao, mango, and pigeon pea. Nature 190: 284.

4. Armstrong, G.M. and J.K. Armstrong. 1948. Non-susceptible hosts as carriers of wilt Fusaria. Phytopathology 38:808-826.

5. ________., B.S. Hawkins, and C.C. Bennett. 1942. Cross inoculations with isolates of Fusaria from cotton, tobacco, and certain other plants subject to wilt. Phytopathology 32:685-689.

6. Arruda, S.C. 1939. Mal do Panama en Bernadino de Campos. Biológico 5:198-199.

7. Aryza, H.C. and G.L. Jain. 1962. Fusarium wilt of eucalyptus. Phytopathology 52:638-642.

8. Ashby, S.F. 1924. Bananas resistant to wilt (Panama disease). Trop. Agr., Trinidad 1:172-173.

9. Baker, R.E.D. and N.W. Simmonds. 1951. Banana research--changes in outlook. Trop. Agr., Trinidad 28:43-45.

10. Beckman, C.H., S. Halmos, and M.E. Mace. 1962. The interaction of host, pathogen and soil temperature in relation to susceptibility to Fusarium wilt of bananas. Phytopathology 52:134-140.

11. ________., M. E. Mace, S. Halmos, and M. W. McGahan. 1961. Physical barriers associated with resistance in Fusarium wilt of bananas. Phytopathology 51:507-515.

12. Brandes, E. W. 1919. Banana wilt. Phytopathology 9: 339-389.

13. (Brazilian Consultant). 1929. O mal da banana nos bananaes do Equador. Conselhos para debellar o mal da banana. Bol. Agr. São Paulo 30:378-381.

14. Buxton, E. W. 1962. Root exudates from banana and their relationship to strains of Fusarium causing Panama wilt. Ann. Appl. Biol. 50:269-282.

15. Calderoni, A. V., M. V. Fernandez Valiela, and M. Bakarcic. 1952. Fusariosis y marchitamiento de la papa. Reunion de Papa Pergamino, Argentina, Mem. 1:43-50.

16. ________., ________., ________. 1958. Marchitamiento de la papa. IDIA (Argentina) 129:21-24.

17. Carrera, C. J. M. 1941. La fusariosis o marchitamiento del lino en la República Argentina debida al Fusarium lini Bolley. Almanaque Min. Agr., Argentina 16:141-142. --. 1954. El genero Fusarium; estudio e identificación de especies de la Republica Argentina y paises limitrofos. Rev. Inv. Agr., Argentina 8:311-456.

18. ________. and M. J. M. De Giberti. 1947. Pruebas experimentales realizzadas con Fusarium gramineum Schw. Argentine Min. Agr., Inst. San. Veg. Publ. 3:1-7.

19. Corden, M. E. and R. A. Young. 1960. The fungicidal activity and sorption of nabam in soil. Phytopathology 50:83.

20. Crowdy, S. H. 1947. Observations on the pathogenicity of Calonectria rigidiuscula (Berk. & Br.) Sacc. on Theobroma cacao L. Ann. Appl. Biol. 34:45-59.

21. Deese, D. C. and M. A. Stahmann. 1962. Pectic

enzymes and cellulase formation by Fusarium oxysporum f. cubense on stem tissues from resistant and susceptible banana plants. Phytopathology 52:247-255.

22. ________., ________. 1962. Pectic enzymes in Fusarium-infected susceptible and resistant tomato plants. Phytopathology 52:255-260.

23. Deslandes, J. 1937. Doenças da bananaeira. Rodriguesia 11:199-206.

24. Desrosiers, R. and A. von Buchwald. 1955. Fungicidal treatments for control of cacao cutting disease in Ecuador. FAO Pl. Prot. Bull. 4:17-21.

25. Dodds, K. S. 1950. The breeding of disease-resistant bananas. World Crops 2:56-59.

26. Dongo D., S. L. 1961. Especies de Fusarium patogenas sobre frijol en el Perú. Estac. Exp. Agr. La Molina, Peru Bulletin 3, 16 p.

27. Drost, A. W. 1912. De surinaamsche panamaziekte in de gros michel bacoven. (Quoted by Stover (89), not seen by the writer.)

28. Duggar, B. M. 1909. Fungous diseases of plants, with chapters on physiology, culture methods and technique. Ginn & Co., Boston, 508 p.

29. Feder, W. A. and J. Feldmesser. 1961. The spreading decline complex: the separate and combined effects of Fusarium spp. and Radopholus similis on the growth of Duncan grapefruit seedlings in the greenhouse. Phytopathology 51:724-726.

30. Ford, E. J., J. A. Bourret and W. C. Snyder. 1967. Biologic specialization in Calonectria (Fusarium) rigidiuscula in relation to green point gall of cacao. Phytopathology 57:710-712.

31. Frezzi, M. J. 1939. La presencia del Fusarium bulbigenum v. blasticola en la República Argentina. Physis (Argentina) 15:86-97.

32. Galli, F., H. Tokeshi, P. de C. Torres de Carvalho,

E. Balmer, H. Kimati, C. O. Nogueira Cardoso, and C. L. Salgado. 1968. Manual de fitopatologia, Doenças das plantas e seu Contrôle. Ceres, São Paulo, 640 p.

33. Garrett, S. D. 1944. Root disease fungi. Chronica Botanica. Waltham, Massachusetts, 177 p.

34. Garcia-Rada, G. 1940. Una nueva enfermedad del cafeto, fusariosis. Est. Exp. Agr. La Molina (Peru). Bol 20, 7 p.

35. Goberdhan L. and D. Ganpat. 1961. Cushion gall of cacao in Trinidad. Inter-Amer. Conf., Trinidad & Tobago Proc. 8:270-280.

36. Gordon, W. L. 1954. The occurrence of Fusarium species in Canada. III. Taxonomy of Fusarium species in the seed of vegetable, forage, and miscellaneous crops. Canadian J. Bot. 32:576-590. --. 1956. The taxonomy and habitats of Fusarium species in Trinidad, B. W. I. Canadian J. Bot. 34:847-864. --. 1960. The taxonomy and habitats of Fusarium species from tropical and temperate regions. Canadian J. Bot. 38: 643-658.

37. Grant, T. J. (1929, 1930, Reports to Department of Plant Pathology, United Fruit Company, Tela, Honduras. Unpublished. See also references in Stover (89). Personal discussions with writer.) --. 1962. The role of plant pathology in tropical plantation agriculture. Phytopathology 52:930-936.

38. Hansen, A. J. 1963. The role of Fusarium decemcellulare and Fusarium roseum in the greenpoint cushion gall complex of cacao. Turrialba 13:80-87.

39. Harper, J. L. 1950. Studies on resistance of certain varieties of banana to Panama disease. I. Internal factors for resistance and antibiotica. II. The rhizosphere. Plant & Soil 2:374-394.

40. Heiser, D. G. 1945. A nectria disease of coffee in western Guatemala. Ann. Missouri Bot. Gard. 32:287-296.

41. Hutchins, L. M. 1959. Transmission of cushion gall of cacao by means of tissue transplantation. Communications Turrialba, Costa Rica 66, 6 p. --. 1961. Transmission tests with cushion gall of cacao. Inter-Amer. Conf., Trinidad & Tobago Proc. 8:255-262. --. 1962. Recent advances in cushion gall research. Phytopathology 52:25. --. 1964. Predisposition to cacao knob gall in wounds where flowery gall has been excised. Phytopathology 54:499. --. 1965. Loss of gall-inducing capacity on cacao, when _Calonectria rigidiuscula_ passes from the conidial (fusarium) stage through the perfect (ascospore) stage. Plant Dis. Reptr. 49:564-565.

42. ________. and L. R. Siller. 1960. Cushion gall types in cacao. Inter-Amer. Cacao Conf., Eighth, 1960:281-287.

43. ________. J. Soria and L. R. Siller. 1964. Partial resistance to greenpoint cushion gall in cacao clones completely resistant to flowery gall. Phytopathology 54:499.

44. Jiménez S., E., A. Llano and L. M. Hutchins. 1965. Effect of nitrogen, potassium, and boron on the growth of green point galls in cacao. (ms. seen.)

45. Krug, H. P. 1941. Cafes duros. IV. Relação entre zonas, qualidade de café e porcentagens de microorganismos. Rev. Inst. Café, Brazil 16: 288-295.

46. LeBeau, F. J., M. A. Flores, and M. R. Hernandez. 1955. El cancer del café causado por _Nectria._ Inst. Agr. Nac., Guatemala Boletin 2, 10 p.

47. Loos, C. A. 1959. Symptom expression of _Fusarium_ wilt disease of the Gros Michel banana in the presence of _Radopholus similis_ (Cobb, 1893) Thorne, 1949 and _Meloidogyne incognita acrita_ Chitwood, 1949. Proc. Helm. Soc. Washington 26:103-111.

48. Malaguti, G. 1958. Primeras observaciones sobre la buba agallas del cacao en Venezuela. Agron. Trop., Venezuela 8:115-120.

49. ________. and L. C. de Reyes. 1964. A gall disease of cacao and mango in Venezuela caused by *Calonectria rigidiuscula*. Phytopathology 54:499.

50. ________. and L. Sirotti. 1952. Podredumbre de la corteza de los bucares (*Erythrina* spp.) y del cacao (*Theobroma cacao*) en Venezuela. Agron. Trop., *Venezuela* 2:41-53.

51. Maloy, O. C. 1960. Physiology of *Fusarium solani* f. *phaseoli* in relation to saprophytic survival in soil. Phytopathology 50:56-61.

52. Marengo, L. V. 1942. Fusariosis del girasol (*Helianthus annus* L.). Rev. Fac. Agr. Vet., Argentina 10:130-145.

53. Marshall, K. C. and M. Alexander. 1960. Competition between soil bacteria and *Fusarium*. Plant & Soil 12:143-153.

54. Meredith, C. H. 1941. The growth of *Fusarium oxysporum cubense* in the soil. Phytopathology 31: 91:93. --. 1942. The effect of chemicals on *Fusarium oxysporum cubense* growing in the soil. Phytopathology 32:182-184. --. 1943. Mercury compounds applied to banana plants in the field. Phytopathology 36:983-987.

55. Messiaen, C. M. 1959. Le systématique du genre *Fusarium* salon Snyder et Hansen. Rev. Path. Veg., Ent. Agr., France 38:253-266.

56. Mitchell, R. 1959. Microbiological changes in submerged soil as related to control of *Fusarium oxysporum* f. *cubense* (E. F. S.) Sny. and Hans. (Not seen; M. S. thesis, Cornell Univ., quoted by Stover.)

57. ________. and M. Alexander. 1961. The mycolytic phenomenon and biological control of *Fusarium* in the soil. Nature 190:109-110.

58. Montealegre, M. R. 1954. Cafetales a pleno sol, versus cafetales a la sombra. Suelo Tico, Costa Rica 7:263-275.

59. Mujica, F. 1944. El genero Fusarium en Chile. Agr. Tec. Chile 4:224-229.

60. Nash, S. M., T. Christou, and W. C. Snyder. 1961. Existence of Fusarium solani f. phaseoli as chlamydospores in soil. Phytopathology 51:308-312.

61. Nash, S. M. and W. C. Snyder. 1962. Quantitative estimation by plate counts of propagules of the bean root rot Fusarium in field soils. Phytopathology 52:567-572.

62. Newcombe, M. 1960. Some effects of water and anaerobic conditions on Fusarium oxysporum f. cubense in soil. Trans. British Mycol. Soc. 43: 51-59.

63. Newhall, A. G. 1958. The incidence of Panama disease of banana in the presence of root knot and burrowing nematodes (Meloidogyne and Radopholus). Plant Dis. Reptr. 42:853-856.

64. Osborne, R. E. 1958. Breeding the immune banana. Jamaica Banana Board Res. Dep., Bulletin 2, 8 p.

65. Owen, H. 1956. Further observations on the pathogenicity of Calonectria rigidiuscula (Berk. & Br.) Sacc. to Theobroma cacao L. Ann. Appl. Biol. 44:307-321.

66. Page, O. T. 1959. Observations on the water economy of Fusarium-infected banana plants. Phytopathology 49:61-65. --. 1961. Variation in the banana-wilt pathogen, Fusarium oxysporum f. cubense. Canadian J. Bot. 39:545-557.

67. Parmeter, Jr., J. R. and J. R. Hood. 1961. The use of Fusarium culture filtrate media in the isolation of Fusaria from soil. Phytopathology 51: 164-168.

68. Picado, T. C. 1932. Fusarium disease of coffee in Costa Rica. J. Dep. Agr., Puerto Rico 16:389-400.

69. Pontis, R. E. 1940. El marchitamiento del pimiento (Capsicum annum) en la provincia de Mendoza. Rev. Argentine Agr. 7:113-127. --. A canker disease of the coffee tree in Colombia and Venezuela. Phytopathology 41:178-184.

70. Reinking, O. A. 1934. Parasitic and other Fusaria counted in Costa Rica and Panama soils. Zentralb. Bact. Parasitenk. Infkt. 90:4-17.

71. ________., and M. Manns. 1933. Parasitic and other Fusaria counted in tropical soils. Zeitschr. Parasitenk. Inf. 6:23-75.

72. Risbeth, J. 1955. Fusarium wilt of bananas in Jamaica. I. Some observations on the epidemiology of the disease. Ann. Bot. 19:293-328. --. 1957. Fusarium wilt of bananas in Jamaica. II. Some aspects of host-parasite relationships. Ann. Bot. 21:215-245.

73. ________., and A. G. Naylor. 1957. Fusarium wilt of bananas in Jamaica. III. Attempted control. Ann. Bot. 21:599-609.

74. Rombouts, J. E. 1953. The microorganisms in the rhizosphere of banana plants in relation to susceptibility or resistance to Panama disease. Plant & Soil 4:276-288.

75. Salas, L. J. A. 1958. Estudios sobre la penetración de Fusarium oxysporum f. cubense causante de la enfermedad de Panama en Gros Michel (Musa sapientum L.) en condiciones de laboratorio. (Thesis for Ing. Agrón. degree, Univ. de Costa Rica; see Stover (89)).

76. Scarseth, G. D. 1944. Growing bananas on acid soil. Agriculture in the Americas 4:188-189.

77. de Segura, C. B. 1965. Enfermedades de Cultivos Tropicales y Subtropicales. Editorial Jurídica, S. A., Lima, Peru, 439 p.

78. Sequeira, L. 1962. Influence of organic amendments on survival of Fusarium oxysporum f. cubense in the soil. Phytopathology 52:976-982.

79. ________., T. A. Steeves, M. W. Steeves, and J. M. Riedhart. 1958. Role of root injury in Panama disease infections. Nature 182:309-311.

80. Siller, L. R. 1962. The influence of cushion gall disease upon yields of cacao. Phytopathology 52:26.

81. Simmonds, N. W. 1959. Bananas. Longmans, Green & Co., London, 466 p.

82. Smith, A. L. and J. B. Dick. 1960. Inheritance of resistance to fusarium wilt in upland and sea island cottons as complicated by nematodes under field conditions. Phytopathology 50:44-48.

83. Smith, C. G. 1953. The incidence of fusarium wilt of tobacco as related to root injury by plant parasitic nematodes. (Thesis for Ph. D. degree, North Carolina State College.)

84. Smith, E. F. 1910. A Cuban banana disease. Science 31:754-755.

85. Snyder, W. C. and H. N. Hansen. 1940. The species concept in _Fusarium._ Amer. J. Bot. 27:64-67. -- and --. 1941. The species concept in _Fusarium_ with special reference to the section _Martiella._ Amer. J. Bot. 28:738-742. -- and --. 1945. The species concept in _Fusarium_ with reference to Discolor and other sections. Amer. J. Bot. 32:657-666.

86. Stevenson, J. A. 1918. A check list of Porto Rican fungi and a host list index. J. Dep. Agr. Porto Rico 2:125-264.

87. Stover, R. H. 1953. The effect of soil moisture on _Fusarium_ species. Canadian J. Bot. 3:693-697. --. 1953. The effect of soil moisture on the growth and survival of _Fusarium oxysporum_ f. _cubense_ in the laboratory. Phytopathology 43: 499-504. --. 1954. Flood fallowing for eradication of _Fusarium oxysporum_ f. _cubense._ II. Some factors involved in fungus survival. Soil Sci. 77:401-414. --. 1955. Flood fallowing for eradication of _Fusarium oxysporum_ f. _cubense._ III. Effect of oxygen on fungus survival. Soil Sci. 80:397-412.

88. ________. 1956. Studies on Fusarium wilt of bananas. I. The behavior of F. oxysporum f. cubense in different soils. Canadian J. Bot. 34:927-942. --. 1958. Studies on Fusarium wilt of bananas. II. Some factors influencing survival and saprophytic multiplication of F. oxysporum f. cubense in soil. Canadian J. Bot. 36:311-324. --. 1958. Studies on Fusarium wilt of bananas. III. Influence of soil fungitoxins on behavior of F. oxysporum f. cubense in soil extracts and diffusates. Canadian J. Bot. 36:439-453. --. 1959. Studies on Fusarium wilt of bananas. IV. Clonal differentiation among wild type isolates of F. oxysporum f. cubense. Canadian J. Bot. 37:245-255. --. 1959. A rapid and simple pathogenicity test for detecting virulent clones of Fusarium oxysporum f. cubense using seedlings of Musa balbisiana. Nature 184:1591-1592. --. 1960. Studies on Fusarium wilt of bananas. V. Pathogenicity and distribution of F. oxysporum f. cubense races 1 and 2. Canadian J. Bot. 38:51-61. (Waite, B. H. and) --. 1960. Studies on Fusarium wilt of banana. VI. Variability and the cultivar concept in Fusarium oxysporum f. cubense. Canadian J. Bot. 38:985-994. --. with R. C. Hildreth and N. C. Thornton. 1961. Studies on Fusarium wilt of bananas. VII. Field control. Canadian J. Bot. 39:197-206.

89. ________. 1962. Fusarial wilt (Panama disease) of bananas and other Musa species. Commonwealth Myc. Inst., Phytopath. Paper 4, 117 p. (358 citations.)

90. ________. and B. H. Waite. 1954. Colonization of banana roots by Fusarium oxysporum f. cubense and other soil fungi. Phytopathology 44:689-693.

91. ________. and S. R. Freiberg. 1958. Effect of carbon dioxide on multiplication of Fusarium in soil. Nature 181:788-789.

92. Timonin, M. J. 1961. The interaction of plant, pathogen, and Scaptocoris talpa Champ. Canadian J. Bot. 39:695-703. --. 1961. Effect of volatile constituents of Scaptocoris talpa Champ on the growth of soil fungi. Plant and Soil 14:323-324.

93. Toussoun, T. A. and W. C. Snyder. 1961. Germination of chlamydospores of Fusarium solani f. phaseoli in unsterilized soils. Phytopathology 51:620-623.

94. Viegas, A. P. 1939. A murcha da bananeira ou mal do Panama. Rev. Agr., Piracicaba 14:225-226. --. 1946. Alguns fungos do Brasil XIII. Hifomicetes. Bragantia 6:353-442. --. 1961. Indice de Fungos da America do Sul. Inst. Agronomico, Campinas, 921 p. --. 1961. Murcha do algodoeiro. Brangantia 20:547-556.

95. Volk, N. J. (1926-1940. Unpublished reports on soil explorations in Central America, and results from laboratory research on chemical soil components related to wilt of banana for the United Fruit Company, Tela, Honduras; also unpublished United Fruit Co., Research Dept. Circulars; personal communications with me; and see references by Stover (89) and his discussion.)

96. Waite, B. H. 1954. Vascular disease of abaca or manila hemp in Central America. Plant Dis. Reptr. 38:575-578. -- and V. C. Dunlap. 1953. Preliminary host range studies with Fusarium oxysporum f. cubense. Plant Dis. Reptr. 37:79-80.

97. Wardlaw, C. W. 1929. Panama disease research. Trop. Agr., Trinidad 6:192-197. --. 1930. The biology of banana wilt (Panama disease). I. Root inoculation experiments. Ann. Bot. 44:741-766. --. 1930. The biology of banana wilt (Panama disease). II. Preliminary observations on sucker infection. Ann. Bot. 44:917-956. --. 1931. The biology of banana wilt (Panama disease). III. An examination of sucker infection through root bases. Ann. Bot. 45:382-399. --. 1961. Banana diseases, including plantains and abaca. Longmans, Green and Co., London, 648 p.

98. ________., and L. P. McGuire. 1929. Panama Disease of Bananas. Reports on scientific visits to the banana growing countries of the West Indies, Central and South America. Empire Mktg. Bd., Pub. 20, 97 p.

99. Wehunt, E. J. and A. Q. Holdeman. 1959. Nematode problems of the banana plant. Soil and Crop Sci. Soc. Florida, Proc. 19:436-442.

100. Wellman, F. L. 1930. Nature of spread of Panama disease (Fusarium cubense) in rhizomes of gros michel banana plants. United Fruit Co., Tropical Res. Lab. Circ. 19, 21 p. --. 1930. Plant disease-free seed pieces to control Panama disease of bananas. United Fruit Co., Tropical Res. Lab. Circ. 21, 30 p. --. 1931. Progress of Fusarium wilt inside the rhizomes of banana plants. Phytopathology 21:121. (--. 1937-1939, 1949-51. Unpublished notes on soil and root isolations of fusaria in Maryland, U. S. A., and Costa Rica.) --. 1942. Differences in pH relations of some pathogenically variable strains of tomato Fusarium. Phytopathology 32:271-287. --. 1943. Increase of pathogenicity in tomato wilt Fusarium. Phytopathology 33:175-193. --. 1943. A new species of Fusarium causing vascular wilt of tomato. Phytopathology 33:956-958. --. 1946. Anotaciones de las enfermedades de las plantas en El Salvador. XX. Decaimento del sistema radicular del tomato. El Cafe de El Salvador 16:889-890. (--. 1946-50, 1950, 1953-1957. Results from laboratory studies with non-technical aides, unpublished miscellaneous notes.) --. 1954. Fungi isolated most frequently from dying coffee fruit branches. Phytopathology 44: 509. --. 1954. The Fusarium phase of the root-disease complex in coffee. Phytopathology 44:509. --. 1955. Coffee diseases, insects, and weeds controlled by chemicals. Advances in chem., Ser. 13:43-63. --. 1961. Coffee Botany, Cultivation, and Utilization (Especially pp. 89, 292-294). Leonard Hill, London; and Interscience Publ., New York, 488 p. --. 1962. A few introductory features of tropical plant pathology. Phytopathology 52:928-930. (Introduction to symposium on tropical plant pathology.)

101. Williams, L. E. and D. D. Kaufman. 1962. Influence of continuous cropping on soil fungi antagonistic to Fusarium roseum. Phytopathology 52:778-781.

102. Wollenweber, H. W. 1913. Studies on the Fusarium

problem. Phytopathology 3:24-50. --. 1917. Fusaria autographice. Delineata. Ann. Mycol. 15:1-5. --. 1926 & 1930. Fusaria autographice delineata. Collectio specierum et ex herbariis variis selectarum etc. Berlin, Plates 1-659 and 660-1100. --. 1931. Fusarium-monographie. Fungi parasitici et saprophytici. Zeitschr. Parasitenk. 3:269-516.

103. ________., C. D. Sherbakoff, O. A. Reinking, H. Johann, and A. A. Bailey. 1925. Fundamentals for taxonomic studies of Fusarium. J. Agr. Res. 30:833-843.

104. ________. and O. A. Reinking. Die Fusarien ihre Beschreibung, Schwadwirkung und Bekaempfung. Paul Parey, Berlin 355 p.

Chapter X

LEAFSPOT PROBLEMS OF THE TROPICS

Summary

Leafspots are lesions bounded by functioning leaf tissues; spots are very common and many are severe in the tropics. They vary widely in colors, shapes, surface textures, and sizes. Weakspot, not completely understood, is probably from fungus infection that did not progress, but may be severely defoliating in effect. Leafspots unless controlled are the cause of continuous drain on crop producers and therefore the economy of all tropical agriculture; in many situations the spots are practically disregarded and accepted as added difficulties in crop growing. Many leafspots when at first encountered give an impression of being minor in effect. However, with years and more intensive husbandry many spots are finally recognized as limiting factors. Ecologic zones have profound effects on leafspots. Under some conditions certain spots are extremely serious, under other conditions certain spots do not cause disease loss.

Introduction

Leaf spots are so common on plants of both wild and cultivated species in the tropics that many people accept them as normal markings. The tropical farmer has always known disease leafspots as part of what is found on his crops, he has learned that the spots apparently do not kill the plants outright or even rapidly, and that either a perennial plantation or a growing annual field crop will be able to live with them. In many cases he needs to do nothing more than carry on some additional horticultural treatment such as shade increase or fertilizing to make the crop grow better without attempting to control the spots. A tropical farmer believes that although spots are more or less of minor importance in comparison with wilts and root rots, leaf blemishes may have only slightly harmful effects. He feels they cannot be avoided and as a consequence are to be treated as part of the scene; there are varieties of certain crops in the tropics that have special spots which do not affect others.

In some instances plant crop species are recognized, almost like botanical characters, by their spots. For example, there is the case of Guinea grass, important and popular for pasturage, which is many times known by the short, streak-like, elongate spots that are caused by its *Cercospora fusimaculans* Atk.; in the same way, volunteer plants of sorghum often are marked with infections of *Puccinia purpurea* Cke.; Bermuda grass is often picked out from among other grasses by its spots caused by *Helminthosporium giganteum* Heald & Wolf; and the sour orange is usually well affected with scab spots from *Elsinoë fawcettii* Bitanc. & Jenkins, although sweet orange is practically immune from it. However, there are many infective leafspots that are not so restrictive, and some are diseases of great economic importance.

Losses every year from plant diseases in tropical crops grown in the Western Hemisphere may be conservatively estimated (55, in 1967 paper) to be between 10 and 80 per cent in different regions (averages 24 to 32 per cent), and this is on only 17 of the important crops of which about 200 are grown. The losses considered are in the field and do not include other damage always being incurred during storage and transportation. The annual cost monetarily, wasted because of disease, is difficult to determine as nowhere are comprehensive data on this being collected and recorded. However, in our tropics, the total wastage every year is probably equal to about $4 billion (in U. S. currency). It is a fair estimate that easily over half of the field disease-losses in the Neotropics are from leafspots that are of communicable, parasitic nature.

Range in Leafspot Symptoms

During 1943 to 1963 I made many diagnoses on leafspots*. This was done in particular at some places in South America, but especially in Central American and West Indies parts of the Neotropics. A very wide range in spot appearances was found. It is believed this range indicates something of what is to be handled in leafspot study and it is therefore presented here. Descriptions are from my personal point of view. In addition to the unaided eye it has been necessary to use hand lense, in some cases microscope or dissecting scope, and in a few instances culture studies. There are apparently some ten kinds or types and these are given below, although there are others. For some citations to leafspots see: 1, 2, 3, 5, 11-14, 17-21, 23, 25, 27, 29, 32, 34, 35, 37, 38, 41, 47, and 52. In connection with Cercosporas alone see mention by many Neotropical workers: 6, 7, 10, 13, 15, 17, 20, 21, 24, 26, 28, 33, 34, 36, 38 collections, 39, 45 publ. 1965 especially, 46, 48-50, 52 and 54-56. The greatest contributions from anywhere on this genus come from our tropics, and will be given more attention in the chapters on control measures.

It is well to understand at the outset that there are very many non-parasitic leaf blemishes and they appear specially plentiful in tropical material. These are called "sterile" leaf spots by plant disease clinicians and mycologists, as no fungus or bacterial remains are scraped off spot surfaces. The sterile spots, some of which may resemble bacterial or fungus infections, may come from many things: the start of infections by fungi or bacteria that did not develop; insect punctures or leaf miner work; sunburn or dry wind effects; and blown sand or young leaves beaten together in storms. Numerous states of abnormal physiology may cause leaf spots: toxicity effects from air pollutants

---

*The term leafspot is used in a slightly restricted sense here and should be explained. A generally mildewed, rusted, rotted or blighted leaf is different from a spotted leaf. Blight, or rots, start with an infection that spreads unrelentingly like watered ink in blotting paper. Leafspots, as referred to here, are delimited leaf blemishes of various sizes, shapes, and appearances, but in the main they are hemmed in and surrounded by healthy leaf tissue. Some spots, when close together or in great numbers, may become confluent.

or sea spray and viruses, under certain conditions; remarkable necrotic spotting comes from certain minor element deficiencies. "Sterile" spots from such causes as the above may at times make up nearly a third of the spots brought to pathologists for identification.

Large numbers of the mild flecking from various causes are found on leaves. These flecks are small yellow markings, sometimes with necrotic specks in them. Some may be from resistance reactions as are seen on maize varieties that, beyond flecking, do not become diseased with either *Puccinia sorghi* Schw., or *P. polysora* Underw. Other colors of flecks are found, as for example red flecks on spider lily leaves, where spores of *Stagnospora curtisii* (Berk.) Sacc. germinate but make only limited infections. There are small stipple-like dark points, flyspeck type of marks, from infections of such fungi as *Leptothyrium* sp. on citrus, and *Leptostroma* sp. on mango. Well confined, non-spreading infections of *Pestalotia* sp. occur as specks on mature foliage of gandul where leaves have been fed upon by leafhoppers.

Spots that are small but no longer flecks are very numerous in the tropics. One type of these is the so-called weakspot (see discussion of it in another part of the present chapter). It is yellow, with a sunken center about the size of a dot made with a dull, soft-lead pencil and appears to be the result of arrested infections by various fungi. It is common under certain conditions on coffee, mango, tea, ligustrum, sugarcane and other crops that have leaves that are retained for many months. It is probable that these spots are on all perennial crops grown in the Neotropics.

There are spots of a size a little larger than these smallest kinds. They are more or less the size of from a quarter to half the area of a small fingernail. These are such as many of the infections by *Septoria* spp. on tomato, celery, and other hosts. The shiny black spots from *Phyllachora* spp. are many of this rather small size, and in the main are common on maize and such gramineae, and on leguminosae. Spots that are about twice or three times the average for such as the *Septoria* or *Phyllachora* lesions, are kinds like that on the palm, the gray leafspot (*Pestalotia* sp.), an example of which is illustrated in Fig. 10-1, or the banana Sigatoka (*Cercospora*).

Other relatively small leafspots are caused by rusts

Fig. 10-1. Parts of African oil palm leaflets diseased with gray leafspot, Pestalotia sp. Photograph taken in Honduras, courtesy of B. H. Waite.

(very much more common in the tropics than the temperate zone) on both monocots and dicots. An example is the harsh Andropagon grass with its numerous, short and elongated, slightly raised dark-colored leafspots that are from the uredial stage of Puccinia sorghi Schw. On infecting its alternate host, the herbaceous and very tender broad-leaved Oxalis, the aecial stage of P. sorghi, appears as a spreading, round leafspot powdered on its surface with a yellow dust.

A distinctly different category among the lesser-sized leafspots is the type not rounded in outline but that is angular in shape. This character is specially exemplified by the severely parasitic (in the tropics) bean angular leafspot, Isariopsis griseola Sacc., and another example, very abundant in the tropics, is on cucurbits, the square cornered and straight sided leafspot from Pseudopernospora cubensis (Berk. & Curt.) Rostow. There is another type in this category, often somewhat larger than those just mentioned, which in addition to being irregular in outline, has raised and roughened, dark and convoluted surfaces. An example is one of the stages of the Hevea leafspot from Dothidella ulei P. Henn., another example is on the grass Aristida, where it causes a scab-like leafspot, Dothichloe aristidae Atk., and a third example is on the ornamental leaves of Alpinia on which is found a scabrous leafspot, Catacauma renalmiae (Rehm) Th. & Syd.

There is a considerable assortment of somewhat larger leafspots that can even be called attractive in appearance. They are about the size of a large fingernail, quite commonly are found to have delimited smooth and dark edges that may be bordered with a yellowish-green halo in the leaf tissue outside the spot, and inside the deeply-colored perimeter may be an ashy center peppered with little black bodies. The coffee browneye leafspot, Cercospora coffeicola Berk. & Cooke, is a typical one of these. Many, but not all, Cercospora leafspots are of this character, an exception being the sweet orange so-called greasy leafspot from Cercospora aurantia Heald & Wolf. In cool mountain gardens in the tropics, cabbage may be afflicted with a large round leafspot, Phoma lingam (Tode) Desm., that is studded with dark kernels on its surface. On tomato there is a large leaf and stem spot that, although not always common in some of the Neotropics, is a typical large leafspot with target-board appearance and is caused by Alternaria solani (Ell. & Mart.) Jones & Grout. In accompanying

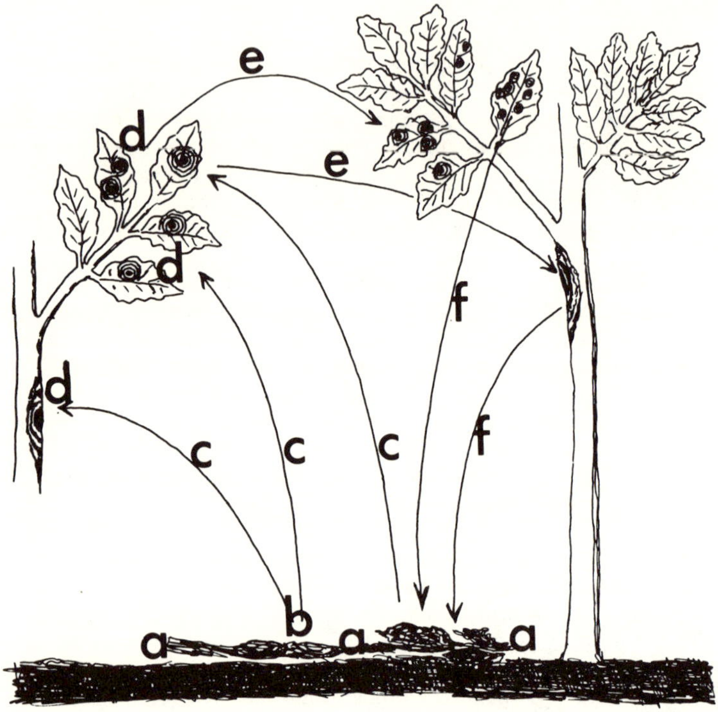

Fig. 10-2. Tomato leafspot, Alternaria solani. Diagrammatic representation: a. the HOLDOVER condition is diseased tomato plant debris on the soil; b. spores are produced on dead debris; c. dispersal up to new foliage; d. from infection by spores, target board-type lesions on leaf and stem; e. spore spread from one plant to another f. inevitable drop of dead, HOLDOVER stage, leaf and stem debris to soil. (Note for temperate zone pathologists: Of course there is no winter in the tropics, but any HOLDOVER stage developed in the tropics is readily adapted to the overwintering stage in the temperate zone.)

Fig. 10-2 is indicated something of the life history of this organism and symptoms of the disease it causes.

There are numbers of leafspots of unusual character in the Neotropics, three of which are as follows. A spectacular large, reddish, target-ringed leafspot, Sclerotium coffeicola Stahel, occurs on coffee, and is specially severe and impressive on the big leaves of the Liberica tree. This leafspot is illustrated and discussed in Chapter XIV. The parasite spreads in concentric rings that are well marked, and each successive ring of infection is of reddish to burnished bronze color. Spots become large, rounded and have gracefully curved, even edges, with the unaltered green color of the uninfected part of the leaf extending to the spot border.

There are several algae that cause tropical leafspots, and the most common in the Americas is Cephaleuros virescens Kunze. It attacks a long list of hosts. Leafspots it causes are rounded they have a velvety and olive-green upper surface on which occur very small but bright orange colored globes. Edges of these algal spots are mossy and described as dendritic.

The most unusual of any leafspot anywhere is the one common on over 500 different species of hosts in moist, cool areas of the Neotropics, and is due to Mycena citricolor (Berk. & Curt.) Sacc. (= Omphalia flavida). Its parchment-like, round and sunken lesions have a slightly raised center. When sufficiently developed the spots have on them slender bristle-like growths tipped with lemon-yellow-colored, glistening bodies that if wet can be picked off between thumb and finger. The most remarkable thing about these spots is that while living they are luminescent in the dark.

Unsolved Leaf Injury Called Weakspot

It is probable that at times a spore of a leafspotting fungus that alights on a leaf and penetrates it to a small extent, cannot develop sufficiently to make a normal, large lesion. This seems to be the case in the coffee orchards in the Orient, Africa, and the Americas (see 55, text publ. 1961, p. 266-267) where the leaf symptom, called weakspot, causes chronic, unseasonal leaf drop that can be controlled by fungicidal applications. The inference from what is seen in those coffee leaves is that the fungus, or fungi, progress slightly beyond the stage of incipiency, enough to form a small amount of toxin but not enough to cause a large area of leaf tissue collapse, and does not produce spores on the spot surfaces. There is always a relatively small blemish, starting as a light green fleck to become transluscent and later in the center to develop a sunken dead part.

From the years 1946 to 1961 isolation and inoculation studies were carried on at intervals respecting special phases of parasitism of *Cercospora coffeicola* on arabica coffee. (For illustration of symptoms see Fig. 10-3.) The fungus grows slowly in pure culture, but if carefully done, isolation of it is successful from bits of surface-sterilized tissue on hard PDA. From researches and observations in Latin America (45, 55, 56) it is clear that conditions for the disease as to moisture, sun exposure, and a mean temperature of 24° C, is good for growth and infection by *C. coffeicola*, and there is always an abundance of continually renewed coffee leaves upon which the parasite can grow and thus multiply its spores. Isolations in 1946 were always at first from obviously vigorous, typical and large, so-called browneye leafspot of foliage on old trees. These isolations in culture invariably produced typical *C. coffeicola*. In later years this organism was repeatedly recovered, over 60 per cent of the time, from weakspot lesions.

In 1950 a lightly shaded, half-manzana-sized field was newly planted with known *Cercospora*-free coffee seedlings that had never been sprayed. This healthy field was within 250 meters and on the leeward side of a large plantation of old trees that were evenly diseased with both weakspot and browneye leafspot. This allowed for breezes to carry ample amounts of inoculum to reach leaves of the young, uninfected trees. And it occurred rather soon, since in six months' time about half the trees in the new planting were *Cercospora*-infected. In nine more months, every tree showed presence of typical browneye leafspot. Counts showed the trees had from one to 15 spots on mature leaves, with an average of a little less than three. A few weakspots (one in every 20 leaves) were also noted.

The trees grew well and after two harvests were again studied for leafspots. The leaf condition in respect to the relatively large numbers of browneye-spots, and the much fewer weakspots, did not change greatly until 1954. By then browneye-spot had increased in severity, there was

Fig. 10-3. Spots on old arabica coffee leaves. Browneye leafspot, Cercospora coffeicola: upper specimen 2 large and 2 smaller spots; middle leaf 1 large spot at tip end and near petiole 1 large from confluence of two infections, while in lamina between are 3 small spots; part of leaf at bottom a large spot on its edge from several spots grown together. Weakspots (see arrows) in some cases so mild as to disappear in printing the picture; weakspots marked with cross showed varying amounts of necrosis. Photograph courtesy Agricultural Experiment Station, University of Puerto Rico, Rio Piedras, Puerto Rico.

considerable defoliation from it, and every tree was pruned and fertilized. In a little more than a year there had been good growth, leaves were well matured, and there appeared about six browneye spots to every 20 leaves; weakspots had greatly increased, and averaged 167 per 20 leaves. Trees that at one time had nothing but browneye-spot had now comparatively few of them but an abundance of weakspots. There were many cases in which browneye-spots and weakspots were on the same leaf, but after a three-month period when defoliation had been notable, the leaves having nothing but weakspot were the most common.

Although coffee-leaf weakspot blemish (the seemingly mild symptoms of this spot are illustrated in Fig. 10-3) from moist, chamber, isolation (see accompanying Table 16) and microscope examinations, showed good evidence of being some kind of arrested infection by Cercospora coffeicola; some series of artificial inoculation studies with that fungus, tried at several times, did not reproduce exact weakspot symptoms. Leaves more than a year old were not susceptible; in rare instances coffee leaves that were recently matured showed small spots from C. coffeicola inoculation and these small abnormal spots, as is the case with weakspot, did not enlarge, nor produce spores. Many times weakspots average about seven spots per leaf. With two or three, or even one located near the main vein, large browneye leafspot causes leaf fall; 8-15 weakspots can be sustained on leaves for months, then suddenly the most seriously weakspot-diseased leaves drop. This latter may occur at the seasonal change following the long, heavy rains.

The weakspot phenomenon is not confined to coffee and has not been adequately studied. It is speculated that this may be a manner in which not only Cercospora but certain other parasitic fungi carry over from season to season, but proof of this still has to be determined. Weakspot has been seen in Florida, which is at the northern border of the Neotropics, on citrus, ligustrum, and a few similar leathery-leaved hosts, although weakspots are not necessarily confined to a leaf that is coriaceous. I observed weakspot on tomato leaves in Colombia, on sweetpotato leaves in Puerto Rico, and on alternanthera leaves in El Salvador.

Weakspot, which is sunken and small, often with a chlorotic halo and no sporulation on its surface, is readily found on leaves in the jungle; the most spectacular blemishes

Table 16. RESULTS OF COFFEE LEAFSPOT ISOLATION STUDIES with blemishes from the browneye leafspot and the weakspot respectively.

Series	Spots used in isolations (no.)	Type of spots	Cercospora spores on surface (% of spots)	Cercospora recovered from tissue (percent)	Other fungi recovered from tissue (percent)
1	150	B(a), 4 weeks old	100	100	0
	150	B, 20 weeks old	6	7	0
2	40	B, typical	0	100	0
	40	W(b), new	0	70	22(c)
	40	W, old	0	30	12(d)
	40	Healthy leaf bits	0	0	0
3	82	W, typical	0	39	55(e)
	100	Healthy leaf bits	0	0	0

(a) Browneye leafspot caused by Cercospora coffeicola.

(b) Weakspot, typical of the small blemishes known by that name. No fungus spores ever found on spot surfaces from either scrapings or seen in cross sections.

(c) Fungi recovered, species of: Pestalotia, Diplodia, Xylaria? and an undetermined organism with brown mycelium.

(d) Fungi recovered, species of: Diplodia, Xylaria, Penicillium.

(e) Nine kinds of fungi recovered, species determined were of genera: Xylaria, Diplodia, Trichoderma, Pleospora and Glomerella; four others of which two were possible Basidiomycetes produced no spores under the conditions in culture.

are the larger, much discolored, vigorously sporulating leafspots. In one hour of looking at the edge of a jungle clearing in Ecuador, weakspot injury was found on species of wild Cinchona (introduced), Theobroma, Philodendron, Cycad, Erythrina, Croton, Balsa, Cyssus, bromeliads, ferns, and Ipomoea. In the characteristic tree growth that is so common in the Neotropics there is continually some leaf drop in progress, and spots of one kind or another are nearly everywhere on leaves. Some leaves appear more susceptible than others, and some trees have a habit of retaining some of their leaves on stems for three or more years. This gives those that live on much greater chance to become diseased, and also to be a source of infective spores over a relatively long period of time. There is no one special season or time, such as the so-called autumn in the temperate zone, when broad leaved bushes and trees drop their leaves.

In the tropics there are leaf flushes and leaf changes, mostly following an irregular rhythm. In some tree species, leaves are shed from only one branch at a time, and the branch may become refoliated before another sheds. In some species, trees standing side by side may drop leaves, but at times a month or so apart. There are some trees that never drop all their foliage, and a few leaves may fall at any time throughout the whole year; in some cases a considerable proportion of leaves may hang on three and four years. Under such conditions of irregular leaf fall and foliage changes as these in and about economically important crops, generally there is close by a continuous supply of leaves with spore-bearing spots on them. This is a principle that must be recognized in connection with work on leafspot diseases in tropical plant pathology. Although fungus spores from many differing spots, either on cultivated or wild-grown leaves, may not always encounter best hosts for their parasitizing, their spore germ tubes may have a chance to invade leaf tissues in a very limited way--and produce weakspots.

Forgotten Loss from Leafspots

Spots are plentiful on leaves in perennial plantation crops, be they broad-leafed trees or bushes, vines, bamboos, sugarcane, musa relatives, short lived herbaceous and seasonally replanted row crops, or in pastures of grasses and legumes. Leaf blemishes usually are not given much attention; indeed, are commonly not recognized as being present. If leaves do die or are cast off, trees recover, or in the wild are naturally replaced. Events of such a nature are accepted. In an increasing fashion, however, growers are more often disturbed by a very spectacular defoliation, but among some the true cause may not always be determined; leafspots in an agricultural region are not likely to receive early publicity, in their effects, or occurrence of new kinds when they appear.

After observations and personal experiences with growers of several crops following the history of their troubles with plant diseases, it is apparent that a regular course of events may be noted. For convenience, the course has been divided into about six stages. After a crop is introduced and accepted, usually the first step in the disease situation is the occurrence of leafspots that appear minor. The second step comes in a season or two when it is evident leafspotting is increasing, but from past experience this step is not feared. The grower has by that time learned to live with it. In the fourth stage spot increase is greatly augmented.

The fourth stage is that in which because of leaf spots, high production of the crops is impossible and a 20 to 30 per cent lower harvest becomes common. The economy of a farm has to be changed and is geared to this condition. Nearby fields may be cleared and planted with good beginning results but will soon give the same lowered production because of leafspot.

A fifth stage is then encountered. By this time it has become certain that foliage spotting does not kill plants down to the ground. Some years in fact, it is less severe than others, though it is always present. However, at about this period a few plants at field edges may contract some virus, or a root disease may kill from three to eight plants or trees in a delimited area. For the first time a disease occurrence is spectacular and causes immediate alarm about the health of the farm or plantation. Study of the situation is carried on, when not only the newly intrusive disease is determined, and control measures are instituted, but also the leafspot, which has been always present, is given long-neglected attention. An example of this is the case of the acerola or West Indies cherry tree in Puerto Rico. It has been grown for centuries, but not until recently when the fruits became of importance as a rich source of vitamin C

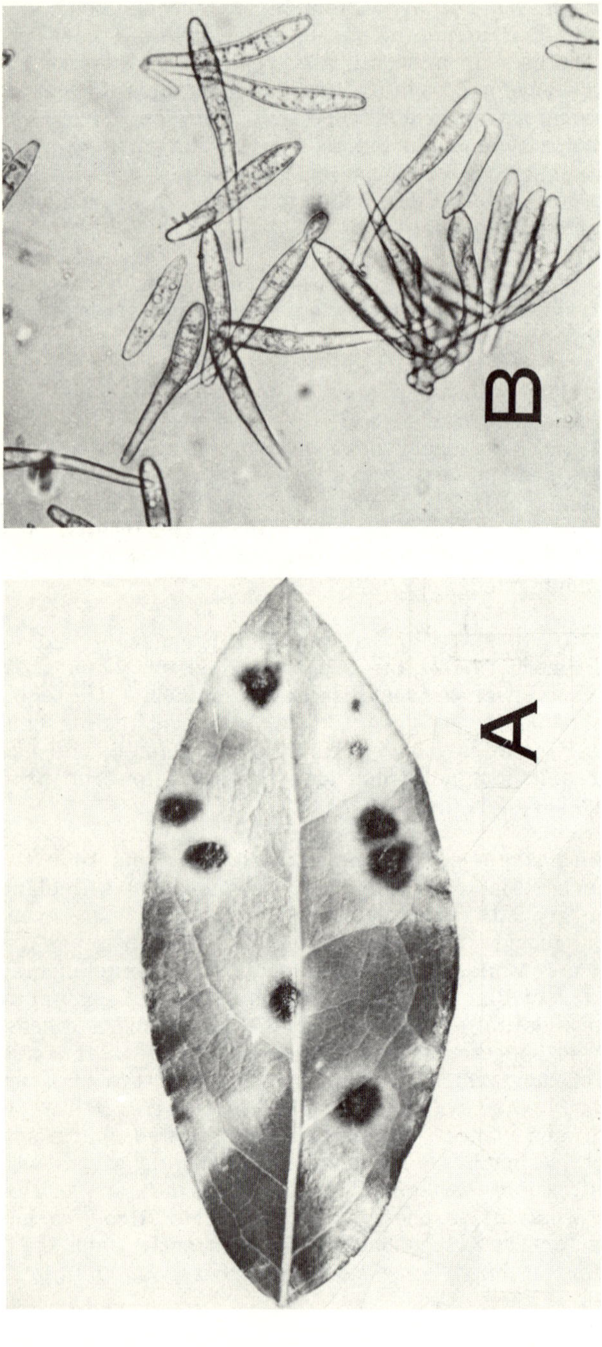

Fig. 10-4. Leafspot of acerola, caused by a typical *Cercospora*. A. Symptoms on leaf consist of black and rounded lesions. B. Photomicrograph of typical spores. Pictures courtesy of Pedro Melendez.

was it thought worthwhile to make a major study of the common leafspot on it. This (Fig. 10-4) is caused by a Cercospora and may be a severe problem under some conditions. Work is in progress on it.

The sixth stage in the course of events, often consists of finding that leafspot severity varies in different field locations. This calls for study by diagnosticians. Suggestions are then sought as to possible ameliorative cultural changes in husbandry, while tests are made for possible chemical control measures and a search is begun for resistant varieties.

During the long period that precedes research to help in leafspot control, the losses--and they at first are mostly either forgotten or unacknowledged--drains away a good proportion of potential productivity. Coffee planters know that the American leafspot, Mycena citricolor (Omphalia flavida) has been a major disease of the crop at least for a century. As an example where coffee once did well in the port area around Santos, Brazil, at one time the trees flourished and were grown thickly. The leafspot appeared, no one knows just exactly when, but the fazendas which had become world renowned for their product developed increasingly serious problems of defoliation. It became so bad that the day came when moist fields had to be abandoned During this period the growers, realizing the questionable future of their crop, had been spreading their holdings into dryer areas, farther away from the original places. They found growing coffee where it was free from the portside heavy dews and rainy periods, and where the crop could develop without shade, there was a condition that meant no longer the severe trouble from the devastation of leafspot. Coffee is not grown around Santos now.

This kind of history in which a crop is driven from an area has occurred in several countries. In addition to the experience in Brazil, I know of coffee areas in Colombia, Costa Rica, Guatemala and Mexico where the trouble was not severely limiting at first, but in later years excess defoliation from the leafspot has led to deserting coffee production. In these places growers know that if they wish to have the crop without expensive spraying, it must be cultured under very light shade. Even with ameliorative methods of reducing the coffee leafspot from Mycena, it is known that, for general purposes, the drain on the coffee crop in several countries from this one disease amounts to about 20 per cent per annum, and in some cases more than that.

Minor Leafspots that Become Major Problems

In Argentina and in Chile, grapes are grown well and the wines are of high quality. At first many vineyardists, with gratification, had harvests year after year without needing to spray. There was occasional leafspotting but not much. However, little by little the leafspot causing mildew, *Plasmopara viticola* (Berk. & Curt.) Berk. & de T., has become established. At first losses were fairly low. Finally the destruction from the disease increased, and nowadays 30 to 50 per cent of the crop may be lost, unless the leafspot is controlled. Unless given early attention the drain goes on for years: from the time the first leaf damage appears until the time the vinyardists, losing heavily, finally turn to spray technology.

In the parts of the Neotropics where the populace is influenced by knowledge regarding the vitamin needs of nations, tomatoes are being grown in more and more intensive fashion. Before tomatoes become a big business crop they are grown in many small plots. The farmers accustom themselves to considering relatively small harvests as normal. They see a number of leafspots such as: *Alternaria solani* (Ell. & Mart.) Jones & Grout, *Cladosporium fulvum* Cke., *Mycosphaerella* (Cercospora) sp., *Ascochyta lycopersici* (Plowr.) Brun., *Phoma destructiva* Plowr., *Stemphylium* sp., *Septoria lycopersici* Speg., and others. All of these are destructive. Taking either the *Septoria* spot, the *Alternaria* spot, or the *Cladosporium* spot, field experiments show that they, in combination or singly, reduce the crop 30-40 per cent.

It is known that an important fig leafspot is from *Glomerella cingulata* (Stonem.) Spauld. & Schr. (*Colletotrichum gloeosporioides* Penz.). It has been seen in numerous countries of Central America and in the West Indies, and at first it is not serious on young trees. However in a few years it becomes severe. Losses from fruit reduction are conservatively estimated as from 10 to 40 per cent. In small plantings these losses are accepted, since the crop is not at present one that has any large industrial interests. The only method of amelioration is to replant new trees. Many people eat less fruit because of fig leafspot.

There is an increasing value given to the papaya in the tropics as a fruit for both airplane shipment to metropolitan parts of the temperate zone and home consumption. On papaya there is a common and spectacular leafspot, caused by Corynespora cassiicola (Berk. & Curt.) Wei. As the leafspot only, it is a bad disease causing defoliation. The fungus produces abundant spores on leaf lesions, and, having much the same size and conformation as spores of a Cercospora these are readily spread. Infections multiply quickly on leaves and the Corynespora also infects the succulent hollow tree trunk, causing dark-colored greasy lesions that in advanced stages extend into the trunk lumen, which results in falling of some trees and decline of plantations.

On banana the Sigatoka leafspot by Mycosphaerella musicola Leach (Cercospora musae Zimm.) attacks so rapidly and vigorously that if it is to be a money-making agricultural enterprise the crop has to be sprayed. On the other hand, for subsistence purposes mostly, the somewhat tolerant cooking bananas or plantains are grown, sometimes in partial shade, without spray control of Sigatoka. I estimate over 20 per cent of food production by plantains is drained away in loss from Cercospora leafspot. Up to the present time, on plantains little has been done for control measures, but this disease is increasingly severe.

Another banana, and abaca hemp, leafspot is caused by Helminthosporium torulosum (Syd.) Ashby. The leaf infections are few and small at first--it is often several years before the disease becomes severe and appears on both the leaf lamina and the succulent leaf sheath in which are buried the sap-carrying vascular elements. Spots on leaves enlarge rapidly, and those on the pseudostems go deeply into the leaf sheath tissues. Injuries are three-fold: reduction of photosynthetic leaf surfaces; poisons introduced into the sap stream that result in hardening and constriction of sheaths; and reducing elongation which results in a common dwarfing of plant growth that may go unnoticed unless Helminthosporium-free plots are closely compared with diseased plots. Growers with plantings thus diseased and who do not spray their fields accept, as an added burden, reduced harvest.

There are hundreds of at first presumably minor Neotropical leafspots that eventually can be shown, as has been done in the few cases given immediately above, to be sources of large continuous drain on the economy of tropical agricultural business. As this is written there have appeared in recent literature frequent announcements of new leafspots and of leafspots that occur but rarely. An example of the latter is the finding in Brazil beans of the so-called floury leafspot, Ovularia (Ramularia) phaseolina (Petrak) Drum. The leafspot is at present not widespread in Brazil. Between 1934 and 1965 it has been collected in that country only three times. The lesions it produces are characterized by a fine-grained farinaceous appearance on the underside of the spot on a leaf. In early stages it causes severe defoliation on bottom leaves of a relatively few plants. Restricted as it is at present it is apparent that the present infrequent occurrence in the state of Minas Gerais, Brazil, may not continue that way. The same leafspot is also known in Nacaragua, and is present in Colombia. It is probably spreading. In the latter country in 1953 it caused minor damage. But in 12 years it increased in severity.

The problem of apparently minor leafspots becoming major limiting factors in agricultural production in the American tropics is an old one. Some suggestive literature on the subject is indicated in citations as follows: 1, 3-6, 8, 9, 13, 15-23, 30, 31, 34, 36-38, 41-47, 52-55, and 58.

Ecology of Leafspot Diseases

The problem of what effect different climates, ecological zones, and other factors have on relative severity of leafspots on crops is more notable in the tropics than in the temperate zone for the reason that very different ecological conditions in the tropics are in many cases close to each other. In the tropics leaf spotting is developing continuously, either on crop plants or on nearby wild growths. It should be repeated that, except for unusual and rare conditions, leafspots if not on a crop in the field are always nearby.

In temperate zone agriculture it is accepted that maintenance of part of the health of most crops must be through spraying technology. In many sections of the tropics, if at all possible, spraying is avoided; generally speaking, use of chemicals on foliage costs much more in the tropics than in temperate zones both in time required for its application and for the chemicals. It becomes a responsibility of plant pathologists to search for evidences of ecological features that help in reducing crop losses in areas where crops are grown; knowledge is also accumulating about weather changes that increase or decrease leaf diseases.

No one has made the much needed, exhaustive review of literature from the Neotropics that deals with ecology and weather effects on plant disease. This could fill a fair sized book, although in many cases observations about this verge on the superficial and these will have to be collated and analyzed along with findings of more objective nature. It seems probable that some day, someone will formulate a system with special reference to disease ecology. For example, certainly there are best zones for: fusarial wilts, black mildews, white mildews, algal leafspots, tar spots, damping off, web blights, Rosellinia root rots, and wood stain diseases. (For bean diseases see 1966 publ. under citation 13, and for more generalized discussion see 1933, 1934, and 1962 publ. under citation 55.) This is in many cases a more intricate kind of study than is expressed in the present tropical forester's ecology (as an example: 51) based on forest zones. Meantime forest ecologists' findings are helpful as they very roughly define differences.

The bean crop grows in most of the arable areas in the ecologic zones of tropical America. It is, however, frost sensitive, it is no xerophyte, and it does not produce well in soils that are very poor or extremely acid or very alkaline. It originated in the Neotropics and for man is the most important single source of protein in all of tropical America, although it is perpetually in short supply. Among leading causes of inadequate bean production are leafspot diseases. It has been shown (see 1966 publ. in citation 13) that ecology is highly significant in this regard. Leafspot diseases vary markedly in severity, depending on field conditions: temperatures, amounts and distribution of rainfall, relative air humidities, cloudiness, and wind exposure in different ecologic zones all have intimate effects on the leafspots--more, probably, on the parasites than on the bean bushes.

There are at least four of the bean leafspots that are examples of fairly clear, ecologically restrictive disease occurrence: It is apparent (13, 38, 41, 47) that the round spot, _Chaetoseptoria wellmanii_ Stevenson, is a potentially serious disease. Symptoms of the leafspot are illustrated in Fig. 10-5. Occasionally it will cause complete defoliation, and I have seen it destroy two small plantings in Guatemala and reduce harvest to less than half full production in a large field in El Salvador. This spot is an example of one fairly

Fig. 10-5. A destructive bean leafspot, *Chaetoseptoria wellmanii*, attacks a wide range of hosts among the *Leguminoseae*. Photograph of diseased common bean leaves, collected in the highlands of Guatemala.

well restricted to moist, moderately cool conditions such as may occur in the ecologic zone that some foresters might designate (51) as the Subtropical Moist Forest type. In the case of bean leafspot from Ascochyta balthshauseri Sacc., Echandi found it in Costa Rica only in cool highlands, within the ecologic zone known as Lower Montane Forest type. It has been identified by me, found in high mountain valleys in Guatemala above the limits of the ecologic zone where it is known in Costa Rica. These two leafspots seem clearly limited to cool-temperature ecology.

In contrast, there are two other bean leafspots that prefer very much warmer conditions. The well known, typically severe tropical bean angular leafspot, Isariopsis griseola Sacc.--although it is seldom if ever of economic importance in the temperate zone--in the warm tropics (17, 20, 21, 34 see 1948 publ., 40, 45, 52) is both common and severe. In Costa Rica (13) it definitely is more severe, largely in fields within the ecology of the so-called Tropical Dry Forest type. In this same somewhat dry, hot zone, beans are additionally afflicted by a leafspot caused by the cosmopolitan mildew, Erysiphe polygoni DC. This disease is not serious or even frequent in occurrence in the wet, cooler ecological zones of the Neotropics.

There are two coffee leafspots that vary remarkably as to disease severity under identical conditions, except for amounts of shade. The plantation owners, or finqueros, in Latin American know that in good volcanic soils, coffee produces well for many years without much if any added fertilization, but that it does best where grown under a medium shade. Sunlight is regulated to be at different intensities by periodic pruning of such shade trees as species of Inga, Erythrina, Cassia, Musa and Ficus. The first three genera are in some ways the most amenable to shade pruning and are much used. Leafspot severity may increase at times in certain locations, which is then a reason for changing shade amounts.

Probably the two most dangerous leafspots of coffee are the browneye leafspot, Cercospora coffeicola, known to be severe under thin shade, which means shade should be allowed to increase, and the American leafspot, Mycena citricolor, known to be severe even under deeper shade. Effects from shade on these two leafspots are exemplified in the accompanying table. Plantation owners are aware of the effects of shade on these two leafspots. These growers

Table 17. COMPARATIVE SEVERITY OF TWO LEAFSPOT DISEASES on year-old Arabica coffee trees under different shade intensities[a].

	Cercospora[b] leafspot study			Mycena[c] leafspot study		
Shade intensity[f]	Plot	Leaves/tree[d]	Spots/leaf[e]	Plot	Leaves/tree[d]	Spots/leaf[e]
Heavy, ca. 80%	1	57	0.1	5	60[g]	19.0
Medium, " 50%	2	42	0.3	6	44[g]	12.0
Light, " 20%	3	34	0.7	7	32	4.8
None	4	34[g]	1.2	8	33	0.1

(a) Experiment at Turrialba, Costa Rica, in 1950. Plots all in the same field were 50 x 50 meters under regularly pruned shade of *Inga vera*, scattered within an area covering two hectares. Young healthy coffee trees were planted near, but not shaded by, old diseased trees. Spots on leaves of these old trees furnished ample inoculum.
(b) *Cercospora coffeicola*.
(c) *Mycena citricolor* = *Omphalia flavida*.
(d) Counts were of the total from all leaves of 10 trees and included dropped leaves.
(e) Average of spots per leaf, from random selection of 20 leaves.
(f) Estimates were made by eye at 2-month intervals.
(g) Over 5 per cent defoliation.

are aware of the effects of shade on these two leafspots. These growers are able sometimes only through shade management and without spraying, even with leafspots present, to secure sufficient spot reduction that good coffee harvests result. The ecological zone in which these spots are of great frequency are in what foresters class as Subtropical Moist Forest and the cooler one above it designated as Lower Montane Forest; although the _Mycena_ spot extends into high Elfin Woodland areas, the _Cercospora_ spot does not occur up there.

It is of note that there is a distinctly different coffee leafspot, the very dark-colored blemish, one called burn, or quema. It is due to _Phoma costarricensis_ Echandi (13). This is confined (55 _publ. 1962)_ to cool high areas, where it is possible it is dependent on a certain amount of insect work and the ample, fairly common misty conditions called garúa or pelo de gato. Another kind of spot, found only in lowland country where it grows at near sea level, is the so-called ragged leafspot. The spots are irregular in shape and variable in appearance, they run together along the edges of old leaves, and are caused by a complex of organisms. Isolations from lesions have yielded mostly species of _Colletotrichum coffeanum_ and _Cladosporium herbarum._ The latter fungus without any other organism present has been repeatedly isolated from lesions that are part of the notable ragged condition in hot-land-grown coffee leaves. The fungus designated as _Colletotrichum_, on being collected as leafspots and incubated in moist chambers at 24° C, developed typical ascocarps of _Glomerella cingulata._ This organism, it was interesting to learn, is widely spread as a secondary invader in leafspots and is found from high, quite cool Elfin Woodland ecology down to moist land at sea level under hot conditions. On the other hand the _Cladosporium_ has been recovered only at warm and dry locations. Algal leafspot from the organism _Cephaleuros virescens_ Kunze occurs in high and low regions, but is more common in isolated coves where a condition is found approaching moderate warmth and moisture.

The above reviews only four leafspots on bean and only four on coffee. These eight are, however, good examples of the great importance of the effect of various environments on leaf-infecting parasites. There are about two hundred plants used in agriculture in the Neotropics and each of these has a number of leaf-spotting organisms common on them. Tropical growers keep watching the diseases as they are affected by crop growing conditions. Sometimes sprays and resistant varieties are used for control. Through changing field locations, however, or varying cultural techniques, some growers often find they are able to greatly ameliorate disease damage and continue in production even with old varieties of crops and with the leafspot organisms present.

Tabulation of Variation in Appearance of Tropical Leafspots

To the reader it is no doubt plain that there may be too much simplification of the subject of this chapter. An attempt has been made to indicate not only the large number of leafspots but also the great diversity in them. In the section entitled "Range in Leafspot Symptoms" some interesting differences are briefly reviewed; the main purpose of the special table below is to give further emphasis to leafspot diversities and to show something more of their multiform appearances supplementing the range in symptoms discussion.

While it is to some extent academic to aim at making an acceptable, artificial organization of leafspots in a separate category from the remainder of leaf blemishes and other non-leaf-infecting diseases, I find it serves a good and practical purpose in consultative and explanatory discussions, to sometimes treat leafspot problems and leafspots themselves as unique. In this table is presented some information relating to only a few of somewhat representative kinds of leafspots. It can be seen that by no means is it possible to enunciate any general classes in which to fit all leafspots. There is no sort of orderly arrangement in the way in which the leafspots are listed. They have been put in as they have been encountered in two weeks of field work answering special calls, and are an example of what the pathologist sees over a short period of time.

Table 18. LIST OF LEAFSPOTS[a] seen in a two-week period of work[b] in Costa Rica during 1953 dry season.

Relative size and shape	Surface and edge	Example on:	Causal organism	Controls suggested
Medium large, angular	Dark, felty, bounded by large veins	Bean	*Isariopsis griseola*	Crop rotation, sprays
Medium small, elongated	Dark brown, on main veins of underside of leaves	Bean	*Colletotrichum lindemuthianum*	Tolerance-resistance sprays clean seed
Medium large, round	Gray surface with black marks in middle, dark edges	Bean	*Chaetoseptoria wellmanii*	Growing the crop at a warmer, lower elevation
Medium large to large, rounded	Dark surface with mildew growth, edges not mildewed	Potato	*Phytophthora infestans*	Spraying, tolerant-resistant variety obtainable
Small, round	Crater-like appearance, raised rim	Coffee, arabica	"Weakspot" probably from *Cercospora*	Spray in early part of rainy season
Medium size, round	Raised straw-colored body in center,	Coffee, arabica	*Mycena citricolor* (= *Omphalia flavida*)	Spray using lead arsenate or

(Relative size and shape)	(Surface and edge)	(Example on:)	(Causal organism)	(Controls suggested)
	slender stalks surmounted with yellow body			Paris Green, reduce shade over trees
Medium size, round	Parchment like, tawny colored center, with dark narrow border	Apple and pear	<u>Sphaeropsis malorum</u> (= <u>Physalospora obtusa</u>)	Spraying
Mostly small but of	White and disfiguring but superficial, can be scraped away with fingernail	Apple	<u>Strigula complanata</u> (fragile combination of alga-fungus; a lichen)	--
Medium size, round	Greenish-brown, felty surface, with occasional slender stalks bearing gold colored globes	Mango	<u>Cephaleuros virescens</u>	Copper containing spray
Large to very large, irregularly shaped but rounded edges	Surface occupied by red and red-brown target rings closely arranged, edges delimited	Coffee, liberica	<u>Sclerotium coffeicola</u>[c]	--

Medium size, irregular	Black and rough surface with small coal black kernels in center	Avocado	_Meliola_ (_amphitricha_?)	--
Small, elliptical and round	Black, shiny surface, entire edges	Maize	_Phyllachora maydis_	
Large, irregular and spreading	Brown with fine dark pubescence (spores seen with hand lens)	Maize	_Helminthosporium turcicum_	Tolerant types available
Medium size to small, elongated, when confluent appears like a "burn"	Relatively smooth surface, specks in ashy center, when elongated bounded by parallel veins, edges dark colored	Musa crops, of banana, plaintain, and abaca	_Cercospora musae_ (= _Mycosphaerella musae_)	Spray, grow in shade
Medium large, rounded	Brown, often spores visible under hand lense, edges delimited	Musa crops banana and plantain	_Cordana musae_	Spraying
Medium small, round	When young white showing through epidermis, with age breaking open	Sweetpotato	_Albugo ipomoeae-pandurata_	Possible resistance

(Relative size and shape)	(Surface and edge)	(Example on:)	(Causal organism)	(Controls suggested)
Medium small, rounded	Black with feathered edges and black cob-webby growth between spots	Sweetpotato	*Meliola malacotricha*	Crop rotation, new plant sett source
Small when young, with age larger, round	Red when small, older gray to parchment yellow, center sprinkled with small dots, edge reddish brown	Lettuce	*Septoria lactucae*	Clean seed, crop rotation

(a) Important leafspots observed, but the list includes only a minor proportion of the leaf diseases found. Those given here are from notes made in connection with identifications and recommendations. Rusts, white mildews, minor element-caused spots, anthracnoses, virus spots, bacterial spots, thread and web blights, and other such affections not included here.

(b) Three growers visited; specimens of which there was uncertainty were taken to the laboratory for microscopic examinations, isolations, and identification.

(c) Noted but not identified at the time. Numerous spots on some older leaves. This disease was first reported for Costa Rica by Echandi (13); working over 10 years earlier I made sketches and collected diseased leaves but did not have positive identification. This spectacular leafspot is illustrated in Chapter XIV.

Chapter X

Literature Cited

(Note: I suggest that for a comprehensive cataloging of a major part of the leafspots of the American tropics up until about 1925, the reader should refer to Saccardo's "Sylloge Fungorum." Additions are continually being made and the valued work by Viegas, "Indice de Fungos da América do Sul," is of special note. A large body of mycological literature is needed to refer to if there is a desire for a complete enumeration of tropical American (Neotropical) leaf infections, spots, and diseases.) I have a list of over 120 names of workers who are dealing or have dealt with this phase of tropical American plant pathology.

1. Abrahão, J. 1961. Combate à ramulose tardia do algodoeiro. Biológico 27:121-124.

2. do Amaral, J. F., C. G. Teixeira and E. D. Pinheiro. 1956. O bacterio causador da mancha aureolada do cafeeiro. Arq. Inst. Biol. (São Paulo) 23: 151-155.

3. Batista, A. C. 1952. Necrose basal das folhas da mangueira e descricão de _Ascochyta mangiferae_ n. sp. e _Didymella mangiferae_ n. sp. Bol. Secr. Agr. Indus. Com., Pernambuco 19:165-171. --. 1952. Algumas novas especies de _Phyllosticta_ e _Phyllostictina._ Bol. Secr. Agr. Indus. Com., Pernambuco 19:212-215. --. 1953. Outras especies de _Phyllosticta_ recolhidas em Pernambuco. Soc. Bot. Brasil, An. Congr. Nac. 4: 84-88. --. 1956. Algumas especies de fungos imperfeitos. Rev. Brasileira Soc. Agron. 12: 53-60.

4. ________. and J. N. da Silva. 1950. Uma nova doença do roseira, causada por _Stagnospora rosarum_ n. sp. Bol. Agr., Pernambuco 17:220-233.

5. Bird, J., A. Krochmal, G. Zentmyer and J. Adsuar. 1966. Fungus diseases of papaya in U.S. Virgin Islands. Univ. Puerto Rico, J. Agr. 50:186-280.

6. Brown, L. R. and J. E. Gonzales. 1956. Cercosporiosis en el algodonero. Inf. Mens. Est. Exp. Agr. La Molina 30, 5 p.

7. Calpouzos, L. 1955. Studies on the sigotoka disease of bananas and its fungus pathogen. Atkins Gard. Publ., 70 p.

8. Campi, M. D. 1939. Helminthosporium turcicum Pass. en la República Argentina. Lilloa 4:5-32.

9. Carvalho, T. and O. Mendes. 1953. A mancha castanha e a mancha preta de amendoim (Arachis hypogea L.). Broteriana 27:156-162.

10. Chupp, C. 1954. A Monograph of the Fungus Genus Cercospora. Publ. author, Ithaca, New York, 667 p.

11. Costa, A. L. 1950. Mancha anular do amendoim causada pelo virus do vira-cabeça. Bragantia 10:67-68.

12. Desrosiers, R. 1954. Cultivo y enfermedades del bananero. El Agro, Ecuador 1:27-29.

13. Echandi, E. 1959. La chasparria de los cafetos causada por el hongo Cercospora coffeicola Berk. & Cooke. Turrialba 9:54-67. --. with A. Salas. 1964. Sclerotium disease of coffee incited by Sclerotium coffeicolum. Phytopathology 54:498. --. 1966. Principales enfermedades del frijol observadas en diferentes zonas ecologicas de Costa Rica. Turrialba 16:359-363.

14. Fawcett, G. L. 1926. Las manchas blancas de las hojas de la caña. Rev. Ind. Agr., Tucumán 17:259-261.

15. Fawcett, H. S. 1936. Citrus diseases and their control. McGraw-Hill, New York, 656 p.

16. Franco, E. 1960. Ocorrência de Curvularia maculans no milho em Sergipe. Rev. Agr., Brasil 14:265-268.

17. Galli, F. et al. (see "Lit. Cit.," chapter 2). 1968. Manual de fitopatologia, doenças das plantas e seu contrôle. Ceres, São Paulo, 640 p.

18. ________., and H. Tokeshi. 1961. Ocorrência de Phoma destructiva Plowr. sôbre tomateiro Lycopersicum esculentum Mill. Rev. Agr., Brasil 36:41-43.

19. Garassini, L. A. 1935. El pasmo del lino Phylctaena? linicola Speg. Rev. Fac. Agr., La Plata 20:170-257.

20. García-Rada, G. 1947. Fitopatología Agrícola del Perú. Agr. Exp. Sta., Lima, 76 p.

21. ________., and J. A. Stevenson. 1942. La flora fungosa Peruana; lista preliminar de hongos que atacan a las plantas en el Peru. Est. Exp. Agr. La Molina, 112 p.

22. Goncalves, R. D. 1938. Principais doenças da videira em São Paulo. Biológico 4:8-10; 76-82; 115; 145-152.

23. Halperín, L. 1935. Informe sobre el ataque del Sphaeropsis malorum Pk. al manzano en el Delta. Bol. Min. Agr., Argentina 37:39-49. --. 1939. Fenómenos a considerarse con respecto al agente etiológico de la mancha angular, del algodonero en nuestro país. Physis 15:99-102.

24. Hodges, C. S. and F. A. Haasis. 1962. Etiology of Cercospora spot of magnolia. Mycologia 54:448-454. -- and --. 1964. The perfect stage of Cercospora magnoliae. Mycologia 56:53-57.

25. Huergo, J. M. 1902. Conferencia sobre la enfermedad de los naranjos y otros enfermedades y parasitos. Bol. Agr. Ganad., Argentina 2:793-832.

26. Issa, E. 1953. A cercosporiose da bananeira. Biologico 19:65-73.

27. Jauch, C. 1937. Las dos especies de Septoria que atacan al apio cultivdo en la República Argentina. Rev. Argentine Agr. 4:258-272.

28. Leach, R. 1941. Banana leaf spot *Mycosphaerella musicola*, the perfect stage of *Cercospora musae* Zimm. Trop. Agr., Trinidad 18:91-95.

29. Lima, A. G., F. A. A. Couto and G. M. Chaves. 1962. Contrôle da mancha púrpura do alho (*Allium sativum* L.) por meio de pulverizações com fungicidas de proteção, combinado com espalhante adesivo. Olericultura 2:203-211.

30. Lindquist, J. C. 1930. Algunos hongos nuevos parásitos de las plantas cultivadas en la República Argentina. Rev. Fac. Agr., La Plata 19:197-210. (See reports of other collections.)

31. Lizer, C. 1914. Quelques notes pour servir de complément au recueil de M. L. Hauman-Merch sur les parasites végétaux des plantes cultifées en Argentina et dans les régiones limitrophes. An. Soc. Sci. Argentina 78:5-17.

32. Lopez, R. and W. Q. Loegering. 1953. Resistencia de variedades de abaca (*Musa textilis* Nee) a la mancha de la hoja y perdidas occasionadad por la enfermedad. Turrialba 3:159-162.

33. McGahan, M. W. and R. H. Fulton. 1965. Leaf spot of banana caused by *Mycosphaerella musicola*: a comparative anatomical study of juvenile and adult leaves in relation to lesion morphology. Phytopathology 55:1179-1182.

34. Marchionatto, J. B. 1926. Fitoparásitos de la Argentina nuevos o poco conocidos. Physis 8:367-372. --. 1928. Fitoparásitos de la Argentina nuevos o poco conocidos. II. Rev. Fac. Agr., La Plata 18:21-29. --. 1941. Epiphytie de septoriose blé. Internat. Pl. Prot. Monthly 15: 113-114. --. 1948. Tratado de Fitopatología. Edit. Sudamericana, Buenos Aires, 537 p. (see Bibl.) (Numerous reports of collections.)

35. Mello, R. F. T. 1960. Observações sôbre a brusone do arroz e seu contrôle. Biológico 26:218-222.

36. Montserin, B. G. 1952. The cultivation of bananas and plantains. J. Agr., Trinidad and Tobago 52:242-268.

37. Mujica Richatt, F. 1942. La septoriosis del apio en Chile. Bol. San. Veg., Chile 2:140-143. (Under "Determinaciones micologicas," published in Agr. Tec., are collections and numerous reports.)

38. Muller, A. S. 1953. A foliar disease of legumes in Central America. FAO. Pl. Prot. Bull. 1:83-84. (Also many collections and reports of them from numerous countries of the Neotropics.)

39. ________., and C. Chupp. 1935. Cercospora de Minas Geraes. Inst. Biol. Veg., Brazil 1:213-220. -- and --. 1942. Las Cercospora de Venezuela. Bol. Soc. Venezolana Cienc. Nat. Bot. 8:35-59. -- and --. 1950. Cercospora in Guatemala. Ceiba 1:171-178.

40. Olave L., C. A. 1958. Resistencia de algunas variedades y lineas de frijol (Phaseolus vulgaris L.) al Isariopsis griseola Sacc. Acta Agron., Colombia 8:197-219.

41. Quiros-Calvo, M. 1950. Chaetoseptoria wellmanii Stev. una enfermedad de las leguminosas. Suelo Tico 4:137-139.

42. Rodríguez, L. R. 1942. La alternariosis de la patata. Univ. Centr. Quito, Bol. Inst. Bot. 1:85-124.

43. Rossetti, V. 1959. Doenças da siringueira. Biologico 25:233-234; 266-272.

44. Sarasola, A. A. 1944. Dos septoriosis de las alamedas argentinas. Rev. Argentine Agr. 11:20-40.

45. de Segura, C. B. 1959. Principales enfermedades de las plantas en el Perú. Soc. Nac. Agr. Com. Def. Técn. Algodon y Ass. Agr. Cañete e Ica (bulletin) 70 p. --. 1965. Enfermedades de Cultivos Tropicales y Subtropicales. Editorial, Jurídica, Lima. 439 p.

46. Spegazzini, C. 1898. Fungi argentinii nova v. critica. An. Mus. Nac., Buenos Aires 6:81-365. --. 1918-1919. Notas micologicas. Physis 4:281-295. (1900-1924, numerous collections and thou-

sands of mycological determinations, many new species, largely from tropical America.)

47. Stevenson, J. A. 1946. Fungi novi denominati. II. Mycologia 38:524-533.

48. Stover, R. H. 1963. Studies on ascospore production by Mycosphaerella musicola in Honduras. Phytopathology 53:890-891. --. 1965. Leaf spot of bananas caused by Mycosphaerella musicola: effect of temperatures on germination, hyphal growth, and conidia production. Trop. Agr., Trinidad 42:351-360.

49. ________., and R. H. Fulton. 1966. Leaf spot of bananas caused by Mycosphaerella musicola: the relation of infection sites to leaf development and spore type. Trop. Agr., Trinidad 43:117-129.

50. Toro, R. A. 1931. Cercosporae of Puerto Rico. J. Dep. Agr. Puerto Rico 15:5-17.

51. Tosi, J. A., Jr. 1960. Zonas de vida natural en el Perú, memoria explicativa sobre el mapa ecológico del Perú. Inst. Interam. Cien. Agr. OEA, Zona Andina. Boletin Tecn. 5, 271 p. (Processed; see also discussion and citations referring to tropical ecology at close of present writer's chapter I. -F. L. W.)

52. Viégas, A. P. 1941. Manchas das fôlhas da mandioca, producidas por Cercosporas. Bragantia 1:233-248. --. 1943. Alguns fungos do Brasil, Cercosporae. Boll. Soc. Brasil Agron. 8:1-160. --. 1955. Manchas do fôlha de Hevea brasiliensis. Bragantia 14:63-69. --. 1961. Indice de Fungos da América do Sul. Inst. Agron., Campinas 919 p.

53. Viera, C. and H. Shands. 1965. A mancha farinhosa do feigoeira commun. Rev. Ceres 12:311-314.

54. Waldlaw, C. W. 1961. Banana diseases including plantain and manila hemp. Longmans, Green & Co., Ltd., London 648 p.

55. Wellman, F. L. 1933. Pathological effect of the east wind in Florida. Plant Dis. Reptr. 17:177-181. --. 1934. Weather conditions associated with seasons of severe and slight celery early blight epidemics in Florida. Phytopathology 24:948-950. --. 1961. Coffee botany, cultivation, and utilization. Leonard Hill, London; and Interscience Publ., New York, 488 p. --. 1962. A few introductory features of tropical plant pathology. pp. 928-930, In Symposium on tropical plant pathology, phytopathology 52:928-953. --. 1967. Impact of plant disease on crop production. (pp. 296-302; 329-330) In Rural development in tropical Latin America (K. L. Turk and L. V. Crowder eds.) Cornell Univ. Press, Ithaca, New York, 446 p. --. 1969. More diseases on crops in the tropics than in the temperate zone. Ceiba 14:1-12.

56. ________., J. C. Walker, A. Cook, and M. Gallegly, Jr. 1951. Effect of temperature on vegetative growth of five coffee disease fungi. Asamblea Latinoamer. Fitoparasit., Mexico Proc. 1:251-257.

57. Wolf, E. A. 1958. Emerald--a new early blight-resistant pascal celery. Florida Agr. Exp. Sta. Circular 8-109, 7 p.

58. Wright, J. 1950. Investigations 1948-1949. Jamaica Dep. Agr. Bulletin 45, 110 p. --. 1951. Investigations 1949-1950. Jamaica Dep. Agr. Bulletin 47, 127 p.

Chapter XI

SCLEROTIAL DISEASES IN THE TROPICS

Summary

There are many hardened structures, the sclerotia or sclerotoid conditions, commonly formed by saprophytic and parasitic fungi in the tropics. Different kinds of diseases, occurring under various kinds of ecology, are caused by the sclerotium-developing, or sclerotoid-body-forming, fungus parasites. Spores spread the diseases, but there are examples in which no perfect stages are found in the fungus' life history. Vigorously multiplying bacteria often consume casually occurring spores and unprotected fungus hyphae on and in soil debris; however, a typical sclerotium or sclerotoid structure resists bacterial dissolution, hyperparasitism, and similar situations.

Introduction

Whenever I have hunted them in the moist tropics, I have had no difficulty in finding many fungus sclerotia or sclerotoid structures. (The Latin word scleroticus means "hardened.") In many parasitic fungi, where the climate is predominantly warm and moist, there develop characteristic resting or holdover structures. These are composed of layers of dead fungus threads adhering to and wrapped around cores of dormant, or at least quiescent, living masses of hyphae, being protected from damage by natural

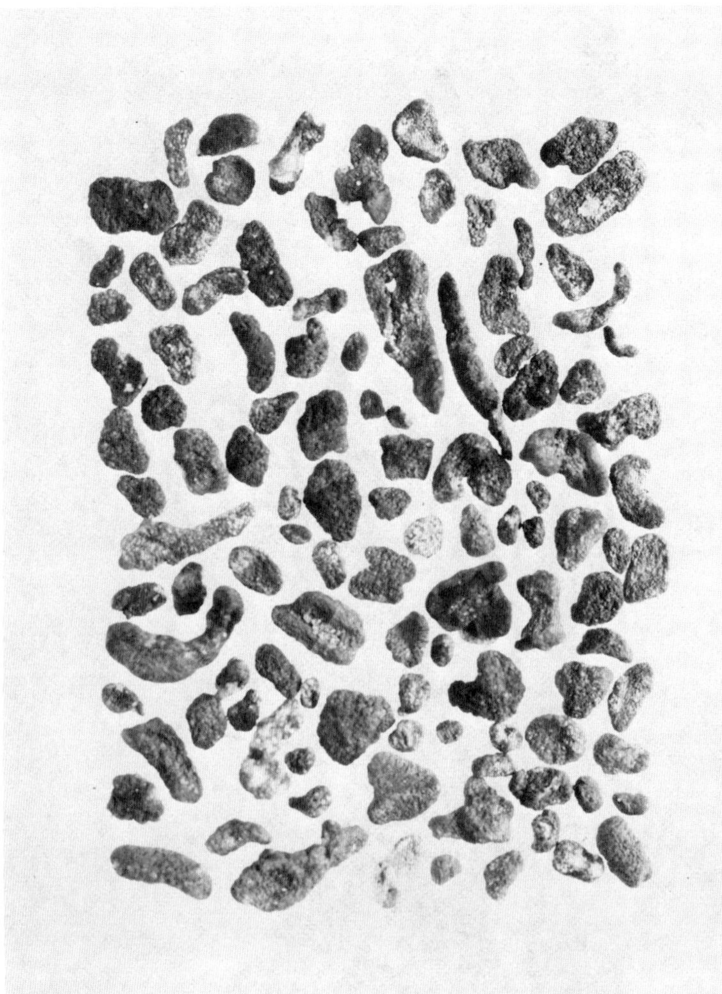

Fig. 11-1. Sclerotia of Sclerotinia sclerotiorum from headrot of cabbage. The fungus HOLDSOVER from season to season as sclerotia in the tropics. Photograph courtesy of Plant Pathology Department, University of Florida, Gainesville. (The same fungus producing nesting disease or cottony rot in baskets of succulent green beans, may be slightly different: often sclerotia are somewhat smaller, dark in color, and rounded to oblong in shape.)

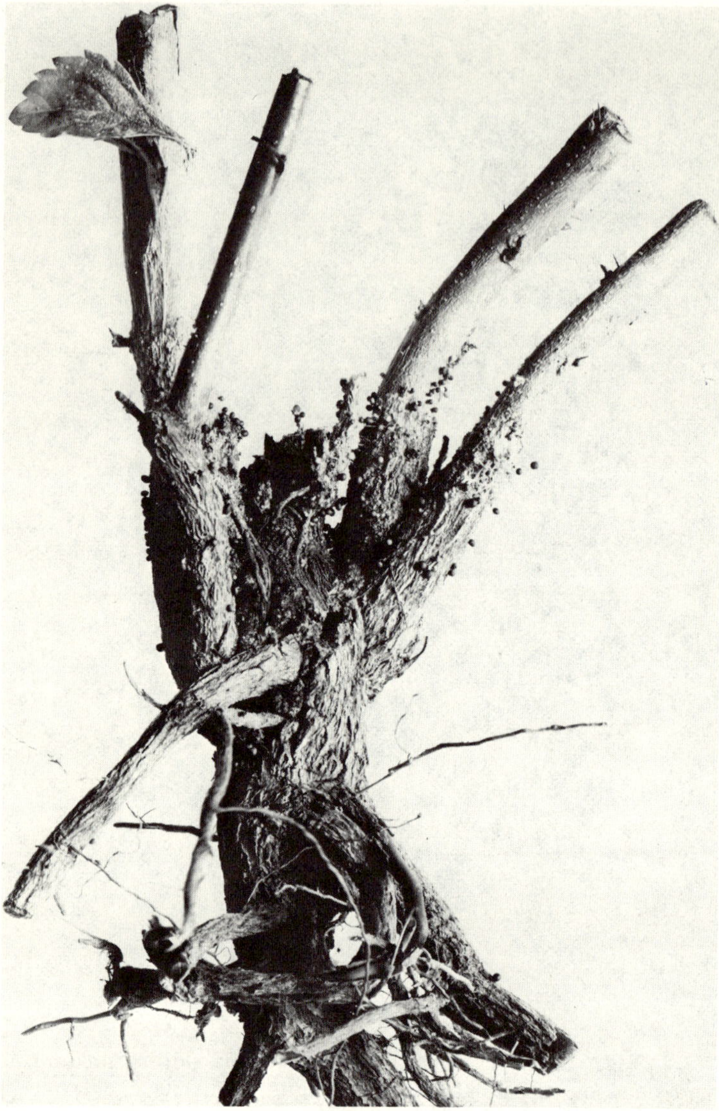

Fig. 11-2. Spectacular, round and small sclerotia typical of Sclerotium rolfsii around the base of an ornamental bush, Duranta repens. Courtesy of Plant Pathology Department, University of Florida, Gainesville.

dangers to the fungus species. Neither excessive dryness nor long periods of soaking rains can readily destroy sclerotia or sclerotoid bodies. In addition, these fungus developments apparently are not favored foods of insects or other fauna of tropical soils, or the extremely large numbers of rapidly multiplying voracious bacteria that live in soil debris and may quickly attack unshielded fungus mycelium. (For examples of spectacular sclerotia see Figs. 11-1 and 11-2.)

A sclerotoid condition develops in bushes or trees with several kinds of branch-end dieback. This is caused, for example, by *Physalospora* (*Diplodia*) in coffee; the killed plant parts harboring the fungus do not drop quickly. In its sclerotic matrix, hyphae of the dieback parasite remain alive (if the dead branches do not fall) during unseasonable conditions. If dieback does drop, soil fauna comminute the tissues and in the presence of moisture, bacteria will complete destruction of the fungus matter thus opened up to attack.

Another, but similar, kind of sclerotic condition occurs with fungus-caused witches' broom diseases. There are a number of such troubles on shade trees used in plantations, on ornamental trees, and on forest trees that furnish lumber. In cacao witches' broom, *Marasmius perniciosus*, the developments are as follows: the hyphae of the parasite become intimately intermingled with hyperplastic host cells in abnormally numerous, short, malformed stems and distended twigs. At a certain point the swelling-growth stops and the broom ceases to grow, wilts, and dies. In cortical parts of the diseased twigs and stems, fungus and host tissues darken, and the interior fungus hyphae remain light in color. Brooms that fall and are covered with wet debris quickly rot. If held aloft on branches, or if fallen but left uncovered on the plantation soil, even though externally making a dead and unsightly mummified appearance, inside the parasite lives on and when the rainy season arrives brooms produce agarics and release basidiospores. This is the final phase of development in the fungus.

When fruits--also called pods--of cacao are infected and then rotted by *Monilia roreri*, the condition terminates in characteristic mummification. This is hardened and compacted; the diseased fruit then finally becomes sclerotic, but spores have already been produced on the surface before it has grown fully solidified. Hence, this co-called mummified fruit, is not a part of the life history preceding the reproductive phase of the latter parasite.

There are, not only multitudes of sclerotoid conditions and situations buried in host tissue as exemplified by types just mentioned, but there are also specialized fungal structures. These are made of mycelium that has massed itself into more or less determinate "resting" bodies, some tuberous and rounded, some fusiform, some flattened and disc-like, some like hard seeds of higher plants, some less dense (and delicate), some irregular and lumpy, and others globular and lens shaped. Such bodies have a tough outer sclera of specialized often dead dermal tissue; this rind surrounds the viable, usually lighter-colored and softer pseudoparenchymatous hyphal material.

Sizes of these sclerotia vary greatly in the tropics. Probably the largest is Poria (Pachyma) cocos (Schw.) Wolf (see Fig. 11-3), which Wolf (54, 55) and Weber (50) describe as sometimes larger than a man's head, down to the dimensions of a hen's egg. It is subterranean, edible, often found deep in the soil, and is parasitic on roots of trees such as oak, eucalyptus, citrus, magnolia and fig and it attacks roots of certain herbaceous hosts like maize. It occurs in the subtropics of the United States and probably much farther south. Very large but somewhat different sclerotia are found occasionally in Brazil (Fig. 11-4). When one of these, about the size of a large grapefruit, was half buried in moist soil and kept for nearly a year it produced sporophores of Polyporous sapurema.

Intermediate to small sclerotia and sclerotoid structures in the Neotropics are reported by numerous authors (for examples, see citations 1-4, 9, 13, 17, 19, 21, 23, 29, 31, 32, 34, 35, 39, 40, 42, 46-48, 53, 57). Examining virgin, jungle-covered soil on the west coast of Guatemala and in a newly cleared Costa Rican jungle, I found an array of sclerotia, from those as large as a walnut to others like a grain of sand. While probably hundreds of saprophytes form sclerotic or sclerotoid growths in the tropics, also a large number of plant parasites do the same thing, and this happens even with hyperparasites (43).

Sclerotial States in Tropical Diseases

The millions of years of fungus evolution that results

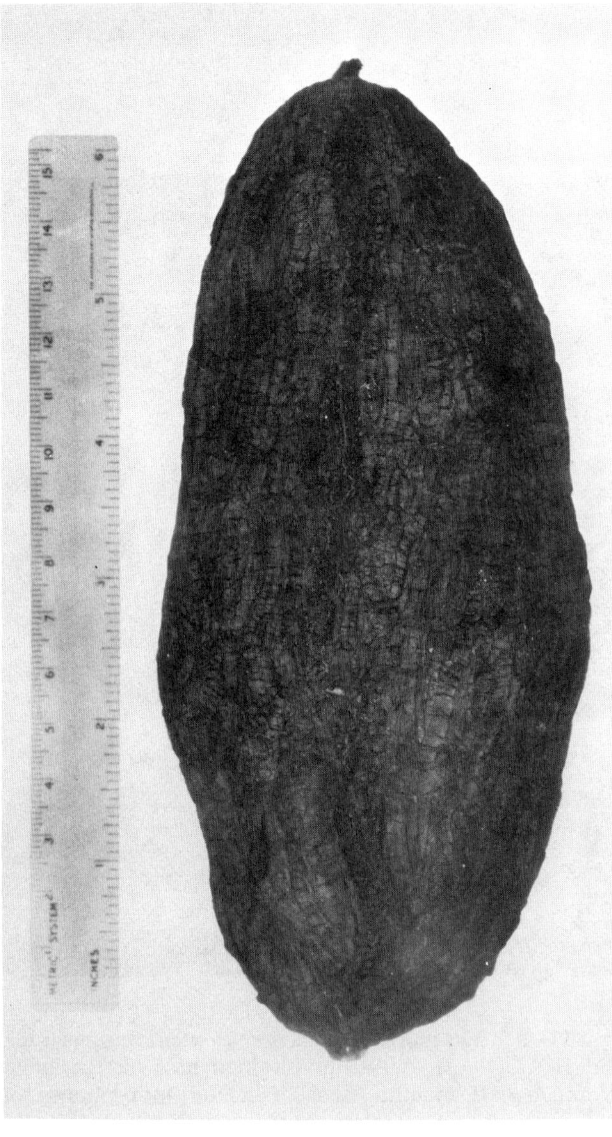

Fig. 11-3. A medium-sized sclerotium ("tuckahoe" or "Indian bread") nearly 18 cm. long; HOLDOVER stage of Poria cocos a parasite of plant roots. Courtesy of Plant Pathology Department, University of Florida, Gainesville.

Fig. 11-4. Example of a large tropical fungus sclerotium, the HOLDOVER stage, maintained half buried in soil for eleven and a half months, that produced sporophores of Polyporus sapurema. Collected in state of Espirito Santo, Brazil. Photograph courtesy of Raul Drummond Gonsalves, Instituto Biologico, São Paulo, Brazil.

in a sclerotial habit in part of the life of an organism is of utmost importance to survival of species in the tropics. In the view of some temperate zone specialists, for examples see works of Whetzel (53), Duggar (13), and Heald (19), formation of a sclerotium is specially for what is termed by them for "overwintering" purpose. This seems particularly logical in lands where summer follows winter and the spore stages from the fungi come to "flower" in the spring. There is no true winter in the tropics and in these conditions survival of fungus species is more difficult than being dependent on remaining alive through a simpler season of cold. The sclerotia and sclerotoid bodies (even though they may be borne aerially (56)) are more useful in the tropics in withstanding long periods of excess drought, much rainfall and floods, and injurious leachates from the abundance of decayed vegetable matter, and withstanding direct action by huge populations of rapidly reproducing bacteria, vigorously feeding insects and other small animals that live in debris, and fungi that live on other exposed fungi.

There is a good deal to be studied respecting the advantages to parasites of developments to save from injury the tender phases of fungus growth. I do not disregard protection in the multitudes of other types of holdover structures of aggregate-cell composition. To point out a few: there are the stromatic creations, from thinly woven pellicles through loosely gathered deposits of mycelium, to hyphae making a thickly and densely structured mat. It is also recognized that numerous parasitic fungi have the habit of collecting parts of their hyphal growths into special sclerotial-like but relatively soft clumps on which spores are borne in a manner that make these sporodochia or synnemata (refer to 28, and special descriptions in chapters of 17, 19, and 35). It seems important to fungi that at times there be for them defense mechanism(s).

It seems, indeed, that some of the sclerotial parasites have the potentialities of being the most persistent disease organisms in the tropics. Examples of two are illustrative. When once introduced into an unsprayed coffee plantation, *Sclerotium coffeicola* Stahel becomes a long-time inhabitant. Descriptions of the sclerotium stage of the organism by Stahel (41), Echandi (14), Aycock (3), Kuyper (22), Martyn (25, 26), Saccas (33), and Icochea and Aycock (20) show it is a hard, resistant body the size of a small pea. From what we know of the continuing field recurrence, it withstands some of the most severe of tropical conditions, that we know are deleterious to fungi.

A somewhat similar fungus is *Sclerotium rolfsii* Sacc., which is common in the tropics. Aycock (3) has indicated its very wide geographical and host range; it has been spread far by man. It is found in the warm temperate zone, in all moist, subtropical American countries, thence south throughout the West Indies, Central America and South America to southern parts of the Argentine and Chile. This phenomenal distribution in itself suggests how highly resistant its sclerotia are to desiccation, decay organisms, vagaries of tropical weather that are so hard on fungi, and the adverse effect of long periods under conditions of transportation. It is occasionally recovered from coffee seedlings which exhibit a killing wilt (Fig. 11-5) but which is not the moist collapsing damping-off so characteristic of *Rhizoctonia* infection.

All is far from settled about certain sclerotial diseases. The occasionally most devastating culm rot, *Sclerotium oryzae* Catt., of rice is an example. This is a common designation for it, but other rice troubles with sclerotia involved (Galli *et al.* (17), Tullis *et al.* (47), Tisdale (46), de Segura (35) and others) associate with it *Helminthosporium sigmoideum* Cav., and *Leptosphaeria salvinii* Catt. In some instances *Septoria nodorum* Berk. is given as the causal organism. A well planned search and collection in our tropics of sclerotia on rice would bring much needed certainty to literature on it.

Something of the wide range of sclerotial diseases is evident from reference to such works as those of Marchionatto (23, 24), Mujica (29), Campi (6), Cano (7), Alvarez and Chupp (1), Jauch (21), and Cook and Toro (9). An organism often thought of as a temperate zone parasite, *Sclerotium sclerotiorum* (Lib.) dBy., may have come originally from the tropics and is frequent in these regions. There it attacks flowers, fruits, roots, stems, growing tips, and the under-bark of trees, and is a serious market problem in fruits and fresh vegetables. A very common, but sometimes not easily seen, sclerotial disease in our tropics is *Sclerotium bataticola* Taub. = *Macrophomina phaseoli* (Maubl.) Ashby that produces masses of black-pepper-size-granules which are small sclerotia inside disease tissue. These sclerotia give to the disease condition the highly descriptive names of ashy stem and charcoal rot.

Long field experience in the tropics with both of the

above sclerotial fungi, S. sclerotiorum and S. bataticola, has shown it is the sclerotial stage that survives either the intensity of the wet season or the severity of the dry season. Fortunately for man, where the worst part of the wet and dry seasons are very limiting, there is the intermediate period for which the grower can time his work, so he may carry on the necessities of husbandry and harvest. However, it is this very period spaced between the heaviest rains and the worst drought (when crops are growing best) that the sclerotia are more likely to be moved around by man, to germinate and to parasitize directly, and if there is a spore stage to be activated to sporulate. With sufficient moisture, hosts are infected and rotted, sclerotia are formed, and the whole cycle of living is ready to repeat itself.

Effect of Drying on Sclerotium

No planned research has been attempted to make a study of what dryness does after the rainy seasons to fungus sclerotia. However, observations have been suggestive in Central America on three of the most readily identified sclerotium-forming parasitic organisms: Sclerotinia sclerotium, which causes stem rot of vainicas or garden beans (Phaseolus vulgaris); Botrytis sp., causing flower collapse and bud rot of amapola or flowering hibiscus (Hibiscus rosa-sinensis); and Sclerotium rolfsii, which brings on basal rot and wilt of higuerilla or castor bean (Ricinus communis) and, as mentioned above, at times wilts seedlings of coffee (Fig. 11-5). Indications are as follows. The organisms of all these three diseases appear to attack most vigorously in about the last month of the rainy season. The organisms make fast-growing mats of hyphae and in the mats are scattered small cottony balls or initials. About them soon grow the protecting covering layer. This stage of development is specially notable at about the time vegetative growths of the parasitizing fungus threads are being slowed down from dry weather. As long as rains are frequent the outer fungus integuments around sclerotia do not attain full hardness. When the main period of drought approaches, sclerotia mature, become characteristically shaped and colored, and are resistant when picked up and pressed between thumb and finger.

During the difficult condition of abnormally long drought, sclerotia or sclerotoid structures are successful in keeping fungi alive. These same sclerotial bodies which

Fig. 11-5. Young typical sclerotia of Sclerotium rolfsii. Developing after wilted coffee seedlings were incubated in moist chamber to determine cause of the disease. Photo made in Guatemala, courtesy of A. S. Muller.

I consider especially important in surviving attack by bacteria and other antibiotic organisms, are apparently equally important in helping the fungus to pass the state of dormancy it goes into by effects from dry season. Some fungi during dry periods, even if subsisting as relatively fragile hyphae or thin-walled conidia, are not always quickly killed by desiccation. They can last, in some cases days or possibly a few weeks. However, if there is time for a fungus to develop chlamydospores or a sclerotium, drought will not kill it for months and often in these stages the organism survives many years.

It is during the rainy seasons that a fungus meets its worst conditions for survival. I observed this in wet seasons three times, on *Fusarium* in Honduras, on *Botrytis* in Costa Rica and on *Sclerotinia* in Puerto Rico. The buried-slide technique was employed in research on the *Fusarium* and *Sclerotinia* pure culture materials, source of which was fungus colonies grown on PDA slants. The sclerotia studied in Costa Rica were of two types: one was a *Botrytis* (species undetermined) that produced sclerotoid structures in tips of withered orchid (*Maxillaria* sp.) flowers or old leaves; and the other was from onion bulbs in the market with sclerotia typical of *Botrytis allii* Munn.

The course of events that occurred underground was determined by working with buried slides, taken out for examination at various intervals. In these cases, fungus mycelium and spores were fixed to slides by four hours of drying on slides before they were buried. Sclerotia were buried pressed firmly against slides. In Puerto Rico the disintegration of mycelium of *Sclerotinia* sp. was seen after nine days, occurring first in old mycelium. During relatively dry conditions, mycelium remained healthy but the cells were smaller in diameter and more densely filled with stain-absorbing oil and other food storage bodies. With rains in the field, or from watering in jelly jars, in six days growing mycelium was larger in diameter than in dry soil. However, with aging of the fungus the bacteria could be seen attacking injured hyphal walls and crowding inside the cell lumen. Once inside, these soil bacteria advanced in both directions and were found forcing openings in septae between fungal cells. This, studied in Honduras, was also seen with *Fusarium.* The *Botrytis* sp. that infects orchids and *B. allii* on onion, exhibited the same results when subjected to underground treatment. In tests using the buried slide technique, mycelial growths from agar cultures were quickly injured and almost completely disintegrated by bacteria in wet soil in three days. Sclerotia of these fungi withstood bacteria in wet soil for 24 days, after which they were germinated on PDA.

Two types of sclerotial-sclerotoid materials were thoroughly dry before being used to test effects of relative air dryness and wet soil on their survival. Well rounded, "fat" sclerotia of *Botrytis* sp. from dead buds of *Hibiscus rosa-sinensis* were collected and air dried. These were then tested for survival under dry and wet conditions with results as follows. These were leathery sclerotia and lived ten months under air dry conditions but did not survive more than two weeks buried in wet soil. Another type of sclerotium was hard but flattened and of another *Botrytis* from seedlings of *Desmodium*. These bits of hardened fungus lived nine months under air dry conditions, but in wet soil survived only three weeks. When the old, matted, leaf tips of orchids developed sclerotoid masses in them, these mats were stored in dry air, and the fungus, *Botrytis* sp., could be isolated after nine months but not at ten months. When buried in wet soil, this orchid organism was still viable after four weeks, but could not be revived after five weeks by planting on agar.

Sclerotial Fungi Without Perfect-Stage Spores

The parasitic, and partially saprophytic, organisms that may be classed in the genus *Botrytis*, of the *Fungi imperfecti*, are well distributed in the Neotropics. Many of the tropical conditions favor such fungi as to parasitism, saprophytism, and survival. Fungi of the genus (see Whetzel (53)) develop sclerotia but are best recognizable by the branched conidiophores with the usually ovoid conidia that are thin walled and short lived; it is these delicate asexual spores that are largely responsible for wide spread occurrence of the organisms.

No one knows as yet of all the members of the genus *Botrytis* that occur in our tropics. Examples of two of little concern in the temperate zone but more serious in the tropics are *B. tulipae* (6) and *B. fabae* (21). Probably all of the relatively few kinds found in the temperate zone are also in the tropics, with possibly more species still to be described by enterprising tropical mycologists. In the winter-free regions the spores encountered are the fragile asexual conidia, although the tougher spores from Ascomycetous stages of Sclerotinia are sometimes found. However, it is in the rind-protected sclerotia and sclerotoid fungus masses that these fungi survive over long periods, and from season to season.

There is a remarkable and unique sclerotial fungus that has vigorous subterranean vegetative growth of cord-like strands. These strings are composed of combined hyphae that permeate alkaline soils in Mexico and adjacent parts of the United States. This especially aggressive parasite is a tropical inhabitant, has been studied since the beginning of this century, and is known in temperate zone literature as the cause of Texas root rot, Phymatotrichum omnivorum (Shear) Duggar. For descriptions, its parasitism and its occurrence, reference may be made to: Duggar (13), Heald (19), Shear (37), Ezekiel (15), Taubenhaus and Ezekiel (44), and Zevada, Yerkes and Niederhauser (57).

The sclerotial stage of the parasite is rounded to a wart-like, sometimes spindle-shaped, brown-colored structure. It withstands extremes of drought, and retains life for the organism in the face of cold, providing that temperatures of below freezing last only for a short time. To infect hosts on the advent of moisture and warmth, the sclerotia "germinate," sending out light-colored, fast-growing mycelial threads in the soil. These gather into strands over host roots, infect and then invade the plant tissues which are thereby severely diseased, causing weakness and death in trees, bushes, herbaceous broad-leaved weeds and crop plants. However, this all but truly omnivorous parasite does not attack grasses, maize, sorghum, and such monocotyledonous hosts. Vegetative mycelium runs through the soil, and, under right conditions of moisture, comes to the soil surface to produce a flat, round, thin mat of growth in which form myriads of small holes lined with basidium-like structures bearing spores. Attempts to germinate these spores have never been successful. These have been given careful research attention by some of the most accurate pathologists and mycologists. While these spores are Moniliaceous in type they appear to be useless since they have lost any ability to infect hosts.

If a naming is attempted of Neotropical fungi classed as belonging to the Monilialies, it will be found that the list is long and the hosts attacked are very numerous. However, as perfect stages are demonstrated now and again

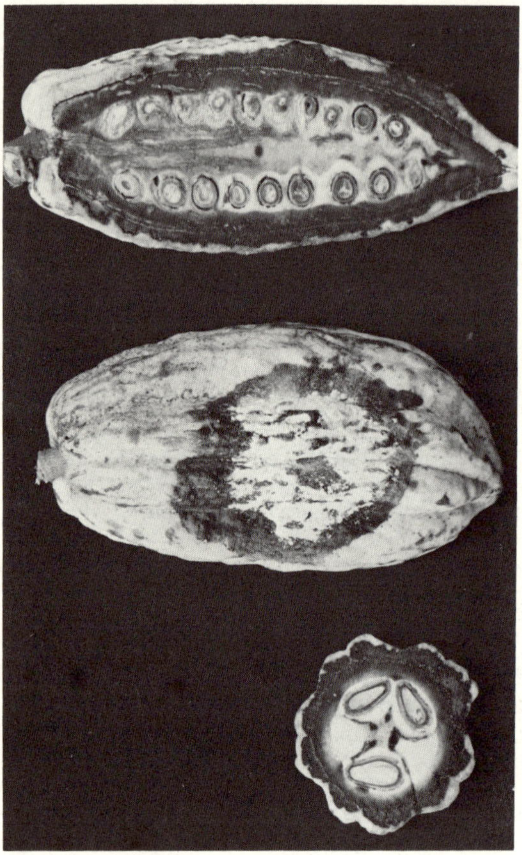

Fig. 11-6. Watery podrot on podredumbre acuosa of cacao: center pod recently developing disease symptoms on its surface including mats of profuse typical monilia spores towards the middle of the decay; pods on either side are cut open to display the hard, dark sclerotoid condition (which may act as a HOLDOVER stage) of the outer layers of the pod. Photograph from Ecuador, courtesy of E. Ampuero, Head of Cacao Research, Guayuquil, Ecuador.

among these, they appear to belong (53) to genera under the Ascomycetes. One Monilia especially well known in South America is the extremely destructive cacao fruit-pod attacking organism which has been under observation for a long time. In spite of work by many, its perfect stage has never been discovered. It is Monilia roreri Cif., first seen by J. B. Rorer in Equador during the years 1916-1917. A photograph of the disease it causes is shown in accompanying Fig. 11-6. In tropical plant pathology this is one of the classical diseases to be noted. A few papers, by no means a comprehensive list, may be seen by reference to citations as follows: 8, 10-12, 16, 18, 30, 35, and 36.

Over two decades after Rorer's first work, when it was described by Ciferi (8) it was still restricted in its distribution to a part of Ecuador. However, in later years it has spread to not only other areas of Ecuador but into Colombia and Venezuela, Panama and to Brazil.

This fungus invades the pods of cacao when they are very young and when it enters it remains as an incipient infection, that only upon later maturation of the pod becomes visible superficially. The growth of the fungus inside the pod is rather slow but after a length of time the diseased area reaches a certain stage of rot and finally a spore-bearing surface appears. This is on the outside of the pod and upon this develops the multitudes of orange- to salmon-colored spores that are then lifted off by breezes and water drops that distribute them to other very young cacao fruits. The rot of this Monilia extends into the middle of the pod or mazorka (see Fig. 11-6), of the cacao and the seeds inside are decayed to a dark and watery consistency, along with the pulp and even the seemingly hard outer surfaces of the pod. Prior to complete rotting the fungus remains viable within the hardening sclerotoid mass in the pod for a long time.

Fortunately this Monilia, which produces such abundant numbers of spores does not attack anything other than cacao crops of tropical America. At least it has not been found on weeds or any other wild plants growing within the cacao regions. This unusual limitation of a parasite to one genus, in this case Theobroma, must be of great significance. The reason for it should be given attention as it is common for tropical parasites to have a wide range of hosts, much wider than is commonly the case with parasitic fungi in the temperate zone. The conidiospores of M. roreri are, although numerous, thin-walled, and are readily killed by sun,

Fig. 11-7. Kernels of maize enveloped in white fungus mycelium, with black stromatic-like, or sclerotium-like bodies under the mycelium. From artificial inoculation and incubation work with Diplodia sp. Photograph courtesy Department of Plant Pathology, University of Florida, Gainesville. (Identical kernel disease symptoms are in moist stored grain, which I found in El Salvador, Guatemala, and Honduras.)

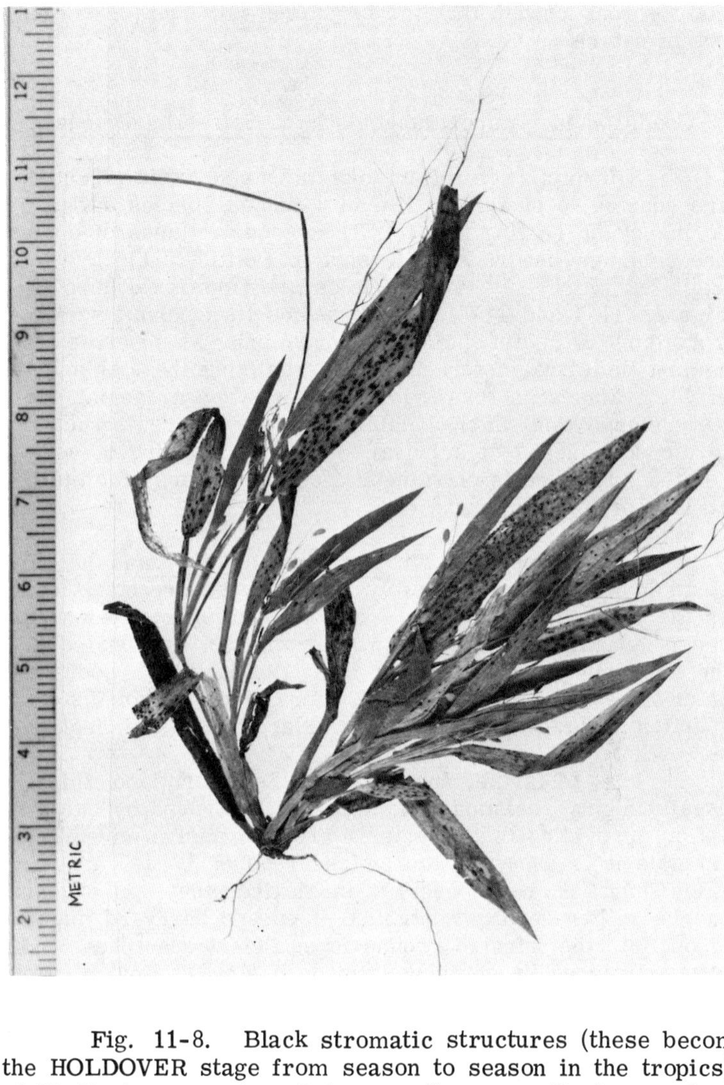

Fig. 11-8. Black stromatic structures (these become the HOLDOVER stage from season to season in the tropics) of *Phyllachora punctum* Orton, on the grass *Panicum xalapense* HBK. Photograph of material from Honduras.

drought and cold. However, the fungus in the sclerotoid, diseased pod hanging on the tree is not quickly killed. In the tropics there is of course the warm wet seasons interspersed by the dry seasons, and it is during the wet seasons that M. roreri flourishes. The dry conditions intervening between one wet season and the next wet season are perhaps not rigorous enough that the organism will ever be stamped out in nature.

Condition in Nature Antagonistic to Sclerotial Diseases

Sclerotia, sclerotium-like, and sclerotoid structures are adapted to sustaining life in a fungus species during dormancy or resting phase. These special kinds of bodies are common (hundreds of tropical parasitic fungi have this habit, of which only two examples are illustrated here in Figures 11-7 and 11-8), which bodies then permit such organisms to retain their life by armoring themselves against biological destruction. In realizing these things the problem arises as to whether there may be a special ecologic complex in which a sclerotium, or kindred structure, is the sole means of survival. An understanding of certain related situations is of value in planning means of either control or amelioration.

In the case of Phymatotrichum omnivorum, the only stage in which it can possibly survive a severe condition is as the sclerotoid structure. No one knows exactly how long under normal situations these sclerotia can live buried in the soil, but they survive for over 12 years in a good state of dryness and in alkaline soil. However, in moist soil with reaction more towards acidity the sclerotia are soon destroyed.

It is of further interest that the sclerotia of this usually highly resistant organism are destroyed by bacterial action, providing these bacteria are in massive amount. For example it is known in the cotton areas of Texas, and the same thing has been observed in Mexico, that this organism can be rather readily controlled if enough barnyard manure is put into the infested ground in sufficient quantities. This is usually done by applying it in deep furrows, close together, which are then plowed closed. It is thus covered and with moisture, as the manure rots, the sclerotia are also destroyed. Knowledge of this sort gives a basis for devising horticultural methods, which will control disease organisms with structures of sclerotia or sclerotoid nature. In warm-moist field ecology, as it occurs in the tropics, where the plentiful and ubiquitous bacterial soil inhabitants can be employed, they will decay sclerotia and, at the very least, greatly reduce infection sources. In the tropics it should be understood as a principle that this can be applied with all sclerotoid fungi in ameliorative, or, where especially effective, complete eradicative treatment. To express this in another way: this may be a means of eliminating expensive chemical disease-control methods. Where there exists a moderate tolerance to a disease in certain crop varieties, the minimal fact of only reducing the amount of a parasite may be sufficient in itself to be practical and allow good harvest in the presence of lessened disease.

Among onion and garlic growers of Peru, Brazil, and Argentina the white rot, Sclerotium cepivorum Berk. (Sphacelia allii Vogl.) is a most troublesome disease (5, 17, 35, 49 and others). It was introduced from Europe, probably about the same time that it came to the United States of America. Wherever found it thrives where soil temperatures are cool. The sclerotium, when found, is the specific diagnostic feature of this parasite. It is the holdover condition; a body that is macroscopic in size, hard and very resistant, and, as in other instances, by temperate zone criteria such a sclerotium is considered as a type of "overwintering" structure. It is certainly the case that in the temperate zone, after winter is over, the white rot sclerotia germinate and infect susceptible plantings. As seen in the mountains of the South American tropics, it is recognized that the typical temperate zone winter season does not exist. However, in some areas at special growing periods temperatures are sufficiently low to allow S. cepivorum to cause disease.

Campacci (5) and others have given attention to temperature relationships of this organism in its cooler tropical haunts. There is considerable difference between the temperature range for vegetative growth of the parasite itself and the temperature range for disease. This is an interesting situation and needs study in relation to the physiology involved. The coldest temperature at which the fungus grows vegetatively is 4° C; it grows most luxuriantly at the range of 20° to 24°, and the warmest temperature at which it will grow vegetatively is 30°. For disease the most rapid and severe rotting is at 10° to 20°, somewhat below the optimum for saprophytic fungus growth. It is weak in attack at 23°, and at 24° or above even in heavily infested soils, onions remain unaffected.

This sclerotial disease occurs in severe form in Brazil, Campacci (5), and in Argentina, Marchionatto (23, 24), and something similar, but not absolutely identified as the same, has been observed in Chile and Colombia. In Peru, according to de Segura (34), it is in gardens at relatively high and therefore cool elevations near Arequipa, but not where grown at low elevations in warm conditions. In Louisiana, also considered in part the Neotropics (see Map A, Chapter I), Tims & Olive (45) have shown that summer-grown crops are free from the disease, whereas fields planted to mature in the cold season are often severely diseased.

This fungus produces sclerotia that are rounded to confluent and almost crust-like under certain conditions. They break up to about the size and shape of more or less scaly seeds and are black, and they withstand extended periods of dry conditions. As contaminants in lots of onion seeds they are capable of being carried long distances for months. The fungus is also distributed at times as invisible, internal sclerotia in onion sets and garlic bulbs used for planting.

It is significant, from the standpoint of dissemination and source of infection, that no spores of _S. cepivorum_, of either the imperfect or perfect stages have been seen in the Neotropics. Reinfestation is apparently always from sclerotia in our lands. If the asexual spore is ever encountered it will probably be the small, catenulately arranged, thin-walled, spherical conidium described by Voglino (49). The perfect stage, if ever discovered, would be an Ascomycete, and, according to Whetzel (53), would be _Stromatinia cepivora_ (Berk.) Whetzel.

Process of Decay and Hyperparasitism of Sclerotia

Either when planted in moist soil, or when stored under moist conditions, the white-rot diseased onion and garlic bulbs are invaded by both soil-borne fungi and by the ubiquitous slimy-soft-rot and other soil bacteria. The amount of bulb decay may be great, and as the bacteria multiply they envelop, act on, and smother the vegetative growths of such fungi as _Sclerotium cepivorum_, _Fusarium_ sp. and _Penicillium_ sp. However, sclerotia themselves of

S. cepivorum are not all decayed although the hyphae of the fungus are eliminated. Sclerotia can be separated from slimy rotted bulbs, and when planted on acidified agar some will germinate. The resistant rind of this sclerotium withstands great amounts of bacterial secretions. Under experimental conditions, I have seen conidia of both Fusarium sp. Penicillium sp. quickly killed by bacteria transferred from the bacterial onion rot that follows sclerotial white-rot.

No research in the American tropics has been carried on to determine which of the soil bacteria are most effective in the continually progressing, rapid decay of plant debris. There is also little information about the specific identity, or the physiology, of bacteria that may be of special value in reducing numbers of plant parasitic fungi. Laboratory isolations from foliage indicate common occurrence of bacteria on aerial parts of plants in the tropics, and the ease with which such organisms are found in the soil. Evidence is that there are indeed extremely large numbers of bacteria in equatorial soils; probably many more in numbers in tropical than in temperate zone locations. It is fairly common in moist warm regions, to see long-lived-on tropical soils being cultivated and growing food with slopes so steep they would have promptly washed away in the temperate zone. On microscopic examination, such tropical soils with such "steep angles of repose" (as agricultural engineers express it) are found to have great abundance of bacteria and fungi. These are in themselves at least partially responsible for holding some soils from erosion.

The "extremely large" numbers of microorganisms that temperate zone workers discover in tropical soils are not to be wondered at. They are in fact normal for warm moist regions, as has been discussed in Chapter I. That this is an old condition is proved by the fact that one of the greatest tropical soil formations is the so-called "laterite." This develops through millions of years of geologic time in which bacteria are breaking down, oxidizing, and transforming parent rock material. Silicon components, so common and important in the temperate zone, are leached away through the effects of bacterial action. And these bacteria, in their nutrient processes, all this while are normally subsisting on both plant debris and other microorganisms, including fungi.

In the tropics, to survive incessant bacterial attack, it must be imperative for fungi to rely on development of sclerotia. It was interesting to me to find fungus sclerotia so small they are invisible to the naked eye and are in truth microsclerotia, best examined under the low power of the microscope. It should be added, that these microsclerotia, fully distinct from chlamydospores, are readily recognized as sclerotia. They have a definite resisting rind of lifeless, thick walled cells, surrounding a solid core of plump hyphae, giving every appearance of being ready to grow and further penetrate the soil.

Bacteria are a main danger to life of fungi, whether parasitizing fruits and foliage, or in the soil. The development of sclerotia is all that protects the organisms against bacterial destruction. However, there is more than bacteria alone to be considered in this regard. In the tropics, where nematodes are so plentiful, it is well known that they also feed on fungi. Numerous times in my laboratory, in researches both by students and by me, saprophagous nematodea have been encountered in moist chamber studies when disease-spotted leaves and dieback twigs were being incubated for development of asexual spores or perfect stages. Nematodes can be so numerous that certain of the fungi will develop not at all, or very poorly, because of the eelworms. However, on the whole, it is not believed that nematodes are a most common limiting factor for existence of soil fungi. On the other hand, this is often the case with bacteria. There are likewise some hyperparasitic fungi that, under conditions still not well understood, invade sclerotia.

I have had special contacts with two fungi of hyperparasitic nature, that have been seen attacking sclerotia. In one case in Costa Rica, a coffee seed bed was under study where young seedlings were being killed by a fungus that had a few small round sclerotia. These bodies were later determined as *Sclerotium rolfsii* and a collection of 19 were planted on agar. Of these, six showed hyperparasitic fungus growth, all of which were the same type, and the organism attacking was determined as *Rhizoctonia* sp. It also invaded sclerotia of this organism growing in pure culture in PDA.

In another case, soil and debris under cacao trees in Ecuador were sifted and small irregular sclerotia were found. On placing them between sheets of clean filter paper they were dried slowly and were taken to my laboratory to be included in material for fungus determinations. In five days the paper was unfolded, and mold of whitish appearance was seen marring the dark color of the sclerotia. When examined under the microscope the mold was diagnosed as *Verticillium* sp. On culturing it was found to act as a hyperparasite on sclerotia of *Sclerotinia sclerotiorum*. With both the *Rhizoctonia* and the *Verticillium*, mycelial threads of the fungi were observed to wrap around the sclerotia they attacked. On mashing the *S. rolfsii* sclerotia it was possible to see, especially clearly with hyphae of *Rhizoctonia*, penetration of the sclerotial rind by the hyperparasite.

Discussion of Certain Climatic Effects on Sclerotia

It has been thought that a type of environment or climate in the tropics may be generally most favorable for sclerotial diseases. However, no unequivocal generalizations seem possible as is seen in the following discussion. On the other hand, in some sclerotial diseases occurrence may be somewhat circumscribed by environment, but there seem to be no over-all parallelisms.

In the case of *Sclerotium cepivorum* in South America (it may also occur in the mountains of Panama since I identified sclerotia of it on garlic bulbs said to come from the hills and on display in a market in Panama City), sclerotia of it are found exclusively in cool country gardens of *Allium* crops. This cool-area occurrence corresponds closely with the temperature relations known to be favorable for infection and disease of garlic and onion. In a second case, *Sclerotium rolfsii* is serious on many vegetable crops and especially on squashes for storage and similar cucurbits grown by indigenous farmers. This sclerotial disease is known to be of greatest frequency in the warm low-country tropics, especially where it is moist. Priests have told me that when the cucurbitaceous crops, called quisquil or chayote, *Sechium edule* Sw., are badly diseased with *S. rolfsii* in clearings on the flat flood plains, Indian growers have moved their plots up into the hills, no more than a quarter day's climb, where it is slightly cooler and the disease has been completely circumvented.

A third kind of sclerotial disease, the carbonosa (charcoal) or pimiento negro (black pepper) of beans is caused by *Macrophomina phaseoli* (Maubl.) Ashby. The myriads of small black sclerotia buried in the bean plant stems are readily identified, and the disease is popularly

Table of Sclerotial Diseases

Table 19. Comparisons found by the writer in the study of selected sclerotial plant diseases which illustrate some of the wide range in characters in types of these disease organism's holdover, or "resting," sclerotoid phase in life history, in the tropics.

Parasite	Host[a]	Type of sclerotoid state	Remarks
Botrytis sp.[b]	*Hibiscus rosa-sinensis* and *H. esculentus*	Dark colored, irregularly shaped, sclerotoid structures buried in diseased buds and flowers	Occurs during end of warm-moist periods in Puerto Rico, seen also in Nicaragua and El Salvador
Diplodia theobromae (= *Physalospora rhodina*)	*Theobroma cacao* and *Coffea arabica*	Hardened and black bark and cortex of "spent" fruiting wood on coffee and small twigs on cacao	Observed wherever there is sufficient moisture, especially following harvest
Macrophomina phaseoli[c]	*Phaseolus vulgaris* and others	Very small and numerous black sclerotia embedded in plant-stem tissues	In West Indies and Panama on low hillsides and flood plains
Monilia roreri	Only on *Theobroma* spp.	Hardened and sclerotic condition of diseased fruit	In Ecuador found in densely shaded, old cacao plantations[d]

Phymatotrichum omnivorum	Gossypium hirsutum, but has wide host range	String-like, fusiform, and irregularly knotted hard subterranean structures	Indigenous to Mexico and contiguous practically frost-free alkaline lands
Poria (Pachyma) cocos	Roots of various trees, also maize	Large sclerotium, the edible "tuckahoe," buried underground	Occasional in subtropical parts of continental U. S. A.
Sclerotinia sclerotiorum	Phaseolus vulgaris but with wide host range	Sclerotia with size, shape, and color of rat droppings scattered in white cottony mold growth	Found in high-, medium-, and low-lying bean fields in Costa Rica
Sclerotium cepivorum	Allium sativum and A. cepa	Dark colored sclerotia, round to flattened in outside scales and sunken in bulbs	From cold-dry fields in Panama and South American countries
Sclerotium coffeicola	Coffea liberica and C. arabica[e]	Typical round, brown sclerotia size of small pea	Described in Suriname and Costa Rica[f]
Sclerotium rolfsii	Cucurbits and many others	Myriads of sclerotia, size and appearance much like mustard seed	Serious for Indian growers in low lands of Guatemala and El Salvador

a) Plant or plants most considered in this work. b) Collected but species were never identified. c) Also called Sclerotium bataticola. d) However, occurrence in Panama is in lightly shaded fields. e) Wild tree host, see Martyn (25), is a Cecropia. f) Not determined by me though seen in Costa Rica, but became of serious localized occurrence and was studied by Echandi (15).

thought of as requiring moisture, or at least being most severe in poorly drained fields. Both in Central America and in the West Indies it has been observed on especially well drained, sloping hillside fields. In the case of a fourth sclerotial disease, Sclerotinia sclerotiorum, on a number of herbaceous vegetable crops the occurrence from the standpoint of ecology may be remarkably widespread. In Central America the disease and the formation of the characteristic sclerotia, are both found at high, cool locations as well as in low, hot areas. In market shipments the disease has been seen causing "nesting" in snap beans from a dry, almost desert, area of Guatemala, and from a moist, practically rainforest type location in Costa Rica. Judging from the four examples discussed, there is no generalization to be made about what climatic conditions favor sclerotia. Specific relationships vary with the diseases.

With respect to moisture, in the sclerotial diseases mentioned above, all require it for development. This is known, and more attention to the problem showed the following. These organisms invade aerial parts and parts at the soil line or just below. It is sometimes felt that, in our tropics, the crop diseases that depend on sclerotia or sclerotoid structures for holdover during unfavorable periods, must be more severe in moist areas. Certainly this is the case in the watery rots of citrus and other fruits, and of leafy and green podded vegetables by Sclerotium sclerotiorum (1, 17, 19, 23, 27, 29, 31, 34, 35, 38), which for the most part require abundant moisture for severe attack. In a different relationship, also in the Neotropics, the sclerotium-forming disease organism, Phymatotrichum omnivorum, is confined to the warm and relatively dry parts of northern Mexico and contiguous parts of the U. S. A.

In the American tropics where shade horticulture is so important in plantation crops, and where it has had much attention, relative amounts of sun and shade are thought of as having fundamental effects on sclerotia and sclerotial diseases. For example in Ecuador the sclerotoid disease of Monilia roreri on cacao is often believed to be more frequent in densely shaded haciendas. However, when the disease is discussed by those knowing it in Panama, M. roreri-infected cacao pods are said to be most common in the fields having lightest shade over them. The sclerotial structures of Botrytis sp. that develop on aging flowers of okra, stachytarpheta, various orchids, amapola (Hibiscus), browalia, and tabernamontana are found both in open cultivation and in shaded locations.

Literature Cited

1. Alvarez-G., L. A. and C. Chupp. 1954. Enfermedades de las hortalizas en Puerto Rico. University of Puerto Rico's Agricultural Experiment Station (Typewritten MS. dated 28 March 1954), 164 p.

2. Andersen, A. L., B. W. Henry and E. C. Tullis. 1947. Factors affecting infectivity, spread, and persistence of Piricularia oryzae Cav. Phytopathology 37:94-110.

3. Aycock, R. 1966. Stem rot and other diseases caused by Sclerotium rolfsii, or the status of Rolf's fungus after 70 years. North Carolina Agr. Exp. Sta. Tech. Bulletin 174, 202 p.

4. Berkeley, M. J. and M. A. Curtis. 1869. Fungi cubensis. J. Linn. Soc. 10:280-392.

5. Campacii, C. A. 1950. A podridao branca do alho e da cebola. Biologico 12:279-281.

6. Campi, M. D. 1941. Botrytis tulipae del tulipan en la República Argentina. Rev. Argentina Agr. 8: 16-18.

7. Cano G., J. 1952. Cultivo del cacahuete en El Salvador. Centro Nac. Agr. (El Salvador). Circ. 58, 3 p.

8. Ciferri, R. and E. Parodi. 1933. Descrizione del fungo che causa la moniliasi del cacao. Phitop. Zeitsch. 6:539-542.

9. Cook, M. T., and R. A. Toro. 1924. A sclerotium disease of yautia. Phytopathology 14:29.

10. Delgado, J. C., E. Ampuero P., and K. D. Doak. 1961. Posible evidencia de resistencia a la Monilia roreri Cif. y Par. en algunos clones de la Estación Experimental Tropical en Pichilingue. Inter-Am. Cacao Conf. (Trinidad & Tobago) Proc. 8:184-192.

11. Desrosiers, R., A. von Buchwald, and C. Bolanos. 1955. Effect of rainfall on the incidence of Monilia pod rot of cacao in Ecuador. FAO Plant Prot. Bull. 3:161-164.

12. Diaz M, J. 1957. Observaciones sobre la incidencia de *Monilia* del cacao en Ecuador. Turrialba 7: 95-99.

13. Duggar, B. M. 1916. The Texas root-rot fungus and its conidial stage. Ann. Missouri Bot. Gard. 3: 11-23.

14. Echandi, E. 1968. (Personal discussion.).

15. Ezekiel, W. N. 1945. Effect of low temperatures on survival of *Phymatotrichium omnivorum*. Phytopathology 35:296-301.

16. de Franco, T. H. 1960. Transmision de la moniliasis del cacao por el *Mecistorhincus tripterus* F. Inter-Am. Cacao Conf. (Palmira, Colombia) Proc. 7:130-136.

17. Galli, F., H. Tokeshi, P. C. Torresde Carvalho, E. Balmer, H. Kimati, C. O. Nogueira Cardoso, & C. L. Salgado. 1968. Manual de Fitopatologia, Doenças das Plantas e seu Controle. Ceres, São Paulo, 640 p.

18. Garcia B., C. and G. Naundorf. 1952. Un año de experimentos sobre control de la Moniliasis en cacao. Cacao en Colombia 1:31-40.

19. Heald, F. D. 1926. Manual of plant diseases. McGraw-Hill, New York, 891 p.

20. Icochea, T. & R. Aycock. 1965. Comparative studies with *Sclerotium coffercolum* and *S. rolfsii*. Phytopathology 55:1063.

21. Jauch, C. 1940. *Botrytis fabae* Sardina en haba. Bol. Inf. Dir. San. Veg. (Argentina), pp. 40-41.

22. Kuyper, J. 1913. Overzicht van de Koffieziekten in Suriname. Bull. Dep. Landb. 31, 5 p.

23. Marchionatto, J. B. 1932-1935. Notas sobre algunos _Sclerotium_ parasitos de las plantas economicas. _Physis_ 11:303-305.

24. ________. 1937. Contribucion al conocimiento del _Botrytis cinerea_ en la Republica Argentina. J. _Agr. Vet._ pp. 179-194.

25. Martyn, E. B. 1929. The _Sclerotium_ disease of coffee and its occurrence in this colony. Agr. J. British Guiana 2:7-10.

26. ________. 1930. The _Sclerotium_ disease of coffee. Some notes on the origin of the disease, its outbreak and control. Agr. J. British Guiana 3:28-34.

27. Miller, P. R. 1953. Lowland rice culture as an economic control of _Sclerotinia_ and rootknot nematodes. FAO. Pl. Prot. Bull. 1:183-187.

28. Morgan, F. and J. Dunleavy. 1966. Brown stem rot of soybeans in Mexico. Plant Dis. Reptr. 50:598-599.

29. Mujica R., F. 1955. Estudios sobre esclerotiniosis. Agr. Tec. (Chile) 15:64-74.

30. Orellana, R. G. 1956. Occurrence of _Monilia_ pod rot and other cacao diseases in eastern Panama. FAO Pl. Prot. Bull. 4:168-169.

31. Rocha Garcia, G. 1956. El cultivo de las vainicas. Peru Dir. Gen. Agr. Bol. 3-4:44-50.

32. Rodrigues Campos, A. 1942. Molestias da mamoneira no estado de Pernambuco. Pernambuco, Brasil, Sec. Agr. Indus. Comer. Bol. 8:55-59.

33. Saccas, A. M. 1958. La maladie des taches zonées de _Coffea excelsa_ en Oubangui-Chari due à _Sclerotium coffeicola_ Stahel. Rev. Mycol. Paris (France) 22 (Suppl. Colon. 2):64-84.

34. de Segura, C. B. 1947. Nuevas enfermedades del lino en el Perú, esclerotiniosis y antracnosis del lino. Est. Exp. Agr. La Molina, Lima,

Peru. Boletin 36, 37 p.

35. ________. 1965. Enfermedades de cultivos tropicales y subtropicales. Editorial Juridica, Lima, Peru, 439 p.

36. Sepulveda L., R. 1955. Biología del Mecistorinhus tripterus F. (Hem. Pentatomidae) y su posible influencia en la transmision de la moniliasis del cacao. Cacao en Colombia 4:15-42.

37. Shear, C. L. 1925. The life-history of the Texas root-rot fungus, Ozonium omnivorum Shear. J. Agr. Res. 30:475-477.

38. Smith, R. E. 1931. The life of Sclerotinia sclerotiorum with reference to the green rot of apricots. Phytopathology 21:407-423.

39. de Sousa, O. F. and E. Abramides. 1952. Ensaios de variedades de amendoim. Bragantia 12:349-358.

40. South, F. W. 1911. Fungus diseases of ground nuts in the West Indies. West Indian Bull. 11:157-160.

41. Stahel, G. 1921. De Sclerotium-ziekte van de Liberia Koffie in Suriname. Dep. Landb. Bulletin 42, 29 p.

42. Stevens, F. L. 1927. New tropical fungi. Mycologia 19:231-238.

43. Stevenson, J. A. 1918. The green muscadine fungus in Porto Rico (Metarrhizium anisopliae (Metsch.) Sorokin.) J. Dep. Agr. Porto Rico 2:19-32.

44. Taubenhaus, J. J. and W. N. Ezekiel. 1930. Studies on the overwintering of Phymatotrichum root rot. Phytopathology 20:761-785.

45. Tims, E. C. and L. S. Olive. 1948. Two interesting leaf spots of fig. Phytopathology 38:707-715.

46. Tisdale, W. H. 1921. Two sclerotium diseases of rice. J. Agr. Res. 21:649-658.

47. Tullis, E. C., J. W. Jones, and L. L. Davis. 1934. The occurrence of stem rot of rice in California. Phytopathology 24:1047.

48. Vergara, C. 1952. La eslerotiniosis del eucalipto. Simiento (Chile) 22:174-176.

49. Voglino, P. 1903. Sur parassitimo e lo sviluppo dello Sclerotium cepivorum Berk. nell Allium sativum L. Staz. Sper. Agr. Italiano 36:89-106.

50. Weber, G. F. 1929. The occurrence of tuckahoes and Poria cocos in Florida. Mycologia 21:113-130.

51. West, E. 1947. Sclerotium rolfsii Sacc. and its perfect stage on climbing fig. Phytopathology 37: 67-69.

52. Wellman, F. L. 1961. Coffee botany, production, and utilization. Leonard Hill, London; Interscience Publ., New York, 488 p.

53. Whetzel, H. H. 1945. A synopsis of the genera and species of the Sclerotiniaceae, a family of stromatic inoperculate discomycetes. Mycologia 37: 648-714.

54. Wolf, F. A. 1922. The fruiting stage of the tuckahoe Pachyma cocos. J. Elisha Mitchell Soc. 38:127-137.

55. ________. 1926. Tuckahoe on maize. J. Elisha Mitchell Soc. 41:288-290.

56. ________. and W. J. Bach. 1927. The thread blight disease caused by Corticium koleroga (Cooke) Hohn., on citrus and pomaceous plants. Phytopathology 17:689-709.

57. Zevada, M. Z., W. D. Yerkes Jr., and J. S. Niederhauser. 1955. Primera lista de hongos de Mexico arreglada por huespedes. Sec. Agr. Ganad., Ofic. Estud. Esp., Foll. Tec. 14, 43 p.

Chapter XII

RUST PROBLEMS OF WARM REGIONS

Summary

History has taught pathologists that true rusts which are abundant in the tropics are probably the most to be feared of any group of plant diseases. *Puccinia* and relatives, evolved in long past geologic time before the era of man, are highly specialized parasites often having 5 spore stages and alternate hosts for completing life histories. *Zea mays* is by far the most widely grown food plant in the American tropics, 3 rusts infect it: *Puccinia polysora, P. sorghi*, and *Physopella zeae.* The first is probably for our tropics the classic example of rust life history complexity: spore stages 0 and I in lesions on *Oxalis* spp.; stage II the main cause of rust epidemics of *Zeae* and *Euchlaena*; stages III and IV following II. *Wheat,* once an exotic in the Neotropics and now a welcome introduced crop, once faced failure from the rusts *Puccinia graminis tritici, P. glumarum,* and *P. rubigo-vera tritici.* The former was much the most serious. Several broad-leafed herbaceous crops such as beans and relatives, tobacco, potato and other solanaceae, and peanuts suffer severely from rusts; very many dicotyledonous trees are rusted seriously in the tropics and some are of prime economic importance. Control of rusts by resistance to physiological races is satisfactory in the temperate zone but has to be redone so frequently in the tropics that something more permanent like tolerance or multilineal resistance is needed.

Introduction

In the history of world plant pathology, the study of rusts holds a prominent place. The recognition of the importance of these diseases is almost human instinct; and it comes from the knowledge of thousands of years of Old World famines due to rust-induced failures in cereal grains (20, 25, 43, 104 in 1971 publ.). For this reason, when a rust is discovered on a crop it strikes a note of extreme foreboding in the hearts of growers. This is in some ways especially appreciated in the tropics, as it has been well publicized that the great coffee growing industry in the tropics of the East, was decimated in some countries because of rust. Coffee rust is still classed as one of the seven great plagues which have caused the most profound changes in economics, populations and histories of men (43, 104 in 1953-1961 publs.).

The reading public and trained biologists, almost all, know what is meant by the phrase "a plant rust epidemic." In the Neotropics, kinds of rusts are much more numerous than in the temperate zone of Canada or northern United States and for lists and occurrences of tropical rusts the reader has only to turn to such works as those indicated in citations: 3, 4, 8, 15, 17, 20, 25, 27, 30, 35, 39-41, 45, 49 especially 1948 publ., 54-56, 61, 66, 68, 70, 75, 76, 83, 85, 89, 90, 93, 99, 106, and 110.

The rust disease symptom most recognized is the specially marked often reddish discoloration. It is a very old kind of disease (early Dawn Man probably knew it) seen as lesions containing multitudes of dust-like particles. These are the microscopic-sized spores which occur in such abundance that with the unaided eye they appear as pulverulent masses. When such masses in sori or pustules are touched they cause yellow, brown, or reddish smudges. In this, it resembles what is found on oxidized iron: discolored powders that are called "rust," and this gave it its popular name.

In determinations of the phylogeny of rust fungi, mycologists agree (see Arthur (4 in 1929 publ.) Gäumann (30), and Bessey (8)) that these organisms are derived from Auriculariaceae, abundant warm temperature fungus saprophytes that palaeontologists know have existed a long time. New forms of these evolved and some became parasitic on the primitive ferns of the warm lands of the Silurian Age, some 350 million years ago, and reached higher specialization later on in the Devonian; times that long preceded appearance of the first amphibians and the dinosaurs. It was from these warm-climate fungus organisms that rusts evolved and the rusts we have today, many of them "living fossils," are similar to rusts that subsisted hundreds of million of years ago in our tropics.

Arthur (4) reviews something of the fossil finds of rusts, reported up to 1929. He and his collaborators accept the extreme age of rusts but have been somewhat skeptical of some of the specific diagnoses of paleontologists. He discussed seven collections of these very ancient rusts; the first two from lower Carboniferous and upper Cretaceous deposits that were laid down under tropical-like warm conditions, and although spores were malformed by pressures and from great age, they are evidently rusts. The remaining five collections were recovered from the Tertiary Period as follows: Aecidites nerii on a fossil oleander, which is a warm country plant from Eocene deposits: Puccinites lanceolatus on a monocotyledonous host from the Oligiocene: Phelonites lignitum on seed of a conifer from the Miocene; Aecidites subcornutus on oak from the Miocene; and A. rhamni tertiariae on buckthorn also from the Miocene.

Pathologists have studied crop plant rusts in the temperate zone much more than in the Neotropics. It is well known that a few Old World plant rusts, the so-called "classic" rusts, in the fields of Europe and North America were long ruinous on wheat, barley, rye, and oats. Varieties of those crops were found that could be grown in specially chosen areas in the Neotropics, and there they were planted but Old World rusts eventually appeared to attack them.

When in the American tropics rusts found their way to the fields of flourishing small grain cereals that were becoming well established in isolated ecologic zones, the disease had serious effect. Documentary evidence and contemporary reports about these early rusts in our tropics are not found. But there is folk story and mention about it. Seeds of cereal grains were brought by early explorers and first colonists; indeed, it was part of the plant introductions program of Ponce de Leon (14) who first lived on the island of Hispañola and established from it first viable European colonies in the New World. From there the tests quickly spread to such other islands as Puerto Rico, Cuba,

Sant'Jago or Jamaica, further afield to a few shore gardens in Florida, and to a few other parts of mainland Central America. Search for relicts of some of these old introductions is now futile. However, there still persist stories about where cereals were once grown. An example is a considerable part of a cool and dry section of the central plateau of Costa Rica where, it was said, once wheat grew well, until an unknown disease destroyed it. For example there is the story told to me in the years 1952 and 1953 by two different farmers who were near the city of Alajuela, Costa Rica, that wheat was an early colonial crop in their area. In 1954 and 1955 the story was reiterated in personal conversations with the late Don Mariano Montealegre, editor of the Costa Rican journal, Revista del Defensa de Cafe de Costa Rica. He believed it fully. So do I.

In the present day, old-type varieties of wheat and most other small grain cereals are known to succumb quickly to rusts in most of the American tropics. However, there are areas well adapted to such cereal production, and, as is shown by recent accomplishments (see such citations as: 10-12, 20, 25, 27, 50. 67, 71, 76, 91, 98), rust resistance or tolerance is all that is necessary to bring back a certain amount of permanent growing of small grains in wider areas in such countries as: Mexico, Costa Rica, Peru, Colombia, Guatemala, Chile, and Argentina.

Some Unique Features of Rusts

A specialist in plant diseases is well aware of the fact that intricate researches demonstrate complex and unique features in rusts: they are generally characterized as obligate parasites (77, 109) although some very specialized techniques have been developed to culture them in artificial media (16, 60, 107); life histories are markedly complex (8, 13, 25, 27, 43, 44, 49 in 1948 publ., 68, 76, 77) requiring in some cases disease on two very different species of plants to perfect the life cycle, although some develop all stages on only one host. It appears rusts are adapted to tropical conditions (20, 25, 72 in 1963 publ., 74, 76, 80, 101 in 1952, 1958 publs., 102, 104 in 1953 publ., 108) and are common in the region of perpetual warmth.

Rust fungi (see Ainsworth's discussion (1) under "Uredinales") are known in general to have five spore types or forms which by international understanding are given

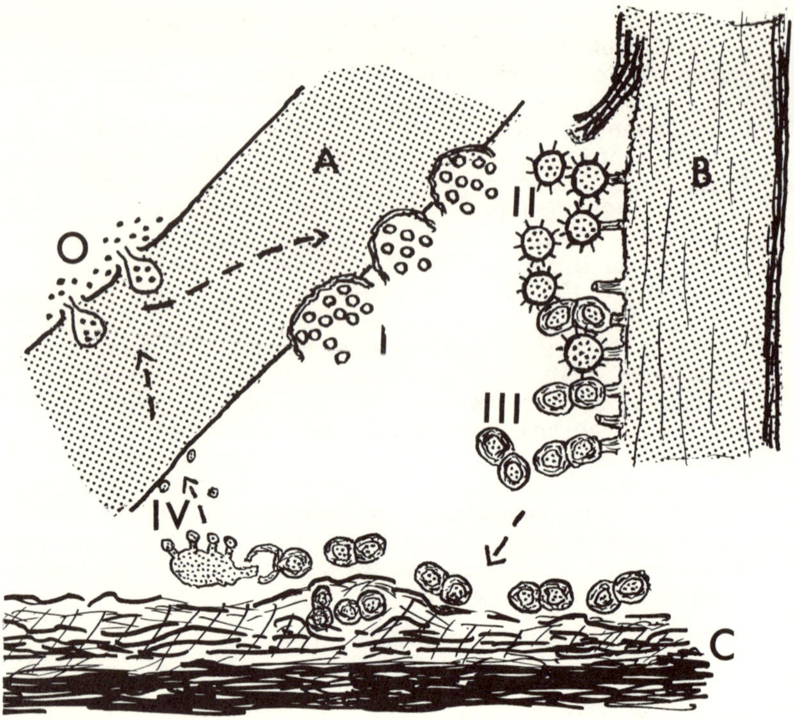

Fig. 12-1. Rust fungus spore types: 0--minute pycniospores, sometimes on the alternate host from, for example, a crop like maize; I--small aeciospores develop following fusion of pycniospores; II--urediospores on, for example, maize of various shapes and appearances, thin walled, light weight with appendages that aid in air dispersal, often main spore in epidemic spread; III--heavy walled teliospores most morphologically stable, where present used for identifications, HOLDOVER stage withstands bacteria in soil and debris as well as extremes of tropical climate; IV--basidiospores from germ tube of teliospore. A (oxalis) and B (maize) are sometimes, not always, two different species of host plants; C indicates soil and debris.

numerals as follows: 0 = pycniospore, I = aeciospore, II = urediospore, III = teliospore, and IV = basidiospore. These spore forms, also called stages, are indicated in accompanying stylized sketches in Fig. 12-1.

In Central America a rust fungus such as Puccinia sorghi often develops on two distinctly different hosts. Host one: spore stages II and III are found on the leaves of the economically important host, the tall and coarse monocotyledonous food-producing, cultivated maize plant. Host two: spore stages 0 and I are found on this alternate host, the relatively inconspicuous oxalis which is a short and tender dicotyledonous weed (Fig. 12-2, A and B). The spore stage IV, or basidiospore, is produced from the teliospore, and in the case of the maize rust fungus mentioned here, is the spore which infects the oxalis.

From the standpoint of the tropical specialist, by far the most common in occurrence and in important effect is the rust spore form known as the urediospore (II). It is the one which gives lesions their "rusty" color; it is readily carried off by winds or air convection currents. It is produced again and again in the same pustule and this gives it the name of repeating spore; this is the stage that is largely responsible for the characteristic rapid and widespread rust epidemics. Such a spore is thin-walled, small, and is studded with spikes and surface markings that increase ratio of spore surface to weight of spore mass. This aids in the spore being lifted up into the air.

As is well known in the tropics, in a large number of cases, rusts there are many times entirely perpetuated by urediospores. In some instances such have been known in the past as fungi imperfecti, causing confusion in mycological determinations (30, 68). Spore and mycelial cells at this stage, II, are haploid, have only one complement of chromosomes in the nucleus, and may be reproduced without nuclear fusion. However, if there does occur a fusion of such simple nuclei, the result is diploidy, and the black and thick-walled teliospore, III, results. In the tropics this evolved condition is a long-ago, holdover-fungus-structure that resists extremes of temperature, moisture, and insolation; it also withstands the characteristically excess numbers of spore-consuming bacteria, fungus hyperparasites, and other kinds of organisms so common in the warm-moist tropical environment. (cont. pg. 470)

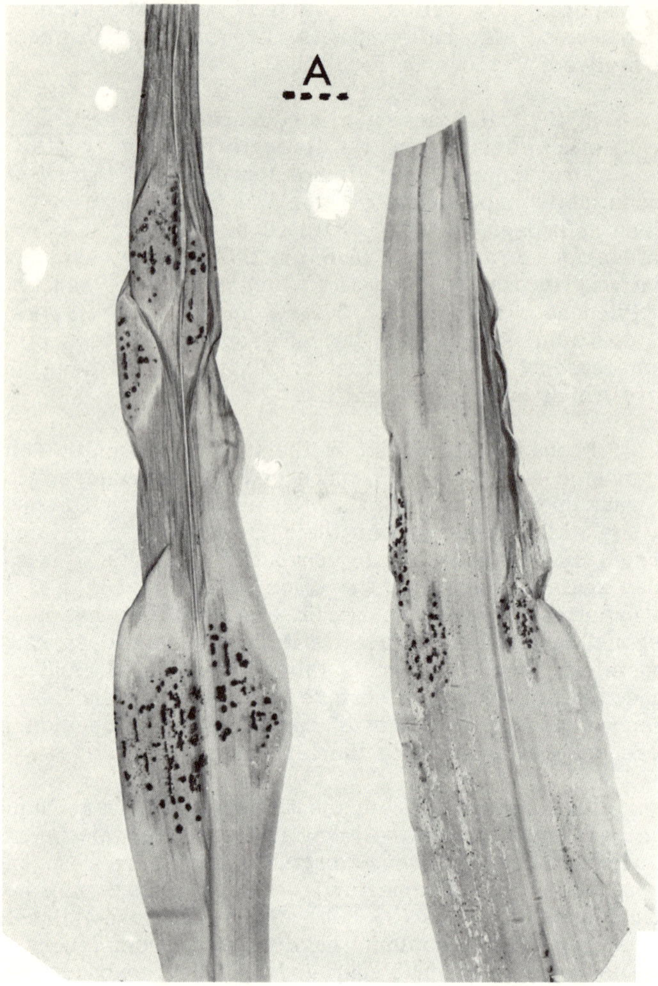

Fig. 12-2. Maize rust, Puccinia sorghi, that has two hosts: A. leaves of the maize crop plant has rust lesions which contain both urediospores (II) and teliospores (III); B (opposite). leaves of the oxalis weed with rust lesions, on leaf upper sides, u, can be seen dark drops of exudation which contain pycniospores (0) while on lower sides, 1, are produced lesions full of aeciospores (I) the result of the previous fusion of pycniospores.

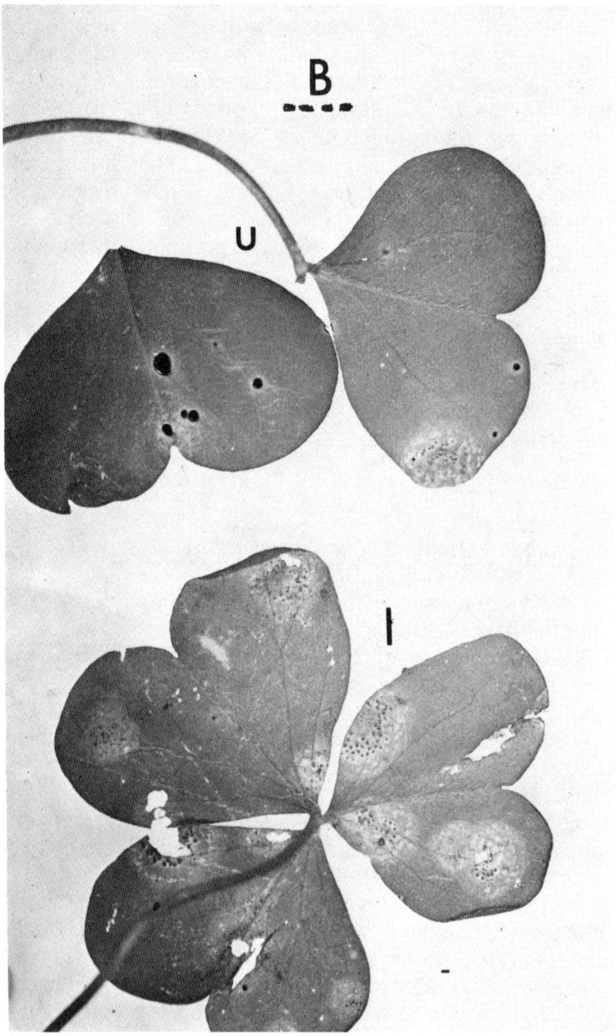

Photographs courtesy of Department of Plant Pathology, University of Florida, Gainesville. (The writer, in his garden in Costa Rica, used aeciospores from locally growing oxalis and produced rust on "sweet corn" seedlings.)

Any rust which develops on two different hosts is known as of heteroecious type, and, as Arthur (4) points out in his classic 1929 book, it is more common in the temperate zone than in the tropics. Where only one host is involved a rust is designated as autoecious, a situation common in the tropics. Moreover, it is of further interest that in some cases rusts that are heteroecious in the temperate zone no longer require, or depend upon, an alternate host, and flourish as autoecious in character in the tropics: a good example of this is the wheat rust, Puccinia graminis tritici, which needs no barberry bushes in the Neotropics for it to be the most serious limiting factor in wheat growing where the conditions there are auspicious for wheat.

May it be emphasized and repeated, in brief: in the tropics it is not necessary for many a rust fungus to pass through all the five spore stages of 0, I, II, III, and IV. It can carry on, for it, a very successful and flourishing parasitic existence dependent on only spore type II with, in a few cases, III developing as lesions age and host tissues die.

Observations About Neotropical Maize

No food crop is so important, so widespread or so popular in the American tropics as is maize, Zea mays. It was called "Indian corn" by early English speaking visitors to the New World, but the great antiquity of it in the Neotropics is in a small way indicated by the numerous, old, indigenous names for it, for example: ixiim, tizate, capulín, pupo, cacapo, malacajaco, malojho, melho, mah-híz, maíz, and maize. Many other names are lost in the history of movements of the crop from one area to another; for example, maize seed from a dark purple-husked and large kernel variety called "capulín" in high cloudy and cold mountain valleys of El Salvador, in 30 or more years was said to become "pushagua," that is a pale-lavender-husked small kernel variety grown near sea level where it is hot and the skies mostly clear. There are literally thousands of folk-selected varieties which have become adapted to very different ecological conditions--and probably many such selections are tolerant to diseases in various special areas. In the tropics, maize grows at elevations of from sea level to over 3500 meters (nearly 12,000 ft.), in all kinds of soils, but always under the watchful eye of man. It cannot survive in the wild, but produces where wheat is a failure.

In El Salvador, a relatively short-stalked, small white kernel maize variety is grown by small farmers for tortillas and fodder. It is planted in a warm, moist but not wet area where sun shines every day, and at an elevation of approximately 500 m. (1640 ft.). The plant develops so rapidly that, in the same field, crops can be sown one following another to produce four harvests a year (three for grain and one for major cattle fodder), and sometimes a fifth short growth for work oxen food. In this area the main rust is Puccinia sorghi, but it seldom is severe as apparently the plants are naturally tolerant to its attack.

In Guatemalan mountains that are about 250 kilometers (180 mi.) northwest, is another maize-growing location with an elevation of about 1500 m. (4921 ft.). There it is cool and there is much rain with some fog nearly every day. The rust, of another kind, Physopella zeae, is the most common leaf infection in this area. (I found it also in highlands of El Salvador.) The maize grown is tall, broad in its leaves, and the grains are large with fine stripe markings. Here the grain is grown for tortillas and animal fodder, but it requires over a year and sometimes up to 18 months for it to reach harvest.

In the Andean highlands of Ecuador, south from Central America about 2000 km. (600 mi.), maize grows on high plateaus at an elevation of about 2000 m. (6562 ft.). There it is relatively dry, cool, and windy, and there are many bright days. The grain is often purple in color and is produced for human food and fermented beverage; the leaves and stout stalks are used for shelter structures, bedding and animal fodder. Leaves brought to me showed pustules of two rusts, Puccinia sorghi and P. polysora, although the latter was not severe in its occurrence. Whether the crop is disease-free or rusted it is used and the peasants harvest one crop a year which requires a long growing season of over seven months.

Almost anywhere--and maize grows anywhere except in extremes of shade, desert, cold, wet, alkaline or acid soil, and perpetual strong wind--where the plant stands, rust disease is found to be almost universal on it in the Neotropics. This remarkable crop with its vigorous root system is fast in plant growth. Its leaf surface is always greater than are its minimum necessities for maturing grain If, from rust disease, 15 or more per cent of leaves are destroyed, a plant still matures its ear (sometimes somewhat degraded in weight and number of kernels, but still usable). From this combination of the innate natural capacities of this plant and with additional help from folk selection, has come superlative rust tolerance and the greatest food producer over the widest range of conditions in the American tropics and subtropics.

As long as maize is grown as a non-mechanized crop in milpas, which is the primitive method still used by most small farmers in underdeveloped areas, scattered plots of maize that are somewhat rust tolerant are able to sustain a limited population. A milpa is a cleared place of small dimensions surrounded by trees, and prepared by the clearing-burning method. Ashes and jungle debris enrich the soil. After two or three years nutrients are mostly gone from the clearing, rust and other diseases have appeared in the maize crop, and habitually the plots are shifted to another virgin location.

When numbers of people to be fed increase substantially, crop growing must be redoubled. This occurred during the heyday of Aztec, Toltec, Arawak, Mayan, Pipil, north Incan and other such large pre-Columbian nations. These had their own types of cultured society. They were highly specialized maize-growers; their food supply, much of their shelter, and their religion were all related and affected by this single crop. Fields were numerous and storage was well done. However, under a greatly intensified cropping, a domesticated plant species such as maize accrues to itself diseases. The maize peoples often must have been seriously plagued by plant diseases, and especially by rusts, on this their main agricultural product. Diseases such as the rusts may have been a great, if not a main contributing factor in the eventual weakening and final subsidence of what was at one time a group of, in their manner, civilized, art- and religion-conscious nations and city-states.

American Tropical Maize Rusts

There are three rusts on maize (or "corn") in the Neotropics. The most common is *Puccinia sorghi* Schw., and it occurs not only in the tropics but also generally in maize-growing parts of South America (19) and in the temperate zone of North America, except in the Pacific Northwest. The second rust, in frequency of occurrence, is *P. polysora* Underw., a common plague in our tropics and found north in our hemisphere in such states as Texas and Georgia. The third rust, verging on rarity, is the long known but at times mycologically confusing (47, 73, 108) Physopella zeae (Mains) Cumm. & Ramachar (= Angiospora zeae), (15), which appears to be a tropical inhabitant of the cool mountains and is found in parts of Central America. In the American tropics the notable effects of elevation on the three rusts is shown by Cummins (15) and by Schieber et al. (73 and 74). They report that in Central America P. sorghi is most common and severe in fields at the higher locations where the weather is cool. Their studies show as well that P. polysora is more often at distinctly lower elevations where temperatures are warmer. On the other hand, the Physopella zeae rust is likely to be irregular in its distribution. Comparative appearances of the teliospores of these rusts is indicated in accompanying sketches in Fig. 12-3, A, B, and C. Vernacular names for the three rust diseases are interchangeable, some of the most common being "roya," "moho," "mbirú," "herrumbre," "chauistle" and "tizón." (Further details are found in citations: 14, 15, 40, 41, 48, 56, 65, 73, 76, 85, 99, 103).

The rusts are considered seasonal in their appearance in some areas. In Jamaica and Trinidad it was found by Dale (17) that P. sorghi is most severe in the mid year, while P. polysora appears later and Physopella zeae is more likely to occur between the seasons for polysora or sorghi. In both El Salvador and Costa Rica P. polysora was always most commonly to be seen after rainy seasons had passed through the canicula, at which later time downpours were less severe, and mean temperatures slightly cooler. I observed P. sorghi all the year round, but it was most prominent towards the first part of the dry season.

Puccinia sorghi, it is interesting to note (36), was given its specific name by the pioneer American mycologist Schweinitz who mistook the leaf on which he found it to be from a sorghum rather than a maize plant. It is known (20) to be confined to the graminicolous hosts maize and teosinté, and to Oxalis in its pathogenicity. At one time it was thought (4, 36) that it was not often found in large quantities outside the temperate zone but it is probable that it originated in the tropics of the western hemisphere. It occurs widely: for example, high in the Andes at Cuzco, Peru, and at near sea level in Sanford, Florida. The disease spots are ordinarily chestnut brown in color, making blisters that break open to release powdery masses of millions of urediospores (II).

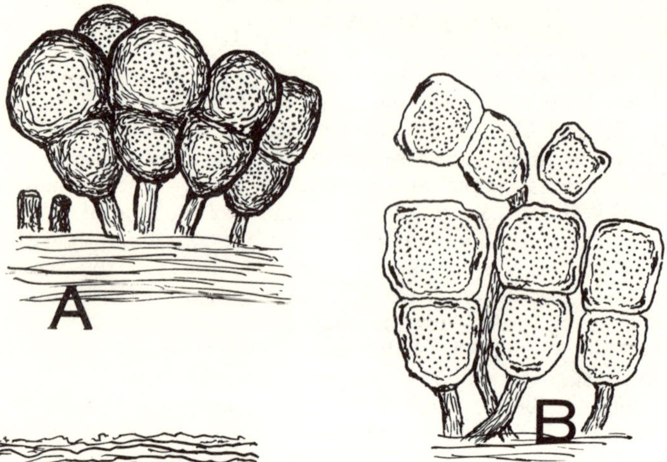
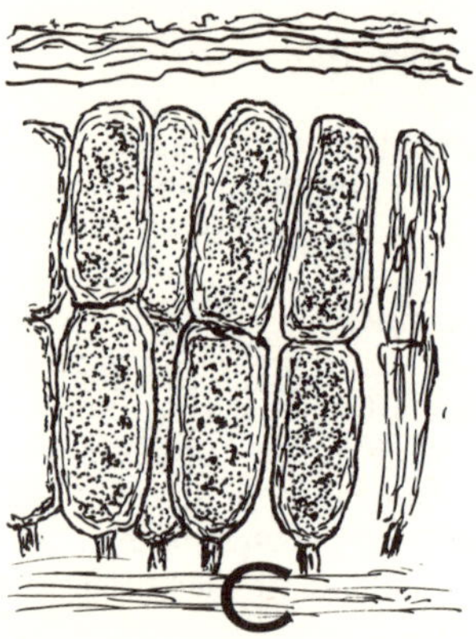

Fig. 12-3. Teliospores of rusts on maize: A. *Puccinia sorghi* in an elongated lesion, spores heavy walled and densely filled with matter are rounded and somewhat short, under a thin sheet of host tissue that has rifted. B. *Puccinia polysora* in relatively small elliptical lesion, spores fairly hyaline tending to be angled, some held on irregular lengths of sporophores and pushing aside lip-like structures of host tissues. C. *Physopella* (*Angiospora*) *zeae* in a relatively small dark lesion, has elongated spores on short sporophores and crowded under unbroken layers of host tissue. (From pencil sketches made from El Salvador material in 1944-1945, of free hand sections under a high-dry lense of the microscope.)

When the spots of <u>P. sorghi</u> are more or less emptied of urediospores--but with the occurrence of proper moisture and warmth--black teliospores (III) develop inside and around the borders of an old so-called "spent" lesion. They are thick-walled and highly resistant to extremes of drought, rain, bacteria, hyperparasitic fungi, nematodes, and other bad conditions. If a teliospore (III) germinates it produces basidiospores (0), followed by aeciospores (I) which on being blown or otherwise moved from the lowly weed, oxalis, on alighting on maize or teosinté causes urediospore (II) stage rust spots on their broad leaves. Thus is the life cycle completed.

It is readily found (e. g., 104 in 1949 publ.) that rust may attack introduced breeding stocks of maize, with great severity. On susceptible plants all ages of leaves are heavily attacked, and the rows of such plants are typical of the disease symptom described as "burned." Dusting and spraying with fungicides are then absolutely necessary to preserve the genetic material. In some imported maize, rusts are not artifically controlled: many experimental lines have been seen in Panama, Nicaragua, Guatemala, Costa Rica, and Puerto Rico so severely diseased with <u>P. sorghi</u>, and some with <u>P. polysora</u>, that maize plants had neither good pollen-filled tassels nor good ears.* On the other hand, growing next to them, the native maizes (that may not be high producers but are tolerant of rust and

*In a letter dated August 8, 1968, Dr. Eugenio Schieber, the Central American plant and tree rust specialist in Guatemala, says as follows: "I am happy to see you believe with me maize (corn) rusts are so important in the American tropics. Numerous maize breeders do not consider them as very serious pathogens. Why I do not know. Under natural conditions where maize has been so long, rusts are found serious in isolated valleys, but sometimes not seemingly widespread. However, when I search I find they occur many places. You should see how destructive <u>P. sorghi</u> is in imported maize varieties brought to the tropics, and how severely new maize lines from the temperate zone are destroyed by <u>P. polysora</u> .... I also agree with you about the prehistoric importance. I believe that a rust like <u>P. polysora</u> on maize was probably a large part of the reason for the sudden and unknown cause for the disappearance of the large populations of Mayan people in, for example, such a great city as Tikal."

have long grown in and are adapted to, local conditions) may have many spots of rusts on leaves but mature their ears of grain. In the worst seasons the lower two leaves of the natives may be severely spotted, but the top few leaves grow well, even with a fair sprinkling of lesions.

There is much need to study ecological effects of maize rusts. Through phytotron manipulations of one kind and another it is certain that much basic information will be learned. The time is ripe to do this, combining attempts at growing the organism in xenoculture which is done successfully with other rusts, and studying the diseases related to temperature, light, physiology of the different varieties of the host, and possible insect relationships. Yarwood (109) has already shown that P. sorghi is considerably increased in disease effect on leaves that were preheated for short periods before inoculation. Techniques of rust inoculation (73) are known, and all three rusts may be given special attention, very profitably, under well controlled conditions of varying environments.

Two of the maize rusts are considerably different in spore survival related to temperature (102). Urediospores of P. sorghi remain alive much longer at relatively cool temperatures, while P. polysora urediospores survive best at higher temperatures. This coincides with the information cited above that in Central America, P. sorghi flourishes most effectively in cool mountainside fields, and P. polysora is more severe in warmer, low-lying parts. Malaguti (48) has shown that similarly in Venezuela, the rust P. sorghi is more severe at higher elevations in fields where temperatures are on the cool side, but that P. polysora is most severe at the warmer, that is the lower, elevations. Physopella zeae (Fig. 12-4) is apparently more irregular in its occurrence and may have a wider temperature range. It is certain that, under phytotron studies, significant and fundamental differences will be discovered, in disease reactions, occurrences, effects on normally field-resistant and on normally field-susceptible maize lines. Such studies may also give an explanation of such an anomalous observation as that on P. polysora in which this rust that is considered best adapted to warm conditions was reported by Schieber (personal communication) as being in a mountain area in Guatemala at around 1750 meters (6000 ft.) where it is cool, and that he saw the same rust producing severe disease on maize in Africa at an equally high elevation. Such observations as this suggest possible variants--

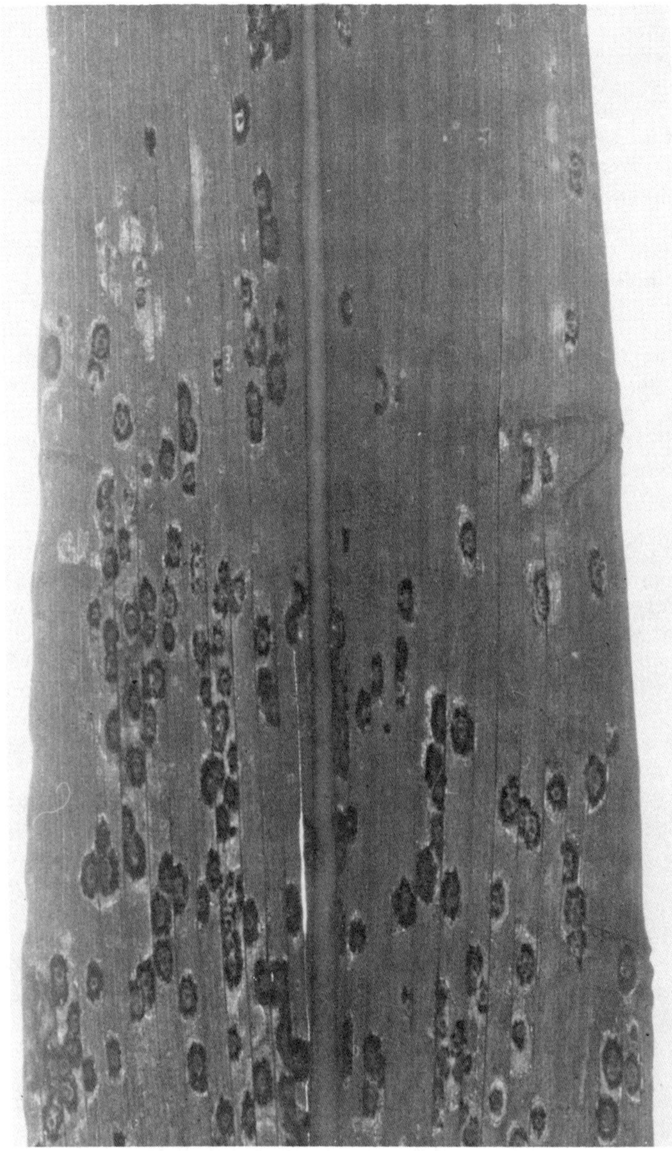

Fig. 12-4. Some characteristic symptoms on maize leaf of rust caused by *Physopella zeae*. Photograph from Guatemala, courtesy of Eugenio Schieber.

physiological race difference in P. polysora--and phytotron studies could uncover any such phenomenon related to an especial type of ecology.

It is to be noted as well that along with normal changes through the shifts due to ever new rust race developments, rusts of maize and other crops are otherwise malleable and are even unstable. The sorghum, Holcus spp., imported from the Old World but by now a long-time valued crop in the Neotropics, recently was noted as the prey in South America of an entirely new rust, Sphaerophragmium sorghi Bat. & Bez. There are numerous rusts on numerous graminicolous hosts in our tropics (36, 47, 65, 70, 94, 99), and spores in the tropics are being released in the winds at all seasons of the year. Not only are there new rust races continuously being developed in the tropics, there are likewise the added possibilities of probable crosses between rust species (38, 59).

When a disease which the Africans and Europeans call "American maize rust," Puccinia polysora, was first found in Africa (15, 58) it was thought of as surprising in occurrence but of little economic concern. This was fortunate at the time since maize, or "mealies," on that continent is one of their most important foods. However, as the disease spread it attacked with severity some of the old maize strains that had been in Africa for centuries and had lost any contact with P. polysora. On these mealies the American rust multiplied rapidly and soon was causing famine. This called for researches which absolutely had to lead to developing control measures (34, 58, 69, 72 in 1965 publ., 86, 87, 108). Realizing the importance of rust variants, western hemisphere maize pathologists have been long aware that their rusts are composed of races (examples of studies: 20, 27, 65, 108). This being appreciated, when breeding results for resistance to American rust on African mealies was reported (23, 34, 69, 72 in 1965 publ., 86, 87), it seemed inevitable that the specialists would determine that there were marked variations in their mealy parasite and they immediately planned to import new host seed, searching for more resistance. Fortunately this was, and is, available in Latin America where the rust is indigenous. There is acceptable tolerance to races of P. polysora in Neotropical maize (see 72 especially 1965 publ.), and thus far in the Americas maize varieties have a wide range of tolerance and, in that, unpurified for resistance there grow good grain crops in spite of the rust; western tropical pathologists have viewed with intense interest any work on physiologic races of P. polysora in Africa and the Oriental tropics. However, it is realized as never before that what is needed now is not so much resistance breeding against separate specific races as securing of more generalized tolerance or resilience that takes into account all manner of the regularly occurring rust variations and mutants.

Exotic Wheat and Its Rust

Wheat or trigo, Triticum aestivum, was a true exotic when first brought to tropical America. In determining some of its history, Dondlinger (22) uncovered reports of Alexander von Homboldt who learned that some time in the 1520's an observant, knowledgeable Negro slave of Cortés found a few loose seeds of wheat in a bag of army ration rice used in the tropics of the New World. These he planted and the progeny was multiplied. Apparently wheat was first introduced to Mexico in about 1529, but it did not become a notable crop there for over a century after that. Wheat seed was also secured from other Old World sources. By about 1547 it was being tried near Cuzco, Peru; in 1769 it was in California; in 1808 it was in Cuba; in 1834-1836 it was grown but not liked in Jamaica and islands of the Bahamas. It had to compete then, and now, with indigenous starch-producers*. After a few hundred years there are still numerous good tropical areas where wheat will not grow; it requires special cool yet sunny locations in our tropics and still remains as an alien plant to many indigenes and peasant folk.

For early settlers, with nostalgia for their old foods, bread was made largely out of precious shipments of wheat grain from their lands of origin far away. For some in the Americas, it is interesting to recall, wheat bread was a gourmet's delight--for others, newly arrived in the strange Occident, it was the taste-connection with homes. The cost increased in shipping, but in the meantime adventurous

*Grain from: Zea mays, Phaseolus vulgaris and other species, Chenopodium spp., Amaranthus spp., and others; roots and tubers of: Xanthosoma spp., Manihot spp., oca, Ipomoea batatas, Dioscorea spp., Calathea allouia, Ullucus tuberosus, Tropaeolum tuberosum, Solanum andigenum-tuberosum, and others.

growers began their harvests. Rust disease appeared, and kindred troubles along with rust drove the plantings to high mountain meadows and table lands; even to the best good soils in dryer and cooler parts of the Antilles, Mexico, Central America, and South America. Since those early days disease controls have been instituted and there are now some nine million hectares of wheat in ecologically suitable parts of the Neotropics.

In this great region, wheat may be attacked by three rusts: stem rust, Puccinia graminis tritici Erikss. & Henn., leaf rust, P. rubigo-vera tritici (Erikss.) Carleton, and stripe rust, P. glumarum (Schum.) Erikss. & Henn.; in tropical America the first is the most widely spread and most dangerous, and it is the main one to be briefly discussed in following paragraphs. Rust has been known since early colonial times in Peru (76) as the epidemic "polvillo" that began to be mentioned after the 1687 earthquake. Wheat fields of part of the coastal area were visited then and after this, by such severe attacks of it that it forced complete abandonment of the crop. There was little scientific study of wheat rust in the American tropics for about two centuries.

The tropical wheat-rust-pathologist works in locations far from the more normal hot and humid lands. Wheat grows best where it is cool to cold and fairly dry. In such countries as Guatemala, Colombia, Venezuela, Brazil, and Peru wheat is grown in the tierra templada and tierra fria, at altitudes that range from about 1500 to over 4000 meters. In the north of Mexico and in the south of the Argentine far from the heat equator this is sometimes at fairly low altitudes.

The stem rust was identified as present in Argentina in 1899 (76) by Spegazzini, and in 1915 it was found by Gassner in Uruguay. It, of course, had been in those and nearby countries, before the dates recorded. Meantime, in Mexico grain planters had learned from hard experience with what they called chauixtle, that if they timed their seed sowing, fields were able to grow during a period that helped escape most severe rust effects, although it was always present and harvests were light. Around 1935 Mexicans were growing wheat on about 540,000 hectares of land. Rust may not have been always severe, but, to the detriment of their national economy, much wheat grain had to be imported at high cost to supplement harvests in Mexico. The disease took continual toll, and a severe epidemic of it occurred in

1951 (10). This was developing elsewhere, too. For example Marchionatto (49) and Fernandez-Valiela (25) indicate that in Argentina there were devastating epidemics of special severity in 1925, 1926, 1929, 1949, and 1950. Since that time, in some localities there have been many other widespread attacks.

It is well known, that in Puccinia graminis tritici there are the same five spore types or stages as have been mentioned above on maize rust. In wheat stem rust, spore stage I, "aeciospore" (single-celled, borne in chains), is found in the temperate zone in the cluster cup structures on leaf lesions of European barberry, Berberis vulgaris L., a plant which does not occur as diseased by it in the Neotropics. However, a close relative of this barberry, B. divaricata Rusly was reported in Peru (see de Segura (76)) and found diseased with an aecidial rust, and in the same country Villanueva and Rojas observed an autoecious rust on a wild, undetermined species of Berberis. Notable as well are listings and citations of reports in Mujica and Vergara (54, 55) in which are given occurrence of several rusts that are not the wheat stem rust. But on Berberis microphylla was reported, among 12 other rusts, the determination in 1911 of P. graminis from Chile on this alternate host. (This is a possibly questionable determination.)

Where no overwintering needs to be considered it is the urediospore, II, that is the common and important spore present in, therefore, the Neotropical wheat growing areas. This, it is to be repeated, is a thin-walled, red-in-mass, single-celled spore and is widely and rapidly wind-disseminated. It lives in the tropics, from wheat-growing season to wheat-growing season, on volunteer or escaped wheat plants or on wild grass hosts. It is also brought long distances on winds sweeping across continents and oceans (25, 27, 43, 45, 49 in 1948 publ., 76, 80, 101, 102). These winds may follow storm tracks (6) or be carrying spores far above in the low layers of stratospheric air streams.

It is well recognized in plant pathological literature that the stem rust of wheat is a preeminent example of a successful wind-borne disease taken hundreds or even thousands of miles. At times it may require some years for a new field planted in an isolated area to become infected. In 1929, I was shown in Guatemala a four manzana field of wheat that had been grown there three years without rust. An officer of the Ministry of Agriculture at that time knew of other such rust-free plots. However, on revisiting the same area in 1952 and in 1961, the rust, Puccinia graminis tritici, was found to be commonly present. As another example: in their 1921-1922 botanical explorations in El Salvador, wheat was found by Standley and Calderon (83) but without rust; the fungus was certainly not overlooked as they were meticulous workers who looked for and reported 28 other species of Puccinia. In 1944-1956 wheat rust was searched for by the writer in fields near the small village of Chalatenango in El Salvador; none was discovered. Some 12 years later Ancalmo (2) reported wheat rust from there.

No region on earth is so well suited as the mountainous tropics to study the results on rusts in easily reached, stabile differences in ecology. Temperatures have marked effects, as is reviewed by de Segura (76) (and see Dickson (20)). Important as temperatures are, they may not be the ultimate answers to all differences in occurrences and reactions of the three wheat rust diseases.

All three rusts have fairly low temperature optimums for germination of the uredo, or epidemic stage spores, although severity of rust damage on hosts depends not alone on temperature. However, elevation and its regulatory effect on temperature is important. This is exemplified by researches of Rafael Villanueva (100) in Peru. He found the stem rust, P. graminis tritici, is most severe at relatively warm altitudes from sea level to 2600 meters, moderate at 2600 to 3000, and from there to 4000 m. it is light in its attack. The stripe rust, P. glumarum, attacked most severely by far at the coldest and therefore the highest elevations, from 2900 to 4000 meters. In not so cold but cool places, from 2900 to 2500, the rust is moderate and light in effect from 2400 meters down to sea level which encompasses the warmer locations. In almost direct contrast to this, leaf rust, P. rubigo-vera, is most severe in the relatively warm sea level to 1000 meter area, medium between 1000 to about 1500, light in attack from there up to about 2900 meters, and above that in the coldest tierra fria, nonexistent.

The profound effect of altitude on wheat rust is recognized in Colombia (63), and there as well as in other tropical countries (3, 9, 25, 27, 39-41, 56, 75, 77, 110) it is taken into account and, as in Mexico (10), used as a tool in the breeding and testing programs for wheat rust control.

Wheat varieties are constantly being bred for the cooler parts of the American tropics. Growing conditions may be favorable for wheat but its permanent establishment is unquestionably dependent on what the rust does. The plant has been kept free from rust by spraying with sulfur and organic compound sprays but thus far in the field in the tropics the method is impractical, except for preserving susceptible but valuable breeding stocks. Fortunately, in addition to being a leading, long-domesticated food crop plant, wheat is among the most genetically variable species in which scientists are interested.

In areas of Mexico, Colombia, Guatemala, Uruguay, Peru, Argentina, and Brazil, wheat varieties of complex inheritances are now being grown. And, among other things, they are mostly rust tolerant or resistant. This dramatic accomplishment in tropical plant pathology is in large measure from transfer and application to cool tropical areas, of north temperate zone methods, findings, and adaptations of breeding materials. These very transfers and applications have had great impact on the tropical public; gaining an appreciation of the dedication it requires for practical food-production results in the presence of disease. At one time the countries from which most if not all rust publications flowed, were such as Russia, Germany, France, Italy, Switzerland, Portugal, northern United States, and Canada, but now there are as many or more research articles appearing from Argentina, Colombia, Uruguay, Brazil, Guatemala, and Mexico.

Rusts of Broad-Leaf Herbaceous Crops

Maize is the greatest stomach filler in the Neotropics, but for protein components in those lands where meats are not plentiful, nothing equals beans. In tropical America, literally uncounted millions of hectares of beans (feijoeiros, frejoles, porotos, habichuelas, alubrias, ejotes, judias, francesas, etc.), *Phaseolus vulgaris* L., are grown every year in plots and fields of all kinds of sizes and in all kinds of ecology. This one crop is of utmost nutritional importance as a protein source of local food in our tropics; it far surpasses in human nutritional importance any crop of a single kind cultivated in the temperate zone. However, on the tropical beans and their relatives, Leguminosae, there is common occurrence of rusts (99) and most belong to such genera as *Ravenelia* and *Uromyces*. It is rusts, often associated in some manner with other parasites, that become annihilating plagues.

Many times bean fields are affected by a blasting combination of diseases, and a usual one encountered by me is a complex of a halo blight, Xanthomonas medicaginis var. phaseolicola (Burk.) Link & Hall associated with the rust Uromyces phaseoli var. typica. This, with at times some insects in addition, has been seen in numerous places to cause complete destruction of many large bean fields. Some years ago I have given (104 in 1967 publ.) as a conservative estimate that annually from disease there is a loss of 20 to 35 per cent of the dry beans grown in tropical America. [These percentages were discussed later with a number of colleagues who considered them low. In some instances, specialists stated that average losses are nearer to 50 per cent, or even more.] This is from the total complex of diseases, but the one most consistently present and serious is the rust. Where it is beginning to be controlled, other disease losses are being reduced. On the whole in the Neotropics this one rust is probably the most serious cause of protein food loss of any disease on a broad leaved crop-plant.

First symptoms of the rust, as the illustration Fig. 12-5 indicates, are on leaves; on pods there are as well the dark, rust-colored, raised spots at first not much larger than a pin head but becoming larger. Depending on the age of the spot, it may be surrounded by a yellow halo. The spores in these spots are urediospores (II) that are quickly released and spread to new leaf tissues, and new plants. Leaves when diseased become tough and almost glassy in texture as a result of toxins, and this is accompanied by stunting and defoliation. In high elevations and cool climates these spots may turn darker in aging leaves that are held onto plants; some of the spores in the old lesions are then teliospores (III). However, in warm lands all spores are usually spore type II, which are the sources of great epidemics. Volunteer bean plants always serve to keep plenty of these spores present in hot tropical climates, and if volunteer plants are not nearby new fields, other old plantings with rust, even hundreds of kilometers distant, will be furnishing windborne urediospores to infect new plantings.

It is of special interest that all spore forms, that is 0, I, II, III and IV, of bean rust, Uromyces phaseoli typica, develop on its bean host (25, 27, 29, 30, 54, 68, 76); it is an

Fig. 12-5. Rust, Uromyces phaseoli typica, on leaflets of common garden bean. Photograph courtesy of Plant Pathology Department, University of Florida, Gainesville.

auteoceous rust, although pycniospore (0) and aeciospore (I) stages in themselves cause only minor spotting while the urediospore (II) is the stage causing the real damage. (Something of its severity and common presence is noted in such publications as 2-5, 25-27, 29, 39-41, 45, 54-56, 68, 75, 76, 93, 99, 106.) That there are physiological races is unquestionable and in Mexico and Brazil over 33 are designated. In Central America (personal discussions with E. Echandi and other researchers) several undescribed races occur in countries of that region.

I have seen well-known rust resistant beans from one Central American area taken to another, and observed how the introduced strains outbear and keep free from rust. This situation may be true for several years, but especially in some highly purified varieties selected for a distinct area, a few rust spots begin to appear and after a while the resistance eventually "breaks down" in the new place. The interpretation of this "break down in resistance" is that physiological races have arisen that are quite new and different from those in the old area where original resistance was bred. It is the new races that destroy what was once the pride of the geneticists and resistant bean selecters.

The broadest leaves on any annual, dicotyledonous, broad-leaf tropical crop is a type of tobacco, and these handsome plants are recently shown to be subject to a rust. At present no one can be certain as to the importance of the disease, caused by _Puccinia substriata_ Ellis & Barth., which was reported first in 1964 in Honduras by Muller and Acuña (57), and is still present there; in 1967 it was reported by Hartill (33) in far distant Rhodesia in south Africa.

The conspicuous stage of this rust on tobacco is the aeciospore (I) in, in some cases numerous, large lesions and mainly on lower leaves. Muller (personal communication) found these spots most severe one season following an unusual five-day period of heavy fog and drizzling rain. The same rust is also known (99) to attack eight known species of _Solanum_, including _torvum_ and _tuberosum_, and an undetermined wild _Solanum_. (On potato this rust is also given the name _P. tubulosa_ (Pat. & Gaill.) Arthur.)

The alternate host(s) of this fungus on which the uredia (II) and telia (III) develop, are Gramineae. In his list of fungi of South America, Viégas (99) shows that _Puccinia substriata_ is found on at least 39 species in 11 genera of the Gramineae which have a range in growth characters from tall, coarse bamboos and similar canes to grasses in pastures and fine textured lawns; Muller (56 in his 1966 MS) reported many collections from Central America and the West Indies and among them three grass genera suscepts not included by Viégas for South America. The potential of such a rust could be great, and as tropical American agriculture intensifies the rust will probably be the subject of more study leading to its control.

Another rust-afflicted dicotyledonous annual herb grown for food in the Neotropics, is peanut, *Arachis hypogaea*. The plant came originally from the American tropics and has been imported as a crop to, and considerably perfected in, other parts of an appreciative world; e.g., in many countries of Africa and in the warmer southern states of the U.S.A. It has been dug by aboriginees in the wild western tropics since long before Columbus, and still keeps starvation at bay here and there in undeveloped areas. For a long time it has not received from tropical growers as much modernization and productivity improvements as it deserves. Perhaps a reason for this is that it bears its pods below the surface of the soil, making its growth in a crop class different from any other. Moreover the unlettered of tropical America have been known to express the superstition that pods of peanut are underground to hide them from rust disease. [During a plant seed collecting trip for the U.S. Government in 1936, I came upon this same superstition among peasants and goatherds in the subtropical foothills of the Taurus Mountains of south Turkey. A wild edible pea (possible a *Pisum*) that had aerial and subterranean pods on the same vine was collected, shown to, and discussed with Turks of the area. They believed, so they said, that the plant "buried half of its seeds to keep them healthy." Certainly peanut pods cannot be rust infected if they are underground. Several wild types of *A. hypogaea* I have seen have a very distinct tolerance--resistance to rust and bear full pods subterraneanly while the leaves above-ground are yellow with powdery lesions of *Puccinia*.]

However, peanut plants are subject to many infections on both pods and leaves and one of the worst is leaf rust, *Puccinia arachidis* Speg. A brief resumé of some highlights of its history is as follows: it was first known in the southern part of South America, although between 1910 to 1940 it was found in many parts of the West Indies and Central

America (see 39-41, 55, 56, 75, 79). In the next 16 years it was identified from almost all the rest of Latin America, where it has existed, probably, forever. It was encountered in the U. S. A. as an itinerant infection of peanuts in Texas in 1941 and 1942, and apparently it had been seen for a short while in Florida in about 1931, but in both cases it died out and never reappeared. More recently, in 1961 and 1962, Smart (78) and Wells (105) diagnosed it as present in Virginia and North Carolina, respectively, in the United States. Its only source of infection in the temperate zone is urediospores (II) blown in from subtropic and tropic areas that may be long distances away. This rust attacks both *Arachis hypogaea* and *A. prostrata* Benth., and has been seen on other unnamed wild species of *Arachis*.

Numerous Rusts of Trees

There are many rusts on wild trees and woody bushes in the Neotropics, but, as thought of by most practical agriculturists, they are of relatively little concern. One reason for this attitude is that we continually see the natural change-of-leaf that goes on in the tropics: sometimes associated with seasonal variations in precipitation patterns, and sometimes leaf fall does not follow the seasonal rhythms. The general observer, seeing partial dropping of leaves at odd times, has come to expect this as not an abnormal phenomenon, and almost never associates with irregular defoliation the inconspicuous leaf infections by fungi, such as is the case in certain rusts. Chlorosis of the oldest leaves, and early leaf cast occurring in wild forest and jungle, does not bring complaints to pathologists--the very existence of many of the miscellaneous leaf rusts on trees is often unsuspected.

Rust occurrence in tropic jungles is on hosts growing in much more scattered fashion that if they were in the temperate zone where there are pure stands of a species. Even highly susceptible kinds in the warm lands are intermingled with many non-suscepts. The situation is such that most of the time a tropical tree rust does not have even a good chance to build up an excessive abundance of spores, and to become a serious menace in the wild.

Important groups of trees are used for shade, timber, charcoal, ornament, and watershed protection in the American tropics. If these trees are growing where the seasonal drought is long, and where in the wet season rains are not too excessive, their leaves often are attacked by rusts. Towards the middle and closing months of the dry season often there may be found mature leaves that have on their undersides light- to dark-brown rust pustules. Leaflets on compound leaves may become irregularly chlorotic and drop singly, and after a few have gone maybe the main or central petiole will dehisce. Or in some instances, when severely diseased, whole compound leaves fall. (There is much mycological and some pathological literature on tropical tree rusts; only a comparatively small number of citations dealing with these are included, such as: 4, 21, 26, 30-32, 37, 39-41, 46, 52, 62, 68, 76, 84, 92, 99, and 104 in 1952, 1053, 1957, 1961 publs. Some citations about coffee rust, *Hemileia vastatrix* Berk. & Br., have been included, from outside the Western Hemisphere. This dread disease has recently become firmly established in the Neotropics (104 publ. 1970). For a short discussion of this see an appendix about HEMILEIA at the close of this book.)

Many rusts that cause these effects are in the genus *Ravenelia*. There are some 150 species of it, of course restricted in the hosts they attack. One rust I found often is *R. cassiaecola* Atk. on the tree *Chamaecrista aeschynomene* (DC.) Greene, a close relative of *Cassia*. The rust occurs in two forms: as numerous, small and discrete twig lesions on which are borne the remarkable appearing telial spore heads, and as well on the leaflets are found the same fungus causing the very dissimilar appearing pustular spots in which develop the uredinial spores. The presence or absence of certain spore stages are different in some of these tree rusts. For example teliospores of *Ravenelia* are not known on *R. ingae* (P. Henn.) Arthur, that occurs frequently on the common agriculturally used shade trees *Inga laurina* (Sw.) Willd. and *Inga vera* Willd. However, at times I have seen these same trees to be so severely infected by rust that leaves are shed at a period in the dry season when coffee or cacao foliage below is burned by exposure to such strong sunlight. Generally in a short while the same shade trees regain new leaves and the plantation trees below may recover from much of the sunlight damage. No rusted *Inga* leaves ever examined disclosed anything but urediospores of *Ravenelia ingae*. On the other hand *R. mesillana* Ell. & Barth. is found to have on it urediospores, teliospores, aeciospores, and pycniospores in its parasitism of *Cassia bauhinioides* A. Gray.

The large tree Pithecalobium unguis-cati (L.) Mart. is a popular landmark and shade tree along avenues, roads, and in pastures of the West Indies and Central America. The leaf rust on it that can cause unexpected partial defoliation towards the latter part of the dry season, is Ravenelia pithecolobii Arthur. There is a flowering bush or small tree known as pride of Barbados or dwarf poinciana, Caesalpinia (Poinciana) pulcherrima (L.) Sw., that is widely planted in the American tropics and on its leaf undersurfaces is often found the rust R. humphreyana P. Henn. Another such tree disease is from R. cebil Speg., that infects leaves of the tree Piptadenia peregrina (L.) Benth., which is found in our tropics from countries of the Caribbean down to Peru and Chile.

There appears to be in several species of leguminous trees a tolerance so that no rust attacks are excessively severe from species of Ravenelia. As is well known--and it bears repeating--there are many more genera and species of trees in the tropics than in the temperate zone. Susceptible individuals, even in dense growths, are commonly separated at some distance from one another, with many non-suscepts intervening. Even when a single tree is diseased, among several non-susceptible members in the mixed stand, huge numbers of spores do not accumulate. The infectable hosts are from seeds of mixed inheritance, therefore such trees may be heterozygous for disease susceptibility. This is a recognized principle in the tropics and effective in reducing chances of a virulently pathogenic race of a fungus from encountering in sufficient numbers a large and permanent supply of a single, highly susceptible tree variety or strain which would then sort out, multiply and perpetuate a new and more severe race of rust or other parasite.

An example of an introduced plantation tree that eventually became severely damaged by a leaf rust is well known in the Delta area of Argentina (see 25, 26, and 49 in 1936 and 1937 publs.). As early as 1905 pathologists had realized a rust was at least one of the reasons for a declining supply of the much valued, even-grained, fast growing wood for boxes Argentina had to have for its overseas containers in its large fruit shipping industry. The logs used for this are from the alamo tree which is the poplar or abele of Europe, Populus nigra L. var. italica Du Roi. It was also known as the criollo, and had been established for so long a time that some accepted and thought of it as a native of the Argentine.

The famed alamo plantations that at first grew so well fell on evil times and did not flourish, and the presence of a rust on the trees, although common, was not feared. Leaves that dropped were regarded as if they were the result of natural season change. The native wild willows, *Salix*, of which there are some six species, also dropped their leaves and are known to be hosts to a *Melampsora* rust that is common but is not critical in importance. However, through the years it became evident that leaf rust of alamo was causing damage not only in Argentina but also in Chile, Uruguay, and Brazil, and specimens of it were sent to Europe. Eventually it was identified as *Melampsora larici-populina* Kleb.; when the rust caused an unusually severe onslaught in 1935 and 1936 it was given more attention than in the past looking towards its control.

A search was made for resistant alamo tree species and *P. canadensis* Moench, was found to be a good tree but about as badly rust damaged as the standard alamo criollo, *P. nigra italica.* However, when a specially vigorous hybrid of these two, called the alamo "A. M.", was tested in the Argentine Delta it proved to be rust resistant and a good replacement for the old alamo criollo. This was most fortunate as another fine available alamo, *Populus alba* L. had been brought in before as a possible substitute tree for the criollo but it proved to be a failure, as *Melampsora* rust diseased it with marked severity.

An Oriental tree rust of which coffee pathologists have a distinct fear is *Hemileia vastatrix* Berk. & Br., which has a close relative *H. coffeicola* Maubl. & Roger. (These are illustrated in Fig. 12-6 and 12-7.) In fact the former, from the history of its complete destruction of the once prestigious and most important coffee producing region in the world, on the large island of Ceylon, as well as other areas, has caused it to be classed as among the 8 or 10 greatest of all crop plagues (43, 104) in agriculture. While these rusts did not occur in the western hemisphere untill 1970, where most of the coffee is grown on earth, *H. vastatrix* was accidentally admitted to Puerto Rico in 1903* on

*An instructive parallel event is in the Papuan history known through personal correspondence with Dr. Dorothy E. Shaw, Chief Pathologist, Department of Agriculture, Stock and Fisheries in Port Moresby, Papua; and her report "Coffee rust outbreaks in Papua from 1892 to 1965 and the 1965 (cont. p. 494)

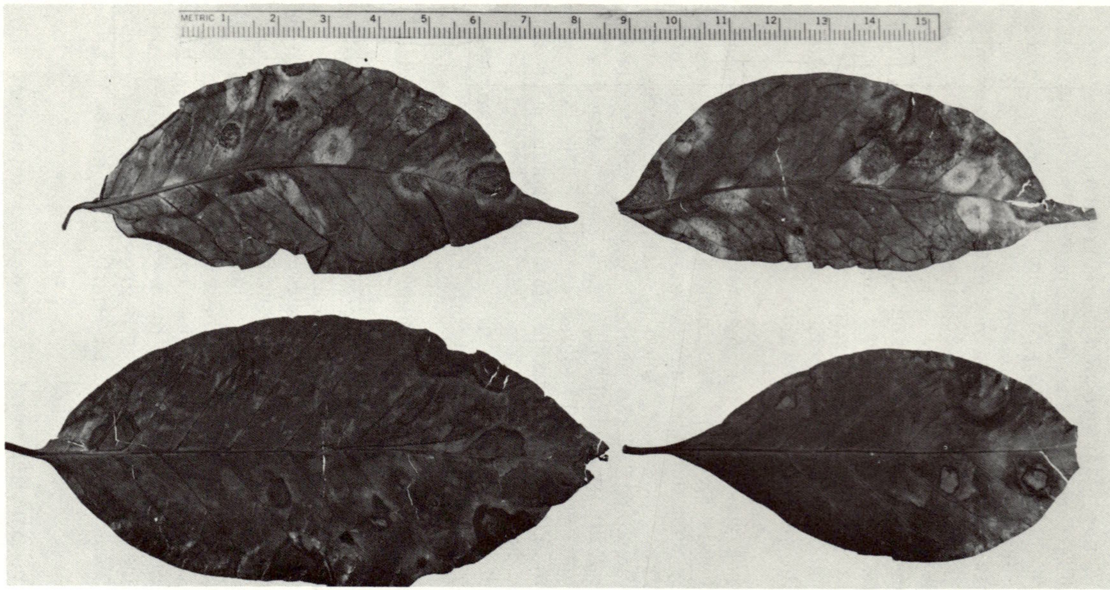

Fig. 12-6. The yellow or common coffee rust, Hemileia vastatrix. Discrete plaque-like lesions; at left on uppersides of leaves lesions appear notably outlined but on this part bear no spores; on leaves at right, undersides exhibit the characteristic yellow to orange powdery masses of spores on lesions. Material for photograph collected in Kenya, but symptoms are identical as seen in Brazil.

Fig. 12-7. The gray rust, Hemileia coffeicola, on coffee (collected in Camaroon). Invasion is systemic and there are no specially localized lesions. At left: upper side of a diseased leaf, darkened but showing no mark from the rust. At right: underside of a leaf, most of which is invaded, and on the host is found scattered, small gray clumps of spores. (This rust is not in the Americas but it can be expected to appear some day.)

experimental coffee seedlings shipped from the Eastern Tropics (27, 75, 104). By quick action, the rust was detected and destroyed, and after one year of careful repeated surveys the western hemisphere was pronounced free from this most dread of all coffee diseases.

Coffee growers and pathologists of our tropics have long been influential (104 see publs. 1952, 1953, 1957, 1961, 1967, 1971) in their backing of strict quarantines against bringing in living coffee plants from outside the hemisphere. However, in spite of great care we now have coffee rust in the Occident. It is evident that *vastatrix* spores were borne by winds over the relatively short stretch of the Atlantic between west Africa and east Brazil.

There are ecologic conditions in Latin America that are entirely satisfactory for rusts of the *Hemileia* genus (see examples in Table 20, at the close of this chapter). All that is necessary for *H. vastatrix* to attack vast amounts of growing coffee in the western tropics is for spores of it to make the right connections under the right conditions. The history of the experience in India with the rust has been given considerable attention by western hemisphere pathologists. Accounts from Mysore State in South India tell that arabica coffee thrived throughout, until rust struck. Then the crop was practically annihilated. Fortunately an arabica coffee found near Chikmagalur showed marked resistance to *H. vastatrix* and soon this coffee, called Old Chik, was multiplied and it replaced the rust destroyed coffee gardens. It was long successful in its resistance, and the Mysore State coffee industry was saved, especially strengthened and increased during 1840 to 1869. In the latter year rust returned, was again devastating, and once remunerative coffee fields were abandoned. Tigers, monkeys

---

eradication program" about *H. vastatrix* on that island. Rust diseased coffee was consigned accidentally during 1892-93 to Papua and by 1903 rust was well spread; it became naturalized and was more serious some years than others. It was in small native dooryard plantings scattered with intervening bush and jungle and had no wild hosts. Of late years, as coffee became of greater economic value on the island, government authorities took all necessary steps and the rust was eradicated. The crop was then replanted and although the eradication program was very costly it has proved successful and very much worth the expenditure.

and deer came back while vines, weeds and bush smothered every sign of cultivation.

Meanwhile, this was not the end. Coffee increased in price and a single tree, resisting rust, was discovered in the Coorg district of India not too far from old forgotten coffee growing centers. Seeds from the tree were planted, retained resistance and became the Coorg variety; it was a good type and most popular in 1880 to 1890. This time resistance was not so long-lived and the Coorg trees showed rust by 1895. Planters were informed by now on these happenings, and an observant grower found a tree free from rust. Its seeds produced resistant trees and, named for the grower, by 1920 the Kent variety was growing in old reconditioned fields that, because of rust, had been allowed at one time to revert to jungle. Rust research men were making a study of H. vastatrix in South India, and from it they learned by 1936 why there were the four repeated "breakdowns in resistance." Each was due to the emergence of a new physiological race of the rust. (See also: 21, 27, 31, 43, 52, 62, 84, 92, 104.)

Tree rusts, before leaves are being cast often escape detection by anyone. Mostly, specimens must be collected by pathologists and examined in detail to see the infections. However, for pure spectacular effect there is no rust in the tropics, or perhaps anywhere, that is so striking in appearance, and so easily seen as one on the pine tree. This is from Cronartium conidenum Hedgc. & Hunt., that involves cones of Pinus montezumae in southern Mexico, in Guatemala and in the West Indies. It is also found on other pines such as P. caribaea, P. oocarpa, P. leiophylla and P. pseudostrobus. The alternate hosts are tropical oaks. If one goes up from the warm flood plains into the cool mountains, one can see forests of tall pines clothed in dark green branches, with many swollen golden bodies glistening in the sun at their tops. This, for the pathologist, is an unforgettable sight.

I experienced this first in Guatemala in 1929 in what was, for the tropics, a remarkably thick stand of pines and estimated about 30 per cent of the cones were diseased. I saw it again in 1962, in the same mountains with a much thinner stand of pines from logging, and made enough counts to estimate that much more than 30 per cent of the cones were rusted. In some parts of Guatemala this cone rust has been more severe and it is said by Schieber (72 in

1967) that it is becoming worse, spreading, and cutting down on seed production. Steps have been taken to work on control measures. The most practical, also the one taking most time, is searching for and selecting disease-tolerant strains of the native pine.

Another tree rust is interesting. From my observations, peach tree rust may be one of the most widely occurring of all tree rusts in the American tropics. This spread is perhaps unusual as ordinarily only here and there are small groups of these exotic (for the Neotropics) trees found around residences in the tropical hills. Often there are hundreds of kilometers of distances between a clump of four or five rust-infected trees to an equally small number of trees in another clump, likewise rusted. The peach rust, _Tranzschelia pruni-spinosa_ Diet., is nearly always on the trees but does not kill them. These trees grow typically from seeds planted by the owners. Fruits from them are in many instances picked green and cooked in a syrupy compote. Insects, birds and curious human fruit watchers make the green harvest habit a requirement if the owners are to eat peaches!

Peaches grown in the American tropics are apparently somewhat rust tolerant. Also for the reason of premature harvest, there is minimum withdrawal of nutrients from the leaves by ripening fruits. The result is that peach leaf rust appears milder than if heavy crops of fruits are allowed to reach ripe-picking stage. The disease is more likely to be found as most severe on trees that are shade-free, and in hills where there is the best air drainage. Some of the most rusted trees seen have been where the sun strikes for most of the day and breezes do not allow long periods of dew or cloudy and drizzly weather.

Physiological Races and Host Tolerance

It was during about 1910 to 1914 that Stakman and co-workers (81) studied wheats inoculated with various collections of rust, _Puccinia graminis tritici._ Not only did some of the rust collections differ in pathogenicities, but the differences were apparently inherited by the fungi as the variations were perpetuated in generation after generation of the parasites. Many other pathologists, working with numerous kinds of disease-causing organisms discovered the same in their collections: whether it was such as

Puccinia, Hemileia, Fusarium, Rhizoctonia, Colletotrichum, Phytophthora, Pseudomonas, or even viruses, multitudes of parasitic variants have been separated and are now known. Because their variant pathogenicities are inherited, and because it makes a difference in reactions to strains of the host they attack, there have grown up designations for the parasitic variants and they may be given letters or numbers. This is how "physiological races" (1) become marked and known; it is a significant concept in plant disease control problems in the tropics.

In the temperate zone, as has been pointed out many times (see 104 in 1971 publ.), the very existence of bread for eating depends upon endless activities keeping track of hundreds of rust races, the movements and changes in them--balanced against alert wheat breeding to produce resistant varieties ready for growers to plant. When it became equally clear to pathologists in the American tropics that wheat rust problems dictated the need for resistant grain, rust race determinations were instituted (for examples of such studies see: 9, 10, 13, 24, 25, 27, 38, 49, 63, 76, 82, 88, 91, 95-98, 101, and 110). As rust researches have progressed it has become plain that tropical races of P. graminis tritici are constantly changing. There are unusual developments in them, including the rise of new and more severe races (10, 24, 27, 28, 96). This is more common in occurrence in the tropics than in the colder parts of the wheat growing temperate zone.

In the temperate zone changes to resistant wheat varieties have been very successful. Sometimes resistance lasts for 20 or more years before physiological rust races disease the old variety and a new wheat is needed. In the tropics, as Borlaug (10) has pointed out, the burdensome changes to new wheat selections may have to be done each four or five years. As is noted in early chapters of the present book, in tropical parasites there are great total numbers of spores that exist, in the tropics there is rapid repeating of generations without hindrance from interventions of dormancy in winter seasons, and variations (mutations) are more commonly expected in the tropics than in the temperate zone. Such things should be reported and there is an Argentine scientific journal "Robigo" published since 1949 devoted exclusively to worldwide rust research reports. Not only is Argentina important in its contributions, but other tropical American countries (in addition to the Europeans, Russians and North Americans) that are making much advance in rust research are Colombia, Peru, Uruguay, Guatemala and of course Mexico.

Mexico, with its large wheat areas so close to the vast cereal acreages in the United States of America, is vitally interested in United States wheat-rust-changes, new races of it, new wheat varieties, and the most modern rust research findings. For reasons of geography and pathology United States and Mexican scientists are aiding, and trading with each other, all information and resistant wheats. During years of effort, one after another race of wheat rust appears, and, by conventional methods an agronomically and commercially valuable wheat variety is bred that not only is good in field and commerce, but likewise is resistant to the latest race or sub-race of *Puccinia graminis tritici*. Through this, the bread grain crop is insured, is again saved for a few years until another change comes in the parasite and the whole process of renewed breeding and selecting of another acceptable strain or variety of the cereal has to be repeated.

There has grown up, among professionals and growers, increasing awareness of the temporary nature of resistance to the changing race of parasites. Among some tropical workers there is the belief that for example, in most, if not all, of the present wheat-growing parts of the tropics, perhaps wheat should be abandoned and another crop substituted. There are those who believe it is hardly worth watching and testing for rust races year after year, and the expense and work involved in securing new resistant wheats is not worth the results. Growers have to be steeled to the expectation that a refined, excellent, popular and pure variety being grown will be pounced upon at almost any time by a hitherto unknown race of the rust. In keeping pace with parasite mutations, there is the never-ending, ever frantic and full time, conventionally carried on testing, breeding, sorting out and multiplying for final release to beleaguered farmers, of the latest new rust resistant wheat variety.

Meanwhile, whether it is in the study of rusts of mandioc, wheat, beans, maize, grass, potatoes, coffee, or in anything else, there is a move by pathologists to learn how to incorporate something better than the conventional type of resistance to combat rust races. This search goes on now in relation to mutations of many other parasites than rusts, but the principle is the same. Caldwell and associates (11, 12, 71, and others, for example 23, 36, 98) have shown much concern about this. In brief, the problem is as follows.

A striking resistance character is genetically transferred into a new crop variety. This resistance is a hypersensitive response to parasite attack, by which the host surrounds the invader with dead or counter-active cells at infections. In cases, thus, with inhospitable tissue, the infection is stopped in its progress. When this resistance character is bred into a new crop, for example as in wheat resistant to Puccinia graminis tritici race 15B, it protects perfectly, so long as there is the expected race population present. However (and this is always to be expected sooner or later--sooner in the tropics), if another parasite race appears, such as P. graminis tritici race 189, the new pathologically different race will, without hindrance, infect, sporulate, be sorted out, reinfect and soon multiply to the extent of being a severe epidemic on the carefully developed, hitherto resistant host variety. The population of the old parasite race or races, starved out, is then replaced by the now flourishing new. This is when "resistance breaks down."

The situation is different with the more desirable "tolerant" or less clean-cut character. This is not to be confused with the commonly understood feature known as "intermediate degree of resistance," but is a definite resilience to disease presence. The inheritance of true tolerance is something still being studied. A tolerant plant is susceptible to parasitic infections, but even if the parasite develops on the host, the crop is capable of relatively unhindered growth, maturation, and producing of good yield and valued quality. On an effective and truly tolerant host variety, new races of the parasite may develop but will not be specially accepted by the host, anymore than the old races. The result is that there is no affinity for an accidental happening of a new parasite race; one of them does not proliferate explosively, and overwhelmingly, in replenishment of the old ones which on the resistant tissue are causing little harm. Theoretically, the tolerance reaction by the host should be far more stable in inheritance than is specialized resistance.

A wild plant in the tropics, whose species flourishes for eons on end in the presence of all the many infecting enemies, is not strikingly resistant to specific races of parasites but it is tolerant. This is a type of lenience that must be the result of a multiple kind of inheritance.

Some pathologists are not satisfied in using the term tolerance to name the complex character just suggested. There are those who denote it as a "multilineal resistance." An outstanding practical example of it is in the Mexican work of Borlaug (10, 70 and others) and associates who have for a long time recognized the necessity of replacing the neater, more easily statistically understood dependence on conventional rust resistance. Studies are now well along leading to something more nearly permanent in tropical wheat varieties.

The Mexican pathologists have developed what they conceive of as the multilineal approach. The more satisfactory Mexican wheat variety is no longer a highly purified strain of the crop, selected for being resistant to special and presently prevailing rust races. Seed sown in more recent years has a much more complex inheritance. A variety originates from many sources, and is composed of a mixture of strains; a variety in which all plants have stabile evenness, desirable agronomic and commercial characters, but also, more importantly, there is in the lines of plants in the variety multiple differences in effects from rust race changes. Such a multi-lineal amount of backcrossing in the breeding work fixes the plant phenotype but at the same time preserves multiple kinds of disease reaction in genotypic variability to races of a parasite.

Table of Neotropical Rusts

Note: rusts are very prevalent in the tropics; there are thousands of them, some not readily seen, many never named and some in the process of having their names changed. Taken as a whole they are of great economic importance. In addition to references used in the main body of the present chapter, it might be well for the more interested worker to go to mycological works. For some of the varied occurrences see citations 3, 5, 25, 27, 29, 30, 39, 42, 49 in 1948 publ., 64, 75, 76, 99, and in a book now in preparation: "A List of Tropical American Plant Diseases." It is significant for coffee plantation owners that there are several species of Hemileia causing rust on several hosts in the Neotropics--examples are marked with an asterisk (*)--and occur under a wide range of conditions.

Finally, an attempt has been made to indicate with a plus sign (+) rusts of considerable economic importance to the tropical agriculturist. Some of them as those attacking peanuts, Arachis spp; canna, Canna spp.; pepper, Capsicum; orchid, Cattleya, Epidendrum, Oncidium; coffee, Coffea; fig, Ficus; cotton, Gossypium spp.; sorghum and relatives, Holcus spp.; sweet potato, Ipomoea; flax, Linum; tomato, Lycopersicon spp.; mandioca, Manihotis spp.; tobacco, Nicotiana spp.; passion fruit, Passiflora; beans, Phaseolus; pine, Pinus spp.; frangipani, Plumeria; alamo or abele, Populus spp.; peaches and relatives, Prunus spp.; guava, Psidium; potato and close relatives, Solanum spp.; ullucus tuber, Ullucus; vanilla, Vanilla; grape, Vitis spp.; and maize, Zea are all of first rate significance. Because of a long-time interest in phanerogamic parasites which are so spectacular in the tropics, I surveyed possible hyperparasites that might keep them in reduced numbers and on such plant genera as Loranthus, Oryctanthus, Phoradendron, and Struthanthus are examples of rusts I found on them.

Table 20. NEOTROPICAL RUSTS

Host	Rust
Achras sp.	Scopella sapotae Mains & Cumm.
"	Uredo sapota Arth. & J. R. Johnst. +
Albizzia spp.	Ravenelia papillosa Speg.
Andira sp.	Ravenelia goyazensis P. Henn.
Annona cherimolia Mill.	Phakopsora cherimoliae Cumm. +
Apium graveoleus L.	Puccinia apii Desm. +
Arachis hypogaea L.	Puccinia arachidis Speg. +
" prostrata Benth.	" " (+?)
Artocarpus spp.	Uredo artocarpi Berk. & Br. (+?)
Asparagus officinalis L.	Puccinia asparagi DC. (+?)

Host	Pathogen
Bauhinia sp.	*Scopella bauhiniicola* Cumm.
"	*Uromyces bauhiniicola* Arth.
"	*U. guatemalensis* Vest.
"	*U. imperfectus* Arth.
"	*U. jamaicensis* Vest.
"	*U. praeterctus* Vest.
Berberis divericata Rusly	(An aecidial rust--undet. in Chile)
" microphylla Forst.	*Puccinia* (possibly *graminis*, in Chile)
Berberis spp.	*Puccinia berberidis* Mont. (in Chile and Mexico)
Beta vulgaris L.	*Uromyces betae* Tul. ex Kickx (+?)
" " var. cicla L.	" " (+?)
Bixa orellana L.	*Uredo bixae* Arth. (+?)
Caesalpinia pulcherrima Sw. (= Poinciana pulcherrima L.)	*Ravenelia hymphreyana* P. Henn.
Cajanus cajan (L.) Millsp.	*Uredo cajanus* Syd. (+?)
" "	*Uromyces dolicholi* Arth. (+?)
Canna coccinea Mill.	*Puccinia cannae* (Wint.) P. Henn. +
" edulis Ker.-Gawl.	" " +
" indica L.	" " +
Capsicum annuum L.	*Puccinia paulensis* Rangel (+?) [for severe leaf malformation and symptoms of rust pustule, see Fig. 12-8.]

Fig. 12-8. As can be seen on garden pepper, the tropical rust Puccinia paulensis causes notable leaf malformations. This type of effect is common in many kinds of rust diseases. Photograph taken in Guatemala courtesy of A. S. Muller.

Cassia sp.	Ravenelia casseicola Atk.
"	R. hymphreyana P. Henn.
"	R. mesillana Ellis & Barth.
"	R. microspora Diet.
"	Uredo cyclogena Speg.
"	U. lutea Arth.
Cattleya dowiana Batem.	Hemileia americana Mass. * +
Chamaecrista aeschinomene (DC.) Greene	Ravenelia cassiaecola Atk.
Cleome spp.	Puccinia cleomis Jacks. & Holw. +
Coffea arabica L.	Hemileia vastatrix Berk. & Br. *(In Puerto Rico in 1903; eradicated but recently finally arrived in American tropics.)
Commelina nudiflora L.	Phakopsora tecta Jacks. & Holw.
" "	Uromyces commelinae Cke.
Desmodium sp.	Phakopsora meibomiae Arth. (+?)
"	Puccinia sp. (+?)
"	Uromyces antiguanus Cumm.
"	U. hedysari-paniculata (Schw.) Farl.
"	U. orbicularis Diet. (+?)
Dolichos lablab L.	Phakopsora pachyrhizae Syd.
" "	Puccinia sp.
" "	Uromyces phaseoli (Pers.) Wint. (+?)

Epidendrum spp.	Uredo epidendri P. Henn.
"	U. guacae Mayor
Erythrina sp.	Ravenelia platensis Speg.
"	Uredo erythrinae P. Henn.
"	Uromyces erythrinae Legh.
Erythroxylon coca Lam.	Bubakia erythroxylonis Cumm.
Ficus sp.	Cerotelium fici (Butl.) Arth. +
"	Physopella fici (Cast.) Arth. +
Gossypium barbadense L.	Phakopsora gossypii (Arth.) Hirst. f. +
" brasiliensis Macfad.	" " +
" hirsutum L.	" " (+?)
Gossypium sp.	Puccinia schedonnardi Keller & Sw. +
Helianthus annuus L.	Puccinia helianthi Schw.
Holcus halepensis L. var. sudanensis L.	Puccinia coronata Cda.
" "	P. purpurea Cda. +
Holcus sorghum L.	Sphaerophragmium sorghi Bat. & Bez.
Indigofera spp.	Ravenelia indigoferae Tranz.
Inga laurina (Sw.) Willd.	Ravenelia ingae (P. Henn.) Arth. (+?)
I. vera Willd.	R. ingae (+?)
I. spp.	Chaconia ingae (Syd.) Cumm. = R. ingae (P. Henn.)

Ipomoea acuminata Roem. & Schult.	*Puccinia crassipes* Berk. & Curt.
" "	*P. opulenta* Speg.
I. batatis L.	*Coleosporium ipomoea* (Schw.) Burr.
Ipomoea sp.	*Aecidium convolvulinum* Speg.
Ixora ferrea Benth.	*Aecidium ixorae* Arth.
Jacaranda sp.	*Aecidium circinatum* Wint.
"	*A. jacarandae* P. Henn.
"	*A. puttemansianum* P. Henn.
Linum usitatissimum L.	*Melampsora lini* Lév. +
Loranthus spp.	*Aecidium goyazense* P. Henn.
"	*A. loranthi* Thuem.
"	*A. struthanthi* Jacks. & Holw.
"	*Puccinia loranthi* Speg.
"	*Uromyces circumscriptus* Neger.
Lycopersicon esculentum Mill.	*Aecidium cantense* Arth. +
" "	*Puccinia pitteriana* P. Henn. +
Lycopersicon sp. (wild)	*Aecidium cantense*
" "	*Puccinia pitteriana*
Malva spp.	*Puccinia heterogenea* Lager.
"	*P. heterospora* Berk. & Curt.
"	*P. malvacearum* Bert. ex Mont.
"	*P. modiolae* Syd.

"	*P. sherardiana* Kaern.
Manihot utilissima Pohl. (= Janipha manihot HBK.)	
" "	*Phakopsora jatrophicola* Cumm. +
" "	*Uromyces cnidoscoli* P. Henn. +
" "	*U. janiphae* (Wint.) Arth.
" "	(Rust, undetermined widespread in Central America +)
Momordica sp.	*Aecidium momordicae* Juel
Montezuma speciosissima DC.	*Phakopsora gossypii* (Arth.) Hirst. +
Nicotiana tabacum L.	*Puccinia substriata* Ellis & Barth. (+ ?)
" tomentosa Ruiz & Pavón	*Puccinia nicotianae* Arth.
Oncidium cavendishianum Batem.	*Hemileia americana* Mass. + *
Oryctanthus botryostachyus Eichl.	*Uromyces urbanianus* P. Henn.
Passiflora edulis Sims	*Puccinia cubensis* Sacc. & Syd. +
" rubra L.	*P. scleriae* (Paz.) Arth. +
Paspalum sp. and numerous other grasses	*Puccinia substriata* Ellis & Barth. (+ ?)
Phaseolus vulgaris L.	*Uromyces phaseoli* var. *typica* Arth. +
Phorodendron spp.	*Aecidium loranthi* Thuem.
Physalis viscosa L.	*Aecidium viscosum* Lindq.
" "	*Puccinia physalidis* Pk.

Pinus leiophylla Sch. & Ch.	Cronartium conigenum Hedgc. & Hunt +
" montezumae Lamb.	" " +
" oocarpa Schiede.	" " +
" peudostrobus Lindl.	" " +
Piptadenia peregrina (L.) Benth.	Ravenelia cebil Speg.
Pithecalobium unguis-cati (L.) Mart.	Ravenelia pithecalobii Arth.
Plumeria spp.	Coleosporium plumeriae Pat. +
"	Puccinia sp.
Populus alba L.	Melampsora larici-populina Kleb. +
" canadensis Moench	" " +
" nigra var. italica Du Roi	" " +
Potentilla heterosepala Fritsch	Phragmidium guatemalense Cumm.
Prunus amygdalus Batsch.	Tranzschelia pruni-spinosae (Pers.) Diet. +
" armeniaca L.	" " +
" cerasus L.	" " +
" domestica L.	" " +
" japonica Thunb.	" " +
" persica Batsch.	" " +
Psidium guajava L.	Puccinia psidii Wint. (= P. sugneurophila Speg.) +
Randia spp.	Phakopsora randiae Kern & Thurston

Salix spp.	Melampsora paradoxa Diet. & Holw. (+?)
Salvia sp.	Puccinia impedita Mains & Holw.
Sechium edule Sw.	Uredo sp. (+?)
Sida spp.	Aecidium odoratum Wint.
"	Puccinia heterospora Berk. & Curt.
"	P. lobata Berk. & Curt.
"	P. malvacearum Bert. ex Mont.
"	P. sherardiana Kern
"	Uredo malvicola Speg.
Solanum (andigena Juz. & Buk. - tuberosum L.)	
"	Aecidium cantense Arth. (+?)
"	Puccinia imitans Syd. (+?)
"	P. pittieriana P. Henn. +
"	P. substriata Ellis & Barth.
"	(= P. tubulosa (Pat. & Gaill.) Arth. +)
Stachytarpheta sp.	Aecidium stachytarphetae P. Henn.
"	Puccinia urbaniana P. Henn.
Struthanthus spp.	Aecidium goyense P. Henn.
"	Uromyces circumscriptus Neger
"	U. socius Arth. & Holw.
Tabernamontana sp.	Aecidium ochraceum Speg.

Tabernamontana sp.	Hemileia juruensis Syd. *
Terminalia spp.	Uredo buchenaviae Kern. & Whet.
	U. terminaliae P. Henn.
Theobroma glauca Karst.	Puccinia thlaspeos-glauco-phylli P. Henn.
Ullucus tuberosus Caldas	Aecidium cantense Arth. +
Triticum aestivum L.	Puccinia glumarum (Sch.) Erikss. & Henn. +
" "	P. graminis-tritici Erikss. +
" "	P. rubigo-vera tritici (Erikss.) Carlet. +
Vanilla planifolia Andr.	Uredo Scabies Cooke (+?)
Viola spp. (wild)	Puccinia cingens Bom. & Rouss.
" "	P. violae (Schum.) DC.
Vitis caribaea DC.	Phakopsora vitis (Thuem.) Syd. (= Angiospora ampelipsidis Thirum. & Kern) +
" vinifera L.	" " " " +
Zanthoxylum spp.	Aecidium xanthoxylinum Speg.
Zea mays L.	Physopella zeae Cumm. & Ram. (= Angiospora zeae Mains) +
	Puccinia polysora Underw. +
	P. sorghi Schw. +

Literature Cited

1. Ainsworth, G. C. 1961. Ainsworth and Bisby's Dic-

tionary of the Fungi. Commonwealth Mycol. Inst., Kew. 547 p.

2. Ancalmo, O. 1959. Lista preliminar de enfermedades parasitarias en las plantas de El Salvador. Serv. Coop. Agr. Salvadoreño Amer., Boletin Tecn. 22, 29 p.

3. Arentsen, S. 1944. Hongos uredinales que atacan a nuestros cultivos. Agr. Tec., Chile 4:54-56.

4. Arthur, J. C. 1904. The aecidium of maize rust. Bot. Gaz. 38:64-67. --. 1922. Uredinales collected by Fred J. Seaver in Trinidad. Mycologia 14:12-24. --. (with collaborators: F. D. Kern, C. R. Orton, F. D. Fromme, H. S. Jackson, E. B. Mains and G. R. Bisby). 1929. The plant rusts (Uredinales). John Wiley, New York, 446 p. --. Las royas de los vegetales (Uredinales) del Perú. Est. Exp. Agr., La Molina, Perú Boletin 2, 14 p.

5. Báez, J. R. 1920. La herrumbre o polvillo de las habas y porotos. Nuestra Terra, pp. 295-296.

6. Bartholomew, J. 1950. The advanced atlas of modern geography. McGraw-Hill, New York, 108 p. + 47 pages of index.

7. Batista, A. C. and J. L. Bezerra. 1960. A Brazilian sorghum rust caused by _Sphaerophragmium sorghi_ n. sp. Nova Hedwigia 2:245-348.

8. Bessey, E. A. 1935. A text-book of mycology. Blakiston's Son, Philadelphia, 495 p.

9. Boasso, C. S. and M. N. Levine. 1952. Roya de la hoja del trigo _Puccinia rubigovera tritici_ en el Uruguay. Arch. Fitotec. Uruguay 5:21-31.

10. Borlaug, N. E. 1954. Mexican wheat production and its role in the epidemiology of stem rust in North America. Phytopathology 44:398-404. --. 1958. The impact of agricultural research on Mexican wheat production. New York Acad. Sci., Ser. 2, Trans. 20:278-295. --. 1959. (pp. 12-24) _In_ Proc. First Internat. Wheat Genetics Symp. Univ. Manitoba, Canada, 268 p.

11. Caldwell, R. M., J. F. Schafer, L. E. Compton and F. L. Patterson. 1957. A mature-plant type of wheat leaf rust resistance of composite origin. Phytopathology 47:690-692.

12. ________., ________., E. C. Leroy and F. L. Patterson. 1958. Tolerance to cereal leaf rusts. Science 128:714-715.

13. Castronovo, A. M. 1943. La biología de _Puccinia rubigo-vera secalis_ en la República Argentina. Rev. Argentina Agr. 10:244-249.

14. Colon, E. D. 1930. Datos Sobre la Agricultura de Puerto Rico Antes de 1898. Cantero, Fernandez & Co., San Juan, Puerto Rico, 302 p.

15. Cummins, G. B. 1941. Identity and distribution of three rusts of corn. Phytopathology 31:856-857. --. 1943. Annotated check list and host index of the rusts of Guatemala. Plant Dis. Reptr., Suppl. 142:87-88. --. and P. Ramachor. 1958. The genus _Physopella_ (Uredinales) replaces _Angiospora_. Mycologia 50:741-744.

16. Cutter, V. M., Jr. 1959. Studies on the isolation and growth of plant rusts in host tissue cultures and upon synthetic media. I. _Gymnosporangium_. Mycologia 51:248-295. --. 1960. Studies on the isolation and growth of plant rusts in host tissue cultures and upon synthetic media. II. _Uromyces ari-triphylli_. Mycologia 52:726-742.

17. Dale, W. T. 1955. A preliminary list of Jamaican Uredinales. Commonwealth Mycol. Inst., Mycol. Paper 60, 21 p.

18. Deighton, F. C. 1951-1954. (Four years of Sierre Leone observations on maize rust. First report outside Americas. Cited by Wood and Lipscomb.)

19. Díaz, H. 1965. Maíz. Estac. Exp. Tucumán, Publicación Miscl. 17, 47 p.

20. Dickson, J. G. 1956. Diseases of Field Crops. McGraw-Hill, New York, 517 p.

21. Diehl, W. W. n. d. ("Solving the Puerto Rican coffee rust mystery." Unpublished, cited in 1961 publ. Wellman.)

22. Dondlinger, P. T. 1912. The book of wheat; an economic history and practical manual of the wheat industry. Orange Judd Co., New York, 369 p.

23. Ellis, R. T. 1954. Tolerance to the maize rust _Puccinia polysora_ Underw. Nature 174:1021.

24. Favret, E. A. 1947. Presencia de la raza 15 de _Puccinia rubigo-vera tritici_ en la Argentina. _Rev. Invest. Agr., Argentina_ 1:63-64.

25. Fernandez Valiela, M. V. 1952. Introducción a la Fitopatología (Ed. 2). Gadola, Buenos Aires, 872 p.

26. Fresa, R. 1941. Royas que atacan al álamo híbrido italiano Arnoldo Mussilini en el delta del Paraná. Rev. Argentine Agr. 8:19-24.

27. Galli, F., _et al._ (see "_Lit. Cit._," Chapter II). Manual de _Fitopatologia, Doenças_ das Plantas e seu Côntrole. Ceres, São Paulo, 640 p.

28. García-Rada, G., J. Vallega, W. Q. Loegering and E. C. Stakman. 1942. An unusually virulent race of wheat stem rust, no. 189. Phytopathology 32:720-726.

29. Gassner, G. 1909. Algunos observaciones sobre el polvillo de las porotes (_Uromyces appendiculatis_). Univ. Montevideo, Rev. _Sec. Agron._ 4:125-129.

30. Gaümann, E. A. (Transl. C. W. Dodge.) 1928. Comparative morphology of fungi. McGraw-Hill, New York, 701 p.

31. Gopalakrishnan, K. S. 1951. Notes on the morphology of the genus _Hemileia._ Mycologia 43:271-283.

32. Gonçalves da Silva, S. 1939. A ferrugem do pimento. Biologico 5:253-254.

33. Hartill, W. F. T. 1967. A rust of tobacco in Rhodesia.

Rhodesia Zambia Malawi J. Agr. Res. 5:189.

34. Hemingway, J. S. 1955. Effects of Puccinia polysora rust on yield of maize. East African Agr. J. 20:191-194.

35. Holway, E. W. D. 1897. Mexican fungi. Bot. Gaz. 24:23-38.

36. Johnston, C. O. and E. B. Mains. 1931. Relative susceptibility of varieties of sorghum to rust, Puccinia purpurea. Phytopathology 21:525-543.

37. Johnston, J. R. 1942. Diseases and insects of pine trees in Guatemala. Amer. Sci. Cong. (Washington, D. C., U. S.) Proc. 3:245-250.

38. Johnson, T. and M. Newton. 1946. Specialization, hybridization, and mutation in cereal rusts. Bot. Rev., United States of America 12:337-392.

39. Kern, F. D. 1932. Supplement to Uredinales. In Botany of Porto Rico and the Virgin Islands. Scientific Survey of Porto Rico and the Virgin Islands. New York Acad. Sci. 8 (part 2):226-231. --. 1938. Additions to the Uredinales of Venezuela. Mycologia 30:537-552.

40. ________., and H. W. Thurston, Jr. 1940. A further report on the Uredinales of Colombia. Mycologia 32:621-629. -- and --. 1943. Additions to the Uredinales of Venezuela-II. Mycologia 35:434-445. 1944. III. Ibid. 38:54-64. 1945. IV. Ibid. 39:503-517. --., --., and H. H. Whetzel. 1933. Annotated index of the rusts of Colombia. Mycologia 25:448-503.

41. Kern, F. D. and H. H. Whetzel. 1926. Some new and interesting Porto Rican rusts. Mycologia 18:39-47. --. and --. 1930. Uredinales. In C. E. Chardon and R. A. Toro, Mycological explorations of Colombia. Porto Rico Dep. Agr. J. 14:301-348.

42. Lamas, P. J. A. 1950. Resultados de las experiencias con lino realizadas en la estacion experimental nacional de Colonia Mascias. Argentina Dir. Gen. Fom. Agr., Bol. 2:2-11.

43. Large, E. C. 1940. The advance of the fungi. Henry Holt & Co., New York, 488 p.

44. (Laudon, G. F. and J. M. Waterston.) 1964. Descriptions of pathogenic fungi and bacteria. (Commonwealth Mycological Institute, set 1, 10 sheets.) Commonwealth Myc. Inst., Kew.

45. Lindquist, J. C. 1951. Uredinales de la provincia de Mendoza. La Plata Univ. Nac., Fac. Agron., Rev. 28:65-114. --. 1952. Notas uredinológicas. Eva Peron Univ. Nac., Fac. Agron., Rev. 28: 213-228. --. 1953. II. Ibid. 29:34-44. --. 1954. Las especies Argentinas de *Ravenelia.* Eva Peron Univ. Nac., Fac. Agron., Rev. 30: 103-128. --. 1956. Uredinales de las Sierras de Cordoba. Cordoba Acad. Nac. Scien., Bol. 39:353-374. --. 1957. Notas uredinológicas IV. La Plata Univ. Nac., Fac. Agron., Rev. 33: 73-82.

46. MacLachlan, J. D. 1938. A rust of the pimento tree in Jamaica, B. W. I. Phytopathology 28:157-170.

47. Mains, E. B. 1934. *Angiospora,* a new genus of rusts on grasses. Mycologia 26:122-132. --. 1938. Two unusual rusts of grasses. Mycologia 30:42-45.

48. Malaguti, G. 1962. Las royas del maíz en Venezuela. Agron. Trop., Venezuela 12:103-111.

49. Marchionatto, J. B. 1926. Polvillo, roya negra, roya lineal del trigo. Rev. Cent. Est. Agr. Vet., Argentina 19:364-371. --. 1932. La roya amarilla del trigo en la zona cerealera. Bol. Min. Agr. Nac., Argentina 30:215-221. --. 1936. Grave attaque de rouille sur peupliers dans le delta du Paraná. Mon. Int. Plant. Prot. 10:145-146. --. 1937. Los efectos de la roya en las alamedas del Delta. Internat. Def. Plant Monthly 11:173-174. --. 1948. Tratado de Fitopatología. Editorial Sudamericana, Buenos Aires, 537 p.

50. Martinez, E. J. 1951. Estimación de los daños causados por las royas de los cereales (*Puccinia*

triticina y P. coronata avenae). Rev. Invest. Agr. Buenos Aires 5:465-482.

51. da Matta, E. A. F. 1954. Ferrugem do maxixe (Puccinia cucumeris P. Henn.). Bahia Inst. Biol., Bol. 1:59-62.

52. Maublanc, A. and L. Roger. 1934. Une nouvelle rouille du cafeier au Cameroun. Soc. Myc. France Bull. 50:193-202.

53. de Menzes, O. B. 1952. Identificacão de 4 racas fisiologicas da ferrugem do feijoeiro. Dusenia 3: 309-312.

54. Mujica Richatt, F. 1952. Determinaciones micologicas V. Agr. Tec., Chile 12:108-109. --. 1954. Roya del frijol. Simiento 24:96-100. --. 1055. Determinaciones micologicas VIII. Agr. Tec., Chile 15:95.

55. ________., and C. Vergara C. 1945. Flora Fungosa Chilena, Indice Preliminar de los Huespedes de los Hongos Chilenos y Sus Referencias Bibliograficas. Imprenta Stanley, Santiago, Chile, 199 p. --. and --. 1961. Addenda a flora fungosa chilena (I). Dep. Invest. Agr., Santiago, Chile, Boletin Tec., 6, 60 p.

56. Muller, A. S. 1950. A preliminary survey of plant diseases in Guatemala. Plant Dis. Reptr. 34:161-164. --. 1952. Plant disease records at Zamarano, Honduras, 1950-1952. Ceiba 3:85-91. --. 1953. Plant disease problems in Central America. FAO Plant Prot. Bull. 1:136-138. --. 1966. List, unpublished, of his own reports and determinations of Puccinia substriata on grasses of Central America. Typewritten, 4 p.

57. ________., and H. Acuña Asegueda. 1964. Rust of tobacco discovered in Honduras. Phytopathology 54:499.

58. Nattras, R. M. 1952. Preliminary notice of the occurrence in Kenya of a rust (Puccinia polysora) on maize. East African Agr. J. 18:39-40.

59. Newton, M., T. Johnson and A. M. Brown. 1931. Hybridization between Puccinia graminis tritici and Puccinia graminis secalis. Phytopathology 21:106-107.

60. Nozzolillo, C. and J. H. Craigie. 1960. Growth of the rust fungus Puccinia helianthi on tissue cultures of its host. Canadian J. Bot. 38:227-233.

61. Olive, E. W. and H. H. Whetzel. 1917. Endophyllum-like rusts of Porto Rico. Amer. J. Bot. 1:44-52.

62. de Oliveira, B. 1954-1955. As ferrugens do cafeeiro. Rev. Café Português 1:5-13; 2:5-12; 6:5-15; 7:9-17; 8:5-18. (Other publ.)

63. Orjuela N., J. 1951. Las royas del trigo en Colombia. Ofic. Estud. Esp. Mexico, Fol. 4:61-70.

64. Presley, J. T. and C. J. King. 1943. A description of the fungus causing cotton rust, and a preliminary survey of its host. Phytopathology 33: 382-389.

65. Robert, A. L. 1962. Host ranges and races of the corn rusts. Phytopathology 52:1010-1012.

66. Rudorf, W. and M. M. Job. 1931. La existencia de Puccinia glumarum tritici (Schmidt) Erik. et Henn. en los países del Río de la Plata. Arch. Soc. Biol., Montevideo 1363-1370 (Quoted and cited by Marchionatto, in 1943 publ.)

67. Rojas M., E. and A. F. Swanson. 1958. El trigo y la roya negra en el Peru. Reunion Interam. Fitog. Fitop. Ent. Edaf. Proc. 3:209-210.

68. Saccardo, P. A. 1882-1931. Sylloge Fungorum. 25 volumes. For Uredinales and Ustilaginales see vols. 7, 9, 11, 14, 16, 17, 21, 23.

69. Saccas, A. M. 1955. La rouille américaine du mais (Zea mays L.) due à Puccinia polysora Underwood au Cameroun et en Afrique Equatoriale Francaise. Agron. Trop., Nogent 10:499-522.

70. Sanchez P., A. 1951. Estudios sobre royas de

varios pastos. Mexico Ofic. Estud. Esp., Fol. 4:107-110.

71. Schafer, H. F., R. M. Caldwell, F. L. Patterson and L. E. Compton. 1959. Combination of resistance to wheat leaf rust. Indiana Acad. Sci., Proc. 68: 89. --[same authors]--. 1963. Wheat leaf rust resistance combination. Phytopathology 53:569-573.

72. Schieber, E. 1962. Estudios preliminares sobre la roya de Guatemala causada por _Physopella zeae._ Reun. Cooper. Interamer. Mejor. Maiz, Proc. 8:76-78. --. 1963. Principales enfermedades del maiz en Guatemala. Reun. Cooper. Centroamer. Mejor. Maiz, Article 7, 5 p. --. 1965. Distribution of _Puccinia polysora_ and _P. sorghi_ in Africa and pathogenicity of these species on corn lines with Latin American plasm. Phytopathology 55:1074-7. --. 1967. Pine cone rust in the highlands of Guatemala. Plant Dis. Reptr. 51:44-46.

73. ________., and J. G. Dickson. 1963. Comparative pathology of three tropical corn rusts. Phytopathology 53:517-521.

74. Schieber H., E., A. E. Rodriguez V. and S. Fuentes F. 1964. Distribution of _Puccinia sorghi_ and _P. polysora_ in Central Mexico. Plant Dis. Reptr. 48:425-427.

75. Seaver, F. J. and C. E. Chardon. 1926. Uredinales (pp. 111-144.) _In_ Mycology, botany of Porto Rico and the Virgin Islands, scientific survey of Porto Rico and the Virgin Islands 8 (Part 1), New York Acad. Sci., 208 p.

76. de Segura, C. B. 1965. Enfermedades de Cultivos Tropicales y Subtropicales. Editorial Juridica, S. A., Perú, 439 p.

77. Shaw, M. 1963. The physiology and host-parasite relations of rusts. Ann. Rev. Phytopath. 1:259-294.

78. Smart, G. C., Jr. 1962. Peanut rust in Virginia. Plant Dis. Reptr. 46:65.

79. South, F. W. 1911. Fungus diseases of ground nut in the West Indies. West Indies Bull. 11:157-160.

80. Stakman, E. C. 1940. Observations on stem rust epidemiology in Mexico. Amer. J. Bot. 27:90-99.

81. ________., J. J. Christensen and H. E. Brewbaker. 1928. Physiologic specialization in *Puccinia sorghi.* Phytopathology 18:345-354.

82. ________., and W. Q. Loegering. 1950. Razas fisiologicas de *Puccinia graminis tritici* en Mexico. Mexico Ofic. Estud. Esp., Fol. 3:24-26.

83. Standley, P. C. and S. Calderón. 1925. Lista Preliminar de las Plantas de El Salvador. Topografía La Unión, San Salvador, El Salvador. 274 p.

84. Stevenson, J. A. and R. Beam. 1955. An annotated bibliography of coffee rust (*Hemileia* spp.) U. S. Dep. Agr., Div. Dis. Survey Publication, 3, 80 p.

85. ________., and F. L. Wellman. 1944. A preliminary account of the plant diseases of El Salvador. Washington Acad. Sci. J. 34:259-268.

86. Storey, H. H. and A. K. Ryland. 1955. The American rust disease of maize caused by *Puccinia polysora* Underw. East. African Agr. Forestry Res. Organ., An. Rept. 1955:43-46.

87. ________., A. K. Howland, and J. S. Hemingway. 1958. East African work on breeding maize resistant to the tropical American rust. Empire J. Exp. Agr. 26:1-17.

88. Straib, W. 1937. Las razas fisiologicas de *Puccinia glumarum* de sudamérica y su comportamiento en la infeccion comparada con el de las formas europeas. Arch. Git. Uruguay 2: 217-233.

89. Sydow, H. 1884-1888. Fungi Guarantici. Hedwigia

17:42-48, 67-96, 119-134; 18:263-286; 19:34-48, 91-96, 241-265; 22:186-224; 26:5-74.

90. Sydow, P. and H. Sydow. 1904. Monographia Uredinearum seu specierum omnium ad hunc usque diens descriptio et adumbratio systematia. I. Genus _Puccinia._ Fratres Borntraeger, Lipsiae 972 p. -- and --. 1909. Fungi Paraenses. Hedwigia, 49:78-87.

91. Teobaldo Llosa, P. 1942. El problema de la roya del trigo en el departmento de Ancash. Dir. Agr. Ganad. Perú, (Bulletin) 21 P.

92. Thirmulachar, M. J. and M. J. Narasimhan. 1947. Studies on the morphology and parasitism of _Hemileia_ species on Rubiaceae in Mysore. Ann. Bot. 11:78-79.

93. Thurston, Jr., H. W. 1940. The rusts of Minas Geraes, Brazil based on collections by A. S. Muller. Mycologia 32:290-309.

94. Tuley, P. 1961. _Puccinia polysora_ on _Tripsacum laxum_ and _Zea mays._ Nature 190:284.

95. Vallega, J. 1940. Especialización fisiológica de _Puccinia graminis tritici_ en la Argentina, Chile y Uruguay. Rev. Argentine Agr. 7:196-219. --. 1942. Especialización fisiológica de _Puccinia coronata avenae_ en Argentina. An. Inst. Fitotec. Santa Catalina 2:53-84. --. 1942. Razas fisiologicas de _Puccinia triticina_ y _P. graminis_ communes en los cultivos de trigo, en Chile, en 1940 y reacción de algunos trigos a las mismas. Min. Agr., Chile, Tec. Boletin 3, 32 p. --. 1944. Especialización fisiológica de _Melampsora lini,_ en Argentina. An. Inst. Fitotec. Santa Catalina Publicación 39, 15 p. --. 1949. Razas de las royas de los cereales halladas en el archipielago de Chiloe. Lilloa 21:169-172.

96. Vallega, J. and G. Stillger. 1948. Mutación de color en _Puccinia graminis tritici._ Rev. Invest. Agr., Argentina 2:93-96.

97. ________., and H. Cenoz. 1958. Physiologic races

isolated. Period: 1938-1958. Robigo 6:32.

98. Vergara C., C. 1961. Ensayo para determinar el tanto por ciento de perdidas causadas por *Puccinia graminis tritici* en trigo. Agr. Tec., Chile 21:58-65.

99. Viégas, A. P. 1961. Indice de Fungos da América do Sul. Instituto Agronômico, Campinas, 921 p.

100. Villanueva, R. (1965, quoted by de Segura, pp. 93-94.)

101. Volosky de Hernandez, D. 1952. Distribución geográfica de las razas fisiológicas de *Puccinia graminis tritici* en Chile. Simiente 22:112-113. --. 1953. Estudios preliminares sobre el *Puccinia glumarum* (Schm.) Erikss. del trigo en Chile. Agr. Tec., Chile 18:159-165. --. 1958. Relación del desarrollo de *Puccinia graminis tritici* en campo e invernadero. Reunion Interamer. Fitog. Fitop. Ent. Edaf., Proc. 3:211-212. --. 1958. Physiologic races isolated Period: 1956-1958. Robigo 6:34.

102. Von Meyer, W. C. 1963. The influence of storage conditions on the longevity of uredospores of *Puccinia polysora* Underw. and *Puccinia sorghi* Schw. Plant Dis. Reptr. 47:614-616.

103. Wehlberg, C. 1962. Leaf diseases of sweet corn in southern Florida. Phytopathology 52:287.

104. Wellman, F. L. 1949. A list of maize diseases from a limited area in Costa Rica. Plant Dis. Reptr. 33:81-85. --. 1952. Danger of introducing coffee rust into the Americas. Inter-Amer. Inst. Agr. Sci., Inf. Bull. 39/41:2-3. --. 1953. Observaciones en *Hemileia vastatrix*. Primera Mesa Redonda de Café, San José, Costa Rica. Documenta pp. 21-26. --. 1953. The Americas face up to the threat of coffee rust. Foreign Agriculture, U. S. D. A., Washington, D. C. 17:47-52; 64. --. 1957. *Hemileia vastatrix*: Investigaciones presentes y pasadas en la herrumbre del café y su importancia en la America Tropical. Federacion Cafet. Amer. (Central America), Secc. Divulg. 23, 68 p. (Transl. to

English by M. K. Visvesvara of Indian Coffee Board in 1963, 66 pages typescript deposited in Plant Pathology Dept. Library, North Carolina State Univ., Raleigh, N. Carolina.); --. 1961. Coffee Botany, Production, and Utilization. Leonard Hill (Books) Inc., London, and Interscience Publ. Co., New York, 488 p. --. 1967. Impact of plant diseases on crop production (pp. 296-302, 329-330). In K. L. Turk and L. V. Crowder eds. Rural Development in Tropical Latin America, Cornell Univ., Ithaca, New York, 446 p. --. 1970. The rust *Hemileia vastatrix* now firmly established on coffee in Brazil. Plant Dis. Reptr. 54:539-541. --. 1971. Plant diseases, an introduction for layman. Natural History Press, Garden City, New York, 130 p.

105. Wells, J. C. 1962. Peanut rust in North Carolina. Plant Dis. Reptr. 46:65.

106. Whetzel, H. H. and H. S. Jackson. 1928. The rusts and smuts of Bermuda. Trans. British Myc. Soc. 13:1-32.

107. Williams, P. G., K. J. Scott and J. L. Kuhl. 1966. Vegetative growth of *Puccinia graminis* f. sp. *tritici* in vitro. Phytopathology 56:1418-1419. -(Same authors)-. 1967. Sporulation and pathogenicity of *Puccinia graminis* f. sp. *tritici* grown on an artificial medium. Phytopathology 57:326-327.

108. Wood, J. I. and R. Lipscomb. 1956. Spread of *Puccinia polysora* with a bibliography on the three rusts of *Zea mays*. U. S. Dept. Agr., Agr. Res. Serv., Spec. Publication 9, 59 p.

109. Yarwood, C. E. 1956. Obligate parasitism. Ann. Rev. Plant Physiol. 7:115-142. --. 1966. Heat predisposition to corn rust. Phytopathology 56:510-511.

110. Zenteno Z., M., W. D. Yerkes Jr. and J. S. Niederhauser. 1955. Primera lista de hongos de México, arreglada por huespedes. Sec. Agr. Ganad., Ofic. Estud. Expeciales, Folleto Tecn. 14, 43 p.

Chapter XIII

TROPICAL SMUT DISEASES

Summary

Smut fungi were present in tropical climates millions of years ago. In general there are four main smut types: kernel smuts, flower smuts, leaf smuts, and stem smuts. Some favor cool (mountainside) areas; some are most severe in warm to hot lowlands; some develop well over a wide range of conditions; all require some but not excessive moisture. Sugarcane smut does best in a warm climate, and is an example of a devastating tropical disease until controlled. It causes stunting, proliferation of slender stalks, and poor stands, and the sporulation is on a "whip"-like organ produced on the host plant's apex. Chlamydospores germinate in five to six hours and slowly infect the cane sett where bract attaches to stalk; from these spores mycelium becomes systemic in a few weeks and in six to ten months, sporulating "whip" is produced. Most temperate zone wheat, barley, rye, and oats smuts occur, now, in our tropics. On rice, the most important smuts are the kernel, leaf, and false smuts. Sorghum and maize crops are widely distributed and smuts develop almost anywhere they grow; these diseases are serious problems. The potato tuber smut occuring in South America up through Mexico is being studied in the Neotropics. A number of unusual tropical smuts are found on grasses and bamboos; most remarkable occurrences are on tillandsias, dicotyledonous herbs (some are important crops), on woody bushes and trees, and on palms.

Introduction

Smuts, or carbónes, the blackening diseases caused by fungi that belong to the family Ustilaginales, are common in the American tropics. The hosts most frequently thought of on which to expect tropical smuts are the grasses, but the smuts occur also on dicotyledonous weeds and herbaceous crops, on ornamental plants, on bushes, and even on trees. I have seen in El Salvador the color of kahki trousers turn grayish and much blackened at the cuffs from tramping through grasses in a smutted pasture (*Paspalum* spp.)--from literally billions of chlamydospores of *Sphacelotheca* sp. In Costa Rica in a dry western flood plain at the edge of the first foothills, a herd of zebu cattle along an old road was seen stirring up a large black and dusty cloud that was found to be from the smutted heads of chuba, *Cyperus esculentus* L., infected with *Cintractia* sp.

However, as a usual thing smuts are not markedly spectacular in their effects in the Neotropics. Although smuts are readily found, there has not been the intensive, decade after decade accumulation of smut disease on a susceptible crop in extremely broad stretches of land. Occasionally, and now more than ever, certain smuts are becoming more troublesome as crop growing stabilizes to feed an enlarging, more modern population. At present it is an acknowledged fact that, except for a few, tropical smuts are considered of interest chiefly because of their mycological significance. However, pathologists cannot rely on passive interest in these organisms, as some of the instances briefly discussed in this chapter will illustrate.

In North America and Europe the reason smuts have become so well known and studied is that they attack many of the smaller cereals and unless controlled they are truly limiting factors. Additionally smut diseases have had historical interest. They stimulated development of techniques in the study of diseases that at one time had never been even contemplated. In the temperate zone observations on cereal smuts has been in progress since almost prehistoric times. It is well known that in the very beginnings of the science of plant pathology, along with rust on grains, one of the first diseases studied was smut and the first treatments towards controls used were in Europe (11-13, 23 publ. 1948, 37) where pioneer workers employed brine, lime, sea water, solutions of copper sulfate, other kinds of crude chemicals, and hot applications for treating seeds to control smuts.

The smut fungi on small grain crops were studied in great detail, under the conventional microscope as it was being perfected in its early days. Examinations dealt especially with morphology, germination of chlamydospores, fusion habits, promycelial developments, production of sporidia, means of entry into hosts, habits of systemic growth, floral infections, and epidemiology. Such matters laid a good part of the basis that showed a scientific world that plant pathology had as much in the line of difficult intricacies in its understanding as was true in much learning about animal and human pathology. For citations of long-known temperate zone studies and a review of some classic works on smuts see Heald (13) and Dickson (18). Since about the first quarter of the present century, many of the most important and valued studies on smuts have dealt with problems of resistance, its genetic understanding, and development of new and practical seed treatments.

Four Common Types of Smut Disease

Nobody can be positive about the origin of the cereal and other kinds of smuts in the Neotropics. However, it is possible that millions of years ago these organisms were indigenous to the tropics, spreading then into the temperate zone. When a collector searches over a Neotropical area and gathers material of diseased and mycological specimens, he inevitably finds numerous smuts. It is probable that in the Neotropics there are some 600-1000 grass species, and over three fourths of these are affected by smuts. There are, besides grasses, many dicotyledonous plants diseased by smuts and in our tropics large numbers of them have not been given much study.

Generally considered, there are at least four different types of important tropical smuts, based on symptomatology, and for teaching purposes at least the types can be listed briefly in about the following manner. One type is the seed smuts, many of which belong to the genus *Tilletia.* Some are characterized by diseased grains that break open and have a black, dusty look. It is this distinctly carbonaceous effect which makes the disease striking in appearance in the field. In other seed smuts the grain is infected and swelled by spores crammed within it, but at first they remain inside an abnormal ballooned seed coat. Inside is powdery, black, carbon-like material. These are often spoken of as covered smuts, the causal fungi belonging usually to the genus *Ustilago.*

A second type are those belonging to the head smuts. With these, the parts that are affected on the host are the flowering structures. The genus Cintractia has in it several species that cause such diseases. There are some of these that are most remarkable in the dicotyletenous hosts and are so specialized that they are only able to invade flowers, some depending on special insect vectors. Among these head smuts are others that belong in a large measure to three other genera as follows: Sorosporium, Sphacelotheca, and in a few cases to Ustilago.

In the third type are found two kinds of leaf smuts and these are spoken of as either blister or stripe kinds. For example in the blister smuts are those which only infect leaves, causing swellings of an abnormal smut-powder-filled spot of convex configuration on the upper surfaces of leaves. These blisters are from attack by species of fungi that belong to the genus Entyloma. They are more common than is ordinarily thought of, and at times can be found in profusion in certain parts of the Neotropics. The other blister type of leaf smut causes the so-called stripe, or sausage lesions symptoms of that shape. Inside all such swellings, of leaf lamina, are black spore masses which fill the blisters. The fungi that cause these stripes and sausage-shaped smut lesions in leaves are of two genera, Urocystis and Ustilago. Urocystis cepulae Frost is an example, and symptoms of attack by it are shown in accompanying Fig. 13-1.

The fourth, remaining type of smuts is the one known as the stem smuts. These usually belong to the genus Ustilago, and are marked often by the development of conspicuous brown or black layers of smut around the internodes of stems. On superficial examination of young diseased plants it is found that the infection does not appear outside but is apparently well within the stem itself. However, in grasses, as the apical meristematic cone matures, the smut then pushes out and appears at the nodes of the haulms. An interesting example of a type of stem smut found in the Neotropics is one that attacks certain grasses, among them the widespread common quack grass. This stem smut, Ustilago spegazinii Hirsch., infects the seeds from internal fungus growth of systemic nature. It may remain internal for maybe two to four years without any outward sign, and when conditions arise that are exactly right, suddenly it produces swellings containing multitudes of spores at the nodal portions of the stems. The tuber of potato is in reality a specialized subterranean stem, which makes it understandable to place the Thecaphora smut of these tubers as among stem smuts.

It is of interest that it is the stem smuts, especially, that become systemic and are perennial within their hosts. As long as the host lives and produces stems the stem smut will then appear again and again, season after season, and produce more spores that then break out, are then lifted into the air, or released into soil in the field where the spores germinate and infect other nearby host plants.

Climatic Features Favoring Certain Smuts

Pathologists are well aware that climate is of utmost importance in problems they face, and, although this phase of plant disease study has not been given all the attention it deserves in most tropical areas, there are several points known in connection with certain smuts. It appears that in our tropics there are smuts in both cool country (such as high valleys, mountain table lands, and areas bordering on the temperate zone), and warm land (for instance, certain low-lying flood plains: protected, and broad, sun-drenched valleys emptying into warm seas). The former appear to be possible chance introductions after they had evolved into organisms adapted to cool parts of the temperate zone, and the latter seem old, long-time, warm tropical inhabitants. Works on ecology of smuts, giving more details than are briefly mentioned and discussed below, are found in, for examples, such citations as: 5, 6, 8, 11-14, 17, 18, 23, 26-29, 32, 35, and 36.

In such a disease as wheat smut (Tilletia caries (DC.) Tul., and T. foetida (Wallr.) Liro), understanding about weather conditions and ranges of seasonal fluctuations in both air and soil temperatures are of much practical value in control and avoidance. It is known from Mexico to Brazil and Argentina that the minimum temperature for fungus growth and infection is a cold 5° C, its optimum is a fresh and cool 18°, and the warmest temperature for growth of the smut is 25°, a condition that is comfortable to a human being but not hot. Above this, where, or when, it is really hot the fungus is inactive. On the other hand, the wheat plant grows very slowly at the low 4° C, its optimum for growth is at about 25° and the high temperature of 30° is near the point where its growth stops. With

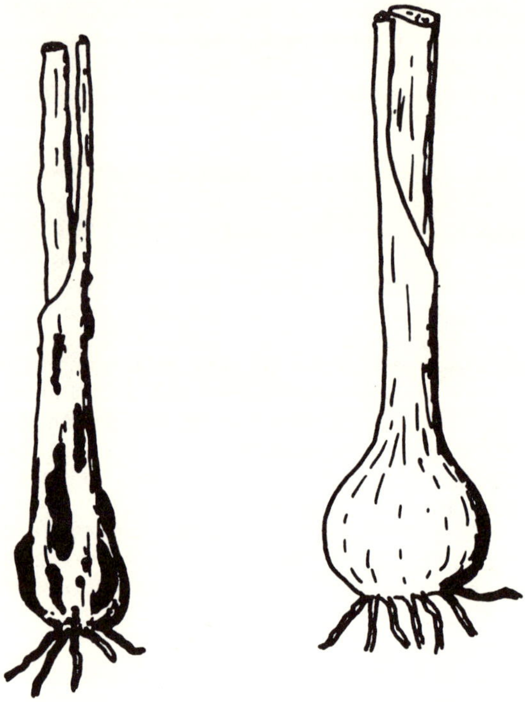

Fig. 13-1. Typical symptoms of onion smut, *Urocystis cepulae*. Plant on right unaffected. (This smut is of comparatively rare occurrence in the warm American tropics, because of its dependence on cool temperatures (36) in infecting its host.)

respect to the disease itself, the temperature range best for it, is from about 5° to 12°; thus if the field production of the crop can be so timed that it mostly escapes this critical 5°-12° range, even with smut spores present in abundance the smut is powerless to infect.

There is another cool country smut, also on wheat, that causes carbón volador, Ustilago tritici (Pers.) Rostr. which may result in much fear by growers because of its spectacular symptoms. However, thus far in our tropics it is, at best, only sporadic in occurrence. This smut is known to be most serious in cool and moist areas. High humidity during wheat flowering is essential, and the best air temperature for flower infection is about 16° to 18° C. At below 15°, and with air relative humidity at 30 per cent and below, this smut is practically eliminated. In the oats smut, Ustilago avenae (Pers.) Rostr., timing field growth to avoid cool temperature is likewise important as this disease is severe at a range of from about 18° to 22°. Vegetable gardeners of South America and Mexico (3, 11, 12, 14, 29) have encountered in almost rare instances the onion smut, Urocystis cepulae Frost (Fig. 13-1). Where found it is only severe if seed planting insures germination and very young seedling growth exposed to soil infections in temperatures at about 10° to 20°. For growth of the fungus itself, the minimum is below 9°, optimum is 13° to 22°, poor growth occurs at 25° to 28°, and at 29° development is stopped.

On oats the smut from Ustilago kolleri Wille is soil-borne, and has temperature optimums for infection that are considerably dependent upon the amount of moisture in the soil: at 15 per cent the optimum is 15°; at from 20 to 25 per cent the optimum temperature is near 20°; at over 26 per cent the optimum is about 25°; and in soil saturated with water its lack of oxygen inhibits action of the fungus at any temperature.

Maize, with a remarkably wide range of conditions where it can grow, is infected almost everywhere with its well known carbón de saco, boba, carvão comum, or smut (Ustilago maydis (DC.) Corda). The fungus chlamydospores germinate at from 9° to 35° C. In fields the disease is usually scattered in occurrence, but in normal fields it is common on the crop causing from a trace to 10 per cent annual losses. Under special conditions I have twice seen small experimental plots of imported maize varieties grown in which no plant escaped smut. In cool highlands the smut causes infections but minor depredations, but in warm lowlands the infected maize plants show severe effects. It may readily be considered a warm-country smut, along with the related smut on sugarcane.

In the Orient, for a long time sugarcane was known to be affected by smut, Ustilago scitaminea Sydow. Probably it was not until about the third decade of the present century that the smut was accidentally introduced into the Neotropics from the eastern tropics. It is known on cane in Argentina, Paraguay, Bolivia, (Peru?), Brazil, and questionably has been reported from the West Indies. It was first identified in Argentina in 1940 and by 1944 and 1946 had spread to cause much concern in Brazil. Researches in both Brazil and Argentina show most ideal conditions in nature for it are a warm and moist climate.

In pure culture U. scitaminea chlamydospores may germinate, but they grow very poorly, at 15° C. The best temperatures both for spore germination and fungus growth are from 24° to 30° and up to four or five more degrees there is little fungus injury. When sugarcane setts (seed pieces) are planted in contaminated soils, if the fields are relatively dry and the season fairly cool, at about 20°, the setts may germinate and grow without disease. However, under more moist and warmer conditions infection takes place in eight to ten days. Penetration of the fungus is at the juncture of bract on cane stalk and from there the fungus proceeds into the meristematic tissues. Work in Brazil shows that it requires approximately one month before the smut becomes fully systemic, after this it may be five to ten months, even longer in cooler areas, before the whip-like disease tip appears. This gives some idea of the period the planter must calculate on in his planting plans.

Sugarcane Smut

Ustilago scitaminea, the stem smut disease of sugarcane, is devastating, as is well known from reports by such as those of Carvalho (4), Cross (7), Fawcett (10), Marchionatto (23) and Mattos (24). It caused great consternation in the western hemisphere tropics when Fawcett reported it as unquestionably present in Argentina at least as early as 1940. It spread rapidly and with extreme severity in the great frost-free Tucumán sugar growing region of that country. Large numbers of hectares of the crop had to be abandoned and bankruptcy was about to occur, before resistant canes were introduced. Since then with application of continuing research on this smut, Tucumán regained its place as a dependable major sugarcane region. It was soon in Brazil and Paraguay. While it has been brought under control and is now of minor concern in these two countries, it spread to Bolivia, and has remained a major sugarcane disease in that country.

Sugarcane smut has a bad name among tropical pathologists, and on intensive West Indies sugarcane-growing islands like Cuba, Puerto Rico, Barbados, the Windwards, and some others, severe smut attack would be disastrous. It is equally feared in Louisiana and Florida in the U. S. A., and in some specially favored sugar growing areas in Central America. When it has been reported at different times from such islands as Trinidad and Hispañola and the country of British Guiana, scientists have been quickly sent to confirm or disavow such findings. Several surveys on this have proved that this dread smut is still absent from the Caribbean region. As late as 1963 from visits and querying new visitors I was able to report that sugarcane smut was not in Puerto Rico, St. John, St. Thomas, St. Croix, Cuba, Martinique, Hispañola, Trinidad, Barbados and the Bahamas.

This is a remarkable smut and it produces three distinct symptoms. The first effect is a general weakening and stunting of the plant; second, there may be a proliferation of slender stalks; and later on occurs the production of the black so-called "whip" at the growing tip. See Fig. 13-2 for this symptom. This is a slender rat-tail shaped growth, characterized by a distinctly black discoloration, and is covered with a delicate, shining skin. When this black structure comes to maturity, the outside silvery tissues die, split apart, and disintegrate, and the carbonaceous dust is freed to contaminate the new plantings. As has been pointed out above, the fungus develops systemically in the sugarcane stool and unless eradicated develops generation after generation of the whip-like, black, chlamydospore-filled structures at any new shoot apex.

Work has been done on culturing of the fungus (10). Fresh chlamydospores germinate rapidly in five to six hours, and on agar surfaces the hyphal growth is best at a warm range of temperatures, from 24° to 30° C. In inoculation studies, the fungus acts comparatively slowly in its invasion

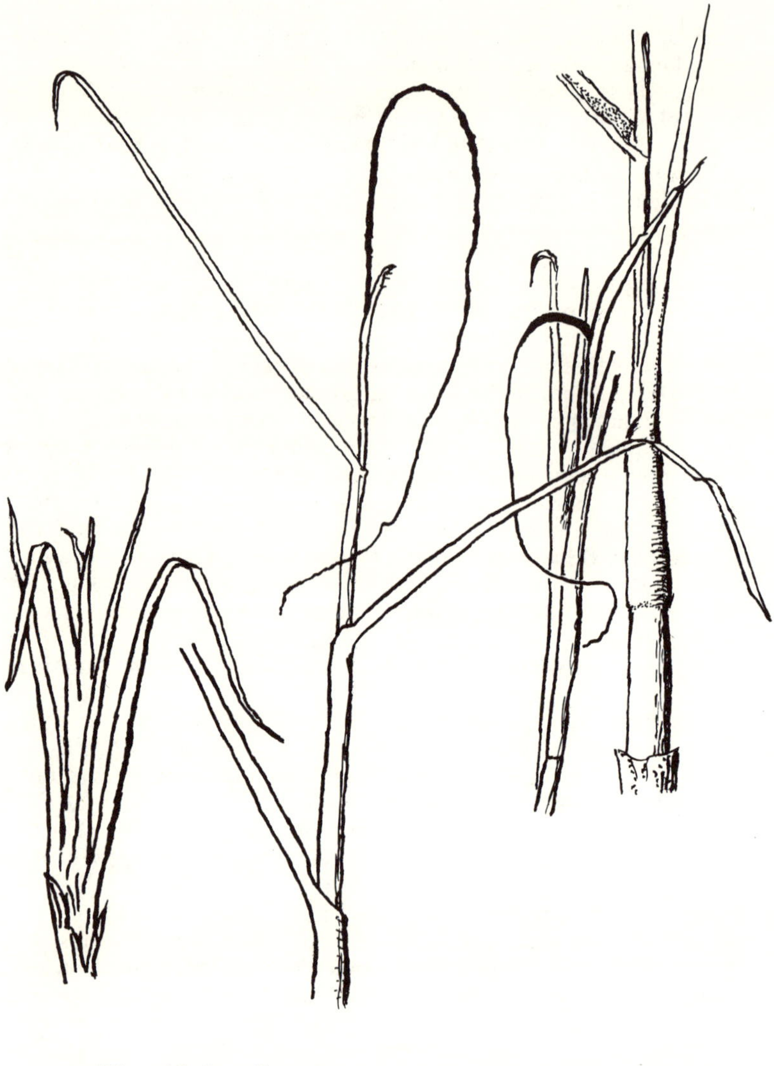

Fig. 13-2. Symptoms of sugarcane smut, Ustilago scitaminea. On left: stunted plant. In center: weak canes with spectacular blackened whip or rat tail, of abnormally stimulated growths of plant tips, that are filled with smut chlamydospores. On right: part of a cane thought to be healthy. (Based on sketch and a color transparency made by a grower from Paraguay who showed these to me while I was in Peru.)

of the meristematic tissues. These are finally found to contain parasitizing mycelium after some eight to ten days following the time when the fungus was first placed in contact with the tender cane tip. In pure culture the fungus does not use starch, which is the form of carbohydrate available in young sugarcane tissues. However, in nature there are large numbers of bacteria present that transform the cane starch into sugars assimilable by <u>U. scitminea.</u> Vegetative growth takes place most vigorously in a host's newly growing tissues. Once established the fungus mycelium spreads out from parenchyma and extends itself then systemically, entering into xylem and phloem elements. Thus, it grows internally, up and down the plant. No sporulation occurs until the infection has reached maturation.

The spores that develop are black binucleate chlamydospores, and are formed in masses in young subepidermal tissues at the growing tip of the cane plant. It is pointed out by Hirschhorn (14 publs. 1949 and 1950) that, differently from other smuts, the nuclei of <u>U. scitaminea</u> double up in hyphal cells at a very young stage of the life history of the organism, prior to chlamydospore development. In other smuts the usual course of development is that chlamydospores do not become binucleate until the time of their maturation.

It is fortunate that the sugarcane organism does not attack all the types of sugarcane. For example the so-called noble types of sugarcane, of the species <u>Saccharum officinarum</u>, are resistant to this smut. In some varieties they are highly tolerant--even almost completely immune--and by using them, sugarcane crops are grown in many parts of the tropics that are thus far free from smut.

However, many of the sugarcanes that are grown in the American tropics and chosen for resistance to mosaic are those that have the genetic characters of some of the other smut-susceptible species of <u>Saccharum</u> in them. For example, <u>S. spontaneum</u> and <u>S. barberi</u> are types that, although resistant or tolerant of certain diseases and conditions, when crossed with the so-called noble canes result in progeny that may be susceptible to the <u>Ustilago scitaminea</u> smut. In many cases there is a good deal of tolerance to it in the presently grown varieties in the Neotropics that have been developed for the region, but in many cases in Florida and Louisiana U.S.A., West Indies, Central America, and South America there are varieties being grown that are also highly susceptible to infection by the organism.

More Than One Sugarcane Smut

It has already been suggested above that, as a general thing, smut diseases are less fearsome in the Neotropics than certain other groups of diseases such as viruses, rusts, fusarial wilts, web blights, cercospora leaf damages, root rots, and some others. However, although this judgment is a popular one, faith in disease status quo is based on the fact that highly technological agriculture is still in its infancy in the American tropics. As more intensified growing in fields and pastures becomes stabilized in practice, it is probable that some diseases, like some of the smuts, will increase in concentration of infective matter. In addition there will always be in our tropics a greater variability in parasites than is found in the temperate zone. Mutations in the tropics are constantly occurring, and sooner or later a race will come forth with greater virulence and attack hosts never before touched.

In the long history of smut studies, it has been found, in Europe and the northern parts of North America, that these parasites have the capacity to vary greatly. Under species there develop subspecies and in these are found races that show marked differences in pathogenicities (8, 15, 16, 23, 29-35). Such differences are of the greatest economic importance.

Upon becoming, as it is, a fear-engendering threat to sugarcane production in the warm regions of the Americas, the mycology of sugarcane smut, _Ustilago scitaminea_, has been given increased attention in the Neotropics. The smut was introduced from the Orient, and the findings in India as to its identity and varieties of it are significant in the western hemisphere. Mycologists describe four sugarcane smuts as follows: the main species _U. scitaminea_ Syd.; a subspecies _U. scitaminea_ Syd. var. _sacchari-officinarum_ Mundkur; and a true second main species _U. consimilis_ Syd. From work by Hirschhorn (14) in South America, it is known that _U. scitaminea_ is divided into six groups and the subspecies _sacchari-barberi_ is approximated in her studies. An important point of note here is that these smuts are not static organisms but are labile and show they are changing. In Africa it has been found that the sugarcane smut attacks not only sugarcane but also several wild grasses including, expecially, _Imperata arundinacea_ and _Erianthus saccharoides_*. In addition smuts are known on uncultivated grass-like sugarcane relatives such as _Sac-_

charum sinensis, S. fuscum, and S. spontaneum.

Cereal Smuts in Our Tropics

Wherever small cereals (wheat, barley, rye, oats) are introduced into the Neotropics, smuts usually can be found. In the past they have been serious locally. However, under more modern situations, cereal grain seed is treated for these diseases. This treatment, in conjunction with selection of fields where the climate favors crop production with minor hindrance from smuts, with proper crop rotation practices, and with alert actions of growers upon expert advice, tends more and more to reduce losses from smuts. Cereal pathologists of such countries as Argentina, Uruguay, Colombia, Mexico, Brazil, and Chile have long had access to a lengthy history of research about smuts in the Old World and in temperate zone North America. In our tropics, intelligent advantage is taken of what others in more advanced lands have discovered. (The bases of knowledge about cereal smuts, and some of the newer findings, are suggested in such citations as: 6, 8, 11-13, 14 especially publs. of 1942 and 1953, 16, 18-20, 22, 23, 26, 29 publ. 1965, 31, 32, 34, 37-39).

In Argentina, in the high mesas of Brazil, and in the cooler parts of Mexico, where wheat is now established as of great importance in food production, an early cause of fear for the future of the new plantings was often the two bunt or smut diseases: Tilletia tritici (DC) Tul. and T. foetida (Wallr.) Liro, and sometimes combined with them Ustilago tritici (Pers.) Rostr. Wheat growers depend on pathologists to devise new or to adapt old measures to control their field diseases; in our tropics this is being done with success. A review of certain of these much used means of control and amelioration in the American tropics is presented in the next-to-last chapter in this book.

When the small cereal grain rye, that is planted in

---

*In connection with future possibilities in the Neotropics, where parasitic fungi are in a more malleable state than in some other parts of the world, pathologists eventually may find other grass or grain smuts attacking sugarcane. For example it has been reported in India that the sorghum loose or head smut, Sphacelotheca cruenta (Kuehn) Potter, attacks flowering parts of sugarcane.

relatively small areas in Chile, Peru, Argentina and a few other countries, ordinary varieties of the crop are often ruined because of the presence of head smut, a Ustilago, and a kernel smut, a Tilletia, unless care is exercised to control these diseases. In many, if not all cases it is believed by local growers and many technicians that these smuts are direct contaminants from the smuts on wheat. There is one smut on rye in the tropics apparently confined to that host. This disease from the fungus Urocystis occulta (Wallr.) Rab. causes dwarfing of plants and aborting of flower spikes.

There are tropical areas where wheat is wanted but where it will not grow well. To produce small cereal grain, it may be that both oats and barley are planted as a substitute. Sometimes these two crops do well, sometimes they may be failures from diseases such as smut. Unless considerable care is exercised, especially in the warmer and slightly more moist locations, oats may be severely attacked by loose smut, Ustilago avenae (Pers.) Rostr. In a few places, I have known of occurrence of the closely related covered smut U. kolleri Wille, which is mostly thought of as one of the rare oats diseases in the tropics.

On barley, possibly one of the more spectacular diseases seen in the tropics on small grains is the semi-loose smut, Ustilago nigra Tapke. It is many times reported as common, but in reality it is seldom in such amounts that it is a severe limiting factor. In Guatemala, a grower found that all he needed to do for control was for laborers to collect disease grain heads one day a year in the field. (He paid a special bounty on given numbers of the diseased spikes.) This smut is recognized almost everywhere that barley is tried as a crop in the confines of our tropics. Reports of it come from the West Indies, from Central America, from Mexico, from the south border states in the U. S. A., and from South America.

It is likewise to be noted that on tropically grown barley, the covered smut, Ustilago hordei (Pers.) Lagerh. is also well known. Occasionally the same smut is seen on oats. However, although it may be widely scattered in these two crops, this smut is not a serious problem where some of the simplest of means of control and amelioration are employed.

Smuts Attacking Rice

Rice, a major stomach-filling world crop, originally from the Orient, has become a profitable farm product in many parts of the western hemisphere tropics. While some of the largest amounts are grown in such countries as Peru, Uruguay, Colombia and Brazil, it is important at the extreme north edge of the Neotropics, in the U. S. in such states as Louisiana, Mississippi, and Texas and the southern parts of Arkansas and California. It is also grown in certain locations of the West Indies and Central America. This cereal crop lends itself well to mechanization. It can be handled in such a manner that no hand labor is needed in soil preparation and fertilization or in irrigation procedures; even weeding and disease/insect spraying has been done successfully by airplane or long boomed spray rigs; harvesting may be by mechanized methods of most modern kind. More and more mechanical means have to be and are being introduced in the tropics. The plant is nearly the most efficient producer of carbohydrate food in the world. Its popularity in the hungry American tropical region has increased a great deal in the last half century. While it has languished as a crop in some places in the Neotropics because of diseases, information about these is accumulating, and numerous practical means of their control are being devised. Smuts need to be watched to be sure they do not develop beyond reach of control measures.

One of the rice smuts is a leaf disease caused by *Entyloma dactylidis* (Pass.) Cif., which is also classified as *E. oryzae* Syd. This parasite invades leaf tissues and produces raised, blister-like lesions (sori) that have a color ranging from cloudy to almost black. These sori, when broken open, are found to be full of dark-colored chlamydospores. Ordinarily it is a minor disease but occasionally fields are found where it is serious.

Where rice is planted year after year in the same ground, and where diseased volunteer plants abound between plantings, the amount of infective spores thus both maintained and multiplied may result in concentrated areas of leaf smut. In very severely infected plants, leaves display yellowing with much weakened stems and lighter kernels. It is in the most severely diseased plants that the leaf blister-lesions appear strongly packed with infections. The result is spectacular and gives a fearsome appearance; but when only the lowest leaves are thus diseased, there seems to be much less reduction in numbers of kernels or in their relative plumpness.

Rice disease specialists in the American tropics find that in practically any established rice growing area this smut, E. dactylidis, is almost certain to be present. Indeed, it is considered pan tropical in occurrence; known from the rice fields about as far north as the crop will grow, in Texas, Arkansas, and Louisiana, and then south in Mexico, in Central America, in the West Indies, and down into Bolivia, Venezuela, Brazil, Colombia, and Peru. There are, also, two other rice smuts, all of which are about equally widespread.

An important rice smut, quite different from the above, in the Neotropics, is kernel smut. The causal fungus Tilletia barclayana Sacc. & Syd. was known in early days as a Neovossia, and for some time as Neovossia horrida; later mycological (33, 34) evidence was established that perhaps it should better be designated as Tilletia horrida, and it finally became evident that this name and Neovossia barclayana were synonymous. The name T. barclayana is now the name most commonly accepted. The kernel smut may attack a high percentage of rice plants in the field, but except in rare cases such as on inoculated plots it has not been severe, and on the whole seemingly as of this writing is of little economic importance in the Neotropics.

Kernel smut is not especially conspicuous in plantings where it occurs in minor quantities, and I have repeatedly visited replanted parts of previously seen slightly affected plots without noting the disease. It was only at harvest that it was finally discovered. Hulls of unmilled but diseased kernels appear gray, and the freshly threshed grains have lost their normal glisten and are considerably dulled. It has been reported as of concern in the U. S. A. (Arkansas, Louisiana and Texas), in Trinidad, in British Guiana, in Peru, in Venezuela, and I have seen it in Panama. I have also received specimens of rice kernel smut from growers in both El Salvador and Costa Rica.

A third rice disease known as a smut, called head smut by some, perhaps best called false smut or carbón falso, is caused by an interesting organism, Ustilaginoides virens (Cooke) Tak. (3, 8, 11, 12, 29, 31, 34). This is not strictly speaking a smut fungus. As a matter of fact, it is an Ascomycete and is closely related to Claviceps sp. U.

virens is one of those organisms (certain lichens, liverworts, and fungi) which form a group found in the tropics that still develop gemmae. These primordial dissemination organs in *U. virens* grow at the surface on the young fungus mass, or *sclerotium*, before it is hardened but after it has usurped the place of the rice grain. These gemmae can be seen with the naked eye as granular particles; they are irregularly rounded, thick-walled and can readily withstand excess rainfall, ward off massive bacterial action, and hold over periods of drought.

It is known (see Dickson's (8) review and citations) that the sclerotia of *U. virens* produce conidia during part of the life cycle. The spores appear greenish in color, which is the reason for the specific name of the fungus. These conidiospores on maturity become studded with short and rounded excrescences. When the spores germinate they produce secondary spores, in a manner very similar in many ways to that carried out by members of the *Ustilaginales*. Dissemination of these sporidia is still to *be fully studied*--there are insect vectors, only a few individual flowers are infected in a rice raceme.

When seen in the field (it is a comparative newcomer to our tropics, for example it was not reported in Brazil (12) until 1946), its presence is now common and the false smut disease is noted for one thing--its scattered occurrence of spherical sclerotia in rice panicles (Fig. 13-3). These are olive drab to greenish in color while the rice crop is in the process of maturing. As drying advances, however, the color of the sclerotia darkens, these small fungus masses shrivel, and altogether they become less conspicuous than while host plants are living. This is specially at harvest when the sclerotia are slightly more covered by the glumes of the rice grains.

This disease, widespread in the Neotropics, although mostly on rice, is found infrequently on maize, for example in Mexico and Panama. It has been collected on rice in all main parts of the American tropics: in the warm south edge of states in the U. S. A. (All Gulf States and California); in Central America (Costa Rica, Nicaragua, Honduras, Guatemala, Panama); in the West Indies (Florida, Puerto Rico, Surinam); and in South America (Colombia, Bolivia, the Guianas, Brazil, Peru). It requires for its most abundant development, conditions of high humidity and warmth (24° to 30° C) for from four to six weeks for pro-

Fig. 13-3. Symptom on rice of black swollen grains of so-called carbón falso or false smut, Ustilaginoides virens, from photograph taken in Venezuela by A. S. Muller.

duction of ascospores, and the disease itself is most severe at 28° with 98 per cent relative air humidity during the period of rice flowering.

Long ago it was introduced to our tropics from Asia and other parts of the Orient, but the trouble has not become a limiting factor up to the present. However, with new and specialized rice varieties being constantly bred and selected, and the "tailoring" of types to be used in localized conditions, pathologists are on the alert as to any of its special developments. Selfings and crossings and selections in rice may result in not only excellent agronomic characters but also in increased susceptibility to some such disease as false smut, ordinarily of little concern. The more modernized are conditions of growth, the more land is planted, and the more refined and perfected are the varieties, the more concern must be kept in mind about how hitherto mild and rare and unexpected diseases will react.

Sorghum and Maize Smuts

In the tierra caliente are by far the most potentially important food-producing lands of Latin America (see Chapters I and II for fuller details). These are in greatest part warm flood plains, from close to sea level to about 500 meters (1640 ft.) above, reaching up into the more broken edges of the foothills. They are relatively flat to mildly rolling lands and are highly suitable for becoming intensive farming country. In these parts, soils are some of the best in the tropics, and the ecology is specially fit for growing the stomach-filling carbohydrate and protein containing tropical crops. Here there is no cold or even cool season, there is abundant sunlight*, there is for the most part reasonably well distributed seasonal rainfall (exceptions are delimited parts where there is excessive rainfall and, vice versa, where there are narrow strips of desert on the Pacific coasts of South American countries), and there is in many of the dry areas sufficient proximity to moist mountains from which irrigation waters can be tapped to alleviate the drought below. However, there are many plant diseases here. In some ways the tierra caliente is not, for human beings, the physically most pleasant part for man to live and work in but with air conditioning and with increasing modernization many discomforts are being overcome.

One of the great difficulties in maintaining the health and vigor of people in some warmer parts of the Neotropics is that at times there is not enough feed grain to produce enough meat to supplement requirements of a tropically-forced vegetarian diet. Grain sorghums are not always eaten by human beings in Latin America. The grain and fodder from them is excellent animal feed, and the plant can grow not only in all the areas where maize produces grain, but also in poor soils and in dry areas that will not support the best growth of maize. Moreover, grain sorghums also will grow on the same land, following the second to fourth final planting of maize in a single year, giving before the dry season a last, good grain and forage harvest for feeding animals.

---

*In a part of the Pacific littoral of northern South America, a cloud or haze blanket is maintained from effects of the Humboldt Current, and annual rainfall may be 7000 mm. (280 in.). Here bright sun is seen only part of an hour after sunrise and part of an hour before sunset. However, the 12-hour daylight is photosynthetically adequate for growing such food crops as beans, plantains, rice, maize, cassava, grain sorghum, and sugarcane, and as well plantation cash crops like rubber, cacao, and coffee.

Sorghum (maicillo) came originally from the Old World tropics and its wild progenitors flourished in dry and windy areas of such countries as Egypt, Sudan, and India. Plants of it in the Neotropics are resistant to dry conditions. They are from one to over six meters tall and grow vigorously in warm or hot climates but do not do well in cold, high mountain fields. Sorghum produces grain and forage in calcareous soils where maize may be a failure. While in the North American temperate zone grain sorghum is grown as an annual, in the Neotropics there are tall "criollo" types that produce year after year as rattoon crops from the same original base and root. Grain sorghums are found in many parts of practically all the countries of the West Indies, Central America, and South America; these grains are most intensively grown, using technological methods, in the middle west and south of the U. S. A. and in the warmer portions of Argentina. Some of the most serious problems regarding the current and necessary expansion of tropical American production of sorghum deal with diseases. The most common complaint made to me about "maíz" and "maicillo" grain disease from growers in Latin America is the disease called "carbón," "enfermedad negra de mazorcas" (black disease of heads), or smuts.

When a blackened and spoiled head of sorghum is examined in the tropics, a number of discoloring fungus parasites may be discovered invading it: *Curvularia* sp., *Cladosporium* (*herbarum*?), *Aspergillus* sp., and in chaffy parts the small spored *Nigrospora oryzae* (Berk. & Br.) Petch is often found. At times the comparatively large spored *N. sphaerica* (Sacc.) Masson appears, coming from irregular black lesions on old flower peduncles, and at bases of kernel receptables. Seeds from such diseased receptacles do not fill out well, and in some cases are aborted. Undetermined species of *Penicillium* and a dark-colored *Fusarium* add further to the blackened appearance. Occasionally stalk and panicle bases of sorghums display a distinct black dry rot. When isolations are made from it and cultured on standard potato or maize meal agar the fungus found is *Diplodia zeae* (Schw.) Lev.; if the black stems are held in moist cloths or in moist chambers, after 30 to 90 days *Physalospora zeae* Stout may be decided upon as the determination for the causal organism. Such disease conditions are black and smut-like in appearance and generally lumped under the one name "carbón" among most growers.

Such blackening and smut-like effects are found also on ears of maize. These are black or nearly black, dry rots of stalk and base of ear caused by *Diplodia zeae* (Schw.) Lév. (*D. natalensis* P. Evans). Working in country where henequen was much injured by definitely known *D. theobromae* (Pat.) Nowell, I found in a nearby patch of maize over 40 per cent of the old maize stalks with black to gray-black dry rot. On isolation and culturing this dry rot proved to be caused by *D. theobromae.* Sorghum was infected upon inoculation with pure cultures of it. The fungus from pure culture grew on old but healthy maize or sorghum stalks causing dry rot, but no perfect stage of *Physalospora* was secured.

A black fungus growth occurring on maize ear husks was repeatedly brought to the experiment station one year in El Salvador as an example of some kind of smut-like infection. It was thought by growers to be stopped in its development because of early onset of the rainy season. Moist chamber studies, and pure cultures grown on laboratory grade filter paper, proved the fungus to be *Chaetomium* sp. Leaves with greenish black smut-like markings have been seen in Ecuador, and these markings had fungus growth in them of *Ustilaginoidea virens* Tak. This is rather rare in occurrence and affects bottom leaves that are nearing death and long past any physiological good for the host. The fungus named has been definitely identified in maize in certain of the Caribbean countries. There is probably no Neotropical region where this crop is grown without charcoal black, smut-like disease attacks somewhere at the base of the plants. This is characterized by lodging and is caused by *Macrophomina phaseoli* (Maubl.) Ashby. A sometimes severe blackening of maize kernels can occur. This destroys the good flavor of the grain, and peasantry fear its effect on the health of those who eat it. This blackening, that has been described as "ojo de negro," from black spotting in kernels, is caused by the fungus *Nigrospora* (mostly *oryzae*).

Among the various types of sorghums, as well as Johnson grass and broom corn, there is a severe smut disease. In the heads of stalks there occur instead of grain, clumps of black, dusty, and sooty masses of useless matter. These masses are open to the air, and in a severely affected field have a characteristic smell. The disease is caused by the loose smut, *Sphacelotheca cruenta* (Kuehn) Potter, and it occurs wherever the crops mentioned above can grow.

Without special search, I have found it in six islands of the West Indies, five countries of Central America and two countries of South America. Although not considered of major importance in the U. S. A. it has been found severe in Central America in attack on certain imported varieties of sorghums from the U. S. and Europe brought in for agronomic tests.

If collected at a rain-free time before S. cruenta has progressed too far, there is found a distinct and characteristic symptom: sorghum grains are replaced with delicate and easily ruptured, long pointed sacks that extend much beyond the chaffy grain receptacles; the sacks rupture easily especially when the dry season comes; inside these sacks are masses of dark brown, almost black, spores. When sacks break open heads of sorghum are sometimes seen that are nothing but a sooty mass. Numbers of spores produced are enormous and they contaminate healthy host seeds. On planting healthy but spore-smeared seeds, they germinate, as do the smut spores, and the fungus attacks and becomes systemic in its infection of young seedlings. A common result of this is stunting of the host, and when diseased plants of sorghums, broom corns or Johnson grass head out, the fungus comes to maturity and spores are produced in the plant's flowering parts. However, systemic infection is not the only mode of attack. Under certain conditions, spores may reach very young flowering heads on healthy plants. In observations I made in El Salvador it appears that if this is at a period of extra heavy dew fall, smut spores cause direct flower infection. In some such cases only a sector or portion of a head will have smut in it.

Another disease, caused by the fungus Sphacelotheca sorghi (Link) Clinton that is closely related to the one just discussed, is found in many parts of the Neotropics. In this smut the sorghum heads appear to have--instead of kernels--swollen, rounded, resistant sacks which are not readily broken. However, when these are mashed between thumb and finger, the smut sack is seen to be filled with a mass of dark-brown spores. These sacks break during harvest and as might be expected the myriad spores are released to contaminate healthy seeds. I have seen it on sorghum in Cuba, El Salvador, Costa Rica and Brazil. At times it causes serious losses. In El Salvador I saw it in 1945 in a native grown field of criollo variety sorghum and it had attacked 38 per cent of the heads. The next year planted next to the criollo field were rows of a sorghum from imported seed and only 1 to 8 per cent of plants became diseased. Seed from this latter field was harvested and replanted in the dry season that year. Grown under irrigation, it produced 18 per cent diseased heads. A year later, the disease percentage was 42 and the seed was then treated by the grower.

The sorghum disease called carbón de la panoja, cabeza negra, or black head smut is caused by Sphacelotheca (Sorosporium) reilana (Kuehn) Clinton. It is never reported as resulting in serious losses but is known in the West Indies and throughout Central America and South America (1, 11, 12, 19, 23, 29, 31, 38, 39). I have seen it once in a few roadside plants growing near Santa Tecla, El Salvador, and once in a small farm plot west of Esquintla, Guatemala. In this disease, the whole sorghum head is involved. The first appearance is a large, spectacular bladder-like gall emerging from the sorghum plant boot at heading time. The disease soon progresses and when the large fungal sack breaks open there remains, around a black fibrous tangle, a dark and sooty spore mass.

Not only is this tropical fungus, S. reilanum, on sorghum but is is seen occasionally on teosinté and on maize. Its occurrence is reported on maize in Mexico and in Argentina. The fungus has been studied in detail by the well-known smut specialist Miss Hirschhorn (14) in Argentina. She found it produced spore-filled swellings that affected not only the inflorescences, both tassel and ear, but also stalks and leaves. She demonstrated considerable variability in Sorosporium reilanum and she named three new species of it, two of which are S. paranensis Hirsch., and S. platensis Hirsch.

Hirschhorn, working with these three smuts, included in her studies the most severe of all maize smuts, known in Argentina by the common names as bolsas and carbón, the well known and widely distributed Ustilago maydis (DC.) Cda. = U. zeae (Schw.) Ung. She secured crosses between her S. reilanum and U. zeae material, and from it there was produced a fungus hybrid with extreme differences in sizes and shapes of resulting chlamydospores. It also is evident that there is much innate variability in fungi of the genus Sphacelotheca in South and Central America. Not only are there head smuts on maize and sorghums and relatives, but also on many other grasses. This suggests the inevitable possibility that these organisms on cultivated crops arose in the Neotropics. It is also apparent that this region has in it many other smuts of economic importance. There is, thus, an eternal threat to be watched for from new smuts that may be coming out of the Neotropics.

The most common smut of maize is Ustilago maydis (DC.) Corda. Numerous reports (see citations in such texts as: 11, 12, 23, and 29) indicate its presence from one end to the other of the Neotropics. Some of its distribution will be indicated in the forthcoming "A List of Tropical American Plant Diseases" in the warmer but not desert parts; in wet and cold areas the disease is comparatively unimportant. This smut is often known as carbón or bolsa in Spanish, although Indian growers have names of their own. For example, in Mexico two such indigenous names are huitlacoche or jit-lecoc. Finding such names, and they occur in other indigenous American languages, is an indication of the recognition of this crop disease before white man appeared.

Symptoms of this smut which is so well known are the presence of spectacular, livid to whitish in color, tightly stretched spore-filled swellings and disease masses. The spores are black and look like ground carbon. These practically take the place of the plant's ears, but are also found at times in smaller swellings on other parts of the plant. Where the crop is grown year after year on the same land, using contaminated seed, and where there is much diseased crop debris and infested soil, losses are severe. In Argentina (9, 11, 23) it is known in every part of the country where maize is grown, and losses are of considerable economic importance especially in dry years. In Peru according to de Segura (29) it is alarmingly severe at times. I made counts in plots of maize in Panama in which 29 per cent of the ears were diseased. On the west coast of El Salvador a plot was seen in 1945 that had 22 per cent ear smut on the third crop of that year and 32 per cent the first crop in the next year. In that same year reports were gathered from 10 other maize fields and losses ranged in them from 5 to over 20 per cent. On the other hand, from wet areas such as in Turrialba, Costa Rica, with rainfall over 3000 mm (120 in.) per year only from 0.10 to 0.26 per cent of maize plants were infected and it was difficult, at times impossible, to find the disease.

Diseases Not Always Recognized as Smuts

The fact goes back into prehistoric ages that there are several blackening plant diseases. Certain of them are so spectacular and so common that the descriptive names given are a permanent part of our languages. In Spanish such are "carbón," "mazorca negra" or "matanegra," and in English there are words "smut" or "brand" which is an early synonym for "burnt." Names like these are used especially in reference to a sooty-mass symptom, characterizing a disease. In tropical America this may result in uncertainty in general discussions of plant diseases.

In the Neotropics some difficulties arise for the pathologist when a whole series of plants, from fast-growing herbaceous kinds to woody bushes and trees, are diagnosed by growers as attacked by carbon or smut and the fungi involved are found to belong to such Ascomycetous genera as *Capnodium* and *Meliola*. This problem of snyonymy is because of common names, for such organisms as the black Capondiaceae and the black Meliolaceae are abundant in the tropics, and observant diagnosticians soon must learn to know them and then readily separate them from the Ustilaginaceae or Tilletiaceae (which are Basidiomycetous).

Additional complications arise when various members of the gramineae are seen in which the heads appear blackened and are called smutted. On making microscopic examinations of scrapings from blackened heads of *Paspalum*, *Holcus*, *Coix*, and *Sorghum* I encountered spores of species of *Curvularia*, *Cladosporium*, *Nigrospora*, *Colletotrichum*, *Fusarium*, and *Helminthosporium*. On host leaves and caulms blackened spots had spores from species of *Phyllachora*, *Physalospora* and *Piricularia*. On heads, kernels were darkened and malformed on *Paspalum* from *Claviceps* sp. and on *Euchlaena* dark brown heads were examined and found parasitized by a combination of a *Rhizoctonia* and *Aspergillus*.

In the West Indies a grass occurs, *Sporobolus indicus* (L.) R. Br., which is known as smut grass for the good reason that its inflorescence is invariably blackened. This blackening, however, is from the fungus *Helminthosporium ravenelii* Curt. as a main inhabitant with other molds that also may appear on it. The *Helminthosporium* seems fairly superficial in its parasitism on the sticky flowers, although no attempt has ever been made to learn how this grass might perform, freed from its fungus incubus. In any case the black fungus growth giving the name smut, is not from any member of the smut-producing fungus family Ustilaginales. However, in Trinidad the Ravenna grass, Erianthus sp. is clearly smutted by the organism Ustilago sp. In nearby Costa Rica and Mexico a surprising smutted host, Tillandsia leiboldiana, is found with black smut sori on its leaves from Ustilago tillandsiae Patterson.

There are in the tropics (and many have invaded the temperate zone) numerous true leaf smut diseases that do not have a dusty, carboniferous, blackening appearance as a symptom. Instead, the lesions are filled with light-colored chlamydospores that are buried internally, with the occurrence of aerial stages of spores that leave on the surface a dull, whitish, powdery deposit. These are white smuts. Determined from chlamydospore examination, these white smuts are classified in the genus Entyloma and if the sporidia, or conidia, are the only spores examined they will be found to belong to either Entylomella or Cylindrosporium. In the immature condition, whitish streaks are caused from several physiological forms of the stripe smut, Ustilago striiformis (West.) Niessl that attack leaves on various small cereals and on grasses, but these forms are not true white smuts.

White smuts of considerable economic importance in the Neotropics are these: on rice, Entyloma oryzae H. & P. Sydow; on water lily, E. nymphaeae (D.D. Cunn.) Setch.; on wild Physalis, E. australe Speg.; on wild Solanum, E. australe; on ragweed, E. polysporum (Pk.) Farl.; on wild lettuce, E. compositarum Farl.; on dahlia, E. dahliae Syd.; on the so-called Spanish needle or bidens, E. guaraniticum Speg.; and on spinach, E. ellisiae Halst. It is notable that E. australe affects not only the host Physalis alkekengi but also a weed, Solanum sp. Although E. oryzae is considered specific to the rice as its host, numerous other grasses, for example species of Panicum, are also diseased with smut and it may become necessary to make studies to determine if some of the white smuts on wild gramineae are types of E. oryzae, or physiological forms of it temporarily harboring on nearby wild or pasture grasses, or specially adapted to other hosts.

Some of the remarkable characteristics of tropical smut fungi are well illustrated by members of the genus Thecaphora in cool parts of the Neotropics. For example on the host Desmodium sp., infection by T. deformans Dur. & Mont. causes symptoms more expected of smuts, in that seeds are swollen and have in them chlamydospores. The same is true of the stem-affecting smut T. pustulata Clint. on bidens which is one of the Compositae. Thecaphora also causes the seed smut of another one of the Leguminosae*, Trifolium tridentatum Lindl. Species of Lupinus and of Vicia are likewise attacked by Thecaphora spp. causing a seed smut. Of all the smuts, one of the most interesting and still not completely studied in all its relationships, is the Peruvian smut, Thecaphora solani Barrus (2, 5, 11, 29, 30) that attacks potatoes of at least the four species Solanum tuberosum, S. andigenum, S. stoloniferum, and S. goniocalix. Chlamydospores develop buried in the tubers, here illustrated in Fig. 13-4, A and B. The fungus causes hyperplasia and from it the tubers develop protuberances and other deformities. Spores are crowded into buried sori reddish to brown in color, and while the tuber is in newly harvested condition it is fairly firm. However, after considerable time in storage, especially if temperatures are not cold and there is normal ventilation, the diseased tubers become leathery and slightly shrunken. If kept long enough the most severely invaded tubers dry out and, when squeezed, there may be ejected a dusty cloud of chlamydospores. T. solani was at one time only in Peru but is known now in Venezuela, Colombia, Ecuador, Panama and Mexico. It can be confidently expected to appear, sooner or later, in all potato growing countries of the western hemisphere. (There is a peduncle diseasing smut on Solanum sp. in Bolivia (31) and the organism has been determined as Urocystis hieronymi Schroet.)

Smuts on Trees

There are a number of Neotropical smuts that are confined to attack on grass leaf material. A well-known one is Ustilago striiformis (West.) Niessl that has

---

*Lest it appear that the species of smuts belonging only to the genus Thecaphora are the more likely to attack leguminous crops, it is of interest that E. Echandi collected some time ago in Costa Rica a leaf smut disease of bean, Phaseolus vulgaris. Examinations show it has typical buried sori, and chlamydospores in them are of such character that John A. Stevenson (personal communication) believes a new genus will have to be erected to describe this new smut organism.

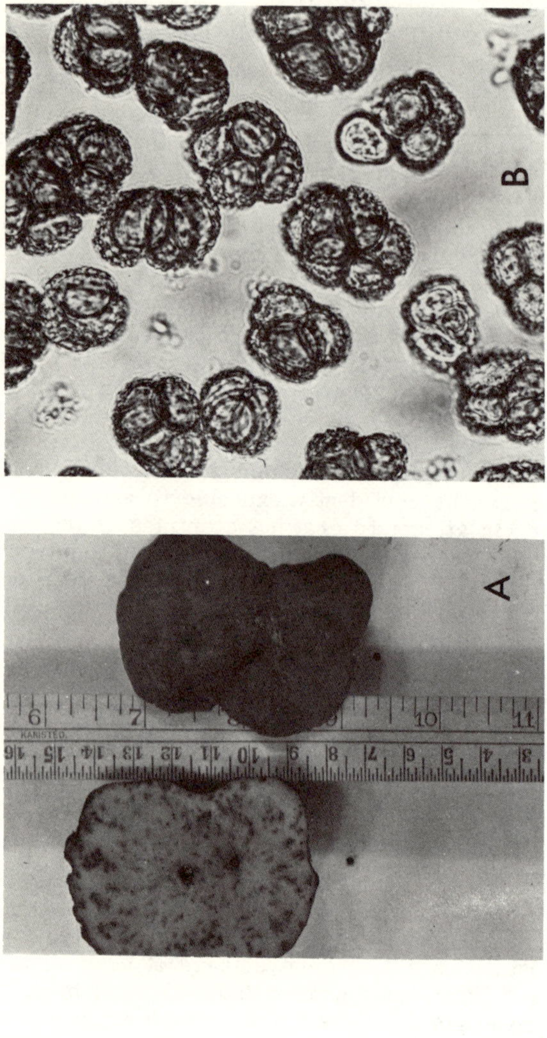

Fig. 13-4. Buba or smut, Thecaphora solani. A. Tuber of Andes potato, Solanum andigenum, malformed by smut; on cutting are seen inside islands of discolored tissue. B. From such diseased potato can be obtained myriads of smut chlamydospores. (Picture A taken in Venezuela, courtesy of A. S. Muller; picture B from work of M. F. Barrus, which see (2), courtesy of H. D. Thurston, Plant Pathology Department, Cornell University, Ithaca, New York.)

many hosts in the gramineae susceptible to one or more physiological forms. It is known as leaf stripe smut or simply stripe disease. Another is U. longissima (Schlecht.) Meyen, and there is also U. macrospora Desm. The striking symptom that characterizes any of these are slightly raised, both long and short, dark-colored stripes on leaves. One of the rather rare smuts is Tolyposporella chrysopogonis Atk. that infects leaves of Sorghastrum nutans L. in parts of Texas that is within the north region edge of the Neotropics. T. sporoboli Jacks., is a true smut that attacks leaves of the smut grass Sporobolus indicus mentioned a few paragraphs above in this section.

Leaf-inhabiting smuts of dicotyledonous plants are not on the whole exceptionally abundant in the Neotropics. It may be there is something in the general construction or physiology of such leaves that does not make them specially good substrates for most smuts. However, there is no rule against this, especially in the tropics: One such smut is on the so-called West Indian cedar or cedro, Cedrella odorato L. a vigorous forest tree that under good conditions grows to be 125 ft. tall. It is attacked by the smut Doassansia sintenisii Bres. In mixed jungle stands the smut is not seriously destructive; it might readily become limiting should the host be grown in plantations or pure stands. Another smut on a tree is Ustilago haesendockii West. found on the mulberries, Morus alba L. and M. rubra L. in South America. An example of a third tropical tree smut is that of Uleilla chilensis Diet. & Neger on the stately South American conifer Araucaria imbricata Pav.; the smut was described in 1897 by Neger (25) on material from Chile, but scattered as it is, to the present day the disease has had minor attention.

In some dryer parts of the northern edge of the Neotropics, such as in California and in Texas, there is a woody bush that may grow to a small tree, known as the bush poppy, planted for its showy flowers. This is attacked by a leaf smut, Entyloma eschscholtziae Harkn. In the whole Western Hemisphere tropics little has been reported on detailed and careful studies of bamboo diseases. Ustilago shiriana P. Henn. is a bamboo smut causing witches' brooming in California and Florida. Some cases have been reported of smuts having been caught by intervention of quarantine officers on Oriental bamboo plant shipments and not allowed to become established in the Americas. But the smut from Shrophira is resident in South American bamboo.

On many palms is an organism related to those that cause smuts, and it is often called "false smut," or even "false rust" in some cases. This fungus is *Graphiola phoenicis* (Moug.) Poit., and in the Neotropics it is widely dispersed. It is of course impossible to know when it first appeared in the Western Hemisphere and it may be that it is an indigenous organism. (See such citations as: 11-13, 21, 23, 31.)

In my experience it was one of the more interesting organisms I came upon when I began my collections. Palms in general are not considered as being able to have the large numbers of leafspots that many plants may. At least to some the palm leafspots that do occur are less notable than in most of the more tender, broad-leaved, dicotyledonous trees and herbs. The common thought in the tropics is that palms are quite resistant to diseases. Wilts that occur on palms are relatively infrequent and this also lends credence to the idea among many people, who are not necessarily specifically interested in plant pathology, that this group of trees is free from disease. However, along with leafspots such as that caused by *Pestalotia* and its many species, Basidiomycetes causing rots, and certain of the species of *Colletotrichum* and many others, the palms are well supplied with leafspot diseases.

When a person picks up leaves of palm that have many a yellowed spot and sees on both sides of it rounded to elongated crusty growths that are about the size of a black-headed pin, he wonders what these might be. In some cases they are on the older leaves of palm trees but do not seem on superficial examination to have caused a great deal of leaf drop or other notable destruction. On the other hand, as newer leaves are examined it is seen that the disease bodies are developing on them and these growths are something different from the ordinary spot. When these crust-like developments were shown first to people who had never seen them before, they asked me if they might not be the droppings from insects.

These dark, crusty excrescences are found to be tightly attached and to have a crater-like depression, in the center of which is a little hole. If young enough when examined, the bodies will be found to have from these center holes long streamers of yellow materials. When examined under the microscope these streamers are found to be a mixture of hyphae and spores. This is characteristic of *Graphiola* spp.

It might be interesting to list here a number of the palms that have been found to have these particular types of infections on them. I have listed what I determined as Graphiola phoenicis on the following: Phoenix dactylifera L., P. canariensis Chaubaud, Washingtonia sp., Pseudophoenix spp., Cocos nucifera L., Elaeis guineensis Jacq., Guilielma gasipaes (HBK.) Bailey, and Corypha umbraculifera L.

While the parasite G. phoenicis is the one species most often considered the causal fungus, it is known that other species of Graphiola occur in the Neotropics. Two others of these are G. congesta Berk. & Rav. and G. thaxteri E. Fisch. A smut closely related that attacks palms is Stylina disticha. (Erenb.) Syd. which infects the Livistona palm in the Orient.

Much is still to be learned about smuts that are to be found on trees. They are an example, again, of features of further interest in plant pathology in the Neotropics leading on to more discovery. For many years mycologists would not consider that such organisms as Graphiola, Shrophira, and Stylina which are known as Graphiolaceae could be included among the smuts. However, this is now generally acknowledged by mycologists who often include under the Ustilaginales the Ustilaginaceae, the Tilletiaceae, and the Graphiolaceae.

Special Table of Some Neotropical Smuts

The special Table 21, below, is a sampling of 56 tropical smuts rather than a complete list. The latter can be obtained from consulting various mycological works. These are numerous and voluminous but are not included in the Literature Cited at the close of this chapter, for lack of space and because many of the multudinous descriptions and reports are still of more academic than plant pathologic importance.

The list in the table is based on my own personal reactions as to those of interest either because of economic concern and because some are unusual and not much referred to in plant disease literature. Ideas regarding relative economic importance of certain of the smuts are only subjective; few numerical data are published and these are so scattered as to be impossible to draw upon to give an overall, objective conclusion.

Grasses seem to be the kinds of plants that are most commonly harboring smuts, but, from what I have seen and from literature consultation, in the Neotropics there are a few grass genera, such as Poa and Tripsacum, that have on their species few if any smuts. On the other hand there are other grass genera such as of Panicum, Paspalum, Pennisetum, and Stipa that have many infections of smuts.

It appears to me that the fact that some plants seem completely immune or at least without any kind of smut in our tropics is reassuring and of economic interest. There are, indeed, even certain rather large families of plants that are important in man's life that have no tropical American smuts reported on any of their genera or species. A partial list of some of the most important families included because of special interest to me, that are not smut-ridden is as follows: no smuts on cactuses; no smuts on Cruciferae; Cucurbitaceae; Euphorbiaceae; Lauraceae; no smuts on Malvaceae except on the genus Sida; Malastomaceae; smut only on the genus Morus in the family Moraceae; none on the Musaceae; Myrtaceae; Orchidaceae; Pinaceae; in the Rosaceae apparently the only smut is Ustilago acaenae Diet. & Neger on the little known plant genus Acaena; Rubiaceae have no smuts; Rutaceae; Salicaceae; Sterculiaceae; in the family Umbelliferae the smut Entyloma eryngii (Corda) de Bary occurs on Eryngium nudicale; Verbenaceae have no smuts; and the Vitaceae seem free.

Table 21. A NUMBER OF PLANT SMUTS that are encountered in the American tropics, selected for apparent importance and from consultation requests.

Name of smut	Remarks
Cintractia caricis (Pers.) Magn., black smut	On Schoenus sp.
" patagonica Cke. & Mass., " "	On Cyperaceae
Sphacelotheca peruviana Zundel smut	On Sporobolus sp.
" pamparum (Speg.) Clinton, smut	On Pennisetum spp.; common and severe in South America

Sphaecelotheca reilana (Kühn) Clinton, head smut	On maize; scattered occurrence; may be confused with common smut
" _sorghi_ (Link.) Clinton, covered kernel smut	On sorghum; common, often causing severe losses in grain
Stylina sp., smut	On washingtonia palm; occasional report
Testicularia leersia Cornu, smut	On _Leersia_ sp.
Thecaphora pustulata Clinton, stem smut	On _Bidens_ sp.
" _solani_ Barrus, internal smut in tubers	On common and wild potatoes; becoming more dangerous
Thecaphora spp., seed smuts	On Leguminosae such as _Trifolium_, _Lupinus_, _Vicia_
" stem smuts	On Convolvulaceae
Tolyposporella chrysopogonis Atk., smut	On _Sorghastrum nutans_
" _sporoboli_ Jacks., black smut	On _Sporobolus indicus_
Tilletia barclayana Sacc. & Syd., kernel smut	On rice; most important smut of this crop
" _caries_ (DC.) Tul., rough spored bunt	On wheat and rye; severe in special locations in American tropics
" _eremophila_ Speg., smut	On _Muehlenbergia_ spp.
" _foetida_ (Wallr.) Liro, smooth spored bunt	On wheat and rye; about same as _T. caries_
" _hypsophila_ Speg., smut	On _Aristida_ and on _Stipa_.

Name of smut	Remarks
Tilletia maclagani (Berk.) Clinton, smut	On Panicum spp.
" ulei Schroet. & P. Henn., smut	On Paspalum spp.
Uleilla chilensis Diet. & Neger., tree smut	On Araucaria imbricata
Cintractia spp., black smut	On Rhynchospora sp.
Doassansia sintenisii Bres., leaf smut	On the tree Cedrella odorata
Entyloma australe Speg., white smut	On species of Solanum and Physalis
" bizzozerianum Sacc., leaf smut	On Digitaria spp.
" dactylidis Pass.) Cif., leaf smut	On rice; common but thus far causing minor damage
" guaraniticum Speg., white smut	On Bidens
" lineatum (Cke.) J. J. Davis, leaf smut	On rice; well-distributed and of some economic importance
" nymphaeae (D. D. Cunn.) Setch., white smut	On water lily
" oryzae H. & P. Sydow, white smut	On rice and certain other grasses; occasionally serious
" sida-rhombifolia Cif., smut	On genus Sida in the Malvaceae
Graphiola phoenicis (Moug.) Poit., "false" smut	On palms: date, canary, washingtonia, coconut, African oil palm, and others; occasionally severe causing notable leaf nec-

	rosis in date palms under moist conditions
Neovossia horrida, syn. *Tilletia barclayana*	On rice, see under *T. barclayana*
Sorosporium chardonianum Zundel, smut	On *Pennisetum* spp.
" *platense* Hirsch., smut	On *Andropogon* sp.
Sphacelotheca braziliensis Zundel, smut	On *Andropogon* sp.
" *cruenta* (Kühn) Potter, loose kernel smut	On sorghum; common, losses up to 30 per cent if uncontrolled (see note*)
" *distruens* (Schl.) Stev. & Johnst., smut	On millet and broomcorn
" *paspali-notati* (P. Henn.) Clinton, smut	On *Paspalum notatum*
Urocystis andina (Speg.) Cif., smut	On *Anemone* spp.
" *cepulae* Frost, bulb smut	On onion; in cool locations
" *occulta* (Wallr.) Rab. dwarfing and spike abortion	On rye; several reports but thus far not severe
" *oxalidis* Paz., smut	On *Oxalis* spp.
Ustilaginoides virens (Cooke) Tak., false smut	On rice and at times on maize; scattered, more common on rice, losses not fully estimated
Ustilago affinis Ell. & Ev., smut	On grasses, especially *Stenotaphrum* spp.
" *avenae* (Pers.) Rostr., loose smut	On oats; at times from appearance very severe

Name of smut	Remarks
Ustilago haesendockii West., leaf smut of tree	On Morus alba and M. rubra
" maydis (DC.) Corda, common smut	On maize; very common, losses from trace to 30 per cent; in special introduced temperate zone stock every plant diseased; also on Euchlaena
" nuda (Jens.) Rostr. loose smut	On barley; not often seen but potentially serious
" sacchari Rab., smut	On Erianthus spp. and reported on Saccharum officinarum and relatives
" scitaminea Sydow, whip smut	On sugarcane; devastating unless control is practiced; reported on Erianthus spp. (possibly this and U. sacchari have been confused)
" shiriana P. Henn., witches' broom smut	On bamboos
" spegazzinii Hirsch., stem smuts	On various grasses
" stipicola Speg., stem smut	On Stipa sp.
" tillandsiae Patters., leaf smut	On Tillandsia sp.
" tritici (Pers.) Rostr., flag smut	On wheat; sporadic, occasionally has caused severe losses

*Sphacelotheca cruenta is known to cause infection of flowering parts on sugarcane in India. Whether this is a physiological race or type that has not reached the Neotropics is not known, but it should be watched for by pathologists in the Western Hemisphere.

Literature Cited

1. Baker, R. E. D. 1943. Notes on some diseases of field crops, vegetables and fruits at the Imperial College of Tropical Agriculture. Trop. Agr. Trinidad 20:28-32.

2. Barrus, M. F. 1944. A Thecaphora smut on potatoes. Phytopathology 37:712-714.

3. Campacci, C. A. 1950. A ferrugem do alho. Bragantia 16:185-187. --. 1951. A carvão verde do arroz. Biológico 17:72-74. --. 1951. O carvão verde do arroz. Biológico 17:72-74.

4. Carvalho, R. S. 1949. Carvão da cana. An. ESALQ, Brazil 6:1-12. (not seen cited by Galli et al.)

5. Christiansen G., J. 1967. El cultivo de la papa en el Perú. Museo Nac., Lima, Peru, 351 p.

6. Costa Neto, J. P. 1963. A cárie do trigo. Correio Agro-Pec., Brasil 3(1):10.

7. Cross, W. E. 1960. Studies on sugar cane smut in Tucumán. Internat. Sugar J. 62:239-242, 271-274.

8. Dickson, J. G. 1947. Diseases of field crops. McGraw-Hill, New York, 429 p.

9. Favret, E. A., A. P. Cercos, and A. Avila. 1954. Sustancia producida por Ustilago maydis con acción sobre el gluten. Rev. Invest. Agr., Argentina 8:187-193.

10. Fawcett, G. L. 1941. El carbón o tizon de la caña de Tucumán. Indus. Azucarera, Argentina 46: 80-81. --. 1944. El carbón de la caña de azúcar (Ustilago scitaminea Syd.). Bol. Est. Agr. Tucumán 47:3-15.

11. Fernandez Valiela, M. V. 1952. Introducción a la fitopatología. 2d ed. Gadola, Buenos Aires, 872 p.

12. Galli, F. et al. (see "Lit. Cit.," Chapter II.) 1968.

Manual de fitopatologia, Doenças das plantas e seu contrôle. Ceres, São Paulo, 640 p.

13. Heald, F. D. 1926. (see especially pp. 662-703.) Manual of plant diseases. McGraw-Hill, New York, 891 p.

14. Hirschhorn, E. 1934. Las especies del género Ustilago en la Argentina. Darwiniana 3:362-363. --. 1941. Notas sobre Ustilago bullata. Rev. Argentina Agr. 8:160-163. --. 1942. Revisión de las especies del Tilletia de la Argentina. Rev. Mus. La Plata, Argentina, Bot. 5:1-20. --. 1942. Nota critica sobre las especies de Urocystis de la Argentina. La Plata Mus., Argentina, Notas 7:81-92. --. 1949. Un nuevo metodo de infección artificial con el carbón de la caña de azúcar. Rev. Invest. Agr., Argentina 3:335-344. --. 1950. Caracteres del ciclo evolutivo del carbón de la caña de azúcar. Rev. Inv. Agr. Argentina 4:317-324. --. 1953. Tipo de germinación de Tilletia eremophila Speg. Rev. Inv. Agr., Argentina 7:69-72.

15. ________. and J. Hirschhorn. 1935. Los carbónes del maíz en la Argentina. Rev. Fac. Agr. Univ. La Plata, Argentina 20:108-140. --. and --. 1939. Formas fisiológicas en Ustilago zeae de diversas localidades de la Argentina. Physis 18: 181-221.

16. Holton, C. S. and J. A. Hoffmann. 1968. Variation in the smut fungi. An. Rev. Phytopath. 6:213-242.

17. Hungerford, C. W. 1922. The relation of soil moisture and soil temperature to bunt infection in wheat. Phytopathology 12:337-352.

18. Jones, E. S. 1923. Influence of temperature, moisture and oxygen on spore germination of Ustilago avenae. J. Agr. Res. 24:577-597.

19. Kern, F. D. and H. H. Whetzel. 1926. Ustilaginales (pp. 107-111). In Mycology. Botany of Porto Rico and the Virgin Islands. Scientific Survey. Porto Rico and the Virgin Islands. Vol. 8, part 1. New York Acad. Sci., New York, 208 p.

20. Lindquist, J. C., A. A. Sarasola, R. S. Cianeros et al. 1954. La caries enana del trigo (Tilletia contraversa), en la Argentina y el Uruguay. Eva Peron Univ., Argentina, Agron. Rev. Ser. III, 30:231-243.

21. Luna Ercilla, C. A. 1950. La palmera productora de datiles. Argentina Bol., Prod. Fomento Agr. 2:2-22.

22. Miccio Peralta, L. R. 1939. El carbón duro Ustilago hordei Pers. Kell. y Sw. de la cebada. An. Rur. Min. Buenos Aires, Argentina 7:89-93.

23. Marchionatto, J. B. 1944. El carbón de la caña de azúcar en Tucumán. Inst. San. Veg., Publicación, 4 p. --. 1948. Tratado de fitopatología (see especially pp. 142-160). Editorial Sudamericana, Buenos Aires, 537 p.

24. Mattos, R. A. 1949. Antecedentes sobre el carbón de la caña de azúcar en el Chaco. Lilloa 21:155-166.

25. Neger, W. 1897. Sobre una enfermedad dañina del pino chileno Araucaria imbricata Pav.) An. Univ. Chile 97:115-116.

26. Nieves, R. 1933. La caries o carbón hediondo del trigo. Bol. Min. Agr., Argentina 32:397-411.

27. Potter, A. A. and L. E. Melchers. 1925. Study of the life history and ecologic relations of the smut of maize. J. Agr. Res., U. S. A., 30:161-173.

28. Reed, G. M. and J. A. Faris. 1924. Influence of environmental factors on the infection of sorghums and oats by smuts. Amer. J. Bot. 11:518-534. --. and --. 1924. Influence of environal factors on the infection of sorghums and oats by smuts. II. Experiments with covered smut of oats and general considerations. Amer. J. Bot. 11:579-599.

29. de Segura, C. B. 1960. Andean disease of potatoes on the coastal plain. FAO Pl. Prot. Bull. 8: 70-71. --. 1965. Enfermedades de cultivos

tropicales y subtropicales. Editorial Jurídica, Lima. 439 p.

30. ________. and A. Quevedo Diaz. 1955. Occurrence of Thecaphora smut on Solanum stoloniferum on the coast. FAO Pl. Prot. Bull. 3:141-143.

31. Stevenson, J. A. 1926. Foreign plant diseases, a manual of economic plant diseases which are new to or not widely distributed in the United States. United States Dep. Agr., Washington, 198 p.

32. Tapke, V. F. 1948. Environment and the cereal smuts. Bot. Rev. 14:359-412.

33. Tullis, E. C. and A. G. Johnson. 1952. Synonyny of Tilletia horrida and Neovossia barclayana. Mycologia 44:773-788.

34. U. S. Department of Agriculture. 1960. Index of plant diseases in the United States. Agr. Handbook 165, 531 p.

35. Vizioli, J. 1953. Fatôres ecológicos e patogenicidade do carvão da cana de azúcar. VI Congr. Internat. Microbiol., Rome 5:258-266.

36. Walker, J. C. and F. L. Wellman. 1926. Relation of temperature to spore germination and growth of Urocystis cepulae. J. Agr. Res. 32:133-146.

37. Wellman, F. L. 1971. Plant diseases. Natural History Press, New York, 130 p.

38. Whetzel, H. H. and F. D. Kern. 1926. The smuts of Porto Rico and the Virgin Islands. Mycologia 18: 114-124.

39. Zeman, V. 1916. Consideraciones sobre el carbón y la caries de los cereales. Rev. Fac. Agr. Vet. Univ. La Plata, Argentina 12:330-340.

Chapter XIV

CORD, THREAD, AND WEB DISEASES

Summary

The numerous blights, caused by recumbent-growing Basidiomycetous parasites are common in the tropics. Those which have a hymenium as a sporulation stage, seem to release their spores almost altogether at night. Some of the most spectacular of the blights, more commonly in the wild, are wiry cord-like fungi that have agarics as the perfect stage. Parasitism is on foliage and branches, causing dieback and decline. Thread blights (kolerogas) also have a travel stage, like the cord fungi, and this thread grows on the underside of branches and attacks the underside of leaves. Effects are severe and are of considerable economic importance on many plantation tree crops. Basidiomycetous fungi are known that have no travel stage, and the infections they cause are delimited leafspots. One of the most widespread of all Basidiomycete blights is the pink disease on very many tree hosts, which has four distinct phases in its life history. The well known damping-off organism (often called Rhizoctonia) is one of the most widely known of any parasite in the tropics.

Introduction

In these chapters it is pointed out repeatedly that there is a greater abundance of parasitic organisms in the Neotropics--the American tropics (it is also true in eastern hemisphere tropics or the Paratropics)--as compared with the temperate zone. Some of the most notable phenomena respecting tropical diseases are the large numbers of parasitic lower Basidiomycetes that crawl, penetrate at soil line, and those on foliage: web, entwine, and hand. The large amount of susceptible wild and cultivated types of plant growths, along with agricultural habits of shifting of fields, sometimes obscures the true importance and need for understanding these diseases. Many times, fortunately, ameliorative changes in husbandry practices give such good results that chemicals need not be used for control in localities marginal for the most favorable development of these diseases.

It has been both convenient and illuminating for me to find there is a series of surface traveling parasites ranging from the coarsest aerial cord growths, in which agaricaceous basidiocarps are in the perfect stage but are seldom seen and saprophytism may be as important as parasitism; to finer strand types of aerial coarse thread blights, where a woven parasitic pellicle of non-infective, spore-bearing hymenium forms on the host in back of the forward traveling threads; to organisms that are neither webs nor cords but induce large leafspots having discrete edges with spores produced centrally; to more delicate aerial foliage web blights, with their non-infective, spore-bearing hymenium within the mass of spider-web-like enwrapped and irregularly diseased leaves; and finally to the soil-inhabiting, fungus, seedling-infection disease that causes stem dissolution (damping-off), in which the fungus invasion occurs usually at or near soil line immediately behind the very fine traveling hyphae.

Wire-like Cord or Horsehair Blights

The most spectacular phase of parasitic fungus cord growth is commonly a surface-borne, macroscopic, wire-like structure. From a cord come the delicate-appearing hyphae that are the absorbing organs which penetrate the host surfaces. It is astonishing on first observation to see these coarse fungus cords where they are overrunning, twisting, looping, and tangling (see Fig. 14-1) among leaves, stems and branches of both woody bushes and giant trees. I have studied these blights at such divergent elevations as sea-level (e.g., Barro Colorado Island in Gatun Lake, Panama Canal Zone), and mountainside locations over 1500

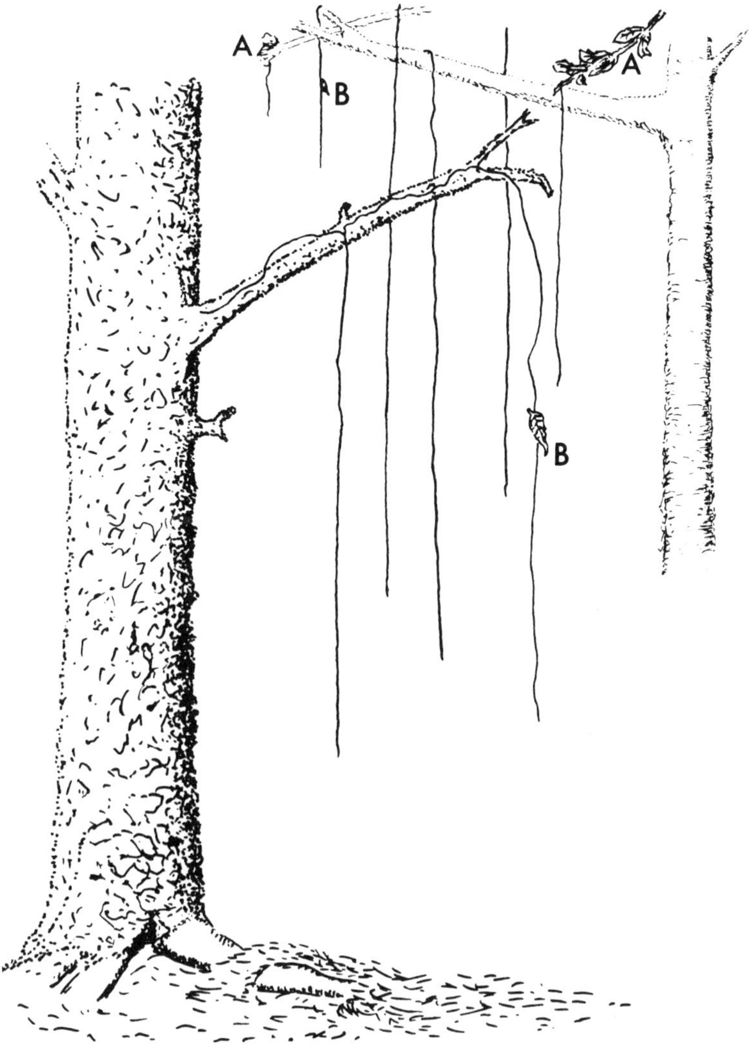

Fig. 14-1. Appearance of cord-blight fungus cords, hanging from leaves and branches of jungle trees in Panama: A, ends of leafy branches with cords growing from them; B, dead leaves attached to cords (not often seen by me with this black cord fungus). Host trees were not determined botanically.

meters above sea-level (e.g., near road on the way up to the summit of Irazú volcano, Costa Rica); undoubtedly there were various different organisms involved. It is obvious the cords are a travel stage of the fungi, as it can be seen that they infect the host in places along the course they pursue; not only are leaves killed and detached from twigs, but also often the twigs that bear them and the minor small branches that are involved by the fungus cords are killed.

The cords are of various macroscopic appearances (as is seen in Fig. 14-2, illustrating the appearance of three different cords). They may be black and about the diameter of horsehair to that of the lead in a pencil; in some they are cylindrical and white; others are coarse and of various colors, tightly pasted to the bark of stems and twigs. When they are freed and drop into the air they turn into the slender cylindrical structures, sometimes meters long, of mycelium.

In certain cases the cords glisten brightly in the light, and in others they are smooth but dull. In these blight fungi the spectacular developments are the cylindrical traveling growth stages. However, the important structure upon which such mycologists as Petch (28), Dennis (10), and Singer (37) base differences in species is the character of the small mushrooms which are the perfect, spore-bearing stage that comes from the parasitizing cord. Agarics are not abundant. They require some search, and are mostly found (37) on cord, disease-killed host branches that have fallen to the ground. In a few instances I have seen sporocarps develop on still attached dead branches. This occurs towards the close of a rainy season.

Thread blights, a distinct group discussed below, are not to be confused with the cord or horsehair blights. These latter are wiry cord, parasitizing fungi apparently belonging exclusively to the genus *Marasmius*. There are non-cord species of jungle *Marasmius* that subsist upon entirely dead substances in moist locations. They are found on such materials as old bark of many kinds of trees, on jungle debris, and on decaying or rotted wood. There are also some species of *Marasmius* that are active causal agents in general root rots. However, the cord blight organisms are aerial in habit.

In observations by Weir (50) in the Amazon jungles

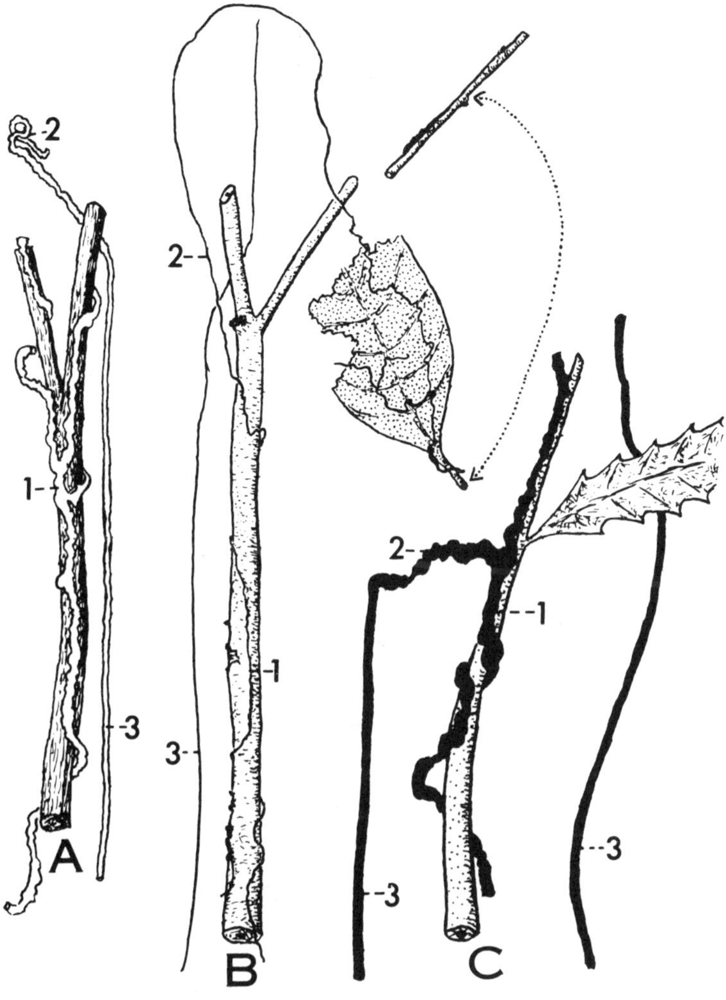

Fig. 14-2. Three different types of cord-blight fungi (none had spore bodies allowing classifications), on twigs of diseased jungle trees of undetermined species. A. Relatively thick, slightly rough, white cord collected in Costa Rican lowlands. B. Slender, shiny black cord from protected cove in foothills of Puerto Rico. C. Relatively thick, dull surfaced, black cord collected in Panama. 1. Irregular, flattened, and knobbed portion of cord at attachments. 2. Area in cord of change from attachment to hanging. 3. Straighter, cylindrical, hanging condition of cord.

it was seen that the wiry cords of M. sarmentosus Berk., and M. equicrinus Muell. appeared to resemble "long, coarse, brown horse-hair." They were overrunning the stems, twigs, and leaves of hevea rubber. He found no actual symptoms of distinct parasitic attack, although he saw that young leaves of rubber seedlings were sometimes deformed from presence of horsehair fungi. Weir considered that the cords probably originated in jungle debris. In swamps he found that on plants "the mycelium was sometimes very conspicuous, forming tangles of considerable extent."

I have at times been much impressed with cord fungus tangles, and have even collected birds' nests (Fig. 14-3) that had been fashioned exclusively from these materials. I have traced parasitizing black and wiry cords from definite, small, dark-colored pads that are the points of origin on small branches, along bark-covered surfaces. The cords then grow out to those twigs with leaves. In an abandoned corner of a Central American plantation a three-year-old coffee leaf was observed, detached from the stem but still hanging by a red cord, which was not koleroga, and the cord had proceeded from a small mat of growth that had established itself among algae, lichens and leafy liverworts on the leaf. In an Ecuador cacao plantation, where shade was furnished by jungle trees, both a cacao tree and four wild trees were seen in which cord blight was starting. In all these cases it was possible to trace points of origin, and these were at branch crotches. Such experiences indicated that cords of parasitism do not necessarily originate in jungle debris.

The slender cord fungi that blight are, in my experience, most commonly black or creamy white in color. The strands they form are not adherent all along their lengths as they grow. However, during rainy seasons and in moist locations, at intervals of about half a hand's span and less, there may be an occurrence of very slightly swollen, almost nodal appearing parts, from which many very fine fibers grow out to form an interlacing pad. The nodal-appearing parts attach themselves with these fibrous pads to moist bark or wet green stem and leaf tissue. These attachments are fairly strong. When I tested them I found that the wire-like cords will often break before the attachments give way. It is at such hold-fasts that infection takes place and if these are on leaf lamina the green color is at first intensified where the parasitism occurs and later the leaf dies.

Fig. 14-3. Bird's nest built of free hanging cords taken by birds from cord-blight fungus growths. (Secured by me in Panama jungle in January 1970.) From appearances probably about four different kinds of fungus cords were used by the nest builders. At the side is a dead twig with cord still attached that had been taken to the nest.

When cords are broken and fall to jungle litter the torn ends germinate as white mycelium.

If a cord breaks while stuck onto a twig or branch it may germinate there. In some cases I have seen these broken cords where they fell to the ground and remained without evidence of germination. A wiry, black-horse-hair blight was found causing severe disease on a mountain rain-forest tree, Micropholis garcinifolia, in Puerto Rico. There the fungus structures were vigorous in the wet and cool conditions and were collected to see if they would produce sporocarps. When taken to the warm laboratory near sea level and held in moist chambers with newly rotted leaves in the bottoms, the broken tips of fungus cords did not grow out, but were quickly invaded by bacteria that were at the same time rotting the debris. In moist chambers with moist filter paper, or old jungle trash in the bottoms, the cords developed mycelial growth at broken ends. In pure culture, growth on potato dextrose agar was strong smelling and built up mounds of mycelium but produced no sporophores. In fact, in the haphazard type of accidental collecting done by me in Puerto Rico and Costa Rica, I did not see sporophores from these cord fungi, but I did find them in Panama where the sporocarp resembled that of *Marasmius equirinus* Muell.

Thread Blighting Phenomena

There is a considerable difference between thread blights and cord organisms. In thread blights, for example, there are characteristic hymenial layers where spores develop, but no sporocarps. Neither the cord nor the thread blights are of much economic concern in the temperate zone but occur along the north edge of the Neotropics, in such U. S. states as Louisiana and Florida (12, 15, 22, 24, 37, 42, 44, 48, 53), and I collected a typical thread blight a little farther north in a sheltered cove in the North Carolina mountains. Thread blights are often devastating in effect in the tropics (4, 10, 11, 15, 22, 26, 27, 34, 35, 37, 39, 43-45, 49-51, 53) on such cultivated trees as many species of coffee, pecan, citrus (lime, lemon, orange, grapefruit, tangerine), Diospyros, Pittosporum, lichi, tung, rambutan, fig, cacao, pomegranate, apple, pear, tree tomato, jungle and shade trees, and it will also attack ferns. The causal organism is the one most commonly known as *Pellicularia koleroga* Cooke (= *Corticium stevensii*). The true thread blight fungus forms for its travel stage a light-colored, flattened, rather delicate thread, loosely adhering to stem and leaf tissue on the host it attacks. In addition it produces a thin and spreading, spiderweb-like pellicle or weft of parasitizing mycelium on the undersides of leaves which are thereby diseased. For illustration of these phases see Fig. 14-4. The toxic effect is partly revealed in the fact that diseased leafy parts become necrotic, turn black, and wither, which for some workers has given the web disease the name black rot or withertip. There are other characteristic symptoms: dead leaves attached by strands of thread are left dangling to tree stems; leaf infection must take some time, for if it

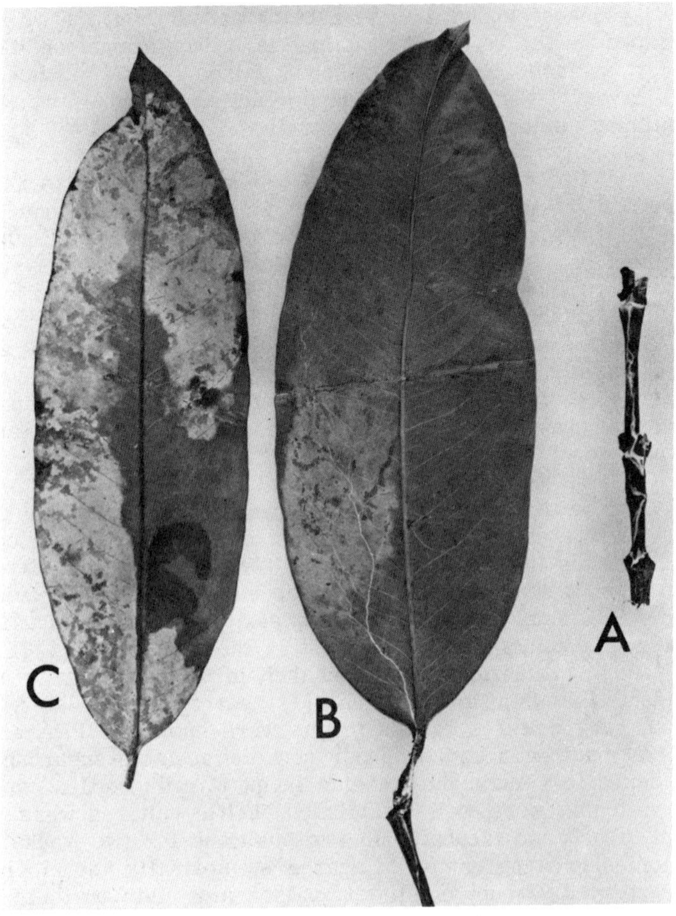

Fig. 14-4. Thread blight (Pellicularia koleroga) disease always occurs on undersides of mangosteen (Garcinia mangostana) or other tree foliage. A. Broad, adhering, travel threads, that on aging become HOLDOVER organs, of the fungus on under half of twig. B. White travel thread spreading like a fan into a sheet or pellicle of nutrient absorbing hyphae on the lower side of a leaf. C. Pellicle, no longer vegetative, transformed to a diffuse hymenial sheet bearing basidia and basidiospores. (Material collected in Panama Canal Zone.)

is old leaves that are attacked they will have time to absciss and will be cast off except for the threads tying them to the host; youngest leaves will turn black and dry out but will be retained on the stem; after a leaf is fully diseased on its under surface, but is still showing signs of life, a sheet-like layer of the fungus begins production of spores which continues as long as the leaf survives.

While working on coffee I was not ever able to accomplish the classic type of exact study on spore production that Carpenter (8) made with the closely related *Pellicularia filamentosa* (Pat.) Rogers on rubber. However, from numerous microscopic examinations it became evident to me that the rhythm of spore production by coffee koleroga at night-time, and complete cessation during the day, was almost exactly like that which Carpenter described for target board spot of rubber. Spores of coffee koleroga germinated well on maize meal agar and in some cases formed strands in culture.

The threads or mycelial strands, in nature, are of special interest from the standpoint of where they develop. The threads themselves when in a strand are in the travel stage, in which condition they proceed rapidly to the undersides of stems, petioles, and leaves. This I studied using different sources of fungus, first in the laboratory in El Salvador in 1945 and 1946, and then in Costa Rica with a student (Bianchini and Wellman (4)) several years later. When pure cultures of this fungus were grown on PDA and maize meal agar and on 2 per cent coffee-stem-sugar agar hardened to remain on one side in quart milk bottles, and the colonies were well established, bottle-cultures were then turned and incubated in two positions for two weeks. Colonies growing on top of agar grew normally, developing an expected amount of slightly raised mycelium on their surfaces. Colonies which grew hanging upside down on agar, extended about the same distance laterally as the "normal position" cultures, but in addition the fungus let fall from its colony streamers of mycelium. When *Rhizoctonia solani* Kuehn, which causes damping-off in coffee, was grown in cultures maintained in an "upside down position," the isolates did not produce mycelium to hang off the colony which developed in any way like the streamers subtending from the kolerga isolate. It was my conclusion (51) that with respect to kolerga: "The fungus has a distinct, positively geotropic character."

One of the questions often asked by planters is, how does the kolerga thread blight spread in the plantation? There have been some who believed that field spread was accomplished entirely by bits of mycelium attached to killed leaves blown about from tree to tree. Observations for years gave me no convincing evidence that this occurred. A field study was made of this in an experimental planting that was well shaded and ten hectares in extent, and started with trees all free from thread blight on the coffee.

In one corner abutting on sugarcane which is never infected, twelve young trees were marked and had put below them several branches with diseased leaves on them. In another corner again next to sugarcane, bits of pure culture of the organism were taken from tubes and placed in tree crotches. Within five days bacteria were all through the culture material, killing the fungus. In neither corner did thread blight develop during six years of observations. In a third corner of the same planting that was not disturbed, but was next to an old plantation severely infested with thread blight, the disease first appeared in the middle of the second year in two trees. In another year, six more showed disease. These were obviously infected by air-borne spores. Detailed examination showed the thread of traveling mycelium started on branches at breast height and above. The travel threads increased in breadth on reaching young stems and petioles, spread a little and appeared to grow the most rapidly when the forward-reaching fringe of tightly appressed mycelium touched the underside of a leaf and deployed, fan-wise, in its most highly destructive, pathogenic phase.

There is in numerous Basidiomycetous blights, a rotting and withering effect that is irregular and progressive. In many the moving forward of the blight is because of superficially occurring and active, eventually infecting, travel mycelia that are structured organs of parasite progress. The development of travel mycelium, is seen at times in certain rhizoctonia diseases even in the temperate zone, where it consists of slender marcoscopically visible hyphal strands (such as Wellman (51) found and illustrated on cabbage) that advance infection to nearby hitherto healthy plant parts.

A notable observation by growers about thread blight organisms is that they can be considerably affected by certain practices. For example, if koleroga blight is unusually serious in part of a coffee finca, a common practice is to abandon the worst part, remove shade trees, cut the coffee down, allow bush to grow five or six years, re-clear, and then return it to coffee or other crop production. Thread blights are not carried over in the soil, but precariously subsist as wild fungus inhabitants on jungle foliage. By breaking the life cycle, a thread blight invasion is removed and has to come again from outside the cultivated field.

In some of these thread blights, the travel organs of the fungi are shortened and almost eliminated. One of these parasitic organisms, *Pellicularis microsclerotia* (Matz) Weber, is found in many parts of the Neotropics. It is frequently on herbaceous leguminous plants and may be especially destructive in fields of beans, *Phaseolus*. It is most severe if the air is highly charged with moisture and the soil of the field never dry. This along with warmth, afternoon rains, and an abundance of inoculum, results in webs of the fungus enfolding and disasterously marching down lush and closely planted rows of a field.

On close examination, when foliage that has been disease-softened and irregularly stuck together is parted, on the injured plants is found the causal wisps of cohering cobweb-like fungus. These are on surfaces and in between stems and leaves. Moreover, in these brownish webs are soft, brown, irregular, small sclerotia like little crumbs of debris caught up in the gossamer growths.

When examined early in the day, before the morning rains set in, wisps of web on the ground and such webs over diseased stems, I found to have the dust-like bloom (on spore-producing hymenium, or pellicle of growth) of basidiospores. Under the microscope many basidia are discernible in mounts made from such material. On material collected later in the day, only in rare instances could a basidiospore be seen. All spores are thin-walled and found to be most numerous in material collected at night. As soon as rains fall during the day it is difficult to find sporulating basidia on parts of the web blighting *Pellicularia*.

Leaf Spotting Without Traveling Mycelium

Knowing of the travel mycelium phenomenon in rotting blights by Basidiomycetous fungi, it is interesting that discrete leafspots occur from closely related organisms, but

Fig. 14-5. Leaf spots, some grow large, and red; there is no thread or other such travel stage; on Liberica coffee from infections by Sclerotium coffeicola. (Material collected by R. D. Rands in Dutch Guiana.)

which do not have these travel mycelium habits. The lesions are often large, and they are rounded, with a central point. Encircling this point, one ring after another develops until a clear target-board pattern is evident, as in Fig. 14-5. In the Neotropics, a number of hosts are susceptible to somewhat similar target spot disease and some common examples are: on Hevea rubber the target leafspot, Pellicularia filamentosa (8, 21, 29, 41); on coffee the red target leafspot, Sclerotium coffeicola (2, 19, 38); and on citrus the yellow leafspot, Corticium areolatum (15, 38). On grasses and sorghum, leaf blades have regular markings in the form of bands, one succeeding another from Pellicularia filamentosa and P. (Corticium) microsclerotia (24, 26, 35, 43, 48).

It may be many years before all details about Basidiomycete target spots are completely known. There are certain steps in life histories of the causal organisms that are similar. Much of this is evident from such works as those by Carpenter (8), Stahel (38), and Echandi (13), and from lesser, but corroborative, observational studies by me. Resistant, quiescent stages of these organisms in the tropics are on surfaces and in the form of bulbils of rounded, smooth, dark-skinned, and seed-like sclerotia, or clumped and crowded-together-masses of fungus bristles, or, in some, heavy walled chlamydospore-type cells. Resistant chlamydospore-like cells are also buried inside the blackened lines of host tissue that concentrically mark a target-board spot; in these, resistant cells are most numerous in the broader band of darkened tissue that forms the perimeter of such a spot.

Quiescent cells are often dark-colored short, thick-walled, but not true spores. In this condition they are adapted to withstand effects from extremes of naturally present fungus-consuming bacteria and other microorganisms. The hyperparasitic attacks become specially numerous in moist and warm environments of the tropics, and the organisms live upon unprotected fungus growth. The quiescent cells of blighting fungi are also adapted to readily survive effects of severe short or long dry periods between wet seasons.

In the field, if a diseased leaf has a large spot located on a central or main vein, or should there be several spots on a leaf it will be dropped. If that fallen leaf becomes covered with wet debris, the parasite is hopelessly smothered. Where a dropped leaf is left uncovered, or where a spotted leaf is held to its stem, and the rains return, the chlamydospores are revivified and they germinate. Germ tubes develop hyphae which appear to take on saprophytic character, and as Sherwood (36) shows for Thanatephorous, they anastomose and so there is probably nuclear exchange and combining in cells. This is the effect apparently that is needed to incite non-parasitic, fruiting growth of the fungus onto healthy parts of the leaf or even onto soil or adjacent dead foreign matter. In the case of Corticium (= Pellicularia or Thanatephorous) there is formed, externally, a crust or papery thin structure. This is a resupinate, non-parasitic hymenium, which is known for example to form an aerial sleeve around the base of perfectly healthy coffee and cabbage (51) seedlings. On the hymenium develop tufts of spore-bearing basidia. The hymenium also forms at the edges of target spots on old coreaceous leaves, at the time the season of long rains returns. Carpenter (8) describes on rubber the sporuliferous growth as whitish "bloom" on the dorsal side of the target spots. Spores are shed from their hymenial source and become air carried to new locations on nearby hosts.

In the typically areolate (target-board) spot of citrus, Stahel (38) showed that the original leaf infection is by a basidiospore. This attaches itself by an appressorium and the germ tube develops irregular swellings which Stahel considered minute sclerotia. These in turn germinate to become parasitizing hyphae. The fungus extension from this grows out in all directions, largely at night. During the day its outer edge is slowed or stopped. Hyphae multiply and radiate, growing apace, and each night they form in citrus leaves a new water-soaked ring about 1 mm. in width. When day comes the so-called "water soaking" collapses, leaving gummed cells at the outer edge of the ring. Rings, in citrus areolate spots, normally number from 10 to 20, although Stahel saw one spot with 47 rings.

This citrus areolate spot, in the manner in which it starts at a point and extends in the leaf in ever widening rings, does not have a suggestion of a special emerging and over-growing traveling mycelium to carry the infection to another area. The spot becomes delimited; when two or a few spots start close together and then come into juxtaposition the spot centers and outlines do not disappear nor do the edges run together. Also, there are no mycelial connections to any satellite spots. In connection with target spot of rubber, the same kind of independence is found in these spore-originated spots.

Observe the spectacular, red, large, target-type spot on Liberica coffee leaves that parallels in most of its habits certain other target spots. The parasite in its "mycelia sterilis" stage is known (2) as _Sclerotium coffeicola_ Stahel. It has been studied by Icochea and Aycock (19), Echandi (unpublished work) and others. Leafspots on coffee in moist seasons bear on dorsal surfaces no recognizable spores, but gemma-like "bristles," which are loosened and airborne. These are the dissemination organs that start the spots. In this target spot, as in such on citrus and rubber, there is no travel mycelium to move the parasite in its life activities, reaching for new and greener pastures that may be nearby. The symptom of leafspot from _S. coffeicola_ is seen in Fig. 14-5.

There is a significant link, between the target-spot disease and web blight. Echandi (13) working under controlled conditions, found that basidiospore infection of common bean leaf by _Corticium microsclerotia_ was first a small, delimited, dark-bordered spot. This spot as seen in the field by me, starts as a central point and spreads outward in all directions; the target-board effect is only slightly suggested in the earliest stages. In a few days of warm moist weather, the spot expands, rotting increases, and from it the travel mycelium organ is soon developed to carry on further, more rapid coalescing, and more complete web blighting and damping-off. Some symptoms of this disease --the webbing, the start of necrosis (rotten tissues), and the myriads of small sclerotia--on pods of green beans are seen in accompanying Fig. 14-6.

Widespread Tropical Pink Disease

A ravenous disease fungus crawls over and kills branches on many tropical trees. Sometimes it appears that infections are not as frequent among weak trees as among vigorous specimens. Coffee and cacao trees maintained in the sun are more susceptible to pink disease than those crops under shade (51). This disease is characterized by a pinkish and yellow-tinged color of mycelium of the causal fungus, _Corticium salmonicolor_ Berk. & Br. The disease has been known for over a century and was a cause of serious concern long before 1873 when it was described (12, 15, 23, 26, 31, 35, 39, 40, 50). It has an extensive host range that includes almost all the tree and woody crop-plants or ornamentals we know in the Neotropics. The

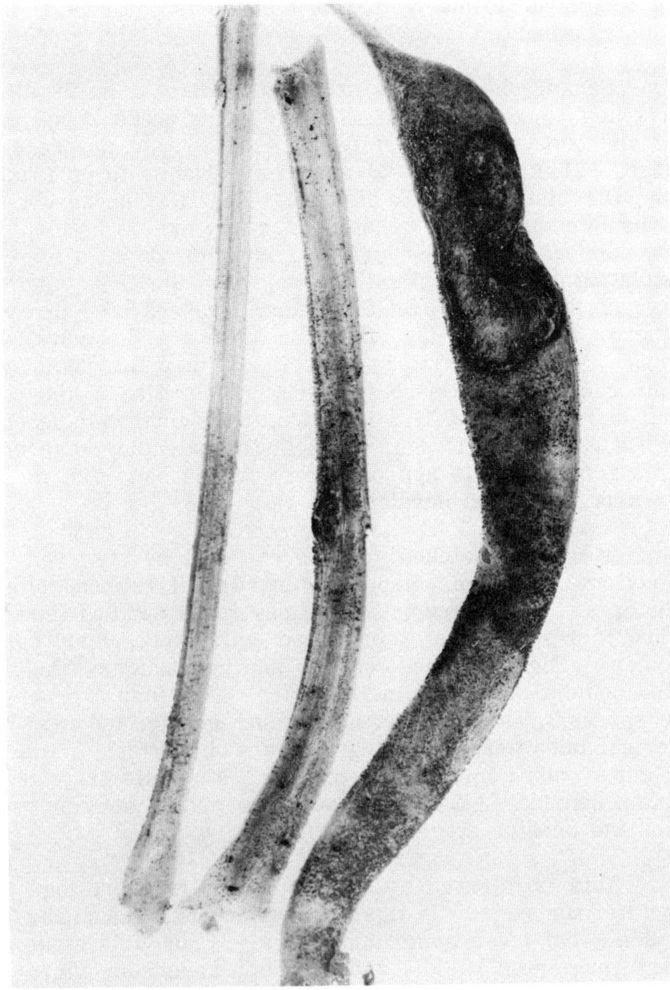

Fig. 14-6. Web blight, Pellicularia microsclerotia, of garden bean. At the right, pod wrapped with webbing liberally supplied with thousands of very small sclerotia. Notable necrosis occurring on diseased pod under the webbing. To the left of the pod, webbing and small sclerotia on petiole and peduncle of bean (leaves rotted). Photograph courtesy of Plant Pathology Department, University of Florida, Gainesville.

fungus also attacks numerous tougher herbaceous plants and even occasional climbers. This disease is pantropical in its distribution.

I made observations on the pink disease on 25 different host species. I had infected coffee trees at my back door and one year made daily notes on disease progress and change. There are several successive stages to be found in the life history of pink disease. The first occurrence of this notable disease on a coffee tree appears as a thin, white, appressed cottony growth. In a few days, a non-sporulating, silky mycelium extends from the first origination. At the periphery of this there are separated threads that grow faster than the others. These are the traveling organs. They appear to be superficial; they lift easily and do not cause actual tissue invasion. Dissection of the material shows that the first infection, the actual cell penetration, is about 5 to 15 mm. back of the extending edge of fungus growth. It is approximately at this band that the first pink coloration develops.

In closely watched growths can next be observed, in the cottony mycelium, clumps or nodules of salmon-colored structures. These are more or less loose bundles (sporodochia?) of spore-like, thin-walled cells, irregularly gathered together. These are not common and appear to be the so-called conidia spoken of in literature. But they are not true spores. They are very irregular in size and may be spherical but where close together are misshapen. These are more truly gemmiferous organs. It is this stage that is often considered the imperfect form and is sometimes called the necator stage.

This is followed by a more nearly true fruiting, or sporuliferous stage. In this, a crustaceous, resupinate and somewhat tough hymenium forms. From this structure emerge numerous basidia and, as seen under the microscope, on each of these are borne four elongated to pear-shaped basidiospores. It is approximately in this condition, if there is sufficient dry weather, that the salmon-colored fungus growth often cracks in characteristic manner at right angles across the branch or twig it is parasitizing.

In studies of diseased coffee and orange in Costa Rica, a certain amount of pink fungus growth was evident, spreading under suberized bark and developing at first small mats. Such mats became the irregularly rounded masses or clumps of thick-walled mycelium, almost chlamydospore-like in appearance. These clumps I watched serve as gemmae that are loosened and germinate promptly in dew. This is the stage that serves as the holdover condition of pink disease on coffee.

While the situation has not been such that any well-planned, long-time research project could be carried out, I have seen the pink disease in 14 tropical countries of America, in Africa and the Orient. One of the problems relating to this disease is its means of dissemination. In personal observations I have not been able to find the necator stage very commonly. It was early concluded that the so-called conidia, at least under the conditions of study, often do not germinate and are not an important means for spread of the fungus.

During studies on the gemmiferous stage of the pink disease, it was found that these bodies are readily detached from the fungal hyphae that bear them. In addition the gemmae even if dried for a day germinate readily in dew. They were moved from one branch to another by rain, and thus spread the disease in the tree. Field observations indicate the gemmae are an important means of distribution from one tree to another. However, there seems every reason to believe that the main method of dissemination is through the basidiospores of C. salmonicolor. These are most abundant at night on mature fruiting hymenia.

There are doubtless different races of this parasite. In one Central American country a large orange grove was visited at the owner's request, because of its decline condition. The difficulty was largely due to pink disease that affected 26 per cent of the best trees. Surprisingly, in the "pink disease" there were some trees which had an otherwise typical but totally white growth of the fungus. A diseased branch with the white growth was secured and when it was examined under the microscope, the fungus was identified as unquestionably C. salmonicolor. I consider it probably a white mutant of the parasite.

One of the characteristics most notable in the literature study of this disease is its wide host range; the fungus seemingly attacks almost anything if it is woody, the fungus has the proper chance, and conditions are favorable. However, in one case, in my laboratory, a typical and morphologically non-varying isolate in pure culture was secured from badly diseased sun-grown coffee. When tests were made using this isolate to infect orange tree twigs, a very scanty, weak, and non-fatal growth on green twigs developed. When used at the same time to infect coffee, the twigs were readily invaded and died.

Rhizoctonia Damping-off

[Note: the startling lack of mycologists--and the serious need for them--in the Neotropics is realized by pathologists studying Rhizoctonias that are blight or rot-causing parasites. A pathologist often finds that isolates are difficult to name with accuracy outside of lengthy and time-consuming application of specialized mycological techniques. Of this the diagnostician and control consultant is painfully aware when he cannot make the special studies but sees in literature relating to his problems, references to such generic fungus names as *Botryobasidium*, *Ceratobasidium*, *Hypochnus*, then *Corticium*, or *Pellicularia*, and of late years *Thanatephorus*, for *Rhizoctonia*.

[In above sections in this chapter I have continued to use certain names which are standard in numerous texts. It is recognized that, mycologically speaking, organism names of some web, thread and damping-off blights are in a state of change. In our tropics conservative practices are such that it is still understandable and practical to retain some of the old names.

[The convenience based on old names, in referring to past and present studies, is not to be understood as in disregard of the painstaking mycological conclusions still in the process of coming to light: see such works as Talbot (43), Donk (12), Flentje (16), Flentje and Stretton (17), and others. It is clear that the diseases with *Rhizoctonia* spp. as the causal agents are varied. However, once the fungi are induced to sporulate they are more and more being brought together under the single binomial *Thanatephorus cucumeris* (Frank) Donk. Nevertheless, the difficulty with investigators in the tropics is still how to provide scientific designations, mycologically perfect, for such a variable group (20) as: the *Thanatephorus cucumeris* so much found as a basal infection on rice and other grasses; the *T. cucumeris* characteristically associated with the target leafspot on Hevea rubber or another areolate (target-patterned) leafspot of citrus; the *T. cucumeris* group that deposits scurfy scabs on such plant storage organs as potato tubers and that will not cause cabbage seedling wirestem disease; or the widely known T. cucumeris organisms that infect seedlings causing wet decay and death. On the mycelial stage this group is still often designated as Rhizoctonia solani. In any case there is a good body of literature from the tropics on effects and limiting factors of the disease it causes (for examples: 1, 3, 5-7, 9, 11, 14, 16-18, 23-25, 29, 32, 43, 46, 47, 49, 51, and 52).]

A soil-inhabiting primitive member of the Basidiomycete parasites related to web and thread blight organisms is easily isolated and often called Rhizoctonia solani (= Corticium (Pellicularia) filamentosa). As found in the Neotropics, and as cultured on PDAgar under ordinary laboratory conditions it is usually reasonably fast-growing, non-sporuliferous, but causing seedling blight collapse. The hyphae of the organism are brownish in color and as is well known, they infect tender seedlings of a large number of crops. The result from this infection is so well known that it hardly needs much description. There is a darkening and water-soaked appearance of seedling stems at the soil line. These diseased stems, where close together, have a very fine tangle of hyphal strands that bind the mass and when torn apart a moist appearance gives it the descriptive name of "soupy" or "damping-off." (The designation "soupy" when used in Latin America, is the basis for one of the common names there, "sancocho," for this disease.) This kind of collapse and rot does not occur on old plants.

To the pathologist, this disease is known from literature to be widely distributed, existing in fact in a large part of the arable lands on earth but more severely in warm moist soils. In the temperate zone there are some crops such as certain trees, and some exotics, and others, on which it has never been reported. In the Neotropics no one knows if there are any crops whose seedlings are resistant, and there is the liklihood that as a crop is more and more intensively grown, sooner or later it will have damping-off reported on it.

So long as crops are scatteringly grown in an extensive manner rather than in an intensive manner, the losses from damping-off are much less than one might expect; so long as the main means of cultivation is by hand, workers can eliminate diseased ("bad-looking") seedlings, and damping-off is taken care of more readily than if the cultivation were by machinery. Another thing that makes it less notable in the tropics is because initially the disease occurs only in occasional seedlings readily removed by laboring fingers. The rest of the seedlings are then free to grow unthreatened. This kind of disease on seedlings may act more as a thinning process than almost anything else in the tropics. However, in fields of repeated plantings, where it has been observed closely in such crops as beans, squash, melons, and peppers, it has been found that field damping-off, the killing of young seedlings not in seed beds but in well-separated rows in the field, is also prevalent. If all were given careful attention and good data secured of those seedlings that were killed, in some areas it would be found to be a very serious and important feature in disease losses.

In several series of isolations in material from Panama, Costa Rica, and El Salvador, surveying the cause of sancocho in seedlings of coffee, tobacco, and tomato, I isolated from these dying seedlings many different kinds of organisms. Some of them were of course typical of what might be expected. It was clear the most common occurrence was of *Rhizoctonia*; next, *Pythium* and *Phytophthora*; and in some cases, *Sclerotium* (*rolfsii?*). Isolations further demonstrated that there were at least four different species of Fusarium involved in some of these sancocho disease areas. Some of the organisms that by themselves caused sancocho were unexpected. For example *Nigrospora* is occasional in seedling collapse of tobacco. A *Penicillium* infects young cotyledons and it progresses and results in damping-off in coffee seedlings. Another organism involved in seedling collapse of both coffee and tomato is a *Pestalotia*. In a case of bean damping-off the organism isolated was a *Phoma*. In two instances *Diplodia* was isolated from dying tomato seedlings. However, in by far the largest number of instances, as indicated above, the most common and important organism is *Rhizoctonia* = *Thanatephorous*.

Where rains are frequent and heavy and there is a poorly defined dry season, the soil-inhabiting *R. solani* does not seem to grow as well and produce as much disease as where alternating with the wet there are long dry seasons. On the other hand, desert conditions do not allow *R. solani* to survive.

There are a number of natural enemies or hyperparasites of the *Rhizoctonia* that causes damping-off or sancocho in the tropics. One hyperparasite isolated with some frequency is *Trichoderma* sp., which in my studies was found mostly in moist regions. Another fungus that is an evident *R. solani* parasite is *Choanephora* sp. Bacteria of undetermined kinds are also inhibitors of *R. solani* in culture and in nature.

In some areas I have seen certain nematodes that have been feeding on *R. solani* in soil and in tobacco and in coffee seedlings. The nematodes observed seemed to be only a very few at certain times. However, after severe sancocho attack, rotted tobacco and coffee seedlings when examined under a dissecting microscope exhibited nematodes (undetermined types) feeding on the fungus in great numbers.

There are times when *R. solani* is evidently a parasite on other fungi. This has been carefully observed in the temperate zone (5). Probably it occurs much more commonly than suspected in the tropics. I have been much impressed in finding *R. solani* in cultures attacking and growing over several types of Basidiomycetous organisms. In two cases *R. solani* from soil wrapped around hyphae of what appeared to be another *R. solani* but which had been recovered from a rotting coffee seedling hypocotyl. *R. solani* has also been seen in my laboratory enveloping hyphae of two other fungi, both undetermined species of *Pythium* and on hyphae of *Phytophthora palmivora.* It was of special interest when a culture of *R. solani,* isolated originally from a seedling of kenaf that was rotted at the soil line, was accidentally placed in a culture plate with a *Pellicularia* non-sporuliferous type of organism that caused bean leaf web blight. In this case the soil organism invaded the more delicately growing leaf organism at its colony edge and eventually grew over the apparently less vigorous culture of the leaf fungus.

Five So-called "Pellicularia" Diseases

In work carried out in Costa Rica by Bianchini and Wellman (4) comparisons were made of five diseases, all considered as incited by species of *Pellicularia.* This genus name was following the criteria given by Rogers (29) and confirmed through comparisons of descriptions in Neotropical literature such as in: 8, 12-16, 27, 41, 43, 44, 48-53. The "Pellicularia" group of fungi have not been well studied

Table 22. CERTAIN AERIAL AND SOIL-LINE BASIDIOMYCETOUS BLIGHTS from recumbent parasitic organisms occurring in the American tropics.

Kind of blight	Typical causal organism(s)*	Part of host attacked	Examples of hosts
Black wiry cord	*Marasmius equirinus* Berk. (A) 37 *M. viegasii* Singer (A) 37	Leaves, twigs and small branches	Cacao, coffee, hevea, jungle trees
White wiry cord	*M. pulcher* (Berk. & Br.) Petch (A) 37	same as above	Cacao, tea, jungle
Thread blight	*Pellicularia koleroga* Cooke (= *Koleroga noxia* Donk) (H) 43	Leaves and petioles and small stems	Coffee, cacao, ferns, other trees
Web blight	*Pellicularia microsclerotia* (Matz) Weber (H) 43	Leaves and petioles	Beans, hibiscus
Targetboard spot	*Sclerotium coffeicola* Stahel (G) 37	Leaf lamina	Liberica and other coffee
Halo spot	*Corticium areolatum* Stahel (H) 37	Leaf lamina	Citrus species
Pink disease	*Corticium salmonicolor* Berk. & Br. (G, H) 15	Twigs and branches	Wide host range

Damping-off	Pellicularia filamentosa (Pat.) Rogers (= Rhizoctonia solani Kuehn) (= Thanatephorous cucumeris (Frank) Donk) (H) 12, 16	Seedling tissue at soil line	On almost all seedlings
Neck rot	Sclerotium rolfsii Sacc. (= Pellicularia rolfsii (Curzi) West (G, H) 2	Ground rot of various aged tissues at ground line	Herbaceous plants, occasional on seedling trees
Rot at base of tree	Clitocybe tabescens Bres. (A) 30	Decay of trunk at crown, at and below ground line	Casaurinas, many other hosts
Rot at base of herb	Rhizoctonia violaceae Tul. 33	Decay at ground line	Alfalfa
General root rot	Marasmius sacchari Wakk. (A) 37	Decay of roots below ground line	Sugarcane, banana

*Organisms that sporulate on hymenium marked (H), those that develop gemmae marked (G), those with sporocarps that are agarics marked (A). The number after each letter gives a citation that may be referred to for something about the mycology of the organism.

in our tropics, for the good reason that both mycologists and pathologists are still in short supply in the region.

The "Pellicularia" diseases are designated here using both the name published by the authors and in the second place sometimes a more recent designation. All the diseases are readily encountered in a wide mountain valley in Costa Rica. They are: 1. Coffee thread blight, *Pellicularia koleroga* Cooke = *Koleroga noxia* Donk; 2. Cacao thread blight, probably *P. koleroga* but designated more often as *Pellicularia filamentosa* = *Thanatephorus cucumeris*; 3. A bean web blight *Pellicularia microsclerotia* (Matz) Weber; 4. Crotalaria web blight *P. filamentosa* = *T. cucumeris*, somewhat more scanty in parasitic growth than the bean disease, but probably the same causal organiam in both cases; and 5. the ornamental bush hibiscus foliar attack, caused by *Pellicularia* sp. that has some similarities to both the thread and web blights.

In culture studies that I have carried on (51) in El Salvador long prior to work in Costa Rica, it was observed that the koleroga organisms on coffee appeared to grow in a positively geotropic manner when growing fungus mats were suspended "upside down." Both physical and physiological phenomena in connection with positive geotropism await study with this organism. This apparent positive geotropism was noted again in the culture studies published by Bianchini and Wellman (4); see above on page 572.

The "Pellicularia" infecting the ornamental hibiscus bush is of special note. Mycelial growth is finer and under the microscope hyphae are about half the width of those either from thread blights or from web blights. It grows well in culture and the isolate studied had a habit of producing numerous anastomoses. These were so typical and common that it was considered during our studies that it might have special genetic significance. An interesting observation on this organism is that it has been seen on three different occasions, acting as a serious parasite of a tree cricket, *Idiarthron (atrispinum?)*, that feeds on coffee and as well on hibiscus foliage.

Table of Basidiomycetous Blights

On pages 586-7, Table 22, of blights that are in the group briefly discussed in the present chapter, illustrates some of the things general pathologists find of special interest in the tropics. In temperate zone work, training, reading, and research deals only rarely with blights caused by Basidiomycete fungi of recumbent and hymenial character. When it does, it is mostly with two diseases, often called "Rhizoc damping-off" and "Rolfsii ground-rot." And these are of unusual severity only under high temperature and moisture conditions; uncommon in much of the temperate zone.

It seems to me that presentation of a whole gamut of differences in this group of organisms, while it may be interesting, may also indicate features of which there is still much to be learned.

Through fortunate circumstances it has been possible to collect, observe, make identifications, and otherwise carry on mostly short-time studies on all the organisms here mentioned, except Marasmius viegasii and Corticium areolatum. Isolations have been made of other Marasmiuses, the Pellicularias, Sclerotium, the Corticiums, and Thanatephorous. Inoculations were made with pure cultures but at no time was a cord secured from isolations from an aerial Marasmius. As anyone would know, infections could be easily obtained under proper conditions from S. rolfsii, the Pellicularias, and C. salmonicolor. Koleroga thread blight was difficult to manipulate so it would develop a thread adapted to living on a coffee branch, but with attention it was accomplished.

Literature Cited

1. Alvarez-G., L. A. 1939. A mahogany seedling blight in Puerto Rico. Caribbean Forester 1:23-24. --. 1946. The control of Rhizoctonia damping-off of pepper and eggplant in Puerto Rico. Univ. Puerto Rico J. Agr. 30:69-96.

2. Aycock, R. 1961. Symposium on Sclerotium rolfsii: summation. Phytopathology 51:107-108. --. 1966. Stem rot and other diseases caused by Sclerotium rolfsii, or the status of Rolfs' fungus after 70 years. North Carolina State Univ. Agr. Exp. Sta. Tech. Bulletin 174, 202 p.

3. Bateman, D. F. and A. W. Dimock. 1959. Influence of

temperature on root rots of poinsettia caused by Thielaviopsis basicola, Rhizoctonia solani, and Pythium ultimum. Phytopathology 49:641-647.

4. Bianchini P., C. L. and F. L. Wellman. 1958. Experimentos en el control de Pellicularia del cafe y ciertas diferencias en pelicularias de cinco huespedes. Turrialba 8:73-92.

5. Butler, E. E. 1957. Rhizoctonia solani as a parasite of fungi. Mycologia 49:354-373.

6. Boosalis, M. G. 1956. Effect of soil temperature and green-manure amendment of unsterilized soil on parasitism of Rhizoctonia solani by Penicillium vermiculatum and Trichoderma sp. Phytopathology 46:473-478.

7. Cardona A., C. 1954. Pudriciones fungosas radiculares del frijoles (Phaseolus vulgaris L.) en el valle de Medellín, Colombia. Rev. Fac. Agron. Medellín, 15:137-209.

8. Carpenter, J. B. 1949. Production and discharge of basidiospores by Pellicularia filamentosa (Pat.) Rogers on hevea rubber. Phytopathology 39:980-985. --. 1951. Target leaf spot of the hevea rubbertree in relation to host development, infection, defoliation, and control. U. S. Dep. Agr., Tec. Bulletin 1028, 34 p.

9. Carrera, C. J. M. 1951. La enfermedad de las almacigas de coniferas en la República Argentina. Fac. Agron. Vet. Buenos Aires, Boletin 30, 51 p.

10. Dennis, R. W. G. 1951. Some Agaricaceae of Trinidad and Venezuela. Leucosporae. Part I. Trans. British Mycol. Soc. 34:411-482.

11. Dongo D., S. L. 1961. Estudio de la chupadera fungosa del frijol. Estac. Exp. Agr. La Molina, Peru, Boletin 1, 15 p.

12. Donk, M. A. 1941. Nomina generica conservanda and confusa for Basidiomycetes (fungi). Bull. Bot. Gard. Buitenzorg. Ser. 3, 17:155-197. --.

1954. Notes on resupinate Hymenomycetes--1. On *Pellicularia* Cooke. Reinwardtia 2:425-434. --. 1957. The generic names proposed for Hymenomycetes--VII. Thelephoraceae. Taxon 6: 103-123. --. 1958. Notes on resupinate Hymenomycetes--V. Fungus 28:16-36.

13. Echandi, E. 1965. Basidiospore infection by *Pellicularia filamentosa* (= *Corticium microsclerotia*) *the incitant of web* blight of common bean. Phytopathology 55:698-699.

14. Exner, B. 1953. Comparative studies of four *Rhizoctonias* occurring in Louisiana. Mycologia 45: 698-719.

15. Fawcett, H. S. 1936. Citrus diseases and their control. McGraw-Hill, New York, 656 p.

16. Flentje, N. T. 1956. Studies on *Pellicularia filamentosa* (Pat.) Rogers. 1. Formation of the perfect stage. Trans. British Myc. Soc. 39:343-356.

17. Flentje, N. T. and H. M. Stretton. 1964. Mechanism of variation in *Thanatephorous cucumeris* and *T. praticolus*. Australian J. Biol. Sci. 17:686-764.

18. Frezzi, M. J. 1947. Contribución al estudio del damping-off o enfermedad de los almacigos en la República Argentina. Min. Agr., Buenos Aires, Publicación Miscl. Ser. A, 3, 40 p.

19. Icochea, T., and R. Aycock. 1965. Comparative studies with *Sclerotium coffeicolum* and *S. rolfsii*. Phytopathology 55:1063.

20. Kerncamp, M. F., D. J. de Zeevw, S. W. Chen, B. C. Ortega, C. T. Tsiang, and A. M. Khan. 1952. Investigations on physiologic specialization and parasitism of *Rhizoctonia solani*. Univ. Minnesota Agr. Exp. Sta., Tec. Bulletin 200, 36 p.

21. Kotila, J. E. 1945. *Rhizoctonia* foliage disease of *Hevea brasiliensis*. Phytopathology 35:739-741.

22. Large, J. R., J. H. Painter, and W. A. Lavis. 1950. Thread blight in tung orchards and its control.

Phytopathology 40:453-459.

23. Marchionatto, J. B. 1940. La *R. solani* y la *R. crocorum* en la República *Argentina*. J. Agr. Vet., Argentina 1939:327-333. --. 1944. Manual de Las Enfermedades de Las Plantas. Editorial Sudamericana, Buenos Aires, 368 p.

24. Matz, J. 1917. A *Rhizoctonia* of the fig. Phytopathology 7:110-118. --. 1921. Una dañina de la habichuela. Puerto Rico Insular Exp. Sta., Circulár 57, 8 p. --. 1921. The Rhizoctonias of Puerto Rico. Puerto Rico J. Dep. Agr. 5:2-30.

25. Nolla, J. A. B. 1932. The damping-off of tobacco and its control in Puerto Rico. J. Dep. Agr. Puerto Rico 16:285-324.

26. Nowell, W. 1916. Pink disease in the West Indies. Agric. News, Trinidad 15:238-239. --. 1926. Diseases of Crop-Plants in the Lesser Antilles. West India Committee, 325 p.

27. Orellana, R. G. 1955. Cacao diseases in Mexico, Nicaragua, Costa Rica and Jamaica. F. A. O. Plant Prot. Bull. 4:35-37.

28. Petch, T. 1947. A revision of Ceylon *Marasmii*. Trans. British Mycol. Soc. 31:19-44.

29. Rogers, D. P. 1943. The genus *Pellicularia* (Thelephoraceae). Farlowia 1:95-118.

30. Rhoads, A. S. 1952. The destructiveness of *Clitocybe* root rot to plantings of casaurinas in Florida. Lloydia 15:161-184.

31. Rorer, J. B. 1910. Pink disease of cacao. Bull. Dep. Agr. Trinidad and Tobago 15:86-89.

32. Ruppel, E. G., D. K. Barnes, and A. Santiago. 1965. Potential of three *Rhizoctonia* species to incite foliar blight and damping-off on *Tephrosia voglii* in Puerto Rico. Phytopathology 55:612-614.

33. Schultz, E. F. 1919. La *Rhizoctonia violacea* en los

alfalfares de Tucumán. Rev. Ind. Agr. Tucumán, Argentina 10:154-162.

34. Seaver, F. J. 1944. The horse-hair fungi. Mycologia 36:340-342.

35. de Segura, C. B. 1965. Enfermedades de Cultivos Tropicales y Subtropicales. Editorial Jurídica S. A., Lima, Peru, 439 p.

36. Sherwood, R. T. 1969. Morphology and physiology in four anastomosis groups of _Thanatephorus cucumeris_. Phytopathology 59:1924-1929.

37. Singer, R. 1946. Type studies on Agarics II. Lloydia 9:114-131. --. 1962. The Agaricales in modern taxonomy (2d. ed.). Hafner Publ. Co., New York, 915 p. --. 1964. Monographic studies on South American Basidiomycetes, especially those of the east slope of the Andes in Brazil. 2. The genus _Marasmius_ in South America. Sydowia 18:106-358.

38. Stahel, G. 1921. De Sclerotium-ziekt van de Liberiakoffie in Suriname. Dep. Landbu. Suriname, Bulletin 42, 29 p. --. 1940. _Corticium aureolatum_, the cause of the aureolate leaf spot of citrus. Phytopathology 30:119-130.

39. Stehlé, H. 1954. Quelques notes sur la botanique et l'écologie végétal d l'Archipel des Caribes. J. Agr. Trop. Bot. Appl., France 1:71-110.

40. Stevenson, J. A. 1916. Pink disease of citrus. _Corticium salmonicolor_. Puerto Rico Progress, April 10, p. 4.

41. ________., and J. B. Carpenter. 1950. Herbarium records of _Pellicularia filamentosa_ on hevea rubber. Plant Dis. Reptr. 34:70.

42. Stroube, W. H. 1954. Host range of the _Rhizoctonia_ aerial blight fungus in Louisiana. Plant Dis. Reptr. 38:789-790.

43. Talbot, P. H. B. 1954. Micromorphology of the lower Hymenomycetes. Bothalia 6:249-299. --. 1958.

Studies of some South African resupinate Hymenomycetes. Part II. Bothalia 7:131-187. --. 1965. Studies of Pellicularia and associated genera of Hymenomycetes. Persoonia 3:371-406.

44. Tims, E.C. and F. Bonner. 1942. Studies of fig leaf blights. Louisiana Acad. Sci., Proc. 6:13-34.

45. Toro, R.A. 1929. Cultivos--una epitofita del café en Cúcuta. Rev. Cafetera Colombia 2:171-173.

46. Tucker, C.M. 1926. La enfermedad de la raíz del cafeto en los semilleros. Rev. Agr. Puerto Rico 13:129-131.

47. Viégas, A.P. 1957. Podridão das raízes do cafeeiro. I, II. Bol. Supt. Serv. Café S. Paulo 32(368):7-16; (369):11-19.

48. Weber, G.F. 1931. Web-blight, a disease of beans caused by Corticium microsclerotia. Phytopathology 29:559-574. --. 1963. Pellicularia spp. pathogenic on sorghum and pearl millet in Florida. Plant Dis. Reptr. 47:654-656.

49. ________., and L. Abrego. 1958. Damping-off and thread blights of coffee in El Salvador. Plant Dis. Reptr. 42:1378-1381.

50. Weir, J.R. 1926. A pathological survey of the para rubber tree (Hevea brasiliensis) in the Amazon Valley. U.S. Dep. Agr., Dep. Bulletin 1380, 129 p.

51. Wellman, F.L. 1932. Rhizoctonia bottom rot and head rot of cabbage. J. Agr. Res. 45:461-469. --. 1953. Some important diseases of coffee. pp. 891-896. In Plant diseases, yearbook of agriculture 1953, Washington, D.C., 940 p. --. 1961. Coffee botany, cultivation and utilization. Leonard Hill, London; Interscience Publ, New York, 488 p.

52. ________., J.C. Walker, A. Cook, and M. Gallegly, Jr. 1951. Effects of temperature on vegetative growth of five coffee disease fungi. Asamblea

Latinoam. Fitopar. Mexico, Proc. 1:251-257.

53. Wolf, F. A., and W. J. Bach. 1927. The thread blight disease caused by Corticium koleroga (Cooke) Höhn. on citrus and pomaceous plants. Phytopathology 17:689-709.

Chapter XV

FOLIAGE DISEASES BY AGARIC AND GEMMIFEROUS FUNGI

Summary

Two of the most spectacular and classic tropical plantation tree diseases are the Marasmius-caused witches' broom of cacao and the Mycena-caused American leafspot of coffee. On cacao the witches' broom causes obnoxious broom-like growths on branches and also infects flowers and fruits; infection is in very young plant tissues by air-borne basidiospores whose germ tubes enter stomata. The perfect stage of the witches' broom fungus is a Basidiomycete. The brooms produce many lavender-colored agarics, and from them come millions of spores. The host range of this disease is narrow; it appears only on a number of species of trees in the family Sterculiaceae. On coffee the American leafspot causes serious leaf spots and defoliation; infection is by rain-splashed gemmae, or infection bodies, which penetrate directly into the epidermises of leaves, of fruits, and in young stems that are not protected with suberized bark. The perfect stage of the fungus is a Basidiomycete, and in special conditions are produced small lemon-colored agarics. Although these are not important in disease spread, the gemmae are. The gemma is many times larger than a spore, is heavy (for its size it is heavier than a hen's egg), and in nature is only splashed-off onto new locations when wet weather arrives. As a well known leaf-spotting organism it is unique in being luminescent. Study of this luminescence disclosed that it is bright when conditions are optimum for growth; its brightness is dulled by cooling or by heating; an extreme brightness is notable when cultures are treated with stimulating chemicals, and conversely reduced by numbing narcotics; it is specially brightened when cultures are contaminated by certain fungi--bacteria will dull luminescence. The fungus has an unusually wide host range, from mosses to asters. The gemma of this fungus is a special body and similar structures occur with numerous other fungi and higher plants; it is probably an archaic means of distribution that may have later evolved into a spore in fungi.

Introduction

When pathologists from the American tropics gather in conventions among themselves for much needed comparisons of results, discussions about the witches' broom of cacao and the American leafspot of coffee are nearly always on the program. If no formal papers on discoveries or control measures are presented about these diseases, there are always those outside the auditorium asking advice and others helping with findings. In the eastern tropics, far from the Americas, there is fear of the cacao witches' broom and the coffee leafspot. In Africa and in the Oriental tropics, strict quarantines, enforced both by governments and by intelligent growers, have thus far restricted these diseases to the Western Hemisphere, i.e., the Neotropics.

The witches' broom and the leafspot that are the subjects of this chapter are, like some other tropical plant diseases, confined to tropical areas where there is much moisture, where temperatures are moderately warm, and where shade conditions obviate effects of excess sunlight and desiccation. In the case of the susceptible crops, the conditions are such that the ideal factors for disease readily parallel conditions that are ideal for the crops. Where infections from these diseases on the favored crops are common, there is no profitable harvest without the employment of adequate disease control measures.

These diseases in the tropics (as important in some cases as rusts and wilts are in the temperate zone), leafspot and broom, where common, have been of such consequence in the development of coffee and cacao plantations, that the growers, and the pathologists keeping the plantations productive, spend their lives in the study of them. In fact, cacao growers and coffee planters always take into account these important diseases; cacao producers realize that if witches' broom is present and uncontrolled, plantations will be complete failures. Coffee men know the repeated histories of the leafspot where it is severe and uncontrolled: driving out fincas and destroying agricultural prospects of great sections of a country. Important as these two diseases are in the tropics, it is a remarkable oversight that they are given so little attention in general world phytopathological literature.

It is recognized among tropical pathologists that foliage diseases (stem and leaf spots and rots, and in a few instances multiple twig developments as well as rare localized mild swellings), are from certain fungi of the Ascomycetes, some of the fungi imperfecti, some of the Basidiomycetes, many rusts, and some of the thread-web blights and their relatives. It is unusual in any part of the world to find fungi that, on causing disease, develop as part of the syndrome the growth of mushrooms on diseased tissues. In this chapter are discussed two such diseases from Basidiomycetous organisms: *Marasmius perniciosus* Stahel, the cause of cacao witches' broom; *Mycena citricolor* Saccardo, the cause of a coffee leafspot. In the former occurs monstrous overgrowths; in the latter is the wierd luminescence of diseased spots. The two diseases are nothing less than unique in the annals of plant pathology.

Foliage Monstrosities and Lavender Agarics

The most disturbing experience a knowledgeable grower of the cacao, or chocolate, tree (*Theobroma cacao* L.) can have in the American tropics is discovery in his plantation of the disease called witches' broom. This disease has been known since 1895 (5, 13). Its natal home is the area of the north-central part of the great drainage basin of the Orinoco and Amazon river systems; the natural habitat is in low wet jungles of South America. At first it spread rather slowly; it required 80 years before it surmounted the Andes with man's help and settled in Ecuador. It is now present in Brazil, Bolivia, Colombia, Surinam, British Guiana, Ecuador, Peru, Venezuela, Trinidad, Tobago, and Grenada. Except for the last three named islands, it does not occur in the West Indies, and it is not in Central

America, or in any of the Old World tropics. For some of the reports and observations on it see citations 3-8, 16-22, 28, 29, 32, 35, 41, 45, and 47.

First symptoms seen by planters are the ugly, obnoxious and tumescent, broom growths (see the accompanying Fig. 15-1, A and B) that are the disease on young branch ends. Instead of the usual, normal slender branch with its large, alternately arranged, broad green leaves there is a discolored broom. This is a monstrosity of multiple stems, there may be 20 or more, crowded together; sickeningly fat, short and flabby twigs that are tawny to brown in color. Leaves on the brooms are haphazard in development, often wholly absent. The broom may have a fairly well grown leaf or two at its base, but in many cases its many swollen twigs retain their stipules, which normally are soon dropped in healthy stem growth. Some stipules develop as strap-like structures unhealthy in color and unhealthy in texture or shape. (All effects of unidentified toxins.)

Witches' broom also attacks cacao flowers. If disease-free, they are pinkish and held on slender nodding peduncles. When infected they become discolored, stiff, tough, and awkward appearing, often star-shaped things that die. If fruit has set and becomes infected, the resulting pod is lumpy and hard, it may be small and carrot-shaped with a blunt tip, perhaps oxheart-shaped--sometimes it is markedly irregular in shape--and it may be a quarter- to half-sized but excessively hard (called indurated); a pod that is invaded late in its growth may have some of its "beans" left inside that are still usable, but this is rare. The diseased fruits (Fig. 15-2) are generally all considered a total loss.

On a lateral branch, a broom appears like the loathsome, upward-curving claws of some animal no one has ever seen. Numbers of brooms on a tree depend upon tree size and how many branch tips it has. I have counted 230 on a single large tree in a well grown plantation. The result is a catastrophic reduction in phytosynthetic leaf surface. I worked in plantings where no crop had been harvested for years because of the disease, and others (18) know it can destroy all productivity.

Sometimes knowing about the changes in scientific names is important, as, especially in plant disease, they may indicate historic steps in better understanding of the

Fig. 15-1. Symptoms of the witches' broom, or escoba de bruja disease of cacao: A. Still growing, ugly tip of an affected branch; the normally dormant adventitious buds push out as sick, fat, and stubby growths, a few misshapened leaves are partially developed and stipules grow excessively. Photographs courtesy of E. Ampuero, Ministry of Agriculture and National Cacao Experiment Station, Guayaquil, Ecuador.

Fig. 15-1: B. A multistemmed branch tip typical of the final development of the disease; necrosis of all tissues has spread into all stems and leafy parts.

Fig. 15-2. *Marasmius perniciosus* attack of cacao pods. Upper row: If infection is just before last fruit meristematic tissues mature, pods are somewhat smaller than normal; as they age they become necrotic and valueless. Lower row: Infection of meristematic tissues of newly forming fruits causes marked stunting, malformation, and hardening; necrosis develops and pods are of no use.

causal organism; the parasite that causes witches' broom of cacao is an example. When the abnormal growth was given detailed study, before and at the beginning of this century, there was no certainty about its incitant. Early mycologists in 1901 and 1902 thought but could not prove it to be due to an *Exoascus.* The work of Van Hall and Drost (43) was hailed as clarifying, for they found one of the imperfect fungi always present, and in 1908 it was named *Colletotrichum luxificum.* To this day--and I know this very well from my own troubled experiences--*Colletotrichum* spp. are common contaminants that easily overgrow slower acting tissue invaders. This is specially true in cacao witches' broom. The organism that certainly was the causation of the cacao brooms, was found by Stahel (41) a few years after Van Hall and Drost, and in 1915 he described the Basidiomycete as *Marasmius perniciosus.* His studies and observations were confirmed in different places. Acceptable as his determination has been, mycologists have had some questions about it (5, 14, 15, 39, 41) and Singer was forced both by considerations of others (14, 15) and his own findings to re-classify this organism as *Crinipellis perniciosus* (39). However, a main body of researchers still retains, for reasons of "nomina conservanda," the name Stahel long ago gave it, and in this book the older designation is used. The agaric stage of the fungus is illustrated in Fig. 15-3.

The curious appearance of witches' broom has attracted the attention of experts who have studied it since well before the beginning of the present century (examples in citations: 3-6, 19, 26, 32, 41, 47). At first there was confusion as to how *Marasmius perniciosus* attacked its host. Isolations can be made from diseased cacao tissue and from parts of an agaric. The fungus, as pure culture, grows on agar media, but bits of the mycelium do not infect the host when artificially placed in the usual manner on wounded and uninjured suberized bark of stems, or on leaves or on dormant fruiting trunk cushions. It is likewise true that even when spores (from agaric caps) are used as inoculum on such tissues that have hardened, they cause no disease.

Using spore suspensions on very young epidermis gives positive results. As shown by several researchers (examples: 3-7, 13, 19, 41), the oval spores which average about 4 to 5 by 10 to 11 microns in size, can be obtained in quantity caught from the little mushroom caps; placed gill-side-down on glass slides in moist chambers. When the season for basidiospore release is at its height, in the shade atmosphere of a cacao plantation, the spore load is great--millions of spores can come from the several agarics that come on only a single broom--the agarics are lavender-colored and produce spores in abundance for a week or more. Under normal conditions of nature, spores are discharged in large numbers from sundown until about midnight, and after that time the numbers subside to a very few until there are practically none by morning.

That seeming, and actual, diurnal habit is not from reaction to light. It is known that regardless of where the sun is, best temperatures for sporulation are from a little more than 16° C. to about 28°. Not much more than a few degrees below and above these temperatures stops all spore release, no matter how dark or how light it is.

It can be seen that the parasite is well adapted to moist-warm, tree-inhabited tropics, for on wet cacao tissues a fresh spore germinates in 30 minutes, but in dry air it is killed in less than an hour. In the tree canopy, wet-season air never dries out and in some areas even dry-season air is never exactly desiccant. A spore sends out a long slender and sinuous germ tube. This penetrates through one of the stomatal openings on the hosts' new growth whether it is the tip of a very young jorcette or a young fruit's skin. It requires several days for the fungus to grow into meristematic tissues, but it is then that the fungitoxic effects develop. (See discussions, also references, in: 3, 5, 17, 19, 41.)

After an inoculation at a growing tip or jorcette, it requires about a month and a half to three months before first visible symptoms are discernible. Once the excess growth (the broom) starts, it expands rapidly. The monstrous development is completed in less than an additional month. Necrosis sets in quite soon after and the broom dies, but is not necessarily separated from the branch. In fact it may remain held up in the air for a year or more, bearing agarics. And whether it remains on the tree or is fallen on jungle debris, a broom needs a dormancy period before agarics form. It is in the matured broom that the fungus holds over from season to season. The time from spore infection to maturation of the coarse broom it causes requires from four to five months. The first rain to wet the dormant broom, even if it is a comparatively small shower, results in raising small, smoothly-shaped fresh mushrooms out of the ugly, ragged, dead broom. These

Fig. 15-3. A sketch of the pale lavender-colored agarics of *Marasmius perniciosus*, on diseased, dead, and spongy cacao twigs from a witches' broom development that has fallen to the moist earth and debris under a cacao tree.

mushrooms look like pinkish or lavender caps, held on top of their silky, white and cream-tinged stalks (Fig. 15-3). Spores released from these innocent-looking caps are airborne, and are carried, in comparatively quiet air as far as 90 meters from the agarics on tree-held brooms. Working in the cacao region in Ecuador, I saw that in one extensive self-shaded area every susceptible cacao seedling was diseased; the clouds of night-released spores, acting in perfect manner as inoculum, that we then relied upon in testing large numbers of seedlings for resistance.

The Narrow Host Range

Wild cacao trees can be found in many areas, from Mexico and Central America to parts far down in South America. Some of these, and I have seen many, may result from seed that escaped from cultivated commercial cacao, *Theobroma cacao* L., which then succeeded in multiplying naturally in tropical jungle surroundings. Many exhibit witches' brooms. There are also numerous relatives of commercial cacao. Desrosiers (16), Viêgas (44), Gonçalves (22) and others indicate over 20 species of the genus *Theobroma*, of which some are suscepts. Among the species it is known that T. cacao, T. bicolor, T. obovatum, T. pentagona, T. grandiflorum, T. sprucianum, and T. speciosa at least are infected with witches' broom. This last species is, perhaps, the most susceptible of any. Using an especially dense spore suspension for inoculum, Van Hall (43) produced the disease in Herrania sp. on a tree from the wild that is in the cacao family, Sterculiaceae. Brooms also occur on a third genus in this family, Sterculia elata, S. pruriens, and S. speciosa (44).

Knowing the hosts affected by a disease, especially if it comes from the jungle or other wild growth, is of great practical importance. As anyone realizes, hosts that are susceptible other than the crop species, are potential sources of more disease infection and must be considered in relation to modes of control. Such workers as Rorer, Pound, Stahel, Garcés, Baker, Desrosiers, and several others have given a good deal of attention to the possibility of Marasmius perniciosus having a wide host range. Various genera and species of trees and plants outside the Sterculaceae have been inoculated and found to be absolutely untouched by the witches' broom fungus. In addition, literally many thousands of kinds of trees and plants in cacao witches broom areas are exposed wet season after wet season to massive deposits of basidiospores from M. perniciosus. Cacao disease specialists have, thus far, never found M. perniciosus infection on any plant or tree outside the cacao family.

A Special Tropical Parasite on Coffee

For two centuries a unique and devastating leaf-spotting and defoliating parasite, Mycena citricolor Sacc., has been seen and studied on coffee (for examples of reports see: 1, 2, 12, 18, 26-28, 30, 33, 35, 38, 42). It has caused a great deal of alarm and years ago in the American coffee lands it was erroneously thought to be the Oriental rust which is so destructive--the worst coffee disease on earth. In all there are over 200 reports and other publications on its appearance on coffee in the American tropics. The fungus has been isolated in pure culture, and some of its characteristics, growing in the laboratory, have been given attention.

Unless controlled, the disease (for symptoms of characteristic leaf-spotting see Fig. 15-4) regularly causes serious--sometimes 20 to 30 per cent--losses (27, 28, 33, 35, 38, 42, 48) and is called by several names: gotera, viruela,

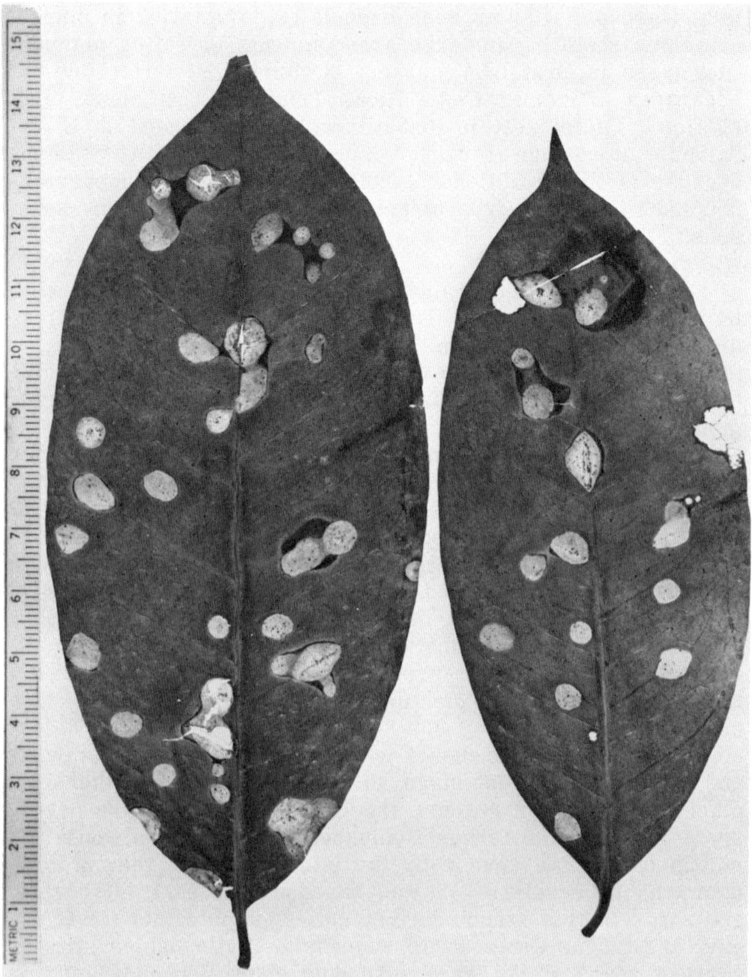

Fig. 15-4. The so-called American leafspot of coffee, caused by Mycena citricolor. A spot is generally a rounded lesion, may be elliptical if on a main vein, gray in the center with a slightly thickened, dark narrow halo of cells. At the very middle of the central dead area there may be left a straw-colored structure developed at the time the gemma infected the leaf tissue.

ojo de gallo, mancha, malamuerte, American leaf spot, false rust, and cat's eye. At times it causes complete tree defoliation. It is found in 27 Neotropical countries and island-groups. It has been especially devastating in some countries resulting in large abandonments of coffee plantation areas in Brazil, Venezuela, Colombia, Puerto Rico, Nicaragua, Panama, Costa Rica, Trinidad, Guatemala, and Mexico. Of interest is that although it is present in El Salvador it has never caused much coffee field abandonment because of the conditions of coffee growing there apparently coupled with short drought-periods breaking into rainy seasons.

Fruits, green branches, leaves and flowers are all infected by the fungus. The main symptom seen from this disease is spotting of leaf lamina, causing extensive defoliation. Sequeira & Steeves (37) have shown that the action of the parasite is such that it inhibits in the leaf the hormone production that keeps a leaf attached to the tree stem. Of course, this results in abnormally rapid leaf fall. There is also notable evidence of toxic effects of a systemic nature. Repeatedly I have observed, on coffee branches with part of the leaves spotted, uninfected leaves that are brittle and tough in appearance. Characteristics of the disease also include the maturation and cessation of growth of terminal leaves before they reach full size, poor flowering, and an abnormal drop of fruit that has set. Often there is a general, slightly lighter-green color of leaves on diseased trees, even in shaded plantings.

The causal fungus is one of the higher Basidiomycetes and has as its perfect stage an agaric. However, there are places in the tropics where the little mushrooms are never seen, although the fungus flourishes, and where its only means of reproduction is by the infection body (gemma, cabiceta, cabecuella). It was based on the early discovery and study of this structure that coffee pathologists first gave the fungus some of its successive Latin names: in earliest years stalks were seen with a swollen end, and under the microscope on it were seen minute spore-like tips (actually hyphal ends, slightly enlarged in preparation for plunging into host tissues), peripherally arranged in stilboid fashion (if those tips had been conidia). This resulted in the name Stilbum flavidum Cooke. In sequence the name was then changed to Stilbella flavida (Cooke) P. Henn., and to Pistillaria flavida (Cooke) Speg. By work in Brazil, the perfect stage was found to be a diminutive lemon-yellow agaric and it was designated Omphalia flavida (Cooke) Maubl. & Rangel. On showing, in 1949, that this genus designation was contestable (39) the name was again changed to Agaricus flavidus (Cooke) Singer. However, there was some question of this generic dispensation, and finally through search back into preserved plant diseased specimens deposited in the famous herbarium in Kew, England, it was discovered by Dennis (14) that long ago this fungus had been given the even more valid binomial of Mycena citricolor (Berk. & Curt.) Sacc. It had been first collected as a small, rather rare agaric on ferns and other plants in Cuba where it was of no economic importance in the highland, chilly and wet "elfin forest" regions. This name was quickly recognized as more properly applicable in Neotropical coffee research circles, and is now widely accepted. The parasite is definitely restricted to the Neotropics and it has a host range of mosses to asters (36, 48), with many more than 550 host species listed.

Gemma and Lemon-Colored Agaric

The appearance of the organism is attractive. On leaf spots are developed some 15 to 50 slender stalks tipped with a glistening yellowish body, shining like a topaz. Such a thing is far from being a spore, it is large enough to see with the naked eye, and with care can be picked off its stalk between the thumb and forefinger. It is many celled, and is shaped like a very tiny door knob (see drawings in Fig. 15-5). This is the infection body already mentioned, and it is also called a gemma, a head, a cabezuelo, a cabecita, and cuerpocilla. When it rains these are easily dislodged from their supporting stalks. I have observed in nature that on falling to the ground or onto decaying vegetable mold and detritus, these infection bodies germinate but do not grow. However, if the bodies or gemmae are planted on sterilized vegetable mold and/or which is carefully moistened and held at the proper temperature, the fungus will grow, but with some difficulty. It only flourishes if the sterilized mold is fortified with a nutrient solution. It also grows well in a culture liquid like Richards solution. It is readily isolated and grows on sterilized grains, PDA and many other nutrient agars, sterile steamed bread, and such materials.

Incubated in a drop of rainwater on a young but full grown coffee leaf, the fresh cabecita germinates quickly and,

Fig. 15-5. Agarics on their stipes (the tallest is 12.5 mm. high), and gemmae on stalks of *Mycena citricolor.* Sketches made of fungus in April, 1948, from pure culture on unhulled rice grain and agar medium in a large petrie dish. Caps of agarics had from 4 to 22 gills. In sketch on right the gemmae (one has fallen top down) are rounded flattened bodies like minute door knobs furnished with a fringe of infection mycelia around their edges.

under a dissecting microscope, the infecting hyphae are seen on its periphery to grow percepitably in an hour's time. In fact if especially well placed, I have seen cabecitas to have anchored on leaves in less than 60 minutes.

The cabecita or gemma deserves a special description. It is a translucent, yellow bit of the organism, with the round part on top of a short neck or collar. The body is partly hollow and if it is not wet the collar tightly clasps the tip of the stalk it grows on. When moistened, the clasping collar loosens from its stalk. After that a splash of rain which touches the cabecita readily takes it away, with the cabecita at the bottom of the drop.

A drop of rain is able to carry one to three cabecitas, and when the drop lands on a leaf the cabecita settles and is immediately in contact with the leaf epidermis. The gemma is fringed with tips of hyphae that are filled with dense protoplasm; the tips appear to have pointed ends that penetrate directly into host epidermis. Under tropical elfin forest or rain forest conditions, host leaves always remain moist long enough that the gemmae of *Mycena citricolor* can germinate and infect. Where such moist conditions are often of short occurrence between droughts, as in many parts of Brazil, especially where coffee is grown without shade--and even in some parts of El Salvador and Guatemala on the Pacific slopes--a comparatively small amount of manipulation of field cultivation methods and proper thinning of shade, and the like, will leave conditions so changed that the leaf-spot becomes a matter of little concern.

While in nature the gemma-producing phase of the parasite develops readily in spots on living leaves, it also grows well in agar cultures. Moreover, I found (48) that exposure to daylight was absolutely essential--a period of somewhat less than 15 minutes being sufficient--for cabecita or agaric formation. In agar cultures the fungus grows more in the dark than in light. From studies under carefully controlled conditions, Rodriguez and Arny (31) published in 1967 evidence that best cabacita production is in near ultraviolet and infrared light; that "nutritional factors in the host cannot replace light requirements of the fungus." It also will not produce agarics in nature unless it has, besides light, a sufficient amount of killed tissue from which to draw its nutrients. I (48) reported in 1952 that contaminants in pure cultures stimulated agaric development (Fig. 15-6); in detailed studies Salas (34) found a *Penicillium* especially superior in its ability to induce basidiocarp production in *Mycena citricolor*.

Production of the small lemon-yellow agarics does not occur in nature unless the coffee tissues that have been killed by the parasite are subjected to washing rains or dews and proper temperature. Using pure cultures and also diseased leaves in moist chambers it was often necessary to irrigate cultures or diseased leaves daily to produce abundant agarics. In addition, while a few agarics would grow at fair warmth (above 26° C), by far the most profuse development came in regularly irrigated illuminated material held at 22° and 24°. This range of temperature almost precisely duplicates what is known (49) about the best temperature for growth of the vegetative stage of the organism in pure culture.

Fallen *Mycena*-spotted leaves around diseased coffee trees are the most common producers of agarics, although sometimes the agarics can be collected in sparse numbers on leaves still hanging to trees. This, of course, only occurs in hill country, when it is cool and moist in the tropics. Many workers know this, and a few (example: 50) indicate

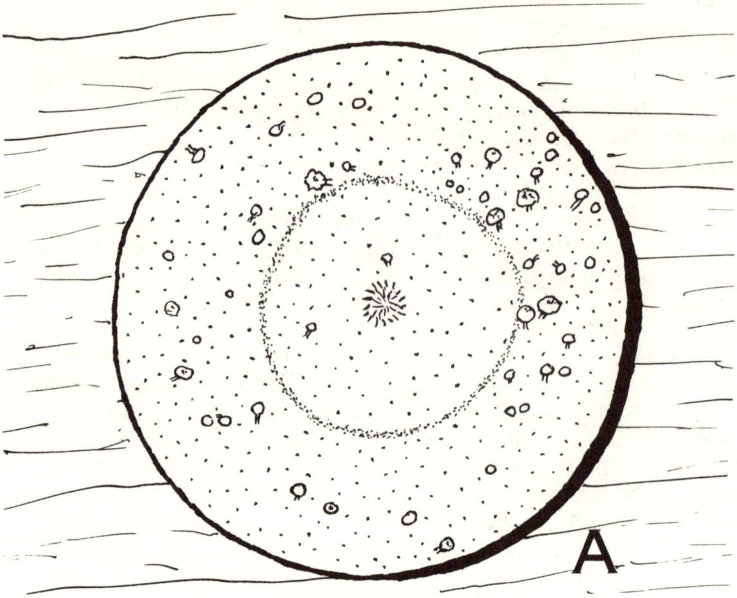
A

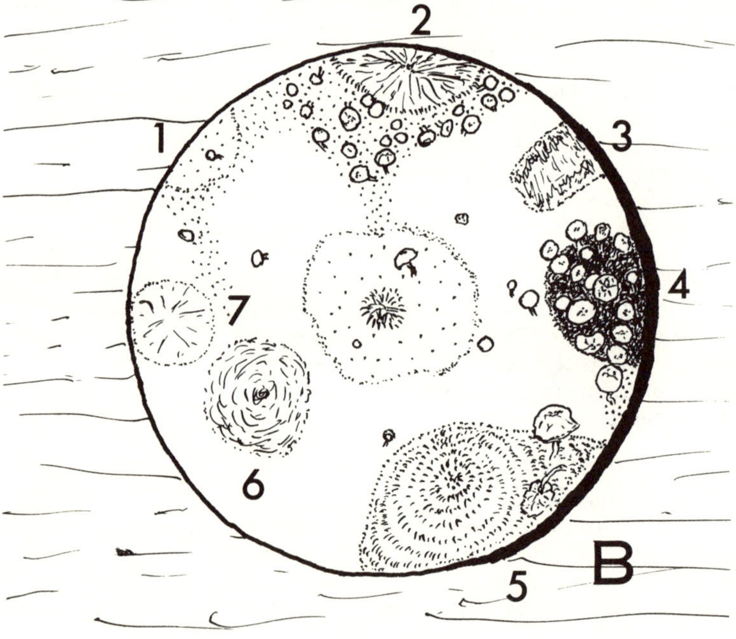
2
1
3
4
7
6
5
B

that there is likewise on hillsides an upper or too cold a limit as well as a low or too warm a limit. In moist chamber cultures, at 24°, the perfect stage is most successfully secured on diseased leaves placed in the chambers with their upper surfaces oriented upwards, but washed off with spray at least once daily.

Findings such as the above, suggested reasons for the repeated observations that the yellow agarics are fewer in number at the close of the wet season in diseased coffee located in a cold western slope of a hill, than in a wet spot or at a slightly lower elevation where air temperatures are at times as high as 26° to 28° but where dews are heavy and there are still a few light rains. Here the cabecitas or infection bodies are common and abundant. However, at a little higher elevation of about 200 additional meters (656 ft. more), where a severely diseased coffee orchard was on an eastern slope and the rains were regular and temperatures were 18° to 22°, there were numerous cabecitas but in addition there was special profusion of agarics.

The agaric caps, illustrated in Fig. 15-5, are small--in some cases so little there is not much room on them for many gills or even basidia. In several instances agarics were collected that produced only 10 to 15 basidiospores. In some cases, however, over 50 and even a few hundred basidiospores were secured from single, especially well-developed agarics. The basidiospores remained alive a short while in rainwater and single spores were planted on wet coffee leaves. Although these spores germinated, in no case did I secure infection from such single basidiospores. However, leaf infection was successful from bits of mycelium from agar-grown colonies of the organism isolated from

---

Fig. 15-6 [on facing page]. *Mycena citricolor* (isolate CR 97) cultures on potato dextrose agar petri plates: A. Pure culture of the organism. B. Contaminated culture, 1. inoculated with one gemma of *M. citricolor* CR 97; 2. colony of *Colletotrichum* sp. from coffee leaf; 3. colony of bacterium from root of coffee seedling; 4. colony of *Diplodia* sp. mixed with *Pestalotia* sp. from twig of coffee dieback; 5. colony of brown colored *Aspergillus* sp. from coffee cherry; 6. and 7. both colonies of fungi undetermined, but secured from rotting coffee leaves (developed black agarics in nature). Production of gemmae indicated by stippling, and agarics by outlines of caps.

agarics or from field-collected leaf spots, and from infection bodies that had been produced in pure culture or picked off disease spots on leaves. Studies by Sequeira (36) on the multi-nucleate condition in cells of M. citricolor suggest that it is only after a sexual fusion takes place between cells from separate basidiospores that the resulting organism is fully pathogenic. I never obtained infections from single spores.

The Luminescence Character

The most remarkable characteristic of lesions from this parasite on coffee or on any plant is that the vigorougly still active moist spots are especially luminescent in the dark. This luminescence has been seen by many, and is found in several species of fungi in the genus Mycena (40). As early as 1878 the luminescence seen at night had given the Spanish name of candelillas* to the coffee leafspot disease (51). It is a subject of recurring interest to biologists and pathologists. One of the classic discussions of it is by Buller (10, 11). Researchers (9-11, 23-25) have studied this kind of luminescence, not only in M. citricolor but other organisms. The reason for its occurrence as a "purposeful" characteristic is difficult to explain. Some believe it is from a physiological reaction somewhat similar to that which results in the so-called "phosphorescent waters" (which is not truly phosphorescent) found in the sea from bacterial action. I have made, over a period of years, several studies on the luminescence in the coffee leafspot fungus and have reported some of the results in 1952 (48).

It is impressive to step at night into a dark laboratory illuminated solely by thousands of steadily shining cultures. The tubes, plates, and flasks of living fungus emit an eerie glow of pale, sky-blue light. Under certain conditions there may be a mottle in it, but the same hue remains constant, no matter from what culture or living lesion on any host. Where chemicals are applied, the hue may be varied by the color of the test solution, but when washed off the same "sky blue" is again visible.

*Candellilas ("little candles") have been regarded by some with superstition. A laborer friend told me he often gathered new spotted coffee leaves which his small daughters played with at night looking at them under a blanket.

I noted that there were in certain cases consistent differences in intensity, and since electronic means of reading luminescence variations were not available, a visual system of estimation was devised. At first this was based on what was seen from test tube cultures of vegetative mycelium. Growing on PDA slants, at 24° C., a culture gained its optimum brightness at between three to five weeks of age; a month-old growth was considered standard. Numerical luminescence ratings were finally decided upon as follows: no luminescence = 0; if barely visible = 1; slight illumination of white paper = 3; brightens fingers = 6; the brightness of a standard month-old PDA test tube culture grown in the dark at 24° and in moist atmosphere = 7; luminescence strong enough to cast pencil shadow on paper = 9; luminous enough to read text-book printing = 10, 11; and very bright, equal to the luminescence of a firefly = 12.

On repeating determinations, the worker becomes more and more certain of his decisions on numerical readings. They are readily demonstrated and taught to colleagues; results coincide well when readings are made and also when test examples are mixed and again read.

The ordinary month-old standard PDA cultures I employed (the reader should see Fig. 15-7 for examples) had a rating of "7," except that the uppermost, thinnest part of the culture slant touching the glass, had a luminescence of "8" or "9." This was disregarded in experiments. Clumps of mycelium grew out in puffs at the center in some slants, and the clumps were sometimes "9" and "10." This greater luminescence was retained and sometimes it was a characteristic all over a new slant surface when the "bright clump" was grown on fresh agar. Also "dull clumps" were found. In one case a strain was isolated that was a dull "4" in brightness, while over-all rating of the mother isolate in new growths was repeatedly scored at a brightness of "7." Differences in luminescence appear to indicate variation in metabolism; Johnson (24) used it as a reaction tool.

Luminescence from common isolates was found to be strong, "8" to "10," on the synthetic agars like Czapek's and Richard's. No cabecitas or agarics developed in these media but vegetative growth was excellent. In this connection, Wellman, R. Segall and E. Echandi (1952, unpublished manuscript) made agar media using Richard's syn-

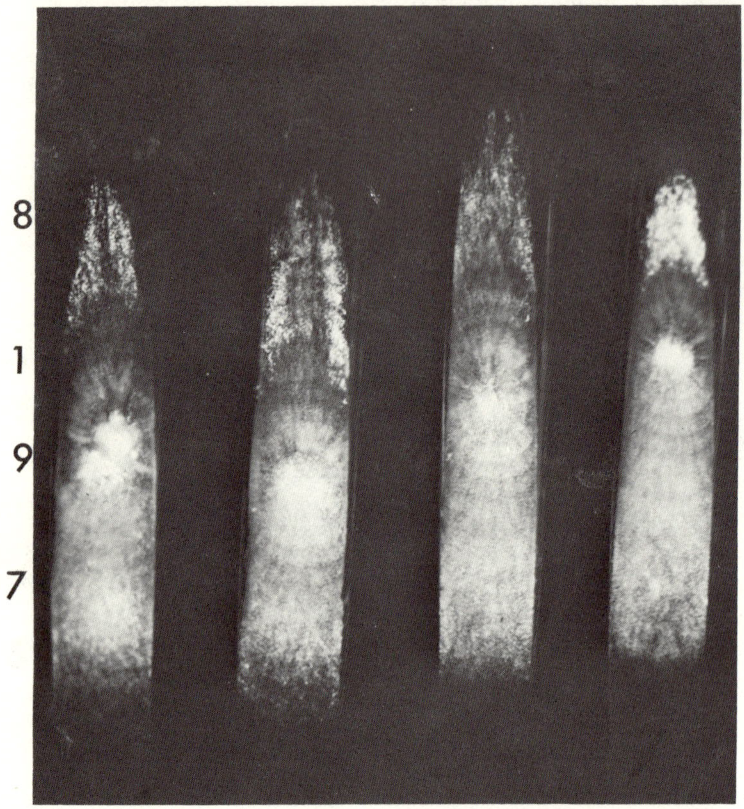

Fig. 15-7. Luminescence from six-weeks old mycelium of *Mycena citricolor* growing in large test tube slant cultures on synthetic medium (no thiamin). Luminescence determinations made at center of the test tube on left; numbers give indication of "readings." To take the picture I used the direct luminescence of the mycelium in total darkness. Cultures were arranged and focused on by a camera, photograph film inserted and lense opened; time of exposure 16 hours in the dark.

thetic as the base and adding certain vitamins. The latter were secured from proprietary preparations sold in drugstores for human consumption. Cultures were in light near a window but protected from sun. Vitamins A, B (Thiamin), D, and C (ascorbic acid) were tested. It was only when thiamin was present that infection bodies and agarics grew.

Table 23. CHANGES IN LUMINESCENCE RATINGS of standard cultures of Mycena citricolor given different temperature treatments. (All cultures rated "7" to begin with, except the last two which were "8." In each test, six matched cultures were used per treatment.)

Temp. (C.)	Min.	Luminescence	Recovery time required to 7
24	60	7	--
2.5	1.5	3	9 hours
37	2	2	48 "
28	3	4	24 "
24	3	7	--
17	3	3	24 "
9	3	2	24 "
30	7	4	4 min.
30	7	5	4 "
36	4	2	4 "
16	4	1	4 "
8	7.5	2	3 "
42	120	0	killed
50	30	0	killed
0	60	0	(After 48 hrs. lum. 2, then death)
24	(1 month)	7	--
24	(1 month)	8	--
28	(1 month)	8	--

An ice-water bath at 2.5° C required 1 min. 28 sec. to reduce luminescence from "7" to a dim "3"; 2 minutes after the cultures were put in 24° water they returned to "7." All readings were made by me in a completely dark room. Work was done at night, after dinner, and after 15 munites were spent in the dark allowing eyes to become adjusted. (This is about what Buller (10) as well as I concluded was the ideal time to make luminescence observations.) All readings were called through a closed door and recorded by my helpful wife.

A warm bath, held at 37°, reduced the luminescence of cultures with "7" to a dimness of "2" in two minutes. These then required about 48 hours held at 24° to return to their old brilliance of "7." After numerous tests it became evident that the temperature of 24° was the most optimum for luminescence. This appears to be a significant finding, as from culture studies it was seen (59) that under controlled conditions the fungus has an optimum vegetative growth at 24°. In one study a set of six matched cultures with "7" luminescence were treated at 30° (6 degrees centigrade above optimum); it required seven minutes to reduce the luminosity to "5"; on returning to 24° the luminescence was brought back to the brightness of "7" in only four minutes. When warmed to 36°, or 12 degrees above optimum, the luminescence fell to "2" in four minutes. When the cultures were immediately changed to 24° the luminescence again came back in 4 minutes, but if held 10 hours at 36° the cultures were killed.

At the same time as the above warm tests were being made, cold treatments were also being studied. A set of cultures that had luminescence of "7" were treated at 16° which is 8 degrees colder than optimum. In four minutes, luminescence had gone down to less than "6," but it returned to "7" within four minutes when the temperature was brought back to 24°. A matched set of cultures cooled to 8° (16° lower than optimum) were reduced in luminescence to "2" in a little less than 8 minutes. When the culture was returned to the optimum temperature, the dim luminescence of "2" returned to normal luminescence in 3 minutes.

Upper and lower temperature limits were determined for luminescence. Heating to 42° and 50° dimmed cultures rapidly, and in less than an hour they were killed. The warm end point was at a temperature of above 36°. At 42° the mycelium was killed. Chilling quickly dimmed luminescence in vigorous cultures, and freezing for one hour obliterated luminescence. However, being frozen for only three minutes, while it was a severe shock, did not kill the fungus; after some hours at 24°, life and luminescence returned. If held at freezing for more than an hour, and certainly overnight, the cultures were killed.

Several experimental matched sets of month-old PDA cultures with a reading of "7" were incubated at 24° C for a total of eight months. This was under subdued light, in a moist atmosphere, and there was no great shrinkage of the agar slants. At the end of this long incubation the fungus growth had piled up considerably and there was some granular, yellow appearance. The luminescence was "3." When these old cultures were subjected to 36° they were killed within less than 10 minutes, and were killed immediately when they were frozen solid.

While many cultures growing on different media exhibited differences in degree of luminescence, it was not possible to get satisfactory proof that the relative brilliance of luminescence indicated the relative growth rapidity of the culture. Among interesting cultures were some in clear liquid Richard's medium. These were fully submerged in half filled 200 ml. flasks. The fungus grew very slowly but was abundantly luminous in its completely submerged mass of hyphae at the bottom of the flask. Moreover, the mycelial webs remained luminous for over 10 months, long after it seemed new growth had ceased. When these cultures died by the end of 11 months, they were not luminescent.

Preliminary tests were made to see if the luminescence could be extracted in some simple manner. Brightly luminescent cultures were used. The mycelium was ground in water using fine emery powder in a mortar with pestle. This was then filtered and it had no luminescence in either the filtered extract or the filtrate. Thorough grinding was fatal to the fungus, and the light it emitted. Extractions, using either alcohol or chloroform, from either moist or rapidly air-dried mycelium showed no luminescence. Drying killed the mycelium. So did leaching of the fungus with kerosene and ether. In no case was luminescence freed from living or dead mycelium.

At numerous times, after my eyes had become fully adjusted to darkness and the detection was easy of fine differences in luminescence, microscopic examinations were made of luminescent culture mycelium under low and high dry objectives. In the kind of set-ups used, the luminescence could never be detected looking through the microscope objective.

Luminescence Increased by Contaminants and Chemicals

The organism M. citricolor grows well in pure culture. As indicated above, it is the vegetative mycelium alone that is luminescent. Neither the gemmae nor the agarics show this character, although it was reported by Buller. It may be possible he had a race of the fungus with such a character. Working literally with many thousands of isolates and observing millions of gemmae and agarics, I have never seen luminescence in M. citricolor except in living, vegetative material; studies of it were made in 11 countries of Central America, the West Indies, and South America.

As long as cultures are grown on agars using certain refined synthetic media (Richard's, Czapek's, modified Crone's), there are no gemmae or agarics. However, when cultures are infected at the side--or accidentally contaminated--by Penicillium spp. and by a brown culture with mycelia having clamp connections but otherwise unidentifiable, both the agarics and the cabecitas (gemmae) develop. There is a strong line of demarcation, what appears to be antagonistic conjunction, between M. citricolor and the other fungi. This is represented in accompanying Fig. 15-8. When examined for luminescence the line is always found to be very

---

Fig. 15-8 [on facing page]. Petri dish cultures of Mycena citricolor isolate CR 97, to show effects of antagonisms (?) of a few other fungi. A. Sketched in daylight. B. Appearance in total darkness. 1. Point of original inoculum, M. citricolor CR 97, with luminescence at point of about "9," remainder of pure Mycena growth about "7." 2. Inoculated with Colletotrichum from coffee, grew intermingled with Mycena, luminescence about "8," inhibited gemmae, stimulated numerous agarics at edge of colony and luminescence about "9." 3. Inoculated with an unknown fungus having white young mycelium blackening on ageing, gemmae inhibited, profuse agaric development, under agarics luminescence about "12." 4. Inoculated with M. citricolor isolate Guate 14, result normal growth and gemmae production, luminescence about "7." 5. Inoculated with Cladosporium sp. from ragged leaf disease of coffee, grew over Mycena, no gemmae or agarics and only slight luminescence of about "1." (There is never any luminescence from agaric caps or stipes, or from gemmae.)

A

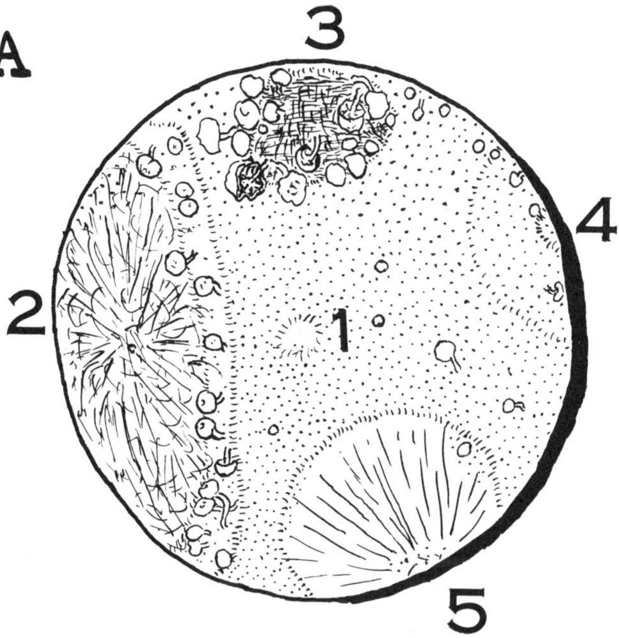

B

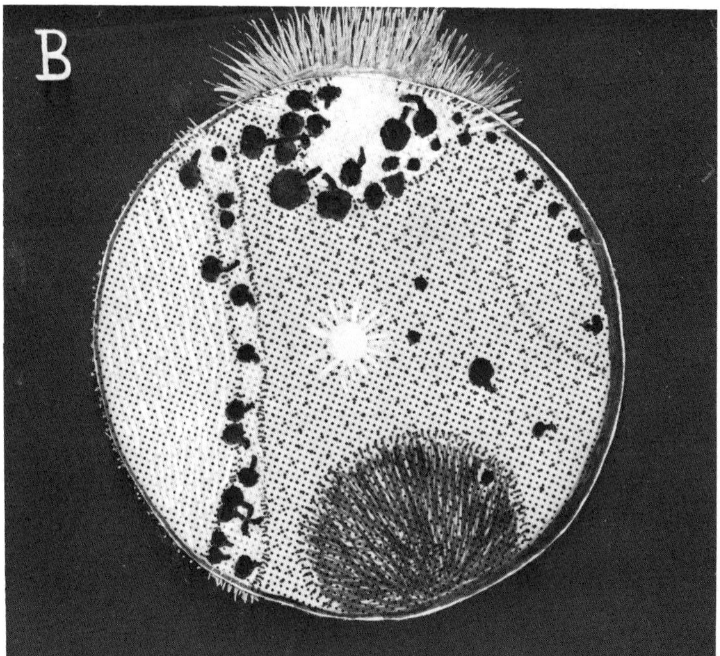

bright, stronger than "12," from the antagonism, and it is near these lines that the agarics and gemmae appear. Normal and vigorous, purely vegetative growth in the conditions described registers at between "5" and "7" in luminescence. Bacterial contaminants all dulled M. citricolor luminescence.

Studies were instituted to determine the effects on luminescence of certain chemicals. Standard glass test tube slants that had grown M. citricolor for a month were used; all had a luminescence of "7." Three slants were used per treatment. Results from these treatments are indicated in the accompanying table.

Table 24. EFFECT OF CHEMICALS ON CHANGING LUMINESCENCE in vigorously growing Mycena citricolor cultures: all treatments triplicated

Chemical	Luminescence ratings[a] after: 1 hour	36 hours
No treatment	7	7
Rainwater (direct from sky)	5	1-2[b]
Distilled water	4	3-7
Lactic acid 0.5%	2	2-3
Ethyl alcohol 0.5%	9[c]	5-7
Sodium biphosphate 3.0%	6	9
Potassium hydroxide 1.0%	1	1
Potassium permanganate 0.5%	1	1
Hydrogen peroxide 0.5%	2.5-3.0	6
Ammonium hydroxide 1.0%	1	1
Aluminum phosphate 5.0%	5	8
Nicotine sulphate 0.1%	7	9
Sucrose 2.0%	8	9
Bordeaux mixture (1-1-50)	1	1
2,4-D (5 ppm.)	9	9
Maltose 2.0%	7	9
Adenosine triphosphate 0.01%	4	14
Acetic acid 0.01%	1	1

a) See preceding discussion for luminescence readings and criteria of ratings. "0"-none; "3" brightens white paper; "7" standard of a month-old normal culture; "14" brighter than flash of the common firefly.
b) Bacterial contaminants found.
c) In another series this alcohol at 0.01% left the culture at "10" after one hour and "9" after 36 hrs. Nembutol and aspirin at 0.1% in an hour reduced luminescence to

"1," from which the cultures did not recover, although they remained alive. When transferred they grew and luminesced normally.

Of interest is the fact that dilutions of acids, alkalis, a fungicide, and such compounds as aspirin and nembutol all caused the luminescence to be dimmed. However, dilutions of certain phosphates, a sugar, 2, 4-D, ethyl alcohol, whisky, gin, dilute tobacco extract, and nicotine sulphate all caused distinct brightening of luminescence. The chemical compound that caused the greatest relative increase in luminescence was adenosine triphosphate, which is known to be an enzyme stimulant.

Judging from tests for luminescence, and the fact that "fruiting"--formation of cabecitas or agarics--is either in so-called "natural" (or tissue) media or in synthetic media (and there only in the presence of contaminants), it appears that this fungus is one that does not synthesize certain needed materials by itself. In nature the organism is an active parasite, never a saprophyte. To study this further, culture work was done with sterilized and unsterilized leaf mold, dead coffee leaves, jungle debris, soils, and sugar and peptone-fortified sand. Any of these materials if sterilized, supported luminescent growth of the fungus. In unsterilized materials the fungus died*, not because of starvation but from the apparent overwhelming action of bacteria.

So far as thus far determined, luminescence has no apparent special value to the fungus. That is, it has no attractive or repulsive character that might be considered a factor of the adaptative needs of a species in evolution. Luminescence seems to be an expression here of the fact that this organism is vegetatively living. It seems to be a mark of the species, such as a certain color or some other character. Obviously, it deserves further study.

---

*Evidence is that active growing vegetative mycelium, the infection bodies or gemmae, and the agarics of the fungus are readily injured in nature; the holdover stage of the parasite appears to be in the quiescent mycelium resting inside the halo of dark-colored tissues around disease spots on broad-leaved evergreen hosts.

Epidemiological Habits of a Rain-Splashed, Heavy Infection Body

Both superficial and detailed observations show that in its natural spread, the parasite *Mycena citricolor* may remain within small areas of plantations several years; in other cases large numbers of previously disease-free trees will show sudden *Mycena* leafspot well-spread over them after three rainy seasons.

Natural spread was studied (48) on seedlings in a moderately wet and cool valley location where there were daily prevailing winds from east to west. Maps were made of 11 plots. Each plot had a large diseased ("mother") tree in it that furnished abundant inoculum, and around each mother tree were growing numerous seedlings. In four of these plots new disease appeared mostly on the leaward seedlings, but in six plots new disease was found on the windward. In two plots disease spread both ways. Another point of note was that in nine months time the spread while considerable, was not rapid, and did not sweep along to affect all seedlings. On the other hand, after five years some plots of seedlings around diseased "mother" trees were 100 per cent infected but others were only 14 per cent diseased. Isolated plots with disease-free "mother" trees grew with no diseased seedlings around them. Thus there is in this disease a unique, very closely localized spread, which is certainly not like that of rusts, Cercospora spots, mildews, cacao witches' brooms, and numerous other diseases that have windborne spores.

Knowing from field researches in one location that seven years went by without agarics appearing, the only possible means of disease spread was through movement of the infection bodies, and these were given special attention (48 publs. 1949, 1950, 1953). It was soon learned that cabecitas, easily seen, occur on hair-like stalks or filaments, arranged sparingly on leafspots. The infection bodies are tightly held onto these stalks, and if tested at a dry time of day the bodies do not fall off, can not be blown off using hard blasts of breath, and if picked off between thumb and finger a body takes with it the tip of its stalk. Dry bodies are not dislodged when stroked with objects of any kind: pencil, needle, a twist of wool, a camel's hair brush, finger tip, shirt sleeve, or coffee leaf. A dry blast of air from an air compressor at 30 lb. pressure will tear off cabecitas, but with part of a stalk attached to each one.

Observations show dry tree crickets, and dry tree snakes do not dislodge cabecitas they crawl over. Careful field observations never disclosed that cabecitas were blown off their stalks in dry windy weather.

On the other hand, presence of rain or heavy dew prepares for ready removal of cabecitas. Once well wetted they are easily dislodged by almost anything after a short period of soaking. While I (48) had not been able to prove spread of the disease by working and brushing through diseased trees with my arms, Sequeira (36) reported in 1958 that wetted cabecitas could be loosened and moved from tree to tree by workmen. It is true that in the laboratory, wetting cabecita-studded cultures loosens them quickly. If these cultures are vigorous and young, over 90 per cent of the infection bodies may be slipped off into water.

Cabecitas quickly sink in water, and tests were made of their buoyancy in solutions of different densities, regulated by adding differing amounts of NaCl. In three trials, using densities (determined with a standardized Baumé densiometer) ranging from 1.000 to 1.150 it was found that cabecitas do not float until the solution reaches the highest density of about 1.115 to 1.123. Fresh hen's eggs float well in solutions with densities of only 1.060 to 1.090. During field studies after rains, cabecitas are often seen splashed away from the spots of their origin and lodged elsewhere. In two separate instances, using a hand lens, cabecitas were watched that first had become loosened on their stalks by moisture, and then at the very moment of ultimate contact a free rain drop was seen when it "picked off" the cabecita and carried it away.

As it was known that cabecitas are moved by water, in still air a laboratory drop tower 250 cm. high was set up, and drops of distilled and rain water were directed under controlled conditions. Leaf spots with gemmae on them were watched for water drop effects, and splashed onto paper covered platforms. Whenever a drop carried off a cabecita the act was seen and a drop with its infection body could be traced; the distance was recorded. Under the conditions farthest distance for travel of a droplet containing an infection body was 564 mm. (approx. 22.5 in.), and the farthest for a cabecita-free droplet to splash was 587 mm. Droplet sizes were measured and it was found from 25 cabecita-carrying droplets that they were from 0.4 to 4.6 mm. in diameter (average 2.95). Most droplets that were carriers contained one cabecita but a few took more and one carried away a phenomenal group of six. This demonstrates the kind of limited action that occurs in the field and certainly does not guarantee rapid spread.

It had been learned in the meantime that leafspot was most severe on coffee that was well shaded. It had also been seen repeatedly in the laboratory that healthy cabecitas germinate quickly in water and will infect only wet leaves. It then became important to know how much time is required for infection bodies to attack leaves. Healthy, vigorous coffee leaves brought to the laboratory were seeded with specific numbers of infection bodies in a well-spread, thin coating of water. At intervals they were then gently washed, turning them on edge in a water container to float off the cabecitas not anchored to the leaves. This was done first at short intervals which were then extended to hours. All such inoculated leaves, with 200 cabecitas per leaf, after loose infection bodies floated off, were incubated in a dimly lighted room for three days in high humidity and at 24°. After four series of these tests it was found that on two leaves in 50, one cabecita on one leaf and two on another appeared to have anchored and caused infection after 20 to 30 minutes of undisturbed exposure before the cabecitas were floated off. There was some uncertainty about those shortest-time apparent infections. However, in the interval from 30 minutes to an hour, two to five cabecitas per leaf did anchor and infect. An hour is adequate in the wet tropics for a few of these "bits of meat," the infection bodies, to germinate and act as the organs of fungus-spread, and insure the species survival in its habitat.

It has been observed many times that the coffee leafspot multiplies after rainy periods, and appears to have no increase in dry seasons. These general conclusions, as Alvarado (1) as well as many others indicate, are acknowledged by both planters and scientists. In rounding out studies it was necessary for me (48) to make objective counts to prove such contention. In a field of trees, weather conditions were watched, and leaf counts were carefully recorded during eight periods. Starting with an average of 6 lesions per tree it was clearly evident that during rainy periods leaf lesions multiplied. In a long rain-free period there were only a few added infections (from heavy dews), and counts of spot increase was only an added 4.5 per tree. However, following that during two moist periods accompanied by rains, the average spot increase was 102.4 and 178.2

per tree respectively. (See citation 48, first 2 papers.)

Spectacularly Inclusive Host Range

In early works (for examples; Saenz (33), Alvarado (1), Nowell (28), Puttemans (30) and Carvajal (12)), the inference was that this organism was possibly in large measure a coffee parasite. As further researches were carried on it became clear that many other plants were also susceptible to attack by this parasite. Such findings as those of Buller (10) showed that the fungus caused disease spots on glass house grown *Bryophyllum* leaves. In connection with mycological problems, Dennis (14) went back in the Kew herbarium to old collections of this tropical leaf-spotting fungus. These had been secured first in "elfin forest" ecology, on top mountains in Cuba where the fungus was found on grasses, ferns and other such low-growing plants.

It remained for Sequeira (36) to have made the most exhaustive study of the remarkably wide host range of *M. citricolor*, which he reported in 1958. He found it causing leaf spots on plants from ferns to asters. In one series I found that the fungus attacks mosses, liverworts, and numerous hosts of higher plants not seen or reported before. Many of the hosts are in a volume entitled *A List of Tropical American Plant Diseases* that I am now preparing. There are more than 550 species of plants known to be susceptible to infection by this fungus. The Neotropical coffee organism does not infect palms, cycads, pandanus, bamboos, banana and its relatives, or species of *Heliconia*. A study which would have extremely interesting repercussions would be why *Mycena* leafspots on some hosts, such as certain Gramineae, never produce cabecitas. Many hosts are in this category.

There are so many hosts that are infected by *M. citricolor* cabecitas that it is now more significant to search for species that cannot be infected than for those which the organism can attack. It is of great interest that a gemmiferous fungus of the Oriental tropics known in East Africa, and described by Wallace (46), is not *M. citricolor* but it has a very large host range.

Archaic Infection Bodies or Gemmae

In plant pathology we are especially fortunate in having for study parasites that produce gemmae and thrive in areas such as the so-called "rain forests" and the wet "tropical elfin forests." These conditions are representative of the climatic situation in the eras long past, when great algal structures, giant club mosses, and other plants grew and produced the tremendous accumulations that became coal deposits. In the wet parts of the Neotropics forest and swamp complexes survive in which plants and fungi retain structure and habits of truly archaic nature. It has been my great fortune to see some of them, and it is my conclusion that fungi that have infection bodies have retained this character that belongs in the category of the truly primordial.

A great number of fungi in their age-old aspects have characteristics that are close to and in many cases almost analogous with the algae. It is probable that fungi "split off" from the algae, evolving into the saprophytes and parasites we know now. Some such fungi have a large part of their existence as vegetative mycelium; ordinarily they do not produce spores or other means of reproduction; they depend on growth-traveling mycelium or pieces of their structure, gemmae, blowing from place to place or being carried from one region to another by winds, water, and other means. Characteristically the simplest form of transferrable life, of fungal life at least, would seem to be a loosened filament or piece of fungus and this is found in some tropical fungi. The piling up of the hyphal mats, not in the form of a hard sclerotium, may be a step in evolving towards higher fungus spore development.

An infection body, it must be understood from the outset, is not a spore. A spore is a highly evolved, microscopic body from one to a few cells in content and a true spore arises from specialized structures. With respect to infection bodies, which are in many cases macroscopic organs, we have a different situation. They are made up of vegetative cells intertwined or otherwise put together for mutual protection and for nutritive purposes. Infection bodies of perhaps most archaic types are relatively tender to the touch and are easily mashed when squeezed. The tender mass of tissue that makes up an infection body as understood here, is almost what might be called "a bit of meat"; in it are stored foods to be mobilized in the growth process so that, if parasitic, the infection can be rapid and vigorous.

It is my belief that the reason that there is rapid germination, vigorous primary infection, and quick invasion of hosts by these organisms is because there is a strong backing of mycelial contents in the infection bodies. These contents are drawn upon to insure the quick germination and rapid direct penetration of the organism.

As this infection body situation is to my mind a unique one, it seems well to list a few types of them. It is well understood that in addition to a few fungi, there are certain algae, liverworts, lichens, and a few ferns that have something similar to infection bodies. They are specialized small structures of tissue often called gemmae, and they act as naked seeds, not as parasitic organs. In reality there are undoubtedly many tropical non-parasitic fungi that have structures like infective bodies.

Possibly the simplest of infection bodies is a type of texturally tender, loose clump of fungus hyphae that is light in color bordering on brown in some cases. These infection bodies may be such as belong to certain species of so-called *Rhizoctonia*. However, these infection mats are not to be confused with the much harder sclerotia that are developed so commonly, and described in connection with scurfy scab on the surface of potato tubers. A second type of infection body is a sporodochial clump that may be orange or salmon in color. This sporodochium is rounded and falls away to be air and water borne. Such is characteristic of the organism *Corticium salmonicolor* that causes the pink disease of many tropical crops. A third type of infection body consists of bundles of hyphae and these bodies are usually small and gray in color. An organism that produces these is one that attacks coffee leaves and is designated as *Sclerotium coffeicola*. A fourth type is round to pear-shaped and is yellow to slightly orange in color. An example of this is the fungus *Mycena citricolor*.

A fifth infection body type, and there are probably others, is oblong to cone-shaped and soft; it may be white but is usually pink and flesh-colored. This organism, so far as I can determine, has never been described, but I collected it several times in moist forests in Central America. It is also in a rainforest area in Puerto Rico where it infects leaves of low-growing palms. The same type of in-

Table 25. SOME DIFFERENCES that may be noted between the cacao witches' broom and the coffee American leafspot diseases*

Item Compared	Witches' broom or escoba de bruja	American leafspot or ojo de gallo
Plantation tree attacked	Cacao (*Theobroma cacao* L.)	Coffee (*Coffea arabica* L.)
Causal organism	*Marasmius perniciosus* Stahel	*Mycena citricolor* Sacc.
Type of fungus	Agaricaceous Basidiomycete	Agaricaceous Basidiomycete
Host range	Narrow (in family Sterculaceae)	Wide (from Bryophytes up to and including Compositae)
Original home of disease	Low lying, warm wet jungles of north central Orinoco-Amazon drainage basin	Mountainous parts of West Indies, Central and Northern South America, in cool, wet and foggy "elfin" (dwarf) forests
World distribution	Neotropics only	Neotropics only
Infection organ	Basidiospore	Gemma
Court of entry	Stomate in young epidermis	Direct epidermal penetration
Vegetative host part infected	Branch tip, involving meristem	Leaf, involving all laminal tissues

Incubation (from infection to first apparent symptom)	Six to twelve weeks, depends upon conditions of infection and host growth	Two or three days, depending on age of leaf
Disease produced	Swollen growths, unhealthy multiple stems, dying branches	Leaf spotting, defoliation, final killing of trees
Perfect stage	Medium sized lavender agaric	Small lemon-yellow agaric
Season of infection	Rainy season	Rainy season
Fungus holdover from season to season	In semidormant hyphae in diseased tissues of old witches' brooms	In semidormant hyphae in dark border at edge of old leaf spots

* Many names have been and are used, those that are common in English and Spanish are included here; Latin binomials for the parasites are those generally accepted by plant pathologists of the Western Hemisphere tropics.

fection body was also found causing disease on such crops as tea and coffee and such ornamentals as Hibiscus rosa-sinensis, a delicate species of Begonia, and Stachytarpheta. It has appeared rather well contained within small areas.

There are distinct advantages in being an infection body in the wet tropics. For one thing these vigorously living bodies will grow almost immediately under conditions that are good for them. When infection bodies are rinsed with sterile water, which is then plated out, bacteria are readily recovered. The bodies are eventually, but not quickly, injured by being in contact on the surface with a mass of bacteria. This is unusually favorable to organisms of parasitic nature in the tropics, because bacteria attacking fungi are so commonly present there.

There are certain disadvantages for an infection body, of course. For example, they are killed by drying and sun exposure. This is true of all infection bodies I have studied. Second, light rains are not in themselves sufficient for distribution. The rains need to extend slightly, from two or three days to maybe a week. Third, the infection bodies can be seen by birds and it has been observed that certain small birds pick up Mycena citricolor infection bodies for food. (Unfortunately, it was impossible to kill the birds and dissect them to corroborate this observation.) A fourth disadvantage is that it requires a long time, and more living material, to make up an infection body than a spore. This is evident to anybody who considers the situation. It may be repeated that infection bodies are primitive and archaic distribution bodies.

Literature Cited

1. Alvarado, J. A. 1935-36. Tratado de caficultura práctica. Tipografía Nacional, Guatemala. 2 vols, 1227 p.

2. Ashby, S. F. 1925. The perfect form of Stilbum flavidum Cke., in pure culture. Kew Bull. Misc. Inf. 1925:325.

3. Baker, R. E. D. 1953. Witches' broom disease of cacao (a review). Trinidad Imper. Coll. Trop. Agr., Rept. Cacao Res. 1945-1951:116-118.

4. ________., and S. H. Crowdy. 1943. Studies in the witches' broom disease of cacao caused by Marasmius perniciosus. I. Imperial Coll. Trop. Agr., Trinidad, Memoir 7, 28 p. -- and --. 1944. (Ditto. II. Ibid. Memoir 8, 28 p.)

5. ________., and P. Holliday. 1957. Witches' broom disease of cacao. Commonwealth Mycol. Inst. Phytopath. Paper 2, 42 p.

6. ________. et al. (W. T. Dale, S. H. Crowdy, R. K. McKee, C. A. Thorold.) 1941-1944. Witches' broom disease investigations. I-XI. Trop. Agr., Trinidad 18:107-116; 19:207-209; 20:5-12, 156-158, 176-181, 188-194, 239-241; 21:170-176, 196-199.

7. Bartley, B. G. D. 1960. Observations on the reaction of young cacao seedlings to artificial inoculation by Marasmius perniciosus Stahel. Inter-Amer. Cacao Conf., Colombia, Proc. 7:104-117.

8. ________., and E. F. Iton. 1959. A note on the infection of Herrania species by Marasmius perniciosus Stahel. Trinidad Imper. Coll. Trop. Agr., Rept. Cacao Res. 1957/58:53-54.

9. Berliner, M. D. 1961. Diurnal periodicity of luminescence in three Basidiomycetes. Science 134:740.

10. Buller, A. H. R. 1934. Omphalia flavida, gemmiferous and luminous leaf-spot fungus (pp. 397-454) In Researches on Fungi, Vol. VI, Longmans Green & Co., London, 513 p.

11. ________., and R. C. Vanterpool. 1926. The bioluminescence of Omphalia flavida, a leaf-spotting fungus. Phytopathology 16:63.

12. Carvajal B., F. 1939. Ojo de gallo (Omphalia flavida). Rev. Inst. Def. Café de Costa Rica 7:535-576.

13. Dale, W. T. 1946. Witches' broom disease investigations. XII. Further studies on the infection of cacao pods by Marasmius perniciosus Stahel. Trop. Agr., Trinidad 23:217-221.

14. Dennis, R. W. G. 1950. An earlier name for Omphalia flavida Maubl. and Rangel. Kew Bull. 1950:434. --. 1951. Some tropical American Agaricaceae referred by Berkeley and Montagne to Marasmius, Collybia or Heliomyces. Kew Bull. 1951:387-410.

15. ________., and D. A. Reid. 1957. Some Marasmoid fungi allegedly parasitic on leaves and twigs in the tropics. Kew Bull. 1957:287-392.

16. Desrosiers, R. 1954. Diversidad genética del cacao como base en la selección de resistencia a la enfermedad de la escoba de bruja. Turrialba 4: 131-134.

17. Dudman, W. F. and R. Nichols. 1959. Examination of culture filtrates of Marasmius perniciosus Stahel for gibberellin-like substances. Trinidad Imper. Coll. Trop. Agr., Rept. Cacao Res. 1957/58:48.

18. Galli, F. et al. (See "Lit. Cit." Chapter II.) 1968. Manual de Fitopatologia, doenças das plantas e seu contrôle. Ceres, São Paulo, 640 p.

19. Garces O., C. 1940. Enfermedades del cacao en Colombia. Min. Econ. Nac., Bogota, Colombia (Bulletin), 99 p. --. 1946. La escoba de bruja (Marasmius perniciosus) del cacao. Univ. Colombia, Fac. Agr., Rev. 6:329-369.

20. Garcia, C. 1956. La escoba de bruja en Colombia 5:19-23.

21. Garcia, C. and G. Naundorf. 1955. Observaciones sobre los cacaotales del Rio Mira, con especial referencia a la enfermedad escoba de bruja. Cacao en Colombia 4:5-9.

22. Conçalves, J. R. C. 1965. Theobroma grandiflorum (Spreng.) Schum. as a source of inoculum of witches' broom disease of Theobroma cacao. Trop. Agr. Trinidad 42:261-263.

23. Harvey, E. N. 1948. The luminescence of living things. Scientific American 178:46-49 --. 1952. Bioluminescence. Academic Press, New York, 650 p.

24. Johnson, F. H. 1948. Bioluminescence: a reaction tool. Scientific Monthly 67:225-235.

25. McElroy, W. D. and H. H. Seliger. 1962. Biological luminescence. Scientific American 207:76-89.

26. Maublanc, A. and E. Rangel. 1914. Le *Stilbum flavidum* Cooke, forme avortée de l'*Omphalia flavida* n. sp. Bull. Soc. Mycol. France 30:41-47. (In Portuguese: 1915. Bo 1. Soc. Agr. Indus. Com., São Paulo 16:69-75.)

27. Montealegre, M. R. 1938. Estudios sobre el café. Rev., Inst. Def. Café de Costa Rica 7:347-357.

28. Nowell, W. 1923. Diseases of crop plants in the Lesser Antilles.

29. Pound, F. J. 1938. Cacao and witch broom disease (*Marasmius perniciosus*) of South America. With notes on other species of *Theobroma*. Youille's Printerie, Port-of-Spain. (Special publ.) 59 p. --. 1940. Search for resistance to withbroom in cacao. Trinidad & Tobago Agr. Soc., Proc. 40:35-37.

30. Puttemans, A. 1907. O *Stilbella flavida* parasita sobre *Tabernamontana*. Rev. Soc. Sci. São Paulo 2:95.

31. Rodriguez, C. J. Jr., and D. C. Arny. 1966. Role of oxidative enzymes in the physiology of the leaf spot of coffee caused by *Mycena citricolor*. Phytopath. Zeitschr. 54:375-384. -- and --. 1967. Effect of light on gemmae formation in *Mycena citricolor*. Canadian J. Bot. 45:675-679.

32. Rorer, J. B. 1913. The Surinam witch-broom disease of cacao. Board Agr. Trinidad & Tobago, Circular 10, 13 p.

33. Saenz, N. 1895. Memoria Sobre el Cultivo del Cafeto. Casa Editorial J. J. Perez, Bogota, Colombia, 185 p.

34. Salas, J. A. 1970. Induction of the perfect stage of *Mycena citricolor* (Berk. & Curt.) Sacc. by

other fungi. Resumenes de Trabajos, Soc. Amer. Phytop., Div. Caribe, Univ. Panama, Enero 4-10, 1970. (Paper No. 9.)

35. de Segura, C. B. 1965. Enfermedades de Cultivos Tropicales y Subtropicales. Editorial Jurídica S. A., Lima, Peru, 439 p.

36. Sequeira, L. 1954. Nuclear phenomena in the basidia and basidiospores of Omphalia flavida. Mycologia 46:470-483. --. 1958. The host range of Mycena citricolor (Berk. and Curt.) Sacc. Turrialba 8:136-147.

37. Sequeira, L. and T. A. Steeves. 1954. Auxin inactivation and its relation to leaf drop caused by the fungus Omphalia flavida. Plant physiol. 29: 11-16.

38. da Silveira, F. 1938. A mancha de ferro do cafeeiro. Rev. Inst. Café, São Paulo 13:1028-1030.

39. Singer, R. 1942. A monographic study of the genera Crinipellis and Chaetocalthus. Lilloa 8:441-534. --. 1949 (1951). The Agaricales (mushrooms) in modern taxonomy. Lilloa 22:1-832.

40. Smith, A. H. 1947. North American species of Mycena. Univ. of Michigan Press, Ann Arbor, 521 p.

41. Stahel, G. 1915. Marasmius perniciosus nov. spec. der Erreger der Krulloten-Krankheit des Kakaos in Surinam. Dep. Landb. Suriname, Bulletin 33, 25 p. --. 1919. Bijdrage de kennis der Krullotenziekte. Dep. Landb. Suriname, Bulletin 39, 34 p. --. 1932. Contribution to the knowledge of witch-broom disease. Trop. Agr., Trinidad 9:167-176. --. De Krullotenziekte in Brasilie. Indies Mercuur 58:71.

42. Uribe Arango, H. 1947. La gotera (Omphalia flavida) del café. Univ. Colombia Fac. Agron., Rev. 7:249-260.

43. Van Hall, C. J. J. and A. W. Drost. 1907. Les balais

de sorciere du cacaoyer provoques par Colletotrichum luxificum n. sp. Rec. Trav. Bot. Neerland 4:243-319.

44. Viégas, A. P. 1961. Indice de Fungos da América do Sul. Inst. Agron., Campinas, 921 p.

45. Vieira, J. T. 1942. Lagartão ou vassoura de bruxa. Soc. Agron. Brasiliero 5:393-400.

46. Wallace, G. B. 1949. Defoliation of crops by a gemmiferous fungus. East African Agr. J. 14: 141-143.

47. Wardlaw, C. W. 1929. Witch-broom in Surinam. Trop. Agr., Trinidad 6:348-349.

48. Wellman, F. L. 1949. Propagación del ojo de gallo por la lluvia en las fincas de café. Suelo Tico, Costa Rica 2:13-17. --. 1950. Dissemination of Omphalia leaf spot of coffee. Turrialba 1:12-27. --. 1951. Tree defoliation for control of Omphalia disease of coffee. Phytopathology 41: 38. --. 1952. The luminescence factor and culture characteristics of the American leaf spot fungus of Coffea. Reunião Lat.-Amer. Fitogen. Fitopar., Campinas, Proc. 2:67-68. --. 1953. Some important diseases of coffee. (pp. 891-869) In Plant diseases, the yearbook of agriculture 1953. U. S. Dep. Agr. Washington, D. C., U. S. A. 940 p. --. 1954. Evidencia de resistencia a las enfermedades en los cafetos. Turrialba 4:52-57. --. 1955. Coffee diseases, insects, and weeds controlled by chemicals. American Chem. Soc., Adv. In Chem. 13:43-63. --. 1962. A few introductory features of tropical plant pathology. (pp. 928-930) In Symposium on tropical plant pathology. Phytopathology 52:928-953.

49. ________., J. C. Walker, A. Cook, and M. Gallegly Jr. 1951. Effects of temperature on vegetative growth of five coffee disease fungi. Asamblea Latinoamer. Fitopar. Mexico, Proc. 1:251-257.

50. Wolf, F. A. 1949. Notes on Venezuelan fungi. Lloydia 12:208-219.

Chapter XVI

ALGAL DISEASE OF PLANTS IN THE TROPICS

Summary

Algae in a primal, moist and warm, perpetually tropical-like earth long ago, were forerunners of the beginnings of plants. At the present time, many kinds of affections occur in the tropics on higher plants from algae: some superficial, some smothering, some internal (both disease causing and benign), some seriously injurious. The commonest plant parasitic alga, *Cephaleuros virescens*, was the subject of much pioneering plant pathological studies in the Oriental-Occidental tropics. It is water-borne in its spread; penetrates as a small thallus down between leaf cells. The host range of this alga is wide, although some plant species appear to remain free from its attack. This alga becomes a lichen on proper combination with a fungus; in certain cases it differs as a parasite depending on whether it is acting as an alga or as a lichen.

Introduction

Algae as photosynthesizing organisms are regarded with special interest by paleontologists. Remains of algae are in stone deposits of extreme age; it is possible that algae and fungi arose at about the same geologic epoch, the Cambian-Ordovician. Algae were the dominant plants on a warm wet earth approximately 400 million years ago. During the Silurian epoch some 130 million years later came a lessened dominance of algae and more early, vascular, land plants, some of which were being parasitized by bacteria, fungi, and algae. Probably all these, plants as well as parasites, were involved in the huge coal deposits of the Carboniferous epoch.

Algae, then, are a sector of organisms of great age and they still grow best in tropical warmth. In temperate zone seas where it is cooler, marine algae develop at depths of from approximately 40 to 60 meters only, and in warm tropical waters they go down to more than twice that far. They may grow free in water and some are attached to solids; some live within--and some live epiphytically outside --other algae and other plants. Some are parasites. (For reviews on these matters, the reader should see such works as: 9, 10, 19, 20 and further citations given by those authors.)

The great algal fauna of thc earth is so extensive that it is studied by specialists (algologists) who spcnd all their lives in fruitful pursuits of scientific information about the algae. While most of these organisms are independent plants growing in brackish and salt waters, absorbing nutrients directly and making their own carbohydrates by photosynthesis, there are also many of them that are inhabitants of fresh waters. There are, moreover, some fresh water algae adapted to aerial growing and a large number of them are common in the tropics. This is true especially in the wet tropics. In many areas both green and red algae occur in felt-like coatings or pubescence on rocks, on the hair of certain animals like the three-toed sloth, on tree trunks and branches, on exposed soil, and on other such surfaces. Among these algae are some that are parasitic; a number are plant disease producing organisms of considerable interest, and some are important.

Various Kinds of Diseases from Algae

In the moist tropics, where conditions are very favorable for algal growth, these organisms are often abundant and at times may be surprisingly troublesome. In my experiences water impounded during rains as a supply reserved for dry season spraying, in a few months can have so much algal growth it has to be discarded. Tanks of drinking water, if left uncovered, even though not contaminated with undesirable bacteria and amoebae, may quickly develop copious algal growths that are unsightly, can be poisonous, and give an unpleasant flavor. Ponds prepared for raising fish soon develop algal growths; under some conditions the "green scum" becomes super exuberant, to the point that the fish may die. At times sweet-water lakes receive massive additions of surface waters that change the balance, and then excessive growth or "blooms" of algae occur which can be deadly poisonous.

When it comes to plant troubles, the difficulties are less forthright, and sometimes more complicated. During years of work in Central America I was taken to, or had sent to me, many examples of what were thought to be plant disease materials of algal nature; such algal materials cause many questions among growers. Using texts and articles such as were available, alga identifications were made as closely as possible. Sometimes only genera could be determined. Many different kinds of effects (see accompanying Table 26) were observed.

Green superficial growth is perhaps the simplest effect and is seen on field soil and on shaded soil in road cuts, on stonework, on tree bark, and in sheltered parts of patios surrounded by walls. It is from a simple terrestrial alga such as Protococcus sp. or Demococcus sp. In tree cavities and wet branch clefts a mixture of algae, along with yeasts and other organisms, are found. Most commonly they are species of Spirogyra or Gleococcus. In such cases, although considered disfiguring by the gardeners, it often occurs that the green and reddish superficial growths, or the accumulations at the edges of cavities or in branch clefts, are actually not of true parasitic nature. Removal of shade and washing with a weak copper sulfate solution controls such algal growth.

When smothering occurs from species of either Spirogyra or Trentepohlia, and this was studied critically, it is always found in long-lasting wet locations. This is ameliorated by pruning away protecting foliage or by improving ground drainage as in the case of low places in fields of sugarcane and lawn or pasture grasses. The algae may not be killed but they no longer flourish and plants then outgrow the algae.

Collected sporophore examples of species of fungi, especially of the genera Fomes, Ganoderma, Poria, Trametes, and Hydnum (if they happen to escape excessive insect ravages), are often so covered and injured in appearance by the erosive effects of algae of species of Chlorella,

Table 26. DIFFERENT PATHOLOGICAL EFFECTS OF ATTACK BY ALGAE on higher plants in the American tropics

Name of algal organism	Occurrence	Effect on host
Anabaena sp.	In roots of *Cycas*	Coralloid root deformation
Cephaleuros minimis	Stems and petioles infected	Penetration of injuries, causes swelling, splitting, strangling, and death
Cephaleuros virescens	Leaves of very numerous hosts, fruits of several	A. Penetration of injuries becomes lesion. B. Penetration of stomata, sometimes with defoliation. C. Excess algal growth shading interferes with leaf photosynthesis. D. With fungi*, develop lichens
Chlorella ellipsoidea, *Chlamydomonas* sp., *Coccomyxa* sp., etc.	On exposed parts of Polyporaceae	Erosion of surfaces
Chlorochytrium lemnae	On leaves of *Lemna*	Mild parasitism, slight stunt
Demococcus sp. (*Pleurococcus* sp.)	Moist ground, rocks, tree bark, etc.	No notable plant infections seen, superficial, develops lichen
Nostoc sp.	In lumen of *Spahgnum* cells	No notable disease
Nostoc sp.	In cavities of *Anthoceros*	No notable disease

Phycopeltis epiphyton	Upper surface of leaves on many hosts	Little or no penetration, some interference with photosynthesis; with fungi becomes lichen*
Rhodochytrium spilanthidis	Red spots on leaves of Artemisia	Mild parasitism, slight stunt
Scytonemeae sp.	Leaves of Acalypha	Penetration of stomata, leaf ultimately dies
Scytonemeae sp.	In pith cells of Gunnera	No notable disease
Spirogyra sp., Gleocapsa sp., etc.	Wet cankers, branch clefts	--
Spirogyra sp.	Bases of sugarcane; lawn forage grasses	Smothering
Stomatochroon sp.	Leaves of various hosts, e.g. Theobroma	Penetration of stomata, becomes systemic; fruits along main vein sides
Trentepohlia sp.	Twigs, both surfaces of leaves	Smothering

* Developments observed by me: lichens formed are classified as Strigula (complanata?); no doubt others of the plant-affecting algae enter into symbiotic relations with fungi, becoming lichens, probably Strigula spp. There must be several Strigulas, all acting as thin white plates and spots on the surfaces of plant parts.

Chlamydomonas and Coccomyxa, that the species and sometimes the genus of the fungus is often hard to determine.

Typical green filaments of species of the alga Nostoc are found in Puerto Rico in openings that occur in thalli of liverworts such as Anthoceros. In material collected in Guatemala, twisted and coiled growths of Nostoc resembling strings of green beads were found in large outside cells of the moss, Sphagnum sp. This also occurs in Costa Rica and no doubt many other countries. In some instances the plants of this moss may be dead, but they appear green from the profuse algal growth inside the dead, transparent cells of their leaves. In such cases as these, at least superficially, there seems to be no adverse pathology involved from Nostoc invasion.

In heavy rainfall regions, in the giant-leaved Gunnera with its soft growth, there is an alga buried in the green pith cells of its rhizomes, and this alga is a species of Scytonemeae. In aerially exposed, coralloid roots of Cycas sp., and Zamia sp., are growths of the alga Anabaena. In all such rhizomes as these, the algae appear parasitic but oddly enough do not seem to have an adverse effect on the aerial growth of the plants in which they are found.

There are numerous trees with tough, smooth and broad leaves of which a few do not dehisce but stay on stems three or four years in the tropics. For examples: mango, codiaeum, certain palms, magnolia, avocado, hevea rubber, coffee, castilloa rubber, and cacao. It is on such old leaves there are found more developments of Phycopeltis epiphyton. Symptoms are flat brown disks, some of which are very reminiscent of certain severe leaf spots and at times cause growers concern. However, this alga, in spite of its frightening aspect at times, is not a serious parasite. It injures the epidermis of leaves, and nutrients from materials are leached out of epidermal leaf cells, but there seems to be no extensive penetration of leaves. On mango and fishtail palm, in a few cases, insect-injured leaves were found in which disks of this alga were seen to grow over and envelop the leaf edges that had been cut by the insect feeding. But even here the alga did not enter the exposed mesophyll on the leaf.

The aquatic plant Lemna sp. is often weakly affected by an alga, and one of the common field weeds, Artemisia sp. has more or less benign red spots on its leaves from an alga. The aquatic plant is apparently infected by Chlorochytrium lemnae and the weed by a red colored alga, Rhodochytrium spilanthidis. These are both examples of distinctly mild parasitism.

The well-known Cephaleuros virescens Kunze is the one most commonly given as an example of a parasitic alga. It is, moreover, quite often encountered as a matter of clinical importance because it has been mistaken for other severe disease producing organisms. Furthermore it is often called rust or red rust (e.g.: 2, 5, 8, 15, 25, 30). In Puerto Rico, 55 years after the coffee rust Hemileia vastatrix had existed for short while in that island but had been successfully eradicated, I was told, seriously, that in special corners the Hemileia rust could still be found on coffee. This was indeed a frightening possibility. A careful search of the areas suggested disclosed to me three narrow mountain valleys in which the alga C. virescens, called red rust by growers, seemed especially frequent on coffee leaves. However, in repeated trips up to 1962, I found nothing on the island of Puerto Rico of any coffee parasite belonging to the Uredinales. (I know Hemileia well from work in Brazil, Africa and the Orient; for a survey of its geographic distribution and in particular its occurrence in Brazil, see Appendix 1.)

In rare instances, for example on leaves of Philodendron hastatum, the alga C. virescens apparently penetrates stomata. Leaf attack is seldom excessive on this host and, as on Allamanda, or avocado (Fig. 16-1), it is only disfiguring in appearance. In some cases on coffees, where there are several algal infections per leaf, defoliation occurs. This is a more serious situation, as the amount of cherry is closely correlated with numbers of leaves that are kept on fruiting branches (29). Where there are only one to three spots per leaf, leaves seem to be retained as well as if uninfected. It is tentatively believed that in some hosts deleterious effects from this alga are probably due both to the disturbing transportation of nutrients and the reduction of photosynthesis. In herbaceous plants like Alternanthera bettzickiana, infections that at times take place through stomata cause more severe effects. A leaf of this lush ornamental, if infected early, can be severely malformed and even one infection on a leaf may cause it to be cast off by the plant. If young growing points of it become infected, the terminal growth is killed.

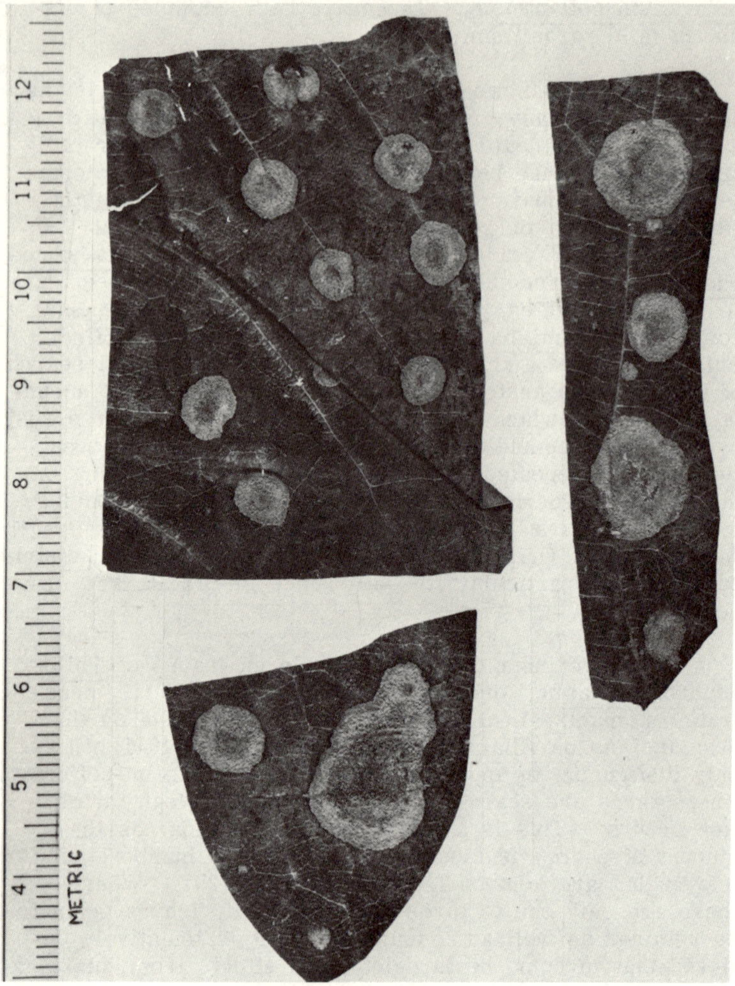

Fig. 16-1. Leafspots from the alga Cephaleuros virescens on avocado. During dry seasons the spot edges, as illustrated here from Ecuador material, appear as more or less entire and smoothly rounded. With coming of the rains the algal colonies grow and the spot edges become mossy and dendritic; the alga filaments elongate and sporangia develop which makes such spots appear so different they may be erroneously called rust.

One kind of infection by a species of the alga _Scytonemeae_ shows a different habit from any other algal attack. This was specially noted on _Acalypha wilkesiana_. The disease produced is known as black mold, symptoms of which are shown in Fig. 16-2. The algal swarm-spore apparently enters a substomatal cavity and soon begins division to make up a thallus. This develops as a filament, and enters natural openings in the mesophyll. From it the dark, olive-brown filaments grow out through stomata into the air on the underside of leaves. Here they form a brownish symptom called black mold and after a time infections cause plants to drop leaves.

The case of _Cephaleuros minimis_ in its invasion and effect on hosts, is somewhat different from its congener _C. virescens_. The latter with its large sporangia appears more distinctly as a leaf-inhabiting organism. With _C. minimis_ which has much smaller structures, the alga pierces twig and petiole tissues, and is not on leaves. Hand-cut cross sections of fresh twig tissue were examined under the microscope; hyperplasia results in cambial host tissues if they are sufficiently young at the time of attack. In old infections on _Theobroma cacao_, after twig swelling, the bark splits apart and the reproductive stages of the parasite emerge in the air from between flakes of cortex. The rifts in bark and cortex are sites of secondary invasion by certain bacteria, fusaria, diplodia, colletotrichum and other weak parasites.

A specialized type of algal disease was studied from the effect of infections by _Stomatochroon_ sp. This was seen in moist jungle conditions, in Costa Rica, and found on leaves from a species of _Inga_, from an artificially inoculated naturalized _Rosa_ sp., on leaves of _Theobroma cacao_, on leaves of _Croton_ sp., and on leaves of a wild, broad-leaved unidentified evergreen bush of the family Rubiaceae. The alga was found only present in moist and well-shaded locations. It infects through stomata and becomes systemic. Once within the mesophyll, the algal thallus grows along the main veins of leaves, and finally the fruiting structures emerge at considerable distances from where the original infection takes place.

A Specially Studied Parasitic Alga

The alga _Cephaleuros virescens_ (= _C. mycoidea_,

Fig. 16-2. The black mold of Acalypha leaves from Scytonameae hofmannii. The parasite is seen as masses of black filaments on undersides of the leaves, and after weeks the leaves drop. (This material collected in Honduras by P. C. Standley.)

Mycoidea parasitica) for a long time has been a disturbingly present plant pathogen in the tropics. It was first studied from a period of before 1827, to 1913 in the Dutch East Indies and in the American tropics (5, 6, 7, 9, 11, 23, 26). In a 1929 review of the development of plant pathology in some 50 years of successful progress up until that date, the then dean of the science, E. F. Smith (18), emphasized researches done on this organism. What had been found about *C. virescens* and the disease it causes in its tropical habitats was one of the early and great milestones in plant pathology. At the present writing, in the Neotropics it is still the commonest of the parasitic algae and is still being given attention (1-4, 7, 8, 11-14, 16, 17, 21, 22, 24, 25, 27-30). It is known in the Oriental as well as Occidental tropics; its geographical distribution in the western hemisphere extends as far north as the southern part of North Carolina, where it is found in well-sheltered locations, and south, far down into Argentina. It is effectively killed by the annual occurrence of a few weeks of freezing weather. From the standpoint of moisture, the parasite requires enough humidity that water drops are held on leaves for several hours at a time.

Symptoms of attack by *C. virescens* are varied. It causes scurf-like types of irregularly edged leafspots. The most serious effect from leaf infection is early leaf fall. Fruits of fig, mango, coffee, guava, and hevea show scabs from infection and are badly blemished and malformed by it. It girdles lushly green plant tips; woody twigs are sometimes split open and the bark roughened from attack. These may then become infected by anthracnose and, later, dieback results.

This alga, it will be remembered, was, and is, known as the red rust of tea in the Eastern tropics. It is also spoken of as a rust in the American tropics not only on tea, but also on coffee, on citrus, on avocado, on guava, on *Lagerstroemia*, and on many other trees and ornamentals. At one time the greatest fear by rubber growers in the Western Hemisphere was that the *Oidium*, the dread mildew disease of hevea, would reach the Occidental tropics. Even as well-known as mildew is by pathologists, rubber leaves with Cephaleuros infections have been brought and sent to me from far corners of Brazil, Colombia, Venezuela and Peru, from growers in countries in the American tropics because the leaves were thought to be mildewed.

Germination and Infection Phenomena

To make controlled inoculation studies, diseased fresh leaves were brought in from various hosts. Work was in a warm laboratory in Puerto Rico, but if the algal material was kept moist in a glass moist chamber, but not wet, the sporangia survived a few days. Under these conditions if the sporangia are then made wet with water they soon germinate. The swarm spores that emerge have flagellae that whip them about which makes them active for a day or more in drops of water. In many instances they do not settle down immediately, but eventually some come to rest.

The swarm spores can readily be watched on glass slides. When sporangia are germinated in distilled water, the spores swim around rapidly with jerky movements. In a good many instances they suddenly cease activity under the scope and literally explode. I observed this same phenomenon and corroborated the observations reported by Wolf (30) and suggested by Ward (28).

When sporangia are placed in rain water on a leaf surface, however, they do not explode. Here they may be moving for many hours but then slow down and sink to the leaf epidermis. The swarm-spores are seen to aggregate into small thalli of a few cells; thus begun, final development after anchoring starts in about 5 days. Up to the time this occurs, if only a single swarm spore is present it does not stick to the leaf. Before they become anchored, even thalli of a few cells are washed off rather easily. They are also killed if a straw decoction full of saprophytic bacteria is introduced in the water drop around young thalli.

In studies I made about one out of approximately 50 of these small thalli grew to a size that could be seen by the naked eye; only about one in 100 of them finally stuck to a leaf and penetrated the epidermis. It appears that where a thallus sticks it is in a place where there is some unevenness of the epidermis, or an injury of some kind, that allows the alga to lodge and begin attack.

The details of histology of the parasitism by *Cephaleuros virescens*, remain about as recorded in drawings and discussion by Wolf (30) and before him by Ward (28); not much has been added by later work. In my own observations, that corroborate older findings, it is apparent that an algal thallus sticks somewhere because it gains some kind of entrance into the leaf tissue, pushing between cells. Free-hand sections reveal that the alga pushes its cells underneath the leaf epidermis; it does not break into these host cells, but it is squeezed beneath them. Cells are evidently under pressure, for they are forced into unnatural shapes and in some cases this stimulates cell division where it would not occur otherwise. In one avocado leaf under one lesion I found four, and under another lesion five, rows of abnormally small mesophyll cells. In a normal part of the leaf there was only one row of cells.

It was impossible for me to be certain whether or not the alga affects the host by poison or enzymatic action. There is no general yellowing, or development of discolored haloes around lesions. In some cases, with enough leaf spots there is unseasonable defoliation, and this seems to be due at least partly to the injury to nutrient absorption by the parasite. It may also come from the strong shading of the host leaf by excessive amount of algal lesions on its upper surface. In some instances when I was studying coffee and mango problems in the West Indies and in Central America, no leaf defoliations took place if only superficial growth of mosses and liverworth occurred on leaves. However, on the same trees, leaf drop was occurring induced solely by numerous infections of the *C. virescens* alga.

It is repeatedly stated by pathologists, and it is given in texts, that *Cephaleuros virescens* will infect hosts that have leaves of coreaceous, that is leathery, character. A classic statement is that given by Ward (28) who said it "occurs on the upper surface of the leaves of so many plants widely separated in affinity and origin, that one must regard the species of the supporting plant as an accident. One general feature, however, is common to all these leaves: they are invariably hard and persistent, differing moreover in degree in this respect." However, I found in El Salvador that distinctly non-leathery tomato leaves were attacked by the organism, and that the tender, lush, brilliantly colored, soft leaves and stems of *Alternanthera* were also infected. In other countries it was found on numerous such thin, rough or smooth, but relatively uncoreaceous leaves as those on plants of *Commelina*, wind lily ("flor de pasqua"), tobacco, mulberry, *Acalypha*, *Antignon* ("coralita"), and *Desmodium* sp.

Host Ranges of the Common Alga

In Puerto Rico, I had the good fortune to be close to algal-diseased, old, mixed plantings of various trees and plants. There were a total of 23 plots; 16 gave most interesting results and these are represented in the final table of this chapter. In each of these, plants and trees of from five to eight species were grown closely together, sometimes accidentally and sometimes by design, but all were located so they had a chance of being well inoculated by *C. virescens* and/or *S. complanata.* At least one host species in each plot was badly diseased and served as an undoubted source, a fountain of abundant inoculum, that could readily reach the other species growing under or with branches mixed in with it. In these special plots, alone, were 34 alga susceptible species. Wider observations were also carried on in visits in Central America and other West Indies islands. Notes were made from 1950 to 1963 in many locations and during that time 158 hosts were noted as susceptible to the algal leafspot. In connection with such a host range study, it is well to recall that Wolf listed over 50 susceptible species (28) and that Ruehle (16) reported he had found 44 species in south Florida that were susceptible to infection by *C. virescens.*

There is an especially interesting list from Brazil of host plants susceptible to *C. virescens.* In looking over plant disease herbarium collections made by a number of workers over several years and deposited in the main herbarium, it was determined by Batista and Lima (3) that hosts attacked by *Cephaleuros virescens* comprised 448 different species. The plants that were diseased ranged from ferns to the Compositae. In character of leaf, the range was from soft types such as *Commelina* sp., through intermediate kinds, to hard-leaved sorts such as *Magnolia* sp. There is no doubt that more attention would demonstrate that in Brazil the host range would be even larger.

It is of interest to look for reports of the occurrence of this parasitic alga in the coldest and the warmest parts of the United States (see U. S. Crops Research Division (25)). These reports are mainly from the temperate zone of the country. In the United States Department of Agriculture's *Index of Plant Diseases,* general disease notices cover a total of 3390 crops, ornamentals, and other plants. Among these species are 767 introduced from the tropics and grown in warm protected places along the northernmost edge of the Neotropics, such areas as gulf-bordered parts of Texas, Louisiana, Alabama, Mississippi, and Florida, and in warm-moist areas of California. On looking through this list, it is evident from what is known that--conservatively speaking--much over half of the tropical plants growing in the continental United States and mentioned in the work are susceptible to algal attack. However, in this _Index_ only 17 are reported as suscepts of these 767 introduced hosts growing in the continental U. S. A. But if reports on the tropical plants did include observations from Puerto Rico, Hawaii, and the Panama Canal zone, 52 hosts were noted as attacked--indicating further the truly tropical nature of the disease.

In researches on a tropical parasite that has demonstrated as wide a host range as the common leafspot alga, it is perhaps of greater importance, as a next step, to know more about those hosts that are not susceptible. In this regard, only a start has been made (see in part the table at the end of this chapter).

In my observations in Puerto Rico, _Codiaeum_ (the greatly variable so-called "croton" of hedges and other tropical plantings) is not attacked by _Cephaleuros virescens._ Repeated visits and search did not disclose such infections on it in the island. However, it had been previously noted that in Honduras and in Costa Rica the alga infects a variety of _Codiaeum_ there, apparently one of its specially different variable forms. It is never seen as severe on this particular garden croton but it was serious enough to be brought to the pathologist's laboratory by gardeners for purposes of diagnosis and control suggestions.

The large ornamental _Lagerstroemia speciosa_ trees, called queen of the flowers, growing in Puerto Rico, are found to be highly susceptible to the alga (Fig. 16-3). An increased amount of the infection had built up in foliage of this tree along a driveway in Rio Piedras. Lesions averaged over 50 per leaf, sometimes as much as 110. There was considerable variation in detailed appearance of vegetative characters among these trees, and some that were badly infected by the alga showed early, unseasonable, leaf fall. Out of over 30 trees in one location, two were seen that seemed specially vigorous and during five years of observation averaged, instead of about 50, some four or five alga lesions on leaves. It was startling to note that bushes of the crepe myrtle, _Lagerstroemia indica_, that is

Fig. 16-3. Leaf infection by Cephaleuros virescens of the tree Lagerstroemia speciosa. The leaves are affected on their upper or dorsal surfaces, and in Puerto Rico as shown here on the largest leaf, there may be so many spots that there is more surface covered with alga than without.

closely related to L. speciosa, did not have alga infections.

It seems to me of great interest that Philodendron hastatum, an herbaceous climber, was observed to become severely attacked by the alga but only on the underside of the leaves. Usually it is the upperside of a leaf that is infected. This P. hastatum was growing intertwined and thoroughly mixed with another species of the vine, P. nepenthites. The leaves of the latter, under and touching leaves of the former that were severely diseased, remained absolutely free from algal infection.

Another observation seemed of special significance to me in some of my work in Puerto Rico. This experience indicated the probably occurrence of varietal resistance. In an old, thoroughly mixed, plot of several kinds of tropical fruit and weed trees, there was a large amount of leafspot from C. virescens. But in the tangle, located in such a position that it was showered year after year with natural inoculum of alga, was growing an unnamed seedling orange tree which had every one of its more than 1200 leaves free from algal leaf spot. Within this same plot, next to the seedling, were two other orange trees of the Parson Brown variety. These had an abundance of algal infection on leaves except the youngest that were then in flush.

Change in Parasitism of an Alga When It Becomes a Lichen

Uncer certain conditions Cephaleuros virescens becomes part of a lichen. It is well known that a lichen is a complex organism composed of fungus and alga in symbiotic union. In this symbiosis, alga and fungus are so intimately mingled that the lichen operates as a different growth and in the case of C. virescens with its fungus, it becomes the well-known lichen, Strigula complanata Née. Although since the days before the beginning of the present century C. virescens has most commonly been considered as the basic algal ingredient of S. complanata, I have observed what I determined as this same lichen occurring on foliage which had no other alga than Cephaleuros minimis attacking twigs bearing lichen-spotted leaves and in numerous cases abundant Strigula growth has been seen and the only accompanying alga found was Phycopeltis sp. In the case of two talipot palm trees with leaves liberally sprinkled with a Strigula lichen, the only alga on the palm trees was Trentepohlia sp.

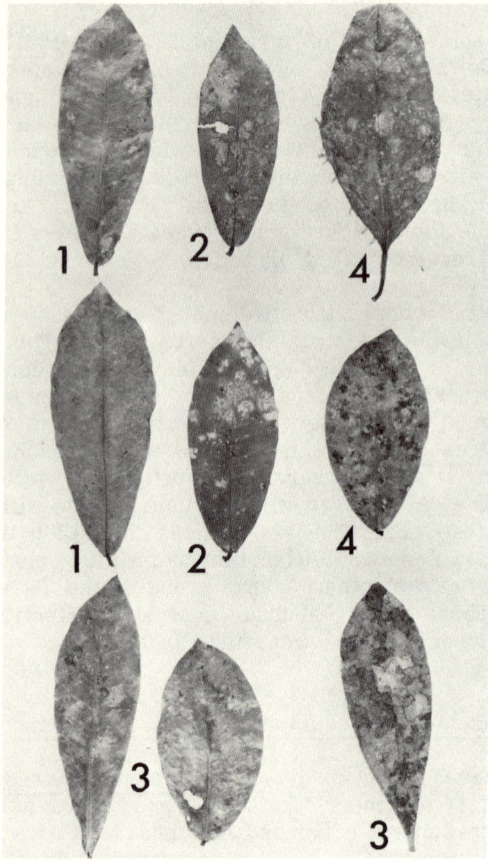

Fig. 16-4. Accumulation of lichens, other thallophytes, epiphytic micro-liverworts on leaves of an ornamental tree, <u>Citrus</u> sp., collected in Panama at different ages: one year old (1), young colonies of <u>Strigula</u> barely visible macroscopically, non-parasitic bacteria and fungi found microscopically; at two years old (2), <u>Strigula</u> colonies more evident; at three years old (3), green and gray lichens at sides of and growing over white <u>Strigula</u>; and at four years old (4), burdened with lichens, bacteria, fungi, a <u>Geococcus</u>-like alga, the alga <u>Trentepohlia</u> sp. (see tufts on left edge of top leaf), and on both oldest leaves plants of micro-liverworts. All leaves alive but after first year coarsely stiff and hardened.

This phenomenon of lichen development is often spoken of as an example of symbiosis; the connotation being there is a mutual benefit. The usual concept is that after the alga-fungus combination is well established the lichen it forms is adapted to reproduction by bits of tissue called gonidia and gemmae, of the lichen body. These are carried away by rain drops. On lodging upon a proper surface, under proper conditions, a gemma then germinates and is able to grow, forming another lichen colony. However, it is possible to see at times the alga growing separate from the fungus, and then at times the process of the combination going on between the two can be detected. Microscopic examinations show the fungus wrapping around and killing some of the algal cells, indicating to me, the alga is in part a parasitized entity, the fungus gaining from its association but not necessarily being seriously dangerous to the alga. (In moist tropics often the lichen *Strigula* may be the first macroscopically observable thallophyte to settle on a long-lived leaf. Microscopic study almost invariably shows scanty threads of fungus and some bacteria on the leaf epidermis between lichen colonies. Soon algae are apparent to the naked eye--see accompanying Figure 16-4--and as well miniature foliaceous liverworts and lichens other than the original *Strigula* are visible, in some cases completely obliterating the first occurring lichens.)

While the first detailed studies on this lichen came from the Oriental tropics, researchers and observers found it abundant in the Occident. It is surprising to some that studies on the evident complexities of the relationships between tropical alga and fungus, first determined nearly a century ago, have not been given more modern attention. The reason of course is that extremely serious pathological problems of other nature have crowded onto the scene in the American tropics. These urgent prospects require the almost complete devotion of workers to matters of crisis in virulent disease epidemics. The tremendous advances in techniques, development of pure culture methods for both algae and fungi, the availability of the electron microscope, handling of tracer elements in physiology, and like things, would yield answers of greatest interest to the understanding both of the parasitism of the algae and to the formation and development of the lichen as it affects plant disease lore.

Scrapings and hand-cut sections of *Strigula complanata*, which I have seen under the microscope, had in them probably mostly Ascomycetous fungi. Under such conditions as during the beginnings of rainy seasons, and under heavy shade where leaves never really dry off, spores were recovered of species of *Pestalotia*, a short-spored *Cercospora*, a *Diplodia*, a *Colletotrichum*, a small-spored *Fusarium*, a large-spored *Curvularia*, a *Phoma* (?), some medium sized elongated double spores resembling ascospores of *Gibberella*, and occasionally large double spores of an undetermined fungus with part of the sporophore attached. I saw no suggestion of a Basidiomycetous fungus as part of the lichen.

Strigula complanata, is considered a serious economic pest in some areas on certain crops. The white and/or gray color, so characteristic of it, is probably an effect from the fungus constituent(s)* more than from the algal elements. It has been given such descriptive names as milky disease, ashes and lime, cenizio, cal enfermedad, lime splatter, white lichen, and cream spot. It appears as a thin layer, and colonies are more or less circular in outline. These spots, markedly different from alga spots (see comparisons in Fig. 16-5), are mostly on the upper surfaces of leaves and seldom on the undersides. They are of such a structure that although closely stuck to leaves they may be scraped off. When thus taken off, and free-hand cross sections of a parasitized leaf are examined under the microscope, it has been found repeatedly that there is some penetration of the leaf epidermis. This results in a type of mild injury; in addition, the closely-stuck white plaques also serve to shade green leaf tissues, and this reduces photosynthesis.

Concluding solely from results of my observations on natural field infections, there are four different kinds of host relationships as to occurrence of or freedom from the parasitic alga or the parasitic lichen. (The reader should realize that some of the hosts may be shifted from certain of these groups if they are artificially inoculated, under most ideal controlled conditions.)

1. Hosts susceptible to *Cephaleuros*, not to lichen: Bougainvillea vine, flowering Hibiscus or common amapola, Boston fern, *Antignon* or coralita, the low growing bulbous windflower, wine plant or Alternanthera, and tomato.

---

*Unfortunately other work never allowed me to make isolations and to grow pure cultures of the organisms involved in the lichen *Strigula complanata*.

Fig. 16-5. On leaves of rubber (*Hevea brasiliensis*), green lesions of the parasitic alga *Cephaleuros virescens* and small white colonies of *Strigula complanata*, the common lichen. In a few spots the alga is evidently in process of combining with a fungus to become lichen--for two examples see arrows. Collected in Honduras.

2. Hosts susceptible to *Strigula*, not to alga: mangosteen, Tabebuia, a species of *Commelina*, coconut palm, African oil palm, Talipot palm, *Calliandra*, loquat, *Diospyros*, *Pandanus*, *Achras*, cupey, wild fern, *Allamanda*, ivory nut palm or *Phytelephas*, and the ornamental reclining Phoenix palm.

3. Hosts susceptible to both *Cephaleuros* and *Strigula*: cacao, ramboutan, coffee, pulusán, hevea rubber, mango, vanilla vine, *Eugenia* or Malay apple, *Dillenia*, *Cryptostegia*, certain species of Citrus, acerola, Sapodilla, Macadamia, *Flacourtia*, *Melaleuca*, bush jessamine, three species of mulberry, tamarind, nutmeg, and guava.

4. Hosts not susceptible to either alga or lichen: at least 11 species of bamboo, banana and its close relatives, heliconia, giant gardenia, Monstera, two species of royal palm, pothos, areca palm, the Montezuma tree, a *Paspalum* grass, crepe myrtle, Terminalia, and one undetermined species of Desmodium.

A special situation occurs with regard to parasitism by the lichen on the spectacular ornamental vine *Congea*. It was studied several times both in the West Indies and in Central America. The most showy part of the plant is the colored flower bracts and these had spotting from the lichen *Strigula* but no infections by the alga *Cephaleuros*. Under ample exposure all the true leaves of the vine are susceptible to algal infection.

It is possible (probable) that there are biologic races of *Cephaleuros* and of *Strigula*. During my work, it appeared many times, in Central America and the West Indies, that the alga does not infect leaves of such trees as coconut palm, African oil palm, and date palm, although in these same regions there is abundant attack by the lichen *Strigula*. However, in Brazil, according to Batista and Lima (3), the palms are definitely diseased by *Cephaleuros*, suggesting a race difference between the alga in the Caribbean and the Brazilian locations.

Wolf (30), in his work on *Cephaleuros virescens*, did not report the occurrence of the lichen *Strigula complanata* on host leaves in the areas bordering the states on the Gulf of Mexico as it affects southern continental United States. This needs further study and more observations. It would be of special significance to determine if a non-lichen-forming race of the alga exists; it may be there are ecological reasons for the lichen only occurring farther south towards the heat equator in the tropics. Certainly the myriad of different fungi in south Florida, Louisiana, and such regions, where C. virescens occurs, should have in them species that might well combine with and parasitize the alga and cause the lichen to be developed.

Field observations in Costa Rica, El Salvador, and Puerto Rico indicate that the lichen can multiply as a lichen, Strigula complanata, without any spots of pure alga near; Strigula has been seen many times to be spreading as a lichen and nothing else, from leaf to leaf, and from one host to another.

Tabular Comparison of Parasitic Occurrences of Alga or Lichen

There is at times an interesting difference in the relative amounts of the alga, Cephaleuros virescens, and of the lichen, Strigula complanata, developing on hosts. Some attention was given to this at various times in the field in moist parts in several locations in Puerto Rico. Final results on some of the observations are represented in Table 27 at the end of this chapter. It is only Plots L, M, and N of this table that are closely comparable. They were near each other on one farm and data on them were secured during part of one morning's visit.

In all plots, groups of trees and plants were regarded as equally inoculated. Over a plot a tree (a vine in one case) was growing that showed ample amounts of long established living inocula. Tree and plant individuals in plots had about the same possibilities of receiving infective materials from the inoculum source. Estimates of spot amounts (see footnote b of Table), of alga and lichen were made at near the closes of the 1958, 1960, and 1961 rainy seasons. This is the best time for final observations on rain-splash carried disease.

The algal spot of Table 27 is one that is readily determined, and there was no doubt as to its presence, its pathogenicity, or its specificity as C. virescens; the lichen spot is always called S. complanata but it may have varied as to some of its fungal components. It has not been determined experimentally, but fungal parts of a lichen complex may be from any one of several of the macroscopically invisible (but commonly present in the tropics) fungi, growing as fine filaments over leaf surfaces. The lichen in the Table is always a moderately thin structure seemingly plastered on a leaf. Under the plaque it forms there can be demonstrated in the host a slight invasion of epidermal cells. Although not easily washed off with water, a lichen spot can be scraped away with edge of a dull knife or a thumb nail.

Observations were made of things of special interest to me in certain Plots:

In Plot B. It is notable that in three species of *Musa* (with regard to the species concept in this genus see Appendix 2 at the end of this book) there were no leaf infections from either alga or lichen. Many observations before and since this time, on bananas and plantains both in plantations and growing scattered in jungle edges, corroborate the conclusion that these hosts are not susceptible to alga or lichen, at least in Puerto Rico.

In Plot D. In this location, old bushes of the commonly grown, red-flowered, and popular ornamental called amapola, *Hibiscus rosa-sinensis*, was liberally spotted with alga. Next to one such bush was growing another, especially large flowered importation from the Orient, also an old *H. rosa-sinensis* bush, whose branches were intermingled with the common amapola. The special large-flowered hibiscus, surprisingly, had no algal spots on its leaves.

In Plot F. One orange tree had on its lower leaves a quite heavy infection of *Strigula* but no alga, though there was much of the latter nearby; it could not have escaped receiving inoculum of alga. In another part of this Plot, a grass, *Paspalum* sp., that was shaded and without question liberally showered with both alga and lichen showed symptoms of neither. The three large seedling hevea rubber trees, sources of abundant inoculum for the Plot, were visited several times to see about spot occurrence on fruits. After three months it was found that fruits showed algal spots and in some cases malformations from alga infections--but there were no lichen symptoms on fruits while lichen and alga spots were numerous on leaves next to fruits.

In Plot K. A severely alga-spotted bougainvillea vine was growing freely and it was draped over a dwarfed and much encumbered, six-year-old *Roystonea* palm. However, no palm fronds were infected. About the palm tree trunk, and climbing through the bougainvillea growth was a vine of an undetermined, wild species of Vanilla. On its leaves and stems were found many spots of both alga and lichen. At the base of the diseased bougainvillea vine--intertwined with the diseased vanille--was a large clump of the low-growing Commelina virginiea that was absolutely free from either alga or lichen.

In Plot M. Under a mango tree, with many alga and lichen leaf spots, at its drip edge was observed a bush of Acalypha with both Cephaleuros and Strigula leaf infections. This was of special note as the only alga I ever saw before on leaves of this bush was Seytonemeae sp. Around the Acalypha and scattered beneath the mango were more than 50 plants of a wild fern, but no leaves of this fern showed either lichen or alga spots.

Table 27. OBSERVATIONS OF ALGA AND LICHEN ATTACK--the alga Cephaleuros virescens and the lichen Strigula complanata (the alga combines with a fungus to produce the lichen)--in Puerto Rico.

Plot	Source of inoculum[a] over plot	Hosts in plot	Amounts[b] of: alga	lichen
A	Nephelium lappaceum	N. lappaceum	5	2
		Mangifera indica	1	2
		Garcinia mangostana	0	2
B	Theobroma cacao	T. cacao	2	2
		Musa sapientum	0	0
		M. cavendishii	0	0
		M. textilis	0	0
		Elaeis guineensis	0	3
C	Coffea arabica	C. arabica	2	2
		Artocarpus communis	1	3
		Lagerstroemia speciosa	3	1
		Vanilla planifolia	3	4
		Citrus sinensis	2	4
D	Eugenia malacensis	E. malaccensis	3	1
		Codiaeum indica[c]	1	2
		Acalypha hispida	1	1
		Tabebuia pentaphylla	0	3

Plot	Source of inoculum[a] over plot	Hosts in plot	Amounts[b] of: alga	lichen
		Cryptostegia grandiflora	1	2
		Recdia edulis	0	4
		Hibiscus rosa-sinensis[c]	0	0
		H. rosa-sinensis	3	0
E	Dillenia indica	D. indica	5	1
		Colocaprum antillum	0	0
		Calliandra portoricensis	0	3
		Mangifera indica	2	3
		Malpighia glabra	1	2
F	Hevea brasiliensis	H. brasiliensis	4	4
		Citrus sinensis[c]	0	4
		C. sinensis	2	4
		Morus alba	4	4
		Tamarindus indica	3	3
		Myristica fragrans	3	3
		Paspalum sp.	0	0
G	Psidium guajava	P. guajava	5	3
		Melaluca leucodendron	2	5
		Noronja emarginata	2	3
		Citrus sinensis	1	3
		C. sinensis[d] hybrid	4	5
H	Artocarpus integrifolia	A. integrifolia	5	1
		Diospyros mobola	0	4
		Citrus sinensis	2	4
		Ocrasia eliptica	0	1
		Psidium guajava	3	4
		Philodendron (pothos)	0	0
I	Tamarindus indica	T. indica	2	4
		Cinnamomum zeylanicum	2	1
		Philodendron (pothos)	0	0
J	Morus nigra	M. nigra	5	4
		Vanilla planifolia	4	3
		Roystonea borinquena	0	0
		Cynodon dactylon	0	0
K	Bougainvillea spectabilis	B. spectabilis	5	0
		Vanilla sp.	3	3
		Roystonea borinquena	0	0
		Hibiscus rosa-sinensis	4	0

Plot	Source of inoculum[a] over plot	Hosts in plot	Amounts[b] of: alga	lichen
		Commelina virginiana	0	0
		Polypodium sp.	2	0
L	Phytelephas macrocarps	P. macrocarpa	0	5
		Polypodium lycopodioides	0	0
		Allamanda violaceae	0	3
		Codiaeum variegatum	0	1
		Cordyline terminalis	0	5
		Phoenix sp.	0	5
		Chrysalidocarpus lutescens	0	0
M	Mangifera indica	M. indica	3	5
		Philodendron hastatum[e]	2	1
		P. nepenthitis	0	0
		Caladium sp.	0	1
		Acalypha godseffiana	1	3
		Polypodium sp.	0	1
N	Morus nigra	M. nigra	5	3
		Phoenix reclinata	0	5
		Montezuma sp.	0	0
		Manilkara nitida	0	0
O	Rourea glabra	R. glabra	5	5
		Phyllanthus urenaria	0	0
		Commelina nudiflora	0[f]	0
		Jesamine pubescens	3	5
		Centrosema sp.	5	2
P	Lagerstroemia speciosa	L. speciosa	5	2
		Codiaeum variegatum	0	0
		Bamboo (large type)	0	0
		Bamboo (small type)	0	0
		Ficus nitida	5	5
		Alternanthera sp.	5	0
		Mangifera indica	3	4

[a]Inoculum always occurred in abundance on host listed as source of inoculum.

[b]Amounts were estimated as follows: 0 = no spot; 1 = a few, one or two spots on a number of but not all mature leaves; 3 = medium amount, from three to four spots on most mature leaves; 5 = very numerous; from five to

twenty or more on most of the large mature leaves.
[c]A special variety of C. variegatum.
[d]Hybrid of C. sinensis and C. grandis.
[e]Alga was on undersurface of leaf, lichen was on uppersurface.
[f]Infections found in other locations.

Literature Cited

1. Albuquerque, F. C. 1964 (pp: 312-314, diseases of black pepper in the Amazon region, report in Portuguese and original not seen). Anais do XIII Congresso da Sociedad Bôtanica do Brasil, Jan. 1962. Univ. Recife, Inst. Mycologia, 512 p. (Rev. Appl. Mycol. 44:118-119. 1965).

2. Alvarado, J. A. 1935. Tratado de Caficultura Practica, Tomo I. Tipografia Nacional, Guatemala, 524 p.

3. Batista, A. C. and D. A. Lima. 1949. Lista de suscetiveis da alga Cephaleuros mycoidea Karst. em Pernambuco. Bol. Sec. Agr. Indust. Comer., Pernambuco, Brazil, 16:32-46.

4. Ciferri, R. 1930. Phytopathological survey of Santo Domingo, 1925-1929. J. Dep. Agr., Porto Rico 14:5-44.

5. Cook, M. T. 1913. The diseases of tropical plants. Macmillan & Co., London, 317 p.

6. Cunningham, D. D. 1879. On Mycoidea parasitica, a new genus of parasitic algae, and the part it plays in the formation of certain lichens. Trans. Linnean Soc. London, Bot. Ser. 1:301-316.

7. Delacroix, G. 1911. Maladies des Plantes Cultivées dans les Pays Chauds. A. Challomel, Paris, 595 p.

8. Fawcett, H. S. 1936. Citrus diseases and their control. McGraw-Hill, New York, 656 p.

9. Fitting, H. 1910. Ueber die Beziehungen Zwischen den Epiphyllen Flechten und den von ihnen bewohnten Blaettern. Ann. Jahrd. Bot. Buitenzorg, Suppl. 3:505-509.

10. Fott, B. 1959. Algenkunde. G. Fischer, Jena, 482 p.

11. (Freeman, W. G.) 1914. An algal disease of cacao, Cephaleuros virescens. Bul. Dep. Agr. Trinidad & Tobago 13:263-264.

12. Marchionatto, J. B. 1926. Fitoparásitos de la Argentina nuevos o poco conocidos. Physis 8:367-372.

13. Nowell, W. 1923. Diseases of crop plants in the Lesser Antilles. West India Committee, London, 383 p.

14. Orjuela Navarete, J. 1965. Indice de enfermedades de plantas cultivadas en Colombia. Ofic. Invest. Esp., Inst. Colombiano Agrop., Boletin Tec. 11, 66 p.

15. Raciborski, M. 1900. Parasitische Algea und Pilze Java's I. Theil. Staatsdruckerei, Batavia, 39 p.

16. Ruehle, G. D. 1941. Algal leaf and fruit spot of guava. Phytopathology 31:95-96.

17. de Segura, C. B. 1965. Enfermedades de Cultivos Tropicales y Subtropicales. Editorial Jurídica, S. A., Lima, Peru, 439 p.

18. Smith, E. F. 1929. Fifty years of pathology. Internat. Plant Sci. Congr., Proc. 1:13-46.

19. Smith, G. M. 1933. The fresh-water algae of the United States. McGraw-Hill, New York, 716 p.

20. (________.) 1963. Algae. Encyclopaedia Brittanica 1:596-607.

21. Stevens, H. E. 1918. Florida citrus diseases. Florida Agr. Exp. Sta. Bulletin 150, 96 p.

22. Stevenson, J. A. 1918. Citrus diseases in Porto Rico. J. Dep. Agr. Porto Rico 2:43-123.

23. Thomas, N. 1913. Notes on Cephaleuros. Ann. Bot. 27:781-792.

24. Toro, R. A. 1929. Plant disease notes from the Central Andes, II. Phytopathology 19:969-974.

25. U. S. Crops Research Division, Plant Disease Surveys. 1960. Index of plant diseases in the United States. U. S. Dep. Agr. Handbook 165, 531 p.

26. Ward, H. M. 1887. On the structure, development, and life-history of a tropical epiphyllus lichen. Trans. Linn. Soc. London Bot. Ser. 2; 88:87-119.

27. Weir, J. R. 1926. A pathological survey of the para rubber tree (*Hevea brasiliensis*) in the Amazon Valley. U. S. Dep. Agr. Bulletin 1380, 129 p.

28. Winston, J. R. 1938. Algal fruit spot of orange. Phytopathology 28:283-286.

29. Wellman, F. L. 1961. Coffee botany, cultivation, and utilization. Leonard Hill; and Interscience Publ., New York, 488 p. --. 1962. A few introductory features of tropical plant pathology (pp. 928-930) *In* Symposium on tropical plant pathology. Phytopathology 52:928-953. --. 1965. Pathogenicity of *Cephaleuros virescens* in the Neotropics. Phytopathology 55:1182.

30. Wolf, F. A. 1930. A parasitic alga *Cephaleuros virescens* Kunze, on citrus and certain other plants. J. Elisha Mitchell Soc. 45:187-205.

Chapter XVII

PARASITISM BY: MISTLETOE, DODDER, TREE

Summary

Non-parasitic plants: orchids, long mosses, strangling figs, bromeliads, many others damage tropical tree growth with tearing weight. The large parasitizing family, Loranthaceae, of flowering, fruit-producing, upright to trailing bushes cause different degrees of disease production from severe tissue irritation and death of host parts, to submergence in host tissue and almost benign development. Parasitism may be from direct primary infection by seedlings, to action by exogenous roots absorbing through sucker-like organs (haustorial pads or hapters). Some Loranthaceae are known that are root-infecting and some are at most semi-parasitic, later becoming free-living bushes or small trees. A large parasitic tree of the family Santalaceae is known in Brazil that attacks other tree roots and kills cacao. Slender voracious parasitic vines (dodders or "plant vampires") are numerous: two most serious in effect are a _Cuscuta_ (family Convolvulaceae) and another "vampire" a _Cassytha_ (family Lauraceae). They both infect in identical manner almost any green growing plant; they do not invade heavily suberized parts, mosses or fungi. The _Cuscuta_ is very vigorous and far outgrows _Cassytha_. The two have never been found together in nature but will infect each other when placed against each other by artificial means. Several other families of plants are parasitic in the tropics. The remarkable features of at least scientific concern are the extended series of parasites on hosts and then parasite-on-parasite-on-parasite. This is further extended by fungi and other thallophytes attacking the parasitic plants.

Introduction

In Neotropical countries there is nothing so important, from the standpoint of agricultural cash income, as plantation products (coffee, oranges, medicinals, spices, cacao, rubber, etc.) from tree crops. As the years progress these are being grown more and more intensively, and lands given over to them are steadily encroaching on wild areas. For generations, untouched jungles have been held as hereditary enemies of struggling man, but now, with newer knowledge, there is increasingly more widespread appreciation of the need for preservation and proper use of tropical trees and woodlands. From the wild comes such direct benefits as harvests of charcoal or fuel wood, structural timber or paper pulp logs, fine woods used in the arts and crafts, certain nuts, and other similar products. More indirectly, but as well recognized, are other values of tree-growing lands such as the containment of erosion, the preservation of much needed future and present water sheds, and the amelioration of climatic conditions.

All tree crops are long-lived; each individual tree of a plantation is specially handled and worked with, occupies its space permanently, and is a costly item that deserves protection. When a tree is threatened with parasites it is much more serious than when herbaceous plants, sown in rows, have their parasitic injuries. The latter, if need be, can be cut down, pulled out and burned, or plowed under. With passage of time, the original land can be re-prepared and a new crop sown. Trees, along with their greater importance in tropical agriculture, although attacked by a full list of microorganisms, also are known to be subject to phanerogamic parasites. [Phanerogamic parasites are plants that have flowers and fruits, but have become so changed in adaptation that they depend on a parasitic existence at the expense of other plants. Botanical families that have parasitic members are: Loranthaceae, Convolvulaceae, Lauraceae, Balanophoraceae, Orobranchaceae, Rafflesiaceae, Santalaceae, and Coniferacese.] These highly organized parasites differ very greatly in their habits, and some attack herbaceous plants rather than trees. Whatever they affect (see review articles by Wellman (78) and references therein) in the American tropics there are branch- and foliage-diseasing mistletoes and dodders, at least one tree

Fig. 17-1. Strangling fig, Ficus sp., in El Salvador that has attached itself to the trunk of a coffee tree; seed of the fig was deposited by a bird in the crotch of the coffee tree at the point indicated by the arrows. Growth is enveloping and overburdening the coffee pushing it over (the coffee tree trunk has lichens on it). As can be seen, roots of the Ficus not only bored into the soil but also crept along the land surface which was on a hillside.

that attacks trees, and root-diseasing species of Scrophulariaceae and Santalaceae. These are in very much larger numbers in the Neotropics than in the American temperate zone.

Non-Parasitic Epiphytes That Destroy

There is a great deal of botanically interesting--and on a popular basis, spectacular--growth of curious plants attached to trees in the tropics. Such plants--strangling fig (Fig. 17-1) is an example--are often spoken of as parasites among visitors and tourists, and it does appear that many trees are very much weighed down by them. These tangled and complex masses consist of such things as bromeliads, orchids, Ericaceae, many moisture-adapted cacti, mosses and ferns, algae, fungi, certain climbing bamboos, and strangling figs. In the Neotropics, trees hold on their branches sometimes actually tons of this kind of epiphytic growth, which is borne on the surface as a burden and bound to trunks and branches which act for the epiphytes as scaffolds.

Main branches on trees have epiphyte "gardens" on them that can be many years old. Judging from vegetative bromeliad and fern developments, and from counts of flower stalk remains of orchids like *Cattleya skinneri* that bloom once a year, on one felled tree in Costa Rica--mass amounts of epiphytes that weighted over half a ton were estimated to be at least 15 years old. Other masses, on neighboring much larger and older trees, were probably more than twice that age.

It is evident, to those with experience in rain forests, that epiphytes often develop in excessive amounts. The ecology of epiphytism has not as yet been fully studied in the western tropics. However, a botanical wonderland exists on occasional old and large tree branches--I have counted on some over 100 kinds of epiphytes. On others there may be a preponderance of only one or two species. Growths of epiphytes become such a burden, during rainy seasons especially, that trees, in such locations as river banks or wet hillsides, are felled by the extra weight. Another situation occurs when limbs, carrying ever increasingly heavy epiphytic growths, bend and some drop off. First the bark is cracked and for a few years the cortex is thus opened up to wet weather enhanced by insect and fungus invaders; then rots start in large limbs; if such decay is near the main trunk it is the beginning of tree death.

If the epiphyte growths do not have in them strangling figs (*Ficus* spp.) or lianas (*Cissus* spp. and others), burdened branches may withstand the strain much longer than when the stranglers are present. Strangling figs at times may be the most impressive of the non-parasitic epiphytes. Seeds of these stranglers are bird-carried, and germinate on branches or at the crotches or some equally advantageous spot on a tree where they are caught. A seedling grows quickly, its roots are directed downwards, and to the sides, around the structure of the tree that supports the strangler. A strangler has no special preference for any special species of tree to support it; it will grow on palms as well as on dicotyledonous types, on cliffs, and even on abandoned stone or brick chimneys. After a period of time, the roots of the strangler reach earth and it becomes an independent tree. In a few years, within the wrappings the growth of the strangler has made, are the completely dead and smothered remains of the original host tree.

It has been observed that many times trees supporting on their limbs profuse growths of such epiphytes as orchids, ferns, and bromeliads are not much attacked by Loranthaceous parasites. On the other hand, *Ficus* spp. are frequently infected by phanerogamic parasites. In all of my observations of many thousands of lianas I have seen only one attacked by a Loranth and that was a species of *Phthirusa*.

Parasitism of Loranthaceae

Because mistletoes or matapalos which belong to the Loranthaceae have such an unusual means of growth--they are plants that flourish without touching the ground--that looks almost otherwordly, they have long been regarded with awe, religious concern, and mystery, and as sources of magic power (23, 29, 40). It has been held (53) that the first time the word which later became "PARASITE" was used, was in the very ancient, informative and understanding reports of Theophrastus (88) written some 2000 years ago.

In the Neotropics, species of parasites belonging to Loranthaceae are very numerous. Linnaeus was sent many specimens from the New World tropics and he (54) described the plant genus *Loranthus* from those tropical materials; it is upon characters he erected for the genus that the botanical family is based. Plants of Loranthaceae are typically warm country organisms (many monographs show this, a few are indicated in citations: 1, 17, 28, 29, 36, 44, 46, 66, 73, 74, 77, 78, 90, 91; some examples of great botanical collections and reports including tropical Loranthaceae are: 4, 9-12, 16, 21, 27, 34, 45, 59, 52, 58, 63, 70, 82-84, 87, 101). In my experience this family of parasites never attacks plants (trees or shrubs) of the Monocotyledonae. Occasionally seeds will fall and germinate on Musa spp. but will not live long. In Australia a mistletoe was found to live for months on an orchid, but it finally died (103). It is well known that temperate-zone-adapted species of the family are harvested as "mistletoe" for the Christmas greens markets of Europe (Viscum spp.) and of North America (Phoradendron spp.). It is likewise recognized, by those in the holiday greens business, that it grows best in warmer regions of the temperate zone.

This great family of phanerogamic parasites is rich in botanical and phytopathological interest. There are some 40 genera alone in the family, and under these the total number of species is between 900 and 1000; about 35 species occur in temperate zone North America (28, 29, 92). There is much variation in these species and in some cases, conditions under which they have grown appear to sometimes exaggerate or at other times reduce plant characters.

For example, in my experience, using both living material and dried specimens, it has not always been easy to separate species in the genera Struthanthus and Phthirusa. Often, uncertainties in these and other genera encountered are enough that species names seem best dropped and numbers used instead. This variability was found in permanently marked, vigorously parasitizing plants watched several years in attacks on the hosts Coffea arabica, Theobroms cacao, Croton reflexifolius, and Spondias mombin. On these the parasite originally determined as Struthanthus orbicularis (HBK.) Blume showed changes in leaf color between yellowish growing for a time on cacao, and dark green upon transfer to coffee. The prehensile character of the parasite leaves noted when growing on cacao did not appear on a plant determined as this same species growing on Croton. When seeds from the Struthanthus without prehensile leaves (i.e., growing on Croton) were germinated and grown on mombin, the young parasitic plants again had normally prehensile leaves.

The parasite S. oerstedii (Oliver) Standl. & Cald. as an old vine had an abundance of ectophytic roots on the host

Croton, but young plants of S. oerstedii examined close by, growing on coffee, had very few to no exogenous roots. In two years' time watching the same plants of these same parasites, these roots then later became numerous. The short inflorescence, said to be typical of S. oerstedii, varied. Some types were found that had abnormally long inflorescence growths. Moreover, in these plants, spikes of flowers occurred in which their bracts were absent. Six of the characteristically small plants of Phthirusa pyrifolia (HBK.) Eichl. attacking a dooryard tree of Croton reflexifolius HBK. had dark reddish brown flower spikes. When the same tree was visited seven years later there were 15 new growths of P. pyrifolia. The oldest was large and vigorous but had lost almost all leaves; on it were many green flower spikes. Other changes in formation such as the shape of leaves and stem appearance, induced me to believe I might then be seeing P. theobromae. All this time, typical P. pyrifolia growing on coffee in the nearby shaded plantation remained stable in its botanical characters and coloration. (A well established parasitic connection of P. pyrifolia on a branch of arabic coffee is shown in Fig. 17-2.)

Much study is still to be done on many problems of distribution and spread of the Loranthaceous, bushy, parasites in the Neotropics. In my work in Central America, (see 98, reported 1964) it was found that seeds of these parasites were carried, and deposited entirely by fruit-eating birds which visit the parasitic bushes and possibly prefer Loranthaceous berries in their diets. This conclusion, although I arrived at it independently, corroborates the older classic researches on this by W. M. Docters Van Leeuwen (18) in the Oriental tropics. More recently it has again been observed by Kuijt (46) in the American tropics. Many of the seeds in the glue-like juices of the curious fruits that contain them have never been adequately described. There are some studies on dwarf mistletoes in the temperate zone (34, 79, 81), but no one knows much about the many differences seen in seeds of tropical American Loranthaceae. What are needed are findings not only on anatomy and classification, but also on the periods such seeds live, modes and conditions for their germination, and infection phenomena.

Among many pathologists there is only superficial interest about phanerogamic parasitism. It seems undoubted that in the tropics lies the most to be learned about diseases from Loranthaceae. Some has been studied (see

Fig. 17-2. An infection by Phthirusa pyrifolia on a branch of arabica coffee. (Ants living in holes at the base of the parasite's roots.)

Fig. 17-3. Sketch of a tree branch upon which have germinated three seeds of Psittacanthus calyculatus, giving appearance of multiple or split-apart cotyledons. Infecting pegs of the parasites are pushing with sharp effect in their initial piercing of branch bark. The seedling on the right, having made successful juncture with its host's sap system has grown its first juvenile leaf.

"Literature Cited") and much more, a very great deal more, remains to be studied. I have been much interested (98, publs. 1953 and 1964) in marked variabilities in parasitic reactions. The following few paragraphs are by no means inclusive of the findings.

The simplest and perhaps one of the most spectacular kinds of Loranthaceous parasitisms may be that of Psittacanthus. One of these parasites' seeds, planted by a bird on a tree branch, germinates and may produce a multiple-split endosperm (see sketches in Fig. 17-2) and quickly develops inside the host a single, massive suctorial organ. This forms the base of a vigorous aerial bush; the point of junction swells like a knob: the mass of parasitizing tissues on top of the hyperplastic mass of tissues enlarged from apparent host irritation reactions. Psittacanthus bushes look like strong feeders; often a host branch beyond the point of infection dies in a few years.

I found, upon dissecting fresh, young diseased branch material, the great extent to which the host wood rays are enlarged, disarranged and in some parts squeezed together. There seem to be no hyphal-like, elongated parasite cells, and no sinker developments that grow down into the host stelar elements. Absorption of nutrients from the host is through an intervening, thin tissue layer between the parasite and the host. Microscopic examination of this layer discloses cells that are irregular in shape and organization. Thousands of absorption elements are closely in contact with this apparent feeding zone. However, there is no true fusion of parasite with wood tissues, nor is there root-like extensions growing into host wood. (In Fig. 17-4 is a view looking up into the foliage of a tree in which grow 14 parasitic bushes.)

Parasitic bushes of Psittacanthus thrive on tropical hard wood trees for a number of years; when a bush parasitizes a trunk (Fig. 17-3) it lives as long as the host tree exists. When the parasite is on a branch that dies the much softer parasite structure is attractive pablum for insects and fungi and after a while it falls away, cleanly, leaving swelled, grooved, often handsomely convoluted, hat- or cup-like host structures (Fig. 17-5) called flor de palo, woodroses, and such names. Woodroses of Central America and the Northern Andes are from Psittacanthus parasites (Fig. 17-6). The closely related Myzodendron is reported (22) to cause the knob-like woodrose growths on the more or less dwarf beech trees, Nothofagus, of Southern South America. It may be that in the Neotropics no other kinds of Loranthaceae produce woodroses, although species of Phoradendron and a Struthanthus are sometimes suggested as the agents of woodrose reaction (1, 46 and others). Thoday (90) found in the Oriental tropics small woodroses formed on the Casuarina tree, from infections by Phrygilanthus spp.

There is a genus displaying remarkable habits that assisted me in connection with understanding different degrees of parasitism among members of the mistletoe family, Loranthaceae. This, genus Gaiadendron (see citations 27, 29, 46, 60, 70, 73, 74, 82, 84, 98), has root parasites in it that in size are small trees to high bushes. In the dry season they bear fragrant yellow flowers and are evergreens, like any typical member of their family. As seen in mountain forests of Costa Rica, it is possible to trace the whole course of the development. I have seen seeds that germinated on the basal part of rough-barked boles of host trees, and on sturdy cane-like trunks of woody bushes. It is possible to find the attachment of the seedling to the host support, and, by dissection, aerial roots are seen to penetrate

Fig. 17-4. A chinaberry tree, *Melia azedarach*, observed on an El Salvador farm roadside was parasitized with fourteen (counted) bushes of Loranthaceae. Upon making collections from these, four different kinds were identified (numbers indicate growths collected): 1. *Phthirusa pyrifolia*; 2. *Phoradendron rensoni*; 3. *Struthanthus costaricensis*; and 4. *Psittacanthus calyculatus*. X. Examples of host leaves.

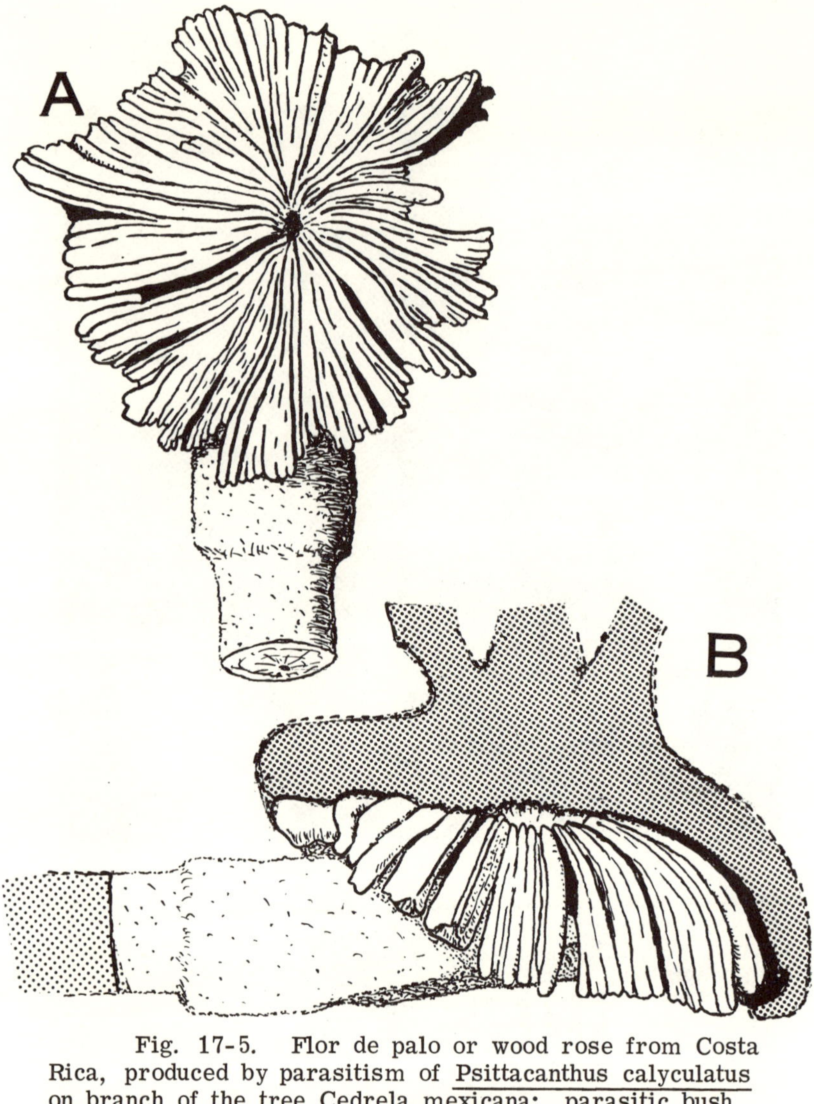

Fig. 17-5. Flor de palo or wood rose from Costa Rica, produced by parasitism of *Psittacanthus calyculatus* on branch of the tree *Cedrela mexicana*; parasitic bush located out towards end of branch grew vigorously several years finally killing the branch tip. Characteristically the parasite eventually died, and its relatively soft bush tissues rotted by diseases and insects fell away from malformed host wood leaving a structure about the diameter of a man's fist. A. Sketched upright and facing the viewer. B. A side or horizontal view with shading indicating parts lacking when sketches were made.

Fig. 17-6. Flor de palos, or woodroses in situ (circled) in a tree on an El Salvador roadside. Host, Spondias purpurea, parasite, Psittacanthus calyculatus; the parasitic bushes grow to be large, but for undetermined reasons all on this tree had died at about the same time. Arrows indicate new Psittacanthus infections that could be seen from the ground as seedlings on branch bark.

in places to the cortex of the host. The rather thick, scaly, positively geotropic aerial roots grow down, sometimes more than two meters, tightly appressed to its host. When such a root finally meets the soil it becomes fibrous, like a normal subterranean root, goes farther downward and then may attach itself in the soil to the host root. (There are also other Loranthaceae that infect roots (2, 5, 60, 62).) It is thus a part-time parasite (98, my publ. 1964, see also Kuijt 46, his publ. 1963); later on it may turn into an autonomous free-living small tree. It develops as a fair sized trunk and wood does not splinter easily. In Central America *Gaiadendron* has been used in constructing articles of furniture and parts of ox-drawn wagons.

Another degree of parasitism occurs that is different, caused by species of *Loranthus*. On the germination of a properly situated seed, its root pierces both bark and cortex on a branch: as the young plant loses its juvenile appearance, growing and becoming a bush, the place of infection on the host is a perfect union; the attachment place is smooth; it gives no special external evidence of injury. Indeed, in many cases it appears benign, it looks as good a connection as if it were from a successful graft made between a compatible stock and healthy scion.

I could not carry out all the much needed studies seeking definitive information about the internal parasitic activities (endophytic developments) of an apparently more intimate and more benign disease from attack by *Loranthus* sp. My concept of this type of parasitism is somewhat based on the results of old temperate zone studies with respect to actions of such as *Viscum album* L. and *Phoradendron flavescens* Nutt. (see for examples of studies: 29 lit., 32, 78 lit., 80, 91 in 1956 publ., 95, 102.)

In dissection studies in 1945 of *Loranthus robustissimum* on the tree *Inga vera*, it was impossible to find evidence of tissue necrosis from disease by this parasite. Seedlings of the Loranthus on penetration, with a specialized peg or foot (Fig. 17-7), develops very special ramifying structures (sinkers that are parasitizing roots) submerged in host. tissues; in some cases it is possible to find in a diseased host branch an almost continuous internal sleeve of parasite growth; it can extend half to 3/4 of a meter up and down inside a branch. The internal sleeve of the parasitic organism appears not to inhibit any absorption of nutrients by the cortex of the affected branch. In addition to

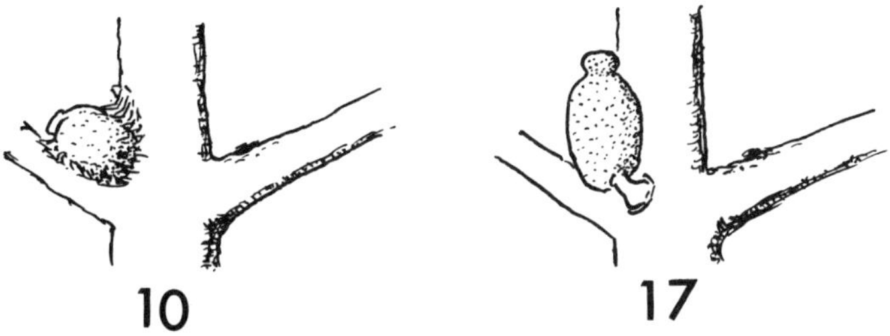

Fig. 17-7. A seed of the parasite Loranthus sp. (?) planted August 10 with its sticky matrix in the crotch of an upright stem of an arabica coffee tree in El Salvador and marked (10). It was sketched that day, watched daily and sketched again a week later, which was August 17 (17), when it had swollen, lost some of its adherent matrix, germinated and its hypocotyl was starting to enter the branch bark. The seed appeared to have changed its orientation.

the internal sleeve of parasitic tissue, when these branches are cut crosswise, sinkers of Loranthus tissues are readily seen between host wood rays. In the old branch, parasitic sinkers are numerous and penetrate like fine nails or coarse pins only a few milimeters apart into the host wood. However, in no case is there swelling from this severe invasion, and no killing of branches. A series of a few different types of loranthaceous connections on hosts is shown in accompanying Fig. 17-8.

A perhaps more complex type of parasitism in the Loranthaceae is that characteristic of the abundant, drooping and vine-like killers of trees, parasites of the genus Struthanthus. This is a low-growing, bush-like, crawling plant that extends over and wraps itself about the host; the characteristic drooping habit is from its slender almost prostrate branches. It starts from germination of a seed that sends a specialized epicotyl into the host branch. This is the primary connection, but it is relatively weak and there develop from the aerial stem numerous true, exogenous roots. These have a scaly bark and are studded with numerous hapters or haustorial pads. ["Hapters'" or "hapterons" (referring to an aerial organ that served to attach plant, lichen, or fungus to a substrate) are more common in

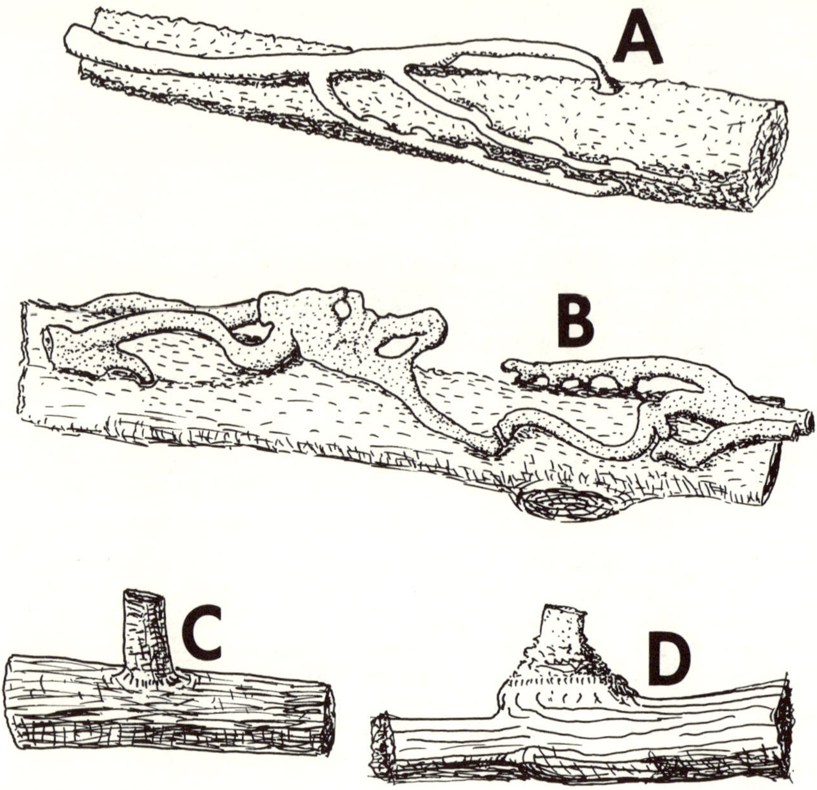

Fig. 17-8. A few different types of connections by higher plant parasites in Central America on tree branches: A. External slender roots (some with typical hapters) of Phthirusa, and an end root that has lost its terminal dominance developing an end sucker, but side roots thereupon becoming dominant; B. external roots, sometimes irregular as well as readily self-grafting (some of these exogenous roots with hapters) of Struthanthus; C. smooth parasitic connection by Loranthus that if part way back from end may not kill tip of branch attacked; D. rough and swelling, evidently irritating connection from infection by year old Psittacanthus, but which on aging swells more, wrinkles, and the tip part on the host branch is commonly killed.

lichens, are encountered among fungi in the Nidulariaceae, and are another character lending itself to the unique qualities found in many phanerogamic parasites.] These hapters soon attach themselves to the host branch, begin absorbing nutrients, and the vine-like growth then extends; without much apparent difficulty, the parasite readily outgrows and escapes from dependence on its primary connection.

The commonly occurring Struthanthus orbicularis (HBK.) Blume, is destructive on many plantation trees. It is taken as one of the best examples of the type of parasitism practiced by the vine-like Loranthaceae. The foliage of this remarkable parasite is yellowish in most locations, although occasionally green under heavy shade. It is a severe pest on such plantation tree crops as coffee, cacao, orange, lime, mango, avocado, and all of the agriculturally employed shade trees. I have never seen it on Hevea rubber but it was found on Castilloa rubber. Palms like coconut, date, and the African oil palm are not infected by it.

From my observations and the studies of Kuijt (46), S. orbicularis has prehensile leaves. This prehensile character is found not only on S. orbicularis but also on S. costaricensis Standl. and Cald. as well as on species of Oryctanthus. These hooking and clasping leaves aid in the capturing and drawing together of host branches which are then subjected to killing aerial root parasitism. The aerial roots (exogenous, extracortical) of these parasites are produced from stem and irregular twining and reversing branches. The stimulus from host contact causes the spiralling branches to change appearance at their tips, where leaves are smaller or drop, and where the vine wraps around its host and sprouts roots. The roots extend outwards and as soon as they come into contact with a host limb they send out side shoots that wrap around host parts, and then hapters form on these.

A significant state in the understanding of parasitism by tropical mistletoes, is illustrated in the intimate connection made with its host by the Struthanthus. The holdfast organ or hapter starts as a pimple on the root where it touches host bark. It grows rapidly, flattens as it expands, and forms, appressed to the host surface, a lip-like growth around the holdfast perimeter. In addition, there appear to be secretions of an adhesive material. From these developments the hapter sticks tightly, and is not easily lifted off, even before actual penetration has begun. This is the veritable "kiss of death." Towards the hapter center, where it is softening and otherwise preparing its victims bark, it sends down penetrating tissue bundles. These, as they develop, are reminiscent of sinker structures which in *Loranthus* are like a deepening wedge, but in *Struthanthus* are a line or ridge of bundles boring in. The entering structures go through suberized cells (bark) and proceed into close association with host xylem elements. That these penetrating structures have force, and can go too far, is clear. I have seen parasitized slender stems of hosts (*Croton* sp. in Costa Rica and *Inga* sp. in El Salvador) that showed rifts in host bark on the opposite side from where hapters attacked. On scraping at the rifts it was found they were ridges of parasitic projections that had gone clear through!

There is a singularly interesting and ironic occurrence in *Loranthaceous* parasitism of trees. Every so often (46, 71, 98) although host trees become so overwhelmed with aerial bushes or vines as parasites that host foliage is completely usurped, the host continues to live for a time. As the parasite acts, it not only may live on its host but can be at the same time self parasitic. Using all the advantage of its parasitic habit it depends on (feeds upon) the host for its life and reproduction. However, as the terminal situation approaches for the host tree, there may be a long time when there is interchange of nutrients between host and parasite. In the end, with all tree leaves gone, the tree may continue to act as the root system for a great mass of parasitic bushes in its branches. I have seen bushes of parasite act as photosynthetic organs of nutrition for the host tree during a considerable time, sometimes two or more years, before the parasite-weighed-down-host finally dies. (See also work of Rediske and Shear (71).)

A Parasitizing Tree That Kills

There has been pointed out in above paragraphs, a fact not often alluded to in plant disease literature: members of the family Loranthaceae are known which act for a while as parasites and after a time spent in a semi-parasitic existence, grow up as small trees with no further reliance on a host. There is an even more rarely published fact, of considerable interest in the field of pathology: in the American tropics there is a tree of fair size that parasitizes and kills other trees. Apparently the first published notice about this particular parasitic tree in the Western Hemisphere is the short article that appeared in 1938 by Reis (72). He found in the state of Bahia, Brazil, a tree that killed the money-making cacao tree. This parasite he called "mata-cacau," or, the "tree that kills cacao."

This discovery was confirmed subsequently, and, as it had a distinct economic importance, special effort was made to identify and prove some kind of pathogenicity. Alvim and Seeschaaf (2) described the securing of material and experimentation under greenhouse conditions. Seedlings of the presumed parasite and of cultivated cacao were planted next to each other in containers. Where wooden partitions were inserted to separate the questioned parasite and cacao, they both grew. Where the cacao and the unknown were not separated by a barrier the cacao trees showed debilitating effects and eventual death, while the other wild parasitic seedlings attacked cacao and the parasites made very good growth at the expense of cacao. An illustration of this is shown in Fig. 17-9.

The parasitic tree was studied botanically and determined as being one of the Santalaceae. It was described and given the name Acanthosyris paulo-alvimii Barroso (5). Its host range is under investigation; on certain occasions it has appeared as a limiting factor in producing cacao. This parasitic tree grows to be 12 to 15 meters tall (about 40 ft.), with its trunk of 20 to 35 cm. (about 12 inches) in diameter. Branches are slightly spiney; leaves are alternately arranged and smooth of acuminate shape; fruits are rounded, drupaceous, 6 to 8 cm. (about 2 1/2 in.) in diameter, with, buried in a reddish pulp, seeds that are oily and with an agreeable flavor.

On digging to study the roots, it is found that the Acanthosyris root swells at its tip when it finds a host root. The swelling enlarges, like a hapter, settles onto the host and sends a central penetrating development to breach the root bark and cortex (Fig. 17-10.) Inside the host, parasitic tissues grow alongside and take nutrients from the phloem and xylem elements of the cacao tree. At the present time there is no complete information about the natural distribution of Acanthosyris in the wild, or its effect on jungle inhabitants.

There are no good reasons to doubt that many trees,

Fig. 17-9. Glasshouse study of effect from root parasitism by young tree of Acanthosyris paulo-alvimii. On left flourishing, eleven-months-old tree of parasite; on right debilitated seedlings of cacao, one devoid of leaves but tip still alive and one with four small and weakened leaves. Uninfected cacao seedlings grown for comparison under identical conditions, had stems more than three times as high and with fifteen or more large and healthy leaves. Photograph taken in Bahia, Brazil courtesy of P. de T. Alvim.

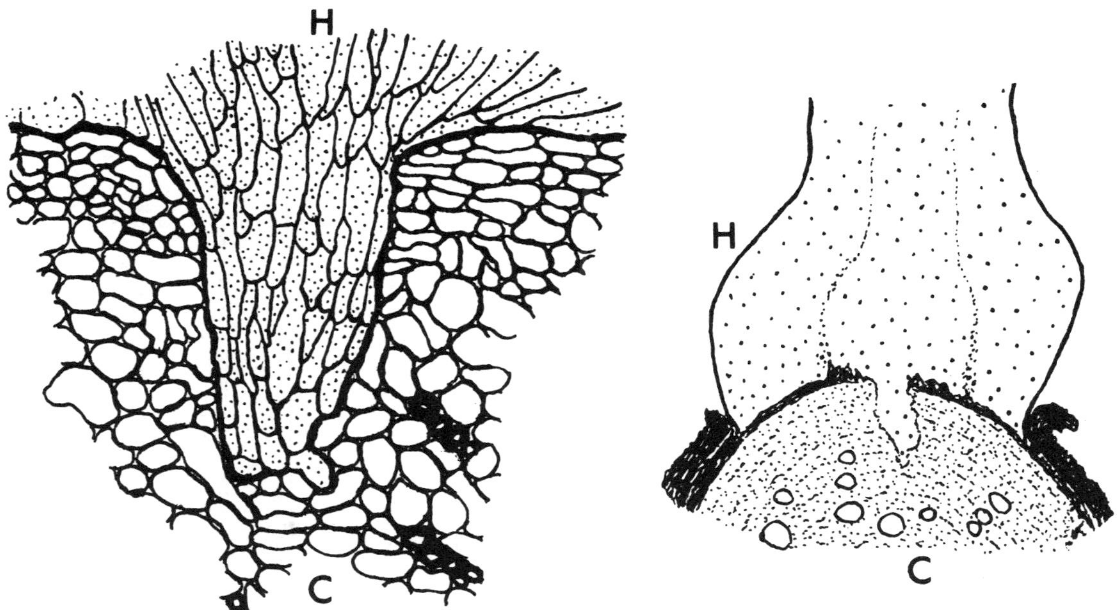

Fig. 17-10. Cross sections of roots of cacao (C) being penetrated under the root-searching, root-invading haustorium (H) of the parasitic tree Acanthosyris paulo-alvimii in Brazil. On left, drawn from microscopic study of a recently penetrating parasite sinker; at right, a less magnified older infection showing swollen part of the haustorium, its pushing aside of superficial, suberized root cells, and the absorbing sinker penetrating towards the center of the host root. Sketches after drawings and photographs courtesy of P. de T. Alvim.

may, in addition to other characters, be parasites on other trees. There is a great wealth of botanical and forestry studies describing, classifying, and giving ecological information on native trees in the Western Hemisphere tropics; numerous genera and species of the Santalaceae are known to be present. However, trees feeding on trees are not phenomena that are discovered by conventional botanical or other studies and observations. Thus far, it seems much probably depends on an economic problem being involved to cause dendrologists, pathologists, or others to dig up roots and prove pathogenicities of trees.

One tree, _Santalum album_ L., is well known in plant disease literature as a parasite of other trees and plants in the Orient, especially in India. Interestingly, decades ago attempts were made to introduce _S. album_ trees into Puerto Rico (11), but they finally died: my searches never discovered any of them in later years. But it is well to know that a tree having parasitic intentions on another tree does not necessarily have to be a member of the Santalaceae. For example, from work of deLaubenfels (50) in New Caledonia, the parasitic activity of a true conifer--far from related to sandal--was proved.

Two Important Tropical Dodders

In the Neotropics are numerous, fast-growing, often-seen, and soft-tissued dodders. One of these belongs to the genus _Cassytha_ with about 15 species, while the other is in

---

Fig. 17-11 [on facing page] Sketches from a special study in Puerto Rico by the writer comparing the two dodders, _Cuscuta americana_ (A) and _Cassytha filiformis_ (B), grown on artificially infected low branches of a tree, _Eugenia_ sp.: a, example of a length of vine where it attacks host showing attachments by suckers or hapters and swellings above hapters; b, self parasitism in _Cuscuta_ occurs also in _Cassytha_; c, travelling vegetative vine (_Cuscuta_ smooth and glistening, color bright orange-yellow, _Cassytha_ striated and dull and slightly pubescent, color green to green-olive); d, tip of vine in _Cuscuta_ smooth and light colored, in _Cassytha_ pubescent and dark in color; e, fruiting habit; f, attacking vine carefully stripped away to display underside of hapters and marks on host (the latter slightly emphasized).

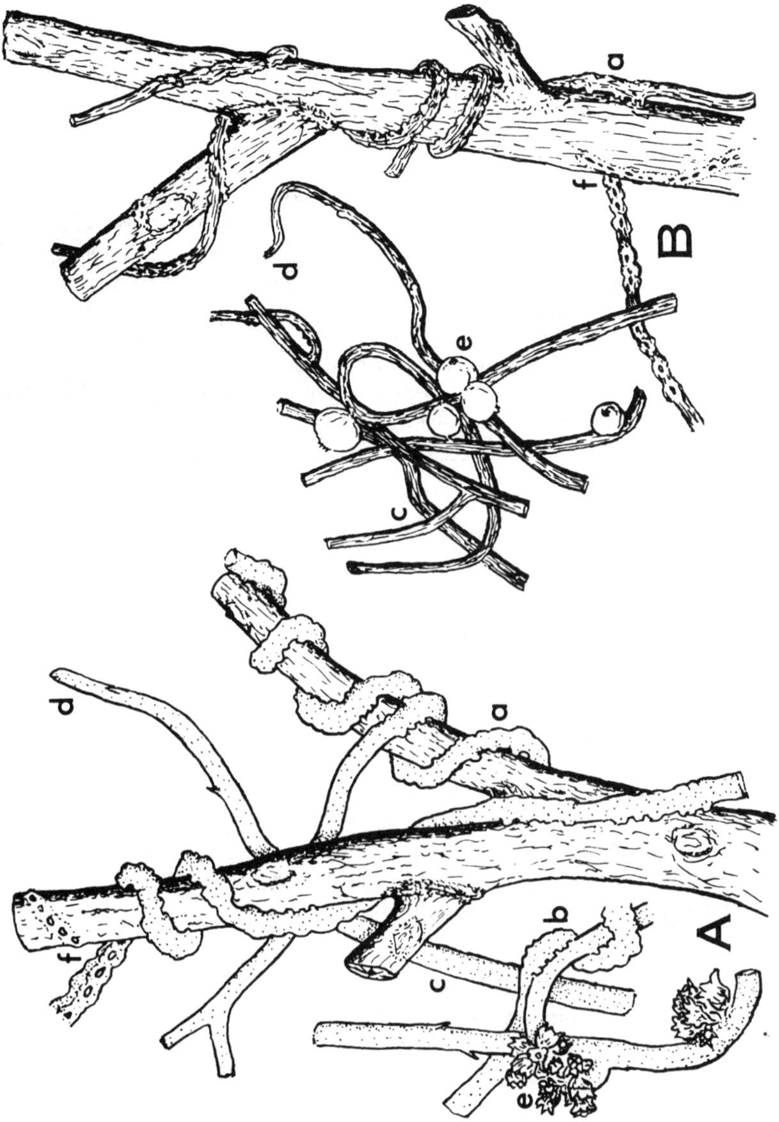
A
B
a
b
c
d
e
f

the genus Cuscuta which has some 160 species (see citations 26, 74, 94, 98, 104 and others). In the tropics, the consistently occurring parasitism of the dodders is at a specially advantageous situation because of the perpetual growing season.

Although there are about 175 species of dodders in the American tropics, in the wild they are seldom spectacular. Natural enemies keep them from excessive development; in heavy jungles they are not commonly seen, but are found mostly in open woodlands, treeless expanses, and edges of forests. On the other hand, where they have escaped from the wild, multiplied in fields, and become important infestations on cultivated crops, they are at times masses of parasitic growth that cannot be ignored; occasionally their evident parasitic enveloping appearance is stunning to the beholder. Of all the Neotropical dodders, the ones perhaps most widespread and most severe, and the ones to which a pathologist may be forced to give most attention, are Cassytha filiformis L. and Cuscuta americana L.

Both of these voracious plants are slender vines that grow fast and vigorously. The Cassytha, whose berries are carried floating in sea water, is often found near tropical beaches as Guppy (31) has shown. [I have frequently found it within the sound of the Caribbean and the Gulf of Mexico, from Venezuela to northern Florida and in the West Indies.] Occasionally wild pigeons feed on fruits and carry seed inland (Wellman and Woodbury (99)). The second dodder expands lushly in warm lowlands but is not confined there, for it is also found well up in mountain valleys and on cool mountain table lands. As is shown in many reports (examples: 10, 11, 16, 19, 21, 22, 24, 25, 38, 43, 49, 52, 55, 56, 58, 61, 63, 65, 68, 70, 74, 82-87, 92, 94, 98, 99, 101, 104), dodders are probably in all regions of the American tropics, their vines enwrapping and living on unsuberized parts of about any kind of host plants.

The two dodders specially discussed here, are widely separated phylogenetically in their taxonomic relationships. The observer is only a little surprised to learn that the Cuscuta vine, a "plant vampire," belongs to the highly evolved family Convolvulaceae. Many members of the family such as morning glory, are twining vines, and the Cuscuta flowers, on close examination with a hand lense, look like miniature morning glory flowers. In addition, seeds of this dodder are black, flattened, and held in a chaffy, dry capsule very similar to that of morning glory. When it comes to the Cassytha dodder, also a "vampire" vine, the observer may be excused for his surprise when he learns it belongs to the supposedly only moderately evolved, family Lauraceae. Members of it are woody shrubs, and various sized trees, such as bay and laurel, that have juicy berries as fruits and leaves within which is a pungent sap. The vine of the Cassytha is, interestingly, also pungent when crushed, and its fruits are white and juice-containing, somewhat like bayberries. (For some similarities and differences in appearance between these two botanically unrelated dodders, see Fig. 17-11.)

In spite of the wide taxonomic divergence between them, both dodders as vines are phenomenally similar in appearance. Superficially they are so alike as vines alone, that side by side, they can be called by the same name. Even trained workers are led astray at times confusing Cuscuta americana with Cassytha filiformis.

Both dodders attack in very similar fashion. Their vines wind around, hold tightly to, and absorb from, the hosts. The voracious parasitic plants are, for their kingdom, veritable vampires that suck the juices from stems, twigs, and leaves of grasses, weeds, trees, and all kinds of crop plants. Certain armored stems are not susceptible: heavily suberized multiple layers of bark on trunks or main branches of trees and woody bushes (but young stems with slight suberization are invaded); leafless bases of two- or three-year-old sugarcane stalks; highly lignified outer tissues of trunks of palms and cycads; strong old, leafless stems of bamboos, pandanus, and large grasses. It is known, however, that leaves of any of these hosts are susceptible to dodder infection (99).

In successful parasitism, bead-like swellings expand at points along the turgid, lushly growing, adhering vine. From these swellings develop, rather rapidly, the holdfasts or hapters (48, 61, 69, 74, 80, 89, 98, 99). As they close down and adhere to a sap-containing source, there come at their centers, growing structures (see: Bennett (8), Fernandez Valiela (22), Lackey (48), Pierce (69), Roger (74), Solms-Laubach (80), Srivastava and Esau (81), and Thoday (89)) and softened outside cortex cells. These parasitic structures penetrate further to become intimately in contact with host xylem and phloem elements. In some instances Cuscutas cause distinct gall-like developments on half-strangled stems

Table 28. COMPARISONS[a] BETWEEN TWO IMPORTANT TROPICAL PLANT DODDERS

	Item of comparison:	Cassytha filiformis	Cuscuta americana
1.	Botanical family	Lauraceae	Convolvulaceae
2.	Coloration of vines	Greenish to olive brown	Nearly orange to light yellow
3.	Texture of vines	Somewhat tough, slightly pubescent	Turgid and brittle, shiny and smooth
4.	Fruits	Juicy white berry with a round seed	Dry capsule with flat black seeds
5.	Some common names	Dodder, love vine, angel hair, cabellos de angel, fideillo, tente en el aire	Dodder, yellow strangling string, bejuco amarillo, bejuco de mona, fideos
6.	Reaction to sea water submergence	Vine alive for over 14 days, fruits remain afloat for over 15 days	Vine collapsed in 1 day, seeds and capsules sink promptly
7.	Rapidity of growth	Fairly slow	4 to 5 times faster than *Cassytha*
8.	Branching	Monopodial	Dichotomous
9.	Time to clasp host	Requires over 2 days	Less than 1 day
10.	First sign of hold-fast (hapter)	Smooth gentle vine swelling	Distinct, almost delimited vine swellings
11.	Shape of hapter	Tends to be elongate	Rounded
12.	Period hapter is easily stripped off young coffee branch	Requires about 8 days before the permanent connection is solid	Enters to make a solid permanent connection in less than 3 days
13.	New hapter effect on	Swelling of cortex tissue under	Swelling of cortex tissue under

	young coffee branch	hapter	hapter
14.	Old hapter effect (when stripped off completed parasitic connection)[b]	Narrow spindle shaped pointed ended scar where penetrating tissues enter	Elongated, round ended ellipse of scar where penetrating tissues enter
15.	Histology of parasite penetration in outer tissues of branch	Slight dissolving of cells under epidermis, parasite strands held back by certain areas of pericycle but finally break through	Distinct and rapid dissolving of groups of epidermal and cortex parts, and very little hindered by pericycle
16.	Histology of parasite association with host phloem	Grows beside phloem, deranging arrangement but without apparent entrance into host cells	Grows beside phloem, deranging arrangement but without apparent entrance into host cells
17.	Histology of parasite association with medullary rays	Pushes between wood-cell elements and goes to central pith	Pushes between wood-cell elements and goes to central pith
18.	Histology of parasite association with xylem	Derangement of normal xylem structure, but no breaks found	Derangement of normal xylem structure, and occasional tracheae found broken
19.	Host range	Probably all green vascular plants	Probably all green vascular plants

[a]Studies were made both in field and laboratory, as can be seen from the text discussion. Histological results from parasitized young Arabica coffee branches, both living material, and from specimens stored in Formal Acetic Acid fixative.

[b]See sketches Fig. 17-11.

(43, 80, 85, 86) and the parasites can easily live over from season to season in parasitized host (3). Host range studies (Wellman and Woodbury (99)) were made in the island of Puerto Rico, where much botanical work has been done and where both of these parasitic dodders flourish.

In field work on host infections by these two dodders, their flowers and fruits are most often present to aid identifications. However, it is equally possible to make absolutely correct diagnoses on nothing but vines with these two dodders. The tendril of *Cuscuta americana* is a conspicuous, almost brittle, odorless, string-like vine that is glaucus, i.e., smooth and shining, with a silky sheen. In color it is characteristically from pinkish-yellow to shades of deep chrome, averaging what might be called a golden or nearly orange in shade. It has occasional, very small scale-like growths (much reduced leaves), that are almost invisible. On the other hand, a length of *Cassytha filiformis* is different. For one thing it is more slender than the other vine. *Cassytha* surfaces are dull; under a hand lense they are striated up and down the vine, and a fine pubescence is seen towards a young tip. Very small, scale-like leaves can be found that are elongate, appressed to the vine, and hairy. Coloration is from a hay green and chartreuse, varying into slightly yellower and grayer shades to greenish and olive browns. As already noted, *Cassytha* vines, when plucked and crushed, give off a pungent aroma.

Positive parasitism, as seen in laboratory and field, is demonstrated when a vine shows distinct permanence in its parasitic connection to the host. In this permanency, there is definite growth of dodder into host stem or leaf. This is tested by lifting the vine at its parasite attachments (hapters). Permanently attached vines cannot be removed without injury to either, or both, parasite and host. In Puerto Rican field work, *Cassytha* was found vigorously parasitic on 82 species in 35 families and *Cuscuta* with equal or more vigor on 83 species in 37 families. Under artificial conditions even more species of plants were parasitized by both dodders.

Both of these important dodders grown on such tough plants as tropical grasses belonging to the Poaceae, occasionally on the relatively tenderest leaves found on trees of young Cycadaceae and Palmaceae, on large sedges of the Cyperaceae, and on leathery leaves of trees in the family Hyperiaceae (Guttifereae). Both parasites find an abundance of plants to live upon of medium hardness such as Sida spp. of the Malvaceae, branches and leaves of Coffea spp. of the Rubiaceae, and other trees. On the other hand both vines are perfectly capable of parasitizing juicy stems of Commelinaceae, tannin-containing leaf sheaths of Musaceae, leaves of tender ferns of the Polypodiaceae, and delicate tendrils of squash, one of the Cururbitaceae. From the standpoint of the phylogenitic system in plant classification, from ferns to asters, there seem to be no reasons to expect any resistant genera or families.

In later laboratory tests* both Cuscuta and Cassytha vines were used to determine if they might parasitize organisms classified as lower than ferns in the botanical scheme. Young and vigorous tips of those phanerogamic parasites were trained to be in close contact with: mosses, Stematophyllum caespitosum Mitt. and Octoblepharum albidum (L.) Hedw.; a leafy liverwort, Plagiochila sp.; four widely differing foliose lichens of undetermined genera; and fungus sporophores of species of Schizophyllum, Polyporus, Nidularia, Poria, Xylaria, and Auricularia. Among these organisms none was observed in which there was the slightest suggestion that either of the dodders were being infective, or even attracted, to it. Both vines were apparently in some way injured by contact with Polyporus sp., although lesions were sterile; but disease lesions developed in both dodders after an extended period under laboratory conditions, and isolations yielded cultures of a small-spored Colletotrichum. There are several reports of diseases carried by or hyperparasitic on dodders; some examples: viruses (8, 15, 37, 93, 96), Septogloeum (20), crown-gall bacteria (41), and Colletotrichum (51).

Growth habits in these vines are of primary importance in carrying on their parasitism. See the accompanying table for more details. They are both extravagant and luxurious growers, and although starting as seedlings in soil, they waste no time in taking up the parasitic life. In one instance of Cuscuta americana, I marked in the morning a seedling not more than 5 cm. high, standing unattached but

---

* From hitherto unpublished manuscript notes of mine. Attempts were made to culture Cuscuta and Cassytha stem tips on agar medium but, although done by Loo (57), they were not successful in my limited experience, but it was not made a major study, which it deserves.

within 12 cm. of the stem of a small leguminous weed. The next day, in the evening the dodder was attached and had three wraps around its host. Cassytha filiformis seedlings showed the same kind of action, although requiring a longer time, both becoming parasites as soon as possible after their seedlings emerged. In Cuscuta seedlings, watched in an artificially wind-sheltered location, the tips swung around slowly in circles until they came into contact with green host tissue, or died of apparent exhaustion.

Movement of the vine tips is important in the understanding of their pathogenicity. Mutation phenomena were given some study. This was done in still air on a well-lighted laboratory table and at a consistent temperature of 27° C., under glass plates with labelled grids. In the plant of Cuscuta americana the movement of the tip of a "searching vine" was recorded at 30-minute intervals. It moved clockwise and made four complete circles in seven hours. On being presented with the stem of a growing bean plant for a host the vine stopped swinging, soon attached itself and overnight twined closely about the bean stem, again in a clockwise direction.

With the plant Cassytha filiformis, the movement in the "search vine tip" was slow and irregular. It changed in position in more or less a counterclockwise direction, and it made a full swing barely once in seven hours. On being given a growing bean plant for a host it stopped its movement and after two days was adhering to the bean stem. It required four more days for it to wrap about the host, and in this case in a clockwise direction.

These same dodder plants were established on coffee seedlings. After seven weeks on coffee, rapidity of growth was found to be very different between them. From seven measurements in seven days, a vine of the Cassytha grew an average of 0.41 cm. in four hours, while a vine of the Cuscuta grew an average of 1.92 cm. in four hours. Under the laboratory conditions and in the field material, too, the Cassytha vine develops a strong leading vine, maintaining apical dominance for a long time. Finally a side branch will take over for itself and after while gain its own apical dominance. In this parasite the branching habit is dichotomous. Studied in parallel conditions with the other dodder the Cuscuta vine grows differently. It branches monopodially and one side of a vine overextends another quickly as it reaches new host matter for food. There seems to be no marked apical dominance in its system of growth, and by this it is more efficient as a parasite and killer. The accompanying Table 28 gives further comparisons between these two parasites.

Growth characters of the *Cassytha* and the *Cuscuta* are dissimilar, and this situation is probably the reason I (98, publ. 1964; and 99) did not find the two parasitic vines attacking each other in nature. This seemed at first incredible for they were observed hundreds of times separately, but never together, and Puerto Rico was searched diligently for this possibility. (Dodders are common in the Oriental tropics, and in India is found both *Cassytha* and *Cuscuta*, but Bor (9) reported he never saw them together.) To study mutual parasitism between them closely, I brought *Cassytha filiformis* and *Cuscuta americana* parasites together. They were established for one experiment on coffee seedlings, and for another on seedlings of an *Acacia* tree.

Vines of these parasites were twined about each other, and also about themselves. *Cuscuta* grows at least four times as fast as the other, and in these studies it consistently killed vine end after vine end of *Cassytha*. This latter will attack *Cuscuta*, but it is soon overwhelmed, and even its attachments are hyperparasitized by the *Cuscuta* it is growing on; the *Cassytha* does not last long in the grip of its adversary. Many *Cuscuta* vines double back on themselves, become immediately self parasitic, and soon after (maybe from this?) develop buds and flowers. This is not as notable on *Cassytha*.

Some Other Neotropical Parasitizing Plants

There are certain parasitic flowering plants in the family Scrophulariaceae of which a few genera and species are parasites. Usually these are low-growing herbs, often in ill-kept fields and at forest edges. This special group of parasites when collected by such botanists as Standley and his coworkers (82, 83, 84), Gentry (27), and Britton and Wilson (11), are reported as attacking sugarcane and some other crops in South America. One of these parasites, *Alectra* sp., is typical of some others. The host roots are attacked by subterranean holdfasts. However, once well started a parasitic plant may have the host connections severed and the *Alectra* will live and produce mature seeds. Another related parasite is *Striga* sp. already known in our tropics (8) which has been introduced to the southern part of the American temperate zone as reported by Garriss and Wells (26) and McGrath et al. (59). As it spreads, control studies (e.g., 64, 75, 92, 100) are going to be of increasing importance. It continually reappears here and there and it may be reasonably expected sooner or later to be in the Neotropics proper.

Others of the herbaceous root parasites belong to the Belanophoraceae. Botanists recognize their short growths that appear bleached and bronze streaked, reminiscent of fungi*. They are widespread, but not common and often somewhat hidden in their locations, see Wellman (99 in 1962 publ.), Lanjouw (49), Standley and Steyermark (84), and others. In our tropics they have had no critical attention as subjects for pathological study.

Because of its historical interest and its value in commerce, along with many other Oriental subjects, trees of *Santalum* (family Santalaceae) have been transplanted to the Occidental tropics. It is said to be struggling in a few locations in the Caribbean region. Seedlings can infect roots of some 250 hosts such as bamboos, legumes, aralias, certain gramineae, and others. Less well-known members of the family are native to South America. Among these are bush-like *Cervantesia* spp. and *Quinchomalium* spp., which Macbride (58) found under varying conditions and at elevations from near sea level to high altitudes. A shorter, shrub, called "macachin" and "matatrigo," *Arjona tuberosa* Cav. var. *tandilensis* (Ck.) Dawson, has been known a long time in Argentina as severe on numerous row and forage crops (see reports such as those by Marchionatto (61) and Dawson (17)). They are all characterized by parasitic connections that invade underground parts of their hosts, and in some their roots develop as swollen storage organs. On the whole, these Santalaceae are considered as hemiparasites, as sometimes many grow for considerable periods without parasitic connections.

---

* It should be mentioned that a so-called false broomrape is becoming more and more of a problem in warm countries (42, 55). This, however, is believed to be a soil borne disease, probably from bacterial or some such causation. The abnormal, bleached, unusual host root growths might be confused with parasitism by one of the Balanophoraceae.

Of the many unusual plants found acting as phanerogamic parasites, perhaps some of the most important from the standpoint of the study of most intimate parasitisms may be among those of the family Rafflesiaceae. In many cases the plants grow entirely embedded within the tissues of its host, and the only thing that is seen by the pathologists is the flower stalk and the subtending fruit or flower. The systemically parasitic Apodanthes caseariae Poit. has its flowers standing often less than 5 mm. high. (These are some of the very smallest flowers known to science, possibly there is a Lemna with a smaller one. The very largest flower that exists is another parasitic member of the Rafflesiaceae. This is the world renowned Oriental root parasite Rafflesia arnoldi R. Br. that has a blossom measuring over a meter in diameter.) A. caseariae Poit. infects leguminous trees, as seen in Central America by me (98 publ. 1964) and invaded wood has been traced in its progress below the bark, up and down trunk and main branches, even into the root region.

Hyperparasitism and Super-Hyperparasitism

In a very few plant disease publications, there appear studies that have dealt with hyperparasitism. This is most often considered a feature somewhat rare, apart, and phenomenal in our science of pathology. Results of researches are still fragmentary on what the pathological physiology is--relating to this matter of mistletoe-on-mistletoe (32, 33). As well, chemical control measures (30, 47) for such parasites are not fully developed. In modern plant pathology there is an obvious necessity to know more about natural means that control disease. Much information about this is still to be unfolded in our tropics. In these tropics is where many types of hyperparasitism are commonly encountered. Weir (97) found instances in the Amazon region, but gave it only passing note. Undoubtedly the most spectacular, and grossest kinds of hyper- and super-hyperparasitism are found in complexes of phanerogamic parasitism. Examples can be found in literature such as: 13, 46, and 98 in 1963, 1964, 1971 publications in which Loranthaceae infect Loranthaceae. Not only are there bacteria, fungi and algae that disease the already parasitizing phanerogams, there are also interacting parasitisms among the thallophytic organisms attacking the parasitizing phanerogams; there is self-parasitism of phanerogamic parasites; there are phanerogams parasitizing those

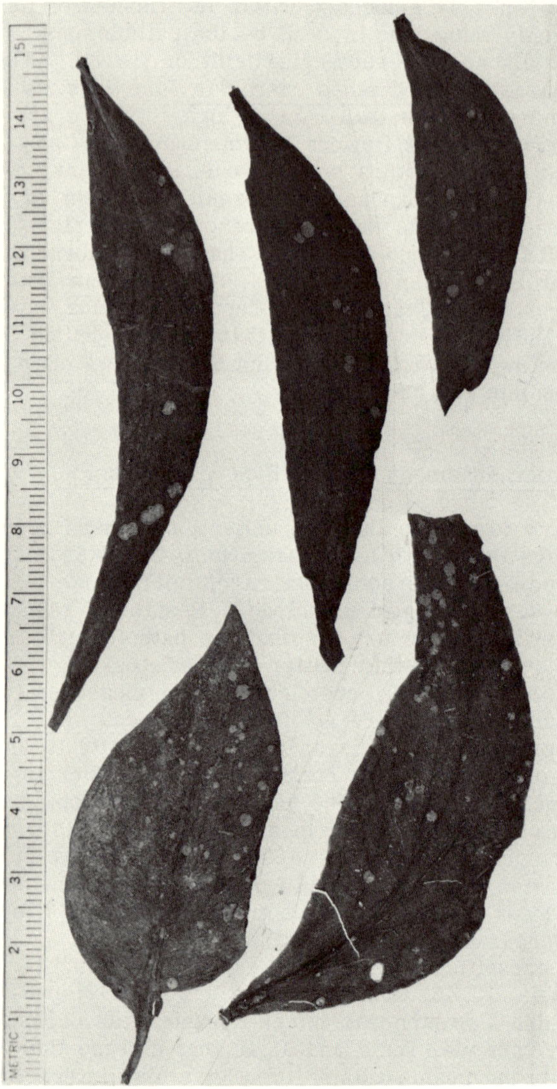

Fig. 17-12. Hyperparasitism. Primary host, tree branch of *Inga laurina* parasitized by *Psittacanthus calyculatus* in El Salvador. Leaves of parasite infected by parasitic alga *Cephaleuros virescens* (light colored, rounded spots) and *Meliola* sp. the black mildew producing reticulated black growths on leaf upper epidermises, and on *Meliola* dead hyphae found invaded by an undetermined bacterium.

Fig. 17-13. Hyperparasitism. Branch of sweet orange with primary parasitism by *Struthanthus* sp. On leaves of parasite: a rust, *Hemileia* sp.; anthracnose spots, *Colletotrichum* = *Glomerella*; carbon or sooty mold, *Capnodium* sp.; an aphid; a scale insect with white tufts; and (on a leaf at the right) a star-shaped seedling of *Psittacanthus calyculatus*. Picture made in Panama.

which are hyperparasitizing other parasitizing phanerogams; and there are examples of phanerogamic parasites at steps towards being, and wholly becoming, independent flowering bushes and small trees.

It is known (e.g., 14, 76, 98 publ. 1963) that parasitic vines of dodder attack growths of Loranthaceae parasitizing woody bushes and trees of various kinds. This is a fairly simple kind of parasite-on-another-parasite. I have also seen it on plants that are not exactly primary parasites but are stranglers (species of *Clusia* and *Ficus*). On bushes of both these genera there were not only *Cuscuta* sp. but also sap-sucking vines of *Cassytha filiformis*. On the host *Ficus* sp. that is a strangling fig, I successfully established a vine of *C. filiformis*, and on this *Cassytha* I put *Cuscuta americana* which it successfully invaded; and then using a known culture, an antracnose fungus, *Glomerella cingulata* was induced to infect the Cuscuta. This dodder is known (67) to be a host of *Colletotrichum* in the temperate zone, but the anthracnose it causes on dodders is still to be studied in the tropics.

Few photographs are ever made of hyperparasitism, and none of super-hyperparasitism. The complexes involved are difficult as camera subjects. Recently, I presented (citation 100, publ. 1971) a diagrammatic drawing of a study I made in El Salvador of a palm tree engulfed by a strangling fig, which was in turn parasitized by hyperparasites, and in addition by super-hyperparasites. However, in accompanying Figure 17-12 are shown diseased leaves of a Loranthaceous parasite, *Psittacanthus*, that was itself parasitizing a tree branch. The parasite leaves are attacked by hyperparasites: *Cephaleuros* and *Meliola*; the latter hyperparasite was found diseased by a bacterium. In Figure 17-13 is an orange tree branch on which is a higher plant parasite, *Struthanthus*, with three fungus hyperparasites on it; another phanerogamic parasite infecting a *Struthanthus* leaf; and insects feeding on *Struthanthus* and *Capnodium* sooty mold starting to form in "super, hyper" fashion on hyperparasites.

During researches carried on in Costa Rica, several instances were found of complexes in series of parasites on top of upright growing parasitic bushes of *Loranthus* spp. on host trees of *Inga* spp. The upright parasites were seen overgrown and parasitically infected by the vine-type woody plants of both *Struthanthus* sp. and *Phthirusa* sp., with, in

Table 29. EXAMPLES[a] OF HYPER- AND SUPER-HYPERPARASITISM of Loranthaceae, in the Caribbean region.

Original tree host[b]	Primary parasite	Hyper- and super-hyperparasitism: Phanerogam(s)	Thallophyte(s)
Pinus	*Psittacanthus*	*Struthanthus*	*Cladosporium*
			Rust (undetermined)
Ficus[c]	*Psittacanthus*	*Phoradendron*	
Inga	*Psittacanthus*	*Psittacanthus*	
Inga	*Psittacanthus*	*Phoradendron*	
		Struthanthus[d]	*Pestalotia*
Erythrina	*Oryctanthus*	*Phthirusa*	
Inga	*Phoradendron*	*Oryctanthus*	
		Struthanthus[d]	*Colletotrichum*
			Mycena
			Capnodium
			Septobasidium
		Cuscuta[e]	*Mycena*
			Colletotrichum
Theobroma	*Struthanthus*	*Oryctanthus*	
Theobroma	*Oryctanthus* sp. 1[f]	*Oryctanthus* sp. 2[g]	
		Phoradendron[d]	
Coffea	*Struthanthus* sp. 1[h]	*Struthanthus* sp. 2[i]	
		Struthanthus sp. 1[d]	*Colletotrichum*
		Struthanthus sp. 2[d]	*Diplodia*
			Fusarium
			Xylaria
Inodes			
Ficus[j]	*Psittacanthus*	*Phoradendron*	*Uleomyces*
			Rosenscheldiella
Hibiscus	*Cassytha*		*Colletotrichum*
Coffea	*Cassytha*	*Cuscuta*[k]	

a. From work by the writer in El Salvador, Costa Rica, Guatemala, and Puerto Rico. See as well examples discussed in the text of this chapter. b. For reasons of simplification and space, only generic names given. c. This strangling fig had killed a tree of undetermined species over which it had grown. d. A parasitic plant hyperparasitizing another. e. *C. americana.* f. *Oryctanthus spicatus.*

g. On *cordifolius.* h. *S. marginatus.* i. *S. orbicularis.* j. This wild strangling fig had recently killed the *Inodes* palm. k. *C. americana* was brought artificially into conjuction with *Cassytha filiformis.*

---

some instances, *Phthirusa* on top of *Struthanthus,* which was already on top of *Loranthus* on an *Inga* tree. In many cases leafspots were seen on *Loranthus* but also they were present on the super-hyperparasites. On leaves of super-hyperparasitizing *Struthanthus* sp., the leafspottings, when spore scrapings from them were examined, were determined as of the fungus *Cercospora* sp., and some of the fruiting structures on spots were *Phoma*-like. In addition, the white spotting of the lichen *Strigula* sp. was found, along with presence of an alga, *Trentepohlia* sp., on old parasite vine stems. A rust was found, probably a *Hemileia,* and a non-parasitic but culturable fungus with clamp connections in its mycelium. Isolations, made from leaf spots and disease-damaged extra cortical roots of *Struthanthus,* disclosed presence of such secondarily parasitic organisms as bacteria (undetermined), *Fusarium* sp. (not *oxysporum*), *Diplodia* sp., *Pestalotia* sp., and a non-sporulating gray-green fungus with markedly fine-grained pubescent mats growing on agar.

What was to me (98 in publ. 1971) a clear prize example of this kind of complex, was studied in El Salvador. [This kind of complexity is known in other parts of the world tropics, as Docters Van Leeuwen (18) has well described in Indonesia.] A palm of the *Inodes* genus, was being engulfed by a strangling fig which was *Ficus* sp. On one of the strangler's branches was a vigorous and well-developed old bush of the parasitic Loranthaceous *Psittacanthus calyculatus.* On this parasitic bush was found another large hyperparasitic bush of *Phoradendron robustissimum.* On this latter super-hyperparasite was found a leafspot, determined by Jenkins and Linder (39) to be *Uleomyces wellmanii,* which in turn was being attacked by another fungus, *Rosenscheldiana phorodendri.* In this case, one can count, including the host, there was a continuous series of six organisms, one upon the other. Microorganisms as parasites on Loranthaceae are becoming known (e.g., 7).

One other instance of a series of parasites superimposed on others (super-super-hyperparasitism) is worthy to relate, indicating that such series are not unusual phenomena. This occurrence was found by me in the 1960 rainy season, on an old *Erythrina* sp. coffee shade tree, felled at the edge of a long neglected coffee finca near Turrialba, Costa Rica. On a large, still living and low hanging branch of the shade tree, was a bushy mass of leafy parasites, the original part of which was *Phoradendron* sp. On its well developed stems was the vine-line parasitic growth of *Oryctanthus* sp. and on top of this was growing *Struthanthus orbicularis* that was several years old. On the *S. orbicularis*, decayed stem parts occurred, and on them Sporocarps were found typical of, *Glomerella cingulata*. Spots of *Mycena citricolor* could readily be identified on the leaves of all four green plants, the shade tree, the primary parasite and the two hyperparasites. On some of the exogenous roots of the *Struthanthus* were seen numerous *Mycena* lesions and a felt organism, *Septobasidium* sp., was sheltering insects and had also covered two *Mycena* infections.

There is probably no place in all biology where hyperparasitism and super-hyperparasitism is so extensive (98 in 1971 publ.) and so remarkable in the series built up, as can be found in connection with the plant pathology of phanerogamic parasitism of plants in the tropics.

Literature Cited

1. Abbiatti, D. 1943. Sinopsis de las Lorantáceceas argentinas. Rev. Argentina Agr. 10:1-25. --. 1946. Las Lorantáceas argentinas. Rev. Mus. La Plata, Sec. Bot. 7:1-110.

2. Alvim, P. de T. and K. W. Seeschaaf. 1968. Dieback and death of cacao trees caused by a new species of parasitic tree. Nature 219:1386-1387.

3. Baker, K. F. 1950. Perennation in relation to control of dodders on native shrubs in Southern California. Phytopathology 40:213-214.

4. Bancroft, C. K. 1914-1950. Report on the botanic gardens and their work. British Guiana Dep. Sci. Agr., Rept. Appendix 2:1-16.

5. Barroso, G. M. 1968. *Acanthosyris paulo-alvimii*--uma nova espécie de Santalaceae. Congresso Brasileiro de Botanica 19 (citation by Alvim).

6. Barry, F. V. 1958. Striga planta parasita de las raices; una nueva amenaza para la caña de azúcar. Com. Defensa Azúcar y Fomento de la Caña 1:1-12.

7. Batista, A. C., J. L. Bezerra and G. E. P. Peres. 1962. Tres novos Polystomellaceae. Broteria 31:41-47.

8. Bennett, C. W. 1944. Studies of dodder transmission of plant viruses. Phytopathology 34:905-932.

9. Bor, N. L. 1953. Manual of Indian forest botany. Oxford Univ. Press, 441 p.

10. Britton, N. L. and C. F. Millspaugh. 1920. The Bahama flora. New Era Printing Co., Lancaster, Penn., 695 p.

11. Britton, N. L. and P. Wilson. 1923-1930. Botany of Porto Rico and the Virgin Islands. Scientific survey of Porto Rico and the Virgin Islands. New York Acad. Sci. 5(pts. 1-4, 626 p.); 6(pts. 1-4, 663 p.).

12. Brown, A. F. 1912. Silviculture in the tropics. Macmillan, London, 309 p.

13. Brown, J. G. 1918. Mistletoe vs. mistletoe. Bot. Gaz. 65:193.

14. Burkhill, I. H. 1906. A parasite upon a parasite--a Viscum apparently V. articulatum Burn., on Loranthus vestitus, Wall, on Quercus incana Roxb. J. Asiatic Soc. Bengal 2:299-301.

15. Costa, A. S. 1944. Multiplication of viruses in the dodder, Cuscuta campestris. Phytopathology 34:151-162.

16. Cufodontis, G. 1934. La piante raccolte durante la Spedizione Biologica Austriaca in Costarica nel 1930. Arch. Bot., Italy 10:235-274.

17. Dawson, G. 1944. Las Santálaceas argentinas. Rev. Mus. La Plata 6:5-80.

18. Docters Van Leeuwen, W. M. 1954. On the biology

of some Javanese Loranthaceae and the role birds play in their life-history. Beaufortia, Misc. Publication 4:105-207.

19. Dutt, B. S. M. 1950. Cassytha L. on Viscum. Sci. Cult. 16:258.

20. Ellis, D. E. 1946. Anthracnose of dwarf mistletoe caused by a new species of Septogloeum. J. Elisha Mitchell Sci. Soc. 62:25-50.

21. Fawcett, W. and A. B. Rendle. 1914. Flora of Jamaica. III. Dicotyledons, families Piperaceae to Connaraceae. British Museum, London, 280 p.

22. Fernandez Valiela, M. V. 1952. Introducción a la Fitopatologia (2d. ed.) Gadola, Buenos Aires, 872 p.

23. Frazer, J. G. 1927. The golden bough, a study in magic and religion. Macmillan, New York, 752 p.

24. Gandara, G. 1920. Enfermedades y plagas del naranjo. Estac. Agr. Centr. Mejico, Bulletin 111, 40 p.

25. Garcia Rada, G. 1942. El control de la Cuscuta en el lino. Dir. Agr. Ganad., Peru, Inf. Agrop. 11/12:333-336. --. 1944. Dodder on flax and its control. Phytopathology 34:704-705.

26. Garriss, H. R., and J. C. Wells. 1956. Parasitic herbaceous annual associated with corn disease in North Carolina. Plant Dis. Reptr. 40:837-839.

27. Gentry, H. S. 1942. Rio Mayo Plants. Carnegie Inst., Washington, D. C., 328 p.

28. Gill, L. S. 1953. Broomrapes, dodders and mistletoes. pp:73-77. In U. S. Dept. Agr. Yrbk. 1953. Washington, D. C., 940 p.

29. ________. and F. G. Hawksworth. 1961. The mistletoes; a literature review. U. S. Dept. Agr., Forest Serv., Tech. Bulletin 1242, 87 p.

30. Greenhorn, C. G. and A. G. Brown. 1957. The control

of mistletoe by trunk injections. Australian Inst. Agr. Sci. J. 23:308-318.

31. Guppy, H. B. 1917. Plants, Seeds, and Currants in the West Indies and Azores. Williams and Norgate, London, 531 p.

32. Harris, J. A., G. J. Harrison, and T. A. Pascoe. 1930. Osmotic concentration and water relations in the mistletoes, with special reference to the occurrence of Phorandendron californicum on Covillea tridentatis. Ecology 11:687-701.

33. ________., and J. V. Lawrence. 1916. On the osmotic pressure of the tissue fluids of Jamaican Loranthaceae parasitic on various hosts. American J. Bot. 3:438-455.

34. Hauman, L. and Irigoyen. 1923. Catalogue des phanérogames de l'Argentine. Mus. Nac., Hist. Nat. Buenos Aires 32:39-58.

35. Hawksworth, F. G. 1961. Dwarf mistletoe of Ponderosa pine in the southwest. U. S. Dep. Agr. Tec. Bul. 1246, 112 p.

36. Hemsley, W. B. 1895. Some remarkable Phanerogamous parasites. J. Linnean Soc. Botany. London 31:306-311.

37. Hosford, R. M. 1967. Transmission of plant viruses by dodder. Bot. Rev. 33:387-406.

38. Hunziger, A. T. 1944. Nota sobre especies de Cuscuta de la flora Argentina. Rev. Argentina Agron. 10:70-73. --. 1947. Sinópsis de las especies argentino-uruguayas del genero Cuscuta. Rev. Argentina Agron. 14:33.

39. Jenkins, A. E. and D. P. Linder. 1952. Two new ascomycetes on Phoradendron. Mycologia 44:557-560.

40. Kanner, L. 1939. Mistletoe, magic medicine. Bull. Hist. Med. 7:875-936.

41. Kelbitsch, H. 1960. Cuscuta als Überträger des

Crowngall-erzengenden Prinzips. Protoplasma 52: 437-445.

42. Kelman, A., G. B. Lucas, and H. R. Garriss. 1955. False brown rape of tobacco in North Carolina. J. Elisha Mitchell Sci. Soc. 71:137-174.

43. Knorr, L. C. 1949. Parasitism of citrus in Florida by various species of dodder, including _Cuscuta boldinghii_ Urb., a species newly reported for the _United States_. Phytopathology 39:411-412. --. 1949. A gall of Tahiti lime and other citrus species caused by dodder. Phytopathology 39:616-620.

44. Koernicke, M. W. 1908. Ueber die Rindenwurzelen tropischer Loranthaceen. Verhand. Ges. Deutsch. Naturfors. Aerzte, Leipzig 80:186-187.

45. Krause, K. 1932-1947. Loranthaceae pp:4-24, _In_ A. A. Pulle. Flora of Suriname 1. Amsterdam, 524 p.

46. Kuijt, J. 1955. Dwarf mistletoe. Bot. Rev. 21: 569-619. --. 1961. Notes on the anatomy of the genus _Oryctanthus_ (Loranthaceae). Canadian J. Bot. 39:1809-1816. --. 1963. On the ecology and parasitism of the Costa Rican tree mistletoe, _Gaiadendron punctatum_ (Ruiz & Pavon) G. Don. Canadian J. Bot. 41:927-938. --. 1964. Critical observations on the parasitism of New World mistletoes. Canadian J. Bot. 42:1243-1278. --. 1964. A revision of the Loranthaceae of Costa Rica. Botan. Jahrb. 83:250-326.

47. Korstian, C. F. and W. H. Long. 1922. The western yellow pine mistletoe. Effect on growth and suggestion for control. U. S. Dep. Agr. Bulletin 1112, 35 p.

48. Lackey, C. F. 1946. Reaction of dodders to stems of other dodders and to their own stems. Phytopathology 36:386-388. --. 1949. Reactions of dodder to host plants used in virus disease investigations. Phytopathology 39:562-567.

49. Lanjouw, J. 1932. Flora of Surinam. J. H. de Bussy Ltd. Amsterdam, 524 p.

50. de Laubenfels, D. J. 1959. Parasitic conifer found in New Caldonia. Science 130:97.

51. Leach, C. M. 1958. A disease of dodder caused by the fungus Colletotrichum destructivum. Plant Dis. Reptr. 42:827-829.

52. Leon, H. (J. S. Sanget) and H. Alain (E. E. Liogier). 1951. Flora de Cuba. Dicotiledoneas: Causarinaceas a Meliaceas. Mus. Hist. Nat. Colegio de LaSalle (Cuba), Contrib. Ocas. 10, 456 p.

53. Link, G. K. K. 1933. Etiological phytopathology. Phytopathology 23:843-862.

54. Linnaei, C. (Linnaeus) 1754. Genera Plantarum Eorumque Characteres Naturales Secundum Numerum, Figuram, Situm, et Proportionem Omnium Fructificationis Partium. Holmiae, Lawrentii Salvii, 521 p. --. 1787. (Translation of last edition by Lichfield Bot. Soc.) The families of plants with their natural characters, according to the number, figure, situation and proportion of all the parts of the fructification. Lichfield Bot. Soc. vols. I & II, 840 p.

55. Lucas, G. B. 1958. Diseases of tobacco. Scarecrow Press, New York, 498 p.

56. Lundell, C. L. 1937. The vegetation of Peten. Carnegie Inst., Washington, Publication 478, 244 p.

57. Loo, S-W. 1946. Cultivation of the excised stem tips of dodder in vitro. American J. Bot. 33:295-300.

58. MacBride, J. F. 1937. Flora of Peru. Field Mus. Nat. Hist., Bot. Ser. 13 (Pt. 2, No. 2):257-661; (Pt. 5, No. 5):1-536.

59. McGrath, H., C. W. Shaw, L. L. Jansen, B. R. Lipscomb, P. R. Miller, and W. B. Ennis. 1957. Witchweed (Striga asiatica)--a new parasitic plant in the United States. U. S. Dept. Agr., Crops Res. Div., Special Publication 10, 146 p.

60. McKee, H. S. 1952. Root parasites in Loranthaceae. Nature 170:40.

61. Marchionatto, J. B. 1948. Tratado de Fitopatología. Editorial Sudamericana, Buenos Aires, 537 p.

62. Menzies, B. P. and H. S. McKee. 1959. Root parasitism in Atkinsonia ligustrina. (A Cunn. ex F. Muell.) F. Muell. Linnean Soc. N. S. Wales, Proc. 84:118-127.

63. Moscoso, R. M. 1943. Catalogus Florae Domingensis. L. & S. Printing Co., New York, 732 p.

64. Nash, S. M. and S. Wilhelm. 1960. Stimulation of broomrape seed germination. Phytopathology 50: 772-774.

65. Nowell, W. 1923. Diseases of Crop-Plants in the Lesser Antilles. West India Committee, London, 383 p.

66. Oliver, D. 1864. Notes on the Loranthaceae, with a synopsis of the genera. J. Linnean Soc. 7:90-106.

67. Parmeter, J. R., J. R. Hood, and R. F. Scharpf. 1959. Colletotrichum blight of dwarf mistletoe. Phytopathology 49:812-815.

68. Parodi, L. R. 1936. Causa probable porque la Cuscuta indecora es poco dañina en los alfalfares argentinos. Rev. Argentine Agr. 3:253-254.

69. Pierce, G. J. 1893. On the structure of the haustoria of some phenerogamic parasites. Ann. Bot. 7:291-327. --. 1894. A contribution to the physiology of the genus Cuscuta. Ann. Bot. 8:53-118.

70. Pittier, H., T. Lasser, L. Schnee, Z. Luces de Febres, and V. Badillo. 1945. Catalogo de la Flora Venezolano. 1. Vargas, Caracas, 423 p.

71. Rediske, J. H. and K. R. Shea. 1961. The production and translocation of photosynthate in dwarfmistletoe and lodgepole pine. American J. Bot. 48:447-452.

72. Reis, A. B. 1938. Mata-cacau. Bahia Rural 5:1976-1977.

73. Rizzini, C. T. 1950. Sôbre Phoradendron fragile. Urb. Rev. Brasil Biol. 10:45-58. --. 1952. Pars generalis prodromi monographie Loranthacearum Brasiliae terrarumque finitarum. Arq. Jard. Bot. Rio de Janeiro 12:39-126. --. 1956. Pars specialis prodromi monographiae Loranthacearum Brasiliae terrarumque finitiniarum. Rodriguésia 18/19:87-264. --. 1960. Loranthaceae. Ann. Missouri Bot. Garden 47:263-290. [Contribution to Flora of Panama by Woodson and Schery.]

74. Roger, L. 1954. Les phanerogames parasites (pp: 2415-2446). In Phytopatologie des Pays Chauds, Paul Lechevalier, Paris, 3 vols., 3154 p.

75. Rogers, W. E. and R. R. Nelson. 1962. Penetration and nutrition of Striga asiatica. Phytopathology 52:1064-1070.

76. Rao, L. N. 1938. A case of triple parasitism? Loranthus vs. Viscum. Current Sci. 7:112.

77. Sanzin, R. 1918. Sobre el Loranthus cuneifolius. Rev. Chile Hist. Nat. 22:126-129.

78. Sanzen-Baker, R. G. 1938. Literature on the mistletoes. Imperial For. Inst. (England), Paper 12, 13 p.

79. Scharfp, R. F. and J. H. Parmeter, Jr. 1962. The collection, storage, and germination of seeds of a dwarf mistletoe. J. Forestry 60:55. --. 1962. Penetration of red fir by the dwarf mistletoe Arceuthobium compylobodum. Phytopathology 52:750.

80. von Solms-Laubach, H. G. 1867-1868. Ueber den Bau und die Entwicklung der Ernahrungsorgane parasitischer Phanerogamen. Jahrb. Wiss. Botan. 6:509-638. --. 1877. Das Haustorium der Loranthacean und der Thallus der Rafflesiaceen und Balanophoreen. Naturf. Ges. Halle 13:239-276.

81. Srivastava, L. M. and K. Esau. 1961. Relation of dwarfmistletoe (Arceuthobium) to the xylem tissue

of conifers. I. Anatomy of parasite sinkers and their connection with host xylem. II. Effect of the parasite on the xylem anatomy of the host. American J. Bot. 48:159-167; 209-215.

82. Standley, P. C. 1937;1938. Flora of Costa Rica. Field Mus. Nat. Hist., Bot. Ser. 18(2):401-780; (4):1137-1616.

83. ________., and S. Calderon. 1941. Lista Preliminar de las Plantas de El Salvador. 2nd. Ed. Imprenta Nacional, San Salvador. 450 p.

84. ________., and J. A. Steyermark. 1946. Flora of Guatemala. Fieldiana: Botany 24(4), 493 p.

85. Stevens, H. E. 1918. Florida citrus diseases. Florida Agr. Dep. Exp. Sta. Bulletin 150, 96 p.

86. Stevenson, J. A. 1918. Citrus diseases of Puerto Rico. J. Dep. Agr. Puerto Rico 2:43-123.

87. Svenson, H. K. 1946. Vegetation of the coast of Ecuador and Peru and its relation to that of the Gallapagos Islands. II. Catalogue of plants. American J. Bot. 33:427-498.

88. Theophrastus (ca. 300 B. C.). Transl. from Greek by A. Hoyt of book by Theophrastus: "Enquiry into plants and minor works on odours and weather signs." W. Heinemann, London; Putnam, New York. 2 vols. total 972 p., publ. in 1916.

89. Thoday, M. G. (Sykes). 1911. On the histological relations between Cuscuta and its host. Ann. Bot. 25:655-682.

90. Thoday, D. 1961. A general survey of the Loranthoideae. Royal Soc. Bot., Proc. 155:1-25. --. 1951. The haustorial system of Viscum album L. J. Exp. Biol. 2:1-19. --. 1956. Modes of union and interaction between parasite and host in the Loranthaceae. I. Viscoideae, not including Phoradendreae. Royal Soc. Bot., Proc. 145:531-548.

91. Trelease, W. 1916. The Genus Phoradendron, a Monographic Revision. Illinois Univ. Press, Urbana, 224 p.

92. United States Crops Research Division, 1957. Witch-weed (Striga asiatica)--a new parasitic plant in the United States. U.S. Dep. Agr., Agr. Res. Serv., Special Publ. 10, 142 p.

93. Valenta, V. 1958. A new yellows virus causing flower proliferations in the dodder, Cuscuta campestris Yunck. Phytop. Zeits. 33:316-318.

94. Viégas, A. P. 1961. Indice de Fungos da América do Sul. Inst. Agron., Campinas, 921 p.

95. Walker, J. C. 1969. Plant pathology (3d. ed.). McGraw-Hill, New York, 819 p.

96. Weathers, L. G., M. K. Harjung, and F. C. Greer. 1965. Dodder as a means of transmitting citrus viruses. California Citrogr. 51:83-84.

97. Weir, J. R. 1926. A pathological survey of the para rubber tree (Hevea brasiliensis) in the Amazon Valley. U. S. Dep. Agr. Bulletin 1380, 129 p.

98. Wellman, F. L. 1953. Various types of parasitisms of tropical American Loranthaceae, Reun. Latino-Amer., Fitogen. Fitopar., Brazil, Proc. 2, p. 50. --. 1961. Coffee botany, cultivation, and utilization. Leonard Hill, London; Interscience Publ., New York, 488 p. --. 1962. A few introductory features of tropical plant pathology: (pp. 928-930). In F. L. Wellman et al., A symposium on tropical plant pathology. Phytopathology 52:928-953. --. 1963. Parasitic habits of phanerogams in the American tropics. Phytopathology 53:26. --. 1964. Parasitism among Neotropical phanerogams. Ann. Rev. Phytopath. 2:43-56. --. 1971. Plant Diseases; An Introduction for the Layman. Natural History Press, New York. 130 p.

99. Wilhelm, S. 1962. History of broomrapes (Orobranche ramosa and O. ludoviciana) and their control by preplant soil injection with methyl bromide solutions. 16th Internat. Hort. Congress 1:112-114.

100. Williams, L. 1942. Exploraciones Botanica en el Guayana Venezolana. Tipografica La Nación, Venezuela, 468 p.

101. York, H. H. 1909. The anatomy and some of the biological aspects of the American mistletoe *Phoradendron flavescens* (Pursh) Nutt. Univ. Texas Bulletin 120, 31 p.

102. Young, H. E. 1947. A mistletoe on an orchid. Australian Orchid Rev. 12:4.

103. Yuncker, T. G. 1932. Genus *Cuscuta*. Torrey Bot. Club Mem. 18:113-331.

Chapter XVIII

CONCERNING TROPICAL NEMATODE DISEASES

Summary

During long periods of geologic time, nematodes, which are minute worm-like animals, have been native to tropical conditions. These organisms vary as to types and cause many tropical plant diseases, e.g., leaf spotting, leaf tip whitening, wilting, internal parasitism in roots--sometimes with host tissue swelling and sometimes without--external root feeding injury, and systemic invasion. Nematodes are widely distributed in the tropics. Considering the total number of these organisms, saprophytes as well as parasites, they are the most numerous of any great group of living things on earth. Some are vectors of plant diseases. Through means of synergistic actions nematodes are known to increase intensities of numerous bacterial and fungus plant diseases; when nemas attack along with the presence of other organisms, they are known to destroy resistance to many crop-plant diseases.

Introduction

Those professionally interested in diseases of tropical crops know that nematodes which parasitize these cultivated plants are, in science, included among the animals. They are microscopic worms (see discussions: Cardeñoso (14), Christie (18), certain authors in the compilation edited by Sasser and Jenkins (55), Thorne (72), Kloss (31), and Lordello (37, 38)), of aquatic habits, that they flourish in

Fig. 18-1. Nematodes, Panagrolaimus sp., found in tomato leaf juice associated with an unidentified bacterium; the two together causing leaf spot. Sketch after preliminary and unpublished drawings by G. Steiner and others made of the worms in situ in 1940.

liquid surroundings whether in the soil or in the sap inside plants. All pathologists are acquainted with the part nematodes play in the field of plant disease study; no specialist in nematology, I have sketched something of the kinds of organisms some of these plant-disease-producing microscopic worms are: Fig. 18-1. They are best in their development in relatively warm conditions, and have been on earth since long before and during the tropically warm-wet eons when dinosaurs were living. Some 9000 species of nematodes have been described, and a good proportion of them are of interest in plant pathology.

It can be seen from looking at the list of literature at the close of this chapter, that there has been considerable study, and for considerable time, on nematodes causing plant diseases in the American tropics. This interest started before 1900 (25, 46, 48) and shortly after (9, 19, 43, 62) in our tropics. In this great region, plant disease nematodes are common, but seemingly not quite so devastating as certain fungi, bacteria, and viruses; nematodes, generally and with reason, are treated as part of the over-all population of parasites that compete for attention in the complex study of plant pathology. Occasional surveys (examples: 4, 9, 17, 31, 37, 45, 57, 67) aid in giving attention to these organisms.

Except for specialized nematode workers in long-established Neotropical institutions in such large countries as Brazil and Argentina (and U. S. state experiment stations in Florida and California), pathologists must handle a large variable list of plant diseases. Extremely close specialization is in the nature of a luxury. Usually, pathologists are so overwhelmed with rusts, spots, rots, blights, mosaics, and wilts that they concentrate on perfecting their experience with these diseases. In recent years nematode diseases have been given more attention in our tropics.

Nematology as a strong component in plant pathology is gaining an accumulation of worldwide results--all plant diseases and their causal organisms are becoming better known. This has been slow with respect to nematodes. However, a few innovators and stimulated, fearless workers have and do pursue profitable and appreciated researches on nematodes causing tropical crop diseases. Sometimes superficial and under laborious conditions, results often begin with the sole business of determining and publishing new names of disease nemas. Sometimes the work stops there; in the best cases research is done on life histories, host ranges, control, and analyses of amelioration practices. (Examples: 2-8, 11, 12, 17, 24, 27, 32-35, 37-40, 50-53, 73-75.)

There are especially favorable conditions in the American tropics for basic discoveries on nematode life and parasitism: there is no winter there. There are areas* in which there is no dry season and perpetually sufficient amounts of water occur naturally, in soils and on surfaces of plants, so these fundamentally aquatic organisms flourish and reproduce there in amounts met nowhere else. Another situation is that many localities for research are available where ideal field temperatures for nemas are forever constant within variations of a few degrees. For comparisons areas can be found of desert places where it is so dry that tradition says it never has rained, in other places it practically rains every day, and these situations blend into parts on either side of normal for crop growing. Whenever agricultural plants grow in the tropics, the soil is not too hot for nematodes to live. On the other hand, there are places where researchers can institute clearing and dry-cultivated fallowing, resulting in sun-baking, usable for both nematode control measures as well as for ecological studies; clearing may be followed by flood- or merely wet-fallow. Another feature, with which the temperate zone cannot compare for pathogen studies, is the fact of the profligacy of host plants, endless chances for variation in the nematodes and the hosts, and in great regions plant life is practically in primeval condition, nearly as untouched now as it was before man ever was upon this earth. In addition there is nowhere that hyperparasitism and other means of natural control of nematodes can be studied with so much success as in the tropics.

It has not been until more recent years that text books and inclusive bulletins (examples: Moreno (45), Thorne (72), Christie (18), Sasser and Jenkins (55), Kloss (31), Steiner and Buhrer (63), Steiner and Ramírez (64), Tarjan (68), Wallace (75), Mai *et al.* (41)) and, as never before, regularly appearing publications, are available with which the tropical nematologist may consult. In agricultural research the tropical scientist, whatever his specialty, much

*Van Gundy (74) must be unacquainted with extensive tropical areas where there is no cold or dry season to induce nematode dormancy.

more than the temperate zone scientist, must be reasonably certain that his efforts will insure results. Many special suggestions and technical methods of nematode study are continually being devised: many citations can be referred to in this regard some of note are 1, 6, 8, 11, 12, 18, 19, 23, 30, 31, 34, 37-40, 50, 51, 53, 55, 56, 59, and in other citations in 64, 69, 72, 75, and 77.

Much attention must be devoted by the nematode student to an extensive background, and there is need of adapting to special research requirements. As with fungi, bacteria, and viruses, the agricultural scientists must learn of the offending nematodes more about their modes of holdover from season to season (recalling there is no winter), means of distribution and spread, and the host ranges involved. To the working pathologist, nematode controls dove-tail well into other regular controls such as disinfestation of nursery or greenhouse soils, spraying, hygiene, quarantine, soil stirring, rotations, mulching practices, searching for and breeding tolerant/resistant crop varieties, and other methods.

Some Types of Nematodes

The nematode population is a variable lot of organisms, and the kinds encountered as plant associates are subjects for life-long studies. For reasons of perspective, a few short statements are all that is attempted here. For wealth of detail, the reader from the Neotropics or elsewhere is referred first to texts (cited in the Introduction to this chapter) and then to publications of more specific interest found at the close of this chapter.

In a manner of speaking, one of the less complex nematodes from the standpoint of food requirements, is the type of free-living, soil inhabiting, saprophyte (saprophagous) eelworm. There are many of these, many if not all are cosmopolitan in soils, and some kinds occur in very wide-ranging situations in different hemispheres of the earth; this was impressively pointed out by Thorne (72). Examples of some of the free-living saprophagous types are certain species of such genera as Rhabditis, Diplogaster, Cephalobus, Longidorus, Panagrolaimus, and Dorylaimus. These, and others like them, are often present and sometimes become troublesome and require attention. Their general characters must be known to avoid confusion in interpreting results of specimens secured from nematode disease recovery studies. Life histories of such free-living nematodes are not given as much study as some others, but is is believed they probably serve a very useful purpose in their consuming of fungi and their comminution and feeding on soil debris, thus hastening its desirable conversion into more available plant nutrients.

Among these free-living nemas are types of occasional predators; these kill parasitic nemas, and they must be of assistance in biologic control. These organisms feed not only upon small animals like rotifers but they also feed upon many saprophytic and plant-attacking nematodes, certain fungi, and other organisms, many of which are involved in plant diseases. As examples of some predator genera can be named Mononchus, Mylonchulus, Nygolaimus, and Sectonema. When examined, some of these voracious organisms are often larger than usual plant parasitizing nemas, and some can be not only carnivorous but also cannibalistic in their habits.

In some contexts, nematodes that disease plants (of which there are over 1000 species: Allen and Sher (1)) are set aside as a type solely on that one character. There are a few workers with these organisms who contend that when it comes to pathology, the problem is more than generic or taxonomically specific differences--the most important thing to these workers is ecology. A few kinds of symptom differences in nematode-caused diseases are as follows.

Leaf spotting is caused by the attack of some nematodes, for example the ectoparasitic Aphelenchoides ritzemabosi (Schwartz) Steiner that has been seen attacking leaves of chrysanthemum (Fig. 18-2) in, for example, Central America and the West Indies. Another nematode leafspot has been seen on both cherry tomato and the large fruited variety as well, caused by Panagrolaimus sp. in Honduras, and which I found also in Guatemala and in El Salvador.

The rice crop in the Orient has long been seriously diseased by the nema Aphelenchus oryzae Yakoo (= Aphelenchoides besseyi Christie) causing the disease called white tip. It also occurs in our tropics. The organisms are like leaf-spotting nemas, they are ectoparasitic, feeding externally on the moist leaf tips; I have seen these parasites in leaf tip guttation. Leaves that are attacked turn white at their terminal ends and if badly diseased the plant produces poorly developed grains or none at all.

There are stem inhabiting (endoparasitic) nematodes in the Neotropics. One example is *Heterodera rostochensis* Wollenw. in its attack on a number of hosts. It is found generally in high altitude locations (10, 60, 72 and others), for example in the Andes of Peru causing disease of quinoa, ulluca, oca, tomato, potato, and certain other Solanaceae. There it invades both stem and root tissue, and causes its most severe effect when roots become attacked by large numbers of the parasite. Another root-attacking, but at the same time ectoparasitic organism is the so-called awl nematode *Dolichodorus heterocephalus* Cobb that inhabits the moist, warm soils of our tropics (e.g., 18, 68, 72) and attacks many hosts. It stings roots in its feeding process, and under moist conditions it often crawls out of the soil and migrates up, on the surface parts and then into the foliage.

Above all, from the standpoint of being widely recognized in the warm countries are the rootknot nematodes. These are mostly of the genus *Meloidogyne* (2, 3, 9, 18, 20, 25, 31, 34, 36-38, 40, 41, 43-49, 53, 58, 67-69, 72). The root en-enlargements they cause (often well illustrated in temperate zone texts) are like the knots in strings and are probably on the largest number of economic hosts of any nematode injury reported in the Neotropics. Rootknots caused by *Meloidogyne* spp.* are well distributed in our tropics, although when present they may not always cause extreme losses; in clinical work, I have had rootknot brought to me year after year often on coffee, especially on cucurbits, and an unlettered peasant showed me a patch of 15 huisquil vines on which rootknot was found from 1943 to 1947 without visible change in aerial vine growth or reduction in fruit production. The *Meloidogyne* nematodes cause many differently shaped swellings and knots. By all odds the giant rootknots (65) seen in Florida and Cuba deserve special mention. On roots of such ornamental vines as *Thunbergia laurifolia* Lindl. and its relative *T. grandiflora* Roxb. galls are often 18 inches to 2 feet (45 cm. to 61 cm.) in diam-

*It is notable in the literature, see Thorne (72), that the nematode genus *Meloidogyne* was first found in the American tropics and studied there by E. A. Goeldi, who worked on it as a coffee parasite before 1888 and published the description of the genus in 1892.

eter. Such huge tropical rootknot galls have also been observed on other plants such as Rheum rhaponticum and on species of Begonia.

Characteristically Tropical Nematodes

It is of note that specialists in nematodes have little to say about the commonly occurring truly tropical nematodes. It remains for a person less specialized, like me for instance, to suggest this. Where they have been looked for in such a country as Peru (57, 60), nematodes are readily found. The nematode problem is geographically very extensive, and is becoming better studied in the tropics. Steiner and Buhrer (63) considered that some of the most difficult tropical crop production problems had to be from the abundant presence of these organisms; these writers believed a reason for developing the well-known, ancient, shifting cultivation method was "avoiding the effect of soil pests, particularly of noxious nematodes." However, many other plant diseases are involved as well. There are good reasons for believing nemas may become increasingly important in tropical agriculture, although some workers feel ill disposed towards dividing groups or types on the basis of geography (see discussions by Thorne (72), Christie (18) and others). Nevertheless certain nematodes are practically confined to warm lands, which is of significance in our Neotropical situation.

There is practical as well as academic interest in recognizing that certain nematodes came from the tropics. Probably here can be found the greatest possible number of variations and related species of nemas. Knowing the tropics gives some idea of the ecology best suited to a nematode, and under what crop conditions it may be the most dangerous. Also, if for example a nematode species seems from circumstantial evidence to be from the tropics, there to the center of it is where scientists must go to search for crop-plant genetic material of most tolerant or resistant character.

One nematode species that seems undoubtedly tropical is the Rhadinaphelenchus cocophilus Cobb, which causes red-ring of coconut. It was found first in the West Indies. It is probable, according to Thorne (72), that symptoms of it were known as early as 1905; he points out that Nowell published the first work about the disease in 1919. Since then it has been of increasing concern in coconut plantations (13, 16, 21, 70). It has been given special attention (work not always published) in Barbados, Puerto Rico, Trinidad, El Salvador, Guatemala, Honduras, British Honduras, Panama, Venezuela, British Guiana, and Brazil. This nematode's effect as a disease appears to be confined to palms, in which it causes infection of certain members in the genera of *Cocos*, *Elaeis*, and *Sabal*. Moreover, although Briton-Jones (13) believed it was transferred by black vultures, the main vector of this nematode appears to be (72) a spectacular tropical insect, the large beetle, *Rhyncophorus palmarum*. It has a rhinocerous-like shaped head, its food is palms and it carries about with it *R. cocophilus*.

For a long time studies on potato troubles have included reference to the golden nematode, *Heterodera rostochiensis* Wollenw. It was first described in northern Europe, and recognized there as adapted to cool conditions where potatoes grew best. Where the nema does not occur there is careful quarantine against it. The trouble does not become serious under hot conditions (Thorne (72)). For many years it was not known where it came from. From the work of Consuelo Bazan de Segura (60)--who discovered it in the Peruvian Andes and pointed out its occurrence in Bolivia--there is reason, as she notes, to conclude it came from that part of the South American tropics which is the place of origin of the potato. At first, in the temperate zone all of its known hosts were in the family Solonaceae with the exception of the snapdragon, *Antirrhinum majus*. In the Andes, de Segura found that it infected all species and kinds of potatoes, as well as the indigenous food plants of oca, ulluca, and quinoa, all long cultivated by aborigines. All indications are that *H. rostochiensis* is a true tropical nematode, that by an accident of evolution it is especially adapted to the cooler conditions present in high Andean valleys and table lands, and that this characteristic has assured its adaptation to temperate zone existence.

The rice white tip nematode, *Aphelenchus oryzae* = *Aphelenchoides besseyi*, has been mentioned in the section above. It seems likely this nema may have been originally from the Oriental tropics. It is seed borne and is therefore readily distributed long distances. One of its habits is to move externally on wet and warm leaf surfaces. Its presence can be noted by the production of white color of leaves. In the Neotropics, this nematode is found at least in the southern and warmest edge of the United States (Louisiana), in Central America (El Salvador, Nicaragua), in the West Indies (Puerto Rico), in a northern part of South America (Venezuela), and in southern South America (Brazil).

A number of nematodes from plants diseased by them in the tropics have been described as new (e.g., 15, 17, 20, 31, 37, 52, 53, 72, 77), and even as this is written other tropically-found nematodes are being studied and described. Some tropical plant pathologists are proportioning much of their time so that a good sector of it includes deeper study of nemas.

That there are tropical nematodes of unusual habits still to be discovered is illustrated by one experience of mine. In 1940 diseased leaves of a small fruited tropical tomato were sent me by T.J. Grant for identification of the causal organism, that had resulted in some unusual and severe leaf spots in Honduras (Fig. 18-2). These lesions could not be satisfactorily identified as anything known by Grant who knew what to expect on tropical tomatoes and who had collected them himself in Lancetilla, Honduras. I made preparations from spot surface scrapings and from cross sections for microscopic examination. No fungus was discovered. There were present, however, bacteria and nematodes. This latter, was of interest--was it a possible scavenger? Isolations resulted in the recovery of a medium slow-growing bacterium, and when isolated in pure culture, it was used for inoculum. Tomato leaves were not infected, whether they had been punctured, bruised or scraped. Meanwhile some crudely crushed leaf spot material also was used as inoculum and the tomato leaves developed spots; in the spots were nematodes. Later that year the leaf spots were given to G. Steiner who studied the nematode present. He found it to be a new species of the genus _Panagrolaimus_ that crawled over plant surfaces, and that caused the leaf spot by injuring leaf tissue and carrying into it bacteria.

The history of this discovery is of a somewhat unfortunate nature. As interesting as the findings were to me, to Dr. Grant and to the late Dr. Steiner, the pressure of other activities by the latter was such that he did not complete his work on this nematode. He made drawings and jotted down information about the nema. In his notes, some of which came to me long after Steiner's death, the name given to the eelworm by him was changed twice and

Fig. 18-2. Severe leaf spotting and defoliation disease of locally grown tomato, Lycopersicon pimpinellifolium, in Honduras. Damage caused by a nematode (Panagrolaimus) in close association with an unusual bacterium never identified. Photograph of original diseased leaflets from a valley in Honduras, collected in August 1940. Courtesy of T. J. Grant.

he finally settled upon _Panagrolaimus hondurae_ n. sp. At present this can only be considered as a tentative name. There was absolutely no question in Steiner's mind that it was a completely new species, occurring in the tropics, having an unusual relationship with a bacterium, and that the genus was _Panagrolaimus_.

Workers in the Neotropics are perpetually studying eelworms that have evolved under tropical conditions, that are specifically adapted to those conditions, and that attack crop plants that grow there. Growers and agricultural scientists in the tropics are very cognizant of the nematode problem, although there is more being done in some crops than in others. It is significant that for many years there has been among plantation owners a common parlance for nematodes that indicate the crop attacked. For example, there are such as "coconut redring nematode," the so-called "banana nematode," "coffee-root nematode," "citrus nematode," "sugarcane nematode," "pineapple nematode," "squash rootknot nematode" and others similarly named. Even recently coffee nematodes, not all the same species, are reported (17) from Guatemala, a melon (49) and a tobacco (22) nema in Brazil, and Dioscorea nemas (58) in Guatemala.

In their review of the nematode problem in the American tropics, Steiner and Buhrer (63) mention only nine specific crops they thought of as being adequately studied for nematode information in this great region. It is believed by others there are more and to my personal knowledge and from literature reviews, over 60 crops have had and are having considerable attention. These are receiving about as much emphasis as available researchers can give them. Considering it in light of the specialized type of work which pathologists have to perform with nematodes as part of the over-all plant disease research, I feel it is a good record, presently reasonably satisfactory in the Neotropics.

Ubiquitousness of Eelworms

There is no great group of living organisms so numerous or so universally present as the eelworm or nematodes in the earth's surface layers, whether high mountain or deep sea, desert or swamp, snow field or sand dune, cultivated or wild land, temperate zone or the tropics. Among the specialists in nematode study, this is a primary principle. Dramatization of it is nowhere so well expressed (see Thorne (72)) as in the classic words of N. A. Cobb: "... if all matter in the universe except nematodes were swept away, our world would still be dimly recognizable .... [We] would find its mountains, hills, valleys, rivers, lakes and oceans represented by a film of nematodes."

It is generally appreciated that when searched for in the wet tropics, eelworms may be found on bark and leaves; at the bases and on the highest tips of tall trees or even telephone poles; in the soil, not only in the rhizosphere but down as far as men dig for deepest ditches; and squirming in moist crevices of palm thatch on roofs of houses and barracoons. In drier parts they may be at times quiescent, but they are easily revived as encysted organisms stuck onto certain seeds or on dead stalks of last year's food and grain plants in the milpa. They are blown around by hurricanes and by rain-drenched wind, are carried in silt-laden flood waters, and, in a dormant condition, they exist attached to dust and crumbs of dirt that can be scraped up from inside Jeeps and Land Rovers and from the hooves of oxen and mules, as well as from all kinds of used agricultural machinery, and from unwashed boots, machetes, and trouser cuffs. They can be easily spread from place to place stuck onto such plant materials as setts of sugarcane, budwood of trees or vines, seedlings of plantation crops or ornamentals, and on or in roots, bulbs or corms. Sometimes nematodes distributed in these ways are harmless, but also they are in many cases the cause of plant diseases. It is obvious that keeping plants in the field free from all nematodes is a virtual impossibility, although efforts at quarantines between countries lengthen for a few years the time before some of most virulent nematodes may cross borders. In the end quarantines are only a stop-gap method; over the centuries they have been an exercise in futility: other means of control must be devised.

Sugarcane nematodes, of which there are several, can be taken as an example of crop-related nemas that have had every chance to become well distributed. When thoroughly studied and determined (this is now in progress, see for examples such citations as: 31, 41, 45, 52, 53, 60, 68, 72), it is probable that sugarcane nematodes will be found to be remarkably similar in regions as far apart geographically as the Everglades area of Florida and the Tucumán of Argentina. Whether in the sheltered, warm, southern border of the continental U. S. A. or in the West Indies-

Central American region, sugarcane nematodes mostly belong to the genera *Hoplolaimus*, *Rotylenchus*, *Radopholus*, *Pratylenchus*, *Heterodera*, *Xiphinema*, and *Meloidogyne*. The "running out" of sugarcane has often been blamed, with good reason, on mosaic and kindred diseases. However, the condition still occurs in places without those virus infections. Often development of roots is meager and lesions on them are common, while roots become stunted from both ecto- and endoparasitic nematodes.

To the already numerous sugarcane nematode population are certain to be added other yet undiscovered kinds. As an example, the works of Román (53) show that on sugarcane there are new types of injury and new species of nematode still being described. It is significant that Steiner, as Thorne (72) discusses it, found in the United States that sugarcane seeds--the true seeds, not cuttings or setts--which had been brought from Turkestan to a quarantined greenhouse for experimental purposes, harbored a new organism, *Anguina spermophaga* Steiner. It is believed that although this nematode was studied and then destroyed under quarantine, the chances are very good for its reappearing, sooner or later, in most any part of the American tropics, carried in by some natural means and in no manner whatsoever connected with its discovery in the quarantine greenhouse.

Not only are sugarcane nematodes being described and named and their occurrences noted, but it is becoming more than ever necessary to study them in ever greater detail. On sugarcane, as well as other grasses and many other plants, there occur *Xiphinema* spp. in many different areas such as in Brazil (31, 37), in Florida (72), in Guatemala (73), and in Puerto Rico (72). These worms are comparatively large with big appetite habits, resulting in much injury to crop plant roots. Not only do these nemas feed on and puncture roots but some are known to be vectors of viruses. For example, *X. index* Thorne & Allen has been shown (72) to carry the soil-borne virus disease fanleaf of grape.

Because *Xiphinema americanum* Cobb is a common ectoparasite of sugarcane roots in Puerto Rico, it was suggestive to some that since the soil-borne chlorotic streak of sugarcane was proved by Bird, Cibes, and Tió (12) to be transmitted under carefully controlled liquid culture conditions by root contacts, so in the field a root injuring nematode also might be involved. Bird (11) then worked intensively to see if there was any such vector relationship with the common *X. americanum* in liquid nutrient cultures, but he secured no positive results of virus infection through the root-feeding activities of this nematode.

Widespread Occurrences

It is impressive to observe the widespread occurrence of certain plant disease nematodes. An illustrative example of pan tropical occurrence can be briefly recounted in the case of one of our most important plant attacking eelworms, the so-called burrowing nematode, *Radopholus similis* (Cobb) Thorne. In the Neotropics this has become specially notorious as the cause of the severe spreading decline (Fig. 18-7) in citrus. It results each year in losses to orange growers. However, it is also of great concern to banana planters of all the American tropics. It is as well a plague that occasionally becomes dangerous in coffee fincas and haciendas. Sugarcane growers know it as one of their much feared pests, and it is known to be bad in its effect on pasture grasses and grazing plants, and on maize and sorghum. It is safe to predict that it will need more study as crop production in the tropics advances into a more technologically exacting profession.

In the history of *Radopholus similis*, reviewed in part in the text by Thorne (72), but from other sources as well, a few highlights are as follows.

Probably the first scientific notice of this organism was in connection with the primarily plant pathological work of Cobb (19) in 1890 and 1891, during his disease studies on severely rotted and injured roots and corms of Fiji Island bananas. In 1898, in some early root disease investigations on coffee in Java, Zimmermann found severe root reduction due to this nematode in areas on that island. Some years later, 1907, still in material from the Oriental tropics, Cobb appears again to have encountered the same nematode, in Hawaii, on sugarcane roots. In 1915, in more of his nematode-oriented pathology, Cobb found what turned out to be the same nematode in the Occidental (Neotropical) tropic region, causing banana root disease in Jamaica. It is not surprising that in about 9 or 10 years a new orange disease was seen in aging Florida groves. This was described to me by older orange men as marked by weakening and eventual dying of trees "and a slowly creeping increase of the problem." There was some argument that spreading decline was virus caused, but exact studies finally resulted in identification and in repeated findings of the nematode R. similis; now proved to be the culprit that causes this disease (Thorne (72)).

Meanwhile, specimens of this eelworm found in connection with a tea-root disease in Java were identified by Steiner as R. similis. A couple of decades later a very old black pepper disease, called yellows, was finally recognized as due to a nematode. In 1950 this was determined by van der Vecht to be R. similis. By this time host range studies were being carried on in Florida (see Thorne (72)); it is now known to attack over 150 different tree and crop plant species, and the probability is that in the future it will be found to be affecting very many more. In the decade of the 1950s it was again recognized that Radopholus nemas were of increasing severity on bananas in Central America. In studies by Newhall (47) and Loos (34, 35) it is now clear that the nemas actually aid in the increase of fusarium wilt in the plantations, and Loos found further that it was undoubtedly the cause of the once controversial black-head disease of bananas; Ruehle and Young (54) in Florida associated R. similis with root disease of mango and avocado.

There are other plant parasitic nematodes in the tropics that show how wide ranging are some occurrences. Perhaps out of several, one more example of such a nema deserves a short review: see Thorne (72) in part for more details, and refer also to his bibliography cited. In early years in the United States Department of Agriculture, Cobb was internationally known as a great plant pathologist. He worked not only on bacteria and fungi but also on nematodes. He had sent to him, or secured, from Florida, California and Alabama specimens of an especially injurious citrus-root nematode. This he described and named as Tylenchus semipenetrans Cobb. Soon it was recognized as occurring also in Spain, Malta, Palestine, Australia, and Argentina. There followed more work on the organism, control methods, and host range studies along with search for resistant types of roots that could be used as grafting stock. It is an insidious disease since often orange trees are not quickly killed but live on in a stunted and wasted condition, bearing crops of small sized fruits. In Argentina, as publications by Marchionatto (42), Thomas (71) and Fresa (23) attest, in the tropics the nematode is a continuing problem. As work has proceeded on it in the Argentine and elsewhere, it is now evident that it occurs not only in the countries mentioned but also in South Africa, the southern edge of Soviet Russia, California, Ceylon, Java, Thailand, China, archipelagoes of the Philippines and Hawaii, in Brazil, and other places.

The pan-tropical (Orient-Occident) and the north-south Western Hemisphere tropical distribution of numerous eelworm pathogens on plants is a recognized nematopathological principle. (It is not unique to nematodes, for it is matched in many instances among fungus, bacterial, and virus disease producers.) While this may be an outstanding but not startling principle, it is very important and helps in the understanding of certain organisms and their complex life along with other parasites; it assists in searching for control measures based on findings of other scientists perhaps thousands of kilometers away. For this reason, plant pathologists in the Neotropics watch the publications from not only African and Asian tropics but also from temperate zone pathologists in the warm southern part of the United States for findings on nematode diseases.

Disease Increase by Combinations of Nemas with Other Pathogens

Nematodes attack plants directly, and they also act in conjunction with other parasites. An early study on this was the work in Brazil by Kramer (32) who found that the nema *Heterodera* was intimately associated with *Fusarium* rot of potato tubers. The occurrence of such a nematode-fungus complex has been more than once the cause of baffling results in pathology. In the temperate zone, working many years and at great expense, good varieties of tobacco have been developed that have high resistance to wilt from *Fusarium oxysporum* f. *nicotianae*. This work is so successful that, under a wide range of conditions, whether in warm temperate zone or tropics, the wilt is well controlled by planting resistant varieties of known reaction.

However, there are times when good wilt resistant varieties contract disease in unexpected amounts. This is at times diagnosed as "running out of resistance." It becomes of mounting concern and is being studied. One example is in reports (27, 61) by researchers who found fusarium-resistant cotton varieties have given uncertain field results when both nematodes and *Fusarium* are together infecting the cotton plants. In another crop the same is known. Planting a fusarium wilt resistant tobacco in naturally or artificially infested soil gives good results--without nemas being present. When soils also contain nematodes, results are irregular (26). In the American tropics the largest proportion of tobacco varieties grown are fusarium resistant, but in recent years where nema infestations have increased there is little to no field resistance. In carnations, studied under highly controlled conditions, a similar *Fusarium*-nematode effect--in the break down of fusarium resistance--has been well demonstrated (59). In the tropics such studies as the above are being given strong attention. As work progresses, for example, on the banana fusarium resistance problem and the selection and breeding of varieties, effects of nematodes (34, 47, 66, 69) are being given paramount attention.

As early as 1906 it was postulated by Cobb that decaying sugarcane roots he dug up in the field must have resulted from feeding of nematodes. Also that the multitudes of root punctures most certainly must be courts of infection for fungus invaders. Conclusive anatomical research in more recent years has shown that without question these types of punctures come from root attack by nematodes*. In 1948, working with *Tylenchus pratensis* (deMan) Filipjev, Jenkins (29) proved not only that these organisms multilated tobacco roots but also that into these damaged roots was initiated the brown rootrot complex. Fungi involved in this were species of five genera: *Thielavia*, *Thielaviopsis*, *Macrophomina*, *Pythium*, and *Fusarium*. For some other studies detailing nematode-and-other-pathogen complications see such works as those by: Alvarez and Lopez (2), Christie (18), Kloss (31), Loos (34, 35), Lordello (37, 38), Lucas *et al.* (39), Newhall (47), Pitcher (50), Powell and Nusbaum (51), Sasser *et al.* (56), Schindler *et al.* (59), Stover (66), and Wallace (75). In such researches are found more details about responses from nematode root

*As examples of different kinds of root reactions from different nematodes, is recognized in a few of the common (symptom description) names used for them: *Criconemoides*, grass-root killing; *Helicotylenchus*, spearing; *Hoplolaimus*, lancing; *Meloidogyne*, root knot; *Pratylenchus*, root lesion; *Radopholus*, burrowing; *Rotylenchus*, spearing; *Trichodorus*, root stub; *Tylenchorhynchus*, stunting; and *Xiphinema*, root lesion, dagger.

attacks. But roots are not the only part of plants attacked. Leaves are spotted from nematodes (72), and Iton (28) has presented evidence that nematodes are involved in spread of Ceratocystis, the sap stain and branch killing fungus, of coffee trees.

There is an exceptionally important feature of some of the nematode-fungus disease studies. Evidence is accumulating that while the disease effects of solely fungus or solely nematode infection are serious enough, when nematode and fungus act together the effects are often significantly and severely increased; more in fact than the expected effect based on what a simple doubling of pathogenicity could do from attack by both organisms at the same time. This marked synergism is well described, for example, by Schindler, Stewart and Semeniuk (59) who proved in Maryland (U. S. A.) that the Fusarium that attacks carnations develops more wilt with the nematode present, and the nematode need be only in small numbers, than if the Fusarium is the sole pathogen present. Indeed, several species of root knot nematodes that infected but caused no injury to carnation without Fusarium, increased notably the virulence of wilting when they were present. In Central America both Loos (34) and Newhall (47) have been able to demonstrate the serious added disease effect of the wilt organism F. oxysporum f. cubense, when the nematodes either Radopholus similis or Meloidogyne incognita were added to it, on banana.

In the southern part of the United States (North Carolina) under summer field conditions and in greenhouse temperatures very similar to tropical effects, the workers Sasser, Lucas, and Powers (56) studied black shank of tobacco as affected by nematode association. The tobacco varieties ordinarily highly resistant to the fungus, Phytophthora parasitica var. nicotianae, became susceptible to black shank in the presence of root knot, Meloidogyne spp. The most significant thing was that the "break down of black shank resistance" was not from mechanical injury but from a deep seated change in plant physiology.

On the other hand, it deserves repeating, fungi are not the only pathogens favored by nematodes. An outstanding example is the work of Lucas, Sasser and Kelman (39) which showed it was entirely the tobacco root injury from Meloidogyne incognita acrita Chitwood that causes increases in infection by the infamous bacterium Pseudomonas solanacearum E. F. Smith. Bacteria and fungi, of course, are not the only kinds of pathogens known to be aided by nematodes. As has already been shown, nemas carry viruses; but up to the present--although considerable effort has been expended on it--the chlorotic streak of sugarcane (11) has not been proved as having a nematode vector, nor does the lethal yellowing of coconut palms (33); field observations and histories of these two virus diseases are most suggestive of nematode vector relations of some sort.

Some Closing Reflections

There is no question that in some places in the American tropics, plant disease nematode study has gained for itself notable impact as a specialty. In some other places there is, equally notable, lessened enthusiasm. There are sophisticated areas in which plant nematology holds a high position in plant disease science, akin to the prestige accorded plant virology and to molecular pathology. However, in some countries, there is recognition, as from the first that more great rapid crop devastations come from such run-of-the-field organisms as those worked with so long: the rusts, the cercosporas, the fusariums, the phytophthoras, the pseudomonases, the rhizoctonias, and the like.

No plant disease organisms are any more interesting biological entities than the nematodes. They squirm under the microscope, they are notable for their many morphological species differences, they have subtle physiological mechanisms, they are remarkable as to cytology, and moreover they present the challenge that studies on them not only adds information to plant pathology but also to helminthology.

Without in any way exposing a prejudiced point of view, agricultural scientists in the tropics have repeatedly stated to me that from their own observations fungi and bacteria speed through soils many times more rapidly than nematodes. These worms wriggle about in soil moisture, but in a year's time have gone very little distance--the farthest may be a hand's breadth. Nemas stuck to agricultural tools, or reproductive plant parts used in planting, have a better chance to cover space. This is more often the occurrence in the temperate zone.

Table 30.
A FEW TROPICAL AMERICAN NEMATODE DISEASES

Host	Plant part invaded	Disease symptom	Nematode genus
Begonia sp.	Root	Giant gall	*Meloidogyne*
Chrysanthemum morifolium	Leaf	Angular spots	*Aphelenchoides*
Cocos nucifera	Root & trunk	Redring	*Rhadinaphelenchus*
Coffea arabica	Root	Rootknot	*Meloidogyne*
Cucurbita spp.	Root	Rootknot	*Meloidogyne*
Lycopersicon pimpinellifolium	Leaf	Raised and rounded spots	*Panagrolaimus*
Musa acuminata	Root	Rootknot Root burrowing	*Meloidogyne* *Radopholus*
Oryza sativa	Leaf tip	Whitening	*Aphelenchus*
Saccharum afficinarum	Root	Feeding lesions	*Hoplolaimus*, *Rotylenchus*, *Pratylenchus*
Solanum andigenum-tuberosum	Tuber-root	Plant	*Heterodera*
Zea mays	Root	Root injury	*Pratylenchus*

In our tropical lands, most crop husbandry does not include vast amounts of mechanical devices moving back and forth, repeatedly, over and in the soil. Largely, tropical crop growing is based on crop rotation and nemas tend to die between cropping changes. Plantation trees, over a decade or more, as a few individuals or as groups may contract and be badly despoiled by nematodes. But, on the whole, it is other disease organisms that cause the greatest damage to trees. There is a lack of fast-spreading ability of nematodes in plantations, and it is not to be wondered at--nematodes travel very slowly and not much soil contamination between trees occurs, in plantation-type growing.

The scrutiny of history, inquiries, and attempts at imagining habits of the pre-Columbian people of the Americas, have not caused me to conclude that the ancient form of agriculture was unusually affected by either the presence or the absence of nematodes. Some have held that milpa or shifting-type field agriculture was instituted because nematodes were so exceedingly destructive. More likely this kind of crop growing was because of the unavailability of tools other than primitive ones using fire; the fragility of tropical soils; insects; and the presence of rusts, leaf spots, blights, and basal rots.

Literature Cited

1. Allen, M.W. and S.A. Sher. 1967. Taxonomic problems concerning the phytoparasitic nematodes. Ann. Rev. Phytopathology 5:247-264.

2. Alvarez G., L.A. and L. Lopez M. 1954. Influence of root-knot nematode on the decline in vigor of the red spanish variety of pineapple in Puerto Rico. J. Agr. Univ. Puerto Rico 38:61-72.

3. Ancalmo, O. 1961. Nematode on Tabebuia pentaphylla in El Salvador. Plant Dis. Reptr. 45:826.

4. Ayala, A. and J. Román. 1963. Distribution and host range of the burrowing nematode in Puerto Rican soils. J. Agr. Puerto Rico 47:28-37.

5. ________., ________., and H. Gandía. 1963. Effect of soil fumigation used in the control of pine-

apple nematodes in Puerto Rico. J. Agr. Puerto Rico 47:76-90.

6. ________., ________., and A. C. Tarjan. 1963. Comparison of four methods for isolating nematodes from soil samples. J. Agr. Puerto Rico 47: 219-225.

7. ________., and C. T. Ramírez. 1964. Host-range, distribution, and bibliography of the reniform nematode, Rotylenchus reniformis, with special reference to Puerto Rico. J. Agr. Puerto Rico 48:140-161.

8. Bain, F. M. and S. A. Fedon C. 1951. Investigaciones sobre anillo rojo del cocotero. Agron. Trop., Venezuela 1:103-130.

9. (Barbados Department of Agriculture). 1907. Eelworms. Agr. News. Barbados 6:123.

10. Bell, F. H. and S. B. Alandia. 1955. Golden nematode and other potato diseases in Bolivia. Plant Dis. Reptr. 39:407-408.

11. Bird, J. 1961. Further studies on transmission of the causal agent of the chlorotic-streak disease of sugarcane. J. Agr. Univ. Puerto Rico 45:8-18.

12. ________., H. Cibes, and M. A. Tió. 1958. Transmission of the causal agent of the chlorotic streak disease of sugarcane disease through the roots of plants grown in nutrient solution. Puerto Rico Univ. Agri. Exp. Sta., Tech. Paper 27, 17 p.

13. Briton-Jones, H. R. (and E. E. Cheesman, revised) 1940. The diseases of the coconut palm. London, 176 p.

14. Cardeñosa, R. 1964. Los nematodos y su control. Agr. Trop., Columbia 10:37-39.

15. Carvalho, J. C. 1959. Descrição do macho de Scutellonema boocki, Nematoda: Tylenchidae. Biológico 26:41-44.

16. Chavez B., A. 1948. O anel vermelho do coqueiro e a fumigação do solo con D-D. Bol. Sec. Comer. (Pernambuco, Brazil) 15:356-387.

17. Chitwood, B. G., and C. A. Berger. 1960. Preliminary report on nemic parasites of coffee in Guatemala, with suggested ad interim control measures. Plant Dis. Reptr. 44:841-847.

18. Christie, J. R. 1959. Plant nematodes, their bionomics and control. H. & W. B. Drew Co., Jacksonville, Florida, 256 p.

19. Cobb, N. A. 1914. Citrus-root nematodes. J. Agr. Res. 2:217-230. --. 1915. Tylenchus similis, the cause of a root disease of sugar cane and banana. J. Agr. Res. 4:561-568. --. 1919. A newly-discovered nematode (Aphelenchus cocophilus n. sp.) connected with a serious disease of the coconut palm. West Indes Bull. 17:203-210.

20. DiEdwardo, A. A. and V. G. Perry. 1963. An undescribed species of Heterodera pathogenic to St. Augustine grass in Florida. Phytopathology 53:874.

21. Eenwick, D. W. 1959. The nematological aspects of red ring disease of coconuts. J. Agr. Soc. Trinidad & Tobago 59:373, 375, 377, 379, 381, 383-387, 389, 391, 393, 395.

22. Figueiredo, M. B. 1958. Algumas observacões sobre os nematódeos que atacam o fumo no est de São Paulo. Rev. Agr., Brazil 33:69-73.

23. Fresa, R. 1940. La presencia del nemátode Tylenchus semipenetrans Cobb en las raicillas de los citrus. Physis 19:348-354.

24. Gil Z., G. 1961. Area de dispersion del nematodo del tallo de la alfalfa en Chile y comparacion de resistencia de variedades. Agr. Tec., Chile 21:100-107.

25. Goeldi, E. A. 1887. Relatorio sobre a molestia do cafeeiro na provincia do Río de Janeiro. (Cited by Puttemans; original not seen by present

author.) --. 1889. Biologische Miscellen aus Brasilien. VII. Der Kaffeenematode Brasiliens (Meloidogyne exigua G.) Zool. Jahr. Syst. 4: 262:267. [--]. 1895. Molestia do cafeeiro na provincia do Rio de Janeiro. Rev. Agr., São Paulo 1:166-168.

26. Holdeman, Q. L. 1956. The effect of the tobacco stunt nematode on the incidence of Fusarium wilt in flue-cured tobacco. Phytopathology 46:129. (see also Phytopathology 44:683-685.)

27. Hollis, J. P. 1958. Relations between root knot and Fusarium vascular discoloration in cotton varieties. Phytopathology 48:661-668.

28. (Iton, E. F.) 1966. (Experimental results), pp. 45-56. In 1965 Annual Report Cacao Research, Imperial Coll. Trop. Agr., Trinidad, 81 p.

29. Jenkins, W. A. 1948. Root rot disease-complexes of tobacco in Virginia. I. Brownroot rot. Phytopathology 38:528-541.

30. Joffily, J. M. 1948. A doenca do anel vermelho do coquiro e sua occorência no Brasil. Ser. Nac. Pesq. Agron., Brazil, Boletin 3, 64 p.

31. Kloss, G. R. 1960. Catálogo de nematóides fitófagos do Brasil. Rio de Janeiro, Boletin Fitossanitario 8, 26 p.

32. Kramer, M. 1944. Verminose (Heterodera) e podridão seca (Fusarium) da batatinha. Biológico 10: 148-149.

33. Latta, R. 1966. Attempts to relate nematodes to lethal yellowing disease of coconut palms in Jamaica. Trop. Agr., Trinidad 43:59-68.

34. Loos, C. A. 1959. Symptom expression of Fusarium wilt disease of the Gros Michel banana in the presence of Radopholus similis (Cobb, 1893) Thorne, 1949, and Meloidogyne incognita acrita Chitwood, 1949. Helm. Soc. Washington, D. C., Proc. 26:103-111. --. 1962. Studies on the life-history and habits of the burrowing nematode,

Radopholus similis, the cause of black-head disease of banana. Helm. Soc. Washington, D. C., Proc. 29:43-52.

35. ________., and S. B. Loos. 1960. The black-head disease of bananas (*Musa acuminata*). Helm. Soc. Washington, D. C., Proc. 27:189-193.

36. Lopez T., L. 1934. Cultivo de la fresa. Bol. Agr., Colombia 195:43-48.

37. Lordello, L. G. E. 1951. *Xiphinema brasiliense*, nova especie de nematoide do Brasil, parasita de *Solanum tuberosum* L. Bragantia 11:87-90. --. 1951. *Xiphinema campinense*, nova especie (Nematoda, Dorylaimidae). Bragantia 11:313-316. --. 1956. Sôbre um nematódeo do gênero *Pratylenchus*, parasito das raizes de *Allium cepa*. Rev. Agr., Brazil 31:181-188. --. 1956. *Meloidogyne inornata* n. sp. a serious pest of soybean in the state of São Paulo, Brazil. Rev. Brasil. Biol. 16:65-70. --. 1956. Nematoides que parasitam a soja na regico de baura. Bragantia 15:55-64. --. 1957. A note on nematode parasites of red Anthurium (*Anthurium andraeanum* Lind.), with a description of *Rotylenchus boocki* n. sp. Nematologica 2:273-276. --. 1958. *Meloidogyne incognita*, a nematode pest of fig orchards at the Malinhos region (State of S. Paulo, Brazil). Rev. Brasil. Biol. 18:375-379. --. 1959. A nematosis of yam in Pernambuco, Brazil, caused by a new species of the genus *Scutellonema*. Rev. Brasil. Biol. 19: 35-41.

38. ________., and A. P. L. Zamith. 1954. Constatacão da moletiade and vermelho do coqueiro no Estado do Rio de Janeiro. Redescricão do agente caudador *Aphelenchoides cocophilus* (Cobb, 1919) Goodey, 1933 (Nematoda: Aphelenchidae). An. Esc. Sup. Agric. Luiz de Quiroz 11:125-132. --. 1956. Novas observaçoes sôbre os nematodos que parasitam a batatinha no Est. de S. Paulo. Rev. Agr. Brazil 31:45-54. --. 1958. Nematodeos que prejudicam as culturas da soja e do algodeiro no Estado de S. Paulo e sau interferencia nos planos de rataçäo. Rev. Agr.

Brazil 33:161-166. --. 1958. On the morphology of the coffee root nematode, Meloidogyne exigua Goeldi, 1887. Proc. Helm. Soc. Washington, D. C. 25:133-137. --. 1958. Nota sobre o genero Trichodorus Cobb, 1913, cum descrição de Trichodorus bucrius sp. n. (Nematoda, Dorylaimoidea). An. Acad. Brasil. Cienc. 30: 103-105. --. 1960. Meloidogyne coffeicola sp. n., a pest of coffee trees in the State of Parana, Brazil, (Nematoda: Heteroderiae). Rev. Brasil. Biol. 20:375-379.

39. Lucas, G. B., J. N. Sasser, and A. Kelman. 1955. The relationship of root-knot nematodes to granville wilt resistance in tobacco. Phytopathology 45:537-540.

40. Madamba, C. P., J. N. Sasser and L. A. Nelson. 1965. Some characteristics of the effects of Meloidogyne spp. on unsuitable host crops. North Carolina Univ. Agr. Exp. Sta., Tech. Bulletin 169, 34 p.

41. Mai, W. F., and H. H. Lyon. 1962. Pictorial key to the genera of plant parasitic nematodes. Unnumbered pages, and 50 plates. (Multilithed). New York State Coll. Agr., Ithaca, New York.

42. Marchionatto, J. B. 1947. La podredumbre de la raicilla de los citrus provocada por el Tylenchus semipenetrans. Argentina Inst. Sanid. Veg., (special publication), Ser. A, 3(36), 6 p.

43. Massee, G. 1909. Coffee diseases of the New World. Kew Bull. Misc. Inform. 1909:337-341.

44. de Mello, L. M. 1958. Nematodeos que parasitiam o melão cantalupe no Estado São Paulo. Rev. Agr., Brazil 33:39-44.

45. Moreno, A. F. 1958. Algunos nematodos parasitos que prejudican cultivos de interes economico. Soc. Cient. Argentina An. 166:115-121.

46. Neal, J. C. 1899. The root disease nematode of peach, orange and other plants in Florida due to the work of Anguillula. U. S. Bur. Ent. (Bulletin) 20, 31 p.

47. Newhall, A. G. 1958. The incidence of Panama disease of banana in the presence of root knot and burrowing nematodes (Meloidogyne and Radopholus). Plant Dis. Reptr. 42:853-856.

48. Noack, F. 1899. A podridās da raiz mestra do cafeeiro. Bol. Inst. Agron. Campinas, São Paulo 10:224-229.

49. Normanha, E. S. 1958. Nematoides em melão. Rev. Agr., Brazil 33:157-159.

50. Pitcher, R. S. 1965. Interrelationships of nematodes and other pathogens of plants. Helm. Abstr. 34: 1-17.

51. Powell, N. T. and C. J. Nusbaum. 1960. The black shank--rootknot complex in flue-cured tobacco. Phytopathology 50:899-906.

52. Ramírez, C. T. 1964. Hoplolaimus puertoricensis n. sp. (Nematoda: Hoplolaimidae). Univ. Puerto Rico J. Agr. 48:127-130.

53. Román, J. 1961. Pathogenicity of five isolates of root-knot nematodes (Meloidogyne spp.) to sugarcane roots. Univ. Puerto Rico J. Agr. 45:55-84. --. 1964. Belonolaimus lineatus n. sp. (Nematoda: Tylenchidae). Univ. Puerto Rico J. Agr. 48:131-134. --. 1964. Immunity of sugarcane to the reniform nematode. Puerto Rico Univ. J. Agr. 48:162-163.

54. Ruehle, G. D. and T. W. Young. 1955. Burrowing and meadow nematodes on avocados and mangos. Florida State Hort. Soc., Proc. 68:288-292. (See also Plant Dis. Reptr. 39:815-817. 1955.)

55. Sasser, J. N. and W. R. Jenkins (eds., 21 contributing authors). 1960. Nematology fundamentals and recent advances with emphasis on plant parasitic and soil forms. Univ. North Carolina Press, Chapel Hill, 480 p.

56. Sasser, J. N., G. B. Lucas and H. R. Powers, Jr. 1955. The relationship of root-knot nematodes to black-shank resistance in tobacco. Phytopathology 45:459-461.

57. Sasser, J. N., O. F. Vargas Gonzales, and A. Martín. 1962. New findings of plant-parasitic nematodes in Peru. Plant Dis. Reptr. 46:171.

58. Schieber, E. 1961. Parasitic nematodes on Dioscorea in Guatemala. Plant Dis. Reptr. 45:425.

59. Schindler, A. F., R. N. Stewart and P. Semeniuk. 1961. A synergistic Fusarium-nematode interaction in carnations. Phytopathology 51:143-146.

60. de Segura, C. B. 1953. More about the golden nematode in Peru. Plant Dis. Reptr. 37:326. --. 1965. Enfermedades de Cultivos Tropicales y Subtropicales. Editorial Jurídica, Lima, Peru, 439 p.

61. Smith, A. L. and J. B. Dick. 1960. Inheritance of resistance to Fusarium-wilt in upland and sea island cottons as complicated by nematodes under field conditions. Phytopathology 50:44-48.

62. Souza, R. 1917. Molestia vermicular nos cafezaes de S. Paulo. Brasil. Agr. 2:88-91.

63. Steiner, G. and E. M. Buhrer. 1964. The plant-nematode problem of the American tropics. Univ. Puerto Rico, J. Agr. 48:69-100.

64. Steiner, G. and C. T. Ramírez. 1964. Bibliography on agro- and plant nematodes of the American tropics. Univ. Puerto Rico J. Agr. 48:101-126.

65. Steiner, G., E. M. Buhrer and A. S. Rhoads. 1934. Giant galls caused by the root-knot nematode. Phytopathology 24:161-163.

66. Stover, R. H. 1966. Fungi associated with nematode and non-nematode lesions on banana roots. Canadian J. Bot. 44:1703-1710.

67. Szkolnik, M. 1950. Nematode root knot of cinchona in the western hemisphere. Plant Dis. Reptr. 34:305.

68. Tarjan, A. C. 1960. Check list of plant and soil nematodes, a nomenclatorial compilation. Univ. Florida Press, Gainesville, 200 p.

69. Taylor, A. L. and W. Q. Loegering. 1953. Nematodes associated with root lesions in abaca. Turrialba, Costa Rica 3:8-13.

70. Tenorio, R. P. 1948. Anel vermelho do coqueiral, em Pernambuco. Bol. Sec. Agr. Comer. Pernambuco, Brazil 15:204-220.

71. Thomas, E. E. 1923. The citrus nematode, Tylenchulus semipenetrans. Univ. California Agr. Exp. Sta. Tech. Paper 2, 35 p.

72. Thorne, G. 1961. Principles of nematology. McGraw-Hill, New York, 533 p. --. 1964. Nematodes of Puerto Rico: Belondiroidea new superfamily, Leptonchidae, Thorne, 1935, and Belonchidae new family (Nemata, Adenophorea, Dorylaimida). Univ. Puerto Rico, Agr. Exp. Sta., Tech. Paper 39, 51 p.

73. ________., and E. Schieber. 1962. American dagger nematode (Xiphinema americanum) on coffee in Guatemala, with suggestions for nematode controls in nurseries. Plant Dis. Reptr. 46:857.

74. Van Gundy, S. D. 1958. The life history of the citrus nematode Tylenchulus semipenetrans Cobb. Nematologica 3:283-294. --. 1965. Factors in survival of nematodes. An. Rev. Phytopath. 3: 43-68.

75. Wallace, H. R. 1964. The biology of plant parasitic nematodes. St. Martin's Press, New York, 280 p.

76. Wellman, F. L. 1961. Coffee botany, cultivation, and utilization. Leonard Hill, London; Interscience Publ., New York, 488 p.

77. Wolff Schoemaker, R. L. P. 1963. Gracilacus peperpotti n. sp. (Nematoda: Paratylenchidae) found in a Surinam coffee plantation soil. Nematologica 9:296-299.

Chapter XIX

MINERAL DEFICIENCIES AND SOIL PROBLEMS

Summary

Plant diseases of abiotic origin occur in the tropics; among them, most attention is given to troubles related to soils. Although many cultivated tropical soils may not erode rapidly, they need special protection because of their fragility as to minor element and other nutrient components. These are readily washed out, resulting in injured soils; many such are sensitive to disease effects. Minor element losses in soils cause distinguishable disease symptoms; such insufficiencies can be significantly limiting factors in crop production. Many agricultural soils are of relatively new volcanic material; deposits may be of varying value: sometimes adding nutrients, sometimes making subsurface hardpans of bad effect. Occasionally, at the north and south extremes of the Neotropics, cold-breakdown disease is encountered on crops; the same occurs in central parts of our tropics at high elevations on mountains. Temperature fluctuations appear to induce other physiological-breakdown troubles such as non-parasitic leafspots. Where crop infection occurs from soil-borne fungus parasites, often there are soil types and conditions either especially favorable or especially unfavorable for such diseases. In the *Fusarium*-wilt disease of the susceptible banana, for instance, research has proved there are short-lived ("susceptible"), medium-lived, and long-lived ("resistant") soils.

Introduction

That diseases of crops show distinct dependence on

environment is perhaps clearer to experienced tropical plant pathologists than to temperate zone workers. Where causal organisms are involved there is no question about their effects, but where, as is often the case in the tropics, one can see marked variations in soils short distances apart coupled with often quickly reached differences in climates, environment in the tropics may be considered to have almost an exaggerated--certainly a leading--role in the presence of absence, or relative severity, of disease. The soils themselves are new in many cases, are easily disturbed, and are being continually influenced by leaching from abundant rainfall coupled with unfailing warm temperatures. Some effects, of course, may be wholly abiotic and not related to soils.

The famous, stable, comparatively little leached, rolling to flat loess soils knows so well in the great plains regions of Europe, Asia, and North America are not duplicated in the American tropics. Of some similarity are the relatively small areas such as the pampas in southern South America and the Mexican continuum of the plains of Texas.

Fragile Tropical Soils

In most tropical soils used in agriculture, it seems (19, 21, 31, 36, 37) they begin as volcanic depositions which are then subjected to the severe weathering-leaching processes inherent in the tropical environment. In spite of the fact that rains and high temperatures are so much more severe in the tropics than elsewhere, there are soils that even as they undergo degradation may not exhibit severe top runoff or erosion (20). It is the permeable, sloping, tropical soils that are most cultivated, and most leached out.

Often in the past, tropical horticulturists, agronomists, and plain land developers have selected areas (without reference to the advice of soils scientists) in which to establish plantations. These areas, picked because of the luxuriant jungle growths can be judged falsely; the logic is only apparent that such successful tree growth in the wild must, without doubt, be a harbinger of equally successful production under domestication.

Clearing is done, bush and felled trees are dried and burned, the soil is given a crude levelling, and then is laid out in fields. When planted it is almost certain that the first sight for the hopeful owner's eyes is that of something good. Well put-in plants, like the giant herbs such as bananas, abacá, and agaves, or the huge grasses such as sugarcane and bamboos, and in lesser fashion the trees such as coffee, cacao, and oil palm, make adequate growth at the first. Maiden crops have a good chance, in any average new land, of from excellent to mediocre productivity. After that it may be a different story if the soil's true capacities cannot equal the optimistic guesses. In poor soil, that was under rich-appearing jungle, plantation decline begins after two to maybe as many as ten rainy seasons have passed (further suggestive discussions: 9, 19, 21, 28, 30, 36, 37, 42, 43, 45, 48).

Soils can be "burned out." (I have seen insidious types of wild morning glory vines tenderly cultured over plantation soils to protect their fragile components. On high mountain slopes I have worked with indigenous farmers (48 in 1945 publ.) watching them use age-old methods. They cut foliage, bushes and weeds to spread over the volcanic soil like a carpet; allowed to wilt and half dry, in the right condition it is then rolled back like a blanket, seeds planted under it in the soil, and the cover unrolled and left flat again over the field.) Without unusual care, many tropical soils cannot support crop growing without the plants exhibiting mineral deficiency effects, and disease.

Mineral Deficiencies

There are numerous effects on growing plants attributable to the lack of certain soil elements. Entirely apart from poor growth and production due to any insufficiency of any or all of the major nutrients of nitrogen, phosphorus and potash, there are minor element deficiencies that are of great concern to tropical agriculturists. To these, the pathologists, with the aid of physiologists, have given a good deal of attention (3, 4, 6-8, 10, 11, 13, 14, 16-18, 23, 24, 31, 32, 34-36, 38-41, 46, 49); in the Neotropics, leading research has been carried on in Florida because of the minor element impoverishment of the sands and mucks used in warm-country crop growing there.

One minor element deficiency symptom that can be quickly diagnosed is that of a bright yellow color on leaves of all ages: a lack of traces of iron. Leaves may have normal size, be reasonably free from other blemish, but the green color is practically absent and what remains is an unhealthy yellow. On dicotyledonous plants and trees the leaf internodes on twigs may be shortened, and in severe cases the end result is dieback; monocotyledonous crop plants are yellow and stunted. On trees of many kinds, this symptom is seen not only as a direct result of the lack of iron. There also are times when iron is demobilized in old branches. This may occur in connection with rejuvenation pruning, when new shoots from adventitious buds have leaves on them characteristically showing the effects of iron deficiency. Leaves on the rest of the tree may be a healthy green.

Another type of loss of green color is that from manganese deficiency. In this, it is the young leaves that first show symptoms. The yellowing is at times almost white with necrosis developing as the condition is prolonged. In a crop like sugarcane, loss of leaf color is in many cases followed by the death of affected mats. Sugarcane men call this such names as manganese-blight or chlorosis-blight. It is found on neutral to alkaline soils. As chlorosis increases, there develop necrotic spots on the leaf blades of dicotyledonous plants.

A minor element deficiency that causes consternation at times is that from lack of traces of zinc in the soil. Plants without it have leaves that are small (the so-called "little leaf" disease) and in some leaves there may be a degree of malformation. In certain plants extra buds may be stimulated to grow, along with the reduction of internode extension (the so-called "rosette" symptom). Lacking a positive determination, the superficial appearance of thus-diseased plants suggest symptoms of virus infections. After the introduction and use of the systemic herbicide 2, 4-D I have seen zinc deficiency symptoms on several kinds of plants closely resembling mild effects of that plant poison. Some sketches of malforming and killing effects from use of 2, 4-D are shown in Fig. 19-1.

A minor element deficiency that usually is not characterized first by a lack of green color, is that from lack of trace amounts of boron. This deficiency in components of the sap stream in plants soon affects the growing tissues. Two types of symptoms can be described as examples. In one, seen in such crops as coffee, the tip bud of a branch dies, and below it in abnormal fashion adventitious buds grow out as stunted twigs, in consequence of which develops a condition called palm frond disease. Flowering occurs on less severely affected parts of the tree, but fruit set is poor and if a good amount does develop there is early drop. In another boron-lack symptom, the vascular strands of the petiole are affected. This is especially notable on the celery plant with its exaggerated leaf petiole; the "strings"

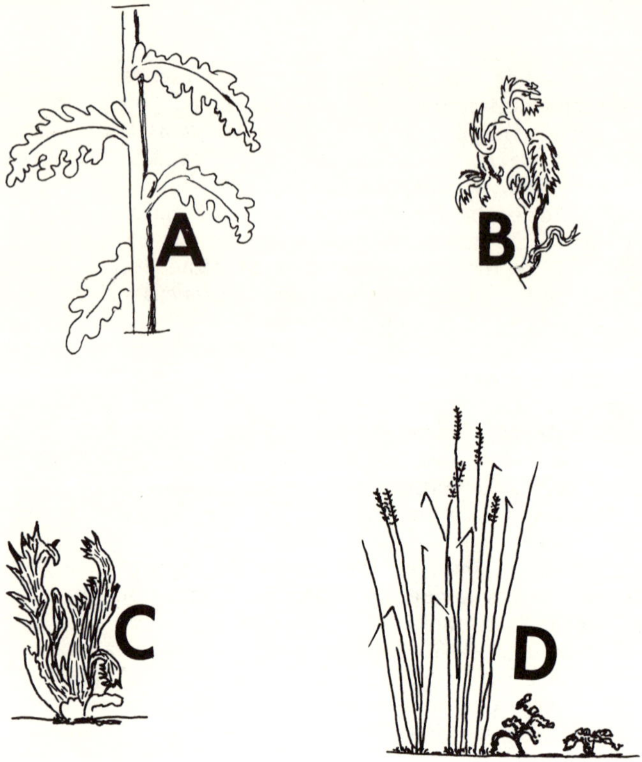

Fig. 19-1. Sketches of plant poisoning effects from use of 2, 4-D as follows: A. healthy leaves of chrysanthemum plant; B. effect of 2, 4-D on accidentally exposed chrysanthemum plant; C. lettuce plant accidentally exposed to 2, 4-D; and D. on left pasture grass treated with 2, 4-D but uninjured, while on right dicotyledonous weeds were being killed by the treatment.

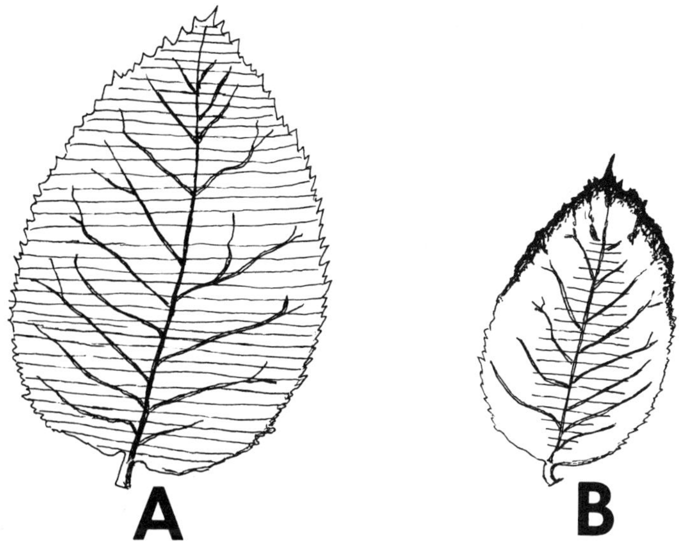

Fig. 19-2. Boron deficiency symptoms on rose leaf. A. Healthy leaf uniformly green and well developed. B. Boron deficiency characterized smaller leaf, green central area, broad yellow margin, and blackened edges.

on it darken, grow more slowly than the pithy more succulent part, and break crosswise causing the crack-stem symptom. Symptoms of these same two kinds, and there are other signs of health disruption, are known from personal experience by me in connection with not only coffee and celery but also mango, cabbage, citrus, rose, maize, and tobacco. These are only a very few of the crops that are susceptible to boron deficiency. [Rose gave a slightly different effect from some hosts. Roses grown in a garden soil known to be boron deficient had small and severely diseased leaves. See Fig. 19-2.]

There are quite a number of minor element deficiencies that can be brought for diagnosis, to plague the tropical pathologist; many are less spectacular and of less easily diagnosed causation than the four given above. Some notable leaf symptoms, (only seven among a larger number of minor element deficiencies) may be described in a sentence each, as follows. In sulfur deficiency, leaf

Table of some minor element deficiencies

Table 31. Symptoms observed in the American tropics from minor element deficiencies affecting certain crops[a]

Minor element deficient	Diseased crop observed[b]		Effects on fruit
Zinc	Coffee	Small leaves, malformations	Fruit set poor, much early drop
Copper	Cacao	Old leaves irregularly chlorotic; young leaves show wilting, some leaves become sickle shaped	Marked reduction in pod harvest, in some pods aborted seeds
Calcium	Tomato	Pale-green leaves, some necrotic leaf tips	Blossom-end rot
Calcium	Avocado	Leaf distortion, dieback on small branches	Darkening of fruits
Boron	Coffee	Tip buds killed on fruit branches, side buds grow abnormally on these branches	Crop reduced up to 90 per cent
Manganese	Sugarcane	Severe chlorosis and stunt, death of plants	--
Manganese	Coffee	Necrotic spotting on leaves, spots die and drop out	--
Phosphorus	White potato	Leaves small and erect	Flowers abort

Phosphorus	Tomato	Leaves small, branch tips erect, foliage often purplish in color	Few if any fruits set
Iron	Coffee	Light yellow chlorosis, same mottling, primary and secondary leaf veins thin green lines	Radical reduction in crop
Magnesium	Maize	Almost white streaks in older leaves, necrotic streaks	Poorly filled kernels
	Coffee	Yellow to white leaves, necrosis if severe	Flowering scarce and little or no fruit set
	Orange	Leaf bronzing	Undetermined

[a]From my notes, made in connection with field diagnoses and answering requests from farmers in Guatemala, Costa Rica, Panama, and Colombia.

[b]Many other crops are affected by these minor element deficiencies.

(N. B. The reader should refer to the researches by specialists making detailed studies on these physiological diseases. As a beginning I suggest attention to the following citations: 3, 4, 6-8, 10, 11, 13, 14, 16-18, 23, 24, 31, 32, 34, 35, 38-41, 46, 49.)

yellowing is usually on younger leaves along with necrotic spotting. In magnesium deficiency, leaf yellowing (sometimes bronzing) is seen on older leaves, but the primary and secondary big veins remain green. In calcium deficiency, young leaves appear a pale green, with necrosis at the growing tip many times resulting in a crook-neck type of growth. In phosphorus deficiency, leaves are usually dark green, held abnormally erect, given a purplish hue, and showing some speckling, necrotic effects. In copper deficiency, there is not uncommonly a wilting of young leaves, while old leaves show a breakdown of laminar tissue; this is often accompanied by one-sided growth, resulting in sickle-shaped leaves. In potash deficiency, chlorosis is many times at the tips of older leaves, where it progresses inwardly causing light streaks in the green; there is some leaf-edge necrosis, and from the dropping away of dead tissues there may result a so-called ragged leaf symptom. In molybdenum deficiency, a strikingly restricted growth of leaf lamina does not inhibit development of the midrib, which results in the rat-tail symptom, accompanied by mottle- and crinkle-effect.

Unfortunately it is not possible from field observations on wild growth for the pathologist to recognize where he is most likely to encounter minor element deficiencies. He has to wait for growers to bring samples to him from their badly developed cultivated crop plants. However, there are some things that are clear: where the volcanic soils are relatively old, where there is good drainage on sloping lands, and where there is abundant rainfall in a well watered area, the minor elements are more likely to be washed out and leached away.

Where rains are frequent and heavy the multiplication of soil microorganisms is profuse. Soluble mineral elements are available and bacteria, algae, fungi, and other microbes live on these nutrients. Where the elements are only slowly released, still bound in rock dust, they are made soluble by biologic action. In many instances, in many soils, minor element deficiencies are not apparent in crops on fields that are new and cleared and planted for the first time. Usually it is only after the passing of several harvests and the carting away of much produce that minor element troubles become acute. Moreover, they are the most severe in tilled land kept free of mulch or protective cover growth. It is believed, and with good reason, that in the soils that lie at a safe distance from but to the leeward of volcanoes, the crops growing there may have less trouble from minor element deficiencies.

Effects of Volcanic Deposits

A large proportion of Neotropical soils being developed agriculturally are, relatively speaking, new volcanic deposits. Some are direct deposits on table lands, some are washed off mountain sides and deposited on slopes and in low flood plains. Of course, how good and how long-lasting these soils are depends wholly on the mother rocks from which the old volcanic explosions came. A good amount of land loses plant-available minor element components quickly.

I visited a coffee finca in Guatemala in 1929 where large trees were called "never bearing" type and had abnormally chlorotic, small, and malformed leaves. Virus causation, which was suggested, was ruled out; the disease had always been on a special slope and it did not spread to nearby trees. Twenty years later the same coffee plantation was revisited: more than a decade previous the "never bearing" trees had been literally half buried by a volcanic deposit from the mountain nearby--in some places the ash was over a meter deep. All the old coffee had had the ash excavated from around them and were long ago recovered from it. They bore heavy crops and in the month of January showed many flower buds ready for the first rains. In addition, no signs could be detected anywhere of any minor element deficiency.

From studies carried on more recently in Costa Rica (see 32, 38, 48), it became clear to me that the trees of the "never bearing" planting seen in 1929 had been suffering for many years from at least zinc and iron deficiencies, and perhaps from lack of other minor elements. The deposit from the volcano had undoubtedly returned to the mountain side trace elements that were not present at the time the plantation that bore no crop was seen long ago.

In connection with fungus disease, volcanic ash deposit can be a bad thing. In several cases it has been reported that ash has weighed down fruit tree foliage and branches to the extent that they have been injured, the wood has cracked, and insects have moved in; with them came dieback from *Diplodia*, *Ceratocystis*, *Colletotrichum* and other fungi. The reports of such things have been word-of-mouth from intelligent growers in El Salvador and in Guatemala. Another kind of physical injury has been seen personally where bananas have been dusted with volcanic ash and the fruit skins, rubbed with the sharp particles, were so badly blemished as to make the crop unsalable. This kind of an effect occurred on bananas growing over 200 kilometers from Volcán Santa Maria of Guatemala and on cooking bananas (platanos) within a radius of 90 kilometers of Volcán Izalco in El Salvador.

At times there may be a layer of the so-called talpitate lying at depths of from an arm's length to twice that under the surface of the ground. This comes from ancient seismic eruptions, and solidifies both through its inherent characteristics and from minerals that have seeped down into it and cemented in it from the soil above; this volcanic hardpan layer may become impervious to good drainage. There is nothing that is more serious to growth of plantation crops than "wet feet," as roots when soaked in water deteriorate rapidly in the tropics, even more quickly than in the temperate zone. For effect of "wet feet" on orange trees see Fig. 19-3. Coffee growing in a medium shallow soil underlain with talpitate, although either mulched in the sun or planted in the shade, appears to be specially susceptible to Rosellinia root rot. The problem of the Fusarium wilt of banana is made more complex, as Stover (43) has shown, when resistant bananas sometimes succumb to wilt, which happens specially in poorly drained Fusarium-infested soil. And this poor drainage is common because the field is underlain with talpitate. Shallow wet soils over talpitate planted to citrus are much more affected by losses from Phytophthora gummosis than well-drained areas.

There is another kind of rootrot that is aided in development by subterranean volcanic deposits that retain moisture. This is seen on sugarcane at mountainous elevations. Ancient volcanic spewings become tuffs in Costa Rica, and I have seen a pure tuff growing good cane that was in a low, warm, and moist field where trash was encouraged to accumulate between rows. The same seed source used in a field farther up (800 meters) was sadly decimated by Pythium root rot. In digging away the secondary top blanket of soil, it was found that rotted sugarcane roots were precisely at the layer of buried volcanic tuff retaining water; in a field 800 meters below, this tuff was not holding water and was the main growing medium in which

Fig. 19-3. In foreground one orange tree of many killed by "wet feet" in a low part of a Florida (U. S. A.) planting subjected to a month of unusually heavy rains; trees in background on better drained land were unhurt. Photograph courtesy of Plant Pathology Department, University of Florida, Gainesville, Florida.

the canefield developed. The cane grew free from rootrot. In a nearby place at 800 meters and at the same rainfall, but where the tuff was mixed in as part of the top soil, only a trace of Pythium rootrot could be found and that was at a point near a road fill that held back drainage water.

Cold- and Physiological-Breakdown Diseases

Generally speaking, if mean temperatures are about 30° C at sea level, the frost line on a mountain is at near 2000 meters, and it can be readily appreciated that temperature effects correlated with altitudes (ecologic zones) are of supreme importance to cold disease. The increase in altitude has the effect of not only cooling air temperature, but there is also more cloud cover, a decrease in light intensity, and on some exposures an increase in relative humidity, any one of which is of great importance to parasitic organisms as well as abiotic troubles.

Observations on abiotic disease from cold on sugarcane showed it was correlated with altitude. Normal low land sugarcane in Guatemala had numerous fungus troubles and a few minor element deficiency troubles. Farmers also had cane cultivated on mountain sides. Going far up to one small field at an elevation of over 1500 meters I found clear evidence of cold injury. It is of the same kind known in such extremes as in the north edge of the Neotropics: Louisiana in the U. S. A., and towards the south edge in the Tucumân of Argentina (12). It is not often publicized, but the part of South America that is visited at intervals by periods of Antarctic chill is subject to occasional serious losses from it. This has been to some extent circumvented in Brazil by a pruning method that ends in developing a thick and low "skirt" of branch growth to protect coffee tree trunks (48).

Not only are such matters as actual frost, low temperature effects, and mineral deficiencies being given attention by pathologists-physiologists. (For want of space it is not possible to attempt more than brief mention of some of the observations and a few findings.) The subject of cold effects includes what might be called cool effects.

It has been pointed out in beginning chapters that air temperatures in agricultural zones are quite stable; they vary much less than is found in the temperate zone.

Fig. 19-4. Examples of what would undoubtedly be determined by mycologists-pathologists as sterile leafspots, are such injuries as shown herewith from undue exposure to sun of grapefruit leaves. Photograph courtesy of Department of Plant Pathology, University of Florida, Gainesville, Florida. I have found and collected such sterile leafspots many times from numerous crops in the tropics. Pruning or removal of natural shade caused it in coffee, cacao, banana, jicama, huisquil, lleren, bean, and other crops.

There are places, for example near sites where the heat equator and the geographical equator are close together, in which at lower altitudes temperatures are close to 30° C (or 86° F), and in these localities there is less than 5° C change between day and night for months at a time. If a plant adapted to an extremely constant shade-moisture-warmth condition is then suddenly taken out of its original habitat and grown elsewhere, it may suffer severe physiological (abiotic) disease. A single example is a rain forest type Philodendron with handsome foliage that is used as an ornamental. Under conditions that are abnormal for it, stomatal exudation sometimes occurs and this leads to seriously disfiguring leaf spots (33). Such spots on other leaves, when collected, dried, and presented to an herbarium mycologist, are recorded as "sterile leafspot" because no spores or other microscopic growth can be demonstrated in connection with the lesions. Examples are shown in Fig. 19-4 of sterile spots of grapefruit leaves. There are times when shade leaves of numerous other genera and species of plants exhibit development of "sterile leafspot" which I have concluded must have developed because of alternate air cooling and air warming. Occasionally spores of definitely non-parasitic fungi fall on leaves. Under ordinary even temperature (only three or four degrees of diurnal fluctuation) such spores are perfectly harmless. However, where the coolness of the night is a distinct change from the warmth of the day, the non-parasitic fungus spores attract moisture, and toxins leach out and cause leaf flecks that may intensify to become necrotic spots.

An impressive effect of sizeable trees is that they regulate and maintain a micro-climate of their own. Bonilla (5) made studies of this in shaded coffee in Colombia, and it has been hinted at in such works as those of Ochse et al. (36), Franco (15), s'Jacob (22), Machicado S. (25), and Sylvain (44). I reached conclusions about it from my reading and personal experiences. One effect is that a body of day-warmed air rises in the evening from the ground upwards, into the zone of the lower branches there, mingling with the foliage; it continues upwards more slowly to the high-held tree canopy. It is in this upper-story that final cooling (sometimes the loss in temperature must be as much as 12° C) takes place in about seven hours. Cool air becomes heavy and then slips downward from the upper foliage of the tall tree tops to the plants below.

This coolness that descends at night from trees is

distinctly felt. During early dry season while working at near sea level in Central America I experienced the following. Sleeping in the middle of an old clearing on a canvas cot protected by a mosquito net, under a thatched roof but with no sides, was a sultry event and I perspired all night; certainly no cover was needed. Five days later, still at sea level, camp was moved to jungle being surveyed for clearing--a 20-minute jeep ride away. The trees were large, and a thatched roof had been erected like the previous one. Bedded on the same cot, protected with the same net, it was warm the first three to four hours after sundown; it was cool by midnight when a sheet was used and in an hour more a light blanket was essential. It is in microclimatic conditions such as described here that appreciably more so-called "sterile leafspots" are found on various crops in the shade: seen especially at times in shaded coffee, cacao, and dwarf banana, the same crops without shade showed more leafspot demonstrably from some parasite causation. Such reaction is confusing for the pathologist and for the growers.

Stress-inducing conditions on leaves need study in relation to abiotic breakdown phenomena. When the natural course of ecology is suddenly changed by something artificial man has brought in, there is reason to expect various kinds of developments such as physiological leaf lesions and discoloration-disruption of tissues in other plant parts, as well. In connection with regular work it was long ago recognized by South Americans (e. g., Marchionatto, 26) that tropical crops are often susceptible to physiological troubles. One of these, seen everywhere I have worked so that I am acquainted with it, is the blossom-end rot of tomatoes. It is under warm conditions that the trouble (17) is more frequent and best studied as to control. Averre (2) has obtained considerable experience with a physiological discoloration of tissues (Fig. 19-5) in the starch-filled roots of cassava that soon makes them unfit for consumption. Alvim (1) has shown that the development of dwarfed, hard, useless fruits of cacao is due to wholly physiological causes. It has been noted that a rice panicle injury (47) results from profound physiological injury associated with sun exposure. These are only a few of the many physiological troubles of plants in the tropics.

Soils and Disease

It is clear, and it needs no detailed presentation as

Fig. 19-5. Physiological breakdown, blackening, of tissues inside the root of cassava or mandioca, from improper storage conditions. Photograph courtesy of C. W. Averre.

it is known to all in the tropics, that there are certain soils which because of varying mixtures of components and because of the conditions in which the soils exist, are poor, medium, or good in their effects on crop plants. It is also true, and not so commonly studied, that there are tropical soils (soils often respond rapidly in the warm countries) that have definitely medium, good, or bad effects on how plant diseases that are in them may react. A few examples of situations as to disease will be briefly mentioned.

The effect of bad drainage is of marked importance, and something of what occurs has been pointed out a few paragraphs above in the section on "Effects of Volcanic Deposits." In old days, before resistant-type root stocks had been discovered, certain poorly permeable clays allowed oranges to grow on their own roots, but when gummosis (*Phytophthora*) appeared the trees were killed. In well-drained soils this is not so serious. This is an obvious situation known by everyone, and related to the maize plant and poor drainage. In spite of the plant's phenomenally wide range in adaptation, and in spite of the fact that a first crop or two may be good even on soil where there is a near-the-surface water table, even so, after the primary appearance of such soil-infesting parasitic fungi as species of *Fusarium*, *Phytophthora*, *Pythium*, and *Macrophomina*, rootrot from one or all of these develops and spreads. The field becomes marked as a severe root-rotting field, and it is impossible to grow a good crop in it. In a well-drained area, where maize roots penetrate deeply and where plant-available soil nutrients are balanced and abundant, good harvests may be secured season after season and year after year, even when there is a small amount of root-rotting continuously present.

A soil that is moist all year and so rich in organic matter that particles of it are clearly seen, is diagnosable as a so-called "Rosellinia soil." In fact, the chances are that the organism may be already scattered in a few points in the land. When susceptible crops are established, once the parasite starts it spreads and causes severe rot in localized areas unless controlled. Soils that have a more mineralized content have much less likelihood of developing this kind of root disease. A bacterial root disease of cassava (mandioca) was found several times in Central American countries over some 20 years. It was invariably on slightly acid soil, in the replace crop, and was especially severe if fields had been cleared and kept fairly free from debris.

The most extensive studies ever carried out, anywhere in the world, on soil in relation to disease, have been those in the banana-growing Neotropics. There, productive field longevity of susceptible bananas in the presence of the wilt Fusarium has long been observed. Some of this is described above in Chapter IX on systemic wilts. Banana growers use the services of (and give jungle experience and training to) pathologists, soils specialists, and agronomists for the purpose of growing the popular gros michel banana as long as possible in their properties. Scientists have put on an objective basis sets of criteria for disease classification of banana lands (43, 17). Some areas can be diagnosed as short-lived (called "susceptible"), some medium-lived, and some long-lived (called "resistant"). The first types grow bananas for from one and a half to about four years, and long-lived soils produce crops for 20 years and longer.

An astute specialist, exploring for best banana soils, first looks over wild vegetation and slope problems. With augurs he draws up what is below and then tests for depths of surface and subsoil layers, and makes determinations on profiles. Using samples extracted, he makes, on the spot, a close approximation to classifying them; through appearance and by smell. Soil collections are then sent to pathological and soil laboratories for other, more detailed, determinations.

As Stover (43) shows in his review paper: "susceptible soils" are those that verge on the acid, with pH ranges of from about 4.0 to 5.6 or 5.9; "Resistant soils" are circum-neutral with readings of from about 6.8 and above; "Semi-resistant soils" have an intermediate range between the acid and the alkaline. In addition to still unapplied but suggestive effects of soil flora antagonisms (see citations: 14; the 1948 publ. in 26; the 1944 and 1946 publs. in 27; and 29, 43), the textures and organic matter contents have marked effects: "susceptible soils" are sandy and sandy loams poorly stocked with nutrients and with high organic contents; "resistant soils" are most likely to be composed of good permeable clays rich in nutrients and with a moderate amount of organic material in them. In my own work in Honduras and in Puerto Rico (unpublished) I found Fusarium oxysporum much the most easily established in the laboratory in sandy soils when compared with clay soils. Stover shows that banana pathologists have secured evidence that their wilt Fusarium, once it is introduced, is more easily recovered from sandy soils. Also, it spreads in banana plantations faster in less fertile sandy loams, as compared with the more fertile and heavier textured soils with good clay components.

In discussing soil-and-disease, examples mentioned above are only a few from many. There is no doubt that studies along these lines are of great economic value to producers. To tropical pathologists there is an awareness that securing for crops something like mutilineal resistance, also known as tolerance or field resistance, has a greater long-time future than the easy more clear-cut and genetically more readily manipulated securing of a high resistance against known parasite races. It is precisely in soils that give the longer-life reactions (or, are "resistant to disease") or in those that are medium long-lived (or, are "intermediately resistant") that tolerant crops do well and even weakly disease-resistant crops will grow for a long time; this may be extremely important in the future for the geneticist-pathologist. Growers will produce field tolerant crop varieties more and more on susceptible soils but will find them most successful on the long-lived and intermediate type soils.

Literature Cited

Note to the reader: There is a large body of literature, a few hundred publications, that could be cited, that would have relevance to mineral deficiency, abiotic disease, and soil problems. At the present writing much is being done with respect to some of these, perhaps more in the temperate zone than in the tropics and the specially interested reader should search them out. I am not attempting more in the following list than to give mostly tropical citations that I referred to and used in the preparation of this chapter. As can be seen, the chapter is no compilation of all that has ever been reported on these subjects from the American tropics. Along with the discussion, the works cited give, it is hoped, highlights that may help in training, and the work entailed in preparatory reading for work on control programs.

1. Alvin, P. de T. 1954. Studies on the cause of cherelle wilt of cacao. Turrialba 4:72-78.

2. Averre, C.W. 1967. Vascular streaking of stored

cassava roots. Internat. Symp. Trop. Root Crops, Univ. W. Indies, Trinidad, Proc.:IV, 31-35.

3. Barnette, R. M. and J. D. Warner. 1935. A response of chlorotic corn plants to the application of zinc sulfate to the soil. Soil Sci. 39:145-159.

4. _______., J. P. Camp, J. D. Warner and O. E. Gall. 1936. The use of zinc sulfate under corn and other field crops. Univ. Florida Agr. Exp. Sta. Bulletin 292, 51 p.

5. Bonilla, G. E. 1952. Estudio comparativo de la distribución de la luz y el viento en dos regiones cafeteras (Caldas y Antioquía). Centro Nac. Inv. Café, Chinchiná, Colombia, Bol. Inform. 3(25):29-35. --. 1952. Estudio de la distribución vertical de la humedad relativa, de la temperatura del aire y del suelo en una plantación de café. Centro Nac. Inv. Café, Chinchiná, Colombia, Bol Inform. 3(34):19-37.

6. Bolle-Jones, E. W. and R. N. Hilton. 1956. Zinc-deficiency of Hevea brasiliensis as a predisposing factor to Oidium infection. Nature 177:619-620.

7. Camp, A. F. 1934. Studies on the effect of zinc and other unusual mineral supplements on the growth of horticultural crops. Univ. Florida Agr. Exp. Sta. Ann. Rept. 1934:67-69.; (with W. Renter, same title, Ibid. 1937:132-135.)

8. Cibes, H. and G. Samuels. 1957. Mineral-deficiency symptoms displayed by tobacco grown in the greenhouse under controlled conditions. Univ. Puerto Rico Agr. Exp. Sta. Tech. Paper 23, 22 p. (Numerous other crops studied over several years, examining effects of minor element deficiencies and results published.)

9. Crane, J. C. and F. L. Wellman. 1950. Edad del henequén en relación con las caracteristicas de la fibre. Turrialba 1:74-77.

10. Dickey, R. D. 1942. A preliminary report on iron deficiency of tung in Florida. Univ. Florida Agr. Exp. Sta. Bulletin 381, 20 p.

11. ________., and G. H. Blackman. 1940. A preliminary report on little-leaf of the peach in Florida--a zinc deficiency. Univ. Florida Agr. Exp. Sta. Bulletin 344, 19 p.

12. Fawcett, G. L. 1923. Algunas notas sobre el efecto del frío sobre las yemas de la caña. Rev. Ind. Agr. Tucumán 14:67-73. --. 1927. Las manchas blancas en las hojas de la caña. Rev. Ind. Agr. Tucumán 17:259-261. --. 1928. La determinación de la temperature en que se hielan las yemas de la caña. Rev. Ind. Agr. Tucumán 19:123-131.

13. Fawcett, H. S. 1936. Citrus diseases and their control. McGraw-Hill, New York, 656 p.

14. Fernandez Valiela, M. V. (2d. ed.) 1952. Introducción a la Fitopatología. Gadola, Buenos Aires, 872 p.

15. Franco, C. M. 1947. O problema do sombreamento dos cafezais em São Paulo. Bol. Supt. Serv. Café. São Paulo 22:708-717. --. 1952. A agua do solo e o sombreamento dos cafezais em São Paulo. Bol. Supt. Serv. Café, São Paulo 27:10-19.

16. ________., and H. C. Mendes. 1949. Síntomas de deficiencias minerais no cafeeíro. Bragantia 9:165-173.

17. Geraldson, C. M. 1957. Cause and control of blossom-end rot of tomatoes. Univ. Florida Agr. Exp. Sta. Circular 101, 8 p.

18. Gonzalez, C. A. and C. Camacho. 1952. Síntomas de la deficiencia del boro en el cafeto, informe preliminar. Costa Rica Min. Agr. Indust., Boletin Tec. 11, 12 p.

19. Gourou, P. 1953. The tropical world; its social and economic conditions, and its future status. Longmans, Green & Co., London, 156 p.

20. Ives, N. C. 1950. Soil and water runoff studies in a tropical soil. Turrialba 1:240-244.

21. Jacks, G.V. 1936. Tropical soils in relation to tropical crops. Bur. Soil Sci., Harpenden, England, Techn. Commun. 34:47-51.

22. s'Jacob, J.C. 1938. Voedings physiologische onderzoekingen bÿ Coffea arabica I. Arch. Koffiecult. Nederl.-Ind. 12e, 1:2-48.

23. Lynch, S.J. and G.D. Ruehle. 1940. Little-leaf of mangos: a zinc deficiency. Florida State Hort. Soc., Proc. 53:167-169.

24. Machicado, M. and P. de T. Alvim. 1954. Sintomatología de las deficiencias minerales de cacao. Turrialba 4:155-163.

25. Machicado S., A. 1946. Influencia del sombrío, el suelo y las practicas de cultivo en el desarrollo del cafeto en sus primeros meses de vida propria; experimento preliminar. Agr. Trop., Colombia, Suplemento Agron. 3, 31 p.

26. Marchionatto, J.B. 1927. El bitter-pit de la manzana. Rev. Centr. Est. Agr. Vet., Argentina 20:325-328. --. 1938. El corazón mohoso de la manzana. Rev. Argentina Agr. 5:179-186. --. 1948. Tratado de Fitopatología. Editorial Sudamericana, Buenos Aires, 537 p.

27. Meredith, C.H. 1941. The growth of Fusarium oxysporum cubense in the soil. Phytopathology 31:91-93. --. 1942. The effect of chemicals on Fusarium oxysporum cubense growing in the soil. Phytopathology 32:182-184. --. 1944. The antagonisms of soil-microorganisms to Fusarium oxysporum cubense. Phytopathology 34:426-429. --. 1946. Soil actinomycetes applied to banana plants in the field. Phytopathology 34:983-987.

28. Milne, G. 1937. Essays in applied pedology. I. Soil type and soil management in relation to plantation agriculture in East Usambara. East African Agr. J. 3:7-20.

29. Mitchell, R. and M. Alexander. 1961. The mycolytic phenomenon and biological control of Fusarium in the soil. Nature 190:109-110.

30. Mohr, E. C. J. (Transl. and notes R. L. Pendleton). 1944. The Soils of Equatorial Regions, With Special Reference to the Netherlands Indies. Edwards Bros., Ann Arbor, Michigan, 766 p.

31. Mowry, H. and A. F. Camp. 1934. A preliminary report on zinc sulfate as a corrective for bronzing of tung trees. Univ. Florida Agr. Exp. Sta. Bulletin 273, 34 p.

32. Muller, L. 1958. Observación y control de las deficiencias de elementos menores en el cafeto. Turrialba 8:126-135.

33. Munnecke, D. E. and P. A. Chandler. 1957. A leaf spot of Philodendron related to stomatal exudation and to temperature. Phytopathology 47:299-303.

34. Norton, K. R. 1965. Boron deficiency in bananas. Trop. Agr., Trinidad 42:361-365.

35. Nutman, F. J. and F. M. Roberts. 1955. Frond drop. A note on an abnormal condition of coconut palms in Jamaica. Empire J. Exp. Agr., England 23: 268-270.

36. Ochse, J. J., M. J. Soule, M. J. Dijkman and C. Wehlberg. 1961. Tropical and subtropical agriculture. 2 vols. Macmillan, New York, 1446 p.

37. Pendleton, R. L. 1950. Agricultural and forestry potentialities of the tropics. Agr. J., United States 42:115-123. --. 1955. El papel de los suelos tropicales en la alimentación del mundo. Turrialba 5:6-16.

38. Perez S., V. M. 1957. Algunas deficiencias minerales del cafeto en Costa Rica. Costa Rica Min. Agr. Indus. (Bulletin) 2, 27 p.

39. Ruehle, G. D. 1940. Zinc deficiency of the avocado. Florida State Hort. Soc., Proc. 1940:150-152.

40. ________., and S. J. Lynch. 1940. Copper sulfate as a corrective for dieback, a new disease of the avocado. Florida State Hort. Soc., Proc. 1940: 152-154.

41. Sánchez Potes, A. 1966. Enfermedades del algodonero, del cocotero y de la palma africane en Colombia. Acta Agron., Palmira 16:1-13.

42. Scarseth, G. D. 1944. Growing bananas in acid soil. Agr. in the Americas 4:188-189.

43. Stover, R. H. 1962. Fusarial wilt (Panama disease) of bananas and other musaceous species. Commonwealth Agr. Inst., Phytopath. Paper 4, 117 p.

44. Sylvain, P. G. 1954. Long-range objectives in studies of the physiology of coffee. Turrialba 4:13-22.

45. Tanada, T. 1946. Utilization of nitrates by the coffee plant under different sunlight intensities. J. Agr. Res., United States, 72:245-258.

46. Tolenaar, D. 1966. Boron deficiency in cacao, bananas and other crops on volcanic soils of Ecuador. Netherlands J. Agr. Sci. 14:138-151. --. 1967. Field symptoms of boron deficiency in cacao. Cacao Grower's Bull. 8: 15-18. (Original not seen. Rev. Appl. Myc. 46:631. 1967.)

47. Tullis, E. C., A. L. Smith, and A. G. Johnson. 1934. Panicles of rexoro rice injured at emergence by sun scald. Phytopathology 24:1043-1044.

48. Wellman, F. L. 1945. Indian gardening in El Salvador. Agricultura in the Americas 5:47-49; 57. --. 1961. Coffee botany, production and utilization. Leonard Hill, London; Interscience Publ., New York, 488 p. --. 1962. A few introductory features of tropical plant pathology. pp. 928-930. _In_ Symposium on tropical plant pathology. Phytopathology 52:928-953.

49. Young, T. W. and J. T. Miner. 1961. Relationship of nitrogen and calcium to soft-nose disease in mango fruits. Amer. Soc. Hort. Sci., Proc. 78:201-208.

Chapter XX

DISEASE CONTROL BY CULTURAL CHANGES

Summary

Plant disease prevention is of special importance in the American tropics; often it is accomplished by means other than chemical applications or disease resistance. Successful plantations progress, and change sometimes because of disease, through several recognizable stages from raw jungle to a highly sophisticated agricultural enterprise. Without scientific help indigenes have long developed their own means of preventing and ameliorating disease losses for their many crops. These means, many of them, are detailed and simple but apparently quite effective. An important broad system for disease control, popular in the tropics, is to move and thus escape old infections. It has been done since aboriginal times to the present; by the smallest of growers to the most extensive of plantation companies.

Much growing occurs among jungle trees and the cultivating of crops under partial shade is a method used by many planters for control of certain diseases. In some cases infections are more serious under shade, but numerous troubles are much reduced by reasonable shading. Antibiotic effects of different sorts appear to be operating in the tropics. These result, over the year, in actual "self healing" effects under some field conditions, and eventual control by antibiosis. There are, naturally, drastic crop manipulations which are used for control of disease: killing parasitic soil organisms by repeatedly plowing fields during hot dry seasons, flooding of diseased fields, and storage of roots in a manner to avoid internal breakdown. Legally unenforceable but needed plant quarantines, given informal and local backing by concerned growers, have been successful in reducing spread and introduction of unwanted crop diseases.

Introduction

> -- There is nothing so important in tropical plant pathology as knowing about all kinds of disease control --

It has been normal and logical in the developing science of tropical plant pathology, that, as it is in general temperate zone plant pathology, the determinative and descriptive phases of the subject are those given first attention (40). In the temperate zone there is a tendency, as sophistication advances, to almost forget disease prevention (32). This tendency in the tropics with its overwhelming problems would be more supremely important. Studies must be designed to lead to ways of managing plants so there is more tolerance--and even in the long view, "self healing"--of field disease problems than ever. Good advances are being made; more is being done to insure crop production relying on a minimum expenditure for disease control in the complexity of conditions in the tropics.

In the tropics, where crop diseases are so numerous and where disease resistant types are often not used (41 publ. 1969) there is much application of so-called "primitive," that is, non-chemical, means of disease control. There is nowhere that this is so important as in the American tropics. Some of the facts plant growers have to face in tropical disease control have been mentioned in previous chapters. For emphasis some need repeating as follows: there are more diseases in the tropics than anywhere else in the world; in the warm tropics there is no seasonal rhythm of the kind caused by annual recurrence of the blasting of winter; where there is no winter the word "overwintering" is out of place, but there is always a holdover of disease organisms from one season to the next; some times there are two and three separate wet seasons a year with dry intervening, in some areas there are no true dry seasons; in many plantation areas leaf surfaces are never truly dry; there are parts where plant life is absent because of extremes of cold (high altitudes) and dryness (desert areas); the holdover of pathogens, where winter never comes, is not to withstand prolonged lifelessness because of freezing, but to protect the parasitic species from the tropically abundant living antibiotic agents (bacteria and fungi and nematodes).

Moreover, it is repeated, that it is in the tropics that there is the most variability in pathogenic organisms. They have been adapted to tropical conditions much longer than in the temperate zone and there are inherent difficulties for the tropical man in the use of some of the art and technology of disease-free crop production systems developed and taught in the temperate zone.

The basis of much tropical plant disease control harks back to the temperate zone but has had its practical effects in the efforts of plantation owners. In many cases they are the ones who have brought science in agriculture to their countries (18). Often coming from important families, planters are also a strong sector of society pushing towards a better country and the advancement of work on plant disease control. The planters form associations around crops and by means of public pressure or even direct hiring, bring plant disease specialists to solve problems in their countries. From such associational activities (e.g., 16, 18, 19, 25, 27, 35) we find that great strides have been made in pathology in our tropics.

Successive Changes During Plantation History and Disease

From discussions with successful plantation owners in several countries, and corroborative personal observations, it can be seen (as I have) that successions of changes occur in plantations, in many cases because of disease situations (3, 8, 10, 14-17, 19, 21, 31, 33, 35, 36, 43). These successions are not confined to any one crop although coffee, bananas, and cacao have been given the more detailed attention by me in this regard. It has also been seen in sugarcane, coconut palm, oil palm, citrus, and Hevea rubber. The changes are fairly consistent in their progressive occurrence.

It can be assumed that the original plantation site is jungle land, of either primordial or secondary growth. This is surveyed and cut through first by brush clearing, next by tree felling, and then by burning the trash. Truly tropical jungle trees, contrary to popular opinion on the whole are not as large nor are their felled trunks as tough-wooded, as the greatest forest trees in the temperate zone. Moreover tropical tree trunks with a few exceptions, usually decay rapidly. The clearing and preparing of soil for planting is more easily done in the tropics than in the temperate zone. After clearing roads are established, fields are laid out, and in the most convenient of them the first plantings are made.

Test plantings look mostly disease-free, as they grow, and even if there are a few leaf spots they seem minor and are disregarded. After the tests, extensive field planting is undertaken. It is usually in the larger fields that severe disease first becomes evident. New plantings thereafter are in directions where little or no disease is found. But attention is given to the simplest and most obvious disease control practices to preserve the older established field growths.

As expanded planting is carried on, more planting material is purchased and with it usually more diseases are brought in. It is about then that first advice is sought from plant disease control specialists. Ameliorative measures, without the use of the expensive and often difficult to transport chemicals and equipment, are most likely to be used.

As plantations gain in size and age their reputations increase and their stabilities are then assured. During this third period, more diseases are seen and they become severe. They may be held back by different kinds of crop manipulations. If successful these measures become routine. If, however, chemicals have to be used, they are given much attention and their applications are made on as sound technical bases as anywhere else in the world, and as part of production. On visiting plantations at this stage, it is evident that in at least some cases great attention is turned to sectors where crops are heaviest, and any means possible is used to intensify cropping there. The effects of the extra attention and forcing the fields is to increase costs, but it also results in a higher quality product.

In larger, older fields, diseases are more important and there are some planters who reach a point where they take pride in disease control even if it is at this period that the most severe effects from disease are experienced. To other planters, diseases of a severe nature are so devastating that rather than conquer the difficulty they cut down the plantation, or stop all care of it, and allow it to return to second growth jungle.

Meanwhile under better planters, established for years, disease control attains more and more expert practice. Of course, the costs of production are also increased. A work force, in addition to those who plant, prune, fertilize, and cultivate, is organized to do nothing more than carry on expert disease controls. These added services are costly, but they become habitual and insure the planter's crop.

With only occasional supervisory attention, the now old planter-owner tends to more and more relegate the plant disease control measures to his subordinates. Often he moves from the plantation dwelling and he may then live at some distance from his holdings, usually in a metropolitan area of some kind such as the country's capital; more and more his vital personal agricultural and plant disease control interests slip away, and maybe the plantation suffers.

At about this juncture the second generation planter appears; modernizing both the houses and the plantings. The weed and bush-choked abandonments, set aside long ago, have grown up almost like original jungle. There are some crop diseases, such as certain blights and leafspots, that disappear under wild growth conditions, and the new generation grower, often a trained young agriculturist, relying on both what he has learned and the experience of others, replants the lost fields. He knows that varieties of his crop are disease resistant, and he is fully aware of spraying and other disease control practices and the use of chemicals.

If replanting can be done, it is more scientifically carried out under the supervision of this new generation. The control of disease is now realized as the only way in which production can be guaranteed. With change and the discarding of old ideas, new ones spawn new systems. Plant pathologists are conferred with to learn about the most recent research results; modern disease control methods are adopted, and these are well inserted into the life and development of the old established plantation.

Such are some of the several general phases in

plantation progress. They result from serious facts: either small farmers or more wealthy planters want stability; disease control is imperative; planters are no longer adventurers counting successes in drastic moves; they know that to escape disease there are no vast, easily-reached virgin jungles ready for the taking; they know, now as never before, that a dependable and trained labor force under an intelligent regimen is indispensable if disease control and special crop husbandry is to be well carried on.

Some Folk Methods Used Against Diseases

The indigenes or so-called "Indians" of tropical America have, in their own old clearings and gardens, worked to prevent extreme losses from plant diseases. It is for their credit, and for our record, to give some examples. The crops with which these peoples (aborigenes) have been most concerned are such as maize (*Zea mays*), beans (*Phaseolus vulgaris* and other spp.), cassava or mandioc (*Manihot esculenta*), small fruited tomato (*Lycopersicon* spp.), hot pepper (*Capsicum baccatum*), sweetpotato (*Ipomoea batatas*), white potato (*Solanum andigenum-S. tuberosum*), squashes and pumpkins, (*Cucurbita* spp.), yautia (*Xanthosoma* spp.), quinoa (*Chenopodium quinoa*), ulluco (*Ullucus tuberosus*), and the like. I have observed the gardening of indigenous populations for some years, and a few studies of it were made in El Salvador (41) and were published in 1951.

The Indian grows his subsistence crops in clearings and often these are in shallow soils. Planting is done with a fire-hardened point on a staff or an iron- or stone-tipped pike. By trials it is found that certain types of tropical soils are less disease-affected, and thus more productive, provided that planting is of mixed species and done so there are no rows. Where possible, soil is kept covered with a light layer of litter. In some cases sunny places are used to grow weeds that are cut, allowed to half dry, and then rolled into an "alfombra" (blanket) which is carried to and unrolled on new planted ground. In other instances, lengths of old stems of such bushes as *Desmodium* sp., or of young *Acacia* sp., and of *Sambucus* sp. are buried under hills of maize, beans, squash, and other kinds of crops; it is believed by Indians that both of these treatments are not only in the nature of fertilizer, but also as an aid against disease.

Fig. 20-1. Seed beds being made: sides reinforced with poles cut from jungle and soil from jungle and from sunken walks dug for drainage, incorporated in the beds. (Photograph taken in El Salvador during acquaintance with this peasant.) Damping-off developed in lowest bed; diseased plantlets were pulled and a mixture of wood ashes and sand was scattered on the bed. Often such field preparation and disease amelioration is used in intensive vegetable and flower growing. The same ancient and serviceable methods are found in many tropical isolated communities.

In addition to inter-species mixing, plantings are handled as small plots and frequently changed. If this is not done, the diseases become severe. A corner or bed is used for one season, harvested, and then even if it is not seriously diseased, replanting is done in another plot. If a severe disease attacks a bed of plants often they are cut down or pulled and allowed to rot where grown, often with weeds piled over them. This land is then "given time to purify" and may not be planted again for a few years.

In El Salvador (41, see 1945 publ.) and in Guatemala, gardens were observed where Indians followed two ways of making seed beds that would most likely keep healthy. In better drained areas, but where seedling rot or damping-off occurs, beds are made on the ground. A sloping spot is cleared of debris and roots, the soil pulverized by rubbing between hands, and a bed is built up slightly by using soil from the sunken walks as is seen in accompanying Fig. 20-1; beds are narrow, about a span high and just wide enough to reach across easily. Sides of beds are firmly supported by poles or bamboo laid along the edges and staked in place. When damping-off appears, cooking-fire ashes or sand may be put around plantlets or the offending seedlings are dug out and discarded at a distance from the bed. Holes from the digging out of diseased soil are filled with virgin dirt from nearby.

The other type of seedbed is actually a slightly elevated platform with low sides to it, made from jungle material. Hand pulverized dirt is used to fill the platform-bed, and seeds are dibbled in with finger and stick. Such a platform-bed is more often found in wet areas, where rains are heavy during seasons. A protecting blanket roll of plant stems or split, flat bamboo pieces tied tightly side by side is laid over these planted platforms to protect against severe rain storms where they occur. Here again, when seedling disease occurs the rotted plantlet is removed along with the soil around it and virgin soil is brought in for replacement. As in ground beds, sands, or ash, by themselves or mixed with road dust, are sometimes distributed around seedlings to reduce damping-off. In some areas devices of flattened bamboo (Fig. 20-2) are used to deflect rain.

I have learned of Indian gardeners, growing flowers and root crops, who pick off mildewed, spotted or rotted

Fig. 20-2. Seed beds in heavy rainfall areas may be wholly destroyed. In some places, as is the case for example in a hill location of Guatemala, the use of protective devices made from flattened bamboo is sufficient to reduce both rain injury and diseases.

leaves by hand*. Indian peasants of Honduras have been known to use on mildewed cucurbit leaves a light powdering of ashes from the wood used in cooking fires. In El Salvador, newly deposited dust from the volcano Izalco was being carefully collected by peasants in my presence, and was said to be used to apply on beans against disease.

A problem that comes to many Indian growers is keeping seed both disease-free and alive from season to season. Such things as fully matured, healthy looking ears of maize in their husks, well dried heads of grain sorghum, pods of beans tied in small bundles, and dried pepper and tomato fruits are often hung in the rafters or under the eaves of peasant dwellings for preservation of the seeds in a sanitary condition. I have also seen another slightly more elaborate system. While it was only observed in maize and beans it was said to be used with some adaptations also with seeds of grain sorghum and lima beans. Unblemished ears, heads, or pods are selected without being opened. In the case of maize the unhusked ears are stacked, loosely, on a newly prepared elevated bed of mixed content; usually such materials as road dust or fresh wood ashes, inter-layered with dry stalks of grasses or maize. Upon this the ears or pods are arranged. This is then again covered with ashes and protected with dry stalks and leaves of such plants as maize, heliconia and sugarcane. Sometimes palm leaves are employed. Such a store is said to be kept for months, located under a roof, and often is in part of the living quarters of the farmer's family (for illustration of this, see Fig. 23 in the author's 1971 publ. (41)).

It is important to note that the labor force, and the plantation foremen recruited from the labor force, come from peasant, often mixed-blood populations, and become experts in plantation work. This is significant for they also often transfer or adopt some of their long-practiced folk methods of disease amelioration or control to their duties in modern-day plantation agriculture. Planters many times

---

*This is not only a Western Hemisphere practice among native gardeners. I (41) saw it, and published remarks about it in 1938, in a subtropical desert area of Asia Minor. Crops in oases were kept almost completely disease free there by the early removal and consumption by men and camels of disease-infected plant parts.

wisely allow old understood methods to be carried on. It may not always be a perfectly successful treatment. But even if only partially so, assiduous application and repeated action give good results in aiding morale and confidence. Meanwhile new techniques are slowly brought in and started along with a training program.

One old folk method, long carried out and by some planters fully acceptable for disease amelioration, is that of changing locations. Planters of major crops in the past, before the advent of plant disease studies, have learned from their unlettered predecessors that often where tree, or other crop diseases accumulate and intensify, sometimes the only answer is to move.

Moving Plantations to Escape Disease

Many centuries before Europeans set foot in the American tropics, expert agriculturists of the time and place carried on movement of their fields to new places. Indications of such early changing occurred during past history of Aztecs, Mayans and other Indians of middle America. There is no doubt plant diseases were rampant then, as now, and when fields of maize and beans were replanted time after time in the same spots, the rusts, root rots, wilts, foliage spots, blights, and other diseases increased, and crop-growing had to be moved.

In the milpa system, relatively small new plots are cleared by having bush and trees chopped down and burned, after which food crops are planted. With maize, disease that often causes drastic losses is one or more of the three rusts*. In a few seasons, when harvests begin to

*Since maize rusts are so important in the Neotropics, personal correspondence and discussions have been carried on about them with the Guatemalan rust researcher Dr. Eugenio Schieber, specialist in maize and tree rusts. It is his belief that rusts of maize have been a serious disease in the American tropics during and since prehistoric times; that ancient crop failures surely occurred because of rusts; and that we have little appreciation of the harmful effects of rusts on the old maize-based civilizations. These rusts were probably the cause of famines and wholesale deaths in many parts of the Pre-columbian tropical America.

give too little, there is the planned-for shift to a new plot, readied nearby and sometimes directly next to the old plot.

This type of moving to escape crop diseases--far away in some cases, at shorter distances in others--has been practiced a long time in our tropics. Recently in Puerto Rico, growers of ñame or yam, *Dioscorea* spp., lost large numbers of roots because of wilt, apparently from a race of *Fusarium oxysporum*. This occurred in fully sun-exposed fields at a fairly low, and therefore, warm location. The amounts of wilt increased, and the growers moved to higher and cooler elevations some half a day's journey by oxcart from the old fields. The crop was better there.

In another case, in ten years' time growers of white potatoes in Costa Rica had moved their fields twice, to find locations where *Phytophthora infestans* was less severe. In El Salvador in 1945 there were seven peasant tomato farmers, in easy walking distance of the capitol, who were bothered by occasional wilt from *Pseudomonas solanacearum*. They were encouraged to plant at a little higher and dryer location, where it was hoped the wilt would be slower in its action. They found land a day's walk up the mountain and tests the first year with two crops were quite successful. Homes were built in their new places, but the second year the leaf mold, *Cladosporium fulvum*, caused such serious losses they returned to their old farms.

Early in the history of sugarcane growing in our tropics, planters used such old varieties as purpureo, crystalina, and rayada. These kinds are among those that belong to the noble types, high in sugar content, with excellent milling qualities, and heavy producers, but susceptible to many diseases. These sugarcanes did exceedingly well in American tropical soils, and took much business from the Oriental growers. After some years, however, dying out commenced, pathologists were brought in, and it is now known (28, 31, 36 and others) that such root disease fungi as species of *Fusarium*, *Rhizoctonia*, *Marasmius*, and *Pythium*, and nematodes were involved. When to these were added leaf injuries from several fungus species like *Cercospora*, *Colletotrichum*, *Leptosphaeria*, and *Helminthosporium*, plantations in many areas declined and had to be abandoned. To carry on their well-known businesses, planters moved long distances, where they established new plantations complete with the costly cane mills and residences.

Before the end of the last century, banana plants originally from the Eastern tropics had become well established as suitable subjects for remunerative plantation culture in the West Indies, and Central, and South America. One variety of fine appearance and with good production and shipping characters, was the gros michel. Around it there grew up large and well-managed plantations; a flourishing industry. Cities developed to take care of it, and international transportation and fruit companies were formed. As may be seen in such accounts as those cited by Stover (35), a wilt fungus, Fusarium oxysporum cubense, appeared and in large plantations there was so much disease it resulted in complete destruction and drastic moves.

The fruit companies sent scientifically trained explorers into new regions far from the old sick areas. The men looked for desirable types of soils and climates and proper conditions for laying new railroads, building new wharves, and developing new cities if necessary. Great jungle areas were tamed again and made productive, sometimes in other countries far from the old; some new areas were most of a continent away from the old places that had become overcome by the wilt. Run by experts trained in old plantations, the new developments were soon a success. The dread banana wilt infection was for a time left behind, and, although exceedingly expensive, the extreme long distance moves circumvented disease and resulted in plantations that for a while paid good returns.

The moving of extensive plantations to avoid coffee plant disease occurs in Brazil. This product has been the very life blood of Brazil's national economy and of all the cash crops in the American tropics, coffee is the most fruitful. [Something of the importance of coffee in the American tropics, its supreme position, and details of its history in Brazil have been reviewed in 1961 by Wellman (41).] Plantations of it did very well in some parts of Brazil, for example in the area neighboring the port of Santos. Metropolitan markets in Europe, the United States, and the Near East liked Brazilian coffee. Using all the advantages of friendly nature, abundant labor, and knowledge about production, "fazendas" (plantations) were planted with millions, now billions, of trees. On the whole, for decades all did well. After a while there appeared the American leaf spot (Mycena citricolor = Omphalia flavida), and then in older plantations came root diseases (Rosellinia spp.), nematode infections, dieback (intensified effects of over-bearing with secondary infection from Colletotrichum coffeanum), probable infections of Fusarium sp., berry diseases, and injury of foliage by the fungus Cercospora coffeicola. A complex of diseases was making a failure of a once highly profitable plantation.

To keep up their business, the Brazilian planters went some kilometers distant from the old area. Here, in new holdings, production was again good and relatively disease free. In a while, however, the old diseases developed, sometimes accidentally brought in from abandoned coffee fazendas. So the growers moved again. This kind of moving has happened a number of times at intervals of about 20 or so years. To this day, flying at a low enough altitude over Brazil it is possible in a most extraordinary manner to see and to recognize different "waves" of coffee tree plantings. There beneath you are the oldest abandoned coffee areas, healed over and now growing pastures or subsistence crops and sometimes in second growth scrub. In some cases there are different grades of abandonments with here and there a few coffee trees still present in sheltered parts; some instances occur in which fazendas have been entirely cleared of old coffee trees and planted to food crops. Next to a dead "wave" are diseased fields with still living, but no longer valuable, trees being allowed to succumb to choking weeds and wild bushes. Out in front is a vigorous "wave" in which are newer fazendas giving in appearance positive evidence of good culture and worthy harvest.

In connection with moving certain crops to new locations, it is becoming evident in our tropics that plant pathologists should give more attention to ecological factors on disease. A body of knowledge is being gradually accumulated about the varying effects of environment. The work is slow, and at present there is no real center where the effects of various temperatures, moistures, light, and wind currents on plant diseases are being systematically investigated. Wherever studies are carried out, on environment and disease in the tropics, results have been of significent practical value.

Use of information on the ecology of disease is more and more advocated; it gives growers of susceptible crops possible indication of where the crops, disease-free, may be produced. Recently, the coffee rust *Hemileia vastatrix* on its appearance in Brazil (41 publ. 1970), has stimulated producers to find out if there is an area in that large country where the popular varieties of susceptible coffee can be grown without attack of rust. Of course, limitations of the crop plant must also be considered (2, 41 publ. 1961, and 42). Combining all rust disease information available from a century of observations by European, African, and Oriental researchers has resulted in hopes that there are parts of the cool, dry uplands in Brazilian states of São Paulo and Paraná, and possibly in some other countries, where this rust might not develop in its worst manner. In such works as that of De Camargo (12) it is clearly evident that in Brazil there are areas at least where coffee will grow well and, even if the disease attacks, rust should be readily dealt with by timely spraying with proper fungicides.

Indications have just been given of major movements of sugarcane, banana, and coffee plantations in escaping disease. Results from such moves are temporarily successful. Cacao growers move from old plantation areas where watery pod rot (*Monilia roreri*) and witches' broom (*Marasmius perniciosus*) are severe. Very early in orange growing in our tropics the footrot or gummosis ("gomosis") (*Phytophthora* spp.) was severe especially in some low-lying soils. But this kind of action has its limits. Planters of the crop change their methods and grow the trees on resistant root stocks and in higher, better drained lands.

The papaya growers of Puerto Rico have been long burdened with a virus disease known as bunchy top. This is carried by a leaf hopper, and this hopper, being a weakly flying insect does not flourish in a part of the island where there is almost continuous trade winds. Some papaya growers find little to cause them to move all their planting to this wind swept area, but planters who do, have trees there with very little to no bunchy top. A number of other plantation crops with limiting diseases have managed, through radical moves, to keep on producing harvests.

However, moving a plantation crop from one area to another far distant is a most expensive activity; the dividends of modern plant pathology study are used more nowadays to solve the many disease problems in plantation crops. The jungles and bush growth can no longer be purchased for a song and easily made ready for large-scale plantation culture.

Fig. 20-3. Illustration of effects from shade removal on coffee in my experimental work in Costa Rica. On left: adequate but not too dense shading; good leaf retention and cherry production. On right: one year following shade removal; defoliation from leaf spots, dieback severe from anthracnose, and no fruit produced. Both pictures taken the same day in plots about 20 meters apart.

Partial Shade-Reducing Disease

In the tropics, it is not accidental nor is it a matter of an "old wive's tale" that there is wide acceptance of the idea that shade in many cases is good in reducing (ameliorating) certain crop diseases. Brief mention of some of this subject is given here. There seems to be no doubt (41 publ. 1961) that regulated moderate shade makes the coffee tree easier to grow (see Fig. 20-3); providing there is sufficient soil moisture that roots of both shade tree and coffee tree can be grown together.

Shade-grown coffee has had so much publicity as being of high quality that it is less well-known that in the Americas other crops also use shade and some of these are cacao, abaca (or manila hemp), the non-shipping types of delicate dessert bananas, certain plantains or cooking bananas, yautia, lleren, and ginger. I believe it will be found that in all cases where shade is advised, it is because there is an amelioration of one or more diseases on the crop through the use of it. A very few examples of references and discussions that serve to suggest problems relative to shade that need much further and more detailed study, are those in works by Garrett (17), Lellis (21), Ochse, Soule, Dijkman, and Wehlburg (29), de Segura (31), Wellman (41 publ. 1961), and Williams and Schmitthenner (43).

Tree shade that "protects" a crop has numerous possible effects, some of which are: shade foliage reduces the amount of light that filters down to strike the lower storey of crop plant leaves; crop plant leaf stomata often are closed longer under a shade canopy than where sun has its full effect; shade growth acts as a windbreak; shade leaves intercept dews above and serve to slow down dew deposit below, so that dew remains for a shorter period on the under storey of the crop plant; shade foliage cools in comparison to the warmth of full sunlight; from shade trees, leaves and twigs drop all year around, acting as an accumulation of mulch and a source of fertility about the roots of crop plants; a certain amount of shade affects the balance between succulent and hard growth, and under proper manipulation it may be used to take the place of applications of certain amounts of nitrogenous fertilizer; shade tree branches and leaves are trimmed off ("lopped away") to collect and lay around the bases of crop plants to protect the soil; shade tree roots absorb water in low locations and act to keep wet soils slightly dryer for some crop plants; shade tree roots extract fertility from deep-lying subsoils, and drop it as leaves on the surface of the planted field as seasons wax and wane; canopies of shade trees and the holding effect from shade tree roots are important in preventing soil erosion as well as aiding in the preservation of rivers and ground water.

I had special responsibility in El Salvador and Costa Rica with regard to coffee disease problems for a number of years, and relative effects of sun and shade were given considerable attention, as in many cases for many growers non-technological means of disease reduction were all that could be applied. Some findings (41 publ. 1961) are that excessive shade is not good and that the leaf spot from *Mycena citricolor* and algal attack by *Cephaleuros virescens* are both increased under excess shade graded as 5. (No shade, grade 0; light shade, grade 1; moderate shade, grade 3; and heavy or excess shade, grade 5). Severe leaf-spotting by *Cercospora coffeicola* may occur in coffee that is poorly shaded, grades 0 to 2, but much reduced if shade is increased to grades 3 and 4. Root disease and tree collapse from *Fusarium oxysporum coffeae*--not common in shade 3 to 5--may be serious in shade 0 and 1. Pink disease, *Corticium salmonicolor*, was searched for during nine years in one coffee area where shade was graded at 3 or 4. In the hundreds of hectares visited, at no time was pink disease found. One year nearly 50 hectares were changed to having no shade and in 18 months over 20 per cent of the sun-exposed coffee trees were pink diseased. All around the demonstration fields, in plantings where shade was maintained, no pink disease was seen. In the case of coffee rust, *Hemileia vastatrix* as observed in Brazil (41 publ. 1970), the disease was seen to occur with about equal severity in both sun and shaded locations.

It was found in one area in Ecuador and one in Costa Rica that cacao growing for years under a heavy shade developed slowly and produced light crops. However, on the whole, tree foliage was good. A plot of a few hundred trees had the shade cut to stimulate crop production. Within four months all cacao trees in full sun had dieback on every branch end, while cacao trees in full shade had little or no dieback. (Dieback twigs were taken to laboratories and isolations made from them, resulting in the recovery of various weakly parasitic and saprophytic fungi and bacteria.)

When susceptible types of abaca or manila hemp, grown in the sun and having severe leaf-lamina spotting and leaf-sheath streak from Helminthosporium sp., are given light shade they show a markedly reduced incidence of disease. Such reductions deserve emphasis as follows.

In the West Indies and Central America, some of the more delicate, sweet, dessert bananas as well as some of the best quality (but poor shipping) cooking bananas, when grown in the full sun, are plagued with serious leaf-spotting from Cercospora musae. However, home owners know this, and where they place plants of these crops clustered under partial shade, such as near a high fence or the side of a high building, or below coconut palm trees, or at the edge of the shade of the falmboyant or other large ornamental tree, the amount of leaf spotting on bananas is markedly reduced.

When exposed to full sunlight in the open, such very short-statured root-producing plants as yautia, lleren, and ginger are subject to various leaf spots and blemishes. Some are from: Colletotrichum sp., Cladosporium (herbarum?), Phyllosticta sp., a Phoma-like organism, Alternaria (tenuis?), Penicillium sp., bacteria, and algae the latter two especially during the wet season. The leafspot complex may be at times so serious that storage roots do not develop well, and crops of them may be practically nil. However, planted under good types of trees, with shade kept at medium density, the plants even without supplementary fertilizer produce good crops. And in doing this in the shade, the leaves of the root-producing crops show few injuries due to microorganism parasite attacks.

Antibiotic Reduction of Parasites

The observer when he returns to a disease-harassed field may see season after season a marked variation in the amount of disease from certain crop parasites. Indeed, although Yarwood (44) holds that as a principle biological control plays a minor part in the maintenance of health in plants, in the tropics one is impressed by what may be antibiotic effects that result in unexpected reduction of disease severity. This takes place, on occasion, in spite of the fact there has been no abnormal fluctuations in temperature or moisture, nor any cultural changes in the field. There are instances when disease organisms may almost disappear, then in a season or two return to unwonted prominence and then subside again into a period of marked reduction. It is well known that there are hyperparasites--an old historical example of which is Metarrhizium (34), seldom heard of in modern times--and they are specially common in the tropics (41 publ. 1969). These, along with other antibiotic effects are always operating and they deserve much greater detailed research, and by many workers, than has been expended on the problems in the past.

A revealing experience regarding antibiosis in the case of disease caused by Sclerotium rolfsii (= Pellicularia rolfsii). In the U. S. it is called "southern blight." In reviewing world knowledge about this disease and the causal organism, Aycock (6) states it is present "in tropical and subtropical countries where high temperatures prevail during the rainy season .... The fungus is common in the southern United States, Central America, the Caribbean Region and South America." Further along he says of the fungus that "Mycelium is usually considered to persist for only short periods. [One worker] found that only 1/3 of the tomato plants killed by southern blight in a dry period produced sclerotia after rains occurred, indicating that the fungus had died out in the other 2/3 of the infected plants. [Another worker] showed that the vegetative stage survived for only a few weeks in the greenhouse. Cowpeas and soybeans planted two weeks after soil infestation suffered little or no infection and the organism survived only when a suitable substrate was provided. [It was] postulated that the organism was kept in check by the presence of other soil organisms or by an accumulation of its own metabolic products."

One cannot read tropical plant pathology literature without encountering suggestions in many instances (e.g., 1, 3, 8, 16, 71, 24, 25, 29-31, 35) that, as is discussed in Chapter XIX, distinct soil types are known that have a profound effect on a number of soil-borne plant diseases. One of such is that which effects Sclerotium rolfsii, for it is known (6) that the disease is most severe in sandy soils, expecially sandy loams which have had incorporated into them much humus and bits of partly decayed debris. Such soils sustain growth of the fungus and are markedly on the "susceptible" side if they are somewhat acid. On the other hand, there are other heavier soils in which the fungus grows poorly, and if the ground has good clay components and is slightly alkaline, it is more "resistant." In these latter soils there is present more antibiosis, or some special type of antagonism, towards the parasite. It is significant that the "resistant" effect is destroyed by sterilization in the laboratory. For after steaming or autoclaving, both "resistant" and "susceptible" soils are readily invaded by the fungus.

Work (35, see for other citations) on gros michel banana wilt, Fusarium oxysporum cubense, has given evidence of "susceptible" and "resistant" soils (see Chapter XIX); this banana grown in a "susceptible" soil is a precarious crop as there is every chance that in a matter of months the plants will be quickly wilted. In "resistant" soil, gros michel banana can be grown for years with no destruction from the fusarial wilt. Such experience as these (and also of other soil-borne troubles) has led to more observations on antagonisms. It is pertinent to remark here that perhaps the most difficult phase of plant pathology may be not so much a matter of molecular physiology, refinements in nematology, or the intricacies of virology, but what exactly happens to tropical plant parasites underground. Special study techniques are being devised (some are indicated in: 6, 7, 9, 11, 17, 20, 25, 30, 38, 43) and much more still has to be accomplished.

A fairly simple, in certain cases readily demonstrated, antagonism may be in the often noted effects between different kinds of Fusarium and certain bacteria (24, 25, 30, 35 see for citations, 38). It is well recognized that in many cases bacteria are certainly inhibitors of Fusarium growth. These, working with other antibiotics, result in a "self healing" effect on fields that are planted to disease susceptible crop varieties. Attempts have been made to insure this and in some way artificially to increase antibiosis and the bacterial content of Fusarium-infested soil (24, 25).

In the case of the widespread seedling damping-off organism Rhizoctonia solani (= Thanatephorous cucumeris), there is often sporadic increase one season followed by dramatic decrease of disease in the years following. Elaborate seedbeds for vegetable-seed germination on the slope of San Salvador volcano were almost completely destroyed by Rhizoctonia damping-off one wet season. To the pathologist's consternation, next season Indians returned, and with little change, reprepared the beds and sowed seeds again and there was practically no damping-off, even during the most intense part of the rainy season. In Costa Rica, places were located where for three years coffee seedling growers lost heavily from Rhizoctonia damping-off; five years later, the same seed beds were almost free from the trouble. On the other hand there are some soils in which this fungus remains a hazard. These soils appear to be from relatively new volcanic deposits.

At one time it was throught that certain apparent discrepancies in disease from _Rhizoctonia solani_ might be from lack of distribution. However, this does not seem to be the case. Where given detailed attention, it is found to be widely spread (17) and is even distributed on some seed (7). A certainty that this latter did occur led to analyses of possibilities of this kind of action in normal management and planting procedures with a number of the important plantation crops. To the best of my knowledge, seed distribution of _R. solani_ does not occur in coffee, hevea rubber, tea, nutmeg, orange, and cinchona. It was found, however, to be carried with seed of coconut, on corms ("heads") of musaceous crops, on ginger roots, and as sclerotia stuck to the uninjured surfaces of seedlings of coffee. Modern usage of the younger top parts of sugar-cane for setts, appears to preclude the fungus' being carried around on that crop. However, it must be that this as well as other soil-borne disease organisms are reduced in potential through abundance of and rapid action by antibiotic organisms.

A feature of remarkable interest in connection with studies on _Rhizoctonia solani_ is that not only is it a parasite but it is also a hyperparasite (11). There are only certain conditions (not thus far determined) in which the organism will parasitize other fungi. On earlier pages I indicate some of my observations on this; I attempted growing it with other fungi in jelly glasses of soil. During the process it was learned that the slender, fast growing, thin-walled cells of _R. solani_ are readily injured growing at the surface of soil that is kept warm and wet. In fact, if the wet period is more than eight days at 28° C the web of mycelium on the soil surface disintegrates into a gelatinous mess. Microscopic examination showed bacteria and possible Actinomycetes in the surface slime.

However, if the _R. solani_ is in the stage of dark-colored loose clumps, or of brown sclerotia (chlamydospore-like cells tightly bound into rounded structures), the parasite is not quickly killed under wet, warm conditions.

Furthermore, the clump-like tawny colored infection mats of *R. solani*, on young hypocotyls of coffee seedlings, lived for over a month in warm wet conditions. In ten days diseased parts of damped-off seedlings were fully rotted and something had killed all colorless cells of the fungus whether inside or outside of the host stem.

There are literally hundreds of fungi that develop subterranean and aerial sclerotia during seasonal changes from wet-to-dry-to-wet conditions in the tropics. These sclerotia, many of which have never been described, seem specially adapted to withstand the extremes of bacterial attack that come about during wet seasons. To me, sclerotia are true and effective holdover stages in tropical fungi (in many cases fungi "skip" a resting spore stage but sclerotia preserve them under adverse circumstances).

In connection with identification problems, I was specially interested in four species of sclerotial structures and made minor, crude studies of them in the laboratory. Sclerotia dealt with were: of *Sclerotinia sclerotiorum* within buds and flowers of hibiscus; of *Macrophomina phaseoli* (= *Sclerotium bataticola*) buried in stems of bean plants; of *Sclerotium rolfsii* stuck at ground line to the surface of diseased tomato stems; and of a *Sclerotium* on diseased leaf surfaces of coffee. In all these cases, sclerotia start as light-colored fungus growths, and, whether whole or macerated, at this early stage they rot quickly after incubation on wet soil. Matured, black or brown sclerotia with their tough "skins" or integuments intact may not rot for several weeks to months when incubated on moist wet soil. However, if broken or cut, the sclerotia will exhibit slow lysis or decay at the surface which is exposed if this surface is lightly buried in wet soil. Miller (26) reports observations indicating that very long inundation eventually decays the hard sclerotia of *Sclerotinia*, apparently even if they are mature, tough, and dark in color.

There are numerous other types of fungus growths that harden against bacterial destruction their outer layers of accumulated hyphal matter. These are in such forms as rhizomorphs and stromata of different kinds; it appears that the outer, tough structures are highly resistant to rot and more often than not this sclerenchymatous tissue is dark to black in color.

In contrast to hard-surfaced fungal developments are

the relatively soft infection bodies, or gemmae (41 publ. 1961), such as those of the round Mycena citricolor (= Omphalia flavida), or the faggot-like clumps of hyphae of Sclerotium coffeicola (6). For reasons still not clear these relatively delicate bodies (they can be smashed between thumb and finger), if uninjured, germinate on leaf surfaces and infect them without being inhibited by the epiphytic bacteria (see 20 and other works) that we know are so common and abundant in the tropics. Broken open, the Mycena infection body is quickly rotted on a wet leaf surface.

Disease Control by Some Extreme Changes in Crop Management

Nematodes have been for a long time an important group of disease producing organisms (in about the same category as viruses, bacteria, and fungi). The nemas can be serious in the tropics. Conscientious citrus pathologists know that nematode injury is often held to a minimum when the trees are kept reasonably pruned, fertilized, mulched, and weeded. The same is more or less true for nematodes in coffee. In many a location in Central and South America, soil removed from around coffee tree roots yields numerous nematodes. However, where the plantations are pruned at the time needed, where adequate fertilizer is used, and where reasonable mulching and weeding is carried on, the presence of nematodes causes little if any detectable reduction in coffee yield. If as in some of the more notable cases, coffee is grown disregarding symptoms of dieback, and there is no history of years of previous fertility, and other field practices are neglected, debility increases and nematodes become a critical problem almost killing the trees.

Nematologists, even those who are mainly nomenclature addicts, know and publicize the fact that their worms are wriggling, moisture-loving creatures. Researchers interested in the ecology of nemas (36 also see citations) have learned that plant nematodes increase in wet soils, and in some cases (example, 39) are serious in fields under waterlogged conditions. Taking advantage of this moisture-loving habit, Thorne and Román (37) controlled an otherwise completely destructive infestation of nematodes on tomatoes in Puerto Rico. The tomato fields were so large that chemical treatments were too costly for application.

Therefore, these workers developed repeated deep plowing and fallow, during the hottest dry part of the dry season, and nematodes were much reduced in numbers. They were so few that normal and very good crops of tomato are now grown during the moist part of the year, where dry season plowing and fallow is a regular treatment in that area.

In connection with the nematode problem in bananas, work was carried on by pathologists (22, 23) freeing the corms from the parasites, and also by study, following the development of nematode disease in the fields. The main organism involved in this work is Radopholus similis, the burrowing nematode, that is known to be serious and destructive on bananas in Florida, Panama, and Jamaica. It was found that where citrus is interplanted with bananas, and in the two crops growing so close that roots are interlacing, the banana nematode disappears. In work on some coffee nematodes, Chitwood and Berger (14) corroborate the findings of Claud L. Horn who pointed out and recommended to them the growing of a crop of Crotalaria voglei on nematode infested soil; this green manure crop is practically an eradicant of plant nematodes in soil. This sort of nematode or other disease control is within the means of people who cannot secure fungicides or nematocides.

There are some other kinds of problems that touch close to the poor consumer who is often unable to grow his own food, let alone produce some for sale. In the much eaten, cheap, stomach-filling yuca or cassava root, is often seen a post-harvest breakdown. This disease, described by Averre (4) and probably of physiological causation, is marked by bluish internal discoloration (accompanied by marked change in texture and taste of the cooked product). Symptoms of it cannot be seen on the root surface, but when it is broken or cut open, which cooks do when they buy these roots in the market, the root flesh shows the dark streaks.

The best of the "normal condition" of marketing yuca is such that the roots get to the cook as rapidly as possible after digging. In four or even two days after digging the internal breakdown may have started. When roots are dug and as soon as possible buried in cool moist soil, internal breakdown is said to be slower in appearance. However, these roots cannot be left for more than a few days before internal breakdown starts. Refrigeration at a little above freezing in a moist atmosphere is a successful storage pro-

Fig. 20-4. Preventing excess loss of henequen or cabuya, Agave sp., fiber by close harvest practice. Located where moist weather insured severe leaf spotting, at a period one cycle of harvest was deemed by labor unfit for cutting. These leaves were left, rotted and drooped around the plant stem, but never removed. Cycles of harvest were reinstituted and by harvesting at short intervals leaves were taken off with no loss from disease. Photograph courtesy of the late Jack Dalton of El Salvador.

cedure for weeks; this immediately increases the cost of the roots so that they become akin to luxuries and no longer have their place in the food pattern of many tropical peoples. Averre in later unpublished work studied methods of manipulating storage conditions for yuca roots and found he could successfully store these roots for over a week at room temperature in a gas tight (polyethylene) bag with a few green leaves mixed in with the roots. Breakdown of the leaves appeared to perhaps consume the oxygen in the bag and thus fully inhibit root vascular breakdown disease.

A distinct manner of disease control by husbandry methods is common in henequen or cabuya (*Agave* sp.) production. In Figure 20-4 is a hill planting with a cycle of leaf harvest starting. Moist climate had resulted at one time in a very severe disease season; harvest was not carried out and leaves drooped and fell about the plant stems. However, harvesting was reinstituted, and if followed out at short intervals, the effects of parasitic fungi are not severe enough to destroy leaf fiber content. No spraying is done in Central American agave fiber plantations.

Perhaps aside from the use of sprays the most spectacular extreme change in field work for crop disease control is the flooding of lands. This is rather a special means and depends on soil type, its under surface deposits, and levelness; it cannot be used everywhere. However, it is known in several instances (e.g., 26, 35 also see citations, and I have seen it in Florida where successful long time inundation of soil was used as a means of eliminating nematodes and fungus damping-off in celery seed beds, and for destroying sclerotia of disease fungi in strawberry fields). Flooding is known in Dutch Guayana as a regular practice: when sugarcane is excessively diseased the field is dyked and water turned in to remain several months. Afterwards it is drained, dried and replanted again to sugarcane.

This kind of treatment has been used (35) in ridding banana soils of *Fusarium* so that a wilt susceptible variety could be replanted in the old location. It is called by some "flood-fallowing" (an illustration of it is in Fig. 9-3) and although not a permanent solution, it was found to be of some success for a few years. All such flooding systems by companies require markedly good engineering skills but are at best palliative and the diseases eventually return.

Success of Informal Plant Quarantine

The agricultural activities of the American tropics are the most important industries of that great region. More cash comes from coffee and cacao than from any other products there. An exceedingly large sector of the population depends on these in spite of all the diseases that attack the plantations.

Planters realize there are many diseases of their crops and one of the ways that can help in control is quarantines. Combining data from just coffee and cacao crops shows there is a total of something like 600 diseases on them. This seems almost a staggering number. However, only about 5 per cent of the total diseases are of major concern in the American tropics.

Anyone knows who has worked with plantation crops, and this is especially true of coffee and cacao, that as such crops are studied every so often a few new diseases are discovered and reported. This continues to the present and is not surprising; fields of cultivated trees are continually being moved into new clearings surrounded by wild trees. These wildlings all have many new kinds of diseases hitherto unknown, and occasionally some move to coffee or cacao or other plantation crop. These new diseases are then often distributed in turn to other new plantings and finally find their way back to even the old plantings where the new parasites had never been before.

Another point of importance is that very good genetic and horticultural improvements are continually being carried on. As new planters take advantage of such improvements, the mechanics of crop-growing change, which make crops appear different. Even their susceptibilities and resistances change. This is specially well illustrated in the case of potato (as was shown by Cavia (13) and others). Along with higher productivity to meet competition, there comes a different more expert planter who finds it necessary to make his fields permanent rather than something he is driven away from easily when disease appears. Part of his problem is to keep most severe diseases from his plantation. This means exclusion, if possible, carried out by quarantines, which is difficult in the tropics.

Where travel is by a good system of surfaced highways, railroads, ocean-going ships, or airplanes, there is some chance to watch and to carry on good quarantine examinations; and through these efforts to exclude plant disease exits and entrances. On the other hand manning a successful quarantine program is a certain impossibility within heavy jungles laced either with pack trails or with tropical rivers traveled by rafts and dugout canoes. Moreover there is little common knowledge about plant disease quarantine needs among the indigenes, and the minimum-reading and illiterate peasantry have even less interest. Coffee and cacao planters have to do the best they can against the odds and there has developed over the years an informal program of strong action of their own, autonomously carried out and of special importance.

It is of considerable interest that there has grown up word-of-mouth quarantine understandings about certain crops. This is particularly true in Latin America where there is so much possibility of spread of some of the more important diseases from the other hemisphere, and from country to country within the Latin American scene. The most basic and ideal situation for prevention of plant disease spread by man would be by direct grower action. This is what is virtually occurring with respect to handling both coffee and cacao in the American tropics. Help from intelligent governmental sources is of inestimable value but quarantines are something the individual grower must be first to carry out.

As just one example: it has been known since the years of the last century that coffee can be wiped out by the dreadful rust, Hemileia vastatrix that was and is so serious in Africa and the Orient. It is known to be best carried 1) by man, 2) as lesions on living foliage of coffee, and 3) on either diseased branches of an old tree or on leaves of a diseased seedling. These, therefore, all planters of the Americas guard against. Growers know the seriousness of this dread disease and that once it reached the Western Hemisphere tropics it would be of utmost importance to coffee production. * This was well appreciated and fully publicized many years in all the American coffee growing communities. Through such an educational effort, it is believed safe to say that no coffee planter smuggled in any

*As is well known by now, Hemileia rust was discovered in Brazil in January 1970 (41 publ. 1970). I am convinced it came in by some natural means and not by human intervention.

coffee plants from any source in the eastern hemisphere for any secret planting on some coffee area in the America. This quarantine has been brought about by grower understanding, and serious advisement at intervals by plant pathologists.

For many decades, from 1903 to 1970 (41 publ. 1961 and publs. 1970), it appears there has been no causal or smuggled introduction of eastern hemisphere-grown coffee seedlings to the American tropics. If they had been coming in very often, there is no doubt that the rust, Hemileia vastatrix, would have appeared in the Americas long ago. The planters, through their well understood and well publicized educational programs against coffee rust introduction, are, it is believed, the reason H. vastatrix has been held away so long from the shores of the Western Hemisphere. When it finally did appear its occurrence follows a natural sequence in interval from the west shores of Africa to the east shores of South America, and suggest natural (wind or storm?) transfer across a relatively narrow width of the Atlantic Ocean.

In contrast to the coffee rust situation might be mentioned the coffee thread blight problem. This latter disease is very common on coffee and is caused by the fungus Pellicularia koleroga (41 publ. 1961). It is now reported from every Western Hemisphere coffee growing region. It is a less acute form of disease than rust, but it is at times severe and has been known in the Americas since 1868. It seems possibly to be of Oriental origin but where it first came from is not absolutely clear. It has a rather wide host range and is found in practically all parts of the tropical world where there is sufficient moisture. There is a possibility that it is indigenous to all these parts. However, it has been known in, for example, the tropics of India since immemorial times, that it first occurred in that country. It is now so well distributed that it can be found in a large number of places where at one time it had never been seen before. Under situations similar to this, with a disease so well distributed, it makes little difference whether there is mutual understanding or any rules or laws made against the possible importation of such a malady. The best that can be done is to work on it and apply control measures when the trouble appears.

The witches' broom disease, Marasmius perniciosus, is one of the most dreaded enemies of cacao. It was first found in Surinam; further attention has shown that it probably has its center of origin in jungle regions along the Amazon and Orinoco rivers. For a long time it was believed to be well confined within a circumscribed region. However, after some years it appeared in plantations in the island of Trinidad, a long distance from Surinam, and it was unexpectedly found by Muller (28) in parts of Venezuela far removed from those centers where it was believed to be confined. Since then it has shown a tendency to be slow in its spread but is now known in northern Brazil, Ecuador, Peru, and in one or two other countries.

One of the reasons this disease has moved so slowly is that cacao specialists know that with respect to long distances it can be carried from one place to another only in living budwood and seedlings. It is apparently not distributed readily in seeds. Realizing these things, not only official government agencies, but especially the cacao producers and growers themselves frown very much on carrying about cacao plant materials; there is virtually no traffic of this sort, except by recognized scientists in cooperation with quarantine work.

There are numerous serious diseases of tropical plants that at present are not in the American tropics and are absent solely because they have thus far been excluded. Some of it may be luck, but in many cases this has been because travelers have been intelligent in what they have chosen to carry with them. It is of interest to indicate a few unwanted diseases, other than those that have been mentioned thus far in this chapter. In addition to the well known dreaded common coffee rust, _Hemileia vastatrix_, there is another species, _Hemileia coffeicola_, that is in the Eastern tropics. The latter, at the moment, is pretty well localized in higher parts of Africa. It is a very serious problem in the finest quality, coolest coffee growing regions of Central Africa. It would be a terrific problem in the American tropics and it is hoped that this will not get to our region. There is also much fear that the _Colletotrichum_ coffee berry disease (CBD) may eventually be introduced into the Western Hemisphere.

In general there is grower and folk understanding about the fact that _Colletotrichum_ or anthracnose diseases of tree crops such as cacao, anona, mango, or coffee are often and easily spread by the action of man. This fungus has a complex life history in the field in the tropics, and as is shown in accompanying Figure 20-5 man has an important

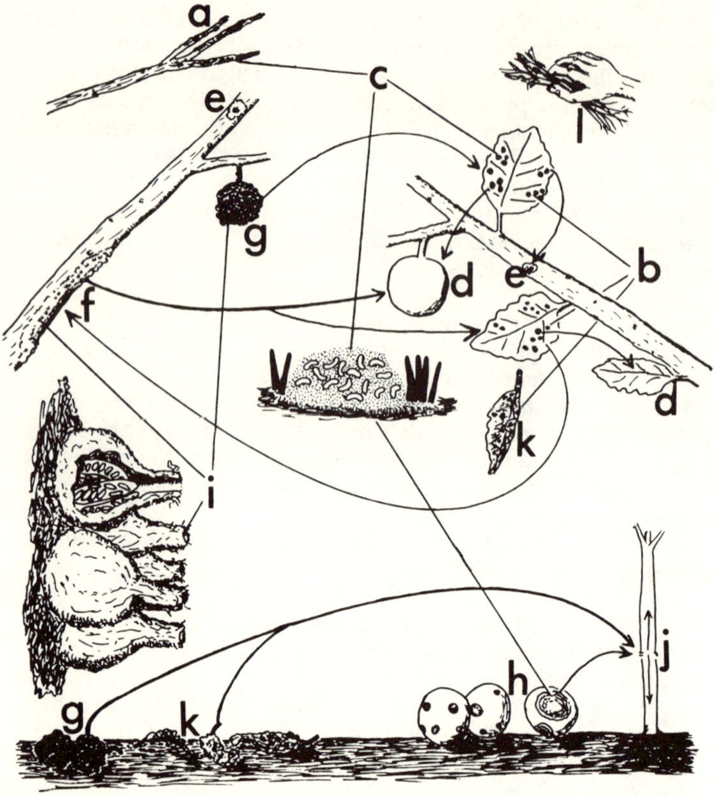

Fig. 20-5. Anthracnose, Colletotrichum = Glomerella, of such tree crops as cacao, mango, coffee, and anona: a. twig dieback; b. leaf spots; c. within spots are found the acervuli or special bodies (here much enlarged) on which develop a sticky drop filled with spores and these spores when dislodged adhere to any surface; d. rain splashes spores to healthy tree parts; e. leaf scars are often infected and the fungus may become systemic in the branch; f. superficial branch cortex wounds are infected and lesions formed that expand slowly; g. dry season fruit mummies are often seen after fruits rot in the rainy season; h. recent rainy season fruit rot causes drop and crop loss; i. the perfect or Glomerella stage (here much enlarged) showing perithecia and spores in sacks or internal asci; j. occasionally seedlings are infected through the cotyledonary connections and the disease may grow slowly inside and cause no outward effect; k. leaf drop from anthracnose; l. often important disease spread by hand of man.

place in its spread and distribution. The severe coffee berry disease of Africa is caused by a *Colletotrichum* and quarantines, it is hoped, will keep it out of the coffee countries of the Neotropics, if not forever at least for a long time.

Another disease of coffee that we have been fortunate enough to escape thus far is a virus that causes stem pitting and a bottleneck trunk of coffee. This is being much studied and is believed to be as much a potential danger to the Western tropics as coffee rusts.

It is well known that the disease which has caused more problems of cacao production than anything else in Africa has been that of the virus that is called swollen shoot. This results in the destruction of millions of trees in an affected region and has been controlled in some places to a considerable extent by rogueing and careful elimination of the insect vectors. However, it is something that cacao people are very much afraid of and is one of their reasons for the closest cooperation with inter-hemisphere quarantine programs. If there are no infected live trees or budwoods that have been secured in another part of the world and brought over to the American tropics, introduction of this disease may not occur.

In the case of hevea rubber everyone knows that the South American leaf disease *Dothidella ulei* is not in the Eastern Hemisphere. Contrary-wise, we have there a serious mildew caused by a species of Oidium that has been found once in Brazil but now does not occur in the West. By a careful system of planter-controlled, autonomous, quarantine these diseases have not been sent from one hemisphere to another so far.

In the case of lettuce we have things in the Orient that we do not want in the Occident. There are some oriental leaf rusts, for example, *Puccinia lactucae*, that are not in the Western Hemisphere. By careful handling of matters it is hoped this may not come to the American tropics for a long time. And this is quite important because lettuce is highly valued as a food crop in the American tropics and a rust would have a very serious effect.

In the Philippines there is a *Cylindrosporium* that attacks sweet potato. This disease is severe and if it were introduced into the Western Hemisphere tropics where the sweet potato is returning to importance as a food crop, it might be a great setback. There is also a sweet potato rust, Uromyces ipomeae, in Africa and every effort is being made to keep it away from our Western tropics. With respect to cotton there are some African leaf diseases we do not want in the Americas, for example, the one caused by Phoma roumii. There are already quite enough diseases on cotton in the American tropics.

In an agricultural region which depends so much on shade culture as we do in the American tropics, if shade trees are attacked by a disease it is necessary that attention be paid to it. In the case of the Grevillea tree there is a serious rust caused by Aecidium warneckenum in the Eastern tropics and it would be a great pity for it to be introduced into the Western Hemisphere.

In the case of cucumbers and melons there is the rust, Puccinia cucumeris, and the smut, Ustilago cucumeris which are great problems in the Old World. They should not be introduced here. There are foreign species of tree rusts that attack Cedrela. One is Triphragmium cedrelae and the other is Uromyces cedrelae. These develop in Japan and Java respectively. Along with these is a mildew on Cedrela, from infections of Uncinlua sp. These are all troubles that we need to keep away from the American tropics and they can well be avoided by careful and well publicized quarantines.

Tea is becoming more and more important as a crop in the West and is being tested in over ten countries. A great deal of attention should be given to the quarantine of tea diseases which might be introduced. One of the most important of these is of course Exobasidium vexans, which can be found in the southern United States but which should not be introduced to the tropics. There are as well some bacterial diseases, viruses, a leaf spot caused by Chateophoma, and a stem blight from Aglosporium on Orient-grown tea. These are all still confined to the African and Oriental tropics, and by careful popularly carried out quarantine systems, can also be kept away from the American tropics.

Literature Cited

1. de Alencar, J. 1941. Podridãe do pé dos citrus. Ceres 2:488-496.

2. Alvarado, J. A. 1940. Las sequías prolongados y su efecto nocivo sobre la fisiología del cafeto. Café de El Salvador 10:219-247.

3. Arentsen S., S. 1945. Enfermedades de los citrus en Chile. Simiente 14(3-4):6-8.

4. Averre, C. W. 1967. Vascular streaking of stored cassava roots. Internat. Symp. Trop. Root Crops, Univ. West Indies, St. Augustine, Trinidad. Proc. IV:31-35.

5. Atkinson, R. E. 1944. Reports on diseases of peanut: southern blight and Diaporthe on peanuts in Virginia. Plant Dis. Reptr. 28:1096-1097

6. Aycock, R. 1966. Stem rot and other diseases caused by Sclerotium rolfsii, or the status of Rolf's fungus after 70 years. North Carolina Agr. Exp. Sta., Tech. Bulletin 174, 202 p.

7. Baker, K. F. 1947. Seed transmission of Rhizoctonia solani in relation to control of seedling damping-off. Phytopathology 37:912-924.

8. Batista, A. C. 1947. Principias doencas das plantas, em o nordeste. Bol. Secr. Agr. Pernambuco 14:5-46.

9. Boosalis, M. G. 1956. Effect of soil temperature and green-manure amendment of unsterilized soil on parasitism of Rhizoctonia solani by Penicillium vermiculatum and Trichoderma sp. Phytopathology 46:473-479.

10. Briton-Jones, H. R. and R. E. D. Baker. 1935. Control of grapefruit diseases by cultural methods in Trinidad. Trop. Agr. Trinidad 12:119-125.

11. Butler, E. E. 1957. Rhizoctonia solani as a parasite of fungi. Mycologia 49:354-373.

12. de Camargo, A. P. 1961. Balanço hídrico no Estado de São Paulo. Instituto Agronomico, Boletim 116, 15 p.

13. Cavia, C. E. 1951. Influencia de subregiones de la

region de Rosario sobre papa Katahdin para semilla. Boll. Prod. Fomento Agr. Argentina 3:22-26.

14. Chitwood, B. G. and C. A. Berger. 1960. Preliminary report on nemic parasites of coffee in Guatemala, with suggested ad interim control measures. Plant Dis. Reptr. 44:841-847.

15. Crane, J. C. and F. L. Wellman. 1950. Edad del henequen en relaciŏn con las caracteristicas de la fibra. Turrialba 1:74-77.

16. Fawcett, H. S. 1936. Citrus diseases and their control. McGraw-Hill, New York, 696 p.

17. Garrett, S. D. 1960. Biology of root-infecting fungi. Univ. Press, Cambridge, England, 293 p.

18. Grant, T. J. 1962. The role of plant pathology in tropical plantation agriculture. Phytopathology 52:930-936.

19. Johnston, J. R. 1913. The relation of cane cultivation to the control of fungous diseases. Porto Rico Sugar Prod. Asoc. Exp. Sta., Circular 3, 13 p.

20. Leben, C. and G. C. Daft. 1965. Influence of an epiphytic bacterium on cucumber anthracnose, early blight of tomato, and northern leaf blight of corn. Phytopathology 55:760-762.

21. Lellis, W. T. 1956. Influencia de la cobertura en la incidencia artificial de Phytophthora palmivora en hojas de plantulas de cacao. Cacao, Costa Rica 3:16.

22. Loos, C. A. 1961. Eradication of the burrowing nematode, Radopholus similis, from bananas. Plant Dis. Reptr. 45:457-461.

23. ________., and S. B. Loos. 1960. Preparing nematode-free banana seed. Phytopathology 50:383-386.

24. Marshall, K. C. and M. Alexander. 1960. Competi-

tion between soil bacteria and Fusarium. Plant & Soil (US) 12:143-153.

25. Meredith, C. H. 1944. The antagonisms of soil-microorganisms to Fusarium oxysporum cubense. Phytopathology 34:426-429. --. 1946. Soil actinomycetes applied to banana plants in the field. Phytopathology 34:983-987.

26. Miller, P. F. 1953. Lowland rice culture as an economic control of Sclerotinia and root-knot nematodes. FAO Plant Prot. Bull. 1:183-187.

27. Moreira, S., A. S. Costa, and T. J. Grant. 1954. Methods of identification and control of tristeza disease of citrus. Bragantia 13:223-236.

28. Muller, A. S. 1945. Plant pathology in Latin America. (pp. 169-171). In F. Verdoorn (ed.), Plants and plant science in Latin America. Chronica Botanica, Waltham, Mass., 349 p.

29. Ochse, J. J., M. J. Soule Jr., M. J. Dijkman, and C. Wehlburg. 1961. Tropical and subtropical agriculture. Macmillan, New York, 2 vols., 1446 p.

30. Rambouts, J. E. 1953. The microorganisms in the rhizosphere of banana plants in relation to susceptibility or resistance to Panama disease. Plant and Soil 4:276-288.

31. de Segura, C. B. 1965. Enfermedades de Cultivos Tropicales y Subtropicales. Editorial Jurídica, S. A., Lima, Peru, 439 p.

32. Stakman, E. C. 1964. Opportunity and obligation in plant pathology. Ann. Rev. Phytopath. 2:1-12.

33. Steiner, G. and E. M. Buhrer. 1964. The plant-nematode problem of the American tropics. Univ. Puerto Rico, J. Agr. 48:69-100.

34. Stevenson, J. A. 1918. The green muscadine fungus in Porto Rico (Metarrhizium anisophilae (Metsch.) Sorokin). J. Dep. Agr. Porto Rico 2:19-32.

35. Stover, R. H. 1962. Fusarial wilt (panama disease)

of bananas and other musa species. Commonwealth Mycol. Inst., Phytopath. Paper 4, 117 p.

36. Thorne, G. 1961. Principles of nematology. McGraw-Hill, New York, 553 p.

37. ________., and J. Román. 1964. Fallow controls nematodes in tomato production. J. Agr. Puerto Rico 48:163-164.

38. Toussoun, T.A. and W.C. Snyder. 1961. Germination of chlamydospores of Fusarium solani f. phaseoli in unsterilized soil. Phytopathology 51: 620-623.

39. Tullis, E.C. 1934. The root-knot nematode on rice. Phytopathology 24:938-942.

40. Walker, J.C. 1963. The future of plant pathology. Ann. Rev. Phytopath. 1:1-4.

41. Wellman, F.L. 1930. Plant disease-free seedpieces to control panama disease of bananas. United Fruit Co., Trop. Res. Lab., Circular 21, 30 p. --. 1930. Nature of spread of panama disease (Fusarium cubense) in rhizomes of gros michel banana plants. United Fruit Co., Trop. Res. Lab., Circular 19, 21 p. --. 1931. Progress of Fusarium wilt inside the rhizomes of banana plants. Phytopathology 21:121. --. 1938. Poverty of human requisites in relation to inhibition of plant diseases. Science 87:64-65. --. 1945. Indian gardening in El Salvador. Agriculture in the Americas 5:47-49; 57. --. 1951. Control del ojo de gallo, Omphalia flavida, por medio de la deshoja de cafetos enfermos. Suelo Tico, Costa Rica 5:42-51. --. 1951. Tree defoliation for control of Omphalia disease of coffee. Phytopathology 41:38. --. 1961. Coffee botany, cultivation and utilization. Leonard Hill, London; Interscience Publ., New York, 488 p. --. 1969. More diseases on crops in the tropics than in the temperate zone. Ceiba, Honduras 14:1-12. --. 1971. Plant diseases: an introduction for the layman. Natural History Press, New York, 130 p.

42. ________., and V. K. Toole. 1961. Coffee seed germination as affected by species, diseases and temperatures. Carib. Reg., Amer. Soc. Hort. Sci., Proc. 4:1-4.

43. Williams, L. E. and A. F. Schmitthenner. 1962. Effect of crop rotation on soil fungus populations. Phytopathology 52:241-247.

44. Yarwood, C. E. 1962. Some principles of plant pathology. Phytopathology 52:166-167.

Chapter XXI

CHEMICAL* DISEASE CONTROL OF TROPICAL CROPS

Summary

The use of chemicals to control tropical plant diseases requires special study. The crops and ecology there are different from those in most of the temperate zone. In warm-moist climates, seed treatments give good results if judiciously carried out. Chemicals of special danger to human beings are inadvisable to use. Compounds suited to the conditions are such as phygon, captan, spergon, and copper-containing materials. Damping-off of seedlings, demonstrated as occurring in the wild, comes from 20 or more organisms, but *Rhizoctonia* and *Pythium* species are

*Chemical names and designations of fungicide compounds are not capitalized herein; wherever a proprietary or trade name is used in this book, it is as an example and its use should be construed as including any other compound with the same content. Certain popular seed treatment and other chemicals have been recently excluded by governments as health dangers: these are not included in this book except in mere mentions in discussion.

Furthermore, I insist that all chemicals, even such common ones as lime or formaldehyde or sulfur or bluestone, are poisons. The grower who has to use any spray or treatment chemical must read and then follow all precautions on the label to prevent hazard to his or his customers' health.

most common on cultivated crops. The recommendations for those are formaldehyde and other soil treatments, with, additionally, fungicides dusted and watered onto the soil around dying seedlings. In caring for cuttings, setts, corms, roots and such, propagating materials are injured with mercurials and some strong fungicides. These are not usable, although ferbam, captan and like organics work well.

What is to be done about spraying and what is a good spray are questions of much concern in our tropics. There is a wide range of diseases and crops involved and of course different diseases require different treatments. Spore presences may be detected by trapping techniques, and then identified. This information, combined with considerations of when tree crops flush and the effects of seasonal fluctuations, determine best times for spraying. The classic bordeaux mixture spray on bananas to control the devastation of sigatoka leafspot has been markedly successful. Its use is based on highly developed, well-engineered spray programs; it was done on bananas on a greater scale than ever before on any other one crop in all the world. The bordeaux mixture, used for a long time, has serious disadvantages. Replacing the laborious methods required with bordeaux are newer air-application techniques; these are well along, combining engineering with scientific study on host and parasite reactions to solve the worst difficulties in spray over large expanses. Orchard oils are used; some have had organic and other spray chemicals added in emulsions. There are problems of special importance in the tropics: extra good from stickers, the role of systemics, the place of an insecticide-fungicide, and the problems found in using eradicants.

Introduction

> -- No matter how far we progress in plant resistance, there will always be need in the tropics for chemical applications in the control of plant diseases --

There are in the tropics large numbers of forever-mutating, varying parasitic organisms. Likewise there is the ever-present certainty that new parasites will appear (221, publ. 1969). In addition, old and untested crop plants are being introduced into virgin areas. No matter how well disease resistance selection is done, the plant geneticist will not be able to stop all disease losses on resistance bases alone.

It is well appreciated also that in tropical countries there are burgeoning fields of many crops that are broadly described as "disease resistant"; some of them have been grown a long time. Examples are: mosaic resistant sugarcane (in the western tropics); fusarial wilt resistant banana (most grown at present in Central America); rust resistant coffee (in the Orient, but being introduced into the Occident since rust has appeared); footrot resistant sweet orange (in numerous parts of the American tropics); rust resistant maize (introduced into Africa from the Central American tropics where the resistance or tolerance is common and ensures harvests now coming back for re-trial in the Neotropics); blast resistant rice (from the Orient, being used in tropical America); and many others. Resistant as these are against the specific disease intended, there are other troubles for which all need chemical treatments.

As more fields of any variety of crop, whether "tolerant," or "resistant," are grown, they inevitably have to be intensively managed. When this happens, the habits of management are changed in the old systems, where once there may have been a non-fertilized weed- or bush-infested field, with largely unpruned crop plants wide-spaced and irregularly arranged, harvest was nearly from the wild. Production was perhaps adequate for the time, sometimes heavy but every other year lighter, which gave a very inconsistent crop history. With modern changes the plants or trees are grown much differently. Closely spaced and well pruned, their cultivated soil is fertilized, mulched, and, when needed, irrigated. Mechanizing is practiced and fields are forced for high and consistent productivity. Under these conditions when infections appear, sprays and other chemicals are a most providential help for the agriculturist in his fight against crop disease.

This book has not been prepared either as a tropical plant disease compendium or as a manual of methods of control. However, some examples of chemical controls in addition to those already given in previous chapters, are indicative of the kinds of diseases combatted and seed treatments or sprays used; these controls are suggestive of lines of further study.

It is not wise for the tropical man to go unhesitantly to temperate zone-tried-and-accepted chemical treatments. There are special problems in the tropics, respecting the effects of such matters as the non-winter, epiphytic bacteria and fungi occurring as natural but invisible growths on leaf surfaces, foliage that never drys, tree crops with no truly annual dormant period, soils with massive numbers of bacteria and fungi that nevertheless change disinfection problems. In some places in the tropics, rains may fall almost every day in the year; there are seeds that need disinfection that are never truly dry nor dormant; and there are difficulties in chemical storage and transportation and spray machine maintenance.

Although bordeaux mixture was used in 1883 (221 in 1961 Publ.) in the tropics (Ceylon) almost since it was first described, as more experience about it was gained in the moist western tropics (221 in 1955 publs.), the transport and storage of it have proved important deterrents to its use far from the temperate zone. Several hundred references have been seen by me on tropical plant spraying with sulfur mixtures and bordeaux mixture. They came out during the years from the 1880s into the 1940s; some may be of historical interest but many constitute mere mentions of these sprays. In some cases, bordeaux had been suggested in reports by individuals who had only read about the spray; these reports have no place in the present chapter. (For anyone wishing sources for more citations on tropical sprays see: Barnes and Allen 19, Brandes 33, Calpouzos 39, Fernandez Valiela 91, Fulton 97, Galli et al. 99, Marchionatto 146, Otero and Cook 169, Ruehle 192-4, de Segura 200, and Wellman 221.) Rather independently of the temperate zone, South American workers used crop sprays with success (e.g., 89, 91, 103, 107, 153, 157, 173, 178) and a good amount was done in the 1930s and 1940s. Some of it had wide acceptance soon after publication, and has had important impact since.

Considering the subject of this chapter, the reader should perhaps be given not only a review of past problems but also those of the present and what is wished for the future. This sort of a review would consist of a large enough series of subjects for a book by itself. This may be done by someone 50 years from now. Meanwhile, the situation is in a great state of change at present, and the challenges for the pathologist are, it seems, almost without limit. The need for more and better chemicals against disease is expanding. And what is learned in the tropics will have important bearing in the temperate zone.

In a recent address entitled "Plant Disease Control, 1970-1980," Dr. D. E. Ellis (85) has made succinct remarks that forcibly comment on chemical therapeutic practices in plant pathology. One strong point he makes is that pathologists perhaps should not use the word "pesticide" for the material employed against a disease and recommends as better "plant disease chemical." Other points of special significance to this chapter are as follows:

> In the future chemicals may play an ever greater role than they do today. Their increased use may be one of the most efficient means of disease control. ... Crop rotations and other cultural measures may not be practical on many of the larger farms of the future. If this is true, then chemical control will of necessity have to be substituted.
>
> We need to have and understand the use of chemicals if we are to be in a position to advise [anyone], to feed expanding populations .... If we as plant pathologists do our job well, both in educating the public and in research, ... I believe that [in the future] we will have newer and better agricultural products, and newer and better disease control measures. Our recent knowledge on epidemiology and effect of environment on disease development, will enable us to use agricultural chemicals more intelligently. ... Newer chemicals are being developed .... The field of systemic fungicides has only recently come over the horizon in a successful manner [and] they will replace many of the protectant fungicides we now use to control plant diseases.

Seed Treatments

While the fundamentals of seed treatment to eliminate diseases in crop plants are the same the world over, in the tropics the disease situation is likely to be intensified. In warm-moist climates, normal in the Neotropics, seeds which are held a long period before planting gather moisture through hygroscopic effects. This results in growth or microorganisms on surfaces, even if they are mainly saprophytes. I have seen in maize, grain sorghum, coffee, and rubber that these act as a serious detriment to seed germination and in pre-emergence damage on seedlings.

Under most small farm and plantation conditions, it is the grower himself who must expend the necessary surveillance to insure a supply of disease-free seed. Where successfully carried on, this means all operations dealing with the seed program are given intensive, personal, and strict scrutiny, if not by the planter-owner, then by the best foreman. Workmen's eyes are exacting and careful about symptoms of disease. With training and specialization, workers can recognize what should be discarded and what should be kept, to have the result of only the most healthy materials used for planting. As modernization increases in more distant localities, agriculturists realize the danger from planting disease-affected seed. Where the seeds are available, they now buy the disease-free-guaranteed product of the seedsman's profession.

With the use of modern developments, the Neotropics becomes less a place of agricultural gamblers and settles into being a needed, dependable, and perpetually productive realm. During the time from about 1955 to 1970 there has been an increasing tendency to test and bring in new seeds, disease-treated, from certified sources of dependable renown developed in Europe, the United States, and Canada. By this importation method many ornamental and florist gardeners and vegetable growers as well have had a guaranteed source of healthy seed and have made progress in establishing businesses. There are also various cereals, legumes, grasses, pasture plants, and other types of crop seeds--all known to be disease free--that are brought in.

In decades gone by, this kind of thing has had such well recognized good effects that certain tropical seed growers now producing seed for and from the tropics, advertise their seeds as known to be disease free, and where necessary treated with protectant chemicals. These specialists are working to take advantage of the growing demand for the best "home grown and dependable seed." However, they find that some of the seed treatments so long a standard in the temperate zone are not always the best for applications in the tropics.

Fresh dry, newly harvested tropic-grown seeds are put in bags or special containers and set aside for storage.

Even tins with friction-held tops, let alone cloth and paper sacks, eventually show moisture effects on seeds. Tradition, or habit, respecting seed storage has been taught and established from temperate zone experiences, where, as in such wide regions as most of Europe and most of the United States, seed viabilities of different crops are established and optimums known. There seeds keep about equally well in stations often a few thousand kilometers apart. However, as an example of tropical effect, there are small countries in the Caribbean region, where in a dry climate station at a western-slope (Pacific side) location, maize and sorghum seeds are readily stored through two years without losing much viability; in a more moist climate less than 100 kilometers away on an east-slope (Atlantic side) station, the seed viability goes down drastically during the first six months.

Such poor germinating seed, wrapped in moistened and sterilized filter paper and incubated for two weeks in a moist chamber, yield growth of such fungi as species of *Aspergillus*, *Penicillium*, *Nigrospora*, *Haplographium*, *Fusarium* (*roseum* (?)) and *Mucor*. Occasional examples of *Streptomyces* have been recovered, as well as *Cephalosporium*, yeast and some bacteria. Washings and scrapings from all of these did not produce seed bed damping-off on newly emerging seedlings of maize or sorghum. Endosperms from dead seeds did not contain any of the organisms recovered from the surfaces of the seed coats. However, treated seeds, partly freed from these microepiphytes, live much longer it appears than if the organisms are left untouched (from unpublished studies untreated seeds are killed twice as fast as treated.)

Disease contaminated seeds have to be treated, especially when it is known they are affected with certain well recognized organisms, either or both on the outside or the inside coverings. For a long time, South Americans (11, 66, 219, 46, 79, 93, 211, 217, 146, 148, and others) and as well workers in Mexico (such as 87, 147) have applied standard control treatments to eliminate cereal smuts. Many times compounds containing mercury are, or have been, employed with success, but other chemicals are replacing (196) the more dangerous compounds. I consider mercury compounds unadvisable in any disease control program for the tropics, and I do not recommend them. North American and European governments are more and more critical of mercury, and mercury uses are specially hazardous with illiterate workers. One year I harvested highly valued selections of a grain sorghum, finding presence of Sphacelotheca holci and S. sorghi, and in my office all seeds were treated with a red colored organic mercury compound and all bags were carefully marked with skull and crossbones. The next morning one 5-kilogram bag was missing, and immediate, urgent word was sent to all workers about the lethal nature of treated seed. The second day the sack reappeared at my office door. No questions were asked, and that was the last time I used that poison, or any other organic mercury.

Some examples of seed treatment recommendations I have made, that were reasonably successful in moist-warm tropics, are as follows.

Maize seed from fields in which seed borne maladies occurred: smut (Ustilago maydis), stalk and ear dryrot (Diplodia spp.), pokka-boeng (Fusarium moniliforme), common leafblight and blast (Helminthosporium spp.), rusts (Puccinia sorghi, P. polysora, Physopella zeae), and cob-rot (Nigrospora oryzae). It was specified that dry seed was to be treated with phygon or captan dust at the rate of 25 grams per 30 kilograms of seed, well mixed, and planted in moist but not wet soil.

In sorghum, grown for grain, seed was from fields in which occurred: surface head disease and infections (organisms recovered were Curvularia sp., Nigrospora spp., Colletotrichum sp., Cladosporium spp., Fusarium roseum-type, and Penicillium or Aspergillus spp., two smuts (Sphacelotheca spp.), the pokka-boeng (Fusarium moniliforme), rust (Puccinia purpurea), and leafblast (Helminthosporium sp.)). Recommended was that the seed be dried thoroughly and treated with amounts of spergon sufficient to coat all surfaces.

Either lima or frijole types of bean seed were secured from fields in which could be found: anthracnose (Colletotrichum lindemuthianum), angular leafspot (Isariopsis griseola), rust (Uromyces phaseoli races), stem rot Sclerotinia sclerotiorum), yeast spot (Nematospora sp.), wet weather pod rots (Rhizopus sp., Cladosporium sp., Fusarium spp.), and infrequently the small leafspot (Cercospora?). Dry seed was dusted with ferbam, at a rate "enough for good coverage."

Tomato seed extracted from fruits was dried on, and the seeds adhered to, strips of new white paper toweling. There were a number of seed-borne diseases in the fields, the most common of which were: leafmold (Cladosporium fulvum), dark fruitrot (Diplodia sp.), peduncle canker (Pestalotia funerea), leaf blight and firm decay of fruit (Phytophthora infestans), stem and branch rot (P. parasitica), bacterial spots on fruits and leaves (Xanthomonas vesicatoria, Erwinia spp., Pseudomonas tomato or Bacterium punctulans), and a yeast-caused fruit rot. The strips of paper with adherent seeds were dusted thickly with either dry spergon or copper oxide, and excess dust was knocked off by light tapping. Seeds were than scraped off the toweling, and stored in envelopes or planted immediately.

With such large and meaty seeds as those of coffee, avocado, cacao, tea, and hevea rubber--where it was desirable to take steps to control seed infestation effects by such organisms as: Colletotrichum or Gloesporium spp., Pestalotia sp., Diplodia sp., Aspergillus sp., Penicillium sp., Cladosporium herbarum, and the like--if being shipped or stored, a good recommendation has been to dust carefully with copper oxychloride allowing only that which sticks to dry seeds to remain, and bury in ground charcoal. Tea and rubber seeds are delicate and specially susceptible to adverse seed treatment chemical poisoning. Wood charcoal is stored in the room under shelter, and after at least two months of standing in the normal, tropical moist-air conditions, it is ground to a crumbly state. Fungicide-dusted seeds of any of the large-seeded bush or tree crops often keep for a few weeks in charcoal. The containers are never sealed as this restricts a gaseous interchange between the large, fast metabolizing seeds and the air.

It was found (221 publ. 1955) that 20 per cent moist charcoal was best for long distance shipment of Hevea rubber. Experiments indicated that coffee seeds could be stored and kept viable for over a year in charcoal in a chamber with a moisture content of air at about 50 per cent, kept at a storage temperature of about 10° C.

It has become of much recent concern in our tropics to know if the coffee rust, Hemileia vastatrix, could be disseminated with coffee seeds. This seems very unlikely. Hemileia rust spores do not survive when planted in soil, and seeds germinate rust free in that soil. Also these rust spores, if under moist and confined conditions of storage, seem short-lived. Tests were carried on and infections failed using abundant numbers of the rust spores from moist material that had been kept less than ten days on diseased lesions on coffee leaves and fruits.

In the case of smuts (*Tilletia* spp.) of small grains, I have had no direct seed treatment experiences. When asked for advice I have suggested liberal dusting with such fungicides as ferbam, spergon, yellow cuprocide and red copper oxide.

Under warm conditions some seed treatments I helped to carry out were harmful to the crop. Dusts of both spergon and semesan on bean and pea seeds caused seedling injury in moist and warm sandy soil, especially during long periods of heavy dews but no rains. Soaks on almost any crop seed, even for as little as 5 or 10 minutes using mercury-containing compounds, did not act well. Although seeds were rinsed and dried, often seedling tissue was killed in warm soil. When phygon and semesan soak was used with maize seed in a field one season, germination was delayed. Ferbam did not injure the seeds. Seeds from the field were dug during slow germination, and were seen in many instances to have distinct phygon injury to coleoptile and to primary root-tip. The penetrating mercury seed treatment did more injury in the tropics than in the temperate zone. This bad effect is specially notable in low-lying districts where it is hot and there is no true dryness during the part of the year when seeds are planted.

The potential values of seed treatments for disease control in the field have been recognized and reported on for many years in the American tropics. Some crops have had special such attention, examples being bean and tobacco (e.g., 10, 129, 199), cotton (e.g., 95), beans and peanuts (e.g., 61, 99, 142, 195), and tomato, pepper and eggplant (e.g., 8, 102).

Applications in Soils

It is well known in plant pathology, and the fact is taught in both tropical and temperate zone universities, that damping-off of young crop plants is serious. Symptoms are described as a soupy rot and are often known literally as "sancocho" or soupy, in Spanish. Little reporting has ever been done about the actual origin--that is, first sources--of damping-off organisms. In one jungle plot in Los Diamontes, Costa Rica, I found seedling damping-off where it seemed impossible to have been introduced by man. A similar area was found near Turrialba, Costa Rica.

When carefully searched for, the disease was seen later in the jungle several times. Revisits showed that some seedlings survived in spite of much disease. This disease as seen in the jungle was in areas of poorly growing and crowded jungle tree seedlings. Some of these soon died, and others as weaklings were later consumed by partly parasitic fungi, by nematodes, and also by the hordes of voracious mites and other animalcules that live on decaying debris. Plants that survive the conditions seem sufficient to carry on the species; those that disappear into the earth are not much thought about; percentages of attack are unknown in the wild: there are no workers counting wild seeds that fall and then comparing those numbers with the numbers of seedlings that grow to healthy trees, shrubs, or herbs. Isolations and microscopic examinations showed the causal fungi to be mostly *Rhizoctonia* or a *Pythium*. A comparison between these two organisms is presented in diagrammatic sketches in accompanying Fig. 21-1.

Perhaps we in the tropics are overwhelmed by our prejudicial knowledge of temperate zone publications. It is possible to make too inclusive a statement such as that by Wellman (221, publ. 1961) inferring that seedling coffee damping-off "is incited [perhaps entirely] by the fungus *Pellicularia filamentosa*. ... Losses may be very high, as much as 60 to 80 per cent." In work since, this organism is certainly the most commonly found, but more intensive attention was carried on subsequently with typical seedling collapse of coffee, from which were isolated other fungi: *Colletotrichum coffeanum*, *Sclerotinia* sp., *Fusarium roseum* (very long spores), *F.* sp. (possibly *oxysporum* but did not invade vascular elements of host), *F.* sp. (with very short and squat spores), *Diplodia theobromae*, *Pythium* sp., *Chaetomium* sp., *Penicillium* sp., and a Basidiomycete which never developed a sporophore. All of these were induced to attack coffee seedlings in glass containers under partially controlled conditions, and all caused seedling collapse of lettuce. In all instances, no matter what the fungus, damping-off was stopped by liberal dusting with ferbam and spergon.

It is to be noted especially that several seedling dis-

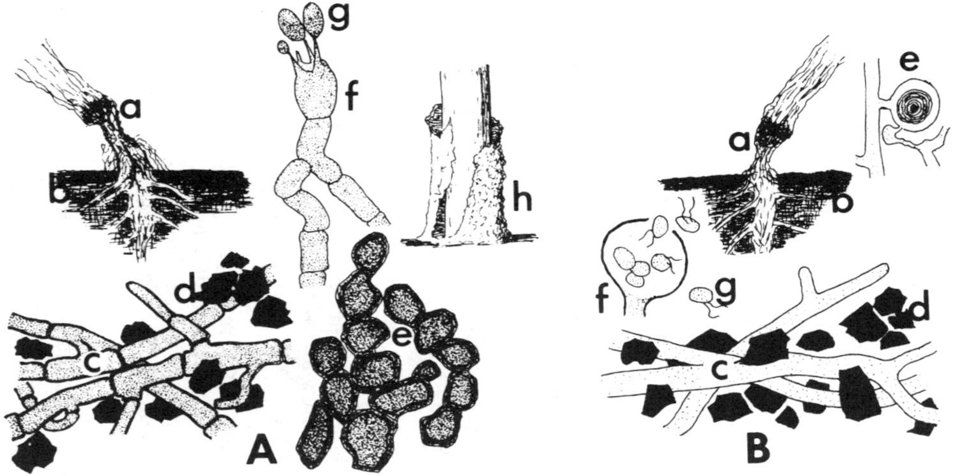

Fig. 21-1. Diagrammatic sketch comparisons, between two different damping-off organisms. A. A Rhizoctonia or Thanatephorus: a. infected base of young collapsed seedling; b. parasite infestation is in top layer of soil; c. magnified aspect of fast-growing, vegetative, and parasitic mycelium; d. soil particles; e. heavy walled resting stage of the fungus, also the HOLDOVER condition; f. basidia on sporulating hymenial growth (see h); g. basidiospores; h. superficial, non-parasitic hymenium, wrapping around a stem. B. A Pythium: a. infected base of collapsed young seedling; b. parasite is in top layer of soil; c. magnified aspect of the non-septate, fast growing, parasitic mycelium of this fungus; d. soil particles; e. reproductive stage where sexual fusion occurs is non-parasitic; f. spore sack; g. motile swarm spores that swim in and are carried by water.

ease fungi are recognized by Marchionatto (146, publ. 1944). He remarks that: "Damping-off disease is very severe.... [It] can destroy 100% of the seedlings attacked ... and can be caused by different fungi (*Pythium*, *Fusarium*, *Rhizoctonia*, etc.) resulting in very similar symptoms."

Parasitic and weakly-parasitic fungi occur in abundance in tropical soils and in the debris at the soil line. At different times many isolations and collections were made and soils and isolates were tested for damping-off. Damping-off from "uniquely virgin" jungle soil was induced by me in many lettuce and radish seedlings.

It was helpful to learn that standard, modern soil treatments that controlled pre-emergence death and damping-off of seedlings from either *Rhizoctonia* or *Pythium* were equally efficacious with 20 other organisms that cause seedling death. Highly sophisticated soil treatment studies (222), although carried on in California, will undoubtedly have much impact in truly tropical countries, far south of the subtropics. Tests are well known in which formaldehyde drenches used in seed beds are good control measures.

A seed bed is soaked deeply with diluted formaldehyde solution: 1 part of liquid formaldehyde to 50 parts of water. The soaked bed, while still wet, is covered to keep in the gas, with three layers of freshly cut banana or plantain leaves, arranged in shingling fashion and held down firmly with stones or other weights. After the third day the leaves are removed and the bed allowed to ventilate for two weeks when it can then be cultivated and seeds sown.

In some cases where such materials as larvacide, chloropicrine, dowfume, methyl bromide and the like are available--liquids which become gas in the soil--are poured into holes made in seed beds. Rates are those manufacturers suggest.

Slow-sprouting seeds, such as old, dry, stored coffee, have been planted in known *Rhizoctonia*-infested soil which after being firmed over the seed was immediately wetted with dilute formaldehyde solution. The bed was not covered and was left to be rained on. There was no injury of the seedlings, and no development of pre-emergence rot or damping-off. Untreated rows in nearby beds showed damping-off. Where damping-off may have already started in seed beds, ferbam, captan and copper fungicide dusts may be sifted along rows, or liquid suspensions of fungicides can be poured on soil near seedlings where the disease has attacked.

Plant pathologists in the American tropics have not been negligent in investigating methods and results from fungicide soil fumigations (8-10, 48, 57, 61, 119, 127, 131, 146, 180, 204), and various means have been shown to be successful in reducing populations of nematodes (22, 23, 31, 32, 54, 88, 90, 138, 156, 174, 189). In certain cases there have been unusual and highly interesting departures: the use of the nematocide DD against tomato *Fusarium* (23); sugar applications eliminating nematodes in soil (90); the employment of lime on coffee-wilt soil to control further spread of the disease (127); the use of amounts of benomyl and thiabenzadole for nematode control (156); the extraction of oil from a local weed, *Chenopodium*, to be used in control of soil nematodes (168); tests of applying nematocides in irrigation water (182, 189); and studies revealing successful killing of fungus sclerotia in soil with drenches of mylone or vapam (204). Further, coffee root disease, *Rosellina*, has been controlled by using soil sprays of fungicides containing copper and by using quintozene (119).

Chemicals on Cuttings or Root Materials

There is sufficient resistance in dormant seeds and enough of a time lapse between application of the chemical and planting that fungicides are used with reduced danger from crop damage. This is not true when the need is for poisons used for disease control on fresh propagating materials. Research by the horticulturist-physiologist can aid in determining when are the best times to dip or wash, or best after-treatments on propagating materials other than seeds. In the tropics cuttings from some crop plants are too tender for strong treatments with some of the usually available fungicides; in warm country propagating benches, chemicals are used with great care (74) and may need to be reduced either as to time of treatment or strength of fungicide used.

In Brazil (213) pathologists were advising growers long before 1940 in the use of potato tuber treatments. It is clear from reading about potato tuber treatments by de Segura (200) or by Chriastensen (58), that there is probably, in these later years, a larger number of tuber-borne diseases in at least part of the Andes than is found in North America or in Europe. Some of the standard 5-minute warm formaldehyde treatments have been regularly employed. With an increasing demand for more food such applications as those of PCNB (99) may require further study. It is evident, from recent experiences and in various pronouncements from various governments, that the use of mercurials will sooner or later be prohibited--perhaps by the time this is published.

Some of the most difficult situations on which to give advice, are those dealing with valued ornamentals. There are, for example, disease problems that are met with on pseudobulbs of orchids (37), and generalizations about their treatment are dangerous to make. One of the problems is that purchasers of plants do not want unsightly residues, and as a consequence various copper-containing fungicides and some carbamates, while they may be good in control of *Colletotrichum* sp., *Rhizoctonia* sp., *Pythium* sp., and other organisms, if used too lavishly or if applied just before plants are to be exhibited, the appearance of chemical stains is a distinct detriment. Some orchid disease cases require careful drying, followed by pruning and surgery, and then a light spray of copper oxychloride on cut parts and on the surface of the medium that is holding the orchid plant.

There are numerous tropical root crops, and under usual tropical management growers know when to dig the roots, when to divide the plants, where to cut them apart, what to do with them before planting, and the amount of dryness that may be required before planting to allow healing of cut places in preparation for starting new feeding roots. With banana and abaca special care is exercised in preparing seed heads or corms; they are washed with water, if possible under pressure (Fig. 21-2), examined for disease, carefully dried, and then planted. Treating with fungicides is not always carried out as, in general, diseased propagating material is discarded. However, under experimental conditions where certain selections must be preserved, fungicidal treatments are needed. It has been learned that strong formaldehyde and mercurial fungicides are unsatisfactory on many fleshy corms and roots because of hindering callus forming on cut areas, and where soils are warm, in some cases such poisons injure the newly emerging feeding roots. I have found this is specially true of corms (heads) of banana and plantain. However, the carbamate fungicides,

Fig. 21-2. Washing with water under pressure, of corms of banana in preparation for planting. They are examined for disease, treated with a non-corrosive fungicide, dried, and taken to the field for planting. Photograph courtesy of Department of Tropical Plant Pathology, United Fruit Company, La Lima, Honduras.

used in dry form as light powders, are effective deterrents to numerous parasites that invade newly cut roots of almost any plant.

Sometimes special types of grafts have to be made, with fungicides applied to insure success. Just one example (6) is the work of Albuquerque in Brazil where tender *Piper nigrum* which is susceptible to both *Fusarium* and *Phytophthora* is grafted onto resistant roots of *P. colubrinum*. Until the graft is healed and growth is well along, it has been found that fungicides must be used or the diseases will get onto the desirable top sector of this multiple black pepper plant.

Another type of propagating material that may require fungicide protection is the kind of cutting or sett used in sugarcane. Upper, or young, parts of selected stalks are cut into lengths having a few nodes; sometimes allowed to dry a short while "to allow for healing" but in other cases they may be planted immediately. In certain old fields there is danger of contamination by various fungi, such as species of *Fusarium*, *Colletotrichum*, *Ceratocystis* (*Ceratostomella*), *Marasmius*, and occasionally *Melanconium*. This is most often if setts are from lower nodes of mother canes. To guard against setts carrying disease externally, fungicides are used (144), and they are dusted or sprayed on lightly using a fixed copper or a carbamate compound. (In some plantations where diseases are prevalent and severe and fungicides are not at hand, brief treatment with hot water from the mill is substituted for chemical treatments.)

A different kind of cutting is that from such an ornamental as rose, or a plantation tree like cacao. In the rose, fairly short stakes are used and they are arranged lying flat on a bed of medium. They are then buried lightly, about 5 cm. deep, with the medium. (Media often are not disinfested, and are such as rotted wood sawdust, partly decayed pergamino from milling of coffee, and sometimes nearby jungle leaf mold.) Cuttings of "stakes" from such trees as cacao, lichi, or mango are cut in span-lengths from a few-months-old stem with about two leaves left at the top. The base is plunged into a moist medium. Sprayed with water daily, and kept under subdued light, difficulties may be encountered on any such cuttings from attack by bacteria and slime mold (unpublished observations) and various fungi and nematodes (74); this is controlled by 2 per cent drenches of carbamates and such material as spergon. Bordeaux mixture, mercury compounds, and copper-containing fungicides are likely to be injurious to cuttings under these conditions.

Still a fourth type of cutting is that of offsets, for example the slips from pineapple. These short shoots grow from the central stalk, and by gloved hand are separated from the mother plant. These offsets become infected at times, and may require dipping (181, 185, 99) in fungicides. Work on this is still in progress using modern fungicides in Brazil. In Central America, I saw, in 1945 and 1946, the use of dips of mercury containing fungicides; results were distinctly harmful. However captan and a weak bordeaux mixture did not injure rooting. Recently, it is said that dips in carbamate and quinone fungicides give better results (I have not seen it personally). Where pineapple diseases are severe, slips are at a premium. Growers often spray field plants bearing offsets, allow them to dry, pick off the slips and then dip them before planting.

Foliage Spray Problems on Various Crops

Although by far the most study and expenditure of money has been given to spray problems in plantation diseases of the American tropics, spray is being used on many other subsistence and ornamental crop plants. As a matter of fact, with increasing need for insuring agricultural production, and with the evolution of special government extension services (not copied from the temperate zone but adapted to thinking and methods for purely tropical conditions and people), there are being developed now more and more habitual spray programs for growers of minor crops.

Numerous miscellaneous crops need sprays and are being sprayed; they gain from it, as references to a varied listing from professional publications shows: orchids (37), white potatoes (9, 14, 163, 26, 206, 212, 213), carnations (44), general home gardens (62, 63, 56, 99, 160), citrus (36, 30, 210, 94, 89), mango (149, 65), green or snap bean (18), cereals (184), tomatoes (59, 183), melons (93), certain temperate zone adapted fruit trees (140. 146), carrots (120), grape (157), beans (21, 203), garlic or onion (139), and petunias (221).

A small list, taken from a few of my consultation

letters to Latin American agricultural experiment stations and to individual farmers developing competence in spraying in relatively small-scale fields, is as follows. From Guatemala and Costa Rica, information was given on sprays other than bordeaux for potato attacked by Phytophthora blight. From Chile, a request for a spray recommendation other than bordeaux mixture for a grape farmer troubled with both Uncinula powdery mildew and Plasmopara downy mildew. From El Salvador, Costa Rica, and Peru came requests for sprays on roses to control Diplocarpon blackspot disease and Sphaerotheca powdery mildew. From Cuba, Puerto Rico, and Colombia, inquiries about a possible non-phytotoxic spray application to control cucumber Erysiphe and Pseudoperonospora mildews. From Panama, a spray control of Oidium on orange. From Ecuador came requests for spray information to control broad bean Botrytis chocolate spot and plant rot. Inquiries were made personally to me while I was in Cuba, Puerto Rico, and Brazil about sprays that are non-toxic to watermelon that would control Colletotrichum anthracnose of leaf and fruits. From Paraguay a letter was received asking for a spray recommendation without poison residue to be used on Cylindrosporium canker of yerba maté. While I was at work in Cuba a spray treatment was suggested to control a complex of tobacco leaf spots, and a top disease of kenaf. From Puerto Rico and the Dominican Republic questions were received from florists shipping ornamental rubber plants for sprays to control Phoma and Phyllosticta leaf spots. Many requests were answered from tomato growers and agricultural agents for spray recommendations to control tomato Cladosporium leaf mold and Phytophthora blight in the countries of El Salvador, Guatemala, Panama, Nicaragua and Costa Rica. This gives a slight idea of the very broad scale of subjects involved in tropical spraying.

A phase of study on spraying in the most intelligent way is that of determining something about the sources of spores, when they are most common in the air, and other matters of epidemiology. Information from special researches of this sort (12, 34, 43, 54, 67, 81, 91, 94, 99, 101, 109, 110, 121-123, 127, 129, 130, 133, 146, 150, 171, 172, 200, 206, 208, 215, 220, 221) are of paramount importance. Results give indications of how spray cycles may be varied, and reduced in numbers per season, depending on when spores are most in evidence. These studies tell also what changes occur in fungal life histories following long periods of spray controls.

In the moist tropic regions, and that is where a large proportion of crops are grown, sprays of generalized nature, such as bordeaux mixture, will keep slightly moist on foliage. There, the hygroscopic character of plant leaf surfaces attracts air moisture, and there are long-lasting dews. Therefore there is always an invisible trace of surface moisture to prevent the best sticking of sprays, and chemicals are more likely to be in partial solution, which leads to toxic effects on the hosts.

There are some special physical difficulties for the spray operator. One problem is that which is due to planting of crops under trees for a shade canopy. Sprays have to be directed either from the ground onto the cash crop below the canopy, or, if from the air, sprays must be made to penetrate that canopy. During part of the year, long rainy seasons represent a severe interference. The locations and contours of fields are variable: some may be broad sweeps of broken hills and quite often planting is on the steep sides of valleys and in the narrow bottoms of them.

It is true that the alternation of wet season and drought gives a chance to time the spraying, as in some diseases sprays are not required in dry season and vegetative growth of a crop is at somewhat of a low ebb. But, when rains do come there is irregularity in their starting; the question arises as to which individual trees will be the first to begin growth. Growth flush results in tender foliage not only susceptible to parasites but also more easily injured by toxic chemicals. And flushing may occur in scattering plantation trees; sometimes on the side of a single tree one branch will flush, after that another, then another. Maybe two weeks later a branch or group of them will flush on another part of the tree. It is not always an easy matter to regulate spray timing or spray strengths under such circumstances.

Warmth stays perpetually and there is a range of temperatures to which fungus parasites are well adapted. In addition, moisture with more warmth may increase toxicity on the host. Insects flourish in tropical warmth, and accidental, not ordinarily suspected, leaf punctures and feeding wounds are more effected by chemicals than normal leaf surfaces sprayed with protectant chemicals.

Spray toxicity is a problem long recognized, and it has stimulated some of the best minds and hardest workers to secure good plant protection materials without excess host injury. References about this in literature (citations may be found: 19, 91, 169; for specific examples see 153, 173, 198, 190) indicate how exceedingly necessary it was in years past in our tropics to develop less host toxic fungicides than bordeaux mixture. Researches on this are more important to the tropical grower than to anyone in any other parts of the world.

A Good Protectant Spray

> Well considered, there are eight requirements for a good protectant spray, and they may be briefly stated in the following sentences. The compound must have good differential toxicity, holding in check or killing the parasite, but being relatively harmless in effect on the host. The spray must be a stable compound that will ship and store well. Its compositions should be relatively non-poisonous to those who use it, and harmless to those who consume the sprayed product. It must stick well to plant surfaces and withstand rain and weathering, but at the same time be sufficiently soluble that it will be slowly released on leaf surfaces to kill infective spores. A good spray requires being standardized by the manufacturer, readily obtainable, and always the same. A compound has to be usable in a practical way, and when run through the appropriate machinery will not cause excessive and recurrent difficulties to mechanisms. Any spray has to be sufficiently inexpensive that it will have a valuable return for the cost, and thus an economic place in crop technology. Lastly, and perhaps somewhat questionably, it would be convenient if the spray could leave a visible residue so that the applier can tell where it has been used.

Although a large part of the "inventing" and actual chemical studies on newer fungicides have been done during the last few decades in the temperate zone, compounds have been sent to or secured by workers in the tropics with avidity and tried (7, 16, 18, 35, 39, 54, 58, 62, 63, 64, 70, 76-78, 82, 83, 107, 112, 117, 143, 179, 184, 210), mostly with great interest; pathologists report good success in many cases.

Applications of sprays on tropical crops are undergoing continuous change and adjustment to take advantage of new advances. Growers who need sprays wisely keep in touch with advisers and pathology institutes. Recommendations differ from decade to decade, often sooner in some cases.

Crops grown at near sea level are sprayed in another manner than at say 1500 meters' elevation. There are even more changes in areas where new crops are being grown and along come newer fungicides and untried combinations of them. It is becoming apparent that the once standard use (99, 146, 200) of bordeaux mixture is almost a thing of the past, no matter how popular it was at one time. There are more and more specific compounds being developed and tested for special diseases.

Some materials are examples of new departures. In Argentina (7) a metal triphenyl hydroxide is good in effect on two celery leaf spots. In Chile (4) lentil rust is sprayed with not only wettable sulfur but also dithane 7-78. In Brazil and Argentina, where the Septoria tomato-leafspot is serious, it is controlled with alternate sprays first of maneb and then of zineb, using detergent additives (16). This alternation of one spray with another is employed with potato Phytophthora in the Andes (58, 84) and elsewhere (55), using such organics as manzate, dithane, or parzate alternately with any one of a few fixed copper fungicides, to which must be added stickers. Systemic fungicides (18) are likewise having their first extended field trials. With every new change, laborers will do better in applications if they can be convinced that their efforts will have an advantage over old treatments.

The Classic Spray on Bananas

It is often repeated in this book that fungus and other foliage-attacking and especially numerous organisms flourish and cause serious problems in the tropics. It has also been pointed out, through the work of many pathologists, that there is rapid and wide distribution of spores that cause diseases of numerous plantation crops. Examples are: several woody hosts infected by conidia of Corticium pink disease blight as discussed by de Segura (200), and Wellman (221, publ. 1955); orange Diaporthe, the cause of black melanose as written of by Fawcett (89) and Brogdon and Lawrence (35); hevea rubber Dothidella leafspot written of by such as de Camargo et al. (43), Galli et al. (99), and Wellman (221 in 1955 publ.); cacao Phytophthora disease as written of by Orellana (167), Befeler (24), Lellis (134-5-6), and Medeiros (152); coffee Hemileia rust (Wellman, citation 221, publ. 1952, 1957, 1958, 1961 and 1970); and banana Mycosphaerella or Cercospora leafspot, also commonly called sigatoka, as written of by such as Lopez (141), de la Hoz (126), Klein (129), Leach (133), Galli et al. (99), Stover (208), de Segura (200), and many others.

The famous banana spraying programs over hundreds of thousands of hectares in West Indian, Central American, and South American countries are truly classic and successful control efforts. Spraying has never before been carried on in such a grand scale, under such great difficulties.

Spraying to control plant disease is the most spectacular of all plant pathology practices. Of all spray programs in any part of the world, the most extensive, the most carefully engineered, and probably the most carefully managed scientifically is banana spraying. Its use is something on which large companies are completely dependent for success in fruit production and the spray operations themselves are sufficient to employ large numbers of workers and technicians who do nothing but carry on its performance.

The causal fungus, Mycosphaerella musicola, has not been in the American tropics for a long time. It was found in Central America in 1934,* and spread rapidly in our tropics after that. Its origin seems to have been in Java where, as Stover (208) points out, Zimmermann found it in 1902. From there it probably was spread by man eastward slowly in the Dutch East Indies (Indonesia) and in 1912 it was observed in large plantations in the banana growing valley of Sigatoka in Fiji. It was from this place the disease got its name. It then soon appeared in Australia. In that country it developed vigorously and probably ascospores

*Several leaf spots occur on banana and some of the fungi are still to be determined. In February of 1930 there was brought to my laboratory a small banana leaf spot on lady finger variety. At first no spores were obtained by spot scrapings, but protected spots on diseased leaves produced growth. Some were of the common Cordana, but others had slender but short multi-celled spores resembling a Cercospora, but not C. musae. (Work on questionable spots stopped.)

were carried into the upper atmosphere where they were swept westward to several parts of Africa. After that, it moved onwards in similar manner across the Atlantic to the western hemisphere. In 1934 it was definitely determined as being in the West Indies and a dozen years later was reported in 29 countries and islands of tropical America.

All commercial varieties of bananas grown in the Americas are susceptible to Sigatoka. By the year 1935, plantation owners could see that the disease was increasing enormously and realized they were in desperate straits from the leafspot. And it was also realized that the very existence of considerable populations in some countries, where banana plantations were the only commercial industry, was contingent on the development of satisfactory spray control (141, 170, 175, 126, 133, 158, 159, 202).

The pathologists selected bordeaux mixture as the best spray possibility. Most first spraying of plants and trees has long been known in the temperate zone with the use of bordeaux. In special cases replacements of bordeaux were tried (216) in the tropics and such materials were used as lime sulfur, iron sulfide, burgundy mixture, and even atomic sulfur with variations of these. However because of its sticking qualities, for a while none was as good as bordeaux.

After feverish but well considered efforts, the value of bordeaux spray was finally proved in control of the sigatoka leafspot. Pathologists, chemists, engineers, agronomists, and economists then formed close collaboration in the next steps. It was learned that the spray was not foolproof; it was a scientist-technologist's province needing proper application at the right time and repetition at proper intervals. Amounts used were about 300 gallons per acre and at first the most popular formula was 5-5-50. The good results led to successful rehabilitation of old abandoned plantations and development of new areas. The best soils free of root disease were selected and intensified methods of banana production were instituted. Leaf disease was no longer a main limitation.

However, controlling banana leafspot on the scale required was the biggest job of plant spraying ever done anywhere in the world. Tropical pathologists are proud of what has been accomplished. Crises in engineering problems were confronted, e. g., the size of spray pipe in relation to pressure requirements; the number of leads of pipes that certain-sized pumping spray centers could handle; the sizes and lengths of pipes that best carried spray underground; the angles and depths at which spray-carrying pipes could be arranged; and the best metals, stopcocks, rubber hoses, and spray guns to use. The securing and transporting throughout the Neotropics of large amounts of copper and lime for the spray, and the guaranteeing of a good supply of clean water, are remarkable feats.

Those in charge of the hundreds of huge spray pumping stations became not only well-trained stationary engineers but knowledgeable spray mixers; they learned exact amounts to use in large tanks; proper dilutions and how and when to combine them; how soon the mixture must be used in the pipes under pressure; what to do when a spray line became plugged (a metal "rat" was invented that was forced through to clean out underground spray pipes).

Plant pathologists also faced numerous crises and knew that the success of the operation was dependent on the mode and time of applications in the field. A uniform piece of work was at first essential before differences were instituted in timing and other changes in routines could be made. The best laborers were recruited for spray gangs and these were trained to carry on a regular round of movements along pre-determined spray paths. Helpers aided with the hoses and men with the spray guns walked along the clumps, or mats, of giant banana plants. At each mat, holding the gun and delivering its spray up at a given angle the sprayer took a certain number of steps in one direction, then a number in another, possibly another, and then returned to his path. Thousands of hectares were protected from leaf-spot, sprayed in this manner for many years, and the banana business recuperated.

However, the use of bordeaux mixture spraying in the tropics has its difficulties. With respect to sigatoka control on banana, there are several prejudicial results from long use of bordeaux mixture. There must be many heavy shipments of special lime that has to be kept dry and used fresh for it deteriorates in the tropics if stored for long periods. Bluestone, or copper sulfate crystals, are required in large amounts, and the copper from which the crystals are made is an important metal of limited supply in peace or war. Ordinary containers used for these materials issuing from the temperate zone are often not well adapted to the tropics.

When suspension-solutions of lime and bluestone are skillfully mixed, they make a fungicide noted for its excellent sticking qualities. The result is that banana fruit, when it comes from the plantation, is liberally covered with "bordeaux blue" which is only matched in its spectacular aspects with the same blue spray residue that incrusts hats, helmets and equipment carried by expert spray gangs and supervisors in the banana fields. Plantation pathologists devised acidified washes to remove the unwanted blue deposits from bananas as well as men. In the extra dipping and type of fruit washing it requires, however, there occurs an increase of fruit bruises. Another result from bordeaux spraying is the bad effects in soil from copper accumulation after many years of sprayings.

Change in Banana Leafspot Spray

Tropical pathologists looked with longing eyes at the great progress made in spraying from the air in plant disease control. In the temperate zone, both fixed-wing airplanes and helicopters are used; spraying is sophisticated and "big business" (221 in Publ. 1971) now as never before. It is the advent of spray oils, their use as fungicide carriers and their transformation into emulsions, that finally turned the trick for air applications on bananas. In accompanying Fig. 21-3 air spraying of bananas with both helicopters and fixed-wing airplanes is shown.

After some 20 years of struggling with the cumbersome but nevertheless successful bordeaux spraying on bananas, there eventually developed the easier, and good, sigatoka leafspot control through air applications, using the so-called orchard or spray oils (see Guyot and Cuille, 114, 115; Desrosiers and Ampuero, 71, 72). An early criticism, from some planters accustomed to the appearance of bordeaux spray on their plants, was that the oil deposits were almost invisible; there was skepticism of the fungicidal efficacy of orchard oils.

However, both in the French West Indies and in Ecuador, repeated trials with oils sprayed on banana, over several seasons, proved its worth. It was shortly being given attention as to refinements, using back-carried mistblowers (187, 214). In other parts of the Neotropics, spray men carried on more detailed studies. It is clear, now, that there is considerable specificity as to oils on banana

Fig. 21-3. Spraying bananas from the air. Top using helicopter, bottom using fixed-wing airplane. Photographs courtesy of Department of Tropical Plant Pathology, United Fruit Company, La Lima, Honduras.

Cercospora leafspot. Calpouzas (42) found that several other Cercosporas on other plants were not so well-controlled as species of that fungus on banana. Fisher (94) shows that oils have been used successfully with Cercospora on citrus. But the main crop that uses this material is the banana. An important value in the tropics for oil sprays is the fact that the amounts of fungicide needed can be so drastically reduced; this is what is called low-gallonage.

Low-gallonage sprays, which have such potential in the tropics, use concentrated spray but apply it in very fine droplets. Such so-called low-gallonage usage with airborne equipment is employed on other crops than bananas (for certain citations see 30, 33, 43, 53, 94, 149). When low-gallonage treatments are carried on with coffee (177, 187, 188) some interesting results are secured: oil emulsions alone stick well on coffee leaves, but when cupric hydroxide in water suspension is used with no oil added, a sticker is required. Airplanes are used successfully to spray coffee with concentrated materials, flying from 5 to 15 feet above the bushes. Droplet sizes most fungicidally effective are 150 to 220 microns in diameter.

In banana work aerial spraying is usually at a little greater distance--not much--from the tops of plants. Droplet sizes that are fungicidally effective are about the same used for coffee or for banana. That fixed-wing airplanes and helicopters are both successfully adapted for low-gallonage spraying over uncertain terrain in the tropics is of great credit to the spirit of technician-scientist teams. An outstanding discussion of some of the basic science required in research on application from the air of low-gallonage spraying is found in the works of Brandes (33) and Fulton (97), and the citations they include of key publications can be referred to for more details.

For reports about precise studies on the science of spray control of banana leafspot see Fulton (97), Desrosiers (71, 72, 74, 75, 84), Calpouzos and co-workers (39-42, 70), Guyot and Cuille (114, 115), and Stover and Dickson (207, 208). Spray oils in fine drops, either alone or as 40 or 50 per cent emulsions, are good controls. Some banana spray men use organic fungicides (such as ferbam) added to oils and have their own private formulas which they use. These formulas are varied during different seasons and the timing of when sprays are to be apllied is of utmost importance.

Exact mycological-physiological-pathological research studies, such as those by Fulton (97), Klein (129), and Stover (208), have been specially needed in assuring the practicing spray (or fog) applicator just when to start after the rains commence and how often he must return to the banana field and repeat his work. After years of study there is good understanding now of why low-gallonage applications, at first put on at 10- to 15-day intervals, after several seasons were reduced to a spray only every three weeks in the warm season, and every four weeks when it was cooler; why under conditions in Ecuador it was best to spray twice a month; and why in some climates trial-and-error tests after repeated studies finally settled as simple rule-of-thumb timing. Sometimes it is more complicated and has become greatly modernized; then spray timing depends on a sophisticated set of observations with tests, giving results on actual leaf coverage, spot appearances and counts along with spore trapping results (For trapping see 150, 171, 172, 121-123).

Amounts of low-gallonage sprays used per spray-cycle vary from about 0. 5 to 0. 75 to 1. 0 gallons of oil per acre. Where bordeaux spray, which of course is carried in water, is used, the rainy season is a serious deterrent to conventional modes of application, as sometimes leaves do not dry for days at a time, and the fungicide is at a distinct disadvantage. With airplane or helicopter applications of oils, control is good in both wet and dry seasons.

Further advances are being made in foliage spray programs. One of the most time- and energy-consuming phases is that of determining by survey where epidemics are occurring, the directions of disease spread, and the relative severity. There is a chance that with more and more perfection of electronic sensing equipment this may be done by mapping of disease from the air (e. g., 34, 206). As experiences progress, the more detailed determinations of the effects of weather changes and climate on crop diseases, as banana leafspot research has so abundantly proved, will inevitably bring the best understanding as to what sprays are best and how they are to be applied, their timing (see such citations as 51, 43, 215, 184, 30), and how they may be combined with leaf absorbing fertilizers and phytohormones (38).

There are a number of newer sprays and some are found to be more specific for some diseases than others.

Several state experiment stations in the U. S. A. have lists of sprays used by their farmers and specialists, and such lists are helpful to researchers in the tropics. A list like that from England (86), but of possibly more international scope, might at some time be published on an at least semiannual basis by such an organization as the United Nations Food and Agricultural Organization. There needs to be a more critical attitude towards fungicides made in the temperate zone for use in the tropics and when, as in Brazil, a group (179) studies the analyses of the dithiocarbamates, for example, enhanced understanding will lead to a more intelligent use of available compounds; the possibilities are increasing for the very desirable prospect of making fungicides in the tropics for the tropics.

Recent researches on problems relating to systemic fungicides have resulted in the use of systemic sprays on tropical crops (18, 39, 112, 117, 210, 127, and others). There are still uncertainties as to just how much leaf tissue transport occurs with copper-containing spray effects. This may be an especially advantageous feature in the tropics, where plant metabolism goes on at high rate over the full 12 months of a year. There is still much to be determined about how such insecticides as lead arsenate or mercurials or copper arsenate act in the tropics as fungicides (49, 82, 83, 176, 186). Learning how these work will lead to replacements and the use of non-toxic substances developed for systemic curative purposes. It will require special spray assay techniques; one example is that studied by Chaves (155).

In connection with spraying extensive plantations in the tropics, there has developed a dependence on delegation of authority in carrying on spray responsibilities. It is a large operation, and requires competent attention from specialists, both of literate, scientifically-trained, and of illiterate, but grower-trained, personnel.

Coffee Leaf Disease Sprays

Coffee plantations ("fincas" or "fazendas"), with their small crowded trees clothed by dense foliage, are a special problem to spray pathologists. Trees are planted close together both in the row and between rows. The branches often whip your face if you are walking between the rows with a knapsack spray machine strapped to your

Fig. 21-4. General view of interior of a coffee plantation shaded with *Inga* trees; the hazards of spraying such a crop are considerable. Passing up and down the rows *with* machine drawn equipment is practically impossible. The result is that spray is applied by light weight equipment carried on men's backs.

back or holding a spray gun directed into the trees. (Anyone who needs glasses, even if he wears a protecting helmet, cannot possibly spray for long in a coffee plantation.)

The spray man has not only to watch what he sprays but to see where he steps. Irregularities of all kinds occur in the finca floor. Shallow holes are frequently dug in some areas to collect organic matter and hold back rain water without allowing excess wash of top soil. Often there are numerous crumbling, 40 and 50 year-old, hardwood tree trunks still remaining from large original forest growth. The placement of shade trees adds to irregularities and must be taken into consideration by the spray man on the ground (Fig. 21-4).

All this is additionally made more difficult because coffee is a crop that is planted upon, and makes money from, what would often be marginal areas: land so steep that no attempts can be made to grow for profit such crops as sugarcane, oranges, hevea rubber, lemon or lime, cacao, bananas, coconut, plantains, or pineapple.

An ample supply of water for spray applications is another problem, particularly difficult if bordeaux mixture, other copper compounds, and certain organic materials carried by water are used. Wells furnishing practically limitless quantities of water are relatively rare in coffee areas. In volcanic deposits, the soils may be rich and if protected by shade almost inexhaustible for tree productivity, but in them water is not easily impounded for long. Springs are almost non-existent in the deep, highly porous soils common in the tropics which have no multiple under-layers of impervious strata to collect and deflect water supplies that eventually come to the surface. Even in older soils--as in the famous, extensive, red coffee-growing fazenda areas of terra roxa in Brazil--rain water is not easily collected, wells are infrequent and limited in yield, and streams dry up after the rains stop. Even only a quarter way through the dry season in the tropics, many rills and small streams are nearly free from running water.

A few more examples of certain special other fungicide application problems in the Neotropics are hereby briefly reviewed. The American leafspot, from the fungus Mycena citricolor (=Omphalia flavida), was early controlled by cycles of applications of conventional bordeaux mixture (see reviews in 47, 218, 221 publ. 1961). With this fungicide there arose difficulties of many sorts: making it is troublesome; it is unpleasant in application; coffee trees in fincas are not planted with spraying in mind; and many times there is a phytotoxic effect. As more modern organic fungicides have been devised, many have been tested on the leafspot. It became evident to Volio (218), Cardona (45), and Wellman and co-workers (28, 105, 221 publs. 1955, 1957, 1958, 1960, and 1961) that such an organic compound as captan was effective provided it could be made to stick. Fungal toxicity is excellent; adherence is poor without a sticker added.

Captan, ferbam, spergon, and phaltan, with stickers, control this special leafspot without causing phytotoxicity. Leaf injury is common at times with some copper-containing sprays. Eradicant fungicides have also been tested and certain ones are excellent in control studies; see reports by Castaño (49), Robles (186), Pereira and Echandi (176), and Echandi and Segall (82, 83). Unfortunately, with human beings involved, there is danger in applying an eradicant which has mercury as the important poisoning component.

Mercury is translocated in the tree and is absorbed into the commercial coffee bean (176). This makes a mercurial fungicide prohibitive in its use on the crop. Arsenic, when the main component, is absorbed into the beans also but the amounts of it are well under acceptable limits in commerce.

In addition to those sprays mentioned immediately above, it is now realized that there is a long list of chemical spray materials usable on coffee (28, 45, 78, 124, 188, 197 and others). (At least in the cases of Cercospora leafspot and Hemileia rust, antibiotics did not control the diseases (127).) It was a considerable dividend, the discovery that certain orchard oils were promising fungicides for coffee. At one time it was thought impossible; oils were believed to kill coffee leaves.

During two separate years, studies I did (unpublished) showed that spray oils applied to the point of run-off were singularly injurious to young flush leaves and to mature not hardened leaves, but were less damaging at first to old mature leaves. The injury is describable as "bronzing" and the old leaves dehisced after about two weeks. Flowers sprayed with oil set fruit without difficulty. When oil was applied in emulsion of 2 per cent strength and "fogged on" in sufficient drops to cover about half of the leaf surface, there was no observable injury on any aged leaf. However, in some cases leaves that had been thus lightly treated were allowed to weather two months and then were brought to the laboratory. There they developed, on inoculation under controlled conditions, a successfully infected-by-*Mycena* leafspot.

Rodriguez and Soto (188) in Costa Rican field studies, where *Cercospora* leafspot of coffee is a severe disease, found that emulsions of spray oil at 2 or 4 per cent were effectively strong in the control of this leafspot. Such use of spray oil further advances the prospects of controlling the *Hemileia vastatrix* rust in its occurrence in the Western Hemisphere. Chaves (54, report 1970) has indicated that oil emulsions appear to have good deterrent effects on coffee rust in Brazil. The fact that there is some precedent for aerial spraying of coffee, although done in the Eastern Hemisphere (177), increases the confidence of pathologists that this rust is controllable using oil sprays with the proper kind of air-application equipment.

It is to be stated as a principle in tropical plant pathology that information obtained from continuing studies on fungicidal sprays should not be limited, even when--as is the case in coffee *Hemileia* rust--disease resistance is obtained; there must always be insurance against that virulent resistant variant of the parasite that evolves and once again attacks. Indeed, in coffee rust this recurring appearance of new parasite races has been seen a number of times in the past (221, publ. 1957 and 1961). A very recent new race of the fungus (205) has appeared in India and was reported in 1968.

In the instance of koleroga (*Pellicularia koleroga*) which becomes common at times in shaded plantations, sprays most often used for control in the past have been bordeaux mixture, or some other copper-containing spray compound, in water (28, 45, and 221 in publ. 1961). Later studies show that stickers are needed with the usual types of water-mixed sprays. Even bordeaux mixture and proprietary copper-containing sprays are improved with added sticker. Koleroga spraying has to be done from the ground. The tough travelling-cords of the fungus are resistant to light droplets of fungicides, whether of inorganic, organic, or of oil content. Bianchini and Wellman (28) found sprays of lead arsenate were promising for koleroga control. This was confirmed by Robles (186) and it is employed in Central America and Colombia for that purpose. If timed properly, three sprays per year give good koleroga control: once at the start of the rainy season, another a month later, and a third at the period of the "canicula" which is a common break in rainy weather during the latter part of the rainy season. These are the usual recommendations given in Central America. In India (221 publ. 1961) the disease is controlled by bordeaux mixture sprayed on twice, first at the start of the monsoon rains and again a little over two weeks later.

Disease control in coffee has an accumulative effect. It requires one or two seasons of spray applications to make a good reduction in the sources from which spores or other propagules arise. It takes considerable experience to adjust spraying to an easier operation after its techniques are learned, and the physical difficulties solved. There is also a need for some time to elapse to convince some growers and their workers that spraying is profitable, as measured by a larger harvest and better or as good taste when the beans produced are cup tested.

Planters know that spraying not only controls a specific disease but also has secondary effects. For example, some fungicides markedly reduce the amounts of mosses, algae, leafy liverworts, and lichens that encumber leaf and limb of many a coffee tree.

Spray for one disease often results in reducing, if not completely curing, a secondary disease. This occurs for example when a fungicide for Mycena leafspot cures weakspot. Although still not perfectly understood (221 publ. 1961) weakspot is the undoubted precursor of the widespread disease known as chronic leafdrop. (Every leaf retained on fruiting branches is of super-importance to production, as there occurs a regular fruit/leaf ratio; varying in different locations and with different coffee varieties; a common count is 11 to 14 cherries for every leaf (see 221 in publ. 1961 and 1970) and it is to be recalled that each cherry contains two grains or beans of commerce.)

A further type of effect from spraying to control coffee disease, is an adverse increase in fumago or other smothering fungi. The sprays that control coffee fungus parasites are in many cases unfortunately also effective against enemies of coffee-residing sucking insects like fulgorids, several scales, mealy bugs, and certain aphids. These thrive on copper-sprayed trees. They excrete drops of so-called "honeydew," and it is on these sugary deposits that there grows the epiphytic, smothering fumago or sooty mold, Capnodium spp.; the presence of insects also adds in some manner to the increased growth of black mildews, Meliola spp., and the alga Cephaleuros, although these two latter are actual parasites on coffee leaves. The disease growth called "felt," Septobasidium spp., is associated with insects on coffee twigs with the fungus growing in protective layers above the insects. In El Salvador, felt loosened like flat chips and dropped away from coffee twigs when it was sprayed with a proprietary ready-made bordeaux mixture with soap and kerosene added. Entomological work (35, 89, 119) shows that insecticides often are the most important part of spray programs and that one of the least persistent, but at the same time most effective insecticides, is malathion.

Leaves of coffee are not the most acceptable objects for retaining spray. They are silky smooth, and have a waxy or oily sheen. Water runs off easily, and falling dew drops can be heard some nights in an undisturbed coffee finca. The leaf surfaces do not retain spray nearly as well as leaves of many other trees. The study of spray stickers on coffee is a specially good subject for research (221 publ. 1955); information gained on this crop will be of fundamental value to spray men all over the world.

Bordeaux mixture sticks on coffee, but in India effects are improved when a sticker is added. Captan, an organic compound of heterocyclic nitrogen composition, proves efficacious against certain fungi in bioassays (54) in the laboratory. However, when I sprayed with a water suspension of captan alone, tropical rains reduced its effectiveness. If coffee leaves are sprayed every three days, even if rained upon daily, they are not infected by Mycena. If sprayed every four days, some infection does take place. Naturally, applied every two weeks captan by itself does not control the leafspot. However, when a sticker is added, captan treated leaves remained practically impervious to Mycena attack, and kept so after a month of weathering. Ferbam is known as a good fungicide, acting well two or three weeks following its being put on some crops in the temperate zone. In the rain-washed and dew-drenched tropics, ferbam on coffee, to control Colletotrichum and Cercospora, lost some of its effectiveness in four days, and in a week's time it was only a third as good against infections as the best combined bordeaux and sticker treatment that had been on for over a month. When a good sticker was added to ferbam it was not so good as bordeaux but was a good protectant after a month of weathering. Two proprietary fungicides were used, one said to contain copper oxide and a sticker, and the other said to be of copper oxychloride with a sticker, and both were made over twice as good as fungicides by the judicious addition of more sticker.

Control of Cacao Fruit Infection

Of all the dicotyledonous fruit-bearing trees of the tropics that are sprayed, the cacao, *Theobroma cacao*, is unique. It was originally an under-story inhabitant of South American jungles, where it was adapted to a warm, moist, moderately shaded condition. In plantations it is readily grown in some shade, and trees are maintained as having medium tall, ungainly sloping trunk or trunks, with a few broadly Y-angled branches. As part of the bark on well-matured larger branches and trunks are the so-called "flower cushions" spaced at intervals; and on these are borne the hanging fruits or pods (mazorcas) set on short peduncles against the tree wood. It is these pods that are infected by disease. In the nature of things, cacao fruits are towards the inside-center of the tree and it is in this location, where often it is wet, that spray must reach to protect the product.

The most widespread disease of cacao is blackpod, *Phythopthora palmivora*, and spraying is a regular and successful treatment against it in many parts of the Neotropics. It has called for considerable study (around 200 reports could be cited on this alone) as losses are occasionally as high as 60 to 80 per cent of the harvest, sometimes a little less; 10 to 20 per cent losses are common. The parasite develops, and is held from season to season, in leaf lesions, in softwood stem cankers, and in the post-harvest remains of diseased pods. There are two special purposes in spraying: one, reduction of the source of inoculum by eliminating or rendering inactive the inoculum sources; the other, protection of growing pods and susceptible foliage. (See such citations as 24, 69, 73-75, 134-137, 152, 161, 162, 165, 201.)

Even where piles of diseased empty pod shells are destroyed after harvest, *Phytophthora* infection from other sources is still a serious menace. One of the best fungicides against it is bordeaux mixture. High volume applications (abundant material sprayed on to at least near the runoff stage) have been successful. Perhaps bordeaux is satisfactory since it has superior sticking qualities; in any case the costs and the type of work involved are serious deterrents to popularity. Low-volume spraying is being tested and improved on other crops; by using copper oxide or other copper-containing spray compounds with stickers on cacao, low-volume applications are found to be good in control.

Spray schedules depend on conditions and are something each cacao grower has to determine for himself. Some apply their chemicals at set intervals of about every three weeks to once a month, throughout the year. Others save on spray costs by using sprays at 10-day or two-week intervals over a month or two while the plantation is in flush and followed by spacing at 30-day intervals another time or two and then no longer spraying until rains and temperature changes again bring on flush growths.

Noting that low volume spraying with copper-containing fungicides has given good results, special attention is drawn to the use of other fungicides. It appears that organic compounds, at least those readily available, are not so effective as inorganics that contain copper. However, in a departure from some of the mixes, oil sprays have appeared effective (165) as a protectant in *Phytophthora* pod disease. With studies and employing results of findings on airplane spray technology, growers are learning about turbulence and drift methods in penetrating with sprays below leafy canopies in cacao plantations.

The witches' broom disease, *Marasmius perniciosus*, presents quite a different problem in cacao spraying. Desrosiers and co-workers have shown this in Ecuador. The fungus infects only tender, meristematic tissues at the youngest tips of either vegetative or reproductive (flower) growths; it is almost impossible in economically acceptable field spraying to keep rapidly developing outside cells perfectly covered with a protectant chemical. A logical help would be the reduction of infective spore sources.

Knowing that the spore-producing agarics of the fungus occur only on dead cacao brooms (see illustration Fig. 15-3), the practical spray control is by use of something to repress sporophore growth. Dead brooms are found from one season to another still held on cacao branches or lying at the floor of the plantation. This means eradicant materials have to be applied by spray men on the ground.

Spray is directed to aerially held brooms and to those fallen, which may be hidden by loose debris and weeds. In their researches, Desrosiers and Bolaños (73) found that there were two specially good eradicant compounds, dinitro-ortho cresol and sodium pentachlorophenate.

A promising series of spray investigations by Desrosiers have given good evidence that low-volume ("low-gallonage") cacao spraying, using zineb in 60 per cent oil emulsion, is effective against witches' broom. It likewise has good results on *Monilia* podrot. However, it has to be applied when the period of *Monilia* infections are most prevalent, and this is during the early part of the rainy season. This is also during flowering, and it appears that oils may have a toxic effect on flowers (75) or at least on fruit set. Until more has been done with oils to secure some formulation not inhibitive to fruiting, generally speaking oil sprays for cacao are still experimental. However, where the *Monilia* disease is very common on susceptible trees, losses are so marked that a spray program is a definite necessity (75). One of the best non-toxic fungicides for this disease is zineb in both low-volume and high-volume applications.

Chemicals and Spray on Various Fruits

There are about 100 special, highly perishable, fleshy fruits native to and popular in the Neotropics. Common names with Latin denominations of a very few are: nispero, *Achras*; granadilla, *Passiflora*; jocote or mombin, *Spondias*; mamey, *Mammea*; anona, guanabana, and chirimoya, *Annona*; and the juicy fruit part of marañon, *Anacardium.* Such fruits have to be consumed without delay and are all subject to infections, bruises, and decay that make them almost impossible to ship long distances for metropolitan stores. One of the most common infections in the lesser tropical fruits is that from *Colletotrichum*, the anthracnose fungus. I identified it over several years, in 80 tropical fruits and in all of those mentioned above.

If the fruit of the banana, *Musa,* is neglected it is much attacked by *Colletotrichum* anthracnose. It is also involved in such problems as crownrot, skin spotting other than anthracnose, scabbing, softening, and miscellaneous infections (17, 103, 107, 108, 154 and others). To reduce harvest bruising, stems of bananas are hooked separately (Fig. 21-5) onto overhead wheels on wires and are hauled from

Fig. 21-5. To reduce harvest bruising--which increases the perishable banana fruits' susceptibility to infection by Colletotrichum (anthracnose) and other diseases--when cut from its plant in the field, a stem of bananas is hooked at separate intervals on a cable. This cable is moved by a tractor, and bananas are thus sent in from the field to post harvest treatment centers. Photograph courtesy of Tropical Plant Pathology Department, United Fruit Company, La Lima, Honduras.

Fig. 21-6. Post-harvest treatment of banana fruits to reduce spoilage in transit and market. Top picture: stems of fruit on right brought in from field; "hands" cut from them and placed in washing and treatment solutions. Bottom picture: after treatment fruit is dried and packaged for shipment. Photographs courtesy of Department of Tropical Plant Pathology, United Fruit Company, La Lima, Honduras.

the field with only minimum initial bruising. Through this and other means pathologists are reducing the problems of crownrot, skin spotting, etc., and special post harvest treatments (Fig. 21-6) have given good results. In plantations sprayed for leafspots, fruit is much nearer to being free from disease than in untreated areas. Another step that means a great deal in producing banana fruits that are free from disease has been the careful bagging of green fruit on plants, while bunches are still growing. But this does not completely eliminate banana fruit troubles. The bunches are harvested with special care, stems are cut into hands, and these are then washed and treated in fungicidal solutions such as that of water and sodium hypochlorite.

Where a fungus such as _Colletotrichum_ causes not only fruit rot but also poor fruit set--as for example where it is abundant on mango--spraying has to be done in a special manner (65, 99). The first phase is to save the newly fertilized flowers, and this is accomplished by two or three sprays at about four-day intervals during flowering. After that fruits hang on a long while and spraying is at monthly intervals during the time when fruits are growing most rapidly. In both Florida and Brazil, not only are cupric sprays advised but also such organics as ferbam and zineb.

In the strawberry (99, 149) which on maturing is susceptible to Colletotrichum, and needs prompt harvesting, spraying is done at least once in the month or six weeks before flowering to reduce the field fruit-attacking anthracnose. After that it is planned that sprays are not applied to these plants until they are just past being fully in flower, but at this time it is necessary immediately. A weak bordeaux mixture or standard zineb is put on plants before flowering. After petal drop there is a weekly spray of thiram or manzate (sometimes alternated). When anthracnose of grape is encountered, a more extended spray schedule is required (151, 80) especially in a climate with a tendency towards being moist while fruits are ripening.

Coffee fruits are known (221) to be attacked by _Colletotrichum_, a _Sclerotinia_, and some other fungi. Bordeaux mixture has been the most common fungicide used to spray the coffee where fruits are diseased. In Africa it has been learned (104) that bordeaux spray reduces coffee fruit set to about half what it should be, so this fungicide must be used with great care on flowers and fruits. Chaves and others (54) have learned that in Brazil, spraying with thiram reduces occurrence of Sclerotinia. This fungicide is much less toxic to flowers and fruit in coffee than is bordeaux.

It is clear to pathologists observing the long time that citrus fruit stays in the field, hanging on trees, that such fruits are not simple to keep clean of infections and free of marks. It is important to preserve healthy appearances, as the market requires it. Fruits that have had sulfur- and copper-containing sprays put on too strong or at the wrong time may become russetted or otherwise blemished. Certain copper sprays have to be used with special judgment (89) or they will kill fungi that are parasitic on scale insects or others. There follows then in many cases an increase of these pests, which is not only bad for the health of the tree, but in addition there develop great deposits of the black Capnodium fungus that is so distinctly detrimental to the appearance of oranges or grapefruit and which spoils limes wanted for squeezing into drinks.

To judge when are the best times to spray for certain fungi, it is becoming more and more necessary to sample the atmosphere content (108) at different periods. The times of recurrent large numbers of airborne spores can then be correlated with the time sprays should be applied adapted to the period of fruit development and the time the fruit is most susceptible to fungus infection.

In Brazil (99) and in Argentina (198) in the cases, for example, of verrugose, Elsinoë = Sphaceloma, or Diaporthe = Phomopsis, some of the best control is by light fogging with orchard oil or use of a weak (1 per cent) bordeaux mixture. Such fungicides are sprayed on when about two-thirds of the petals have withered and fallen. (See also such citations as 60, 89, 99, 112, 130, 200.)

With regard to post-harvest disease of oranges, one of the worst culprits is Penicillium. This and a few others of the same type of fungi yield to washings and dips with some of the newer fungicides (117, 112) such as benomyl, thiabendazole, and sodium orthophenylphenate.

Crops Seldom If Ever Sprayed

It is believed of interest and practical value to point out here that there are some tropical crops that are never, or at least seldom, sprayed or otherwise treated with chemicals to control disease. The spraying of some tropical trees or lesser plants for the crops they produce may be definitely not worthwhile. This kind of a situation is given little attention in standard texts on plant pathology. However, it may be rather important in connection with planning future agricultural programs in the circum-equator parts of the world.

Where lands are wild and yet harvests of nuts, spices or woods are secured by following roadways and trails leading into the bush and jungle, there is no sense nor reason to spray or otherwise use chemicals for disease control. Crops that come under this classification are such as dye woods, the so-called balsam of Peru, Brazil nuts, and like things.

Although the disease spots on leaves of some minor crops, for example lleren, may be numerous, they are not severe under the usual conditions where they are grown. This is also true of the tuber-bearing oxalis, and yautia, as well as chayote. There are some diseases attacking plants like these that are never allowed to develop to severe proportions simply because plantings are moved to fresh locations.

The coconut, and other palm fruits, have valued crops, but the height of the trees do not allow of much spraying or other leaf application.

It is perhaps needless to point out, but it should be recorded, that most shade and roadside trees planted or permitted to grow for fruit are not sprayed. Nature is allowed to take its course with them.

Interestingly enough one of the most important and most technologically studied of all tropical crops, the sugarcane, except very rarely has never been sprayed. Partly this is because of the length of time from planting to making good growth; and then, too, the cycles of harvest that serve to interfere with leaf disease organisms by stopping development through the completion of life histories and also carrying away much disease material. The same thing is true of lemon and citronella grasses.

Maize (corn) should also be included. This is the greatest producer in the largest number of places and over the widest range of ecological conditions, of any of the grasses, Gramineae, in the American tropics. It is not sprayed, not only because spray ordinarily would not increase harvest enough to pay for the expense and troubles*, but also because it is a crop that has so much malleability in its genetic makeup. Inheritable resistances to disease are almost always being found in maize, which means sprays do not have to be depended upon to insure productivity. Sometimes a new disease, or a virulent race of an old one, appears and for a while causes severe losses. But over the years these things do not last long.

Some Unusual Problems in Spray Applications

Growers most of the time grow trees that have some epiphytes but are not excessively burdened with them. In various locations, the ecology is such that epiphytes thrive, and they consist largely of lichens, algae, mosses, liverworts, bromeliads, weed-like orchids, small bushes of loranthaceae, climbing grasses, crawling bamboo, and ferns. It is these that flourish and exist as increasing burdens on living bark of plantation trees; epiphytes become a serious detriment. Examples of the problem of three affected crops are as follows.

On trunks and branches of cacao trees, epiphyte growth can be seen so thick that flower cushions cannot push out their buds. On trees of coffee, epiphytic growth can be found so heavy that it bends over the thicker, upward-growing (orthotropic) leader branches on which the slender horizontal fruit branches develop. This bending results in stimulation of adventitious, ordinarily quiescent, non-fruiting orthotropic buds--and fruit branches stop growth. On trees of hevea rubber, mosses and algae and lichens grow on tapping panels and can be so thick that costly, special trunk cleaning is needed.

With a large supply of hand labor, epiphytes are pulled off and fairly well reduced in quantity. This kind of work is not always thorough, and in the case of cacao the

*An exception to this is in what is called a frost-free part of Florida, where special varieties of "sweet corn" are grown for the winter market, and the fields are sprayed against leaf diseases.

flower cushions are sometimes injured. Spray work has shown that control is also possible by the application of proper chemicals (25, 75) such as weak creoline emulsions, copper-containing proprietary compounds, and wettable sulfur. I have successfully used strong formulas of bordeaux mixture, but an orchard oil put on as a light fog had little effect.

The *Mycena citricolor* fungus with its large and heavy propagules, flourishes under wet-forest to water-drenched rain forest ecological conditions (221 publ. 1961). Under these wet conditions, spread is by water drops although at a more reduced pace than what appears to occur in certain other water-drop dispersals (110) of small-spored parasites.

Use of spray for the *Mycena* on coffee has long been from the standpoint of protection (47, 105, 221 in 1961, and many others). Fungicides most often used in former days have been of varying bordeaux mixture formulas. In some places, under a special kind of plantation growing, coffee is sometimes never free from *Mycena.* The fungus also attacks cinchona and some 500 other kinds of plants, and in some areas sources of infection are plentiful. The ever-moist conditions during times of spray applications require special attention to increasing the sticking characteristics of sprays. Even bordeaux suffers rapid attrition from weathering. Attempts are made to replace standard bordeaux with other copper-containing fungicides, (see e.g., 145, 162, 187) but these fall far short of the adhesiveness displayed by bordeaux mixture.

It is to be repeated here that every spray tried, even bordeaux, has been significantly improved in results when a sticker is included in the spray formula (e.g., 105, 132, 187). Moreover, although this is only with respect to the special coffee *Mycena,* in the tropics, sticker additives in sprays are almost invariably notable in giving added good response in disease control (24, 27, 50, 58, 69, 164, 187, 221 in 1955 publ.). Of interest has been research reported from Colombia (50) on stickers used in sprays against *Phytophthora* on potato. Employing stickers alone on non-treated plants gave a disease amount of about an index of 98; treated with maneb alone, disease was 36; but when maneb was added to sticker, the disease was at an index of only 7 and below. Tests in other parts of the Andes (58) also resulted in proving the value of stickers added to potato sprays.

It should be expected in the tropics that an insecticide also exhibiting fungicidal value will gain special attention. This has been found by coffee pathologists when they learned, first from workers in Colombia, that needed insecticide sprays of lead arsenate and paris green also eliminated *Mycena* leafspot (49, 78, 82, 83, 186) on coffee. From special attention, it is now known that these sprays act as eradicants of the *Mycena* organism.

The poisons definitely are absorbed and translocated in the host and even some of it goes to the seed or bean that is used in commerce. Fortunately the poison in beans is in such a small amount (176) that it in no way inhibits use or sale of the coffee.

Findings that lead arsenate is absorbed in plants and is good in a coffee disease spurred its being tested on some other things, and work in Ecuador (77, 78) has uncovered, for instance, that *Puccinia* rust of potato is in no way controlled by the insecticide. The bare fact that a fungicide may be absorbed in coffee does not necessarily make of it a cure-all. For example, n-dodecylguanadine acetate that is well absorbed in coffee has no fungicidal effect on *Mycena.*

Insect-disease relationships are very common in our tropics. It should be noted here, from a few examples, that the use of various means to control insects is directly responsible in the tropics for the control of those diseases that the insects carry. A well publicized disease-insect relationship encountered by tropical planters is the branch dieback or woodstain diseases from *Ceratocystis* (= *Ceratostomella*) that occurs in cacao, mango and coffee. It is closely involved with bark and wood beetles or weevils (2, 3, 99, 200 and others) and both pruning to excise infested wood and judicial use of insecticides aids in keeping these three crops reasonably free from woodstain.

Another weevil-carried infection of a very different character is the redring disease of coconut and oil palm. Nematodes, *Radinophelenchus cocophilus* are carried by the weevil or beetle *Rhynchophorus palmarum* (111, 125). Combining rogueing processes with palm-tree poisoning results in considerable success.

With the recent occurrence of rust on coffee in Brazil, some special interest has been turned to sprays for its control. Until the year 1970 no spraying for this rust has ever been done in the Neotropics, because it has not been present. However, a multitude of other rusts are, and some have good spray controls. A few miscellaneous examples: rust on potato from Puccinia pitteriana is known to be controllable by spraying (58, 77, 78) using not only micronized sulfur, but certain other organic compounds as well, such as zimate, phaltan, thylate, and manzate. It is of interest that in Ecuador the organic fermate, the fungicide cupravit, and the inorganics lead arsenate and ditox had no effect on this rust. The common guava Puccinia can be well controlled by organic fungicides (13, 99). While leadership in the spraying of cereals to control rusts is going on in the temperate zone (191), it is notable that this is not being neglected in the tropics. Disease-resistant varieties are a long-time development, and, for example in Colombia (184), the farmers spray the crops both with maneb alone and also with maneb combined with both nickel and zinc. In 1970 a systemic fungicide has been found effective (18) on bean rust, also maneb combined with either nickel or zinc salts.

Coffee rust spreads rapidly (in India spores have been recovered (81) from traps 200 to 500 yards in the air above diseased trees) and for the first time in the Neotropics coffee is being sprayed for the disease (54 report 1970). Copper-containing sprays are being tested and so also are numerous organics and systemics. Some of the first tests with orchard oils seem promising. Even if in the end some good resistant varieties are grown, sprays will probably always be necessary to take care of areas not using the resistant types. Also it should never be forgotten that new races may appear suddenly (221 publ. 1961; 205) and without warning and infect resistant kinds.

White mildews, the Erysiphaceae, are common in certain parts of the Neotropics during dry seasons, if temperatures are reasonably cool and dews are frequent. Many crops are attacked, such as potato, tomato, numerous cucurbits, ornamentals, grape, pea, peach and rosaceous fruits, and others. Often sulfur by itself is dusted on in dry powdered form, or in combination with lime (80 per cent sulfur, 20 lime), and various liquid sulfur-containing sprays (58, 68, 98, 69, 99, 128, 200).

Literature Cited

[When it came time to write Chapter XXI I had more than 1000 citations that could have been used. Weeding out duplicate reports and mere mentions of sprays that were never tried reduced this to a more manageable number. Also, some of the older works have been superseded by newer, and some things of wholly academic discussion have been eliminated. The final list of literature contains only those items I worked with in the final preparation of this chapter.]

1. Abiusso, N. G. 1947. Preparación y propriedades del acido penetatiónico: su acción fungicida. Rev. Quim, LaPlata Univ. 20:17-79. --. 1949. Compuestos fungicidas organicos. Almanaque Min. Agr. Nac., Argentina 24:357-358. --. 1952. Compuesto del azufre: su comportamiento respecto a los hongos fitopatógenos. Rev. Inv. Agr., Argentina 6:131-156.

2. Abrahão, J. 1964. Mangueira com antracnose e murcha. Biológico 30:210.

3. ________., V. Rossetti, and R. Elias. 1966. Contrôle da sêca on murcha da mangueira, causada por Ceratocystis fimbriata. Sym., Dep. Prod. Veg., Secr. Agr. São Paulo 4:12.

4. Accatino, P. 1963-64. Ensayo de control químico de la enfermedad roya o polvillo de la lenteja. Agr. Tec., Chile 23-24:7-14.

5. Adams, P. G., G. C. Papavizas, and J. A. Lewis. 1968. Mechanism of control of fusarium root rot of beans with spent coffee grounds. Phytopathology 58:883.

6. Albuquerque, F. C. 1968. Piper colubrinum, a grafting root stock for Piper nigrum resistant to diseases caused by Phytophthora palmivora and Fusarium solani. Pesq. Agrop. Brasil 3:141-145.

7. Alippi, H. E. 1965. Tizón temprano o cercosporiosis del apio (Cercospora apii Fres.). Rev. Fac. Agron., Univ. Nac. LaPlata 41:51-56.

8. Alvarez-G., L. A. 1946. The control of Rhizoctonia damping-off of pepper and eggplant in Puerto Rico. J. Agr. Univ. Puerto Rico 30:69-96.

9. ________., J. Adsuar, and P. Rodriguez. 1953. Fungicidal control of early and late blight of the potato in Puerto Rico. J. Agr. Univ. Puerto Rico 37:288-297.

10. ________., and L. Lopez. 1953. Results of tests with new fungicides for the control of damping-off of tobacco in Puerto Rico. J. Agr. Univ. Puerto Rico 37:283-287.

11. Amoral, R. E. D. M., and A. F. Cintra. 1968. Health conditions of rice seeds in the state of São Paulo Brazil. Biológico 34:175-178.

12. Andersen, A. A. 1958. New sampler for the collection, sizing and enumeration of viable airborne particles. J. Bact. 76:471-484.

13. de Andrade, A. C. 1948. Progressos no estudo dos fungicidas organicos. Biologico 14:57-64. --. 1951. O contrôle da ferrugem da goiabeira por meio de pulverizações. Agr. Inst. Biológico. 20:127-146.

14. ________., and J. M. Salles. 1949. Pulverização da batatinha. Biologico 15:187-188.

15. Arango V., A. 1945. Uso de la creolina. Vida Rurál, Colombia 6:12.

16. Argentine Research. 1967. Ensayos comparativos para control de viruela del tomate. Fitosanitarias, Argentina 5(10):6-14.

17. Badger, A. M. 1965. Influence of relative humidity on fungi causing crown rot of boxed bananas. Phytopathology 55:688-692.

18. Baldwin, R. E. 1970. Control of snapbean rust with systemic and nonsystemic fungicides. Phytopathology 60:1013.

19. Barnes, H. V. and J. M. Allen. 1951. A Bibliography

of Plant Pathology in the Tropics and in Latin America. U. S. Dep. Agr., Bibl. Bulletin 14, 78 p.

20. Barros, O. 1966. A new form of Fusarium roseum associated with root fungus in the Colombian bean. Rev. Inst. Agrop., Colombiano 1(2):79-86.

21. ________., C. Cardona, R. Cardenosa, and R. L. Skiles. 1958. The control of angular leaf spot of bean in Colombia. FAO Plant Prot. Bull. 6:97-101.

22. Batista, A. C. 1948. O anel vermelho do conqueiro e a fumigação do solo com DD. Bol. Sec. Agr. Indus. Comer., Pernambuco 15:356-387.

23. ________., and H. Carneiro. 1947. A ação do DD, comprovada eficiente contra Fusarium bulbigenum var. lycopersici (Brushi) Wr. e R. Bol. Sec. Agr. Indus. Comer., Pernambuco 14:62-71. --. and --. 1947. DD como nematicida, para a horticultura. Bol. Sec. Agr. Indus. Comer., Pernambuco 14:147-157.

24. Befeler F., P. E. 1960. Investigacion sobre el uso de adherentes en fungicidas en el control de Phytophthora palmivora Butl. Inter-Amer. Conf. Cacao., Proc. 7:118-129.

25. Bernal E., G. and J. J. Castano A. 1953. Consideraciones sobre la manera de combatir los musgos del cafeto. Rev. Cafetera Colombia 11:4004-4009.

26. Berta, J. A. and H. D. Tobler Bottini. 1945. Ensayos del caldo bordeles contra la Phytophthora infestans (Mont.) de Bary en Solanum tuberosum. Asoc. Ing. Agron. Rev., Uruguay 17:3-11.

27. Bertelli, L. K. 1947. Preparación del caldo bordeles adherentes. Min. Ganad. Agr. Uruguay, Bol. Inform. 4:170.

28. Bianchini, C. L. and F. L. Wellman. 1958. Experimentos en el control de Pellicularia del café y ciertas differencias en el pelicularias de cinco huespedes. Turrialba 8:73-92.

29. Birchfield, W. 1969. Nematocides for control of plant parasitic nematodes on sugarcane. Plant Dis. Reptr. 53:530-533.

30. Bluffand-de Konig, M. H. 1964. Jaarverslag 1963 (pp. 71-75). Meded. Landpr. Suriname 34, 93 p.

31. Boock, O. J. 1949. O fumigante dowfume W-10 no contrôle aos nematoides da batatinha. Rev. Agr., Piracicaba 24:25-42. --. 1951. Combate aos nematoides pela aplicação de fumigantes no solo. Efeito do D-D e dowfume W-40 no combate aos de batatinha. Bragantia 11:13-18.

32. ________., and L. G. E. Lordello. 1952. Tratamento dos tuberculos-semente de batatinha com brometo de metilo no combate aos menatoides das galhas. Bragantia 12:363-365.

33. Brandes, G. A. 1971. Advances in fungicide application technique. Ann. Rev. Phytopatho. 9:(in press.)

34. Brenchley, G. H. 1965. The early stages of potato blight epidemics. European Potato J. 8:187.

35. Brogdon, J. E. and F. P. Lawrence. 1962. Control of insects and diseases of dooryard citrus trees. Univ. Florida Circ. 139A, 16 p.

36. Brooks, R. F. 1966. Comparison of a helicopter with a conventional airblast sprayer in the application of a citrus spray program. Florida State Hort. Sco. Proc. 78:69-73.

37. Burnett, H. C. 1965. Orchid diseases. Florida State Dept. Agr. (Bulletin) (3), 57 p.

38. Cabrera C., J. 1953. Mezclas entre fungicidas, fertilizantes y fitohormonas en aspersiones sobre de cacao. Acta Agron., Colombia 3:229-250.

39. Calpouzos, L. 1966. Action of oil in the control of plant disease. Ann. Rev. Phytopath. 4:369-390-

40. ________., W. A. Brun, T. Theis, and C. A. Colberg. 1960. A precision spray technique for evaluting oils for sigatoka disease control on individual

banana leaves in the field. Phytopathology 50: 69-72.

41. ________., C. Colberg, and T. Theis. 1962. Oil-spray field trials on Cercospora diseases of tobacco, peanut, zinnia, and gramalote grass. Plant Dis. Reptr. 46:105-106.

42. ________., T. Theis, C. M. Rivera, and C. Colberg. 1959. Studies on action of oil in the control of Mycosphaerella musicola on banana leaves. Phytopathology 49:119-122.

43. de Camargo, A. P., R. M. G. Cardoso, and N. C. Schmidt. 1967. Behavior and ecology of the South American leaf blight of rubber tree under climatic conditions of the São Paulo state plateau. Bragantia 26:1-18.

44. Campi, M. D. 1946. Ensayo comparativo de la eficacia de tres fungicidas sobre la enfermedad del clavel producida por Heterosporium echinulatum (Berk.) Cke. Inst. San. Veg., Argentina (Bulletin) 2, 12 p.

45. Cardona S., H. 1949. Aplicación de los productos químicos para proteges la caficulturo. J. Cafe de El Salvador 19:1299-1303.

46. Carrera, C. 1948. Tratamiento o cura de las semillas de los cereales. Bolsa de Cereales, Buenos Aires 35:19-21; 28-37.

47. Carvajal B., F. 1939. Ojo de gallo (Omphalia flavida). Rev. Inst. Defensa del Café, Costa Rica 7:535-576.

48. Carvalho, J. C. 1958. Influencia de alguns fatóres na fumigação do solo. Rev. Agr., Brazil 33: 79-86.

49. Castano A., J. J. 1957. El arseniato de plomo (DuPont Nu Rexform) en el control de la gotera del cafeto. Rev. Cafetera Colombia 13(130):36-44.

50. ________., and H. D. Thurston. 1968. Efecto de adherentes aplicados aisladamente y combinados

con maneb en el control de Phytophthora infestans de la papa. Agr. Trop., Colombia 24:223-225.

51. Castillo, A. U. 1967. Enfermedades importantes del hule Hevea brasiliensis en Mexico. Fitopat. Mexicana 1(2):16-23.

52. Cercos, A. P. 1950. Antibiotico DINR. 49-I (fungocina) producido por Bacillus subtilis con actividad sobre hongos patogenos de animales y vegetales. Rev. Invest. Agr., Argentina 4:325-337.

53. (Ceylon Department of Agriculture). 1954. Helicopter spraying for the control of leaf disease of rubber in Ceylon. World Crops 6:117-118.

54. Chaves, G. M. 1954. Metodo bioanalítico para evaluacion comparative de adherentes en fungicides orgánicos. Turrialba 4:39-42. --. 1964. Estudos sobre Sclerotinia sclerotiorum (Lib.) DeBary. Experientiae, Univ. Vicosa, Brasil 4:69-133. --. 1970. [Personal letter.]

55. ________., and F. A. A. Couto. 1958. Ensaios comparativos sobre a eficiencia de fungicidas organicos e a base de cobre no contrôle da mela (Phytophthora infestans (Mont.) de Bary) e da septoriose (Septoria lycopersici Speg.) do tomateiro. Ceres 10:303-316.

56. Childers, N. F., H. F. Winters, P. Sequinot, and H. K. Plank. 1950. Vegetable gardening in the tropics. U. S. Dep. Agr., Puerto Rico Fed. Exp. Sta. Circ. 32, 144 p.

57. Chinn, S. H. F. and R. J. Ledigham. 1962. A laboratory method for testing the fungicidal effect of chemicals on fungal spores in soil. Phytopathology 52:1041-1044.

58. Christiansen G., J. 1967. El Cultivo de la Papa en El Peru. Museo Nocional, Lima, Peru, 351 p. (Diseases pp. 196-232).

59. Coelho de Andrade, A. 1951. O contrôle da ferrugem da goiabeira por meio de pulverizações. Arq. Inst. Biol. S. Paulo 20:127-146. --. 1952.

Fungicidas modernas para controlar à requeima do tomateiro. Biologico 18:6-14.

60. ________., and D. Puzzi. 1953. Ensaio de fungicidas para controlar as manchas fungidas dos citrus. Biologico 19:144-148.

61. Compacci, C. A. and M. B. Figueiredo. 1964. Ocorrencia de rhizoctoniose em amendoim. Biologico 30:267-268.

62. ________., and L. O. C. Rezende. 1964. Fungicidas para o contrôle das doenças da folhagem da batatinha. Reun. Fitossan. Brasil, An. 8:49-55.

63. ________., ________., and H. V. Arruba. 1963. Contrôle das doenças da folhagem de pepino. Arq. Inst. Biologico 10:214-216.

64. ________., ________., and J. Nakamoto. 1966. Ensaio de campo com fungicidas para o contrôle das doenças do folha da cebola. Biologico 32: 184-188.

65. Conover, R. A. 1966. Results of recent experiments on control of mango anthracnose. Florida State Hort. Soc. Proc. 78:364-369.

66. Costa, J. J. 1957. Estudio comparativo de los metodos de laboratorio y de campo para la valoracion de anticaries comerciales; sus valores reales y practicos. Rev. Univ. Agron. Vet., Buenos Aires 14:22-42.

67. Crowe, T. J. 1963. Possible insect vectors of the uredospores of _Hemileia vastatrix_ in Kenya. Trans. British Mycol. Soc. 46:24-26.

68. Danies L., C. A. and F. Penaranda Canal. 1941. Mildeu de la vid. Rev. Facult. Agron. Univ. Nac. Colombia 4:1009-1036.

69. Dantas, B. 1951. O lucro de pulverização na lavoura cacaueira. Teoria e practica. Pernambuco, Brasil Sec. Agr. Indus. Comer., Boletin 18:218-256.

70. Delfel, N. E., L. Calpouzos, and C. Colberg. 1962. Measurement of spray-oil volatility and its relation to sigatoka fungus disease control and phytotoxicity on banana leaves. Phytopathology 52: 913-917.

71. Desrosiers, R. 1957. El uso del helicoptero para la atomizacion del banano. Asoc. Nac. Bananeros, Ecuador 2:31-32. --. 1958. The control of sigatoka disease on the gros michel banana by low volume spraying in Ecuador. Min. Fom. Producc. Dir. Gen. Agr., Ecuador Tec. Boletin 1, 58 p. --. 1960. Ultimos avances en el control de enfermedades del cacao en el campo. Inter-Amer. Cacao Conf., Colombia, Proc. 7:99-103.

72. ________., and E. Ampuero. 1957. Control de sigatoka en bananos usando la neblindora protatil. Serv. Coop. Interam. Agr., Quito, Ecuador, Misc. (publication) 2, 6 p.

73. ________., and C. W. Bolanos. 1955. Inhibition of sporophore development of Marasmius perniciosus (stahel) on cacao. Turrialba 5:28-32.

74. ________., and A. von Buchwald. 1955. Fungicidal treatments for control of cacao cutting disease in Ecuador. FAO. Plant Prot. Bull. 4:17-21.

75. ________., and J. Diaz M. 1956. Efecto de diversos fungicidas en el control de la Monilia. Turrialba 6:19-22.

76. Dianese, J. C. 1966. Fungicidas no contrôle da mela da batatinha (Solanum tuberosum) causada por Phytophthora infestans (Mont.) de Bary. Ceres 8(72):1-8. --. 1966. Doenças do abacaxi. Bol. Campo, Brazil 21:5-12.

77. Diaz Moreno, J. and J. Echeverria. 1963. La roya de la papa causada por Puccinia pitteriana y su combate quimico en Ecuador. Turrialba 13:152-158.

78. ________., and V. M. Napoles. 1961. Comportamiento de various fungicidas en el combate de Mycena

citricolor del cafeto en Ecuador. Turrialba 11: 93-97.

79. Drummond, O. A. 1940. Doenças do milho. Ceres 1:478-482.

80. Drummond-Goncalves, R. 1943. Meios de combate a antracnose da videira. Sitios e Fazendas, Brazil 8(3):36-37.

81. D'Souza, G. I. 1965. Mycology. Res. Dep. Indian Coffee Board Rept. 1964-1965:73-84.

82. Echandi, E. 1956. Inhibition of gemmae (cabecitas) production of Mycena citricolor on coffee trees. Plant Dis. Reptr. 40:775.

83. ________., and R. H. Segall. 1958. The effectiveness of certain eradicant fungicides on inhibition of gemmae of Mycena citricolor. Phytopathology 48: 11-14.

84. Ecuador. Experimental Station. 1967. Programa de fitopatologia. Estac. Exp. Santa Catalina, Ecuador. Inst. Inves. Agrop., Ecuador 1965:47-49.

85. Ellis, D. E. 1970. Plant disease control 1970-1980. Address at Univ. Florida, Gainesville, Florida (Processed), 11 p.

86. England, Ministry of Agriculture. 1968. Agricultural Chemical Approved Scheme. 1968. List of Approved Products and Their Use for Farmers and Growers. London, Min. Agr. Fish Food, 149 p.

87. Escobar, R. 1943. El chapete. Agr. Mexicana 59 (12):6-7.

88. Estrada R., N. and F. Villamil G. 1948. Experimento de campo sobre el control de nematodos a base de D-D. Agr. Trop. Colombia 4:37-40.

89. Fawcett, H. S. 1936. Citrus diseases and their control. McGraw-Hill, New York, 656 p.

90. Feder, W. A. 1960. Osmotic destruction of plant

parasitic and saprophytic nematodes by the addition of sugars to soil. Plant Dis. Reptr. 44: 883-885.

91. Fernandez Valiela, M. V. 1952 (2nd ed.). Introducción a la Fitopatología. Gadola, Buenos Aires, 872 p.

92. Fiqueiredo, M. B., R. M. G. Cardoso, and J. Abrahao. 1966. Perdas causadas pelo fungo Mycosphaerella melonis (Pass.) Chiu & J. C. Walker em aboboreiro de moita. Biológico 32:203-205.

93. Fischer, G. L., R. Santoro, and M. Aznarez. 1938. Ensayos de germinación de trigos sometidos a tratamientos anticriptogamicos. Arch. Fitotéc. Uruguay 3:69-85.

94. Fisher, F. E. 1968. Timing of spray applications for control of Cercospora diseases of citrus. Phytopathology 58:553.

95. di Fonzo, M. A. 1941. La desinfestación de la semilla de algodonero y otros factores relacionados con su poder germinativo. Junta Nac. Algodón, Argentina, Publicación 53, 15 p. --. 1943. Parasitos que debe conocer el agricultor algodonero. Suelo Argentino 2:783-784.

96. Fresa, R. 1941. Aplicación del auto-hervido (azufre-cal) en la fruticultura de la zona del delta del Paraná. Min. Agr., Argentina, Almanaque 16: 399-401.

97. Fulton, R. H. 1965. Low-volume spraying. Ann. Rev. Phytopath. 3:175-196.

98. Galano, R. M. J. 1939. Una enfermedad del rosal; el blanco. Rev. Hort., Buenos Aires 5:9-11.

99. Galli, F., H. Tokeshi, P. de C. Torres de Carvalho, E. Balmer, H. Kimati, C. O. Nogueira Cardoso, and C. Lima Salgado. 1968. (Fungicides, pp. 111-142). Manual de Fitopatologia. Ceres, São Paulo, 640 p.

100. Garcia Rada, G. 1947. Fitopatologia Agricola del

Peru. Estac. Exp. Agr., La Molina, Peru, 423 p. --. 1948. Control of peach powdery mildew (_Sphaerotheca pannosa persicae_) in Peru. Plant Dis. Reptr. 32:319-320.

101. Gonçalves da Silva, S. 1944. Noções Sôbre Doenças e Pragas das Plantas e seu Combate. Serv. Inform., Min. Agr. Rio de Janeiro, 140 p.

102. Gonzales M., F. 1941. Informe especiál sobre el control de las bacterias y modo de combatir la pérdida premature del tomate, por medio del uso de compuestos a base de cobre. Rev. Com. Exterior, Mexico 6:91-96.

103. Goos, R. D., E. A. Cox, and G. Stotzky. 1961. _Botryodiplodia theobromae_ and its association with _Musa_ species. Mycologia 53:262-277.

104. Gowda, B. S. P. and J. C. Joshi. 1968. Effect of bordeaux mixture spray on fruit set in coffee (_Coffea arabica_). J. Mysore Agr. Sci. 12:247-249.

105. Grainger, Jr., A. and F. L. Wellman. 1955. Algunos fungicidas usados con adherentes en aspersiones para cafetos jovenes. Turrialba 5:123-131.

106. Gray, R. A. 1958. The downward translocation of antibiotics in plants. Phytopathology 48:71-78.

107. Greene, G. L. 1966. Response of conidia and appressoria of _Gloeosporium musarum_ to hypochlorous acid. Phytopathology 56:1201-1203.

108. ________., and R. D. Goos. 1963. Fungi associated with crown rot of boxed bananas. Phytopathology 53:271-275.

109. Gregory, P. H. 1961. The microbiology of the atmosphere. Leonard Hill, London, 251 p.

110. ________., E. J. Guthrie, and M. E. Bunce. 1959. Experiments on splash dispersal of fungus spores. J. Gen. Microbiol. 20:328-354.

111. Griffith, R. 1968. Relationship between the red ring

nematode and the palm weevil. J. Agr. Soc. Trinidad Tobago 68:342-356.

112. Gutter, Y. 1968. Screening trials for control of green mold in valencia oranges. Plant Dis. Reptr. 52:777-779. --. 1969. Comparative effectiveness of benomyl thiobendazole and other fungal compounds for post harvest control of penicillium decay in shamouti and valencia oranges. Plant Dis. Reptr. 53:474-478. --. 1969. Effectiveness of the pre inoculation and post inoculation treatments with sodium orthophenyl phenate thiabendazole and benomyl for green mold *Penicillium digitatum* control in artifically inoculated eureka lemons. Plant Dis. Reptr. 53:479-482.

113. Guevara, J. M. 1945. La *Peronospora* de la vid. Prejuicio que ocasiona. Su evolucion y tratamientos preventivas. Argentina Sec. Ind. Comer. Bol. 1:259-264.

114. Guyot, H. 1953. La lutte contre *Cercospora musae* dans les bananerais de Guadeloupe. Essais de nebulisation (fogging). Fruits Outre Mer 8:525-532. --. 1954. 9:293-296.

115. Guyot, H. and J. Cuillé. 1954. Les formules fongicides huileuses pour le traitement des bananerais. Fruits Outre Mer 9:289-292. --. and --. 1955. Les traitements fongicides des bananerais. II. Efficacités des differents modes de traitements. Rôle de huile. Fruits Outre Mer 10:101-107.

116. Hainsworth, J. G. E. 1952. Tea pests and diseases and their control. W. Heffner & Sons, Cambridge, England, 130 p.

117. Harding, P. R. Jr. 1968. Comparison of fungicide thiabendazole and sodium orthophenylphenate for control of penicillium molds of post harvest citrus. Plant Dis. Reptr. 52:623-625.

118. Hearn, C. J. and J. F. L. Childs. 1969. A systemic fungicide effective against sour orange scab disease. Plant Dis. Reptr. 53:103-105.

119. Hernandez Paz, M. 1967. Pudrición de la raíz.

Rev. Cafetalera, Guatemala 69:16-17. --. 1967. Dos enfermedades del cafe. Rev. Cafetalera, Guatemala 69:22-23.

120. Hiroce, R. and C. M. Duarte. 1962. Ensaio comparativo sobre à eficiencia de alguns fungicidas no contrôle da queima das folhas (Alternaria dauci (Kuehn) Groves & Skolko) da cenoura (Daucus carota L.). Olericultura 1:96-103.

121. Hirst, J. M. 1952. An automatic volumetric spore trap. Ann. Appl. Biol. 39:257-265. --. 1959. Spore liberation and dispersal (pp 529-538), In Plant pathology problems and progress, Univ. Wisconsin Press, 588 p.

122. ________., and O. J. Stedman. 1963. Dry liberation of fungus spores by rain drops. J. Gen. Microbiol. 33:335-344.

123. ________., O. J. Stedman, and G. W. Hurst. 1967. Long-distance spore transport: vertical sections of spore clouds over the sea. J. Gen. Microbial. 48:329-355; 357-377.

124. Hocking, D. 1967. Fungicides for arabica coffee. Brief notice of candidate fungicides for Colletotrichum coffeanum. Turrialba 17:351-353.

125. Hoyle, J. C. 1968. A simple method for poisoning coconut trees. J. Agr. Soc. Trinidad Tobago 68:459-465.

126. de la Hoz, C. 1940. El problema bananera. Tratamiento de la sigatoka. Rev. Agr. Guatemala 17:98-101.

127. India. Coffee Research. 1968. Mycology. Res. Dep. Indian Coffee Board, Rept. 1967-68:41-45 (processed).

128. Infante, G. 1943. Azufradura en las vinas. Asoc. Rual, Uruguay 70(4):51-52.

129. Klein, H. H. 1960. Control of cercospora leaf spot of bananas with applications of oil sprays based on the disease cycle. Phytopathology 50:488-490.

130. Knorr, L. C., R. F. Suit, and E. P. Ducharme. 1957. Handbook of citrus diseases in Florida. Univ. Florida Agr. Exp. Sta. Bulletin 587, 157 p.

131. Kolbezen, M. J., D. E. Munnecke, and L. H. Stolzy. 1968. Fumigating soils for oak root fungus control. California Citrog. 53:439; 449-450.

132. Kramer, M. and R. C. de Andrade. 1943. Estudos sobre adesivos da calda bordaleza. Biologico 9:317-330.

133. Leach, R. 1946. Banana leaf spot (Mycosphaerella musicola) on the gros michel variety in Jamaica; investigations on the aetiology of the disease and the principles of control by spraying. Kingston, Jamaica (special bulletin publication), 118 p.

134. Lellis, W. T. 1954. La pudrición parda de los frutos del cacao en Bahia y algunos trabajos realizados para su combate. Cacao 3:13.

135. ________., and M. E. A. F. de Mata. 1956. Competencia de fungicidas en el control de la pudrición parda de los frutos del cacao. Cacao 3:16.

136. ________., and O. Peixoto Jr. 1960. Pulverização-versus polvilhamento no contrôle da podridão parda dos frutos do cacaueiro. Inter-Amer. Cacao. Conf., Colombia, Proc. 7:152-158.

137. Lems, G. 1963. Resultaten van een cacao-klonen experiment. Suriname Landb. 11:183+.

138. Lepage, H. S., O. Giannotti, and A. Orlando. 1947. Notas sobre o combate ao nematoide da raiz (Herterodera marioni) pelo fumigante DD. Biologico 13:123-124.

139. Lima, A. G., F. A. A. Couto, and G. M. Chaves. 1962. Ensaio comparativo da eficiencia de fungicidas associados a espalhantes adesivos no contrôle à mancha púrpura da cebola. Olericultura 2:168-175. --., --. and --. 1962. Contrôle da mancha púrpurea do alho (Allium sativum L.) por meio de pulverizações com fungicidas de proteção, combinados com espalhante adesivo. Olericultura 2:203-211.

140. Llosa P., T. 1939. El caldo sulfocálcico y su aplicación como fungicida. Agron. La Molina, Peru 4(16):43-48.

141. Lopes, N. F. C. 1960. A cercosporiose da bananeira. Dep. Def. San. Veg., Min. Agr., Rio de Janeiro, Brasil (Bulletin), 34 p.

142. Lopez M., A. 1957. Efectividad de varios fungicidas en la represión del damping-off y la pudrición de semillas de frijol (Phaseolus vulgaris L.). Acta Agron., Colombia 7:141-163.

143. McCormack, A. A. and G. E. Brown. 1967. Thiabendazole an experimental fungicide for fresh citrus fruits. Florida State Hort. Soc. Proc. 80:230-237.

144. Malaguti, G. and A. V. Rodriguez. 1954. La pudrición de las esquejas de la caña de azucar y posibilidad de su control con fungicidas. Agr. Trop. Venezuela 4:61-67.

145. Maravalhas, N. 1941. Nova sustancia anti-criptogamica sugestoes par o estudo do aproveitamento, na lavoura, de nova sal de cobre. Rev. Agr., Piracicaba, Brasil 16:182-184.

146. Marchionatto, J. B. 1929. Investigaciones sobre el tratamiento seco en la lucha contra la caries del trigo. Univ. La Plata, Argentina, Rev. Fac. Agr. 16:88-92. --. 1929. La lucha contra el carbón volador del trigo. Bol. Min. Agr., Argentina 28:229-231. --. 1943. Enfermedades de Los Frutales y Procedimientos para Combatirlas. Edit. Sudamericana, Buenos Aires, 196 p. --. 1944. La enfermedad de los almácigos y procedimientos para combatirla. Suelo Argentina 3:14. --. 1944. Manual de Las Enfermedades de Las Plantas. Edit. Sudamericana, Buenos Aires, 368 p. --. 1946. Reseña de la sanidad vegetal en la República Argentina. Univ. Buenos Aires, Fac. Agron. Vet., Publicacion 2, 117 p. --. 1947. El control del damping-off o enfermedad de los almacigos. Cienc. Invest., Argentina 3:164-165. --. 1948. Las enfermedades de las plantas, sus causas y remedios. Soc. Cient.

Argentina An. 146:417-428. --. 1948. Tratado de Fitopatologia. Edit. Sudamericana, Buenos Aires, 537 p.

147. Martinez, J. 1951. Tratamientos de las semillas en los cultivas de maiz, hortalizas y sorgos. Ofic. Estud. Esp. Mexico, Fol. Mis. 4:162-165.

148. Martinez, J.A. and P.B. Cruz. 1961. Contrôle quimico da murcha do amendoim. Biológico 27: 191-193.

149. ________., L.O.C. Rezende, and G.M. de Fazio. 1962. Experiencia para o contrôle da podridão da antracnose do fruto do morangueiro. Biológico 28:296-299.

150. May, K.R. 1967. Physical aspects of sampling air-borne microbes (pp:60-80). In Gregory, P.H. and J.L. Monteith (eds.) Airborne Microbes (Symposium). Cambridge Univ. Press, 385 p.

151. Mazzei P., J. 1950. Notas fitopatológicas. Control de la antracnosis de vid. Uruguay Dir. Agron., Publicación 100, 6 p.

152. Medeiros, A.G. 1964. Pesquisas sôbre à podridão parda. Cent. Pesq. Cacau, Tec. Communicado 8, 26 p. --. 1965. Combate à pordidão porda. Cacau Actualidades, Brazil 2:3-4.

153. Medina, G. 1943. Da calda bordaleza ao flordo. Rev. Agron., Porto Alegre, Brazil 7:334-335.

154. Meredith, D.S. 1960. Studies on Gloeosporium musarum Cke. & Massee causing storage rots of Jamaica bananas. 1. Anthracnose and its chemical control. Ann. Appl. Biol. 48:279-290.

155. Merny, M. 1954. Recherche d'une technique de microessais de produits fungicides contre Cercospora musae a la Guadeloupe. J. Agron. Trop. Bot. Appl., France 1:61-70.

156. Miller, P.M. 1969. Suppression by benomyl and thiabendazole of root invasion by Heterodera tabacum. Plant Dis. Reptr. 53:963-966.

157. Montano, A., and M. Webster. 1938. Consideracões sobre a peronospora e antracnose e seus tratamentos. Rio Grande do Sul, Brazil, Rev. Agron. 2:937-941; 1039-1042.

158. Monteiro, M. V. M., J. P. Stupiello, and M. Maeji. 1964. Algumas notas sobre o mal de sigatoka e seu contrôle no litoral paulista. Reun. Fitossan. Brasil, An. 8:31-40.

159. Moreau, B. 1967. La culture bananière en Colombia (règion de Santa Marta). Fruits Outre Mer 22: 557-577.

160. Neto, J. P. da C. 1943. Doenças da cevada no Rio Grande do Sul. Sec. Estad. Negoc. Agr. Indus. Comer., Rio Grande do Sul, Boletim 69, 18 p.
--. 1943. Doenças do tomatevio. Rev. Agron. Rio Grande do Sul 7:261-262.

161. Newhall, A. G. 1968. To increase efficiency or lower the cost of the control of Phytophthora pod rot of cacao. Turrialba 18:79-82.

162. ________., A. Paredes, and L. G. Salazar. 1968. Control of phytophthora pod rot on cacao with fungicides. Results of three experiments. Cacao. 13(1):3-13.

163. Nieto Soto, L. A. 1945. Control de la lancha o gota serena de la patatas y el tomate. Cámera Agr., Ecuador Rev. 7:712-716.

164. Obregon Botero, R. 1940. Experimentos sobre adherentes para fungicidas. Rev. Cafet. Colombia 8:2605-2606.

165. Ocaña, G. and A. J. Hansen. 1960. Preliminary studies on the use of agricultural spray oil for the protection of cacao against *Phytophthora palmivora*. Inter-Am. Cacao Conf., Trinidad Tobago, Proc. 8:196-199.

166. Orellana, R. G. and C. Bianchini. 1956. Combate de la *Pellicularia* del cacao con fungicidas. Cacao 3:15.

167. ________., and L. R. Siller. 1956. Bases para el control quimico de Phytophthora del cacao en relación a la epidemiológia de la enfermedad. Cacao 3:15.

168. Orlando, A. 1951. O valor do oleo de quenopodio e das plantas de erva de Santa Maria Chenopodium ambrosioides L. var. anthelminticum (L.) Gray no contrôle a nematoides que atacam raizes de vegetais cultivados. São Paulo Inst. Biol., Arq. 24:69-80.

169. Otero, J. I. and M. T. Cook. 1937. A bibliography of mycology and phytopathology of Central America, Mexico and the West Indies. Univ. Puerto Rico, J. Agr. 21:249-486.

170. Pacheco H., M. 1937. Estudio sobre la enfermedad de los bananales. Rev. Agr. Guatemala 14:423-429.

171. Pady, S. M. 1959. A continuous spore sampler. Phytopathology 49:757-760.

172. ________., C. L. Kramer, and B. J. Wiley. 1962. Kansas aeromycology. XII. Materials, methods and general results of diurnal studies 1959-1960. Mycologia 54:168-180.

173. Pardo Navarro, L. 1945. Observaciones sobre el caldo bordelés. Agr. Trop. Venezuela 1:43-44.

174. Peachey, J. E. (ed.) 1969. Nematodes of Tropical Crops. Caribbean Symposium in Trinidad. Commonwealth Agr. Bureau, England 355 p.

175. Pereira, H. F., E. R. de Figueiredo Jr., and J. Hussni. 1958. A cercosporiose da bananeira. Sua ocorrência no estado de São Paulo. Experiências e combate. Arq. Inst. Biológico 25:161-184.

176. Pereira, J. F. and E. Echandi. 1964. Residuo de arsénico en hojas y granos de plantas de café asperjadas con arseniato de plomo. Turrialba 14:85-90.

177. Pereira, J. L. 1970. Aerial spraying of coffee for disease control. Agr. Aviation 12:17-20.

178. Pickel, B. 1939. A influencia da calda bordalesa sôbre a folhagem das plantas. Biológico 5:167-171.

179. Piedade, J. R., H. Yoneda, and D. A. Souza. 1967. Análize quimicados ditiocarbamatos fungicidas. Arq. Inst. Biológico 34:175-177.

180. Revilla, V. A. 1952. Control de la chupadera del algodon mediante fungicidas. Min. Agr., Peru, Exp. Agr. Bol. 1:9-11.

181. Rezende, L. O. C., C. A. Compacci, and M. Meiji. 1966. Tratamento de mudas de abacaxi. Biológico 32:55-57.

182. Rhoades, H. L. and J. F. Beerman. 1967. Efficacy of some experimental nematocides applied in the row on vegetables. Florida State Hort. Soc. Proc. 80:156-161.

183. Rios G., M. 1951. Aspersiones en el jitomate en Chapingo, Mexico 1949. Ofic. Estud. Esp., Mexico, Fol. Misc. 4:154-157.

184. Robayo, G., J. Orjuela, and H. D. Thurston. 1967. Evaluación y uso de fungicidas en cultivos comerciales de cebada en Colombia. Agr. Trop., Colombia 23:455-464.

185. Robbs, C. F., M. Amaral, and J. C. Dianese. 1965. A resinose fungica do abacaxi (Ananas sativus Schult.) e a sua ocorrência nos estodos de São Paulo e Minas Gerais. Reun. Fitossan. Brasil An. 9:71-79.

186. Robles C., J. 1967. El arañero y ojo de gallo en los cafetales de Tingo Maria. Agr. Ganad. Trop., Peru 1:15-16.

187. Rodriguez, R. A. 1968. The spraying of agricultural oil on giant cavendish banana. Turrialba 18:373-381. --. 1968. Variation of fungicide deposits resulting from spraying bananas with the mist-

blower machine. Phytopathology 58:553-554.

188. Rodriguez, R. A. and C. A. Soto. 1968. Cupric emulsion for the control of Cercospora disease of coffee. Phytopathology 58:554.

189. Román, J. and J. Badillo. 1965. Response of a sugarcane nematode population to the addition of nematocides in irrigation water. Univ. Puerto Rico, J. Agr. 49:325-330.

190. Romero Brambila, A. 1943. El empleo de las sales de cobre en el combate de las enfermedades fungosas de las plantas. Fitofilo 2:12-19.

191. Rowell, J. B. 1968. Chemical control of the cereal rusts. Ann. Rev. Phytopath. 6:243-262.

192. Ruehle, G. D. 1941. Experiments for control of potato late blight in an epidemic year. Florida State Hort. Soc. Proc. 1941:106-108. --. 1942. The use of zinc salts with copper fungicides on tomatoes in Dade County. Florida State Hort. Soc., Proc. 1942:133-134. --. 1943. The cause and control of avocado scab. Univ. Florida Agr. Exp. Sta. Bulletin 580, 4 p. --. 1943. Cause and control of Cercospora spot and anthracnose of the avocado. Univ. Florida Agr. Exp. Sta. Bulletin 583, 4 p. --. 1944. Outstanding potato late blight control in Florida with a new organic fungicide combined with zinc sulfate. Plant Dis. Rptr. 28:242-245. --. 1953. Organic fungicides for control of anthracnose of mango. Florida Mango Forum, Proc. 13:12-14.

193. ________., and W. A. Kuntz. 1940. Melanose of citrus and its commercial control. Florida Agr. Exp. Sta. Bulletin 349, 54 p.

194. ________., and W. L. Thompson. 1939. Commercial control of citrus scab in Florida. Univ. Florida Agr. Exp. Sta. Bulletin 334, 47 p.

195. Sanchez P., A. 1956. Efectividad de varios fungicidas usadas solos y en combinación para el control del damping-off y la pudrición de semillas en arvejas y frijoles. Acta Agron. Colombia 6: 1-35.

196. Sarli, C. A. 1967. Ensayo de una nueva formulación para la disinfestación de semillas de trigo contra el carbón hediondo (Tilletia spp.). Fitossanitarias 5(10):18-22.

197. Schieber, E. and G. A. Zentmyer. 1968. Myrothecium stem necrosis and leaf spot: important coffee disease in Guatemala. Plant Dis. Reptr. 52:115-117.

198. Schiel, E. 1942. La lucha contra Sphaceloma australis del mandarino en la provincia da Santa Fé. Rev. Agron., Argentina 9:19-27.

199. Segall, R. H. and L. Lopez-Matos. 1956. Control of damping-off of tobacco in seedbeds by the use of gaseous soil fumigants. Puerto Rico Univ. J. Agr. 40:62-66.

200. de Segura, C. B. 1948. Algunos fungicidas, su preparación y aplicaciónés. Rev. Agr., Peru 20: 245-259. --. 1965. (Fungicides, pp. 16-29) Enfermedades de Cultivos Tropiales y Subtropicales. Edit. Jurídica, Lima, Peru, 439 p.

201. Siller, L. R. 1954. Efecto de tres fungicidas en el combate de Phytophthora palmivora en arboles de cacao. Cacao 3:12.

202. Simmonds, N. W. 1959. Bananas. Longmans, Green and Co., New York, 466 p.

203. Skiles, R. L., Barros N., and C. Cardona A. 1958. Control de enfermedades de frijol mediante el uso de varias combinaciones de fungicidas. Reun. Interam. Fitogen. Fitop. Ent. Edafo., 1955, Bogota Colombia, Proc. 3:222-223.

204. Sleeth, B. 1968. Comparative fungicidal effectiveness of chemicals on Phymatotrichum omnivorum in laboratory soil cultures. Plant Dis. Reptr. 52: 232-234.

205. Sreenivasan, M. S. and C. C. Chinnappa. 1968. Occurrence of a new physiological race of coffee rust. Indian Coffee 32:94; 99.

206. Steiner, D. 1967. Aerial photography for land use mapping, cattle inventories, yield forecasting, and crop disease determination. III. Internat. Agr. Aviation Conf., Arnheim, Netherlands March 1966:334-361.

207. Stover, R. H. 1968. Leaf spot of bananas caused by Mycosphaerella musicola: perithecia and sporodochia production in different climates. Trop. Agr., Trinidad 45:1-12.

208. ________., and J. D. Dickson. 1968. Leaf spot of bananas caused by Mycosphaerella, action of oil on the life cycle of the pathogen. Canadian J. Bot. 46:1495-1505.

209. Tartakowsky H., S. and A. A. Garcia. 1942. Ensayos preliminares sobre control del damping-off del tabacco. Bol. San Veg., Chile 2:20-24.

210. Thanasoulopoulos, C. 1969. Some preliminary investigations on the possible control of mal seco disease of citrus with the systemic chemical benlate. Ann. Phytopath. (special number):249-250.

211. Tobler Bottini, H. D. 1946. Estudio del controlar del carbón cubierto, hediondo, caries o tizón del trigo (Tilletia spp.). Uruguay Asoc. Ing. Agron. Rev. 18(72):17-27.

212. ________., and J. A. Berta. 1945. Ensayo del caldo bordeles contra la peronospora de la papa (Phytophthora infestans Bary). Min. Ganad. Agr., Uruguay, Bol. Inf. 2:401.

213. Tocchetto, A. 1942. Tratamento da semente de batata. Rev. Agron. Rio Grande do Sul, Brasil 6:111-112. --. 1947. Mildiú da batata e contrôle. Rev. Agron. Porto Alegre, Brasil 11:176-178.

214. Tollenaar, D. 1955. The control of sigatoka (banana leaf spot disease) by mist-blowing with banacobresandoz. Netherlands J. Agr. Sci. 3:70-83.

215. Van Der Plank, J. E. 1947. Plant Diseases: Epidemics and Control. Academic Press, London,

349 p. --. 1967. Spread of plant pathogens in space and time. pp. 227-246. *In* Gregory, P. H. and J. L. Monteith (eds.) Airborne Microbes (Symposium). Cambridge Univ. Press., 385 p.

216. Vera P., L. 1947. El lacco copro 50, una promesa para controlar el chamusco. Tierra, Mexico 2: 269-272.

217. Viegas, G. P. 1945. Tratamentos de sementes de milho. Bragantia 5:145-151.

218. Volio G., C. A. 1953. Control del ojo de gallo (*Omphalia flavida*) con fungicidas. La Hacienda 48:43, 72.

219. Voloskey De Hernandez, D. 1945. Desinfestantes del carbón de la cebada. Agr. Tec., Chile 5:37-47.

220. Wardlaw, C. W. 1941. The banana in Central America. Nature 147:313-316. --. 1961. Banana diseases, including plantains and abaca. Longmans, Green, London, 648 p.

221. Wellman, F. L. 1946. Aspersiones contra las enfermedades de las plantas. J. El Café de El Salvador 16:894-895. --. 1949. Como disinfectar sus semillas. Suelo Tico, Costa Rica 5:259. --. 1949. Successful spray control of *Alternaria* blight of petunias grown for seed in Costa Rica. Plant Dis. Reptr. 33:69-72. --. 1952. The danger of introducing *Hemileia* of coffee into the Americas, and problems of its exclusion and control. Segundo Asambl. Latino. Genat. Fitop., Brazil, 6 p. --. 1955. Control of rubber diseases by chemicals. Adv. Chem. Ser. 13:31-42. --. 1955. Coffee diseases, insects, and weeds controlled by chemicals. Adv. Chem. Ser. 13: 43-63. --. 1957. *Hemileia vastatrix:* Investigaciones presentes y pasadas en la herrumbre del café y su importancia en la America Tropical. FEDECAME, Sección Divulgacion (Booklet), 67 p. --. 1958. Recent spraying for control of coffee diseases. Coffee & Tea Ind. 81:106-111. --. 1960. Recomendaciones para mejorar el cultivo del café en Puerto Rico. Puerto Rico Univ.,

Estac. Exp. Agr. Boletin 153, 113 p. --. 1961. Enfermedades de algunas plantas ornamentales de Puerto Rico, con recomendaciones para contrarrestarlas. Rev. Agr. Puerto Rico 58:179-213. --. 1961. Coffee botany, cultivation and utilization. Leonard Hill, London; Interscience Publ., New York, 488 p. --. 1969. More diseases on crops in the tropics than in the temperate zone. Ceiba 14:1-12. --. 1971. Plant diseases; an introduction for the layman. Natural History Press, New York, 130 p.

222. Wells, J. M. and J. K. Stewart. 1968. Heat pasteurization and chemical fungicides for control of Fusarium rot of California cantaloupe. Plant Dis. Reptr. 52:262-264.

Chapter XXII

DISEASE RESISTANCE OF TROPICAL CROPS

Summary

Disease resistance is of special value to the Neotropics. It may well be the most important disease-control measure possessed by tropical American pathologists because it lasts, and does not have to be bought and applied, and is specially good when crop breeding includes the incorporation of resistance. Normally resistance is a common feature in nature, more than susceptibility; most plants are resistant to most microorganisms--only specialized parasites attack.

In cultivated crops with disease, a broad-type resistance is encountered, classified variously as field resistance, horizontal resistance, multilineal resistance, and general resistance. This is sometimes difficult to find and perpetuate, but once on hand is invaluable. A more easily isolated and manipulated resistance type is called by such names as simple resistance, specific resistance, monogenic resistance, and vertical resistance.

Many crops have been and are being saved through the development by pathologists of disease resistance. It is of supremely important use in the tropics where parasitic microorganisms are so much more common and vary so much more radically than they do in the temperate zone.

Sources of resistance often are wild plants, from which the crop originated, that have evolved in the presence of disease parasites for eons of time. Such long-ago imported tropical American crops as orange, coffee, banana, sugarcane, rice, are what probably have the best kinds of resistance in wild types in the Eastern Hemisphere. Such native Neotropical crops as maize, beans, hevea rubber, tomato, potato, and tobacco will most likely be found to have resistance in wild individuals still growing in Central and South America.

Solving plant disease resistance problems is a lofty calling. It stimulates the best talent and thinking in tropical phytopathologists. Some of the very greatest satisfactions for any of these researchers have come while dealing with control problems in resistance. There are many involved in this field, and for them our profession has a deep gratitude. Indeed, one worker in Mexico has been tireless and so successful in cereal disease resistance researches that because of his world-wide contribution towards lessened hunger he was accorded the Nobel Peace Prize.

Introduction

As never before, plant varieties in the tropics must be "tailored"--that is, changed and developed by breeding along with selection--to meet modern necessities. Varieties not only must be healthiest, but must be best in quality and economically produce the most. They must be broadly adaptable for many special natural conditions and horticultural-agronomical management. To meet the ever-increasing demands for more food, agriculture is yearly more intensified by both large individual planters, and by groups of smaller operators. Above all, however, harvests have to be insured against failure from disease. Breeding must therefore include a character that gives a degree of resistance--at least (and perhaps preferably) a more or less changeable, multiple tolerance character rather than the fixed simpler type of resistance which sorts out ever more virulent parasites, which then multiply and require a whole new breeding program.

Far into the future there will never be a time in the American tropics when phytopathologists, along with geneticists and agronomist-horticulturists, can possibly relax work in crop breeding for resistance-tolerance to common diseases. Since in the tropics it is more difficult to produce a crop than in the temperate zone, it is in the tropics that there is more to be gained from disease resistance than in any other part of the world. Finding and using disease-resistant crop varieties such as sugarcane (1, 51, 56) has been in progress in the tropics since long before the beginning of the present century.

From early history to the present moment the greatest danger to people in the business of field husbandry and production has been devastating disease in much-needed crops. However, when it comes, as it apparently must at some time and place to every crop, catastrophe begets attention from trained pathologists. No disease is uncontrollable if enough research is done on the problem. In regard to catastrophe much would be accomplished: a full botanical reassessment of the crop plant; a clearer than ever consideration of the current importance of the crop in local and/or world economy; a fuller understanding of the life processes of the host; much more information completed on the genetics and cytogenetics of the crop and its improvement; and, not the least, much more will actually be studied, written, and published, so that lack of information will no longer be the most limiting factor.

Resistance in Nature Is Basic and Common

It is a popular belief that there must be a few kinds of plants that are never diseased by microorganisms. However, the studying pathologist knows this to be so unlikely he would guarantee that every species has its diseases. While relatively speaking only a few microorganisms are parasites (most of them are saprophytes), it is because of these parasites that the evolution of plants has included some kind of native resistance. If this were not so, there would be no chance for the existence of tropical pastures, forests, green valleys, jungles, or wooded hills, let alone cultivated fields and plantations. It appears that one of the most commonly inherited characters of plants in nature is a broad-scale type of disease resistance.

In no doubt all cases, microorganisms are present around plants, but most fungi and bacteria have no effect on leaf tissues. Several times I have had occasion to make examinations of uninjured leaf surfaces on which there were abundant microorganisms. These observations of micro-epiphytes are on collected leaves secured in moist habitats in El Salvador, Costa Rica, Ecuador, and Puerto Rico including materials from grasses, species of *Cynodon*, *Paspalum*, and *Chloris*; a cucurbit, *Sechium edule*; six trees selecting only old leathery leaves on species of *Montezuma*, *Theobroma*, *Ficus*, *Lucuma*, *Coffea*, and *Spondias*; and three palms, species of *Pseudophoenix*, *Cocos*, and *Caryota*.

On leaves of every genus, I observed actively multiplying bacteria that were not causing infections. I saw cells of algae sticking to the surfaces of, but not penetrating, foliage of *Ficus*, *Caryota*, *Spondias*, and *Lucuma*. On apparently healthy leaves, of all of the 13 species, a search was made for fungal remains. Epidermis was stripped off and examined before and after staining in cotton blue solution. Fungus spores of several kinds were seen but no positive identifications were made. If leaves had been retained on the tree during two rainy seasons, numerous superficially attached, growing fungus hyphae were evident, some of which had septae; many of these superficial hyphae had broken and torn ends. There were no infections.

There are many easily found plant species that show frequent evidence of many infections which have no severe effect, thus indicating host resistance. The tropical plant pathologist observes this occurrence of apparent broad-type resistance in cultivated plants. A few random examples are as follows. The palm *Pseudophoenix* sp. is found year after year with fronds liberally studded with spore-bearing bodies of the fungus *Graphiola* without observable injury to the vigor of the tree. Grown by indigenous farmers in the mountains of El Salvador, strains of peanuts occur that have well distributed on their leaves spots of *Cercospora*, but no reduction of the crop is noticeable. Many times small, fruited, hot pepper plantings are seen with most of the oldest leaves spotted by *Cercospora*, but the infection does not interfere with productivity. The cenicero tree, *Pithecalobium dulce* has leaves often mildewed with *Microstroma*, with no discernable injury to the tree. Field after field of the old style, most common maize grown by peasants has much brown-spot, *Physoderma zea-maydis*, but this fungus most often causes no reduction in harvest at all of the crop. Ends of old leaflets on thousands of coconut palms, *Cocos nucifera*, are invaded by a combination of *Leptosphaeria* sp., *Pestalotia palmarum*, and *Glomerella* sp. without danger to developing nuts. Such a list is only a beginning.

It is fully evident that there is broad-type resistance in nature, and it has been studied more and more all over the world in connection with commercially valued crops. This kind of resistance is given various names such as partial resistance, slow-infection resistance, field tolerance, field resistance, tolerance, horizontal resistance, multigenic resistance, broad resistance, and generalized resistance (19, 58, 63, 64, 68, 70, 78, 99, 106, 118, 119, 129, 139). It is significant that this kind of resistance seems so permanent in nature.

Horizontal, or broad, resistance is notable for the fact that it provides no advantage to a new parasite mutant. It is extremely difficult, verging on the impossible, for a parasite to build up a new mutant because the host with this broad or horizontal resistance accepts all races of the parasite while slowing their action. There may be some suppression of the sporulation of the parasite and there may be a reduction of the growth of the parasite in host tissues. In some cases, the infective tissues in the host are more restricted in their occurrence in resistant strains. In other cases, the horizontal resistance of a host confers such a character upon its even abundant infections, that a tree or plant may fructify and produce often as good a harvest as if no parasite were present. Host plants become infected should proper conditions prevail, but if there are epidemics they are slow in developing and a crop with horizontal resistance survives with little to no deleterious effect.

A drawback in relation to field tolerance or horizontal resistance is that it occurs from inheritance by a combination of several genes not simple to find and isolate. The study of its effects through successive host generations does not give the clear cut results that the other, more recognized monogenic type of resistance does.

I do not intend a condemnation of the more easily secured, readily workable, specific-type of resistance. This saves, and has saved, much of our field cropping from utter failure. It is called not only specific-type resistance, but also other names such as racial resistance, single gene (monogenic) resistance, vertical resistance, and perpendicular resistance. Work with the other broader type resistance is being and should be increased, but the more specific resistance is, though less permanent, sometimes the only thing that saves a crop at present.

Vertical resistance is much longer known in plant pathology than the so-called horizontal resistance. It is characterized by host tissues being super-sensitive and repellant to a parasite's attack. In some manner, the offensive organism is unable to subsist in unfettered fashion on resistant host cells. This kind of rejection, it would seem, is clean-cut, and might be the realization of a grower's fondest dream except for the habit of variability in parasites. Through recombining genes or mutations, they sooner or later produce off-types that can once more come back to infect and cause disease on accepted resistant host varieties.

This defeat requires that plant pathologists re-develop a whole new host variety or strain, that is again resistant to a differently virulent race of the parasite. And it must be done at intervals (19, 106, 118, 143): the vertical type resistance requires re-sorting and re-placing by use of newly bred resistant varieties about every 20 years in the temperate zone, whereas it may need change in as few as four to six years in the tropics.

A Number of Resistant Crops in Use

There follow below brief discussions of important developments in a few tropical crops resistant to disease. Principles and broad aspects vary considerably in different crop diseases and each situation is necessarily a special problem in itself, but bases are given below for researching and developing resistant characters. It is plain to professional plant men that there is nothing more important in Neotropical plant pathology than increasingly effective disease resistance in the great crops of the region. This is, and forever will be, a considerable challenge (2, 5, 8, 12, 16, 20, 26, 28, 31, 35, 44, 51, 54, 56, 59, 64, 71, 89-91, 99, 107, 110, 113, 120, 121, 127, 136, 137, 143, 149).

During the last decades of the past century, flourishing sugarcane, *Saccharum officinarum*, of the Dutch East Indies was showing the fearsome mosaic disease marked by stunting and leaf chlorosis (1, 56). It did not kill plants; was not cured by extra fertilization; caused final abandoning of what were once best fields; and the end was disaster. This abnormal disease condition originated in the Pacific Islands and was first reported in Java in 1892.

With little time for refined researches on this disease, Dutch scientists searched for and found wild sugarcane, *S. spontaneum*, resistant to it. Hybrids of the wild and the cultivated canes were resistant and they were accepted by planters for sugar production, and, for names, varieties were given "POJ" numbers (the famous three letters were for the Java Experiment Station where the work was done).

Sugarcane setts from the Orient that had been taken to the Americas grew well. This was fortunate, for Oriental sugar shipments to world markets lagged because of disease, and sugar loaves were brought in ships from the West Indies, which in market parlance included not only the West Indies-Antilles Islands but also protected parts of the mainland like some of Florida and southern Louisiana, and Central and South America. Productions from these our tropics were so good that soon shipping and plantation developments made rapid strides bringing much unused jungle and virgin lands under man's control. Sugarcane growing business was good and in the countries where the crop did well the people connected with it prospered. It was a successful and principal crop until the stunting-yellowing mosaic disease struck it.

This occurrence was reported in Argentina, Puerto Rico and Louisiana during the years 1910 to 1919 and soon after that in Peru. A little later the disease had increased, and during the years 1923 to 1927 it resulted in such grave destruction that in some areas there loomed the end of sugarcane plantations. In Puerto Rico and Louisiana, losses were at least $200, 000, 000 per year. The disease had been accidentally brought in (1, 51, 12) to Argentina in setts from an unknown source in the East Indies. It developed as a severe condition in the famous sugarcane plantations of Argentine Tucumán and spread widely in our tropics by interchange of "seed," or setts. The eventual destruction of all the tropical American cane plantations seemed assured. Fortunately, the POJ resistant canes were known about, and Dutch scientists unselfishly, quietly, and with pride aided in the early shipment of POJ 36 and POJ 213 canes to the Neotropics; these resistant canes, and subsequent "numbers," saved the day. From breeding with these and other sources, sugarcane specialists have developed hundreds of varieties, not only resistant to mosaic and other viruses and other maladies, but better adapted to many different ecological conditions. (This matter of adaptation, is likewise important in various other crops.)

Of all fruit trees found growing on farms in the American tropics the sweet oranges are the most popular. Almost every farmyard, "huerta" or planter's residencia has one to several such orange trees. In addition great national and internationally important agricultural businesses and corporations are based on oranges. Among the world's most notable orange producing lands are parts of the U. S. A., Mexico, Brazil, Paraguay, and Argentina.

As in every history of a good, established crop, orange diseases began to be of much concern. Of them all, a gum-exuding rot in bark of trees at the ground level was the most destructive (49, 51, 56, 115). Almost invariably the first several years of a plantation or grove production is good, but as trees age they are attacked by an abnormality variously-called footrot, root gumming, gomosis, or mal-de-goma. As seen at first many believe the footrot kills the most profitable trees at the most profitable period of their productive life.

In the Neotropics, according to Fawcett (49) the first published notice of footrot was an observation in 1709 by a travelling French scientist, L. Feuillee, who saw gomosis of orange in Lima, Peru. After many other observations over the next 50 or 60 years, this most serious, sweet, orange tree-killing disease, Phytophthora parasitica and other Phytophthora species, was in the groves of Florida and California. It was not long before it was determined as present in Mexico, Brazil, Peru, Paraguay, Argentina and West Indies islands. The mal-de-goma increased in spread and severity, and groves died.

However, although trees of sweet orange were most ruinously affected, the next most popular citrus trees, grapefruit and lemon, were also diseased. Pathologists began detailed studies, using newly devised methods of isolations of microorganisms, and long before 1890, various bacteria and fungi were found in mal-de-goma tissues. In addition, where an occasional sour orange had been planted it was noted by folk and planter observers that it did not succumb. This looked like a resistance effect and technicians began the studies which culminated (49) in works by pathologists from many tropical countries. They gave objective and measurable proofs that resistant sour orange (and certain other citrus) roots could be used, and on them would grow healthy tops of trees bearing sweet oranges.

A uniquely tropical product is chocolate, from the cold-sensitive "cacao" tree Theobroma cacao. It was known in prehistoric times by indigenous populations who secured it for their priests and kings, and in 1502 was taken on a return trip by Columbus to the Imperial Court in Spain.

For decades of the ensuing century, the use of chocolate was confined to emperors, princes, and courtiers of Europe. A drink or a sweetmeat from it for a wider distribution, had not only excellence, but it gained a distinct aura for the splendor it represented. Increased shipments from so-called "West Indies trade," came in sailing craft to special markets. Considered at first as an almost totally magic product gathered under difficulties from widely scattered trees, it was brought by jungle-dwelling men to ships plying the tropics. In the natural course of events, its money-making value was realized, trees were transplanted and grown in a crude form of agriculture, the culinary and manufacturing aspects of it were advanced, and by 1653 the economics of it were such that it became a money-making and popular product.

It is a serious shock when the spectacular witches' broom, Marasmius perniciosus, disease attacks and sweeps through cacao plantations. It is known to destroy 80 and 90 per cent of the annual harvests in many areas.

Studies show (see Chapter XXI) that witches' broom can be controlled by applying protectant sprays at frequent intervals; that the miniature, spore-bearing agarics can be held in check by spraying or manual removal. When word about the disease difficulties became known in chocolate manufacturing circles, prices of the commodity increased radically, and the future of it in international commerce dimmed. This has led to the possibility of developing tree resistance as an alternative to sprays.

Searches in wild country for resistant cacao resulted in some fortunate finds in the Amazon basin (e.g., 7, 39-42, 109, 121). This was followed by further determinations and it is established that both the disease and the host and the host resistance to the disease come from natural stands of cacao areas along the Amazon river.

After persistent work, a number of broom-resistant clones of cacao are now being grown in various locations in tropical America. Examples from Trinidad of a few such clones are ICS 6, ICS 45, ICS 91, ICS 95, and ICS 98, reported some years ago by Holliday (71). Later Ecuadorean workers, Ampuero and Alvarado (6), reported that they had selected from field-collected trees six clones of good broom resistance. In 1963 Lems (78) was able to conclude, after ten years of testing in Surinam, that the two most valued cacao clones he could recommend were ICS 95 and ICS 100, which were not immune but highly tolerant to witches' broom and bore good crops along with scattered infections by *M. perniciosus*.

Banana is another crop with a notable disease history. Corms of banana were first brought to the Western Hemisphere tropics by colonists, missionaries and slavers during from about 1516 to 1750. The crop flourished, soon produced good food for struggling people, and fed cheaply and well large labor populations. However, after some years, inevitable disease troubles appeared, and fields had to be moved from one place to another to escape losses. Meanwhile small and occasional shipments of the fruit went to the temperate zone by short trips in sailing craft, and in Cuba, for example, this developed into a limited but remunerative trade with the relatively close eastern seaboard of the United States.

From the standpoint of Neotropical plant pathology, one of the first important occurrences was the discovery by planters of Cuba of a serious banana disease. Examples of discolored, partly rotted plant material were sent to the United States for study. It resembled a bacterial trouble in some ways but was found to be a fungus. (On a working visit in 1933 to the Cuban Agricultural Experiment Station at Santiago de las Vegas, I discussed with the late Dr. Stephen S. Bruner and two elderly Cubans the history of banana wilt in Cuba. These three told me that the disease was first seen on the island as early as 1905 and probably it had been there some time before. For many years Bruner was interested in this history and had studied banana borer as possible bacterial disease vectors, but believed they aided in fungus, not bacterial, infections.)

A brief review will give something of the importance of banana plantations to agriculture in certain countries of the Neotropics. (For corroborations and details see works cited in the *Musa* appendix at the close of this book.) The banana, and preeminently the gros michel variety, was for a long time the greatest money-making crop in many tropical American areas. Examples of what occurred in certain countries in one year, 1955, are as follows: in Guatemala only 1 per cent of all crop lands utilized were growing bananas, but shipments of it nevertheless aggregated 18 per cent of the total of agricultural exports. In Costa Rica bananas growing in 7 per cent of all crop lands produced 40 per cent of all exports. In Honduras 4 per cent of all crop lands utilized grew bananas, from which were realized 50 per cent of the total of that country's exports. In Ecuador, 12 per cent of all crop lands were utilized for bananas, and from it came 55 per cent of the total exports. In Panama, bananas were grown on only 6 per cent of all crop lands but produced 75 per cent of its total exports.

It was a severe shock to tropical fruit companies--all growing the old disease-susceptible gros michel variety--when frightening, spectacular, and sudden out-breaks of wilt, *Fusarium oxysporum cubense*, occurred in soils of the best plantations. Maps and histories were kept of the march of disease as inevitable encroachment meant ruin. An example is in Surinam, where gros michel banana shipping started about 1904. By 1910 the country sent out 670 "stems" or bunches; when wilt appeared, in two years' time there was no more to ship. In 1870, the first gros michel fruit was marketed out of Jamaica, from pilot trials that were a great success in spite of scattered wilt. In 1937 the island sold 27 million stems; then the disease started a rapid increase and in ten years shipping had practically ceased. In an area of of good soil in Panama this banana since 1890 was being harvested annually. But on account of wilt it was suddenly stopped in 1937.

The Caribbean side of Costa Rica was growing bananas in small amounts as early as 1878; later for decades it was a stable source of income, and from about 1905 to 1913 production was about 11 million bunches a year. Then wilt became common in 1935 to 1941 and only about a tenth of the crop could be harvested. From one part of Honduras the banana was first exported in 1921 and much fruit was sent out, at a rate of some 8 million stems in good years; when the disease came it spread rapidly and by 1939, because of wilt, major banana shipments were drastically reduced.

Never let it be said, however, that fruit growers were not aware of the need of scientific studies on the problem. Research departments of plantations found that certain "long life" soils had less wilt than others, and they made detailed determinations about differences. In some places wilt occurred on soils where it was not expected. Attempts were made to stop wilt spread: spraying, roguing, several kinds of soil treatments including water fallow, digging isolation trenches, and other means. None of them were of long time practical benefit. Meanwhile pathologists found from among the billions of plants in wilted fields occasional "escapes" which they tested, hoping for a wilt-resisting mutant, but these escapes always finally succumbed to wilt. To the pathologists, it has long been evident that the susceptible variety gros michel would some day have to make way for a wilt-resistant banana, and in 1971 this is now in progress.

There was a report in 1912 from Surinam that the congo banana was evidently resistant to the wilt. In the next decade bananas of the cavendish group, both dwarf and giant types, were found to be wilt resistant. Numerous others have been obtained from world collections. But wherever banana men and their engineers and pathologists search, they do not find resistant bananas that have all the commercial characters of wilt-susceptible gros michel. Banana breeding was started as a long-time effort, in 1922 in Trinidad and in 1924 in Jamaica, and is still being pursued (see the appendix on Musa in this book, and citations 12, 56, 121, 127). Using knowledge of the Musa genus, its cytogenetics, and what was needed in the field, gros michel was at first crossed with disease-resistant Musa acuminata malaccensis. Of a number of such seedlings obtained there are two (I. C. 2 and S. 19) that are disease resistant and fairly close in appearance to gros michel. However, they have bunch irregularities and fruit sizes more nearly like their wild male parentage; growing, shipping and marketing trials discouraged commercial use.

There followed more similar crosses, selfings, some back- and inter-crosses, and various cross combinations. In some cases there are unexpected pollen sterilities so the wanted crosses fail. The plant that is successful has to have very high female sterility, for, to be sure, fruits with even only an occasional seed are at a serious commercial disadvantage. This means that in addition to dependable and inheritable wilt resistance, and good commercial character, breeding progeny has to be highly female sterile. This has to come by way of a parthenocarpic male parent, obtainable through breeding, or some lucky choice of a collection out of introductions. Progress is being made on this as cytogenetics discloses the true diploid, triploid, and tetraploid natures known of breeding parents.

The mode of heredity to *Fusarium* wilt resistance for bananas is not fully settled, as some examples will show. Crossing the resistant (possibly immune) pisang lilin from Malaya with susceptible gros michel gives seedlings from pisang lilin that are about 60 per cent wilt-susceptible. When these and others are given special seedling and older plant root-inoculation tests with *Fusarium oxysporum cubense* races 1 and 2, in seedlings of *Musa acuminata* subspecies *errans*, *banksii* and *microcarps*, there occurs differential resistances. The *M. acuminata* "variety lidi" gives reactions in its progenies that indicate it possesses a single dominant inherited factor for resistance to the race 1 of the *Fusarium* wilt organism. There is a suggestion of complication in this as several *Musa* species, including *M. balbisiana*, while growing as young seedlings are clearly susceptible to *Fusarium* wilt, but as mature plants are resistant. However, there seems good reason to begin to believe that resistance is probably a dominant character. There are studies which show that about 50 per cent of the progenies from a resistant male crossed with a susceptible female give about 50 per cent resistant. However, there is much still to be done in understanding this, as a few clones from progenies originally classes as "resistant" show some susceptibility to *Fusarium oxysporum cubense.*

Until something better is developed, commercial shippers, although in certain soils holding onto some gros michel, more and more are using resistant kinds like lacatan, giant cavendish, and dwarf cavendish. These latter three kinds grow well in the excellent banana soils that are at the same time thoroughly saturated with the wilt *Fusarium.* In some ways the most successful of these three resistant collections is a selection of wilt-resistant dwarf cavendish, and especially a variety from it called "valery."

Even if discussed but briefly in the paragraphs immediately above, the reader can see how resistance has been developed in the crops of sugarcane, sweet orange, cacao, and banana. There is clear evidence, also, of the inestimable economic good which comes to our parasite-occupied tropics from obtaining crops that are disease resistant. Tropical growers, both literate and illiterate, realize the unquestionably better results from using plant varieties with disease resistance in them; from increasing publications, the public is apprised ever more fully of what pathologists have to encounter in their work. There are certain general citations (8, 51, 77, 79, 107, 118, 121, 139, 143, 144 as well as other works) on these problems that are worth consultation.

As indicative of the practical and further wide contribution of how much is obtained from disease-resistance-breeding, reference is made to work in countries from Florida in the north to Pernambuco, Argentina in the south. These are of greatest satisfaction in their results to agricultural producers, and let it be stated here that resistance is known in--these are not all--specific crop cultures of the following:

Rice (see for examples, citations 10, 57, 65, 90, 135); coffee (16, 50, 101, 113, 124, 128, 147); cacao (6, 7, 13, 31, 38-42, 71, 74, 78, 84, 93, 109); dry bean (17, 34, 43, 45, 55, 89, 95, 96, 103, 140, 149); white potato (18, 23, 27, 28, 35, 37, 64, 69, 70, 91, 92, 99, 120, 122, 129); wheat (19, 24, 29, 32, 60, 61, 88, 98, 104, 106, 111, 116, 136-138, 141, 148); maize (22, 30, 83, 102, 112, 117, 119, 125, 146); tomato (26, 108, 132, 133, 142); citrus (49, 52, 68, 76, 94, 97, 115); banana (81, 82, 105, 123, 127, 147); sugarcane (1, 21, 51, 57, 86); cotton (53, 85, 130); rye and oats (48, 137); hevea rubber (20, 59); avocado (110, 150); the black pepper (3); vanilla (4); onions and garlic (9); roselle and kenaf (5); garbanzo (72); melons (36); pine tree (37); pangola grass (80); peanut (87); amaryllis (100); sunflower (114); cabbage (131); flax (2, 137); lettuce (145); and tobacco (8). There are numerous cases of lines of certain crops in which it is evident that there is disease resistance which originated from folk selection.

Sources of Resistance

Shifting cultivation, or the milpa kind of growing, is an invaluable method to preserve crops for human consumption in spite of disease. However, because of population pressures, situations develop where little or no shifting is possible. In those areas agriculture is a settled thing. As pathologists are aware, observant growers find plant individuals, or groups of plants, that survive diseases; seed from these sources is kept to be handed down from generation to generation. It is with material like this that the appreciative plant pathologist deals, and using scientific methods he tests, isolates, and fixes better resistant varieties. A plant pathologist working in a country where a crop has been introduced from afar and is then found diseased must go back to its centers of origin (79) to recover long-lost disease resistances.

Using this principle, specialists in the American tropics prove many times that their best chances for obtaining resistant selections of crops that were first native in the Eastern Hemisphere is by securing introductions of plant materials out of that great region. A few examples are: with wheat plasm--for ecologically feasible zones in such countries as Mexico, the highland parts of Guatemala, Chile, Uruguay, certain sections of Brazil, Argentina, and some others--pathologists have obtained resistance to disease, especially rusts (24, 32, 98, 111, 116, 144), by use of material from Europe and the Near East. In South America, rice plasm that is resistant to pasmo comes from far eastern sources (65, 135). The mal-de-goma resistant orange rootstalks are found (49, 68, 76, 94, 115) in tree species mostly out of China. With Western Hemisphere coffee, leafrust and other disease resistant lines come (16, 101, 113, 124, 147) from Africa. The fusarial wilt-resistant bananas obtained for and now being used (105, 123, 127) in the Neotropics, originate in such far off places as the Pacific islands.

However, tropical American pathologists have centers of origin of some of their other more important crops more closely at hand. These, it is known, originate in the Western Hemisphere. It is here their natural resistances have evolved. Plant disease men can thus go into their own, backyard areas in some cases, and from wild plants secure superior and practical sources of resistance. Mention is here made of only eight out of many of these crops; resistant plasms of such perhaps need not be imported from afar. They are, in the Americas, maize (30, 102, 112, 117, 119, 125, 146); bean of dry or frijole type (58, 87, 95, 96, 103, 149); white potato (28, 35, 47, 55, 64, 69, 70, 99, 122); cacao (7, 31, 38-40, 71, 109); hevea rubber (59); peanut (87); tomato (108); vanilla (4); and tobacco (8).

Plant pathologists, breeding highly valued crops for resistance, are after much variability as to horticultural adaptability. An instructive example is cacao, with the attention given it by Ampuero (6, 7), Cheesman (31), Desrosiers (39), Pound (109), and others, searching out clone diversity that will fit it for such widely varying areas as the low and rolling northeast on the Atlantic side of Brazil, the wide and flat stretches of the great littoral in the Pacific plain of Ecuador, the irregularly lying Caribbean country of Nicaragua, and both the low and medium elevated valleys of Costa Rica. At the same time there must be in the clones, or varieties, resistances to at least three diseases: witches' broom, Marasmius perniciosus; dieback from woodstain, Ceratocystis fimbriata; and cushion gall, Fusarium rigidiscula.

Natural variants are being watched for by plant pathologists, and there is an occasional mutant. There have been reports, not so long ago, from cereal disease men: Favret and Rodriguez (48) and Vallega (137) have used spontaneous mutations in their crops in Argentina, and from these they obtain new chances for securing resistance to disease and adaptability to growing conditions. Just when the Brazil tomato crop varieties of best resistance to Fusarium oxysporum lycopersici were well along in selection, Tokeshi, Galli, and Kurozawa (133) found that an entirely new race of the Fusarium had appeared, which made new breeding imperative. Orjuela (104), working with Puccinia graminis tritici has been discovering change in its races in Colombia and race numbers 302 and 319 have necessitated search for different resistant wheat genes to control old disease perpetuated by new parasite races.

This matter of occurrence of parasite races and resistances to them, is well known in our tropics; for details and discussions, reference is made to such South American plant pathology texts as those by Fernandez Valiela (51), Galli et al. (56), and de Segura (121). After finding races of pathogens, the practical result is developing programs of plant breeding against them. This requires much effort and time, but in the end effects are gratifying--amply demonstrated in our tropics by, for example, Borlaug (19) in his classic work on cereal rusts from Puccinia in Mexico; the work on Phytophthora blight organism of potatoes by Castronovo (28) in the Argentine; by Niederhauser, Graham and others in Mexico (63), and by Guzman Naranjo, Thurston, and Heidrick in Colombia (70); researches on the races of Brazilian tomato wilt Fusarium by Tokeshi, Galli, and Kurozawa (133); the work of Schieber on rusts of maize in Guatemala (119); and Erysiphe wheat mildew race studies by Vallega and Cenoz in Argentina (138). Such works as these just mentioned, and there are many others, bring prestige and confidence both to those actually carrying out the exacting studies and to the far-seeing governments and universities through which the pathologists are given facilities and help.

The Stimulus to Pathology from Adversity

It is axiomatic that a good plant pathologist enjoys

his profession, is dedicated to it, and earns his keep. With the needed aid from his administrator he accepts with interest and enthusiasm the challenge of a problem that leads toward disease control. This is his reason for being. Of all the means of control that can be developed, the one from which accrues the most nearly permanent effect, and the one that after a long time gains the widest and deepest attention by the world of pathologists, as well as the laymen growers of the crop for whom he works, is the production of a satisfying disease resistant crop variety, or varieties.

Pathologists are fully aware that no time is to be lost in starting the sometimes slow moving and at first unspectacular researches that lead to help for a crop facing a serious and increasingly menacing disease. The plant disease specialist, if a scholar in the science, takes any possible advantage of what others may have learned about his crop and the disease involved; using anything in any language, nation, or other type of differential. His strong stimuli are the thoughts of not only the scientific endeavors involved--how this increases his position and what this adds to his role in the history of his profession--but also what the success of disease resistance will mean to growers in dire need of harvests from crops cut down by disease. It is the realization and the keeping in perspective of the important juxtaposition of crop/disease-control/humans which livens the blood of disease researchers. (Note for a few examples authors such as: Ampuero and Desrosiers with their co-workers in Ecuador (6, 7, 39-42); Cruz (35, 36), and Moreira, Costa, and Grant (94) in Brazil; Graham and Niederhauser and co-workers (63, 64), and Borlaug (19) in Mexico; Castronovo (28), Di Fonzo (53), and Vallega (137) in Argentina; Schieber (119) in Guatemala; Cortazar (32), and Gonzalez (60, 61) in Chile; and Guzman Naranjo and Thurston and their co-workers (70, 129) in Colombia.) What they and others do is inspiring, requires skill and uncommon zeal, but above all takes years of sustained effort.

Looking back on successful experiences of other tropical researchers, one finds that shorter-term experiments on controls by sprays or soil and seed treatments (see Chapter XXI, for examples and citations) give quick and usable results in some cases after a single season of work. The controls are gratefully used by farmers and also professionally acknowledged; they are in no way at all underestimated by scientists. In fact studies on chemical disease controls play a great and sound part in the scientific permanence and advancement of plant pathologists in research stations.

Indeed, to emphasize this point, it is by the short-term studies with their practical results that the administrators and those in charge of financial matters are assisted in keeping their technical workers happily busy. Each new discovery adds to the proper image of what a man and the institution within which he works presents to the public. The population must have ready proof of the good returns that come at intervals. Meanwhile, many master plant pathologists have under-girding their years scientific advances toward what must eventually develop, deliberately, to be disease resistance.

A severe and devastating disease of a crop is a worthy, natural adversary for a pathologist to conquer by disease resistance. It requires a program that is well planned, yet allows for innovations and new turns in research, always having several years of attention in view. Disease resistance projects, often complex, take into account various phases, including problems of understanding host growth habits, and epidemiology of the parasite and its variations. It obviously requires applications of inoculation, field study, and other techniques that insure adequate parasite attack, to test the reactions of host progenies under various conditions. And besides it is a prominent requirement that the worker look as far into the future as possible and introduce into the new crop variety he is assisting to evolve, the most desirable horticultural-agronomical characters. Breeding of the resistant plant must be such that it has in it inheritance that not only controls disease but that gives good reactions to intensive management.

Research Refinements on Disease Resistance

The first step of a pathologist looking for resistance in a crop is to find healthy plants among a diseased population. In the tropics, where variability induced in plant-inherited characters is more common than in the temperate zone, there are good chances of finding disease resistance. Neotropical pathologists obtain as many samples of materials from as many different reputable sources as they can; recommendable sources are regularly supplemented by pathologists who themselves go to the wild for collecting wild types of the crop where host and disease have existed together for eons of time.

Any pathologist knows about the first steps that follow finding apparently resistant plants. When grown along with susceptibles under disease conditions in field, shade house, screened house, or greenhouse, first reactions are noted. Outstanding individuals are used for seed. Progenies are next employed in planned plantings, and here is where a most practical and valuable refinement of research is used. Tests are so designed that results obtained can be given accurate statistical analyses. The chosen lots giving distinct and practical resistance results can thus be picked out and used in more testing, breeding, selecting and reselecting, and perhaps ultimately become the answer to the wish for a resistant crop variety.

This kind of series of developments in numerous crops, carefully carried out during many years, is the basis of the success in securing resistant varieties of maize, as is indicated in reports of such as Schieber (119), von Bulow (22), and Wellhausen *et al.* (146); this kind of activity is what has given to cacao growers excellent clones resistant to disease, as indicated in reports of Ampuero and Desrosiers (7), Desrosiers (39), and Pound (109). Crops of dry beans are now grown--as reports of Zaumeyer and Thomas (149), Melendez de la Garza (89), Munoz and Cardenas (96), and Olave (103) testify--where certain diseases eliminate all susceptible types. Where strongly developed fruit companies and a host of small growers had long been forced to abandon banana growing because of wilt, works of such as Osborne (105), Shepherd (123), and Stover and Richardson (127) give information enabling resistant varieties to be replanted, making once ruined, *Fusarium*-infested lands again profitable. The growing of wheat, much needed in hungry Latin America, was weakened and nearly stopped by rust, but as can be seen from such reports as those of Vallega (137), Orjuela (104), Borlaug (19), and others, the resistant varieties now produce more and better wheat than was ever known before. Returns from resistance research on these and 20 or more other crops have given hugh benefits to human beings. (I repeat much, for emphasis.)

There is excitement in bringing to needy farmers a resistant crop variety, and it stimulates thinking about why there is resistance and what it is. Comparative determinations between resistant and susceptible tissues require dealings with highly complex physico- and bio-chemical matters in connection with molecular physiology. Often these efforts employ electron microscopy, electrophoresis equipment, and the means to make and interpret paper- and gas-chromatography readings. Most research of this kind is successfully done where there is a great deal of others' experience to call upon, in highly specialized institutional laboratories in the temperate zone. Generally, the needed trained help, the required time, and the laboratory are not available to the tropical man. However, researchers involved with some of the basic features of plant disease resistance have not been wholly neglected in the Neotropics. A number of examples illustrate this.

Two teams of research workers (29, 46) in Venezuela and Costa Rica, have been at work on minute chemical differences in tissues between resistant and susceptible cacao. This has been in connection with the woodstain and branch dieback caused by Ceratocystis fimbriata. Both teams have found that tissue of the highly resistant clone has more chlorogenic acid than the susceptible. How chlorogenic acid changes the action of, or neutralizes, toxins (73) that are produced by the parasite is still to be determined.

When internationally operating banana companies were required by increasing disease to change to fusarial wilt-resistant varieties, they made extensive plantation tests. A kind of banana clone selected as good was one that under many conditions demonstrated adequate wilt resistance. However, the change to a new kind, after 100 years or more of producing the old type, made it imperative as a guarantee for pathologists to determine what objective physiological differences between the resistant and the susceptible could be found in tissues. It is significant that this study was fully subsidized by the commercial banana industry. Reassurance by it that resistance is more than a short-time field effect was essential. Findings (14, 15, 81, 82 are examples; other citations are in items included at the end of Chapter IX) make it plain that there are distinct and dependable differences: in the greater quantity of phenols in resistant banana root tissues--in that there develop gels and tyloses or plug-like growths in the lumina of Fusarium-infected vascular elements of resistant roots--and in the different rapidity with which toxic injury causes resistant root tissues to hasten the healing processes.

Unpredicted Onward Gains from Resistance Study

If this were a manual, or a text prepared by a group,

some one undoubtedly would have been assigned to prepare statements and approximations about objective values tropical pathologists have contributed to harassed growers in the Americas. The monetary values of harvest increases would be found immense, unbelievable. In any case, the best procedure is to accept the good and dispense with detail. But, something happens to plant pathology researchers as they drive towards well-defined disease control. They recognize more to be done than the sole devising of a specific spray or resistance measure. They see its ramifications, and enter prophetically into the resulting chains of reactions, indicated here by a few spare examples.

Tropical American relatively soft, high-yielding, short-lived sugarcanes (with the now almost unheard of names like crystallina, borbon, and otaheite) were found mosaic-susceptible. These had to be replaced with harder-skinned, new, resistant types, at first mostly "POJ" kinds. The machinery used to squeeze juice from soft canes was light and simple, and the Latin American trapiche was moved with little difficulty to keep up with short-term land use. The POJ and such resistant canes were of pedigree and more expensive, but could grow a long time where the old, soft types failed. Pathologists of the West Indies and Tucumán in Argentina were quick to stimulate agronomic science that greatly lengthened the lives of fields, to pay for and to serve many years the much heavier and permanently installed mills required for the newer resistant canes (1, 21, 51, 56, 121).

Tropical pathologists encounter unexpected and scientifically important discoveries in studies originally designed to lead towards securing disease resistance. Of fairly recent occurrence are two outstandingly interesting and new-found predisposition phenomena. Costa (33) shows that a plant that may ordinarily be fully resistant to a virus may succumb to it if it is invaded first by another type virus. In studies on *Fusarium*-resistant vanilla, Alconero and Santiago (4) find otherwise matured and toughened old root tissues--when the natural external fungus *Rhizoctonia* grows around them in usual mycorrhizal fashion--susceptible to invasion by the wilt-causing *Fusarium oxysporum vanillae.*

It seems probable that no one, not even a pathologist, would have had the audacity to foretell to banana fruit companies all the researches and changes that would have to come to them, when they changed to the much better wilt disease-resistant banana variety or varieties. However, the susceptible plant, three times as tall as a man with bunches high in the air, is replaced in many areas now by the resistant plant with its bunch almost sweeping the ground. The old type banana was rugged, took lots of field room, and from it came crops under difficult conditions. The new resistant kind responds well to care. It readily withstands field crowding and intensive husbandry.

It can no longer be half-wild culture. Areas are cleared and levelled in an entirely new manner. There are different and new, planting, growing, and harvesting regimes. Securing the "seed" or "heads" (corms) of the new resistant pedigreed banana is costly and exacting work. It is known that it gives best harvests when treated to field management far beyond the traditional lot of the old susceptible but tough and hardy kind. As a matter of fact, growing such a banana as the more apparently refined but resistant plant results in fundamental, and long overdue, changes to modern banana crop technology. Much of the newer method adapted, moreover, has been due to past findings in the researches on fusarial wilt.

Since pedigreed wilt resistant banana growing is costly, engineers have worked out methods and machines to do many of the operations. For instance instead of sweating men digging holes for planting seed heads in raw, rough soil, the field is now plowed, levelled, and holes dug with power tools. Other power equipment for weed control, fertilizing, and other operations, are common. Spraying is done from the ground in some special instances, but mostly from the air using helicopters and fixed-wing planes.

Resistant bananas are grown with confidence many years in the same field, and because of this permanence of production, agronomists develop new and modern programs. There are now twice or more the number of plants on a given area of land than formerly. There is no fear of wilt spread effects. Heavy fertilizing once thought prohibitive is part of the modern system, irrigation is common, and more rapid harvesting cycles result. Many cultural and fertilization effects are known through basic agronomic researches performed long ago on such matters as effects of soil movement, type, fertility, and wash, in relation to *Fusarium* disease activity.

Producing resistant bananas, as they are now no

longer a gamble against nature, has resulted in settled communities and brought in such comforts as air conditioning and abundant refrigerated foods. The removal of the threat of failure from Fusarium has made it imperative to change shipping habits. For over 100 years bananas were handled as heavy bunches only strong men carried. Cut from high up, the weighty bunch lowered to a man's back was taken to a mule, then to a spur siding on a narrow gauge railroad track, then to a railroad car to go to port. Finally into the hold of a ship, stacked in bulk.

In plantations of low growing resistant bananas, returns have been good and continuous. A bunch is now surrounded with a plastic bag as it matures, when ready it is cut still shrouded, and hooked onto a cable that leads to the fruit-treatment center. There the banana "fingers" are cut away from the bunch in "hands," washed with fungicides and dried, wrapped, and then packed in boxes of corrugated board scientifically designed to reduce injury and infections and yet to allow of even better aeration in overseas shipment than anything available before (see Fig. 21-6). As in no other tropical American crop, pathologists of bananas are at work on many fungi never before incriminated as causing a fruit rot. They are studying likewise not only the fundamentals but the most practical means of sending the largest amount of produce to market from those millions of hectares of modern disease-resistant bananas (11, 62, 66, 67, 75, 126, 127).

A young phytopathologist with a West Indies tobacco growers' institute was hired away from it on a temporary basis. Those who secured him were a group from South America. He consulted and brought to their growers many advances in resistance which have been used ever since. His success, widely recognized, was appreciated in his home environment; he returned to work there and took back some resistant tobaccos. The perspective gained from all this incited in him a new purpose. From it he changed, and he stimulated new thinking in his old directorship and his administrators; through his unceasing efforts, the privately-run institute working on only one crop subsidized by growers, became a government research station that now helps all the farmers on all their problems of ailing crops.

A different thing happened in coffee that has been so long free from rust in the Western Hemisphere. A pathologist was sent around the tropical world to collect rust resistant seeds and to learn how African and Oriental coffee men produced in the presence of rust--how they "lived with it" and how they used resistance. This trip had some unexpected effects. There were obtained the planned-for, various disease-resistant collections, but in addition numbers of other tropical American scientists developed interest in coffee rust disease resistance collections; there was additional effort developed in connection with inter-hemispherical plant quarantine problems; work on coffee botany then at a standstill was given renewed impetus; new information about coffee growing was secured from very widely different ecological regions of the world; and dangers and prospective kinds of needed studies were readily determined when rust finally did appear in the American tropics.

There is no doubt that effects from resistance research on wheat rust in the Mexican tropics are of unique value on grain crops throughout the world. The immediate and intensifying cause of the detailed work on *Puccinia graminis tritici*, was that rust race changes were shown (19, and see Chapter XII in this book) to require new resistances in wheat three or more times as fast in the tropics as in the temperate zone. The average commercial lifetime of an improved resistant wheat variety is about four years in the tropics, whereas there may be 15 to 20 years in the useful life of a variety in the temperate zone.

It is possible that it could have been foretold that specialists in Mexico would be dissatisfied with purified resistant wheat lines. There is little variation in them, for the wheat plant is known to be self-pollinating, and a fixed line has a minimum of disease-resistance resiliency; when a sudden mutant race of rust finds the fixed wheat line compatible with its new disease capacity, it attacks, multiplies, and develops an epidemic. Tropical pathologists are now working with something different: a variety made up of individuals with standard height, appearance and agronomic adaptability, but with great inheritable variability in genetic composition that is wide-ranging in its multilineal resistances to many rust races.

An unpredictable course of events is that the multilineally resistant wheats inspired their creators towards a fresh attention to agronomic problems. New varieties clung to tall straw types that suffered when crowded in the field, but that grew well in poorer soils. With fertilizers added, wheat heads gained weight, causing the tall and thin straws to fall; lodged grain resulted. Realizing this way a crucial weakness, Dr. N. E. Borlaug and his co-workers bred stiff, dwarf straw characters along with long lasting multilineal resistance. These wheat plants can be grown close together on land, can use economically and well extra fertility added to the soil, are sufficiently vigorous and resistant that even if there is a light occurrence of rust it has no effect on the harvest, and with the world-wide reputation gained all this has inspired scientific agriculturists to speak of the "dawn of the green revolution"--the fight to feed an increasingly hungry world in which population outstrips food productivity.

These same principles respecting the solving of wheat rust resistance with correlated growing and production problems in the Mexican tropics are being applied successfully on more cereals in numerous other countries of the Western Hemisphere and the Orient. These concepts of solving disease resistance difficulties are brilliantly used on paddy rice in the Eastern Hemisphere and new varieties more resistant, shorter in stature, and more productive are now being grown pantropically.

It is with gratitude that world leaders view the promise of more food for their populations. There is growing confidence that famines caused by plant diseases can be eliminated and that resistant strains of crops will help bring to a minimum political instabilities in the tropical world from pressures by hungry people, materially aiding steps towards a better and more peaceful world. On this hope was based the decision of the Norwegian parliament to accord in 1970 the highest award that can be made, the world-recognized Nobel Peace Prize, to the plant pathologist of Mexico who did most of his work on disease resistance, Dr. N. E. Borlaug.

Literature Cited

1. Abbott, E. V. 1953. Sugarcane and its diseases (pp. 526-539). In Plant disease. U. S. Department of Agriculture Yearbook Plant Diseases, 940 p.

2. Acosta, P. P. 1967. Rancazua INTA. Nuevo aporte al mejoramiento del lino en Pergamino. Estac. Exp. Agrop. Pergamino, Argentina. Tec. Informe 61, 8 p.

3. Albuquerque, F. C. 1968. Nota prévia sôbre a enzertia da pimenta do reino (Piper nigrum L.). Inst. IPEA Norte, Belem, Brazil, Bulletin 14, 25 p.

4. Alconero, R. and A. G. Santiago. 1969. Mycorrhizal infections of mature portions of vanilla roots by Rhizoctonia solani as a predisposing factor to infection by Fusarium oxysporum f. sp. vanillae. Phytopathology 59:1521-1524.

5. ________., and E. Stone. 1969. Phytophthora parasitica nicotianae in roselle and kenaf in Puerto Rico. Plant Dis. Reptr. 53:702-705.

6. Ampuero P., E. and R. Alvarado. 1961. Variación en resistencia a la escoba de bruja, rendimiento e indice de mazorca en la progenie hibrida interclonal. Inter-Amer. Cacao Conf., Trinidad & Tobago Proc. 8:174-183.

7. ________., and R. Desrosiers. 1961. Comportamiento en resistencia a la escoba de bruja de los clones Silecia-1 y 5 triados del oriente ecuatoriano. Inter-Amer. Cacao Conf., Trinidad & Tobago Proc. 8:156-165.

8. Apple, J. L. 1967. Occurrence of race 1 of Phytophthora parasitica var. nicotianae in North Carolina and its implications in breeding for disease resistance. Tobacco Science 11:79-83.

9. Aquino, M. de la N. and L. J. da G. Wanderley. 1966. O mal das sete voltas nos cebolais do São Francisco. Inst. Pesq. Agron. Pernambuco, Téc. Boletim 16, 42 p.

10. Atkins, J. G., A. L. Robert, C. R. Adair, K. Goto, T. Kozaka, R. Yanagida, M. Yamada, and S. Matsumoto. 1967. An international set of rice varieties for differentiating races of Piricularia oryzae. Phytopathology 57:297-301.

11. Badger, A. M. 1965. Influence of relative humidity on fungi causing crown rot of boxed bananas. Phytopathology 55:688-692.

12. Baker, R. E. D. and N. W. Simmonds. 1951. Banana research--changes in outlook. Trop. Agr. Trinidad 28:43-45.

13. Bartley, B. G. D. 1953. Outlines of the cacao-breeding programme at I. C. T. A. Trinidad Imper. Coll. Trop. Agr., Cacao Res. 1945-51:24-25. --. 1959. The efficiency of a test of the resistance of cacao seedlings to Marasmius perniciosus Stahel. Trinidad Imper. Coll. Trop. Agr., Cacao Res. 1957-58:49-52.

14. Beckman, C. H. 1964. Host responses to vascular infection. Ann. Rev. Phytopath. 2:231-252.

15. ________., and W. C. Mueller. 1970. Distribution of phenols in specialized cells of banana roots. Phytopathology 60:79-82.

16. Bettencourt, A. J. and A. Carvalho. 1968. Melhoramento visando a resistência do cafeeiro à ferrugem. Bragantia 27:35-68.

17. Bocanegra S., S. 1958. Variedad mejorada de frijol canarie para la costa central del peru. Reunion Interam. Fitog. Fitop. Ent. Edaf., Proc. 3rd: 168-169.

18. Boock, O. J., A. Scivittaro, and S. A. Nobrego. 1965. Comportamento de setenta e uma variedades holandesas de batatinha sob as condições do estado de São Paulo. Campinas Inst. Agr., Boletim, 199 p.

19. Borlaug, N. E. 1958. The use of multilineal or composite varieties to control airborne epidemic diseases of self-pollenated crop plants (pp. 12-27). In Proc. First Internat. Wheat Genetics Symp., Univ. Manitoba, Winnipeg, Canada, 268 p. --. 1965. Wheat, rust, and people. Phytopathology 55:1088-1098.

20. Bos, H. and K. G. McIndoe. 1965. Breeding of hevea for resistance against Dothidella ulei P. Henn. J. Rubber Inst. Malaya 19;98-107.

21. Boyd, H. and F. Galli. 1966. Efeito do armazen-

amento da cana-de-azucar sôbre a germinação e sôbre a incidência da podridão-abacaxi, causada por Ceratocystis paradoxa (de Seynes) Moreau. An. Esc. Sup. Agr. Luiz Queiroz 23:229-245.

22. von Bulow, J. F. W. 1966. Ferrugens da milho. Pesq. Agropec. Brasil 1:249-262.

23. Calderoni, A. V. and C. J. Induni. 1954. Resistencia de variedades y clones de papa a la sarna comun (Streptomyces scabies). Rev. Invest. Agr. Argentina 8:195-206.

24. Campos T., A. 1951. Lineas y variedades Mexicanas de trigo y su resistencia a la roya del tallo. Ofic. Estud. Esp. Mexico, Fol. 4:70-81.

25. Capriles de Reyes, L., E. Schulz Schomburgk, and A. Muñoz. 1966. El contenido de ácido chlorogénico en differentes variedades de cacao y su relación con la resisténcia contra el hongo Ceratocystis fimbriata. Agron. Trop. Venezuela 16:273-284.

26. Casseres, E. H. and P. J. Linares. 1950. Producción de variedades de tomatoes para los tropicos humedos. Turrialba 1:7-11.

27. ________., L. C. Peterson and D. Reddick. 1953. Tres nuevas variedades de papas resistentes al tizon tardio. Turrialba 3:86-90.

28. Castronovo, A. 1950. Fuentes de resistencia a Phytophthora infestans. Rev. Inv. Agr., Argentina 4:245-275. --. 1954. Studies on the inheritance of resistance in Phytophthora infestans in potatoes. American Potato J. 31:397-403.

29. Cenoz, H. P. 1952. Resistencia al carbón volador del trigo. Rev. Invest. Agr. Argentina 6:29-88.

30. Cervantes, R., A. Rodriguez V., and J. S. Niederhauser. 1958. Resistencia al virus causante del achaparramiento del maíz. Ofic. Estud. Esp. Mexico, Folletto Tec. 29, 18 p.

31. Cheesman, E. E. 1944. Notes on nomenclature,

classification and possible relationships of cacao populations. Trop. Agr. Trinidad 21:144-159.

32. Cortazar S., R. 1950. Producción de trigos resistentes a los polvillos colorados (Puccinias) en Chile. Ofic. Estud. Esp. Mexico, Fol. 3: 220-232.

33. Costa, A. S. 1969. Conditioning of the plant by one virus necessary for systemic invasion of another. Phytopath. Zeitsch. 65:219-230.

34. Crispin, M., A., W. D. Yerkes Jr., and G. Patino M. 1957. Mejores variedades de frijol. México Sec. Agr. Ganad. Boletin 297, 9 p.

35. Cruz, B. P. B., O. J. Book, and V. Rossetti. 1961. Reações de clones de batatinha IAC à infecção por racas fisiologicas de Phytophthora infestans no campo. Arq. Inst. Biológico 28:7-11.

36. ________., A. P. de Silveira, and N. T. de Mendonça. 1968. Determinação da resistência de algumas variedades de melancia (Citrullus vulgaris Schrad.) à murcha de fusarium (Fusarium oxysporum (Schlecht.), F. niveum (E. F. Sm.) Snyder & Hansen). Arq. Inst. Biól. 35:67-76.

37. de la Puente E., J. M. 1959. Algunos notas sobre el mejoramiento y la genetica de los pinos. Escuela Nac. Agr. Mexico, Tec. Publicación 1, 41 p.

38. Delgado, J. C. and E. Echandi. 1965. Evaluación de la resistencia de especies y clones de cacao al mal del machete provocado por Ceratocystis fimbriata. Turrialba 15:286-289.

39. Desrosiers, R. 1954. Diversidad genetica del cacao como base en la selección de resistencia a la enfermedad de la escoba de bruja. Turrialba 4: 131-134.

40. ________., C. W. Bolaños, and J. Vargas. 1955. Evaluación de clones de cacao en relacion con su resistencia a la escoba de bruja (Marasmius perniciosus Stahel). Turrialba 5:78-82.

41. ________., A. von Buchwald, and C. W. Bolaños. 1954. Selección masal de plantas de cacao de semilla para resistencia a la escoba de bruja. Cacao (Costa Rica) 3:11-12.

42. ________., ________., ________. 1955. Selección en massa de plantulas de cacao con base a su resistencia a la escoba de bruja. Turrialba 5: 134-138.

43. Díaz Polanco, C. and E. de Armas. 1964. Relaciones entre exudados de cuatro variedades de caraotas y el crecimiento de Phytophthora parasitica, Pythium aphanidermatum, e Isariopsis griseola. Agron. Trop. Venezuela 14:169-174.

44. Dropkin, V. H. 1969. Cellular responses of plants to nematode infections. Ann. Rev. Phytopath. 7: 101-122.

45. Echandi, E. 1967. Amarillamiento del frijol (Phaseolus vulgaris L.) provocado por Fusarium oxysporum f. phaseoli. Turrialba 17:409-410.

46. ________., and C. E. Fernandez. 1962. Relación entre el contenido de ácido chlorogénico y la resistencia a las llaga macana o cancer de los cafetos causada por Ceratocystis fimbriata. Turrialba 12:87-90.

47. Estrada R., N. 1958. Mejoramiento de la papa en Colombia, para la obtencion de variedades resistentes a la gota (P. infestans). Inter-Amer. Fitog. Fitop. Ent. Edafo., Reunion 3:178-179.

48. Favret, E. A. and A. A. Rodriguez. 1957. Mutabilidad espontanea en cebada. Rev. Invest. Agr., Argentina 11:309-311.

49. Fawcett, H. S. 1936. Citrus diseases and their control. McGraw-Hill, New York, 656 p.

50. Fernández-Borrero, O. 1964. Patogenicidad del Ceratocystis fimbriata (Ell. Halst.) Hunt y posible resistencia en Coffea arabica L. var bourbon. Cenicafé, Colombia 15:3-17.

51. Fernandez Valiela, M. V. (2d. ed.). Introducción a la Fitopatología. Gadola, Buenos Aires, 872 p.

52. Fisher, F. E. 1969. Resistance in Florida citrus to Cercospora citri-grisea. Plant Dis. Reptr. 53: 112-114.

53. di Fonzo, M. A. 1938. Informe sobre la marchitez del algodonero en la República Argentina. Junta Nac. Algodón, Argentina, Tec. Boletin 5, 11 p.

54. Ford, H. W. 1967. Burrowing nematode resistant rootstocks as biological barriers in citrus groves. Proc. Florida State Hort. Soc. 80:56-59.

55. Gallegos, B. C. 1966. Estudio genético de la resistencia del frijol a Pseudomonas phaseolicola (Burk.) Dows. Agrociencia 1:56-66.

56. Galli, F., H. Tokeshi, P. de C. Torres de Carvalho, E. Balmer, H. Kimati, C. O. N. Cardoso, and C. L. Salgado. 1968. Manual de Fitopatologia, Doenças das Plantas e seu Contrôle. Ceres, São Paulo, 640 p. (see especially pp. 99-110.)

57. Galvez, G. E. and J. C. Lozano. 1967. Identification of physiological races of Piricularia oryzae in Colombia rice. Agr. Trop. Trinidad 23:233-239.

58. Gentry, H. S. 1969. The ancestor of the common bean and its potential as a breeding resource. U. S. Dep. Agr. Res. Serv. Report 1969:74-50.

59. Goncalves, J. R. C. 1968. The resistance of FX and IAN rubber (hevea) clones to leaf diseases in Brazil. Trop. Agr. Trinidad 45:331-336.

60. Gonzalez B., R. 1966. Efecto del ataque del polvillo de la hoja (Puccinia recondita Rob. ex Desm.) en el rendimiento de variedades del trigo. Agr. Tec., Chile 26:16-21.

61. ________., P. Parodi P., O. Moreno M., H. Wulf M., and D. Granger M. 1966. Centrifén, collafén, yafén y alifén, nuevas variedades de trigo para la zona central del pais. Agr. Tec. Chile 26:78-84.

62. Goos, R. D., E. A. Cox, and G. Stotzky. 1961. Botryodiplodia theobromae and its association with Musa species. Mycologia 53:262-277.

63. Graham, K. M., J. S. Niederhauser, and S. Romero. 1959. Observations on races of Phytophthora infestans in Mexico during 1956-1967. American Potato J. 36:196-203.

64. ________., ________., and L. Servin. 1959. Studies on fertility and late blight resistance in Solanum bulbocastanum Dun. in Mexico. Canadian J. Bot. 37:41-49. (See also other works by Niederhauser before and after this date.)

65. Granados R., G. and A. Ortega C. 1966. Hospederas de las vectores del virus de la hoja blanca. Pruebas de transmissiÓn y reacciÓn de differentes variedades de arroz al virus. Agr. Téc. Mexicana 2:276-284.

66. Greene, G. L. 1966. Response of conidia and appressoria of Gloeosporium musarum to hypochlorous acid. Phytopathology 56:1201-1203.

67. ________., and R. D. Goos. 1963. Fungi associated with crown rot of boxed bananas. Phytopathology 53:271-275.

68. Grimm, G. R. and S. M. Garnsey. 1968. Foot rot and tristeza tolerance of smooth seville orange from two sources. Proc. Florida State Hort. Soc. 81:84-90.

69. Guimarães, F., N. K. Luz, and M. T. Liberal. 1958. Trabalhos de melhoramento em batata americana (Solanum tuberosum L.) visando obtenção de resistência ao fungo Phytophthora infestans (Mont.) DeBary. Agron. Riograndense, Brazil 3:35-38.

70. Guzman Naranjo, J., H. D. Thurston, and L. E. Heidrick. 1960. Naturaleza de la resistencia parcial de tres clones de papa al Phytophthora infestans. Agr. Trop. Colombia 16:89-99.

71. Holliday, P. C. 1954. The susceptibility of some Imperial College selections to witches' broom

disease. Trinidad Imper. Coll. Trop. Agr. Rept., Cacao Res. 1953:58-63.

72. Huitrón, M. M. E. and T. A. Campos. 1966. Pruebas de resistencia de lineas de garbanzo a enfermedades de la raiz. Agrociencia 1:67-76.

73. Idrobo M., S. and G. Naundorf. 1956. Producción de toxinas por el Ceratostomella fimbriata en cacao. Cacao en Colombia 5:37-39.

74. Jolly, A. L. and L. L. de Verteuil. 1961. Clonal and fertilizer experiments in Trinidad and Tobago. Inter-Amer. Cacao Conf. (Trinidad & Tobago) Proc. 8:390-416.

75. Kaiser, W. J. and F. L. Lukezic. 1966. Influence of certain environmental conditions on spore dispersal and survival of Cercospora hayi from babana. Phytopathology 56:1290-1293.

76. Klotz, L., T. A. DeWolfe, and R. C. Baines. 1969. Resistance of trifoliate orange stocks to gummosis. Citrograph 54:259-260.

77. Kuc, J. 1966. Resistance of plants to infection agents. Ann. Rev. Microbiol. 20:337-370.

78. Lems, G. 1963. Resultaten van een Cacao-klonen Experiment. Surinaam. Landb. 11:138-145.

79. Leppik, E. E. 1968. Relation of centers of origin of cultivated plants to sources of disease resistance. U. S. Dep. Agr. Plant Introd. Invest. Paper 13, 7 p.

80. Liu, J. L. and J. E. Pérez. 1969. Occurrence of rust on pangola grass in Puerto Rico. Phytopathology 59:1348.

81. Mace, M. E. 1963. Histochemical localization of phenols in healthy and diseased banana roots. Physiol. Plantarum 16:915-925.

82. ________., and E. Solit. 1966. Interactions of 3-indoleacitic acid and 3-hydroxytyramine in fusarium wilt of bananas. Phytopathology 56:245-247.

83. Malaguti B., G. and A. Ordosgoitty F. 1969. El achaparramiento del maíz en Venezuela. Agron. Trop. Venezuela 19:85-97.

84. Manotas A., L. E. 1953. El contenido en azucares y acidos organicos de la mazorca de cacao y su probable relación con ataques de enfermedades. Acta Agron., Colombia 3:177-188.

85. Mathre, D. E., D. C. Erwin, A. O. Paulus, and A. V. Ravenscroft. 1966. Comparison of the virulence of isolates of Verticillium albo-atrum from several of the cotton-growing regions in the United States, Mexico, and Peru. Plant Dis. Reptr. 50:930-933.

86. Mattos, R. A. 1949. Antecedentes sobre el carbón de la caña de azucar en el Chaco. Lilloa 21:155-166.

87. Mazzani, B., and S. Hingosa. 1961. Differencias varietales de susceptibilidad a la roya del maní en Venezuela. Agron. Trop. Venezuela 11:41-45.

88. Mazzei, I. 1949. Ensayo de especificos y hormones en trigo literal preciz infestado con carbón hediendo. Bol. Inform. San Veg., Uruguay 1(3):16-20.

89. Melendez de la Garza, M. 1951. Resistencia y susceptibilidad de las variedades Mexicanos de frijol a la antracnose (Colletotrichum lindemuthianum (Sacc. y Magn.) Bri. y Cav.). Ofic. Estud. Esp. Mexico, Fol. Misc. 4, 118 p.

90. Mello, R. F. T. 1960. Observações sôbre a brusone do arroz e seu contrôle. Biológico 26:218-222.

91. Montaldo, A. 1951. Fitomejoramiento para resistencia a la nematosis de la papa. Agr. Tec., Chile 11: 64-85.

92. Montaldo B., A. and E. de Rojas Pena. 1950. Fitomejoramiento para resistencia al tizon de la papa. I. Prueba de la reacción al Phytophthora infestans. Agr. Tec. Chile 10:66-73.

93. Montserin, B. G., and L. L. de Verteuil. 1956. Trabajo de hibridacion en el Departmento de Agricultura (Trinidad y Tobago) para el combate la escoba de bruja del cacao. Cacao, Costa Rica 3:14.

94. Moreira, S., A. S. Costa, and T. J. Grant. 1954. Metodos para identificação e contrôle da tristeza dos citras. Bragantia 13:223-236. (Numerous other publications by these authors.)

95. Mujica, R. F. 1952. Razas fisiológicas y susceptibilidad varietál de los frijoles Chilenos a la antracnosis. Agr. Tec., Chile 12:37-45.

96. Muñoz, J. M. and F. Cardenas R. 1950. Variación genética de las frijoles de Mexico. Ofic. Estud. Espec. (Mexico), Fol. Misc. 3:62-71.

97. Namekata, T. and V. Rossetti. 1968. Observações sôbre a tristeza dos citrus. Biologico 35:289-290.

98. Nieves, R. 1930. Ensayos comparativos de resistencia a la _Tilletia laevis_ (Kühn) en trigos argentinos e importados, comunes y de pedigree. Bol. Min. Agr. Nac., Argentina 29:297-314.

99. Niederhauser, J. S. 1961. Genetic studies of _Phytophthora infestans_ and _Solanum_ species in relation to late-blight resistance in the potato. Recent Adv. Bot. 163:498-499. --. 1962. Evaluation of multigenic field resistance of the potato to _Phytophthora infestans_ in 10 years of trials at Toluca, Mexico. Phytopathology 52: 746. (See other numerous publications by this author.)

100. Nolla, J. A. B. 1969. A new leaf spot of amaryllis. Phytopathology 59:1348.

101. Noronha-Wagner, M. and A. J. Bettencourt. 1967. Genetic study of the resistance of _Coffea_ spp. to leaf rust. 1. Identification and behaviour of four factors conditioning disease reaction in _Coffea arabica_ to twelve physiologic races of _Hemileia vastatrix_. Canadian J. Bot. 45:2021-2031.

102. Ochoa A., M. 1966. Reacción de líneas y variedades de maíz a la fusariosis. Fitopatologia, Chile 1:14-18.

103. Olave L., C.A. 1958. Resistencia de algunas variedades y lineas de frijol (Phaseolus vulgaris L.) al Isariopsis griseola. Acta Agron., Colombia 8: 197-219.

104. Orjuela, J. 1967. Razas fisiológicas de Puccinia graminis tritici en Colombia y sus implicaciones. Agr. Trop., Colombia 23:465.

105. Osborne, R.E. 1958. Breeding the immune banana. Jamaica Banana Bd. Res. Dep., Bulletin 2, 8 p.

106. Parodi P., P. and R.M. Caldwell. 1966. Herencia de la resistencia a los rezas fisiológicas 1, 2 y 15 de Puccinia recondita f. sp. tritici Rob. ex Desm. en cinco variedades de trigo. Agr. Tec., Chile 26:26-33. (See also numerous other publications by R.M. Caldwell.)

107. Peachey, J.E. (ed.) 1969. Nematodes of tropical crops. Proc. Caribbean Symp. Held in Trinidad. Commonwealth Agr. Bureau, England, 355 p.

108. Porte, W.S. and F.L. Wellman. 1941. Development of interspecific tomato hybrids of horticultural value and highly resistant to fusarium wilt. U.S. Dep. Agr. Circular 584, 18 p.

109. Pound, F.J. 1940. Witches' broom resistance in cacao. Trop. Agr. Trinidad 17:6-8. --. 1940. Search for resistance to witch broom in cacao. Agr. Soc. Trinidad and Tobago, Proc. 40:35-37.

110. de Ramallo, N.E.V. 1969. Sarna o verrugosis del palto en Tucumán. Rev. Indus. Agr. Tucumán 46(2):27-30.

111. Reyes N., R. and L. Penna Acosta. 1958. Fuentes de resistencia a las royas de trigo en Colombia. Reunion Interam. Fitogen. Fitop. Ent. Edaf., Proc. 3:139-140.

112. Roberts, L.M., U.J. Grant, and R. Ramirez E.

1957. Razas de maíz en Colombia. Colombia Min. Agr., Ofic. Invest. Espec., Bol. Tec. 2, 159 p.

113. Rodriguez, C. J. 1956. Nota sobre a resistencia de algumas espécies de Coffea a Hemileia coffeicola Maubl. et Rog. Rev. Café Português 3:48-71.

114. Rojas M., E. 1958. Hacia la creación de tipos de girasol resistentes a la roya. Reunion Interam. Fitog. Fitop. Ent. Edaf. Proc. 3:166-167.

115. Rossetti, V. 1947. Estudos sôbre a gomose de phytophthora dos citros. I. Suscetibilidade de diversas espécies cítricas a algumas espécies de Phytophthora. Arq. Inst. Biológico 18:97-124.

116. Rupert, J. A. 1950. El plasma germinal disponible y necesidad para el majoramiento del trigo en Mexico. Ofic. Estud. Esp. Mexico, Fol. Misc. 3:54-62.

117. Sarria Villano, S. 1954. Estudio descriptivo de cuatro razas de maíz en Colombia. Acta Agron., Colombia 4:161-174.

118. Schafer, J. F. 1971. Disease tolerance. Ann. Rev. Phytopath. 9:(in press.)

119. Schieber, E. 1962. Estudios preliminares sobre la roya de Guatemala causada por Physopella zeae. Reunión Cooper. Interamer. Mejor. Maíz, Proc. 8:76-78. --. 1963. Principales enfermedades del maíz en Guatemala. Reunión Cooper. Centroamer. Mejor. Maíz, (article 7) 7 p. --. 1965. Distribution of Puccinia polysora and P. sorghi in Africa and pathogenicity of these species on corn lines with Latin American plasm. Phytopathology 43:1074-1075. --. 1968. Preliminary studies on Phyllachora maydis affecting corn in Central America. Phytopathology 58:554.

120. Schochaj, R. 1954. Evaluación de las variedades de papa segun su resistencia a Phytophthora infestans (Mont.) de Bary, en plantaciones de primavera y verano en la estación experimental El Sancho, Catamarca. Rev. Agron. Noreste Argentina 1: 135-155.

121. de Segura, C. B. 1965. Enfermedades de Cultivos Tropicales y Subtropicales. Edit. Jurídica S. A., Lima, Peru, 439 p.

122. Sequeira, L. and P. R. Rowe. 1969. Selection and utilization of Solanum phureja clones with high resistance to different strains of Pseudomonas solanacearum. American Potato J. 46:451-462.

123. Shepherd, K. 1968. Banana breeding in the West Indies. P. A. N. S., Sec. B., 14:370-379.

124. Sreenivasan, M. S. 1968. Progress report on screening coffee selections and collections with races of leaf rust fungus at Central Coffee Research Institute. Indian Coffee 32:299-304.

125. Storey, H. H. and A. K. Audric. 1967. Transfer of resistance to the streak virus in East African maize. East African Agr. For. J. 33:131-135.

126. Stover, R. H. 1969. The effect of benomyl on Mycosphaerella musicola. Plant Dis. Reptr. 53: 830-833.

127. ________., and D. L. Richardson. 1968. Pelipita, an ABB bluggoe type plantain resistant to bacterial and fusarial wilts. Plant Dis. Reptr. 52:901-903.

128. (Subramanian, S.) 1969. Plant pathology (report to) Indian Coffee Board 1968-69:77-89 (processed).

129. Thurston, H. D. and J. C. Lozano T. 1968. Resistance to bacterial wilt of potatoes in Colombian clones of Solanum phureja. American Potato J. 45(2): 51-55.

130. Toffano, W. B. 1963. Estudos preliminares sôbre a resistência de variedades à ramulose de algodoeiro. Biológico 29:69-71.

131. Tokeshi, H. 1964. Resistência a Fusarium oxysporum f. conglutinans (Wr.) Sny. & Hans. em algumas variedades da Brassica oleracea L., cultivadas no estado de São Paulo. An. Esc. Sup. Agr. Luis Queiroz 21:3-5.

132. ________., and F. Galli. 1966. Variabilidade de Fusarium oxysporum f. lycopersici (Wr.) Sny. & Hans. em São Paulo. An. Esc. Sup. Agr. Luis Queiroz 23:195-209.

133. ________., ________., and C. Kurozawa. 1966. Nova raça de Fusarium do tomateiro em São Paulo. An. Esc. Sup. Agr. Luis Queiroz 23:195-204; 211-215; 217-227.

134. Trujillo, G. 1969. A possibility for selection of rice varieties resistant to hoja blanca. Plant Dis. Reptr. 53:440-443.

135. Valbuena T., L. A. 1953. Lineas isogeneticas. Reunion Interamer. Fitog. Fitop. Ent. Edaf. (Proc.) 3:144.

136. Vallega, J. 1938. Observaciones sobre la resistencia a la roya de algunas linos ensayados en el Instituto Fitotecnico de Llavallol. Rev. Argentine Agr. 5:25-54. --. 1944. Reacción de algunos trigos con respecto a las razas fisiologicos de Puccinia rubigo-vera tritici communes en Argentina. Rev. Fac. Agr. Vet., La Plata Univ. 11: 1-27. --. 1947. Reacción de algunas especies espontaneas de Hordeum con respecto a las royas que efectan al trigo. Rev. Invest. Agr., Argentina 1:52-62. --. 1950. Resultados obtenidos en las investigaciones--y lo que estos conocimientos significan para la selección de variedades resistentes a las enfermedades. Mexico, Ofic. Estud. Esp., Fol. 3:214-219. --. 1951. Herencia de la resistencia a Puccinia coronata y P. graminis avenae. Rev. Invest. Agr., Argentina 5:523-539. --. 1955. Variación en la población de Puccinia graminis avenae y sus consecuencias fitotecnicas. Rev. Nac. Fac. Agron. Vet., Buenos Aires Univ. 13:459-469. (See also other publications by J. Vallega.)

137. ________., and H. Cenoz. 1943. Reaccion de algunos trigos a las razas fisiológicas de Erysiphe graminis tritici en Argentina. An. Inst. Fit. Sta. Catalina, Argentina 3:45-58.

138. Van der Plank, J. E. 1963. Plant diseases: epi-

demics and controls. Academic Press, New York, 349 p. --. 1969. Pathogenic races, host resistance, and an analysis of pathogenicity. Netherlands J. Plant Path. 75:45-52.

139. Vieira, C. 1959. Rico-23, nova variedad de feijão prêto para a Zona da Mata, Minas Gerais. Ceres, Brazil 11:22-26.

140. Villaneuva Navoa, R. 1958. Las hibridos de trigo de las Estacion Experimental Agricola de la Molina y su valor para la agricultura peruana. Reunion Interam. Fitog. Fitop. Ent. Edafol. (Proc.) 3: 145-147. --. 1969. Herencia de la necrosis marrón y de otros caracteres del trigo en el Peru. Turrialba 19:161-175.

141. Von der Pahlen, A. 1967. Herencia de la resistencia al virus del mosaico del tabaco en el tomato en la Argentina. Rev. Invest. Agrop., Argentina Ser. 2, 4:231-236.

142. Walker, J. C. 1959. Enfermedades de las Hortalizas. Salvat Editores, Barcelona, 624 p.

143. ________., and M. A. Stahmann. 1955. Chemical nature of disease resistance in plants. Ann. Rev. Plant Phys. 6:351-366.

144. Wehlberg, C. and R. W. Meyer. 1966. Bacterial soft rot of iceberg (great lakes) lettuce in the Florida everglades. Plant Dis. Reptr. 50:938-941.

145. Wellhausen, E. J., L. M. Roberts, and E. Hernández. 1951. Razas de maíz en Mexico. Ofic. Estud. Esp., Mexico Secr. Agr. Ganad. (special bulletin), 211 p.

146. Wellman, F. L. 1946. Reporte acerca del adelanto de los estudios sobre las podredumbres radiculares del cafeto en El Salvador. Cafe de El Salvador 16:105-117. --. 1949. Estudio de plantas de café con el objeto de tener mejoras por selección. Suelo Tico, Costa Rica 1:364-365. --. 1952. The danger of introducing Hemileia of coffee into the Americas, and prob-

lems of its exclusion and control. Asamb. Latinoamer. (Segundo), Genet. Fitop., Brazil, 6 p. --. 1954. Evidencia de resistencia a las enfermedades en los cafetos. Turrialba 4: 52-57. --. 1954. The fusarium phase of the root-disease complex in coffee. Phytopathology 44:509. --. 1957. Hemileia vastatrix. Investigaciones presentes y pasadas en la herrumbre del cafe y su importacia en la América tropical. FEDECAME, Sección Divulg. (Booklet), 67 p. --. 1961. Coffee botany, cultivation, and utilization. Leonard Hill, London; Interscience Pub., New York, 488 p. (Unpublished reports on banana Fusarium wilt in 1929 and 1930 for private circulation, from work with the United Fruit Company in Central America.)

147. Zapata Balcazar, M. and L. A. García. 1958. Variedades sinteticas de trigo para Colombia. Reunion Interam. Fitog. Fitop. Ent. Edafo. (Proc.) 3: 152-154.

148. Zaumeyer, W. J. and H. R. Thomas. 1957. A monographic study of bean diseases and methods for control. U. S. Dep. Agr., Tech. Bulletin 868, 255 p.

149. Zentmyer, G. A. and C. A. Schroeder. 1958. Resistance of species of Persea to avocado root rot. California Avocado Soc. Yrbk. 42:106-107.

Appendix 1

HEMILEIA
(Occurrence of Coffee Rust in Brazil)

For nearly 275 years coffee, that is *Coffea arabica*, has been grown in the American tropics free from the rust *Hemileia vastatrix*. This coffee leaf disease (see accompanying illustration) is very well known in the annals of tropical plant pathology. It has long been considered one of the eight greatest plant disease scourges of agricultural man. Until scientists had spent a century of study on it, it was thought almost a certainty that once the rust appeared in a country, the coffee crop there was sooner or later a doomed thing.

The rust doubtlessly originated as a specialized parasite of wild coffee that grows as a small tree in the forests on mountain slopes of northeastern Africa. When coffee was domesticated, the rust spread to it and developed on it strongly. Since 1869, when the fungus that caused it was first described, it has many times proved to be the most devastating disease of coffee in the world. One experience after another is that whole plantations and groups of plantations have been destroyed by *Hemileia*.

The list of rust-despoiled areas is extensive in the Eastern Hemisphere, but a few names of countries where coffee rust has been extreme are as follows: Ceylon, South India (Mysore, Coorg), Java, Tanzania (Tanganyika), Kenya, Uganda, Malaya, Philippine Islands, Sao Tomé, parts of Cameroon, Angola, Dahomey, Togo, Ivory Coast, Ghana, Nigeria, Liberia, and Guinea. From east to west the rust's deliberate march across the Oriental tropical world has been relentless; the hope was that it could not succeed in crossing the Atlantic Ocean to the Americas for in the fields of the Western tropics all commercial coffees are susceptible to the rust.

Neotropical arabica coffee came to the Neotropics during a few years from 1712 to 1718 originating from a

Leafspot symptoms of rust caused by attack of *Hemileia vastatrix*, viewed from underside of a leaf of arabica coffee. Photograph courtesy of A. Branquinho d'Oliveira, Coffee Rust Research Centre, Oeiras, Portugal.

very restricted region, and therefore of narrow inheritance. Yields and commercial acceptability were good, but that is only if crops are reasonably disease free. It is well known that in the 1970s the billions of coffee trees growing in the important 15 or more coffee producing countries of the American tropics are all highly susceptible to H. vastatrix. The prospects are most discouraging for what might follow a complete rust induced failure of the great coffee industry of the Americas.

It is freedom from rust that has assured the success of coffee growing in Latin America. It is this commodity that is now the greatest money-making product of anything from the Neotropics. Moreover, in that great region is Brazil, the largest producer of coffee in the world, and Brazil is the largest country in the Western tropics. It counts on over half of its export revenues to come from the coffee trade. The discovery in January 1970 of Hemileia vastatrix on coffee in Brazil has been of major concern to tropical American plant disease men.

How it arrived in Brazil will perhaps never be known. Coffee growers are completely informed about the dangers of importing live coffee plants from the Eastern Hemisphere and I fully accept that the rust did not come in on smuggled coffee seedlings. In addition, throughout several decades there have been established in the Americas a large number of quarantine laws and programs, expressly dealing with coffee rust. (The reader can refer to many notices of coffee plant quarantines, as noted in issues of the journal Review of Applied Mycology, and also in publications of the United Nations Food and Agricultural Organization.) It has had much attention.

However, in spite of all care, the rust entered Brazil. On 17 January 1970 it was found and determined by the cacao pathologist Arnoldo Gomes Medeiros, in a fazenda where coffee was grown under the shade of cacao trees. This was near the city of Itabuna, in the state of Bahia in Brazil. He obtained immediate confirmation of his discovery from highly respected specialists, and work began in the hopes of isolating the infestation and if possible wiping it out. It was soon discovered to be widely disseminated in not only Bahia but also in two other states, Espirito Santo and Minas Gerais, of Brazil. Its occurrence was so extensive that it was realized it could not be extirpated.

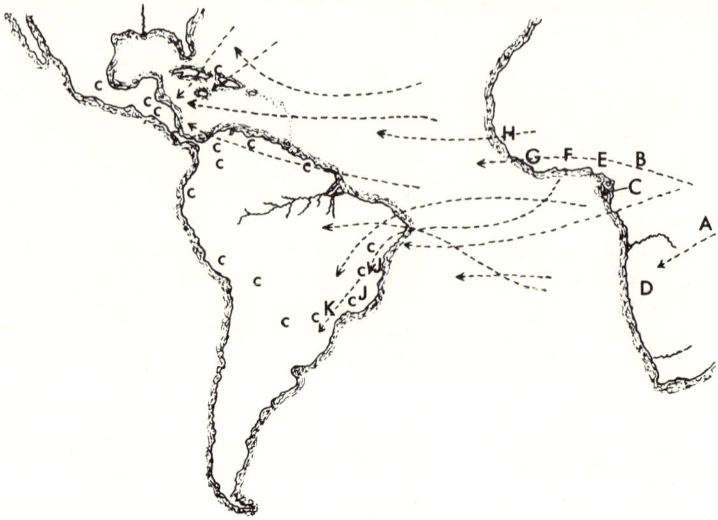

Sketch map for coffee rust, Hemileia vastatrix, appendix. Dashes with arrow points give courses and directions of some common winds and storms; winds in troposphere a few miles aloft, go from Africa to the Americas. Capital letters indicate countries and rust occurrences: A, Congo 1929-1942; B, Cameroons 1931-1952; C, Sao Tomé 1938; D, Angola 1930-1950s; E, Nigeria 1962-1963; F, Gold Coast and Ivory Coast and Dahomey-Togo 1952-1954; G, Liberia 1955; H, Guinea and Sierre Leon 1962-1963; J, States of Bahia and Espirito Santo and Minas Gerais of Brazil 1970; K, State of São Paulo of Brazil in 1971. The "c" lower case, some general locations of coffee growing in the Americas. (Distances: Dakar w. coast Africa to Recife e. coast Brazil by airplane 1717 miles; Dakar to Rio de Janeiro 2763 miles. Shortest direct distance Africa to Brazil about 1226 miles.)

In the accompanying sketch map (capital letters refer to countries and dates of occurrences are given in the caption) can be seen the distribution of Hemileia vastatrix in the western half of the continent of Africa--and then the inevitable crossing of the Atlantic Ocean to the eastern states of Brazil; from there, as might be expected, it is spreading south in 1971 and is now in the States of São Paulo and Paraná.

Since microscopic examination shows spores of Hemileia are evidently adapted to being airborne (it is known from studies that they are both air-carried and water-droplet dispersed), I have attempted to indicate on the map wind and storm movements. It seems reasonable that the rust was brought over to Brazil by air movements. Moreover, it seems definite to me that with fair certainty there will be further spread of the rust to the rest of the Americas.

Some References

Amaral, M. and C. Deruiz Beduin. 1970. A ferrugem alaranjada do cafèeiro, historico e distribuição geografico. Min. Agr. Brasil (Special publ.), 11 p.

Anon. 1970. Ferrugem do cafèeiro. Inst. Bras., Cafe-Inst. Biológ., Estado São Paulo (March 1970), Folha Colorida 11, 2 p.

Anon. 1970. Café. Guerra à ferrugem. Veja e Leia, No. 85 (22 April), pp. 38-39.

Bettencourt, A. J. and A. Carvalho. 1968. Melhoramento visando a resistência do cafèeiro à ferrugem. Bragantia 27:35-68.

Bitancourt, A. A. 1958. A perigo da introdução de doenças do cafèeiro no Brasil. Biológico 24:35-40.

_______. 1970. A ferrugem do cafèeiro. Biológico 36: 31-34.

Chaves, G. M., J. da Cruz Filho, M. G. de Carvalho, K. Matsuoka, D. Teixeiro Coelho, and C. Shimoya. 1970. A ferrugem do cafèeiro (Hemileia vastatrix Berk. & Br.): revisão de literature con observações e comentários sôbre a enfermidade no Brasil. Seiva 30: (Ed. Esp.), 75 p.

Colombia Federación Nacional de Cafeteros. 1970. Esta es la roya: ruina de los cafetos. Centro Nac. Inv. Café (CENICAFE), Bol. Extens. 42, 18 p.

Compacci, C. A. 1970. Resultados preliminares sôbre o contrôle químico da ferrugem do cafèeiro (*Hemileia vastatrix*) em Caratinga, estado de Minas Gerais, Brasil. Inst. Biológ. São Paulo (Typescript), 9 p.

D'Souza, G. I. 1965. (Report of research department) V. Mycology. Indian Coffee Bd., Res. Dept. Rept. 1964-1965:73-84. (Spore trap studies: *H. vastatrix* dry spores airborne at least 200-400 yards.)

Instituto Brasilero do Café. 1970. A ferrugem do cafèeiro no Brasil. (Special publ.) 75 p., 1 map.

Instituto Salvadoreño de Investigaciones del Cafe. 1970. La roya del cafeto. Inst. Salvadoreño Inv. Cafe, Bol. Inform. 93, 5 p.

Mayne, W. W. 1969. A century of coffee leaf disease, 1869-1969. Biologist (J. Inst. Biol.) 16:58-60.

Medeiros, A. G. 1970. (Observing on 17 January, occurrence of *Hemileia vastatrix* on coffee in Fazenda Asuncião, Municipio Aurelino Leal, Bahia.) One page (processed, for President of CEPLAC) dated 26, Jan. 1970, Rio de Janeiro, Brazil.

Noronha-Wagner, M. and A. J. Bettencourt. 1967. Genetic study of the resistance of *Coffea* spp. to leaf rust. I. Identification and behavior of four factors conditioning disease reaction in *Coffea arabica* to twelve physiological races of *Hemileia vastatrix*. Canadian J. Bot. 45:2021-2031.

Nutman, F. J. and F. M. Roberts. 1963. Studies on biology of *Hemileia vastatrix* Berk. & Br. Trans. British Mycol. Soc. 46:27-48. (Important means of dispersal by water drops.)

d'Oliveira, A. B. 1965. Progress report 1960-1965, Coffee Rust Research Centre, Oeiras, Portugal. (Processed) 144 p. (Also other reports from Dr. d'Oliveira.)

________. 1970. Identificación de razas de la *Hemileia* y

tipos de resistencia. (Unpublished.)

Orillo, F. T. and R. B. Valdez. 1962. Selection of coffee species and varieties resistant to rust. Coffee & Cacao J. 5:155-156. (All Central American selections susceptible; 8 African and 4 Indian sel. highly resistant; Brazil hybrid 387 highly resistant; resistance dominant.)

Rourk, J. P. and J. L. Creech. 1970. Coffee rust and countermeasures in the Americas. For. Agr. 8 (no. 33):2-5.

Schieber, E. 1960. La herrumbre del café y la cooperación internacional. El Informador (Guatemala) Agr., Foll. Especial, 3 p.

________. 1963. La herrumbre del café y el Centro de Investigaciones en Portugal. Rev. Cafetalera, Guatemala 1963:21-22.

________. 1970. Observaciones comparativos sobre la roya del cafeto en Brazil y Kenya, Africa. (Unpublished.)

Stevenson, J. A. and R. Beam. 1953. An annotated bibliography of coffee rust (<u>Hemileia</u> spp.). Div. Mycol. Dis. Surv., United States Dept. Agr., Special Publ. 3, 80 p. (Inclusive to that date and very valuable source.)

Sylvain, P. G. 1970. Lista de las principales introducciones de café del IICA-CEI presentando un grado apreciable de resistencia a la <u>Hemileia vastatrix</u>. (Unpublished.)

________., and J. J. Cordoba. 1967. Lista de las introducciones de cafe del Departamento de Fitotecnia y Suelos. Inst. Interamer. Cien. Agr. OEA, Costa Rica, Publ. Misc. 47, 96 p.

Time Magazine, Science Department. 1970. Coffee nerves in Brazil. Time 96 (no. 16):47.

Wallis, J. A. N. 1970. Coffee leaf rust in South America, a report to the International Coffee Organization. London (special publ. processed) 49 p., 1 map.

Wellman, F. L. 1952. Peligro de introducción de la

Hemileia del café a las Americas. Turrialba 2:47-50.

________. 1953. The Americas face up to the threat of coffee rust. For. Agr. 17:47-52; 64.

________. 1957. Hemileia vastatrix. Investigaciones presentes y pasados en la herrumbre del cafe y su importancia en el America tropical. FEDECAME, Secc. Divulg. (San Salvador, El Salvador.) Special publ., 67 p.

________. 1961. Coffee botany, cultivation, and utilization. Leonard Hill, London; Interscience Publ., New York, 488 p. (many references to rust and numerous citations given, see especially pp. 71-83, 120-121, 250-260, 345-349, and plates 36 and 37.)

________. 1970. The rust Hemileia vastatrix now firmly established on coffee in Brazil. Plant Dis. Reptr. 54:539-541.

________., R. Desrosiers, and E. Schieber. 1970. The Hemileia vastatrix coffee rust disease established in the American tropics. Phytopathology 60:139.

Appendix 2

GENUS *MUSA*

Some Botanical and Pathological Differences in Species

Bananas and plantains, under modern study, are morphologically confusing. They have been grown by aboriginal peoples since prehistoric times; these fruits have kept and are keeping millions from starving and are permanently important in present day tropical agriculture. In connection with plant pathological studies these crops have so many diseases (non-parasitic, viruses, bacteria, fungi, nematodes) that the many banana and plantain clones require more definitive and modern scientific classifications. Some assistance towards this end may be served by a brief summary that follows below. From reference to such works included at the end of this Appendix, and to research reports continuing to appear from stations in several tropical American countries, it will be seen that the dessert banana is sometimes no longer classed as *Musa sapientum*, some other botanical features are unsettled, and breeding and cytogenetic studies are still in progress. For citations respecting pathological and botanical problems of *Musa* see Chapters III, IV, IX, XXI and XXII.

Different scientific names are used for banana and plantain. One, *Musa sapientum* L., is for banana, clones of which are characterized by their tender, relatively sweet fruits that are eaten raw, and the other is *M. paradisiaca* L. for plantain, which has fruits characterized by coarseness and more starch and which are cooked before eating. From historical and cytogenetic studies it appears that the giant growing banana plant that Linnaeus saw was of the clone now often called silk, silk fig, seda and similar names. As we now know, it is not a true or pure species that Linnaeus treated as *Musa sapientum*.

Breeding and cytogenetic studies reveal the variety called silk is a triploid of a natural hybrid between the wild, seed-bearing *M. acuminata* L. and the wild seed-

REACTION OF *Musa* spp. AND CULTIVARS (BANANAS AND PLANTAINS) TO *Fusarium oxysporum cubense*[a]

Highly resistant[b]	Moderately resistant	Moderately susceptible	Most susceptible
Chamaluco A B B[c]	Bluggoe A A A	Awak legore A B B	Bluggoe "selection" A A A
Congo or Robusta A A A	Bluggoe "selection" A A A	Golden beauty or I. C. 2 A A A A	Gros michel A A A
Dwarf cavendish A A A	Gros michel x *M. acuminata* A A A	Manzana A A B	Gros michel x *M. acuminata*
Giant cavendish A A A	Lacatan "selection" A A A	*M. acuminata* A A	Gros michel x *M. acuminata*
Green cavendish A A A	*M. acuminata banksii*	*M. acuminata* clone 1877 A A A A	Highgate[d] A A A
Highgate "cross" A A A A	*M. acuminata errans* A A	*M. acuminata* x Gros michel	Lady finger A A B
Lacatan A A A	*M. acuminata microcarpa* A A	Paka x gros michel	*M. balbisiana* (seedling) B B
M. acuminata burmanniana A A	Paka A A	Pisang awak "selection" A B B	Payan A B B

M. acuminata lidi A A	Pisang lilin x gros michel A A A	Pisang lilin x gros michel	Pisang awak A B B
M. balbisiana B B (adult)	Platano enano A A B	Red A A A	Seda A A B
Pisang lilin[e] A A	Rio or Brazil A A A	Red green A A A	Silk A A B
Sikuzani A A	(Manguindanao, abaca)[f]		(Bungalanon, abaca)[f]
Tongat A A			
Williams "hybrid" A A A			
(Tangongon, abaca)[f]			

---

a Race 1.
b Resistance thought by some to be a single dominant character.
c Capital letters indicate ploidy: acuminata (A), balbisiana (B).
d Dwarf mutant of gros michel
e Thought to be nearest to original banana eaten by man.
f Abaca is Manila hemp, Musa textilis Nee.

bearing *M. balbisiana* L. (the diploid condition of the former is designated as AA and of the latter BB). The clone called silk has in it two genomes of *acuminata* and one of *balbisiana.* This clone, the type Linnaeus called *M. sapientum,* under the present circumstances is better represented as "*Musa* silk AAB." In naming another *Musa,* the plantain or cooking banana Linnaeus is believed to have used is the clone commonly called french plantain. Modern study shows it is also a natural hybrid between *acuminata* and *balbisiana,* which resulted in a triploid with AAB composition. This special plantain clone is therefore designatable as "*Musa* french plantain AAB." Both of these great clones have the same genome content but are much different as to edibility and appearance.

The typical giant banana clone called gros michel, so important in tropical American economy, was once classified as *M. sapientum.* Cytogenetic investigation shows it came probably from a naturally occurring triploid of *M. acuminata,* AAA (being a seedless mutant it would have long ago disappeared in nature without prehistoric man protecting it and growing it for food so that gros michel might correctly be given the Latin binomial *M. acuminata* with some varietal derivation attached; but until botanists satisfy themselves about this, and arrive at something better, it is more understandable for pathologists to give this banana the designation "*Musa* gros michel AAA"; also *Musa* AAA. It is interesting that a dwarf type but tetraploid mutant of gros michel is now grown, named highgate. Its designation would be "*Musa* highgate AAAA"; also *Musa* AAAA.

From another tall growing banana there appears to have come a dwarf mutant, and it was studied by three botanists. Using the old accepted system the names for it may be given as either *M. cavendishii* Lambert, or *M. sinensis* Sweet ex Sagot, or *M. nana* Lour. This is grown commercially in the Orient, in Africa, in Brazil, and more recently in Central America. On cytogenic analyses, both the tall form and the dwarf have been proved to be triploids of AAA composition. Clones are with better understanding now given, for the tall cavendish type, as "*Musa* giant cavendish AAA," and for the short cavendish as "*Musa* dwarf cavendish AAA."

To the pathologist, designations such as those above, are much more exact than older Latin binomial or trinomial usages and indicate clearly some of the genetic makeup of the crop under study. However, in disease work there is something much more that is involved. For examples: Musa gros michel AAA and its tetraploid dwarf Musa highgate tetraploid AAAA, are both distinctly susceptible to fusarial wilt (Fusarium oxysporum cubense), while Musa giant cavendish AAA and Musa dwarf cavendish AAA are, generally speaking, highly resistant to this devastating Fusarium disease. All four of these clones are distinctly susceptible to leafspot (Mycosphaerella musicola). However, when gros michel was crossed with a wild Musa acuminata the progeny proved to be somewhat resistant, but not immune, to fusarial wilt and highly resistant to leaf spot. It is notable that Musa 2390 AAAA, derived from highgate, is a dwarf plant and is resistant to both the wilt and the leafspot. Musa pelipita ABB, which is a bluggoe-type plantain, is resistant not only to common fusarial wilt, but also to bacterial wilt from Pseudomonas solanacearum.

Some Musa Plants Being Used in Tropical America

A partial list, from discussions with tropical American researchers, of Musa species and clones is given herewith, indicating the genome, or "set" or "sets" of chromosome populations per cell. A simplified means of designating content of Musa clones is as follows: a single cell, as seen in a pollen grain or in an oogonium, has one "set" of chromosomes and is indicated by a single capital letter. On pollination, the seed is started from which the plant comes, and it then has two sets of chromosomes in its cells and is the diploid condition indicated by a double capital letter. Triploid condition is indicated by three such letters and tetraploid by four letters. The A indicates a component here of M. acuminata, and the B is a component of M. balbisiana.

Group 1 DIPLOIDS	Group 2 TRIPLOIDS (all three parts acuminata)
M. acuminata AA	
M. acuminata subspecies AA	Giant cavendish AAA
Pisang lilin AA	Dwarf cavendish AAA
Paka AA	Dwarf cavendish variety AAA
Sikuzani AA	Gros michel AAA
Tongat AA	Highgate AAA
M. balbisiana BB	Congo or robusta AAA
M. balbisiana subspecies BB	Green banana AAA

Group 2 TRIPLOIDS (cont.)

Lacatan AAA
Pisang serendah AAA
Pisang rajah udang AAA
Rio or Brazil AAA
Williams hybrid AAA
Bluggoe banana AAA

Group 3 TRIPLOIDS
(mixed: 2 parts acuminata, 1 part balbisiana)

French plantain AAB
Silk fig banana AAB
Platano enano AAB
Green french plantain AAB
Giant plantain AAB
Horn plantain AAB
Maiden plantain AAB
Mysore AAB
Manzana AAB
Seda AAB
Lady finger AAB
Sugar silk AAB

Group 4 TRIPLOIDS
(mixed: 1 part acuminata, 2 parts balbisiana)

Chameluco ABB
Bluggoe plantain ABB
Dwarf bluggoe ABB
Pisang awak ABB
Awak legor ABB
Payan ABB

Group 5 TETRAPLOIDS
(all 4 parts acuminata)

I. C. 2 or Golden Beauty AAAA
Clone 1877 AAAA
Clone 2390 AAAA
Bodles altafort AAAA
Highgate tetraploid AAAA

Some Literature on MUSA

Baker, R. E. D. and N. W. Simmonds. 1951. Banana research--changes in outlook. Trop. Agr. Trinidad 28:43-45.

Cheesman, R. E. 1931. Banana breeding at the Imperial College of Tropical Agriculture. Report of Empire Marketing Board No. 47.

________. 1948. Classification of the bananas 3, C. Kew Bull. 1948:145-153.

Dodds, K. S. 1950. The breeding of disease resistant bananas. World Crops 2:56-59.

Purss, G. S. 1953. A disease in williams hybrid banana produced by Fusarium sp. Queensland J. Agr. Sci. 10:126.

Shepherd, K. 1960. Seed fertility of gros michel bananas. Trop. Agr. Trinidad 37:211-221.

Simmonds, N. W. 1957. Botanical results of banana collecting expedition 1954-1955. Kew Bulletin 463.

________. 1959. Bananas. Longmans, Green and Co., London, 466 p.

Stover, R. H. 1962. Fusarial wilt (panama disease) of bananas and other Musa species. Commonw. Mycol. Inst., Phytopath. Paper 4, 117 p.

________., and D. L. Richardson. 1968. Pelipita, an A B bluggoe type plantain resistant to bacterial and fusarial wilts. Plant Dis. Reptr. 52:901-903.

Tomlinson, P. B. 1959. An anatomical approach to the classification to the Musaceae. J. Linnean Soc. Botany 55:799-809.

Wardlaw, C. W. 1961. Banana Diseases, including plantains and abaca. Longmans, Green and Co., London, 648 p.

Appendix 3

MEASUREMENTS AND CONVERSIONS

A Geographical Degree

The length of 1 degree earth latitude is about 111 kilometers, or 69 miles. (The importance of this measurement in the tropics is dealt with in Chapter II.) Every 69 miles north or south of the so-called "thermal" equator it gets a little cooler. For every up-hill kilometer in the tropics it is also cooler. Average temperatures are 0.3° to 0.9° Centigrade cooler for every 100 meters elevation. It will be seen (Map A, p. 26) that the sinous thermal equator does not follow the straight drawn geographical equator, but goes through it only once in its extension from points farthest north to south in the western tropics. In approximating changes in temperature from elevation, the base is the line of the thermal equator.

Useful Conversions and Measurements

There are conversions and measurements of considerable convenience in American tropical plant disease research work. Among scientists the metric system is preferred over the old and local ways of expressing units; some of the latter come from colonial adaptations of ancient Indian usages, others from European folk words, some forgotten in lands where they came from.

It is necessary to know what is used in the United States, as these are employed in the tropics by some personnel of companies, consultants with U.S. backgrounds, and others with their home bases in the northern part of the hemisphere. However, in the tropics where research results are published, it becomes increasingly apparent that effective and respected work is most likely to be reported in metric terms.

Weights (lesser)

Kilograms	Pounds (U. S.)
1	2. 20
2	4. 41
4	8. 82
6	13. 22
8	17. 65
10	22. 04
12	26. 45
14	30. 86
16	35. 27
18	39. 68
20	44. 09
25	55. 11

Weights (greater)

Kilograms	Pounds (U. S.)
100	220
110	243
120	265
130	287
200	441
250	551
300	661

Short lengths

Meters	Yards (U. S.)	Varas (Span.)
1	1. 09	1. 20
2	3. 28	2. 39
4	4. 37	4. 79
6	6. 56	7. 18
8	8. 75	9. 57
10	10. 93	11. 96
12	13. 12	14. 35
14	15. 31	16. 75
16	17. 49	19. 14
18	19. 68	21. 53
20	21. 87	23. 92
25	27. 34	29. 91

Distances

Km.	Miles (U. S.)
1	0. 62
2	1. 24
4	2. 49
6	3. 73
8	4. 97
10	6. 21
12	7. 46
14	8. 70
16	9. 94
18	11. 18
20	12. 42
25	15. 53

Land Measures

Hectares	Manzanas
1	1. 43
2	2. 86
4	5. 72
6	8. 59
8	11. 45
10	14. 31
12	17. 17
14	20. 03
16	22. 90
18	25. 76
20	28. 62
25	35. 78

Surfaces (such as seedbeds)

Sq. m.	Sq. feet	Sq. varas
1	10. 76	1. 43
2	21. 52	2. 86
4	43. 05	5. 72
6	64. 58	8. 59
8	86. 11	11. 45
10	107. 6	14. 31
12	129. 1	17. 17
14	150. 6	20. 03
16	172. 2	22. 90
18	193. 7	25. 76
20	215. 2	28. 62
25	269. 1	35. 78

Rainfall Measurements, per Unit Area

Unit	cubic inches	liters	metric tons
1 inch/acre	6, 272, 640	102, 788	102
1 millimeter/hectare	610, 234	10, 000	10
1 millimeter/acre	246, 954	4, 076	4

Short Measures

Cm.	Ft. & in.
10	0 - 3 7/8
20	0 - 7 7/8
30	0 - 11 3/4
40	1 - 3 3/4
50	1 - 7 5/8
60	1 - 11 5/8

Liquid Quantities (U. S.)

Liters	Pints	Gals.
10	21. 13	2. 64
20	42. 27	5. 28
30	63. 40	7. 93
40	84. 54	10. 57
50	105. 67	13. 21
60	126. 81	15. 85
70	147. 94	18. 49
80	169. 07	21. 13
90	190. 21	23. 78

Temperatures (*H_2O)

Centigrade (Celsius)		Fahrenheit	Centigrade	Fahrenheit
0	*freezes	32. 0	19	66. 2
1		33. 8	20	68. 0
2		35. 6	21	69. 8
3		37. 4	22	71. 6
4		39. 2	23	73. 4
5		41. 0	24	75. 2
6		42. 8	25	77. 0
7		44. 6	26	78. 8
8		46. 4	27	80. 6
9		48. 2	28	82. 4
10		50. 0	29	84. 2
11		51. 8	30	86. 0
12		53. 6	31	87. 8
13		55. 4	32	89. 6
14		57. 2	33	91. 4
15		59. 0	34	93. 2
16		60. 8	35	95. 0
17		62. 6	36	96. 8
18		64. 4	37	98. 6

Centigrade	Fahrenheit	Centigrade		Fahrenheit
38	100.4	70		158.0
39	102.2	80		176.0
40	104.0	90		194.0
41	105.8	100	*boils	212.0
42	107.6	110		230.0
43	109.4	120		248.0
44	111.2	130		266.0
45	113.0	140		284.0
46	114.8	150		302.0
47	116.6	160		320.0
48	118.4	170		338.0
49	120.2	180		356.0
50	122.0	190		374.0
60	140.0	200		392.0

Some Convenient Common Experiment Station Measures

1 Quintal	= 100 kilograms
1 Vara	= 0.836 meters = 0.914 yard
1 Meter	= 100 cm. = 3.28 ft (39.37 inches) = 1.093 yards
1.0 Manzana	= 0.698 hectare = 1.726 acres
1 Hectare	= 10,000 sq. mi. (2.471 acres)
1 Kilometer	= 5/8 of one U.S. mile, approx. (1100 yards)
15 Milliliters	= 1 tablespoonfull (1/2 fluid ounce)
1 Pine (U.S.)	= 0.473 liters
1 Quart	= 0.946 liters
1 Gallon	= 3.785 liters
1.0 Botella*	= 0.757 liter = 1/5 gallon (U.S.)
1.0 Liter	= 1/4 gallon (U.S.), approx. = 1 1/3 botella*
1.0 Ounce	= 0.028 kilogram
1.0 Almud*	= 11 liters, approximately
1.0 Acre (U.S.)	= 0.579 Manzanas

*Varies in different countries.

Descriptives for Soil Reactions

"Intensely acid"	pH 3.0-4.0	(extreme; not good for crops)
"Markedly acid"	pH 4.5-5.9	(Favors a few bacteria but more fungi)
"Circum neutral-acid"	pH 6.0-6.9	(Favors good growth of fungi and some bacteria)
"Circum neutral-alkaline"	pH 7.0-7.5	(Favors fungi and many bacteria)
"Markedly alkaline	pH 7.5-7.9	(Favors certain fungi but more bacteria)
"Highly alkaline"	pH 8.0-9.0	(Extreme; not good for crops)

There occur times when parts per million (p.p.m. or ppm.) are suggested for use in disease treatment applications. Examples of some dilutions with certain antibiotics may indicate proportions:

Agrimycin 17: 21.3% W.P., add 1 1/2 tsp. to 1 gallon of water to make 100 ppm.

Phytomycin: 20% liquid, add 1/2 tsp. to 1 gallon of water to make 100 ppm.

Agri-Strep. Type D: 21.2% W.P., add 1 1/3 tsp. to 1 gallon of water to make 100 ppm.

Ortho Streptomycin Spray: 17% W.P., add 1 1/2 tsp. to 1 gallon of water to make 100 ppm.

Streptomycin Antibiotic Spray Powder (Miller): 8.5% W.P., add 2 1/2 tsp. to 1 gallon of water to make 100 ppm.

(Examples of one part per million: (1) weight of a postage stamp is equal to one part per million of the weight of a person; (2) weight of one gram needle in one ton of rice straw is one part per million; (3) one minute in two years is one part per million.)

Proportions* for Certain Fungicides in Water Suspensions

Fungicide	Amount often added to: 1 gallon	100 gals.
Actidione P. M.	5 tbsp.	6 lb.
Bordeaux mixture		
CuSO4 crystals	2 tbsp.	2 lb.
Lime, hydrated	6 tbsp.	6 lb.
Captan 50% W. P.	1 1/2 tbsp.	2 lb.
Ferbam 76% W. P.	2 1/2 tbsp.	2 lb.
Fixed copper 50% W. P.	2 tbsp.	4 lb.
Karathane W. D., 22.5% W. P.	1 tbsp.	1/2 lb.
Lime-sulfur, liquid	5 tbsp.	2 gal.
Phaltan 75% W. P.	1 1/2 tbsp.	1 1/2 lb.
Puratized Agricultural spray	1 tsp.	1 pint
Terrachlor 75% W. P.	1 tbsp.	1 lb.
Thiram 65% W. P. (Thylate)	2 tsp.	1 lb.
Wettable sulfur (dry) 95%	1 tbsp.	2 lb.
Zineb 65% or 75% W. P.	1 1/2 tsp.	1 1/2 lb.

*Standard: tbsp. = tablespoon measured level; tsp. = teaspoon measured level; lb. = pound (U. S.); gal. = gallon (U. S.); and pint also U. S.

Equivalent Quantities of Wettable Powders (W. P.) for Various Quantities of Water

Amount of Wettable Powder						Water (gals.)
1 lb.	2 lb.	3 lb.	4 lb.	5 lb.	6 lb.	100
8 oz.	1 lb.	1.5 lb.	2 lb.	2.5 lb.	3 lb	50
3 tbsp.	1.5 oz.	2.5 oz.	3.25 oz.	4 oz.	5 oz.	5
2 tsp.	3 tsp.	1.5 tbsp.	2 tbsp.	3 tbsp.	3 tbsp.	1

Example, for a recommended mixture of 4 lbs. W. P. to 100 gals. water, 3 1/4 oz. or approximately 6 1/2 tsp., added to 5 gal. of water would give 5 gallons of spray mixture of the recommended strength.

Weights (U. S.)

28.35 grams = 1 ounce	1 cu. ft. water = 62.4 lbs.
16 ounces = 1 pound	1 gallon fuel oil = 7 lbs.
1 gal. water = 8.34 lbs.	1 gallon kerosene = 6.7 lbs.

Measures (U. S.)

3 teaspoons = 1 tablespoon
3 tablespoons = 1 fl. oz.
8 fl. ozs. = 16 tbsp. = 1 cup
2 cups = 32 tbsp. = 1 pint
2 pints = 64 tbsp. = 1 quart
4 quarts = 256 tbsp. = 1 gal.
128 fl. oz. = 1 gallon

Equivalent Quantities of Liquid Materials When Mixed

To make dilutions 1 part of chemical to water			Add to this amount
Dilution 1:400	Dilution 1:800	Dilution 1:1600	of water:
1 qt.	1 pt.	1 cup	100 gals.
1 pt.	1 cup	1/2 cup	50 gals.
1 tbsp.	5 tsp.	2 1/2 tsp.	5 gals.
2 tsp.	1 tsp.	1/2 tsp.	1 gal.

Example, for 1 part of chemical to 800 parts water, add 5 tsp. of chemical to 5 gals. water, giving 5 gals. of 1:800 mixture.

Relative Humidity of Air Determinations

Dry bulb temp:	Differences in degrees centigrade between dry bulb and wet bulb														
	1	2	3	4	5	6	7	8	9	10	11	12	13	14	15
1	97	66	49	33	17										
2	93	84	52	37	22										
5	86	72	58	45	33	20	7								
7	87	74	62	50	38	26	15								
10	88	77	66	55	44	34	24	15	6						
15	90	80	71	61	53	44	36	27	20	13	6				
20	91	83	74	66	59	51	44	37	31	24	18	12	6		
25	92	84	77	70	63	57	50	44	39	33	28	22	17	12	8
30	93	86	79	73	67	61	55	50	44	39	35	30	25	21	17
35	94	87	81	75	69	64	59	54	49	44	40	36	32	28	24
40	94	88	82	77	72	67	62	57	53	48	44	40	36	33	29

[Relative humidity, or RH, given as percentage]

For use in field situations without exact electronic or

other equipment. Requirements: 2 calibrated centigrade thermometers; strip of thin linen cloth (handkerchief weight) to be wet and attached (one thickness) around bulb of one thermometer; shallow water-holding container in which end of wet cloth will dip; small supply of clean water; leaf for use as a fan to stir air. Dry bulb held in the air within about 15 cm. of wet bulb, air kept lightly moving by fanning and after five minutes temperature observations from both bulbs are made and recorded three times during a 10-minute period, and the closest readings used. Interpolations may be made in above table for intermediate points.

Users should be cautioned that these percentages will be approximate, but determinations will give, nevertheless, highly illuminating results of reasons for disease differences for example in sun and shade, at various elevations or wind exposures, and the like. To be absolutely exact in RH recording requires taking into account not only geographical elevation but as well atmospheric pressure at the moment. However, approximations will be revealing.

Suggested procedure is to prepare a data page with four columns for: "dry," "wet," "difference," and "per cent." If, for example at 9:00 in the morning "dry" shows 30 and "wet" is 25, "difference" would be 5, and "per cent" would read 67; at noon "dry" might show 35 with "wet" at 25, giving the "difference" of 10, and "per cent" of 44; at midafternoon with "dry" still 35, the "wet" might be 21 and a "difference" of 14 or a "per cent" of 28; at midnight with "dry" at 25 and "wet" at 24 the "difference" would be 1 or a "per cent" of 92.

INDEX

Disease items, including controls, are herein prefixed with *, and host items with †.